Handbook of
VADOSE ZONE
CHARACTERIZATION
&
MONITORING

Edited by

L.G. Wilson, Lorne G. Everett, Stephen J. Cullen

CRC Press
Taylor & Francis Group
Boca Raton London New York

CRC Press is an imprint of the
Taylor & Francis Group, an **informa** business

CRC Press
Taylor & Francis Group
6000 Broken Sound Parkway NW, Suite 300
Boca Raton, FL 33487-2742

First issued in paperback 2019

ISBN-13: 978-0-367-40193-1

Series Editor: Fred L. Troise

Library of Congress Cataloging-in-Publication Data

Handbook of vadose zone characterization & monitoring / L. G.
   Wilson, Lorne G. Everett, Stephen J. Cullen, editors.
       p.   cm. — (Geraghty & Miller environmental science and engineering series)
   Includes bibliographical references and index.
       1. Zone of aeration — Congresses. 2. Water quality — Measurement — Congresses. I.
   Wilson, L. G. (Lorne Graham), 1929-   .
   II. Everett, Lorne G. III. Cullen, Stephen J. IV. Series.
   TD403.H36   1994
   628.1'61 — dc20   94-29079

Visit the Taylor & Francis Web site at
http://www.taylorandfrancis.com

and the CRC Press Web site at
http://www.crcpress.com

Editors

Lorne Graham Wilson is a Hydrologist in the Department of Hydrology and Water Resources, University of Arizona, Tucson, Arizona. He received his PhD in soil science from the University of California at Davis, California, in 1962. Following graduation from Davis, Dr. Wilson joined the faculty of the University of Arizona. His research activities at the university have focused on evaluating natural and artificial ground-water recharge under southwestern conditions, vadose-zone and ground-water monitoring at waste disposal facilities (e.g., sanitary landfills and dry wells), and soil aquifer treatment of sewage effluent. In addition to publishing on these topics in the technical literature, he is coauthor, with Lorne G. Everett and Edward Hoylman, of the book *Vadose Zone Monitoring for Hazardous Waste Sites.*

Lorne G. Everett is the Chief Research Hydrologist and Vice President for Geraghty & Miller, Inc., Santa Barbara, and the Director of the Vadose Zone Research Laboratory at the University of California at Santa Barbara. He has a PhD in Hydrology from the University of Arizona in Tucson. Dr. Everett conducts research on subsurface characterization and remediation. He is Chairman of the American Society for Testing and Materials (ASTM) Task Committee on Groundwater and Vadose Zone Monitoring, and is responsible for writing nine new National ASTM Standards on Vadose Zone Investigations. He chaired the Remediation Session of the First USSR/USA Conference on Environmental Hydrogeology (Leningrad, 1990). Dr. Everett has authored several books including: *Groundwater Monitoring, Vadose Zone Monitoring for Hazardous Waste Sites*, and *Subsurface Migration of Hazardous Waste*. His book entitled *Groundwater Monitoring* was endorsed by the Environmental Protection Agency as establishing "the state-of-the-art used by industry today" and is recommended by the World Health Organization for all developing countries.

Stephen J. Cullen is a Principal Scientist with Geraghty & Miller, with over 15 years experience as a soil scientist, vadose zone hydrologist, and in the development and evaluation of state-of-the-art vadose zone characterization and monitoring techniques. He is also a faculty member and Research Hydrologic Specialist with the Institute for Crustal Studies at the University of California, Santa Barbara (UCSB). He has participated in the solution of a wide range of vadose zone characterization, monitoring, and remediation problems including hazardous and solid waste landfill monitoring system design, monitoring system design at a hazardous waste land treatment and low level radioactive waste disposal facilities, and design of a vapor extraction system for petroleum contaminated soils and sediments. Mr. Cullen has published numerous scientific articles on various aspects of vadose zone monitoring and lectures in the Department of Geology, Engineering, and Geography at UCSB on the subjects of soil physics, hydraulics, vadose zone hydrogeology, land pollution, and subsurface contaminant remediation.

Acknowledgments

The editors acknowledge the invaluable assistance of the contributing authors to this text who endured our browbeating and begging for a period of almost four years. We also thank the anonymous peer reviewers who assisted in clarifying, correcting, and supplementing individual sections of text. Finally, we gratefully acknowledge Kathy Walters and Vivian Collier of Lewis Publishers for their excellent editorial assistance and guidance, and Brian Lewis, Publisher, for his support and encouragement throughout this project.

Preface

Vadose zone monitoring is now recognized as a necessary component of a comprehensive subsurface monitoring system. As stated by Cullen et al. in Chapter 1 of this book, ground-water monitoring without an equal emphasis on vadose zone monitoring is illogical. Fortunately, regulatory agencies have come to accept and encourage vadose zone monitoring at hazardous and other waste disposal facilities. In Chapter 2 of this book, Durant and Myers review the evolution of regulations promulgated by the United States Environmental Protection Agency (EPA) and methodologies at Resource Conservation and Recovery Act (RCRA) facilities. Reflecting these developments, the 1980s saw an upsurge of interest and application of vadose zone monitoring methodologies not only for monitoring per se but also as a component of remediation programs. Correspondingly, there has been a growth in the number of practitioners of vadose zone monitoring. Many of these practitioners have had little or no exposure to vadose zone hydrology, much less monitoring.

During the mid 1980s, the American Society for Testing and Materials (ASTM), in cooperation with the EPA, became actively involved through Subcommittee D-18.21.02 (Vadose Zone Investigations) in developing state-of-the-art guidance documents on standardized vadose zone characterization and monitoring approaches. The purpose of this book is to expand upon and consolidate the useful but succinct information included in various ASTM documents, EPA manuals, and other similar texts. This book will be an aid to the new practitioner. Seasoned practitioners will also find the book to be a useful reference source.

Just as vadose zone monitoring should be supplementary and complementary to ground-water monitoring, this book is the analogous complement of the text *Practical Handbook of Ground-Water Monitoring*, edited by D.M. Nielsen and published by Lewis Publishers in 1991. As is Nielsen's book, this volume is a compilation of chapters and sections written by experienced practitioners. Topics covered include basic principles of vadose zone hydrology and prevalent monitoring techniques. Case studies are also included to benefit the reader from field experiences, good and bad. The charge to the authors when preparing their chapters was as follows:

> . . . our intent is to prepare a document that will be useful to the end users of vadose zone monitoring techniques. Our feeling is that many end users do not have the background and experience needed to fully understand the principles, advantages, and limitations of the methods they are selecting. Accordingly, your section should be written in a simple, straightforward manner. Please

avoid complicated mathematical derivations and limit the use of equations to those necessary to explain principles. Think of your section as a primer on the topic.

The authors are to be commended for complying with this request.

The first two chapters (Cullen et al. and Durant and Myers, respectively) review the technical and regulatory rationale for vadose zone monitoring. Chapters in the subsequent two Parts review basic characteristics of vadose zone hydrogeology to serve as the foundation for selecting and understanding the operating principles and site-specific requirements of vadose zone monitoring techniques. Thus, Chapter 3, by Stanley Davis, reviews the vadose zone hydrogeology of the United States, to our knowledge an assessment not previously reported. Similarly, Chapters 4 and 5 in Part III summarize basic hydrologic processes and structural considerations when characterizing vadoze zones.

Following the background covered in the first five chapters, the chapters in Part IV review basic contaminant fate and transport processes, aiding in selecting appropriate methods for site-specific contaminants. Part V, Preliminary Monitoring-Related Activities, expands on information in previous chapters, contributing to the selection of direct and indirect monitoring methods. For example, based on geological and other physical properties of the vadose zone and potential pollutants to be monitored, certain monitoring methods may be favored over others. Part VI reviews the applicability (and limitations) of information gained from Part IV and Part V for modeling pollutant fate and transport in the vadose zone. Part VII covers indirect monitoring methods, i.e., those inferring the presence and movement of pollutants (e.g., neutron logging, time domain reflectometry, etc.). Part VIII covers direct methods (e.g., pore-liquid samplers) for collecting samples of pollutants in unsaturated regions of the vadose zone. Inasmuch as saturated regions occur in vadose zones, Part IX covers saturated sampling techniques (e.g., free-drainage samplers). Although the emphasis in these last three Parts is on monitoring for pollutant movement, the methods and approaches discussed have been used for many years in agricultural applications, e.g., monitoring water and solute movement beneath agricultural fields. The methods are also suitable for monitoring nonhazardous fluids such as percolating water from artificial recharge facilities. Finally, as a link to possible revisions of this book, Part X summarizes emerging technologies in the rapidly-expanding field of vadoze zone monitoring.

L.G. Wilson
L.G. Everett
S.J. Cullen

Contents

PART V
PRELIMINARY MONITORING-RELATED ACTIVITIES

PART VI
MODELING

PART VII
INDIRECT METHODS FOR DETECTING CONTAMINANT MOVEMENT

PART VIII
DIRECT METHODS FOR SAMPLING CHEMICAL AND MICROBIAL POLLUTANTS IN UNSATURATED REGIONS OF THE VADOSE ZONE

PART IX
DIRECT METHODS FOR SAMPLING CHEMICAL AND MICROBIAL POLLUTANTS IN SATURATED REGIONS OF THE VADOSE ZONE

PART X
EMERGING TECHNOLOGIES

Handbook of
VADOSE ZONE CHARACTERIZATION & MONITORING

Is Our Ground-Water Monitoring Strategy Illogical?

Stephen J. Cullen, John H. Kramer, Lorne G. Everett,
and Lawrence A. Eccles

INTRODUCTION

Traditional ground-water monitoring strategies have relied on a process of collecting samples from wells and determining if there is evidence of contamination. This approach is arguably illogical, because if the purpose of monitoring is preventing aquifer contamination, the evidence comes only after contaminants have reached the resource we are trying to protect. Monitoring in the vadose zone, on the other hand, can provide early detection of contaminant migration and the opportunity to actually prevent ground-water contamination. Because this fact has been increasingly recognized in the regulatory arena, ground-water students and professionals need training in vadose zone hydrogeology. While the relative hydrologic importance of the vadose zone is much greater in the western United States, it remains a largely misunderstood portion of the geologic profile across the country. The vadose zone, sometimes saturated but more often not, is complicated by textural heterogeneities, flow instabilities, and preferred pathways. Reliable prediction of contaminant migration is lacking. Thus, ongoing monitoring is the only viable option and development of innovative monitoring techniques should be encouraged in both the public and private sectors. For example, deployment of monitoring techniques such as neutron moderation, dielectric measurements, and soil gas screening represent efficient, low-cost methods for improving monitoring coverage in space and time.

WHAT IS GROUND-WATER MONITORING REALLY?

Attempting to prevent aquifer contamination by monitoring ground water is akin to monitoring a patient's heartbeat to determine when a heart attack will occur. By the time the heartbeat becomes irregular or stops, the patient is typically experiencing a serious trauma. Likewise, by the time significant contamination shows up in a traditional ground-water sampling program, the ground water is already contaminated. Prudence requires attention to monitoring the parameters symptomatic of the condition. In the case of the patient, breathlessness and chest pains are symptomatic of heart problems. In the case of ground-water contamination, the vadose zone is the logical place to look for symptoms because a contaminant release to the vadose zone always precedes ground-water contamination, with the notable exception of ground water in direct hydraulic communication with surface water.

Ground-water monitoring programs have long been used to detect, observe, regulate, and control aquifer water quality. In earlier years, this approach seemed adequate because demand on ground-water supplies was lower, industrial sources of contaminants fewer, and the perceived public health risks low. In today's technological society, agricultural and urban demands placed on ground water are explosive, and the variety of primary or secondary toxic compounds produced is steadily increasing. Improved analytic instrumentation now provides measurement capability of water sample toxics in the parts-per-billion range, while our awareness of the toxicological effects of various contaminants has widened. This has increased the chance of detecting contamination, and elevated our evaluation of the risk. Once aquifer contamination is detected, a series of undesirable and costly events are irrevocably placed in motion, driven by regulatory mandates at the federal, state, or local levels. These events can include more intensive compliance monitoring, corrective action to remove the source of contamination, aquifer remediation, and shutdown of facilities. Moreover, the National Research Council (1990) noted that "Cleanup objectives may not be achievable at all, especially with dense nonaqueous phase liquids." For these reasons, prevention, not mere detection, is the desirable goal.

As the result of vigorous federal legislative activity, the U.S. Environmental Protection Agency (EPA) adopted a Ground Water Protection Strategy in 1984, subsequently revised in July 1991. In clear and concise language, EPA stated that " . . . the overall goal of EPA's Ground Water Policy is to prevent adverse effects to human health and the environment, and to protect the environmental integrity of the nation's ground-water resources . . . ," (U.S. EPA, 1991). A significant component of the EPA's newly revised and rapidly evolving policy clearly places an increased emphasis on prevention of ground-water contamination, and on efforts to achieve a greater balance between prevention and remediation activities.

THE VADOSE ZONE

The vadose zone is that portion of the geologic profile beneath the earth's surface and above the first principal water-bearing aquifer. Flow in the vadose zone is dynamic and characterized by periods of unsaturated flow at varying degrees of partial saturation punctuated by episodes of preferential, saturated flow in response to hydrologic events or releases of liquids. As Bouwer (1978) pointed out, it is inappropriate to refer to this zone as the "unsaturated zone" or to refer to all flow in this zone as "unsaturated flow."

It is time to consider the possibility that we have painted ourselves into a terminology corner. The term "ground water" has historically come to mean water beneath the land surface contained in interconnected pores in the saturated zone that is under hydrostatic pressure (D.M. Nielsen, 1991; Bouwer, 1978; Freeze and Cherry, 1979; Driscoll, 1986). Unfortunately, this leads to an artificial separation in the minds of many between water above and below the water table. In fact, all subsurface water is ultimately connected hydraulically via intergranular films. Water naturally flows to and from the ground-water table in response to gravitational, pressure, and matric potential gradients. To avoid misconceptions about vadose zone flow processes, it would be preferable for all water in the subsurface environment, including vadose zone water, to be considered ground water. Since it is not practical to rewrite the textbooks, we promote the use of the term "subsurface water" as used by Driscoll (1986), and remind readers that ground water is a component of the larger category of subsurface water, which also includes hydraulically connected vadose zone water.

WHY VADOSE ZONE MONITORING?

The vadose zone is an intrinsically complicated three-phase system consisting of interconnected matrices of solid, liquid, and gas phases and is a complex environment in which to characterize and predict water flow. Vadose zone contaminant flow is confounded by partitioning into nonaqueous, dissolved, gaseous, and sorbed phases. Downward saturated flow through the "filter" of a porous medium driven primarily by gravity and hydrostatic head, as described by Darcy and refined by others (Green and Ampt, 1911), is intuitive. Such flow is sometimes diverted in the vadose zone by barriers causing lateral transport, or accentuated by preferred pathways promoting rapid downward transport (Germann, 1988; Glass et al., 1988). Less intuitive is unsaturated flow driven by the attraction of water for soil particle surfaces. Matric forces can be stronger than gravity and result in water flowing uphill! Predictive modeling of such a complex environment is, at best, frustrating. Nielsen et al. (1990), reviewing the state-of-the-art of predicting contaminant transport in the vadose zone, stated that, "The efficacy of accurately predicting the attenuation and eventual location of solutes or constituents in the vadose zone remains undeveloped." They further concluded, "Because the reliability of models for contaminant transport has not been established even for site-specific conditions, it appears that direct monitoring of the constituents in the vadose zone remains a necessity into the foreseeable future."

Beyond the pure logical value of detecting contamination before it reaches ground-water supplies, vadose zone monitoring can decrease the expense and technical difficulties associated with ground-water remediation. In the event of a release, early detection by vadose zone monitoring can result in faster and less costly remediation because of the generally reduced contaminant concentrations, reduced volume of contaminated subsurface material, and higher levels of oxygen to promote degradation and/or recovery.

VADOSE ZONE REGULATORY EVOLUTION

Since the National Environmental Policy Act became law in 1969, society's collective consciousness has been raised regarding the important role the vadose zone plays as a buffer against ground-water contamination. Just after its formation in 1970, the EPA funded the first major national ground-water monitoring study involving numerous noted hydrogeologists such as Banks, Geraghty, Schmidt, Tinlin, Todd, Warner, and others, which was later published by Everett (1980). Supported by the EPA's Environmental Monitoring Systems Laboratory (EMSL) at Las Vegas, this initial investigation recognized the need for early alert or vadose zone monitoring as part of an overall ground-water monitoring strategy. Wilson (1980) first described approaches to monitoring the vadose zone in a report to the EPA.

The Resource Conservation and Recovery Act (RCRA), passed in 1976, identified land treatment as a nationally permitted method for hazardous waste disposal. Because land treatment operates as an open system, vadose zone monitoring was required within 30 cm of the bottom of the treatment zone. As a result of the first federally mandated regulations (45 FR 33248 and 47 FR 32363) covering the vadose zone, *Permit Guidance Manual on Unsaturated Zone Monitoring for Hazardous Waste Land Treatment Units* (U.S. EPA, 1986), was published by EMSL-Las Vegas. This manual, based on a report to the EPA by Everett and Wilson (1986), has since been widely referenced in other federal and state legislative applications.

The RCRA was amended in 1984 and the EPA, under Subtitle I, required vadose zone monitoring at Underground Storage Tanks (UST) sites where sufficient depth to the water table existed. Subsequently, numerous state approaches required vadose zone monitoring when water table depth was shallower than that dictated by the federal position. Thus, vadose zone monitoring has been accepted nationally as a viable part of a monitoring strategy.

In 1974, Congress enacted the Safe Drinking Water Act (public law 93–523) including provisions to protect existing and future underground sources of drinking water. Under Part C of the act, the EPA developed regulations to protect underground sources of drinking water from contamination by the subsurface injection or emplacement of fluid through wells. Although this act allowed Class V wells (dry wells or vadose zone wells) to inject nonhazardous fluids into or above underground sources of drinking water, some states and regions have recognized that this practice may be hazardous. In fact, some have discontinued the use of dry wells associated with gas stations and industrial facilities. Clearly, the use of Class V wells associated with sites where petroleum or chlorinated hydrocarbons are used should be discontinued nationally. Procedures to close and remediate these Class V wells are in preparation (Everett, 1992).

In 1994, amendments to the RCRA are under consideration. Fundamental to the reauthorization will be the inclusion that vadose zone monitoring can be required as part of Subtitle C activities related to Part B permits at hazardous waste treatment, storage, and disposal facilities. Based on the decision of EPA Regional Administrators, vadose zone monitoring will therefore be a key component of our national hazardous waste disposal monitoring program.

A clear indication of regulatory changes can be seen in California. Title 14, Chapter 5 of the California Code of Regulations (CCR) requires that reports from every landfill (operating or closed) contain a chemical characterization of the soil-pore liquid from areas likely to be affected by leachate leaks. In addition, under CCR Title 23, Chapter 15, California requires that every operating and closed landfill have a vadose zone monitoring program in operation during detection, characterization, and remediation of a release from a waste management unit. It is encouraging to note that all California landfills must have a vadose zone monitoring strategy to help prevent these solid waste sites from evolving into Superfund sites. Recognizing that the nation has thousands of solid waste sites, the EPA is developing a vadose zone closure guidance document for solid waste disposal sites in cooperation with the Vadose Zone Monitoring Laboratory at the University of California at Santa Barbara.

VADOSE ZONE MONITORING PROBLEMS

While the concept of vadose zone monitoring as a ground-water contamination prevention strategy began developing about 20 years ago, the theoretical basis was pioneered earlier by agriculturalists such as Buckingham, Gardner, and Richards (D.R. Nielsen, 1991). Before that time the vadose zone was considered the subsurface hydrologic "no man's land" (Meinzer, 1942). Flanked above the soil zone studied intensively by agriculturalists, and below by aquifers investigated by hydrogeologists, this intermediate zone suffered for many years from a lack of interest by researchers who considered it too unimportant or too complicated for study.

As interest grows, environmental scientists from numerous disciplines are challenged by the intricacies of vadose zone phenomena. Unfortunately, many of these well-qualified professionals simply did not receive the specific training necessary to deal with counterintuitive subsurface flow regimes encountered in the vadose zone. Thus, when confronted with a subsurface monitoring problem, investigators have typically utilized what they were taught in school or on the job: traditional ground-water monitoring.

Lack of vadose zone education is exemplified by a common erroneous criticism of vadose zone monitoring when attempting to collect pore-liquid samples using vacuum lysimeters: "the samplers don't work." Inability to collect a sample does not typically indicate a failure of the instruments to operate properly. The majority of so-called failures are actually a failure to understand the principles governing unsaturated flow in the vadose zone. For example, no vacuum lysimeter can extract a pore-liquid sample from a soil with a matric potential greater than one atmosphere. From a practical standpoint, it is not possible to obtain a pore-liquid sample from most soils at matric potentials above 60 centibars. When the soil is so dry and the unsaturated hydraulic conductivity so low that pore-liquid samples cannot be collected, aqueous liquid migration is very close to nil. For this reason, "don't work" can often be interpreted as "don't worry." Suspect lysimeters can be field tested in situ to determine if they have failed to function properly.

Inherited from the early disinterest in the vadose zone is a regulatory bias for ground-water monitoring, the currently preferred approach to "protecting" aquifers. This regulatory incentive to not detect contamination before it reaches an aquifer has an unfortunate ripple effect in the environmental monitoring instrumentation industry. Manufacturers and suppliers of monitoring equipment respond to market demand and the market is driven by the regulatory environment. Because the demand for vadose zone monitoring equipment remains relatively low, the economic incentive to research, develop, and bring new products to the market also remains low. Thus, the innovative new instrumentation required to develop tomorrow's sophisticated vadose zone monitoring networks is not being brought to the marketplace fast enough. For example, there is a need to reduce monitoring network space and time gaps to detect small or short-lived contaminant transport events. Currently available direct pore-liquid samplers have the drawback of a limited radius of sample collection and require a sampler spacing so close as not to be economically feasible for complete monitoring coverage. More frequent monitoring events and closer spacing of monitored positions in space and time require cost-efficient, indirect monitoring techniques such as automated neutron logging, thus far unavailable.

The vadose zone rarely conforms to layer cake stratigraphy; one is continually faced with the problem of how to fully characterize it. The vadose zone should be considered more than homogeneous overburden above an aquifer. Successful design of a vadose zone monitoring network is predicated upon a solid conceptual understanding of the hydrogeologic conditions at the site to be monitored. In order to adequately understand potential contaminant flow and transport pathways, one is required to have a reasonably close estimate of the hydrologic properties. Vadose zone parameters governing contaminant movement must be systematically measured and/or estimated. While the authors do not propose a "Swiss cheese" approach to a vadose zone characterization drilling program, we recognize that too often consulting scientists are asked to infer hydrologic properties of the vadose zone based on sparse boring logs that consist entirely of soil color, moisture content, and particle size estimations by the feel-and-appearance method rather than measured parameters such as particle size distributions and soil water characteristic curves. Because of inadequate site characterization upon which to base a design, many vadose zone monitoring efforts are doomed from the outset.

WHAT ARE THE SOLUTIONS?

Vadose zone hydrology should be considered an essential topic requiring complete coverage in any college or university hydrologic or hydrogeologic curriculum. Currently active ground-water professionals should thoroughly familiarize themselves with vadose zone flow and transport principles and vadose zone monitoring techniques. The resulting increase in professionals equipped to deal with vadose zone complexities can only improve our ability to protect ground water, prevent contamination, and respond appropriately to regulatory trends that clearly point toward a more aggressive stance with respect to vadose zone monitoring.

While our conceptual knowledge of the vadose zone exceeds our technological ability to measure vadose zone phenomena, new and exciting instrumentation is appearing in the marketplace. The time domain reflectometer (TDR) and the frequency domain capacitance (FDC) probe represent two examples of the kind of bold instrumentation development that is needed to fully characterize and more effectively monitor pore-liquid movement in both the saturated and unsaturated regimes of the vadose zone. The fundamental research upon which both instruments was based was publicly funded and conducted in university programs. Just as in other areas of our economy, government needs to find creative ways to work with the academic and private sectors for the betterment of society as a whole. The EPA, particularly EMSL-Las Vegas, looks to contaminant prevention to play a major role in assuring the quality of our ground-water supplies. Some creative, albeit small in scope, projects are being conducted at universities to solve the hydrologic and engineering problems of automated, early alert monitoring (Kramer et al., 1991). It is now imperative that funding of research programs be encouraged to find solutions to the impediments that prevent us from implementing a vadose zone monitoring strategy to its fullest extent.

Characterization efforts can be improved by utilizing a "tool box" approach which incorporates any number of useful and developing evaluation techniques. These include several borehole geophysical techniques that are very sensitive to textural changes frequently undetectable in boring logs (e.g., Kramer et al., 1992), continuous core recovery, surface geophysical techniques, geologic data analysis such as hydrogeologic facies maps (Anderson, 1989; Poeter and Gaylord, 1990) applied to the vadose zone, and use of geostatistical techniques for parameter estimation (EPA, 1988; David, 1977).

Standards development is needed at the national level to bring continuity to subsurface water data collection and reporting techniques. The American Society of Testing and Materials (ASTM) has been active in pursuing this goal and has historically proved its merit in the fields of geotechnical and soil engineering to assure quality control of investigation techniques and quality assurance of the reported results. Standards applicable to vadose zone investigations are already in place, and more are currently being developed. These standards are developed by consensus of the technical membership and are often referenced in contractual agreements and regulatory statutes and guidelines. Standards specific to vadose zone investigations are developed in the ASTM Subcommittee D-18.21.02 Section on Vadose Zone Investigations and in the Subcommittee D-18.04.02 Task Group on Hydrologic Properties of Unsaturated Soils—Laboratory Methods. Additional input and participation is always welcome.

CONCLUSIONS

Whenever possible, prevention of aquifer contamination should be attempted above the water table in the vadose zone. Increasingly, the economic benefits of contaminant prevention and remediation in the vadose zone are being mandated by regulatory agencies and realized by environmental professionals. The EPA has taken the position on the federal level, and in its oversight of state programs, that emphasis will be placed on early detection and monitoring of potential ground-water contaminants in order to expeditiously control and remediate ground-water contamination. Vadose zone monitoring represents an emerging ground-water investigative and pollution-prevention strategy which provides early warning and detection of potential ground-water contamination. Protection of ground water quantity and quality is logically dependent, in part, upon our ability to exploit the vadose zone as the critical buffer zone between human activities at the surface of the earth and the life-sustaining water supplies in aquifers.

REFERENCES

Anderson, M.P., "Hydrogeologic Facies Models to Delineate Large-Scale Spatial Trends in Glacial and Glaciofluvial Sediments," *Geol. Soc. Am. Bull.*, 101, 501–511 (1989).

Bouwer, H., *Groundwater Hydrology* (New York: McGraw Hill, 1978).

David, M., *Geostatistical Ore Reserve Estimation*, (Amsterdam: Elsevier Scientific Publishing Co., 1977).

Driscoll, F.G., *Groundwater and Wells*, 2nd Edition, (St. Paul, MN: Johnson Division, 1986).

Everett, L.G., *Groundwater Monitoring*, (Schenectady, NY: Genium Publishing Corporation, 1980).

Everett, L.G., and L.G. Wilson, Permit Guidance Manual on Unsaturated Zone Monitoring for Hazardous Waste Land Treatment Units, EPA contract 68–03–3090. USEPA, Environmental Monitoring Systems Laboratory, Las Vegas, NV, 1986.

Everett, L.G., L.G. Wilson, and E.W. Hoylman, *Vadose Zone Monitoring for Hazardous Waste Sites*, (Parkridge, NJ: Noyes Data Corporation, 1984).

Everett, L.G., E.W. Hoylman, L.G. Wilson, and L.G. McMillion, "Constraints and Categories of Vadose Zone Monitoring Devices," *Ground Water Monitoring Review*, 1984.

Everett, L. G., Personal communication (1992).

Germann, P. F., "Approaches to Rapid and Far-Reaching Hydrologic Processes in the Vadose Zone," *J. Contam. Hydrology*, 3:115–127 (1988).

Glass, R.J., T.S. Steenhuis, and J.Y. Parlange, "Wetting Front Instability as a Rapid and Far-Reaching Hydrologic Process in the Vadose Zone," *J. Contam. Hydrology*, 3:207–226 (1988).

Green, W.H., and G.A. Ampt, "Studies on Soil Physics: I. Flow of Air and Water Through Soils," *J. Agric. Sci.*, 4:1–24 (1911).

Kramer, J. H., L.G. Everett, and S.J. Cullen, "Innovative Vadose Zone Monitoring at a Landfill Using the Neutron Probe," in *Proceedings of the Fifth National Outdoor Action Conference on Aquifer Restoration, Ground Water Monitoring and Geophysical Methods*, (Dublin, OH: NWWA, 1991).

Kramer, J. H., S.J. Cullen, and L.G. Everett, "Vadose Zone Monitoring with the Neutron Moisture Probe," *Ground Water Monitoring Review*, 12:177–187 (1992).

Meinzer, O. E., *Hydrology*, (New York, NY: Dover Publications, 1942).

National Research Council, Water Science and Technology Board, December 16, 1990, *Groundwater Newsletter*, Water Information Center, Plainview, NY, 1990.

Nielsen, D. M., ed., *A Practical Handbook of Ground-Water Monitoring*, (Chelsea, MI: Lewis Publishers, Inc., 1991), p. 660.

Nielsen, D. R., D. Shibberu, G. E. Fogg, and D. R. Rolston, A Review of the State of the Art: Predicting Contaminant Transport in the Vadose Zone, Report (90–17 CWP) to the Water Resources Control Board of the State of California. Sacramento, CA, 1990.

Nielsen, D. R., Eighth Memorial Chester C. Kisiel Lecture. Presented March, 1989, "A Challenging Frontier in Hydrology—The Vadose Zone," University of Arizona, Department of Hydrology and Water Resources, Tucson, AZ, 1991.

Poeter, E., and D.R. Gaylord, "Influence of Aquifer Heterogeneity on Aquifer Transport at the Hanford Site," *Ground Water*, 28:900–909 (1990).

U.S. Environmental Protection Agency (EPA), Permit Guidance Manual on Unsaturated Zone Monitoring for Hazardous Waste Land Treatment Units, EPA/530-SW-86-040, Office of Solid Waste and Emergency Response, Washington, DC, 1986.

U.S. Environmental Protection Agency (EPA), GEO-EAS (Geostatistical Environmental Assessment Software) User's Guide, EPA/600/4-88/033a, Washington, DC, 1988.

U.S. Environmental Protection Agency (EPA), Protecting the Nation's Ground Water: EPA's Strategy For the 1990's, Final Report of the EPA Ground-Water Task Force. WH-550G. Office of the Administrator, Washington, DC, 1991.

Wilson, L. G., Monitoring in the Vadose Zone: A Review of Technical Elements and Methods, Environmental Monitoring Systems Laboratory, Office of Research and Technology, USEPA, Las Vegas, NV, EPA-600/7-80-134, 1980, p. 168.

2

EPA's Approach to Vadose Zone Monitoring at RCRA Facilities

Neal D. Durant and Vernon B. Myers

INTRODUCTION

U.S. Environmental Protection Agency (EPA) regulations under the Resource Conservation and Recovery Act (RCRA) have required soil-pore liquid and soil-core monitoring at hazardous waste land treatment facilities since the land disposal regulations were promulgated in 1980 and 1982 (U.S EPA, 1980, 1982). To date, the regulations have not explicitly required vadose zone monitoring for RCRA hazardous waste landfills, surface impoundments, and waste piles. In 1988, however, the EPA proposed to amend the ground-water monitoring regulations to give RCRA permit writers (federal and state regulators) discretion to require vadose zone monitoring at landfills, surface impoundments, and waste piles if such monitoring would aid in the early detection of migrating contaminants (U.S. EPA, 1988).

This chapter examines the evolution of the EPA's position on vadose zone monitoring in the RCRA program, and presents the rationale that led the EPA to advocate the use of vadose zone monitoring at certain land-based RCRA facilities. While the chapter does not offer a critical evaluation on the performance of state-of-the-art vadose zone monitoring techniques, it does contain a discussion of techniques that have been successfully used to detect vadose zone contamination.

This paper first appeared as an article in the Winter 1993 issue of the quarterly journal *Ground Water Monitoring and Remediation*, which holds the copyright. GWMR is published by Ground Water Publishing Company, 6375 Riverside Drive, Dublin, Ohio.

REGULATORY HISTORY OF VADOSE ZONE MONITORING IN THE RCRA PROGRAM

Table 2.1 contains a chronology of EPA regulatory actions concerning vadose zone monitoring. As shown in that table, the EPA proposed in 1978 to require vadose zone monitoring for all RCRA hazardous waste landfills, surface impoundments, and land treatment facilities. The proposal explained that vadose zone monitoring is beneficial because it can potentially provide an early indication that ground water is threatened by leaking hazardous constituents (U.S. EPA, 1978). This early warning is critical because it indicates the need for corrective action before extensive contamination occurs in the saturated zone. In the proposal, the EPA maintained that saturated zone monitoring alone does not sufficiently protect ground-water resources because it doesn't provide the capacity to detect vadose zone contamination that may spread widely before it is detected in the saturated zone. This contamination can act as a continuous source, affecting ground-water quality far into the future. In addition, the proposal noted that vadose zone monitoring would not be required at landfills and surface impoundments if the unit was designed with an effective leak detection system (U.S. EPA, 1978). The EPA recognized that many vadose zone monitoring technologies required further testing and research (U.S. EPA, 1980).

The proposed vadose zone monitoring standards for land treatment facilities were

Table 2.1. Chronology of EPA Actions Concerning Vadose Zone Monitoring in the RCRA Program

Year	EPA Action	Comments
1978	Proposed vadose zone and saturated zone monitoring for all hazardous waste land disposal facilities.	Vadose zone monitoring exemption proposed for landfills and surface impoundments equipped with leak detection systems.
1980	Promulgated requirements for vadose zone monitoring at hazardous waste land treatment facilities only.	Vadose zone monitoring not required at landfills, surface impoundments, and waste piles because monitoring technology required further development.
1986	Issued comprehensive guidance manual on design and operation of vadose zone monitoring networks at land treatment facilities.	Guidance designed to resolve vadose zone monitoring technology difficulties experienced by regulated community.
1988	Proposed to require vadose zone monitoring on a case-by-case basis at hazardous waste landfills, surface impoundments, and waste piles.	Impetus for proposal based on improvements in design and application of vadose zone monitoring technology.

significantly more detailed than those proposed for landfills and surface impoundments because the proposal did not contain requirements for saturated zone monitoring at land treatment facilities. The EPA explained that the environmental performance of land treatment facilities was to be evaluated solely on the basis of soil-core monitoring (i.e., analyzing soil cores from below the treatment zone for the presence of contaminants). The EPA requested public comment, however, on whether saturated zone monitoring would be necessary at land treatment facilities in addition to soil-core monitoring (U.S. EPA, 1980).

Public Comments on 1978 Proposed Regulations

With regard to landfills and surface impoundments, some of the public comments received on the proposal favored vadose zone monitoring in lieu of ground-water monitoring in certain instances. Many commentors, however, argued that it was prohibitively difficult to install vadose zone monitoring systems retroactively at existing landfills and surface impoundments (U.S. EPA, 1980). These commentors felt that saturated zone monitoring alone would provide sufficient ground-water protection.

With regard to land treatment facilities, many commentors argued that saturated zone monitoring was necessary in addition to soil-core monitoring. These commentors contended that soil monitoring had certain limitations (e.g., samples are only representative of quality at a single point, whereas saturated zone monitoring yields samples that are often representative of a larger formation). The EPA received several comments, however, suggesting that saturated zone monitoring was unnecessary in light of soil core monitoring.

Regulations for Landfills and Surface Impoundments

The regulations proposed in 1978 were finalized in two sets, the first set being published on May 1, 1980, and the second being published July 26, 1982 (U.S. EPA, 1980, 1982). These regulations, which contained detailed requirements for saturated zone monitoring, did not require vadose zone monitoring at permitted landfills and surface impoundments. The EPA based this decision, in part, on the technical infeasibility of retroactively installing vadose zone monitoring equipment at landfills and surface impoundments. The EPA felt it was impracticable to require vadose zone monitoring at landfills because monitoring equipment could not be installed through the unit without removing the waste, redesigning the facility, or disturbing the integrity of any liner systems present (U.S. EPA, 1980).

The EPA noted that it would continue to evaluate the appropriateness of requiring vadose zone monitoring at new landfills and surface impoundments (U.S. EPA, 1980). The EPA had not yet tested the utility and feasibility of installing vadose zone monitoring technology at the periphery of landfills and surface impoundments. The EPA also was not convinced that applications of pressure-vacuum lysimeters (pore-liquid samplers), geophysical techniques, and soil-gas sampling to monitoring at the periphery of these units could provide an effective means for monitoring of contaminants migrating horizontally through the vadose zone. Discussion of vadose zone monitoring in the 1978 proposal and the 1980 and 1982 final rules instead focused on monitoring the vertical component of contaminants migrating through the vadose zone.

Regulations for Land Treatment Facilities

The final regulations for land treatment facilities required saturated zone monitoring in addition to soil-core monitoring *and* soil-pore liquid monitoring (i.e., sampling and subsequent analysis of liquids obtained from the vadose zone). Saturated zone monitoring is necessary because vadose zone monitoring can only detect the presence of contaminants migrating through the vadose zone and cannot accurately detect the presence and degree of saturated zone contamination (U.S. EPA, 1980). The EPA maintained that soil-pore liquid monitoring at land facilities is necessary as a backup system to ensure that the absence of a hazardous waste constituent in the soil core sample indicates a true breakdown of the waste rather than merely the rapid migration of the waste material through the soil. In addition, the EPA required soil-pore liquid monitoring at land treatment sites and not landfills or surface impoundments because lysimeters at land treatment facilities can be installed directly into the area where the waste had been applied. The Agency also explained that maintenance and repair of lysimeters would be more feasible at land treatment facilities than at landfills and surface impoundments because of the relatively shallow depth of lysimeter installation at land treatment facilities.

VADOSE ZONE MONITORING AS PRACTICED AT LAND TREATMENT FACILITIES

Soil-core and soil-pore liquid monitoring have been required at RCRA land treatment facilities since the regulations were promulgated in 1982. When the EPA promulgated the regulations, it emphasized that vadose zone monitoring at land treatment facilities is not a substitute for saturated zone monitoring (U.S. EPA, 1982). The Agency explained that because liners are not present at these facilities, vadose zone monitoring serves in a capacity similar to a leak detection system. As indicated above, one purpose of vadose monitoring at this type of facility is to provide data on the success of treatment and potential leaks to the subsurface. The information obtained from this monitoring can be used to adjust the operating conditions (e.g., quantity, types, and frequence of loading) at the unit in order to maximize degradation, transformation, and immobilization of hazardous constituents in the treatment zone. For example, if a significant increase of a hazardous constituent is detected in the vadose zone, the facility operator may examine the waste characteristics and application procedures that significantly affect the mobility and persistence of the constituent.

Soil-core and soil-pore liquid monitoring are vadose zone monitoring techniques that are intended to complement one another. Soil-core monitoring provides information primarily on the movement of less mobile constituents such as heavy metals, whereas soil-pore liquid monitoring can provide data on the migration of more mobile, nonvolatile contaminants. Soil-pore liquid monitoring offers the advantage of monitoring liquid quality at a single point over time without repeated disturbance to the media. Soil-core monitoring, on the other hand, allows for sampling when conditions are too dry to support soil-pore liquid sampling.

Historically, concern has been raised that the required placement of lysimeters on the active portion of a land treatment unit can hinder site operations. The EPA has responded by offering lysimeter installation designs that permit monitoring below the treatment zone without interfering with facility operations (Everett and Wilson,

1986). The EPA has recommended that pressure-vacuum lysimeters be installed beneath the active portion of the treatment unit (the portion receiving waste) using lysimeter access tubes laid in trenches that extend from the active portion of the facility to a sampling station at the edge of the facility where the pressure-vacuum instrumentation can be easily operated (Everett and Wilson, 1986). As shown in Figure 2.1, the access tube houses the pressure-vacuum and pore-liquid discharge lines, protecting them from physical or chemical degradation. The EPA documents *Permit Guidance Manual on Unsaturated Zone Monitoring for Hazardous Waste Land Treatment Units* (Everett and Wilson, 1986), and *Criteria for Selecting Monitoring Devices and Indicator Parameters for Direct Pore-Liquid Sampling of Petroleum Hydrocarbon Contaminated Sites* (Everett et al., 1990) contain detailed instruction on the installation of pressure-vacuum lysimeters at land treatment facilities.

In the past, site owners and operators have also experienced problems with plugging and deterioration of pore-liquid monitoring equipment. To address these problems, the EPA conducted extensive research on lysimeters, (especially pressure-vacuum lysimeters) during the 1980s (Everett and Wilson, 1986; Everett et al., 1984; Everett and McMillion, 1985). It was found that the majority of lysimeter problems experienced at land treatment facilities are the result of improper installation and/or operation (Everett and McMillion, 1985).

One of the more common characteristics of improper installation of pressure-vacuum lysimeters is plugging of the sampling cup. The goals of installing lysimeters properly are to ensure good contact between the sampling cup and the surrounding soil, and to minimize vertical migration of pore-liquid along the annulus of the installation hole. Figure 2.2 presents an EPA-recommended approach to lysimeter installation that is designed to meet these goals. As illustrated in Figure 2.2, a bentonite plug should be placed immediately above the lysimeter to isolate the sampling zone and minimize leakage into the annulus, and the sampling cup should be packed-off using a silica flour slurry. This ensures good hydraulic continuity between the cup and the surrounding soil, and acts as a filter to particulates which may clog the pores of the cup (Everett and Wilson, 1986).

Another common problem the EPA found concerning lysimeter use at land treat-

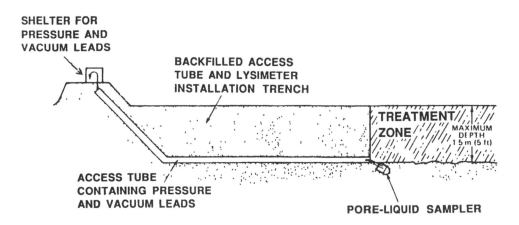

Figure 1.   Profile of a pressure-vacuum lysimeter and access tube installed at a land treatment facility. (Modified from Everett and Wilson, 1986).

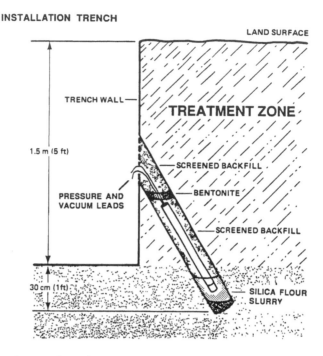

Figure 2. Detailed profile of a pressure-vacuum lysimeter installation within land treatment zone. (Modified from Everett and Wilson, 1986).

ment facilities was that facility operators would often attempt to install and operate pressure-vacuum lysimeters without also monitoring the soil moisture levels. Most ceramic pressure-vacuum lysimeters are only operative at soil moisture levels between 0 and 60 centibars (Everett and McMillion, 1985). In many cases, lysimeters installed at land treatment facilities would not yield samples because the soil was too dry.

In addition, many facility operators installed pressure-vacuum lysimeters with sampling cups constructed of polytetrafluorethylene (PTFE). EPA research has shown that lysimeters made of this material are hydrophobic and only operative at soil moisture levels ranging between 0 and 7 centibars. This research also showed that PTFE lysimeters cannot hold a vacuum greater than 7 centibars (Everett and McMillion, 1985).

Lack of leak testing of lysimeters just prior to installation was also found to be a problem. Unless all leaks in a lysimeter are sealed prior to installation, obtaining a sample will be difficult because the leaks will prevent development of sufficient vacuum for sample collection.

The EPA's experience has shown that if used by technicians who are familiar with the limitations of the technology, vacuum-pressure lysimeters can be an effective means for monitoring soil-pore liquid. A combined subsurface monitoring approach using soil-core monitoring, pore-liquid monitoring, and saturated zone monitoring is necessary to protect ground water, especially at land treatment facilities where the waste might migrate freely from land surface to the vadose zone.

RESEARCH ON VADOSE ZONE MONITORING TECHNOLOGY

The EPA Environmental Systems Monitoring Laboratory (EMSL) has conducted research on vadose zone monitoring strategies and techniques throughout the development of the RCRA land disposal regulations. Although vadose zone monitoring was not required at landfills and surface impoundments in the 1980 and 1982 regulations, EMSL continued to evaluate and develop the feasibility of vadose zone monitoring at these type of facilities. In the early 1980s, EMSL conducted and sponsored comprehensive reviews of vadose zone monitoring techniques (Wilson, 1980; Wilson, 1983; Everett et al., 1983). Research by others outside the EPA began to show that pressure-vacuum lysimeters could be used, in some cases, to successfully detect vadose zone pollution (inorganic contaminants) at the periphery of lined landfills (Johnson et al., 1981).

Since the promulgation of the 1980 and 1982 regulations, the literature has reported an abundance of research on the feasibility of both direct and indirect vadose zone monitoring methods. The EMSL and others have conducted research on soil gas monitoring, surface geophysical methods including complex resistivity, ground-penetrating radar, and downhole geophysical methods such as neutron moderation. Research on these methods indicates that under the appropriate conditions, each method can provide valuable information about the presence and quality of liquids and vapors in the vadose zone.

The EPA has worked closely with the American Society for Testing and Materials (ASTM) toward the development of standard vadose zone monitoring and geophysical monitoring methods. To date, the ASTM Subcommittee on Vadose Zone Monitoring has finalized several standards, including standards for pore-liquid sampling, soil-core sampling, soil-gas monitoring, and neutron moderation applications. Once vadose zone monitoring is implemented in the RCRA program, these standards will become an important component in the design and operation of monitoring networks.

Soil-Gas Monitoring

EPA research in the RCRA, Superfund, and Underground Storage Tank programs has shown that soil-gas monitoring can provide an effective means for detecting and monitoring vapor contamination in the soil gas at sites polluted with volatile organic compounds. Soil-gas monitoring is particularly useful in defining the distribution of volatile chemicals in source-area soils, and can be used to distinguish between soil and ground-water contamination (Marrin and Kerfoot, 1988). In addition, soil-gas sampling can be useful in detecting volatile chemicals in shallow ground water under proper conditions.

One EPA study examined application of active soil-gas monitoring and passive soil-gas sampling techniques for detecting organic contamination at four sites. The results of the soil-gas monitoring were compared to known organic contaminant concentration in the ground water at the sites. The study found good correlation between active soil-gas monitoring and ground-water data, and concluded that given the proper hydrogeologic conditions, active soil-gas sampling can be an effective means for detecting ground water contaminated with organic solvents (Pitchford et al., 1988). The study also found that passive soil-gas monitoring using absorbent charcoal badges (dosimeters) is a less effective means for detecting organic contamination in ground water, but is successful in some cases. Other studies have found, however, that dosimeters for soil gas monitoring can be used at some sites to delineate plumes of certain organic contaminants in ground water (Kerfoot and Mayer, 1988).

Gaseous phase contamination in the vadose zone at some sites may comprise a significant portion of the mass of subsurface contaminant. An EPA-funded investigation at a Superfund site suggests that volatile organic vapors in the vadose zone are the primary source of subsurface contamination at the site. The investigation successfully mapped plumes of carbon tetrachloride and 1,1,1-trichloroethane vapor using active soil gas samples obtained in boreholes at various depths in the vadose zone (Hahne and Thomsen, 1991).

Although soil-gas investigations have been used to effectively delineate volatile organic contamination in the vadose zone and the saturated zone, scientists still debate the quantitative significance of contaminant vapors in soil gas. It has been proposed that soil-gas monitoring be used for qualitative purposes only (Ripp and Robertson, 1990). Whether soil-gas data is used quantitatively or qualitatively, soil and ground-water contamination suggested by soil-gas data should be confirmed by analysis of soil and fluid samples taken at depth (Marrin and Kerfoot, 1988). In the vadose zone, the presence of contamination can be determined from the laboratory analysis of soil cores and soil-pore liquid samples obtained using lysimeters. More detailed discussions of soil-gas sampling and analysis are provided by Devitt et al. (1987), Marrin and Kerfoot (1988), and Kerfoot and Mayer (1988).

Geophysical Methods

The literature contains a growing array of geophysical techniques that may be applied to the problem of detecting vadose zone contaminants at RCRA facilities. However, these techniques only provide an indirect measurement of contamination, and should be used in conjunction with direct techniques such as soil coring and soil-pore liquid sampling.

Electrical resistivity and electromagnetic induction (EM) are techniques that can be used to map electrical conductivity in the subsurface. Contaminants in the subsurface often tend to significantly alter the natural conductivity, and this alteration may be detected using electrical resistivity and EM. The literature contains several examples of waste disposal sites where surface electrical resistivity and EM have been used to detect and delineate ground-water pollution (Mazac et al., 1989; Olhoeft, 1986; and Urish, 1983). Research by Olhoeft (1986) and Walther et al. (1983) has shown that these techniques can also be used to successfully map inorganic contamination within the vadose zone. Electrical resistivity and EM are best suited to use at sites where the pollution shows an elevated level of ionic contaminants (Pitchford et al., 1988; Urish, 1983).

Complex resistivity (also known as induced polarization) measures essentially the same information as electrical resistivity and EM, but also provides information on the presence of active chemical processes in the sediment being tested (Olhoeft, 1985). Pitchford et al. (1988) and Olhoeft (1986) have successfully used this technique to detect organic contamination in the subsurface. Time domain reflectometry, a geophysical technique analogous to induced polarization, is another method that is emerging as an effective means for detecting organic contamination without disturbing the subsurface (Redman et al., 1991; Yong and Hoppe, 1989).

Ground-penetrating radar (GPR) is a geophysical method that entails the surface deployment of radio waves and the subsequent measure of their reflection to map subsurface structure. Research by the EPA and others has found that GPR is capable of detecting hydrocarbon contamination in both the vadose zone and the saturated zone (Pitchford et al., 1988; Olhoeft, 1986; Redman et al., 1991).

Research has shown that neutron moderation can be used to effectively monitor moisture changes in the vadose zone (Wilson, 1981; Wilson and DeCook, 1968). The neutron moderation method is a borehole logging technique that entails the emission of high energy neutrons from a radioactive source. The trajectories of the neutrons are modified by collisions with light atomic nuclei, particularly those in water molecules. This modification, which is detectable by the neutron logger, can be used to locate liquids (including hydrocarbons) in the vadose zone (Kramer et al., 1992).

Brose and Shatz (1987) present a case study of a 30-acre wastewater impoundment facility where neutron logging is used to effectively monitor soil moisture changes to provide leak detection coverage below the facility. The neutron logging system, which was installed prior to construction of the impoundment liner/leak detection system, consists of both horizontal and vertical neutron logging access tubes in order to monitor spatial moisture changes. The literature contains other examples of new waste disposal facilities designed with neutron moderation logging systems that effectively monitor liquids in the vadose zone (Kramer et al., 1991; Unruh et al., 1990).

Neutron moderation can only provide indirect evidence of contamination (moisture changes) in the vadose zone. Soil-gas, soil cores, or pore-liquid samples should be obtained to confirm the presence of contaminants when neutron logging data detects a contaminant moisture front. Neutron moderation has been used beneath a solid waste landfill in conjunction with deep (250+ feet) installation of high pressure-vacuum lysimeters hung from a neutron access casing by Cullen et al. (1991).

VADOSE ZONE MONITORING AT RCRA LANDFILLS, SURFACE IMPOUNDMENTS, AND WASTE PILES

The scientific community's understanding of the uses and limitations of vadose zone monitoring technologies improved significantly during the 1980s. Based on advancement of this technology, the EPA proposed regulatory amendments on July 26, 1988 that would explicitly give RCRA permit writers discretion to require vadose zone monitoring at RCRA hazardous waste landfills, surface impoundments, and waste piles on a case-by-case basis. The proposal is consistent with the EPA's long-standing goal of early detection of leaked contaminants. Vadose zone monitoring can be used to detect contamination close to the source before it reaches the uppermost aquifer. When used in conjunction with saturated zone monitoring, vadose zone monitoring can be used to differentiate between vadose and saturated zone contamination. Early detection is economically advantageous because it provides an early warning for corrective action, thus enabling the facility operator to take action before the plume becomes unmanageable and costly to clean up.

Implementation of Vadose Zone Monitoring Programs

Vadose zone monitoring will likely become an important component in the EPA's effort to characterize and remediate contamination associated with solid waste management units (SWMUs) regulated under the RCRA corrective action program.

Many SWMUs are unlined and would be readily amenable to investigation in the vadose zone immediately beneath the waste.

The EPA also expects vadose zone monitoring will become particularly useful at permitted RCRA facilities in arid portions of the country where the depth to the perennial water table is substantial. The proposal would allow the EPA and the facility operator to select a vadose zone monitoring method that is appropriate for the hydrologic and geologic conditions at a site. For example, if the soil moisture levels at a site were too low to yield a sample to a vacuum-pressure lysimeter, the EPA or the facility operator might select electrical resistivity used in conjunction with soil coring monitoring as the appropriate vadose zone monitoring strategy.

Where possible, vadose zone monitoring should be used in conjunction with saturated zone monitoring to provide an early alert that contamination may be imminent. In many cases, monitoring of the saturated zone without monitoring the vadose zone is illogical because subsurface contamination cannot be detected until the water supply is contaminated. If a vadose zone monitoring system can provide effective monitoring beneath a facility, the EPA may decide that it is appropriate to reduce the scope of saturated zone monitoring at the site. RCRA policy-makers are currently considering a reduction in saturated zone monitoring requirements for such circumstances.

Vadose zone monitoring is not necessary at all RCRA facilities. For example, in hydrogeologic environments where the perennial water table is near the land surface, the vadose zone may be too thin to allow a vadose zone monitoring system. The proposed requirement would grant the EPA discretion to determine if the saturated zone monitoring system at the facility sufficiently meets the goal of early detection of contaminant releases to the subsurface.

Vadose zone monitoring is also appropriate for many facilities equipped with liners and leachate collection systems. Although liner and leachate collection systems may reduce the escape of leachate into the subsurface, EPA experience has found that these systems are often prone to failure. Laboratory studies indicate that the synergistic effects of chemical attack on geomembrane materials, temperature flux in-surface, stress-stretching due to subsidence, animal burrowing, and a host of other degradation mechanisms create a high probability that synthetic liner and leachate collection systems will fail in the long term (Koerner et al., 1990; U.S. EPA, 1983). The majority of studies promoting the efficacy of geomembrane liners only evaluate the short-term effects of single or dual failure mechanisms. Thus, the EPA continues to require saturated zone monitoring at most lined hazardous waste disposal facilities, and may require vadose zone monitoring at these facilities in the future.

Future EPA Action

Since proposing the vadose zone monitoring requirement in 1988, the EPA has reevaluated its approach toward regulating vadose zone monitoring at RCRA facilities. An extensive review of the RCRA statute and existing regulations applicable to hazardous waste facilities has led policy makers to decide that RCRA already provides sufficient authority to require vadose zone monitoring without explicitly amending the regulations. Therefore, the EPA will not promulgate the proposed vadose zone requirement in the near term, and instead will begin to implement vadose zone monitoring programs at facilities on a case-by-case basis.

To facilitate this process, the EPA will develop guidance manuals and policy directives that will prompt permit writers to evaluate the need for vadose zone monitoring on a site-specific basis, and to implement such monitoring where necessary. At present, the EPA is developing a comprehensive guidance manual that will provide RCRA permit writers and facility owners and operators with detailed criteria for selecting, designing, and implementing site-specific vadose zone monitoring programs. The manual will explain sampling procedures, frequency, and sample analysis for a long list of techniques including soil gas methods, pressure-vacuum lysimeters, and neutron moderation. In addition, the manual will provide guidance on actions to be taken in the event that vadose zone contamination is detected. Criteria discussed in the manual will relate closely with the vadose zone monitoring standards being developed by ASTM. As of June 1994, ASTM has finalized standards on soil sampling, pore liquid sampling, soil gas monitoring, tensiometer use, neutron probe use, and measurement of hydraulic conductivity.

SUMMARY

The EPA's proposal to require vadose zone monitoring on a case-by-case basis at RCRA hazardous waste landfills, surface impoundments, and waste piles is based on demonstrated advances in vadose zone monitoring technology, and improved understanding of the benefits derived from its application. EPA and vadose zone monitoring experts have developed a better understanding of the benefits and limitations of technologies such as pressure-vacuum lysimeters, active soil-gas samplers, and neutron probes, and have had the opportunity to evaluate a variety of other vadose zone monitoring methods.

Research has shown that vadose zone monitoring equipment can be used effectively at new facilities, where equipment may be installed directly below the unit prior to construction, and at existing facilities, where equipment installed at the periphery of the unit can be used to detect contamination migrating horizontally in the subsurface. At unlined disposal units, such as those awaiting closure and SWMUs in the corrective action program, vadose zone monitoring networks can be established directly beneath the unit.

While vadose zone monitoring is not appropriate for all RCRA facilities, it could become an integral component of monitoring programs at facilities in areas of the country where the water table is relatively deep. At some facilities where effective vadose zone monitoring systems are installed, it may be appropriate to reduce the scope of saturated zone monitoring efforts.

DISCLAIMER

The contents of this chapter were developed based on EPA policy as articulated in the *Federal Register*, and on the research of the EPA and others. The chapter has not been subjected to the Agency's peer and policy review, and therefore, does not necessarily reflect the views of the Agency.

REFERENCES

Brose, R.J., and R.W. Shatz. "Neutron Monitoring in the Unsaturated Zone," *Proceedings of the First National Outdoor Action Conference on Aquifer Restoration, Ground-Water Monitoring and Geophysical Methods*, National Water Well Association, Dublin, OH, 1987, p. 455.

Cullen, S.J., W.F. Allman, and B.K. Keller. *China Grade Sanitary Landfill: Vadose Zone Monitoring Program, Report to County of Kern, Department of Public Works.* Metcalf & Eddy, Inc. Santa Barbara, CA, 1991.

Devitt, D.A., R.B. Evans, W.A. Jury, T.H. Starks, B. Eklund, and A. Gnolson. "Soil Gas Sensing for Detection and Mapping of Volatile Organics," U.S. EPA Office of Research and Development Report No. EPA/600/8-87/036, 1987.

Everett, L.G., and L.G. McMillion. "Operational Ranges for Suction Lysimeters," *Ground Water Monitoring Review*, 5:51 (1985).

Everett, L.G., E.W. Hoylman, and L.G. Wilson. "Vadose Zone Monitoring at Hazardous Waste Sites," U.S. EPA Office of Research and Development Report No. EPA/600/X-83/064, 1983.

Everett, L.G., E.W. Hoylman, L.G. Wilson, and L.G. McMillion. "Constraints and Categories of Vadose Zone Monitoring Devices," *Ground Water Monitoring Review*, 4:26 (1984).

Everett, L.G., S.J. Cullen, R.G. Fessler, D.W. Dorrance, and L.G. Wilson. "Criteria for Selecting Monitoring Devices and Indicator Parameters for Direct Pore-Liquid Sampling of Petroleum Hydrocarbon Contaminated Sites," U.S. EPA Office of Research and Development, Report No. EPA/600/4-90/035, 1990.

Everett, L.G., and L.G. Wilson. "Permit Guidance Manual on Unsaturated Zone Monitoring for Hazardous Waste Land Treatment Units," Office of Solid Waste Report No. EPA/530-SW-86-040, 87, 1986.

Hahne, T.W., and K.O. Thomsen. "VOC Distribution in Vadose Zone Soil Gas Resulting From a Fumigant Release," in *Proceedings of the Fifth National Outdoor Action Conference on Aquifer Restoration, Ground-Water Monitoring, and Geophysical Methods*, National Water Well Association, Dublin, OH, 1991, p. 699.

Johnson, T.M., K. Cartwright, R.M. Schuller. "Monitoring of Leachate Migration in the Unsaturated Zone in the Vicinity of Sanitary Landfills," *Ground Water Monitoring Review*, 1:55 (1981).

Kerfoot, H.B., and C.L. Mayer. "The Use of Industrial Hygiene Samplers for Soil Gas Measurement," U.S. EPA Office of Research and Development Project Report, 1988.

Koerner, R.M., Y.H. Halse, and A.E. Lord. "Long-Term Durability and Aging of Geomembranes," in *Waste Containment Systems: Construction, Regulation, and Performance. Proceedings of a Symposium at the Amer. Soc. of Civil Eng. National Convention*, San Francisco, CA, 1990, p. 106.

Kramer, J.H., L.G. Everett, and S. Cullen. 1991. "Innovative Vadose Zone Monitoring at Landfill Using the Neutron Probe," in *Proceedings of the Fifth National Outdoor Action Conference on Aquifer Restoration, Ground-Water Monitoring, and Geophysical Methods*, National Water Well Association, Dublin, OH, 1991, p. 135.

Kramer, J.H., S.J. Cullen, and L.G. Everett. "Vadose Zone Monitoring with the Neutron Moisture Probe," *Ground Water Monitoring Review*, 12:177 (1992).

Marrin, D.L., and H.B. Kerfoot. "Soil-Gas Surveying Techniques," *Environ. Sci. Technol.*, 22:740 (1988).

Mazac, O., I. Landa, and W.E. Kelly. "Surface Geoelectrics for the Study of Ground-water Pollution—Survey Design," *J. Hydrol.*, 111:163 (1989).

Olhoeft, G.R. "Direct Detection of Hydrocarbon and Organic Chemicals with Ground Penetrating Radar and Complex Resistivity," in *Proceedings of the National Water Well Association Conference on Hydrocarbons and Organic Chemicals in Ground Water*, National Water Well Association, Dublin, OH, 1986, p. 284.

Olhoeft, G.R. "Low-Frequency Electrical Properties," *Geophysics*, 50:2492 (1985).

Pitchford, A.M., A.T. Mazzella, and K.R. Scarbrough. "Soil-Gas and Geophysical Techniques for Detection of Subsurface Organic Contamination," EPA Office of Research and Development Project Summary No. EPA/600/S4–88/019, 1988, p. 7.

Redman, J.D., B.H. Kueper, A.P. Annan. "Dielectric Stratigraphy of a DNAPL Spill and Implications for Detection with Ground Penetrating Radar," in *Proceedings of the Fifth National Outdoor Action Conference on Aquifer Restoration, Ground-Water Monitoring, and Geophysical Methods*, National Water Well Association, Dublin, OH, 1991, p. 1017.

Ripp, J.A., and C. Robertson. "Soil Gas Sampling Techniques: Problems and Benefits," in *Proceedings: Environmental Research Conference on Groundwater Quality and Waste Disposal*, Electric Power Research Institute, Palo Alto, CA, 1990, p. 13.

U.S. EPA. "Hazardous Waste: Proposed Guidelines and Regulations and Proposal on Identification and Listing," *Federal Register*, 43:58986 (1978).

U.S. EPA. "Hazardous Waste Management System: Standards for Owners and Operators of Hazardous Waste Treatment Storage, and Disposal Facilities," *Federal Register*, 45:33191 (1980).

U.S. EPA. "Hazardous Waste Management System; Permitting Requirements for Land Disposal Facilities," *Federal Register*, 47:32328 (1982).

U.S. EPA. "Synthetic Cap and Liner Systems," Office of Solid Waste draft final report, 1983.

U.S. EPA. "Ground-Water Monitoring at Hazardous Waste Facilities; Proposed Amendment to Rule," *Federal Register*, 53:28160 (1988).

Unruh, M.E., C. Corey, and J.M. Robertson. "Vadose Zone Monitoring By Fast Neutron Thermalization (Neutron Probe)—A 2-Year Case Study," in *Proceedings of the Fourth National Outdoor Action Conference on Aquifer Restoration, Ground-Water Monitoring and Geophysical Methods*, National Water Well Association, Dublin, OH, 1990, p. 431.

Urish, D.W. "The Practical Application of Surface Electrical Resistivity to Detection of Ground-Water Pollution," *Ground Water*, 21:144 (1983).

Walther, E.G., D. LaBrecque, D.D. Weber, R.B. Evans, and J.J. Van Ee. "Study of Subsurface Contamination with Geophysical Monitoring Methods at Henderson, Nevada," *Proceedings of the National Conference on Management of Uncontrolled Hazardous Waste Sites*, Washington, DC, 1983, p. 28.

Wilson, L.G., Monitoring in the Vadose Zone: A Review of Technical Elements and Methods, Environmental Monitoring Systems Laboratory, Office of Research and Technology, USEPA, Las Vegas, NV, EPA-600/7-80-134, 1980, p. 168.

Wilson, L.G. "Monitoring in the Vadose Zone, Part I: Storage Changes," *Ground Water Monitoring Review*, 1:32 (1981).

Wilson, L.G. "Monitoring in the Vadose Zone: Part III." *Ground Water Monitoring Review*, 3:155 (1983).

Wilson, L.G., and K.J. DeCook. "Field Observation on Changes in the Subsurface Water Regime During Influent Seepage in the Santa Cruz River," *Water Resour. Res.*, 4:1219 (1968).

Yong, R.N., and E.J. Hoppe. "Application of Electric Polarization to Contaminant Detection in Soils," *Can. Geotech. J.*, 26:536 (1989).

3

Vadose Zone Hydrogeology in the United States

Stanley N. Davis

INTRODUCTION

In this chapter we will consider briefly the general geologic aspects of the vadose zone in the United States. As discussed in Chapter 1, the vadose zone is a long-neglected portion of the hydrologic system. This neglect has been particularly true of the geologic aspects of the vadose zone. As a consequence, publications summarizing the extent and nature of the vadose zone in different geologic settings are virtually nonexistent. Numerous depth-to-water maps, nevertheless, exist which will suggest, but not define in detail, the thickness of the vadose zone in many regions. Specifics of moisture content, composition of soil gas, chemistry of vadose water, mineralogy of solid particles, velocity of downward moving fluids, and standard hydrogeologic measures such as permeability, porosity, and dispersivity are rarely determined for the vadose zone except for the upper two or three meters which are of direct interest to agriculture.

Owing to several modern developments, the long period of neglect of research into the hydrogeologic aspects of the vadose zone is virtually over. Requirements for subsurface monitoring of liquid as well as gaseous contaminants, the necessity of evaluating hazards from proposed burial of radioactive and other dangerous materials (Evans and Nicholson, 1987), the need for precise estimates of ground-water recharge, requirements for radon surveys, and the desire to understand the geochemical evolution of subsurface water (Domenico and Schwartz, 1990) have all stimulated research with a resultant explosion of knowledge concerning the vadose zone. Even the seemingly esoteric topic of the vadose zone in fractured rock is currently the subject of intense research directed in part toward the evaluation of a proposed repository for radioactive waste in southern Nevada (Evans and Nicholson, 1987). As important and fruitful as modern studies have been, nevertheless, the work has been too scattered and too site-specific to allow more than a few geological generalizations to be made.

Like all attempts to attach labels to natural features, the hypothetical separation of the subsurface into two well-defined hydrogeologic zones called the vadose and phreatic zones commonly causes more confusion than enlightenment. A simple example is shown in Figure 3.1 where no significant geologic complexity exists, yet horizontal correlation of phreatic or vadose zones would be very difficult unless an unusually large number of test holes existed. If faulting, folding, and irregular unconformities are present along with significant fluctuations of the water table, the three-dimensional, time-dependent representation of the vadose zone would be almost impossible to achieve. Despite this pessimistic note, many regions do have extensive and laterally continuous vadose zones. In this chapter, we will start with relatively simple geologic settings and then progress to the impossibly complex.

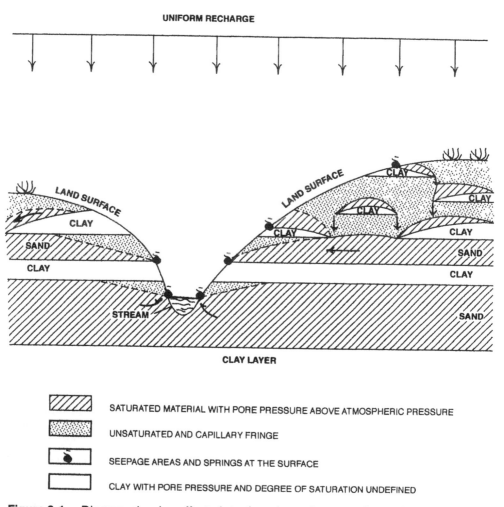

Figure 3.1.   Diagram showing effect of stratigraphy and topography on the configuration of the vadose zone.

PRACTICAL DIFFICULTIES IN RECOGNIZING THE VADOSE ZONE

Most driller's logs of boreholes are not reliable sources of information concerning the vadose zone. Water levels in holes are commonly influenced by large volumes of fluids introduced into the holes by the drilling process. On the other hand, drillers may report extensive "dry" zones in the subsurface below the water table. In deep holes most of the "dry" zones are simply materials of low permeability where little if any ground water flows into the hole. In actual fact, the rocks penetrated are almost always fully saturated and will yield water, but only under very high hydraulic gradients. Nevertheless, a few exceptions exist. Permanently frozen ground in the high latitudes may extend several hundred meters below the surface (Sloan and van Everdingen, 1988). Except for small pockets of brine, these materials are dry with respect to water in the liquid state. Also, thick beds of anhydrite will be dry because fresh and brackish water, if present, would react with the anhydrite to form gypsum under normal conditions of temperature and pressure. Some anhydrite beds are very thick. For example, more than 1,500 meters of anhydrite are reported to be present in the subsurface near Eloy, Arizona (Peirce, 1976). Zones of high subsurface temperatures in regions of volcanic activity may also be dry in the sense that water in the liquid state is absent.

Subsurface geophysical logging can help define the extent of the vadose zone in some regions (Keys and MacCary, 1971; Daily et al., 1992). Resistivity, neutron, and sonic logs have all been used to determine the position of the water table in nonindurated materials, but a number of factors may make geophysical determinations uncertain. With fine-grained materials of high porosity such as silt the problems are made difficult because geophysical properties of 95% saturated silt can rarely be distinguished from those of 100% saturation. On the other end of the porosity scale, such as fractured granite with an effective porosity of 0.05%, small variations in the degree of saturation are even more difficult to quantify.

Surface geophysical methods have also been used to try to map the water table and hence the thickness of the vadose zone. Seismic methods have, perhaps, been the most useful where coarse granular materials are present and where structural and stratigraphic complexities are minimal. The coarse, permeable materials will not retain much moisture in the vadose zone, so seismic velocities will be much slower than below the water table. This contrast allows the use of seismic refraction to locate the water table. Under ideal conditions, electrical resistivity may also be employed to map the water table. Again, coarse sediments which have a well-defined and sharp contrast in water saturation above and below the water are necessary for successful use of resistivity methods.

Information available from petroleum exploration and development can, theoretically, be used to determine the position of the water table. Besides geophysical logging, drill-stem tests, records of losses of drilling fluids, and data from initial seismic exploration could yield some information. Uncertainties are so great, however, that the necessary field and analytical work is rarely undertaken for the sole purpose of locating the water table.

Given enough money, instrumentation developed primarily by soil scientists could probably be adapted to give reliable data on the total thickness of the vadose zone where the water table is below 10 or 20 meters. Such work, however, is expensive and beyond the scope of most routine surveys.

In deep alluvial basins, water levels in water wells are most commonly the only source of information concerning the thickness of the vadose zone. However, water-bearing zones may be effectively confined so that water levels may reflect artesian pressures and have little relationship to the position of the local water table. In other wells, several water-bearing zones may be hydraulically connected with the well so that the so-called static water level which is assumed incorrectly to be the water table is actually some average of the various hydraulic heads in the different water-bearing zones (Figure 3.2).

HYDROGEOLOGIC CONTROLS OF THE VADOSE ZONE

The thickness of the vadose zone probably exceeds 2,000 m in several places in the United States. Excluding geothermal regions, thick sequences of rock containing anhydrite, and other special geologic settings already mentioned, the extreme thicknesses are not extensive and are generally confined to permeable materials in arid and semiarid regions of high relief. In contrast, vast regions which aggregate considerably more than half the land area of the United States had vadose zones less than 50 m thick prior to human development of ground-water resources.

The thickness and distribution of the vadose zone is a function of many hydrologic, topographic, and lithologic variables. Runoff and evapotranspiration generally remove most of the precipitation reaching the land surface. However, the small amount which reaches the water table and becomes ground-water recharge will maintain the elevation of the water table, and hence the amount of recharge will have some complicated inverse relationship to the thickness of the vadose zone.

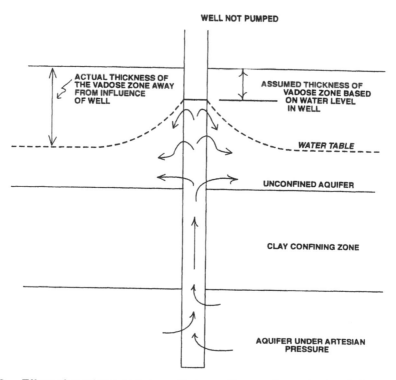

Figure 3.2.   Effect of varying heads in multiple aquifers on the water level observed in a well.

Lithology, which includes the texture and fabric of nonindurated sediments, will control the porosity and permeability of rocks and sediments and their heterogeneity within the water-bearing zones. These properties, in turn, control the ease and rate with which the water table responds to discharge and recharge events in the water-bearing zones. For example, low permeability and high porosities, as found in glacial till, would favor, theoretically, a fairly stable water table, because ground water would not drain rapidly from the system, and a large storage capacity produced by the pore space within the till would accommodate recharge water without causing much increase in the elevation of the water table. Actually, because of permeable fractures in most tills, initial water-table fluctuations in response to recharge are somewhat larger than one might suppose. In contrast with rocks of high porosity and low permeability, a system dominated by low effective porosities but localized zones of high permeabilities, such as in many karst systems, will favor a thick vadose zone which fluctuates rapidly in response to recharge or discharge of ground water.

Topography will help determine the locations of springs and areas of diffuse ground-water discharge which in turn will control the elevation of the water table and the thickness of the vadose zone (Figure 3.3).

THE PROBLEM OF PERCHED WATER

Zones of phreatic water within the vadose zone are called bodies of perched water. They are very important because they greatly complicate the geometry of the vadose zone and can lead to the misinterpretation of field data. Most bodies of perched water are small and involve the storage of a few thousand cubic meters or less of

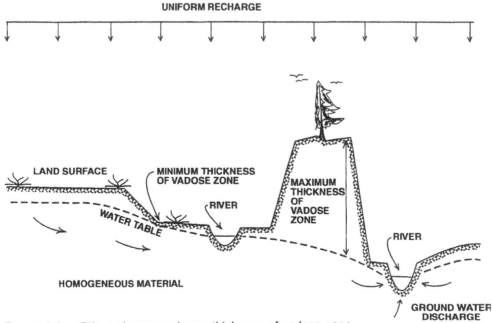

Figure 3.3.   Effect of topography on thickness of vadose zone.

ground water. Also, many of these zones are associated with transient hydrogeologic conditions which persist for only a few days to a few months.

Heterogeneities such as thin layers of clay interbedded with permeable sand can form local perching zones in all types of nonindurated materials. Because permeability is a function of the square of the diameter of the pores which conduct subsurface water, fine-grained clays with pore openings measured in microns will have permeabilities of at least four orders of magnitude less than medium to coarse sand. Stable zones of perched water can, therefore, develop over zones of clay only a few centimeters thick and short-term perched zones over films of clay less than a millimeter thick. Obviously, the identification of such thin layers of clay in the process of drilling a hole would be next to impossible unless a continuous core of the sedimentary material were made.

Human activity, particularly in the arid portions of the United States, has greatly increased the complexity and number of perched zones through, first, a rapid drawdown of ground-water levels due to overpumping local aquifers which leaves water perched on clay layers and, second, through application of excess irrigation water which drains downward through the vadose zone until it piles up on top of zones of low permeability.

One interesting phenomenon associated with larger perched zones is cascading water in nonpumping wells together with a continuous drawing of air into the well (Figure 3.4). Although commonly less than a liter per second, such cascading water may sound like a major waterfall owing to the noise reverberating within the open well casing.

THE VADOSE ZONE IN NONINDURATED GRANULAR MATERIAL

The largest continuous body of nonindurated sediments in the United States underlies the vast coastal plain which extends from Long Island in the northeast, southward along the eastern portion of the country, across the upper Florida peninsula, then westward through Louisiana, and finally southwestward through Texas and ultimately into Mexico (Figure 3.5) (Heath, 1988). Where water levels in wells have not been lowered substantially by pumping, the vadose zone is generally no more than 50 m thick under uplands and less than 1.0 m in swampy lowlands. Perched zones may be present, particularly where water levels have been lowered by excessive pumping. Temporary perching of subsurface water is, of course, present in many areas during periods of active infiltration following heavy rainfall.

The second most extensive and almost continuous body of nonindurated sediments in the United States is made up of till and other closely associated glacial deposits which blanket the northern Midwest and extend eastward into New York and the New England states (Figure 3.5) (Heath, 1988). Till, in general, has a very low hydraulic conductivity, so subsurface drainage is poor. The vadose zone in extensive till plains under natural conditions is rarely more than 20 m thick and is commonly less than 5 m thick. Because of high water tables in many till plains, drainage ditches and tile drains are required to provide a vadose zone of sufficient thickness for general agriculture. Along margins of modern valleys and in terminal moraines, fractures induced by weathering (Williams and Farvolden, 1967) of the till together with lenses of sand and gravel in the till may provide enough permeability to effectively drain the sediments and thereby increase the vadose zone thickness to more than 20 m.

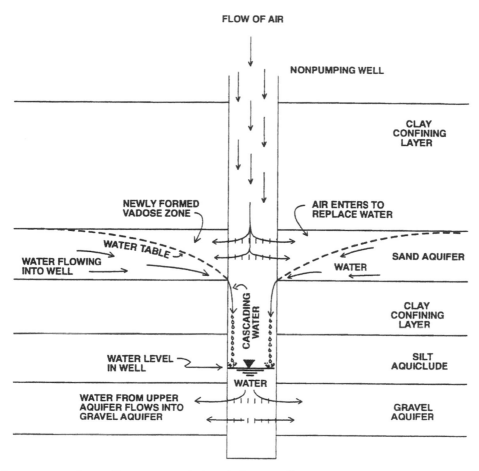

Figure 3.4. Cascading water and air entering a well.

The third largest, more or less continuous body of nonindurated sediments is the High Plains aquifer which is primarily composed of the Ogallala Formation and extends from Wyoming southward into Texas and New Mexico (Figure 3.5) (Heath, 1988). Numerous valleys are incised into the aquifer which induces drainage of ground water with the subsequent thickening of the vadose zone. Prior to extensive pumping of ground water from the aquifer, the vadose zone was generally less than 60 m thick beneath broad extensive uplands and increased to 70 or 80 m along the upper margins of major valleys. Today, extensive overextraction of water from this aquifer has increased the thickness of the vadose zone to more than 100 m in many areas.

The region of western tectonic valleys (Figure 3.5) (Heath, 1988), although made up of almost countless separate sedimentary basins, has many unifying characteristics. Virtually all of the basins are fringed by coalescing alluvial fans which are underlain by heterogeneous mixtures of sand, gravel, silt, and clay. These alluvial deposits interfinger basinward with lacustrine silts and clays. In California and western Nevada some of the basins are cut by active faults which produce ground-water barriers which, prior to artificial ground-water discharge, forced the water to the surface on the upgradient side of the faults. Many of the valleys form closed

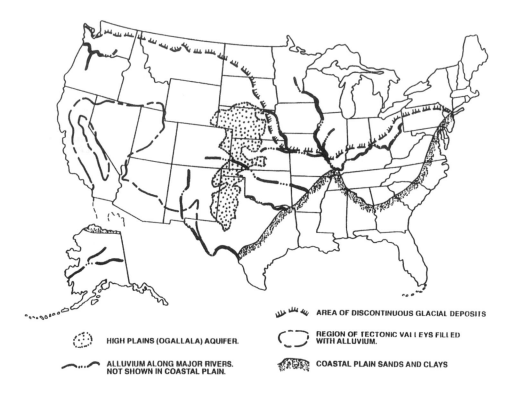

Figure 3.5. Extent of various nonindurated units in the United States. Coastal Plain, glacial deposits, Ogallala Formation (High Plains aquifer), western tectonic valleys, and alluvium along major rivers in the Midwest.

topographic basins without historical surface-water discharge, although some of these closed basins have very effective subsurface drainage from one basin to the next (Eakin, 1966).

Vadose zones within the tectonic valleys have highly variable thicknesses. Some of the tectonic valleys in Nevada that have deep systems of regional ground-water drainage have vadose zones in excess of 300 m in thickness. Vadose zones on the upper slopes of alluvial fans which do not have faults that act as ground-water barriers are commonly more than 100 m thick. In contrast, the vadose zone is less than 1.0 m thick along bands of effluent seepage bordering playas and deeply incised drainage channels.

Major rivers in central United States flow within geologically young valleys which are from 0.5 to 10.0 km wide and filled with 20 to 80 m of nonindurated alluvium (Figure 3.5) (Sharp, 1988). Typically, the alluvium consists of coarse sand and gravel that grades upward into fine-grained silt with some clay. The less permeable silt and clay is commonly 2 to 10 meters thick. Although the total area underlain by alluvium from major rivers is small, its importance can hardly be overemphasized. Municipalities and industries in and near Kansas City, Louisville, St. Louis, and numerous other midwestern cities pump vast amounts of ground water from the alluvial aquifers. Under natural conditions, vadose zones in the floodplains were generally less than 7 m thick except along the banks of very large rivers where ground-water levels are generally adjusted to fluctuating river stages. In places where Pleistocene terraces are present along the rivers, the thickness of the vadose zone may have been as much as 50 m prior to pumping of ground water.

THE VADOSE ZONE IN SANDSTONE AND SHALE

Sandstone and shale are at or near the land surface in broad areas within most major regions within the United States (Figure 3.6) (Heath, 1988). In these rocks, topography, precipitation, rock structure, and rock lithology exert such a complex influence on the geometry of the vadose zone that generalizations which follow have numerous exceptions. This is particularly true in the folded Appalachian rocks where localized zones of perched water caused by lithologic and structural barriers to movement of water make the geometry of the vadose zone almost impossible to define.

Shale below the zone of open fractures is so impermeable that water may not flow into open test holes even after several months. Expensive packer devices to measure pore pressures of fluids are needed to determine the state of saturation of the shale. Such devices are rarely available in routine studies, so the extent of the vadose zone in thick shale formations is largely unknown, although such formations are most commonly assumed to be saturated if water saturates open fractures which normally exist in the upper 10 to 50 m of the shale.

Where the vadose zone can be defined, thicknesses in the eastern part of the United States are usually in the range of 1 to 10 m in valleys and 5 to 50 m at ridge crests and broad uplands. As one moves westward into dry regions, vadose zones

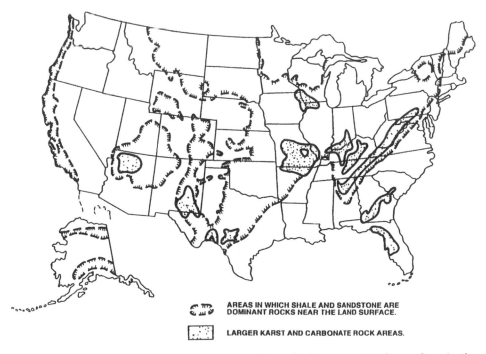

AREAS IN WHICH SHALE AND SANDSTONE ARE DOMINANT ROCKS NEAR THE LAND SURFACE.

LARGER KARST AND CARBONATE ROCK AREAS.

Figure 3.6. Extent of sandstone and shale aquifers which crop out at the surface in the United States. Major karst areas are shown also.

may range in thickness from 10 to 50 m in lowlands and more than 100 m on plateaus and other uplands. Within the region of the Colorado Plateau, vadose zones in upland regions tend to be thicker because of almost negligible ground-water recharge coupled with efficient regional ground-water drainage induced by the deeply dissected topography. These thicknesses are commonly more than 200 m and in places may exceed 500 m, particularly in the region south and east of the Grand Canyon in Arizona.

THE VADOSE ZONE IN CARBONATE ROCKS

Perhaps next to volcanic rocks, carbonate rocks in a karst region (Figure 3.6) present the most complex picture of the vadose zone (Brahana et al., 1988). Most of the complication arises from extreme differences in local permeabilities which may range from less than 10^{-4} darcy for the matrix of a fine-grained, dense limestone to almost infinity for large cavernous solution passages. Numerous caves in karst regions extend far below regional water tables and represent branching, finger-like extensions of the vadose zone into otherwise saturated rocks.

Time-related changes in the vadose zone within carbonate rocks add to the complexity of already complicated geometric features of this zone. Rapid ground-water recharge of carbonate rocks during storms can often cause water levels in solution openings in the rocks to rise more than 10 m during a period of less than an hour. Rises of 30 m are not unknown.

As a consequence of the complex temporal and spatial variations, the characterization of karst hydrology for the purpose of monitoring both the phreatic as well as the vadose zones remains one of the most vexing problems in contaminant hydrology.

THE VADOSE ZONE IN METAMORPHIC AND PLUTONIC IGNEOUS ROCKS

More than 100 different names have been used for the bewildering array of metamorphic and plutonic igneous rocks encountered in nature. Fortunately, most of these rocks can be grouped together as far as their hydrogeologic properties are concerned (Figure 3.7). In general, movement of subsurface water is localized along fractures except in highly weathered rock where water moves between particles of disaggregated rock (Trainer, 1988). Although water-bearing fractures can be found at any depth reached by modern drilling, these fractures are generally less frequent with depth, and their apertures also decrease rapidly with depth. Consequently, from a broad statistical perspective, permeability must also decrease with depth (Figure 3.8). As far as the vadose zone is concerned, it is almost always confined to the weathered zone in the metamorphic and plutonic igneous rocks except in arid regions having significant relief. The thickness of the vadose zone is typically between 2 and 20 meters in humid regions of the Piedmont along the eastern states. In some of the western mountains, dry mines indicate thicknesses in excess of 100 m.

Figure 3.7. Extent of metamorphic and plutonic igneous rocks in the United States.

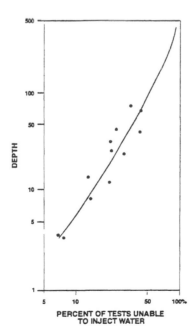

Figure 3.8. Success of water injection as a function of depth in metamorphic and plutonic igneous rocks of the Sierra Nevada, California. Each point which is plotted represents the median of 20 or more individual measurements (Trainer, 1988).

THE VADOSE ZONE IN
VOLCANIC ROCKS

The hydrogeologic complexity of recent volcanic rocks stems from variable types of magma; highly variable mechanisms of intrusion, passive extrusions, and explosive eruptions; and rapid erosion in highlands and deposition in valleys between eruptive events (Stearns and Macdonald, 1946). The resulting complexity nevertheless tends to be erased slowly with time. Porous material will be removed by erosion or be compacted and cemented. Later intrusions of dikes and sills may take place. Most volcanic rocks which are more than about 10^8 years old have hydrogeologic characteristics similar to metamorphic or plutonic igneous rocks.

The understanding of the hydrogeology of young basaltic rocks that originate from volcanic eruptions focuses more on zones of low permeability than on the influence of permeable zones as is the case with metamorphic and plutonic igneous rocks. This focus is owing to the fact that large masses of the young basalt commonly have such high permeabilities that water drains from the rocks rapidly and accumulates only where zones of low permeability such as dikes and buried soils block the drainage. The presence of saturated zones in many thick sequences of young basalt are so scattered that some hydrogeologists picture the system as one with isolated pockets of ground water that are perched within a vast vadose zone. The distribution of these pockets would be determined by geologic vagaries which, due to their complexities, are very difficult to decipher. Some regions have been explored in vain without even finding significant ground water. One of the best known areas is along the flanks of the Humuula Saddle between Mauna Loa and Mauna Kea on the island of Hawaii. Here, available evidence suggests that the vadose zone is at least 2,000 m thick (Stearns and MacDonald, 1946). Elsewhere in the Hawaiian Islands as well as in the western United States, vadose zones in excess of 100 m prevail except in local zones of perched water and next to the seacoast or major rivers and lakes where the vadose zones may be less than 2 m thick.

The hydrogeology of flood basalts in the Northwest (Figure 3.9), although still quite complex, is much less complicated than the hydrogeology of young basalts originating on the flanks of volcanoes. In the Northwest, the individual layers of dense flood basalt are generally 10 to 100 m thick and can extend laterally for distances in excess of 100 km. Vadose zones which originally were from 10 to 100 m thick have been reduced drastically in areas of irrigation because excess irrigation water has rapidly filled narrow fractures in the dense rock. In one year alone, water levels rose 50 m in a newly irrigated area near Moses Lake, Washington. Prior to irrigation, the vadose zone in the region of flood basalts was generally between 20 and 200 m thick.

Volcanic tuff is a granular pyroclastic rock which may be very porous and almost nonindurated except in the central parts of thick beds where welded tuff is common (Winograd, 1971). Fractures in the dense welded portions provide sufficient permeability so that they constitute important aquifers where they are saturated with water. The nonwelded tuff generally has a permeability measured in the millidarcy range or less (Davis, 1969). In mountainous desert areas of the West, the vadose zone may exceed 500 m in thickness in tuff.

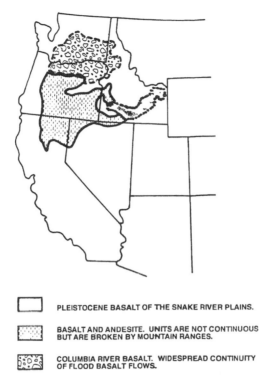

PLEISTOCENE BASALT OF THE SNAKE RIVER PLAINS.

BASALT AND ANDESITE. UNITS ARE NOT CONTINUOUS BUT ARE BROKEN BY MOUNTAIN RANGES.

COLUMBIA RIVER BASALT. WIDESPREAD CONTINUITY OF FLOOD BASALT FLOWS.

Figure 3.9. Extent of basaltic aquifers in northwestern United States.

RESEARCH NEEDED

This brief chapter has been largely restricted to a discussion of some of the broad aspects of the distribution and thickness of the vadose zone in various geologic materials. Owing to a scarcity of information, little more can be said in a general way concerning the scientific aspects of the deeper vadose zones. Nevertheless, I will list some of the research which is needed together with references to a few of the very fine site-specific studies which have begun to fill the void in our generalized hydrogeologic knowledge of the deeper parts of the vadose zone.

The Chemistry of Gases in the Vadose Zone

The gas phase in the vadose zone, particularly the CO_2 content (Domenico and Schwartz, 1990), is one of the most important factors controlling the chemistry of vadose water and, subsequently, ground water which has been recharged by vadose water. Radon, a natural gas, is an item of health concern. Gases from landfills and artificial sources can pose both a threat of explosions as well as a threat to human health.

The Movement of Soil Gases

The rate of movement of soil gases will govern the migration of and subsequent attempts at the removal of noxious gases in the subsurface. Considerable speculation exists concerning large-scale, natural convective movement of gases in thick vadose zones. These large-scale systems would pose obvious questions concerning the artificial isolation of hazardous volatile wastes (Weeks et al., 1982; Weeks, 1987; Massman and Farrier, 1992; Nilson et al., 1992).

The Chemistry of Water in the Vadose Zone

Representative samples of vadose water extracted from deep samples of sediments and rocks are very difficult to obtain. Nevertheless, as already mentioned, the chemical nature of the vadose water must be understood in order to understand the genesis of the chemistry of the underlying ground water (Domenico and Schwartz, 1990). Also, the composition of natural vadose water must be known in order to serve as a baseline against which pollution can be measured. For example, the natural trace organic compounds present in deep vadose water are virtually unstudied.

The Movement of Liquid Water in the Vadose Zone

The importance of very fine-grained materials as well as large gas-filled pores as barriers to unsaturated flow has been long recognized but not carefully measured in deep systems. These barriers, coupled with an uneven availability of water at the land surface, probably means that movement of vadose water downward is highly localized in arid regions and perhaps to some extent even in humid regions. The implications of these highly localized zones are numerous. For one, estimates of deep ground-water recharge based on neutron probe measurements in certain settings in arid regions may be meaningless unless an inordinate number of holes were drilled. Another implication is that water-rock interaction which helps determine water quality may take place in restricted zones which would be leached of gypsum and other minerals but which would not be characterized, except fortuitously, by test drilling.

Nonvertical Movement of Vadose Water

Most geologic models of ground-water recharge picture vadose water as moving vertically downward until it reaches the water table. This must be a misleading oversimplification in many geologic settings. Vadose water can actually build up to saturation on top of an impermeable zone and flow under an almost lateral component of gravity for a long distance. It is important to note that saturation is reached initially with a pore pressure less than atmospheric, so that the water on top of the impermeable zone is not always, technically speaking, perched ground water. The impermeable barrier effect is coupled with the effect of a lateral capillary potential which, due to moisture differences, can act also with a strong lateral component. The direction and velocity of movement of water in a vapor as well as liquid state may also be affected by chemical, thermal, and natural electrical potentials. Effects of the above factors in movement of moisture in shallow vadose systems have been studied for more than 70 years. Very few deep systems have been studied.

Statistical Characterization of Hydrogeologic Properties of the Vadose Zone

Local changes in porosity, permeability, and diffusivity cannot all be quantified and accounted for in modern models describing movement of subsurface water as well as the transport of chemicals. Statistically sound representations of media having equivalent properties must be substituted for the real systems (Durlofsky, 1992; Dykaar and Kitanidis, 1992). Unfortunately, well-controlled studies of deep vadose systems using statistically significant numbers of actual measurements are available for only a few selected areas. Much more work involving actual field measurements is needed. As a simple example, a log-normal distribution of permeability is commonly assumed for given hydrogeologic units. This assumption may be useful for many purposes, but extreme values, which have important regulatory implications, may depart significantly from a log-normal distribution.

Microbiology of the Deep Vadose Zone

As in most of the proposed research topics given above, work with microbiology is well advanced as it relates to the upper 5 meters of the vadose zone. Only during the past 15 years, however, have we seen a widespread scientific interest in the microbiology of the deeper vadose zone (Matthess, 1982). Owing to the complexity of the media, the difficulty of sampling, and the necessity of close collaboration among microbiologists, soil scientists, geologists, and drilling technologists, the work has been slow, albeit very fruitful. The part played by natural microorganisms in the vadose zone in the degradation of organic contaminants is of special current interest (Lee et al., 1987; Hadley and Armstrong, 1991).

CONCLUSION

In conclusion, I will repeat an observation that I made 25 years ago concerning the deep vadose zone. "The cost of drilling and instrumentation, the slow rates of (some) water migration, and the remoteness of some areas that need to be studied all suggest that this topic of investigation will remain a fruitful one for many years, particularly for those with patience, imaginative thinking, and a fat budget" (Davis, 1967).

ACKNOWLEDGMENT

Figures by Helen A. Wilson.

REFERENCES

Brahana, J.V., J. Thrailkill, T. Freeman, and W.C. Ward, "Carbonate Rocks" in *Hydrogeology*, W. Back, J.S. Rosenshein, and P.R. Seaber, eds., Geological Society of America, Vol. 0-2 of the Geology of North America, 1988, pp. 333–352.

Daily, W., A. Ramirez, D. LaBrecque, and Nitao, "Electrical Resistivity Tomography of Vadose Water Movement," *Water Resour. Res.*, 28(5):1429–1442 (1992).

Davis, S.N., "Occurrence of Ground Water in Different Geologic Environments: A General

Statement," in *Proceedings of the National Symposium on Ground-Water Hydrology*, (San Francisco, CA: American Water Resources Association, 1967). pp. 56–71.

Davis, S.N., "Porosity and Permeability of Natural Materials," in *Flow Through Porous Media*, R.M. De Wiest, ed., (New York, Academic Press, 1969), pp. 53–89.

Domenico, P.A., and F.W. Schwartz, *Physical and Chemical Hydrogeology*, (New York, John Wiley & Sons, 1990), pp. 492–498.

Durlofsky, L.J., "Representation of Grid Block Permeability in Coarse Scale Models of Randomly Heterogeneous Porous Media," *Water Resour. Res.*, 28(7):1791–1800 (1992).

Dykaar, B.B., and P.K. Kitanidis, "Determination of the Effective Hydraulic Conductivity for Heterogeneous Porous Media Using a Numerical Spectral Approach, 1. Method," *Water Resour. Res.*, 28(4):1155–1166 (1992).

Eakin, T.A., "A Regional Interbasin Groundwater System in the White River Area, Southeastern Nevada," *Water Resour. Res.*, 2(2):251–271 (1966).

Evans, D.D., and T.J. Nicholson, eds., "Flow and Transport Through Unsaturated Fractured Rock," Geophysical Monograph 42, American Geophysical Union, 1987.

Hadley, P.W., and R. Armstrong, "Where's the Benzene? – Examining California Ground-Water Quality Surveys," *Ground Water*, 21(9):35–40 (1991).

Heath, R.C., "Hydrogeologic Setting of Regions in Hydrogeology," W. Back, J.S. Rosenhein, and P.R. Seaber, eds., Geological Society of America, Vol. 0-2 of the Geology of North America, 1988, pp. 15–23.

Keys, W.S., and L.M. MacCary, "Application of Borehole Geophysics to Water-Resources Investigations," *U.S. Geological Survey Techniques in Water Resources Investigations*, Book 2, Chap. E1, 1971.

Lee, M.D., V.W. Jamison, and R.L. Raymond, "Applicability of In-Situ Bioreclamation as Remedial Action Alternative," *NWWA/API Conference on Petroleum Hydrocarbons and Organic Chemicals in Ground Water – Prevention, Detection, and Restoration, Dublin, Ohio, Proceedings*, National Water Well Association, 1987, pp. 167–185.

Massman, J., and D.F. Farrier, "Effects of Atmospheric Pressures on Gas Transport in the Vadose Zone," *Water Resour. Res.* 28(3):777–791 (1992).

Matthess, G., *The Properties of Groundwater*, (New York:John Wiley & Sons, 1982) pp. 119–139.

Nilson, R.H., E.W. Peterson, K.H. Lie, N.R. Burkhard, and J.R. Hearst, "Atmospheric Pumping: A Mechanism Causing Vertical Transport of Contaminated Gases Through Fractured Permeable Media," *J. Geophys. Res., Earth*, 96(B13):21,933–21,948 (1992).

Peirce, H.W., "Tectonic Significance of Basin and Range Thick Evaporite Deposits," *Arizona Geological Soc. Digest*, 10:325–339 (1976).

Sharp, J.M., Jr., "Alluvial Aquifers Along Major Rivers," in *Hydrogeology*, W. Back, J.S. Rosenshein, and P.R. Seaber, eds., Geological Society of America, Vol. 0-2 of the Geology of North America, 1988, pp. 273–282.

Sloan, C.E., and R.O. van Everdingen, "Region 28, Permafrost Region," in *Hydrogeology*, W. Back, J.W. Rosenshein, and P.R. Seaber, eds., Geological Society of America, Vol. 0-2 of the Geology of North America, 1988, pp. 263–270.

Stearns, H.T., and G.A. Macdonald, Geology and Ground-Water Resources of the Island of Hawaii, Hawaii Division of Hydrography Bulletin 9, 1946.

Trainer, F.W., "Hydrogeology of the Plutonic and Metamorphic Rocks," in *Hydrogeology*, W. Back, J.S. Rosenshein, and P.R. Seaber, eds., Geological Society of America, Vol. 0–2 of the Geology of North America, 1988, pp. 367–380.

Weeks, E.P., D.E. Earp, and G.M. Thompson, "Use of Atmospheric Fluorocarbons F-11 and F-12 to Determine the Diffusion Parameters of the Unsaturated Zone in the Southern High Plains of Texas," *Water Resour. Res.*, 18:1365–1378 (1982).

Weeks, E.P., "Effect of Topography on Gas Flow in Unsaturated Rock: Concepts and Observations," in *Flow and Transport Through Unsaturated Fractured Rock*, D.D. Evans, and T.J. Nicholson, eds., American Geophysical Union, Geophys. Monograph 42, 1987, pp. 165–170.

Williams, R.E., and R.N. Farvolden, "The Influence of Joints on the Movement of Ground Water Through Glacial Till," *J. Hydrology*, 5:163–170 (1967).

Winograd, I.J., "Hydrogeology of Ash-Flow Tuff: A Preliminary Statement," *Water Resour. Res.*, 7(4):994–1006 (1971).

4

Water and Solute Transport and Storage

Peter J. Wierenga

INTRODUCTION

Present interest in the vadose zone is largely a result of concern by the general public about the quality of ground water. People are beginning to understand that drinking water obtained from ground water can easily become polluted, and that great care is needed to prevent such pollution.

Ground water gets polluted by leakage of contaminants from the soil surface to ground water. Examples of pollutants are agricultural chemicals, road salts, and municipal and industrial wastes which are intentionally or accidentally applied to soil from where they leach to the subsoil and into ground water. Contamination of ground water may also occur by contaminants leaking from landfills, disposal sites, leach fields, underground storage tanks, nuclear waste disposal facilities, holding ponds, etc. The rate at which contaminants move to ground water depends on the physical and chemical properties of the soil and of the contaminants, on prevailing rainfall or irrigation rates, and on the depth to ground water. The depth to ground water varies greatly between different regions of the country and the world, from a meter or less to several hundreds of meters or more.

In areas with deep ground water, the unsaturated zone above the water table, also called the vadose zone, may serve as a temporary buffer between the soil surface and ground water. The effectiveness of this buffer in protecting ground water for drinking water, even in areas with deep vadose zones, is often questionable.

In this chapter, we will discuss the basic principles of water storage, the factors affecting water storage, and water and solute movement through unsaturated soil. These principles will be discussed relative to characterization of soil hydraulic properties, and their effects on monitoring water and contaminants in the vadose zone.

0-87371-610-8/95/$0.00 + $.50

WATER STORAGE

Water Content

For most common soils, the soil solid phase occupies from 40% to 70% of the total space. The remaining 60% to 30% is pore space filled with water and air. If the pore space is completely filled with water (no air present), then the soil is called saturated. If the pore space is not completely filled with water, as is the case in the vadose zone, but contains air as well as water, then the soil is said to be unsaturated.

There are two methods to express the status of water in soil. The most common method is to determine the amount, i.e., the volume or the mass of water in soil. The other method is to determine the energy status of water in soil.

The standard method for determining the volume or mass water content of soil is taking samples, drying the samples in an oven, and calculating water contents by differences. (For details see Klute, 1986). Where repetitive measurements are needed, indirect methods may be used advantageously. The most common indirect method for determining water contents in the vadose zone is the neutron probe method. This method involves lowering a radioactive source emitting fast neutrons through an access tube installed in soil. The presence of neutrons slowed by hydrogen is counted, and is an indirect measure of the soil-water content (Klute, 1986). Newer methods for measuring soil water in the field employ the dielectric constant of water which is much larger (i.e., 60–80) than the dielectric constant of dry soil (i.e., 2–5). Most promising among these is the Time Domain Reflectometry method (TDR) which measures the soil-water content in layers of 0–5 cm thickness (Brisco et al., 1992). For a useful review of the TDR for measuring water content see Reeves and Elgezawi (1992).

Energy Relationships

Knowing the amount of water in soil is often not enough. For example, the water content by itself gives no indication whether a soil is saturated or unsaturated. Water content also provides no measure of availability for plant uptake, and is of no value in determining the direction of water flow. For many purposes percent saturation, i.e., the volumetric water content divided by the water content at saturation, is a more useful parameter. However, a complete evaluation of water in soil requires knowledge of the energy status of water in soil.

The energy status of water is expressed through the soil-water potential. The total soil-water potential is formally defined as the mechanical work required to transfer water from soil to a standard reference state, where the total soil-water potential is zero by definition. The units of soil-water potential are those of the reference state. In soil science and hydrology, soil-water potential is mostly expressed in energy/volume (J/m^3; bar) or energy/weight (head, cm or m) (Jury et al., 1991).

The total soil-water potential (ψ_T, in energy/volume units; H, in head units) is more easily understood if it is broken down into its component potentials. For the vadose zone three important component potentials are: gravitational potential ψ_z (with energy/volume units) or z (with head units); solute potential ψ_s (energy/volume units); and (tensiometer) pressure potential ψ_p (energy/volume units) or h (head units).

The gravitational potential is the potential resulting from the gravitational force field. The solute potential results from the difference in potential energy between soil solution and pure free water. Soil solution contains a variety of dissolved minerals and thus its energy status is decreased relative to pure water. For example, a soil whose soil solution contains 1000 ppm of dissolved salts (not uncommon) has a solute potential of about −0.6 bar.

The tensiometric pressure potential (or pressure potential, ψ_p) results from the capillary and other forces in unsaturated soils. Its value is always negative in unsaturated soils, and mostly expressed in bars, cm, or equivalent units. In saturated soils capillary forces are zero, but water below the water table surface is subject to pressure from the overlying water. This latter hydrostatic pressure is often expressed in head units, and is also called pressure head.

Because solute potential does not contribute much as a driving force for water flow (except at the soil-root interface where solute potential differences are most important for water uptake into roots), its contribution is often ignored. Adding the two remaining potentials, and expressing these in terms of head units, one obtains:

$$H = h + z \tag{1}$$

Thus in unsaturated soil the total pressure head H equals the sum of the pressure head (h) and gravity or elevation head (z). The pressure potential or pressure head in unsaturated soils is always negative, becomes zero at the water table and positive below the water table. The absolute value of the negative pressure potential in unsaturated soils is also called tension.

There is a relationship between the amount of water held in soil and its energy status. This relationship is characteristic for each soil and is called soil-water characteristic curve or soil-water retention curve. Examples of water retention curves for a loamy fine sand and for a clay loam (Hills et al., 1989) are shown in Figure 4.1. It shows that as soil dries its tension increases (pressure-potential becomes more negative). Differences between the two curves in Figure 4.1 are significant. The clay loam has much higher water contents at saturation, and at intermediate tensions. The water retention curve for the fine sand is very steep at tensions above 10^3 cm.

Two points on the water retention curve are of special interest to soil scientists and hydrologists. They are Field Capacity (FC) and Permanent Wilting Point (PWP). Field capacity is the water content of soil after two days of redistribution following a heavy infiltration of water. In the older literature, field capacity is often defined as

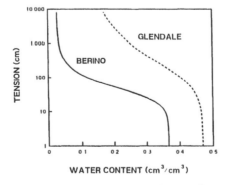

Figure 4.1. Soil-water retention curves for Berino loamy fine sand and Glendale clay loam.

the water held at $^1/_3$ bar tension potential. In reality, careful measurements of tension in well-watered field soils show that the tension at FC varies from 0.04 to 0.06 bar in coarse textured soils, and from 0.06 to 0.1 bar in fine textured soils. The permanent wilting point is the soil-water condition at which movement of soil water to plant roots is too slow to keep up with transpiration losses, and plants do not recover upon wilting. The permanent wilting point is different for different soils and is generally taken around 15 bar, but may vary from 10 to 40 bar tension. Its actual value is of little significance because the water available from soil with a tension greater than 10 bar (see Figure 4.1) is usually small, although of sufficient magnitude for survival of many drought resistant plants.

Water retention curves are frequently determined in the laboratory on intact cores or using disturbed soil samples (Klute, 1986). The methodology allows for starting with wet or dry soil. Unfortunately, this makes a difference, and the final retention curves are different depending on whether the procedure was started with wet or dry soil. This behavior is called hysteresis in the water retention curve (Figure 4.2). In field studies, hysteresis is often ignored because water retention data, with hysteresis measured in the field, is generally not available.

Water storage in soils is often estimated by subtracting the water contents at PWP from the water contents at FC, and multiplying by the depth of interest. For the soils in Figure 4.1, the water stored in a 1 m deep profile is given in Table 4.1. It should be noted that field capacity moisture content values available from the literature often do not represent field measured values, but rather laboratory values which may be too low.

Water retention curves can be described by a number of empirical relationships (Brooks and Corey, 1964; van Genuchten, 1980). Recently the van Genuchten approach has found wide usage in soil and hydrological studies.

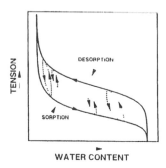

Figure 4.2. Soil-water retention curves with hysteresis. The dotted lines represent scanning loops.

Table 4.1. Volumetric Water Content (cm³/cm³) and Water Storage (cm per 100 cm soil profile)

	FC (0.1 Bar)	PWP (15 Bar)	Storage (cm H₂O)
Berino fine sand	0.116	0.027	8.9
Glendale clay loam	0.404	0.156	24.8

The relationship is

$$S_e(h) = \frac{\theta - \theta_r}{\theta_s - \theta_r} = [1 + (\alpha h)^n]^{-m} \qquad (2)$$

where S_e is the effective saturation, θ is the volumetric water content ($cm^3 cm^{-3}$), θ_r is the residual water content ($cm^3 cm^{-3}$), θ_s is the saturated water content ($cm^3 cm^{-3}$), and h is the tension; α (cm^{-1}), n (dimensionless), and m (dimensionless) are empirical parameters (van Genuchten, 1980). This function fits observed data well for a wide range of soils (van Genuchten and Nielsen, 1985). The curves in Figure 4.1 were calculated with Equation 2, using parameter values presented in Table 4.2 (Hills et al., 1989).

WATER MOVEMENT

Steady Flow

Water flow through soil may be viewed microscopically by looking at water flow through individual pores, or macroscopically by looking at flow through the entire pore body. This latter approach is the more common.

Macroscopic flow of water through soil is described with Darcy's equation

$$q = -K(\theta) \frac{\partial H}{\partial z} \qquad (3)$$

In this equation q is the Darcy flux or volume of water flowing through a unit surface area per unit time ($cm^3/cm^2 day$), $K(\theta)$ is the hydraulic conductivity (cm/day), H the total pressure head (cm) and z is distance (cm). For saturated soils the hydraulic conductivity is assumed constant.

Hydraulic Conductivity

In unsaturated soils, the hydraulic conductivity decreases sharply with decreasing water content. This is because as soil loses water (drains) pores empty, and the cross sectional area for flow decreases. Thus, for soils that drain rapidly such as sands, the hydraulic conductivity decreases much more sharply with decreasing water content than for clay soils in which there are more intermediate size pores. These drain more slowly, and thus the hydraulic conductivity of clay soils decreases less sharply with decreasing water content. An example of changes in hydraulic conductivity with tension for the two soils of Table 4.2 is shown in Figure 4.3. Because water content is related to soil-water tension through the soil-water retention curve (Figure

Table 4.2.   van Genuchten Parameters for Berino Loamy Fine Sand and Glendale Clay Loam (It is assumed m = 1 − 1/n)

	θ_r cm^3/cm^3	θ_s cm^3/cm^3	α cm^{-1}	n –	K_s cm/day
Berino	0.027	0.366	0.028	2.239	541
Glendale	0.106	0.469	0.010	1.395	13

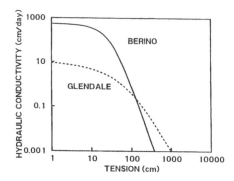

Figure 4.3.   Hydraulic conductivity versus tension for Berino loamy fine sand and Glendale clay loam.

4.1), the hydraulic conductivity is often shown as a function of soil-water tension. Note the exponential decrease in hydraulic conductivity with increasing tension.

There have been many efforts to relate the unsaturated hydraulic conductivity to water content, pressure head, or tension (Gardner, 1958; Brooks and Corey, 1964; Campbell, 1974; van Genuchten, 1980). Because of its simplicity, the Gardner equation has been used quite frequently,

$$K(h) = K_{sat}e^{\alpha h} \tag{4}$$

where K_{sat} is the saturated hydraulic conductivity (cm/day), h pressure head (cm), and α (cm^{-1}) a parameter, equal to the slope of the unsaturated log conductivity function. Values for K_{sat} and α must be determined by fitting Equation 4 to measured values of unsaturated conductivity. By plotting ln K(h) versus h one obtains α and K_{sat} from the slope and intercept, respectively. The empirical value of K_{sat} so obtained is not necessarily equal to the measured saturated conductivity. In practice, the log-linear relationship of Equation 4 is most applicable to the moist soil-water content range (e.g., $-100 < h(cm) < 0$).

A more versatile equation was proposed by van Genuchten (1980). The equation can be derived from the water content-pressure head relationship presented by Equation 2. The equation is:

$$K(S_e) = K_{sat}(S_e)^{\gamma}[1 - (1 - S_e^{1/m})^m]^2 \tag{5}$$

The symbols in Equation 5 were defined before, and are the same as for Equation 2. The parameter γ is often taken to be 0.5 (Mualem 1976), although Hudson (1992) found that γ was greater than 2 for the Berino soil depicted in Figure 4.1. For Equation 5 to be valid it was also assumed that m = 1–1/n. The solid and dashed lines in Figure 4.3 were computed with Equation 5, using parameter values from Table 4.2.

Equations 4 and 5 show a strong dependence of K(h) and K(θ) on tension and water content, respectively. As a result of this dependency, a relatively small decrease in water content may reduce the hydraulic conductivity 10 to 100 fold, or more. This fact is important to consider when one wants to extract soil solution from an unsaturated soil. A high vacuum applied suddenly to a porous sampler may dry out the soil immediately around the sampler, and strongly reduce flow into the sampler.

Transient Flow

Equation 3 is satisfactory for steady flow through soil. For transient conditions, i.e., when water contents or pressure heads change over time, Equation 3 needs to be combined with the equation of continuity. This equation is:

$$\frac{\partial \theta}{\partial t} = - \frac{\partial q}{\partial z} \tag{6}$$

where t is time (days or seconds) and z is elevation (cm). The other symbols are as defined before. Combining Equations 3 and 6 yields the Richards equation:

$$\frac{\partial \theta}{\partial t} = \frac{\partial}{\partial z} \left[K(h) \left(\frac{\partial h}{\partial z} + 1 \right) \right] \tag{7}$$

Equation 7 is for vertical flow without sources or sinks (plant roots). By defining the slope of the water retention function as the water capacity function,

$$C(h) = \frac{d\theta}{dh} \tag{8}$$

one may write Equation 7 in its pressure head equivalent. Thus:

$$C(h) \frac{\partial h}{\partial t} = \frac{\partial}{\partial z} \left[K(h) \left(\frac{\partial h}{\partial z} + 1 \right) \right] \tag{9}$$

Analytical solutions of Equations 7 and 9 are available for some well defined conditions. However, given the complexity of most practical situations (variable upper boundary conditions, layered soils, uptake of water by roots), numerical solutions are used most commonly. Such solutions are available for one, two, or three dimensions, with several of these coupled to solute transport models. A large number of the models have been described in the literature, but unfortunately, very few well-documented models are available in the public domain. Examples of public domain models are SWIM (CSIRO, Div. of Soils; Private Mail Bag, P.P. Aitkenvale, Queensland, Australia 4814), HYDRUS (Kool and van Genuchten, 1991, U.S. Salinity Laboratory, 4500 Glenwood Drive, Riverside, CA 92502) and SWMS (Šimůnek et al., 1994).

Infiltration

The rate at which water enters soil during rainfall or irrigation events has long been the subject of theoretical analyses and field and laboratory studies.

For many years the following equation was used, i.e.,

$$i = At^{-1/2} \tag{10}$$

where i is infiltration rate (cm/day), A is a parameter and t is time (day). Following a series of theoretical analyses, Philip (1969) suggested

$$i = St^{-1/2} + Kt \qquad (11)$$

The first part on the right-hand side of Equation 11 represents early time infiltration while the second part represents long time infiltration behavior.

Equation 11 was derived by Philip (1958) by solving Equation 7 for given boundary conditions. Therefore, the parameters S and K can be computed from known hydraulic conductivity and water retention functions (Haverkamp et al., 1977). However, the procedure is cumbersome, and also requires knowledge of K(h) and θ(h) for the soil of interest. Because determining K(h) and θ(h) is time-consuming, it is often more efficient to determine S and K in Equation 11 directly by matching the equation to measured infiltration data.

Infiltration of water into soil may also be computed with a numerical solution of Equation 7. An example of computed water contents during infiltration of 10 cm of water into a 6-m deep soil column of Berino loamy fine sand (Table 4.2) is presented in Figure 4.4. The example is presented for two initial water contents, e.g., 0.08 and 0.16 cm³/cm³, respectively. HYDRUS (Kool and van Genuchten, 1991) was used for the calculations. Evapotranspiration was assumed to be zero. The results clearly demonstrate the effects of initial water content on the rate at which water moves downward. The soil with the higher initial water content (0.16 cm³/cm³) has a higher hydraulic conductivity and lower storage capacity for the infiltrating water, causing the water front to advance much more rapidly. Thus, if one needs to monitor subsoil moisture movement, one may expect much more rapid responses following rain in moist soil as opposed to dry soil. This is true even though the initial infiltration rate into dry soil is higher than in moist soil.

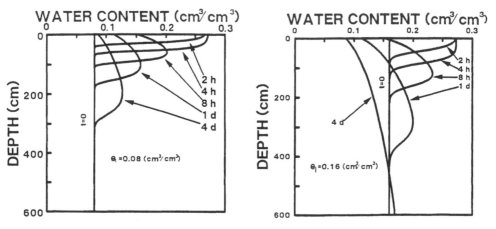

Figure 4.4.   Calculated water content distributions during and after infiltration of 10 cm H₂O in Berino loamy fine sand, at initial water contents of 0.08 cm³/cm³ (left half) and 0.16 cm³/cm³ (right half), respectively. The 10 cm water was added over a 4 hour period.

Redistribution

Redistribution of water in soil commences when infiltration stops. Figures 4.4 and 4.5 show computed water-content distributions. Calculations were again made assuming no evaporation or root-water uptake. Note that decreases in water content are initially most pronounced near the soil surface. As time goes on, water contents decrease over the full length of the profile. Note that even though the initial water contents were quite different for the two simulations (8% and 16%), the water content distributions after 50 and 100 days are practically identical.

The amounts of water draining from the 6-m profiles are plotted versus time in Figure 4.6. This figure again demonstrates the importance of initial water contents

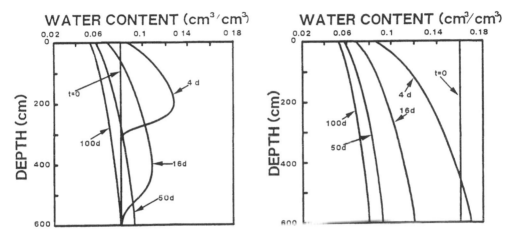

Figure 4.5. Calculated water content distributions after infiltration of 10 cm H$_2$O in Berino loamy fine sand, at initial water contents of 0.08 cm^3/cm^3 (left half) and 0.16 cm^3/cm^3 (right half), respectively.

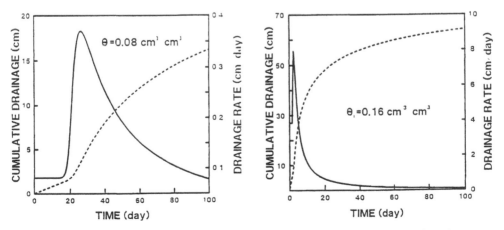

Figure 4.6. Drainage rates (solid lines) and cumulative drainage (dashed lines) versus time after 10 cm water infiltration. Berino loamy fine sand, at initial water contents of 0.08 cm^3/cm^3 (left half) and 0.16 cm^3/cm^3 (right half), respectively.

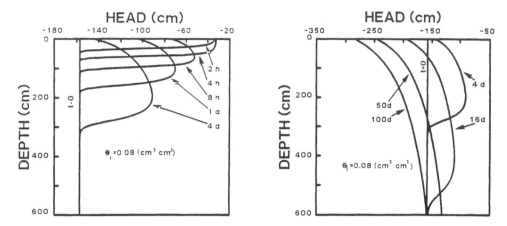

Figure 4.7. Pressure head distributions before and after irrigation with 10 cm water. Berino loamy fine sand at initial water content of 0.08 cm³/cm³.

on percolation of water to deeper depths. Soils at greater initial water content will yield more recharge quicker than drier soils for similar conditions. Of course, evaporation and root-water uptake may greatly reduce the rates and amounts of recharge. After 100 days, the cumulative drainage from the 0.16 cm³/cm³ profile is 64 cm versus 16 cm for the 0.08 cm³/cm³ profile or a difference of 48 cm. This difference is entirely due to the greater initial water storage in the wet versus dry soil (600 × 0.08 = 48 cm). In Figure 4.7, the pressure head distributions were plotted before, during, and after infiltration of 10 cm water. Note that gradients in pressure head in the moist subsoil are not large, even though a significant amount of water passes through. For much of the time the pressure gradients are close to one and do not change much. However, even though pressure head gradients are relatively constant, this does not imply absence of percolation. Downward flow can still be significant.

In terms of monitoring it is important to note that accurate determination of the hydraulic gradient is necessary for calculation of recharge with Equation 3, if $K(\theta)$ is known, or for calculation of $K(\theta)$ if q(t) is known. Pressure heads should be measured with a resolution of 1 mbar or 1 cm H_2O. A resolution of 1 cbar (10 cm H_2O) gives an error of at least 10% over a 1 m depth interval.

FLOW AND TRANSPORT

Equations

Solute transport through soil is the result of three simultaneous processes, e.g., convective or mass transport, diffusion, and dispersion. Convective transport (J_m) refers to the passive movement of a dissolved solute with water. The equation for convective transport in one direction is (van Genuchten and Wierenga, 1986):

$$J_m = qC \tag{12}$$

where q is the Darcy flux and C is the solution concentration.

Diffusive transport (J_D) results from the natural thermal motion of dissolved ions and molecules. This process can be described with Fick's law of diffusion as follows:

$$J_D = -\theta D_m \frac{\partial C}{\partial x} \tag{13}$$

where θ is the volumetric water content of the medium, D_m the porous medium diffusion coefficient, and x the distance. The porous medium diffusion coefficient, D_m, is always smaller than the diffusion coefficient in water as a result of the tortuous nature of the porous medium.

The porous nature of soils and vadose zone materials results from the existence of pores, fissures, wormholes and other features, as well as from the loose stacking of the soil material itself. The different shapes, sizes, and orientation of the open spaces and pores result in fluid velocities which differ from place to place. In addition, fluid velocities differ in an individual pore, based on where the velocity is determined with respect to the pore wall. Velocities are lowest near the pore wall. The velocity variations cause solute to be transported at different rates, leading to mixing, macroscopically similar to molecular diffusion. This mechanical spreading by local velocity variations is called mechanical dispersion (Bear, 1972). The process can be described by an equation similar to Equation 13, where J_h is dispersive transport, and D_h is the mechanical dispersion coefficient. This coefficient is strongly, and often linearly, related to fluid velocity. Thus:

$$J_h = -\theta D_h \frac{\partial C}{\partial z} \tag{14}$$

$$D_h = \alpha \, v \tag{15}$$

where v is the average pore-water velocity. The coefficient is frequently referred to as dispersivity.

Because D_m and D_h are macroscopically similar, these terms are usually added. Thus,

$$D = D_m + D_h \tag{16}$$

where D is the longitudinal hydrodynamic dispersion coefficient (Bear, 1972).

Combining Equations 12, 13, 14, and 16 yields the solute flux J_s:

$$J_s = -\theta D \frac{\partial C}{\partial z} + qC \tag{17}$$

The equation of continuity for soil solution is:

$$\frac{\partial \theta C}{\partial t} = \frac{-\partial J_s}{\partial z} \tag{18}$$

where t is time. Combining Equations 17 and 18 yields the general transport equation for non-sorbing chemicals in one dimension:

$$\frac{\partial \theta C}{\partial t} = \frac{\partial}{\partial z} \left[\left(\theta D \frac{\partial C}{\partial z} \right) - qC \right] \tag{19}$$

Many chemicals partially sorb to soil with the remainder staying in solution. The relationship between sorbed and solution concentrations is:

$$S = K_f C^n \qquad (20)$$

where S is the sorbed concentration and C is the solution concentration, K_f is the Freundlich partition coefficient, and n is a dimensionless parameter. In many cases n may be assumed to be 1. For sorbing chemicals the equation of continuity, 18, is now written as:

$$\frac{\partial}{\partial t} (\theta C + \rho S) = \frac{-\partial J_s}{\partial z} \qquad (21)$$

where ρ is the soil bulk density. Combining Equations 17 and 21 yields:

$$\frac{\partial}{\partial t} (\theta C + \rho S) = \frac{\partial}{\partial z} \left(\theta D \frac{\partial C}{\partial z} - qC \right) \qquad (22)$$

Differentiation of (20), with n = 1, and substitution of the resulting equation in (22), eventually yields:

$$R \frac{\partial \theta C}{\partial t} = \frac{\partial}{\partial z} \left(\theta D \frac{\partial C}{\partial z} - qC \right) \qquad (23)$$

where R is the retardation factor equal to:

$$R = 1 + \frac{\rho K_f}{\theta} \qquad (24)$$

Equation 23 is the general, one-dimensional convective-dispersive equation (CDE) for a sorbing chemical. Defining pore water velocity v as q/θ, and assuming a uniform and constant water content, Equation 23 may be rearranged as follows:

$$\frac{\partial C}{\partial t} = \frac{D}{R} \frac{\partial^2 C}{\partial z^2} - \frac{v}{R} \frac{\partial C}{\partial z} \qquad (25)$$

Equation 25 shows that solute transport is driven by diffusion and dispersion (the first term on the right-hand side of Equation 25), and mass transport (the second term on the right-hand side of Equation 25). When pore water velocity is high, mass transport is dominating the process, but when velocities are low, diffusion becomes increasingly important.

Equation 25 also shows that the retardation factor can be defined in terms of relative velocities of water and solute. The retardation factor is the distance traveled by a solute front when R = 1, divided by the distance traveled by the solute front when R has the assigned value. Or, more simply, R = velocity of water divided by velocity of tracer. Except for negatively charged ions such as chloride or bromide, R values are generally greater than 1.

The flux or velocity need to be known before Equation 23 can be used. If q or v change over time as, for example, during infiltration of water into an unsaturated soil, the equation for water flow (Equation 7 or 9) needs to be solved along with Equation 23. This is most conveniently accomplished numerically, using one of several available computer models (Kool and van Genuchten, 1991).

Transport in Uniform Soil: Steady Flow

Figure 4.8 is an example of solute flow through soil irrigated at a constant rate of 2.06 cm/day (Porro et al., 1993). A 6-m deep by 1-m diameter column was filled with alternate 20-cm thick layers of Berino loamy fine sand and Glendale clay loam. Water was added for several months at a slow but steady rate of 2.06 cm/day. The percentages of saturation once steady flow was reached were approximately 43% for the sand and 81% for the clay loam. Tritiated water was added with the water during the first 11 days of the experiment only. The symbols in Figure 4.8 are measured relative tritium concentration distributions at the various soil depths. The solid lines were obtained by fitting an analytical solution of Equation 23 with $R = 1$, $\theta = 0.31$ and $q = 2.06$ to the data. Measured and fitted data in Figure 4.8 show what is typically observed in these situations, i.e., a decrease in the peak concentration and a spreading of the pulse as the tritiated water moves down the 6 m deep layered soil profile. The curves were predicted with the same values for q, θ, and R for the entire column. Using constant values for q, θ, and R resulted in excellent fits between measured and observed curves at some depths (e.g., 2.05 m) but poorer fits at other depths (e.g., 5.0 m). This is the result of the variability of soil physical properties inside the column. Thus, even though the soil was hand packed with great care, there still is considerable variation in transport with depth. Even greater variation may be expected under field conditions. Except for the 5 m depth, the analytical solution of Equation 23 described the arrival times of the solute fronts quite well, indicating that simplified analytical solutions may be used for first-order estimates of arrival times. However, as is shown in the next example, a good estimate of the retardation factor is most important for such predictions.

The second example of computed solute transport with steady flow of water is shown in Figure 4.9. In this case, water was added continuously to a soil profile at a

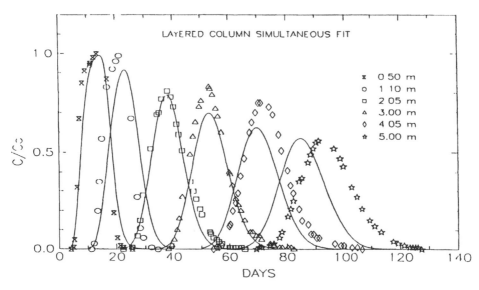

Figure 4.8. Relative tritium concentration distributions in a layered soil column. The solid lines were obtained by fitting an analytical solution of Equation 25 to the data.

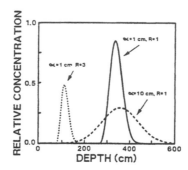

Figure 4.9. Predicted solute concentration/depth distributions for different combinations of dispersivity and retardation factor (R). Flux 3.74 cm/day; water content 0.16 cm^3/cm^3; tension 70 cm H_2O; time 15 days.

rate of 3.74 cm/day, but solute was added during the first two days only. The figure shows the relative solution concentrations with depth after 15 days for two dispersivity values (1 and 10 cm) and two retardation values (1 and 3). The effect of chemical interaction (retardation) on solute transport is quite clear. A retardation factor of 3 instead of 1 significantly delays the downward movement. Thus, high retardation factors cause solute to move through soil or vadose zone material at much reduced rates. An R value of 3 represents a K_f value of 0.41 cm^3/g, assuming a bulk density of 1.4 and a water content of 20% by volume. R values greater than 3, and in many cases much greater than 3, are common for many of the contaminants introduced in the environment. Thus, when monitoring for solutes in the subsoil, one can expect arrival of noninteracting chemicals (R = 1) much before arrival of sorbing chemicals. This difference in arrival times can be very large, i.e., years. Figure 4.9 also shows that dispersivity, α, mostly affects the shape of the concentration/depth distribution, but not the location of the peak concentration. Increased dispersivity values cause greater spreading and decrease the peak concentration, but they barely affect the depth of the maximum concentration.

Transport in Uniform Soil: Transient Flow

Water flow is seldom constant in the field, but changes over time due to variations in boundary conditions (e.g., rainfall, irrigation events). This has a significant effect on solute movement which occurs rapidly during periods of fast water flow. When the rate of water movement is slow, such as during the latter phases of the redistribution process (see Figure 4.5), solute transport is mainly by diffusion, and very slow. Thus in the absence of significant water movement, solutes tend to stay where they are.

Figure 4.10 shows the advance of a solute front through the 6 m profile of Figures 4.4 and 4.5. No chemical interaction was assumed between the solute and the soil (i.e., R = 1). Calculations were again made with **HYDRUS**, and the physical properties of the soil were the same as before (Table 4.2). The distributions of water and solute in Figure 4.10 show that the solute fronts lag behind the water fronts. The magnitude of this lag is dependent on the initial water present in the soil, being greater for soil having a higher initial water content. The model results suggest that the initial water ("old water") is pushed ahead by the traced water ("new water").

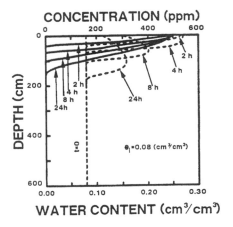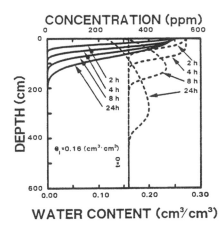

Figure 4.10. Water content (dashed lines) and solute concentration (solid lines) distributions in Berino loamy fine sand irrigated with 10 cm water in 4 hours. Initial water contents were 0.08 cm^3/cm^3 (left figure) and 0.16 cm^3/cm^3 (right figure), respectively.

This phenomenon has been observed in several studies, and is most clearly demonstrated in a recent paper by Porro and Wierenga (1993). These authors applied water containing tritium and bromide to a 6 m deep column. They measured the advance of the wetting front with tensiometers and a neutron probe. The advance of the solute fronts was determined by taking soil solution samples which were analyzed in the laboratory. Figure 4.11 shows the water and solute fronts passing by the 4 m depth. As the water front passed by the 4 m depth on day 18, the water content increased (dashed line), while the soil water tension decreased simultaneously (solid line). However, it took nearly nine more days for bromide, and fourteen more days for tritium to arrive at this depth. This implies that the "old water," containing no tracers, but clearly changing the tension and water content readings at 400 cm, is pushed ahead of the "new water." This has significant implications for monitoring, in that water content monitoring can serve as an early warning signal for contaminant flow. Arrival of a contaminant front will be preceded by increases in water content or decreases in tension at the monitoring depth unless there is preferential or bypass flow. Thus, measuring water contents and/or tensions below disposal areas, tanks, ponds, etc. does serve a useful purpose when monitoring for solutes. Note that the arrival of bromide ahead of tritium at 4 m is due to anion exclusion. Bromide ions are negatively charged. They are repelled by the negatively charged soil particles, and move in the center of soil pores where velocities are greatest. This causes the bromide front to move ahead of the tritium front.

Transport in Field Soils: Heterogeneity

The solute transport examples shown in Figures 4.8 through 4.11 represent close to ideal conditions, such as may be found in carefully packed laboratory or field size columns.

In the field, conditions are not nearly as ideal and one may expect different behavior. This is often the result of heterogeneity in soil properties and of preferential flow.

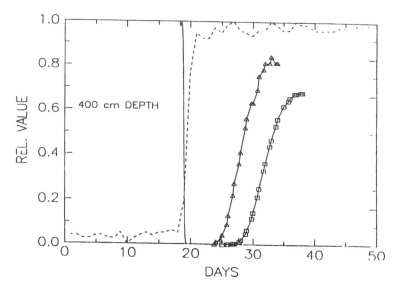

Figure 4.11. Tritiated water (squares), bromide (triangles), water content (dashes), and soil moisture tension (solid line) changes at the 400 cm depth following irrigation of a 6 m deep column with traced water.

One of the main reasons for deviations from ideal transport behavior in the field are heterogeneities in the physical, chemical, and biological properties of subsurface material. Soils and geologic materials are inherently nonuniform. This is a direct result of the formation of these materials and their exposure to various events during past geologic times. Thus, properties which affect solute transport can be expected to vary spatially. This is particularly true for soil hydraulic properties, such as the unsaturated hydraulic conductivity and the water retention properties.

Many studies have been conducted to determine the variability of hydraulic properties in the field. Examples are studies by Nielsen et al., 1973; Warrick and Nielsen, 1980; Gajem et al., 1981; and Nash et al., 1989. Because the main driving force for solute transport in the field is convection (see Equation 23), variation in hydraulic properties and thus convection has a large effect on contaminant transport in the subsoil.

Jury et al. (1987) reviewed chemical parameters in addition to hydraulic parameters affecting solute transport. While these authors paid special attention to dispersion coefficients, less is known about spatial variability of other chemical parameters such as adsorption coefficients, or retardation factors. Even less is known about field heterogeneity of microbial properties.

Heterogeneity occurs laterally as well as vertically. However, soils tend to be more similar in the horizontal direction than in the vertical direction. In the horizontal plane, changes in properties occur more gradual. Extensive layering is observed at most sites, and this causes sudden changes in texture and hydraulic properties. It is not clear how this layering affects solute transport in the vadose zone. However, one would expect that local differences in vertical fluid velocities are evened out by the layering. The result may be that over longer vertical distances, across a number of distinct horizons, solute transport patterns are beginning to look more like those observed in uniform soils. Some evidence for this was shown experimentally by

Porro and Wierenga (1993) who measured dispersivities during unsaturated transport through 6 m deep columns. They found that the dispersivity in a column uniformly filled with soil was twice as high as in a layered column. Similar observations were made by Roth et al. (1991). The implication is that heterogeneity in the vertical direction may have an averaging effect on solute transport in the vadose zone, and may allow for application of simplified models based on Equation 23 for prediction purposes. The parameters in these models would have to represent the average behavior, such as average monthly or annual recharge for q, and an average retardation factor R calculated from values measured for each individual horizon (See Roth et al., 1991).

Transport in Field Soils: Preferential Flow

Solute transport in the vadose zone can also be affected by the structure of the soil or geologic material. In structured soils and in soils containing root holes, worm holes, cracks, and other large pores, water and solute may move preferentially through these pathways. This phenomenon is variously called macropore flow, short-circuiting, bypassing, preferential flow, partial displacement, or transport with "mobile" and "immobile" domains (Brusseau and Rao, 1990).

Preferential flow is a well documented phenomenon known for over 100 years, and many publications, mainly in the soil sciences, deal with it (Beven and Germann, 1982; Elrick and French, 1966; Biggar and Nielsen, 1961). A recent publication by the American Society of Agricultural Engineers summarizes the role of preferential flow on ground water quality, processes affecting it, methods of measuring, and modeling for predictions (ASAE, 1991). According to Beven (1990) the important characteristic of preferential flow is that during wetting, local wetting fronts propagate into the soil to significant depths, thus bypassing intervening matrix pore space. The result is movement of water and solutes to greater depths at faster speeds than would be predicted on the basis of Equation 23. However, at the same time, there is diffusive mass transfer of solutes from the preferential pathways into the surrounding soil matrix and vice versa. The result is that in structured soil one observes early breakthrough and tailing (van Genuchten and Wierenga, 1976). In the field, breakthrough curves may take a variety of forms with early or delayed breakthrough common, as well as extensive tailing (van de Pol et al., 1977; Steenhuis and Parlange, 1990). Although field observations on preferential flow are well documented for relatively shallow soils, the importance of preferential flow on transport through deep unsaturated profiles has not been documented. Because of layering in deep unsaturated profiles, preferential flow may be less important in such profiles than in surface soils.

A great deal of attention has recently been paid to modeling transport through soils with preferential flow. In most models, the porous medium is thought to consist of two domains; a "mobile" domain, where solute transport occurs by advection and dispersion, and an "immobile" domain in which diffusion is dominant (Brusseau and Rao, 1990; Sardin et al., 1991). Transfer of solutes between the two domains is usually presented as a diffusive mass transfer. Details of such modeling efforts can be found in the relevant literature.

For the purpose of this chapter, it suffices to say that as the understanding of the physics, chemistry, and biology of vadose zone transport grows, models will get increasingly complex, containing many more parameters which will be difficult to evaluate independently. At the same time, averaging processes in field soils may allow the use of relatively simple models, with only a few bulk parameters. No matter which route is taken, it is important that physical, chemical, and biological aspects of solute transport are taken into account. This applies to predictions of solute ransport through unsaturated soils, as well as to characterization of vadose zone materials.

ACKNOWLEDGMENTS

The author wants to acknowledge Mr. Lehua Pan, Research Assistant, Department of Soil and Water Science, University of Arizona, who performed the Hydrus modeling and prepared the figures (except Figures 4.8 and 4.11).

REFERENCES

American Society of Agricultural Engineers, *Preferential Flow*, Proceedings of the National Symposium, Chicago, IL, 1991, p. 408.

Bear, J., *Dynamics of Fluids in Porous Media*, (New York: American Elsevier Publishing Co., 1972).

Beven, K., "Modeling Preferential Flow: An Uncertain Future?" in *Proceedings of the National Symposium on Preferential Flow*, T.J. Gish, ed., (St. Joseph, MI: American Society of Agricultural Engineers, 1990), p. 1–11.

Beven, K. and P. Germann, "Macropores and Water Flow in Soil," *Water Resour. Res.*, 18:1311–1325 (1982).

Biggar, J.W., and D.R. Nielsen, "Miscible Displacement: II. Behavior of Tracers," *Soil Sci. Soc. Am. Proc.*, 125 (1961).

Brisco, B., T.J. Pultz, R.J. Brown, G.C. Topp, M.A. Hares, and W.D. Zebchuk, "Soil Moisture Measurement Using Portable Dielectric Probes and Time Domain Reflectometry," *Water Resour. Res.*, 28:1339–1346 (1992).

Brooks, R.H., and A.T. Corey, "Hydraulic Properties of Porous Media," Hydrol. Paper No. 3, Colorado State University, Fort Collins, 1964.

Brusseau, M.L., and P.S.C. Rao, "Modeling Solute Transport in Structured Soils. A Review," *Geoderma*, 46:169–192 (1990).

Campbell, G.S., "A Simple Method for Determining Unsaturated Hydraulic Conductivity from Moisture Retention Data," *Soil Sci.*, 1117:311–314 (1974).

Elrick, D.E., and L.K. French, "Miscible Displacement Patterns on Disturbed and Undisturbed Soil Cores," *Soil Sci. Soc. Amer. Proc.*, 30:153 (1966).

Gajem, Y.M., A.W. Warrick, and D.E. Myers, "Spatial Dependence of Physical Properties of a Typic Torrifluvential Soil," *Soil Sci. Soc. Amer. J.*, 45:709–715 (1981).

Gardner, W.R., "Some Steady-State Solutions of the Unsaturated Moisture Flow Equation with Application to Evaporation from a Water Table," *Soil Sci.*, 85:228–232 (1958).

Haverkamp, R., M. Vauclin, J. Touma, P.J. Wierenga, and G. Vachaud, "A Comparison of Numerical Simulation Models for One-Dimensional Infiltration," *Soil Sci. Soc. Amer. J.*, 41:285–293 (1977).

Hills, R.G., I. Porro, D.B. Hudson, and P.J. Wierenga, "Modelling One-Dimensional Infiltration into Very Dry Soils. Part 1. Water Content Versus Pressure Head Based Algorithms," *Water Resour. Res.*, 25:1259–1269 (1989).

Hudson, D.B., A Transient Method for Estimating the Hydraulic Properties of Dry Unsaturated Soil Cores, unpublished PhD thesis, New Mexico State University, Las Cruces, NM, 1992.

Jury, W.A., W.R. Gardner, and W.H. Gardner, *Soil Physics*, 5th ed. (New York: John Wiley and Sons, Inc., 1991), p. 328.

Jury, W.A., D. Russo, G. Sposito, and H. Elabd, *Hilgardia*, 55:1–32 (1987).

Kool, J.B., and M. T. van Genuchten, Hydrus: One-Dimensional Variably Saturated Flow and Transport Model, Including Hysteresis and Root Water Uptake, U.S. Salinity Laboratory, Riverside, CA, 1991.

Klute, A., Methods of Soil Analysis, Part 1. Monograph 9, American Society of Agronomy, Madison, WI, 1986.

Mualem, Y., "A New Model for Predicting the Hydraulic Conductivity of Unsaturated Porous Media," *Water Resour. Res.*, 12:513–522 (1976).

Nash, M.S., P.J. Wierenga, and A. Butler-Nance, "Variation in Tension, Water-Content, and Drainage Rate Along a 91 m Transect," *Soil Sci.*, 148:94–101 (1989).

Nielsen, D.R., J.W. Biggar, and K.T. Erh, "Spatial Variability of Field-Measured Soil-Water Properties," *Hilgardia*, 42:215–260 (1973).

Philip, J.R., "The Theory of Infiltration," *Soil Sci.*, 85:278–286 (1958).

Philip, J.R., "Theory of Infiltration," in *Adv. in Hydrosci.*, V.T. Chow, ed., (New York: Academic Press, 1969), pp. 215-305.

Porro, I., and P.J. Wierenga, "Transient and Steady-State Solute Transport Through a Large Unsaturated Soil Column," *Ground Water*, 31:193-200 (1993).

Porro, I., P.J. Wierenga, and R.G. Hills, "Solute Transport Through Large Uniform and Layered Soil Columns," *Water Resour. Res.*, 29:1321-1330 (1993).

Reeves, T.L., and S.M. Elgezawi, "Time Domain Reflectometry for Measuring Volumetric Water Content in Processed Oil Shale Waste," *Water Resour. Res.*, 28:769-776 (1992).

Roth, K., W.A. Jury, H. Fluhler, and W. Attinger, "Transport of Chloride Through an Unsaturated Field Soil," *Water Resour. Res.*, 27:2533-2541 (1991).

Sardin, M., D. Schweich, F.L. Ley, and M.T. van Genuchten, "Modeling the Nonequilibrium Transport of Linearly Interacting Solutes in Porous Media. A Review," *Water Resour. Res.*, 27:2287-2307 (1991).

Šimůnek, J., T. Vogel, and M.T. van Genuchten. The SWMS-2D Code for Simulating Water Flow and Solute Transport in Two-Dimensional Variably Saturated Media. V.1.2. Research Report No. 132, U.S. Salinity Laboratory, USDA, ARS, Riverside, CA, 1994, p. 196.

Steenhuis, T.S., and J.Y. Parlange, "Preferential Flow in Structured and Sandy Soils," in *Proceedings of the National Symposium on Preferential Flow*, T.J. Gish, ed., American Society of Agricultural Engineers, St. Joseph, MI, 1990, pp. 11-21.

van de Pol, R.M., P.J. Wierenga, and D.R. Nielsen, "Solute Movement in a Field Soil," *Soil Sci. Soc. Am. J.*, 41:10-13 (1977).

van Genuchten, M.T., and D.R. Nielsen, "On Describing and Predicting the Hydraulic Properties of Unsaturated Soils," *Annals Geophysicae*, 3:616-628 (1985).

van Genuchten, M.T., and P.J. Wierenga, "Mass Transfer Studies in Sorbing Porous Media: Analytical Solutions," *Soil Sci. Soc. Am. J.*, 40:473–480 (1976).

van Genuchten, M.T., "A Closed-Form Equation for Predicting the Hydraulic Conductivity of Unsaturated Soils," *Soil Sci. Soc. Am. J.*, 44:892–898 (1980).

van Genuchten, M.T., and P.J. Wierenga, "Solute Dispersion Coefficients and Retardation Factors," in *Methods of Soil Analysis, Part 1. Physical and Mineralogical Methods*, A. Klute, ed., Agron. Monograph No. 9, 1986, pp. 1025–1054.

Warrick, A.W., and D.R. Nielsen, "Spatial Variability of Soil Physical Properties in the Field," in *Applications of Soil Physics*, D. Hillel, ed., (New York: Academic Press, 1980), pp. 319–343.

Preferential Flow in Structured and Sandy Soils: Consequences for Modeling and Monitoring

Tammo S. Steenhuis, J.-Yves Parlange, and Sunnie A. Aburime

INTRODUCTION

Preferential flow is the process whereby water and solutes move by preferred pathways through a porous medium (Helling and Gish, 1991). During preferential flow, local wetting fronts may propagate to considerable depths in a soil profile, essentially bypassing the matrix pore space (Beven, 1991). Under such conditions, classical methods, such as the convective-dispersive equation, for quantifying flow of water and solutes in uniform soils, may not be valid. Additionally, the presence of local flow channels complicates the problem of locating monitoring devices in the vadose zone. For example, random installation of suction lysimeters, used as point samplers of pore liquids, may result in some units being located within macropores, while other units are within the soil matrix. Kung et al. (1991) demonstrated that the concentrations of water-soluble constituents collected from macropores differ from concentrations within the soil matrix. Accordingly, by randomly installing sampling units, measurements of pore liquid chemistry will be highly variable and not a true representation of the spatial average (Kung et al., 1991).

Although the term preferential flow does not imply any given mechanism (Helling and Gish, 1991), it usually refers to macropore flow, fingering (unstable flow), and funneled flow. Macropore flow involves flow through noncapillary cracks or channels within a profile, reflecting soil structure, root decay, wormholes, ant tunnels, etc. Soil structure refers to the aggregation of the textural units into blocks. A well-

Partly based on a paper in *Preferential Flow*, T.J. Gish and A. Shirmohammadi, Eds., American Society of Agricultural Engineers, St. Joseph, MI, 1991, pp. 12–21. Used by permission of publisher.

structured soil has two distinct flow regions for liquids applied at the land surface:
(1) through the cracks between blocks (interpedal flow), and (2) through the finer
pore sequences inside the blocks (intrapedal, or matrix flow). Simpson and Cunningham (1982) have shown that rapid flow of wastewater through interpedal cracks
may lessen the renovating capacity of the soil because of reduced surface area and
contact time. Similarly, because of the rapid flushing of pollutants through larger
interconnected soil pores, the movement of such pollutants into the finer pores of
the soil blocks may be limited.

Fingering occurs as a result of wetting front instability. Fingering may cause water
and solutes to move in columnar structures through the vadose zone at velocities
approaching saturated pore velocity (Glass et al., 1988). Fingering may occur for a
number of reasons, including changes in hydraulic conductivity with depth and
compression of air ahead of the wetting front (Helling and Gish, 1991). Laboratory
soil-column studies by Glass et al. (1988) presented a qualitative insight into fingering during unstable flow. These studies were conducted to evaluate the effect of
initial water content and its nonuniformity on wetting front stability in a two-
layered system with a fine layer overlaying a coarse layer. Glass et al. (1988) summarized the results of their laboratory studies as follows:

> First, the textural interface provides a very strong perturbation, discretizing
> the uniform flow at a number of discrete point sources. Second, under dry
> initial conditions, fingers that form have a dramatic moisture content
> structure—very wet inside the finger and very dry outside the finger. Third,
> over time, slow sideways diffusion of moisture from finger core areas takes
> place resulting in a steady flow field with finger core areas persisting and
> continuing to conduct most of the flow. Fourth, on subsequent infiltration
> cycles, flow remains concentrated in the same core areas as first delineated
> fingers, thus emphasizing the importance of slight variation of initial moisture
> content between past fringe and core areas on the heterogeneous moisture
> movement and the persistence of fingers. Fifth, when the initial moisture
> content is artificially uniform and high, fingers widen and can coalesce giving
> the appearance of an almost uniform flow field but with horizontal variation
> in flux still apparent.

Funneled flow occurs when sloping geological layers cause pore water to flow
laterally, accumulating at a low region. If the underlaying region is coarser, finger
flow may also occur.

The goals of this section are to describe the limitations of using the classical
convective-dispersive approach for predicting solute movement when preferential
flow occurs; to illustrate preferential flow in structured and sandy soils; and to
present case studies illustrating the effect of preferential flow during vadose zone
monitoring.

PREDICTING SOLUTE DISTRIBUTION USING THE
CONVECTIVE-DISPERSIVE APPROACH

The approach used by many soil scientists to simulate water and solute flow in
unsaturated soils is based on a theory by Richards (1931). van der Molen (1956)
combined aspects of this model with the theory of dispersive movement to predict

the course of desalinization of land in the Netherlands that had been inundated by seawater. The resulting convective-dispersive equation assumes that water and solutes follow an average path through the soil — which is to say that a given molecule, starting at the surface, is equally likely to follow any one of a multitude of available paths.

Typical of models based on the convective-dispersive equation is MOUSE (Model of Underground Solute Evaluation) (Steenhuis et al., 1987), which was developed in an attempt to predict pesticide movement in soils. It assumes an average flow path for water moving through the soil, and includes a statistical description of soil heterogeneities. MOUSE and similar models have proved quite satisfactory for predicting the overall flow of water and the transport of nitrates, as can be seen from Figures 5.1 and 5.2. In Figure 5.1, the observed and predicted recharge amounts are compared for a sandy Long Island soil (Steenhuis and van der Molen, 1986). In this case, the simplistic Thornthwaite-Mather water budget procedure was used, resulting in a remarkably good fit, without any calibration, using only measured parameters. The nitrate concentration in the tile line for a waste application site was predicted with the Cornell Nutrient Model (the predecessor of MOUSE) (Steenhuis and Walter, 1980). Again, a reasonable fit is in evidence, but clearly not as good as for the water flow (Figure 5.2). In contrast, the predictions of pesticide movement with the convective-dispersive equation, except for disturbed soil columns (Figure 5.3), fail to represent the observed values. In Figure 5.4, the aldicarb concentrations of two deep 32 m long cores are compared to simulated values of the MOUSE model. The cores were taken on Long Island for a similar soil and

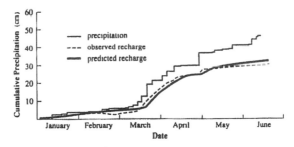

Figure 5.1.   Comparison of predicted and observed recharge for a grass covered plot near Riverhead on Long Island.

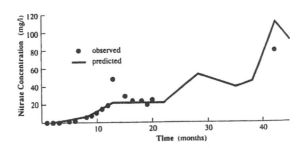

Figure 5.2.   Comparison of predicted and observed nitrate concentrations in tile lines for a waste application site in Illinois.

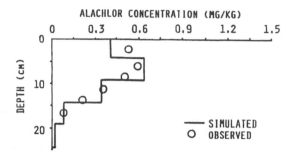

Figure 5.3.   Comparison of predicted and observed distribution of alachlor in laboratory columns.

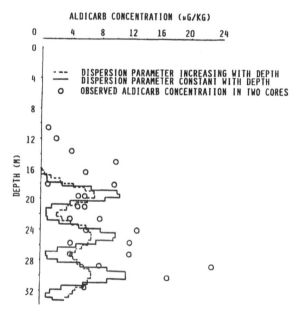

Figure 5.4.   Comparison of simulated and measured aldicarb concentrations for a potato field near Wading River, Long Island.

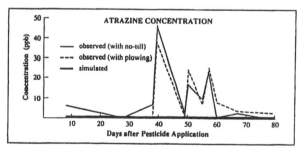

Figure 5.5.   Comparison of predicted and observed atrazine concentrations in artificial tile lines in Northern New York.

near the site where we were able to predict the recharge so well. However, unlike the recharge case, the fit between predicted and observed values remained poor, even after calibration. Figure 5.5 shows an even larger discrepancy between observed and predicted values obtained for atrazine concentration in tile lines for a sandy loam in northern New York. The atrazine was detected in the tile water within a month, while the predicted breakthrough was approximately 1000 days later.

The difference in predictive accuracy between the pesticides, on one hand, and nutrient and water, on the other hand, results, in part, in a wide disparity in the amount of chemicals that will cause health standards to be exceeded. According to limits set by the Environmental Protection Agency, nitrates are tolerable in concentrations up to 10,000 parts per billion, which are not attained unless more than 40% of the amount commonly applied reaches ground water. In contrast, the maximum permissible pesticide concentration is usually below 10 parts per billion. On Long Island, a concentration of 2.5 parts per billion may be attained if just one quarter of one percent of the amount usually applied reaches ground water. Since this limit can be exceeded only if a small amount of solute follows a nonaverage flow path, averaging models based on the convective-dispersive equation are too simplistic. A more sophisticated model is needed that takes preferential flow into account.

PREFERENTIAL FLOW IN STRUCTURED SOILS

In clay and loam soils, most of the soil is of relatively low permeability with usually less than 1% of the area consisting of cracks partially filled with sand and small stones, as well as passages formed by roots and earthworms. When it rains, water infiltrating the ground follows these channels in preference to the surrounding matrix, whose small pores are penetrated comparatively slowly.

The first overview of solute flow through a structured soil was carried out more than 100 years ago by Lawes et al. (1882). Interestingly, they noted firsthand that it is a mistake to regard ordinary soil as an uniform porous media. They theorized that water collected in drains can be separated into two constituents: "direct" and "general" drainage. Direct drainage (or preferential flow in today's terminology) is precipitation that passes with little modification through the soil channels consisting, near the surface, of cracks which are partly filled with sand and small stones that have fallen in from the surface during dry periods. Deeper channels are formed by either roots or earthworms. General drainage (or matrix flow) is water discharged from the pores of the saturated soil which has used, to some degree, the same open channels as the direct drainage. The relative importance of the two forms of drainage water is dependent on the soil type and rainfall intensity. In light soils, rain is absorbed directly by the "main body of soil" and channel flow is insignificant. In heavy soils, preferential flow occurs even before the soil is fully saturated. Only after the soil is fully saturated near the drain does matrix flow commence.

Water in preferential and matrix flow paths may have different solute content. The chemical composition of preferential flow reflects the concentration of the water near the surface, while matrix flow represents the concentration of water around the drain. Thus, when a salt is equally distributed throughout the soil, the salt content of matrix flow is higher than for water in preferential flow paths, characterized by the rainfall composition. The opposite was true (i.e., preferential flow had a higher solute content than the matrix flow) when a fertilizer or tracer were recently surface-applied. Consequently, the solute concentration of drainage water depends on the ratio of water in preferential and matrix flow paths.

Interestingly, the following quotation from Lawes et al. (1882) highlights the importance of preferential flow on the composition of water discharging from a drain:

> The admixture of rain water will be most considerable during heavy rain, when water accumulates on the surface of the land, as water will stand over the heads of all existing channels. Drainage from the surface will cease soon after rain has stopped, the upper layer of the soil being the first to lose its supersaturated condition. As the running at the pipe diminishes, the drainage water will be successfully furnished by lower and yet lower layers of soil, until the soil is no longer in a supersaturated condition above the drain pipe. Under these conditions assumed, it is clear that the drainage water collected at the commencement of a running will be much weaker than that collected at the end.

The assumption of drainage water from the surface reaching the drain before that of the lower layers is contradictory to current water and solute transport theory based on the convective-dispersive equation.

Investigations by Hegert et al. (1981), Steenhuis and Muck (1988), van Es et al. (1991), Shalit et al. (1992), Aburime (1986), and Richard and Steenhuis (1988), are consistent with the findings of Lawes et al. (1882) in that, when the tile outflow rate is high, the water composition resembles that of the surface soil solution flowing through the larger pores toward the tile line. They also confirm that the distribution of the solutes is highly variable, even over a very short distance.

Many of the preferential flow findings of the historical paper by Lawes et al. (1882), despite its prominence, have not, up to now, been incorporated into models for simulating solute flow. New models that try to incorporate preferential flow through cracks and macropores are by Ajuha (1991), Tsjang and Tsjang (1989), Stagnitti et al. (1991), Steenhuis et al. (1991), and Corwin et al. (1991). The advantage of these models is that the concentration of the percolating water is dependent on the rate at which the water is applied. Also, the concentration of the recharge water is sensitive to the time period between rainfall and chemical application.

PREFERENTIAL FLOW IN SANDY SOILS

Even in homogeneous soils, water does not necessarily take an average flow path. Hill and Parlange (1972) were among the first to document preferential flow in homogeneous sand at low infiltration rates. Further research by others and us has led to an understanding of many of the mechanisms involved in this phenomena (Hill and Parlange, 1972; Raats, 1973; Philip, 1975; Hillel and Baker, 1988; Glass et al., 1989a,b). Of special importance in understanding the phenomena is the technique developed originally by Hoa (1981), and adapted by Glass et al. (1989c) for instantaneously determining the moisture content of a 1 cm thick sand layer by measuring the intensity of the transmitted light.

This line of research has shown that water can find its way down through sandy soils in a number of channels that we call *fingers* (Hill and Parlange, 1972; Raats, 1973). A poorly conducting layer of topsoil at the surface over a more conductive subsoil produces a wetting front instability. Gravity drives the instability and surface tension has a contrary, stabilizing effect. The balance between gravity, expressed as $(k_w - q)$ (cm/sec), and surface tension, expressed as $S_w^2/(\theta_w - \theta_i)$ (cm²/sec) gives the diameter, d (cm), of the fingers. In accordance with dimensional analysis

$$d = \lambda \frac{S_w^2}{(\theta_w - \theta_i)(k_w - q)} \tag{1}$$

where λ is a universal number, q is the entering flux, θ_w and θ_i are the average water contents behind and ahead of the wetting front, S_w is the sorptivity, and $k_w = k(\theta_w)$ – $k(\theta_i)$, where $k(\theta_w)$ and $k(\theta_i)$ are the soil water conductivities at their respective water contents. θ_w can be called the water-entry value.

To solve the transport equation for the fastest growing disturbance, λ is explicitly calculated with Bessel functions for three-dimensional fingers as $\lambda = 4.8$ (Glass et al., 1991). For two-dimensional fingers, $\lambda = \pi$ (Parlange and Hill, 1976; Glass et al., 1989a,b). In the field, k_w tends to be much larger than q, and, for most sandy soils, the quantity $S_w^2/(\theta_w - \theta_i)k_w$ is only a slightly increasing function of θ_w. These fundamental properties imply that d is essentially independent of q, and the amount of water carried by the fingers, θ_w, is a property of the soil. By using the Brooks and Corey relationships, Equation 1 can be simplified for small infiltration rates to (Liu et al., 1994)

$$d = \frac{2 \lambda \theta_f \left(\frac{dh}{d\theta}\right)}{\eta + 1.5} \tag{2}$$

where η is the exponent in the Brooks and Corey conductivity function and usually has a value between 3 and 4, θ_f is the moisture content in the tip of the finger, and $dh/d\theta$ is the slope of the wetting curve at θ_f.

The distribution of water within a given finger is not uniform. Not only is there a sharp gradient at the finger's cylindrical surface, responsible for lateral diffusion, but, even more strikingly, there is an increase of water content with depth. This variation with depth can easily be described by solving the transport equation for a constant velocity of finger growth, v, yielding (Selker et al., 1992a,b)

$$\frac{k(\theta)}{\theta} = \frac{v}{1 - \dfrac{\partial h}{v \partial t}} \tag{3}$$

Selker et al. (1992c) showed in the laboratory that, in addition to layered soils, instabilities would also occur by raining, at a much lower rate than the saturated conductivity. Thus, fingering may occur widely when soils are coarse or water-repellent. Indeed, this has been confirmed by several field and laboratory studies (Hendrickx et al., 1988; van Dam et al., 1990; Hagerman et al., 1989; Glass et al., 1988; Hillel and Baker, 1988).

In sand that is quite homogeneous, the fingers are nearly vertical and do not merge with one another. If the sand is less homogeneous, however, the fingers deviate from a strictly vertical path and can merge. When this happens, they do not come together on equal terms, like the arms of a Y. Instead, one finger continues on its course, while the other donates water as a tributary entering from the side. In the field, where heterogeneities in the soil are far more pronounced than in laboratory experiments, most merging takes place in a shallow layer near the surface called the *induction zone*. Below the induction zone, merging is less common, as long as the water flows through a relatively homogeneous medium. But, further down in the soil, sloping interfaces of compositionally different layers may act as funnels (Kung, 1990a,b). Such structures can concentrate the water from a large area into a single finger. Contrariwise, fingers may split. This is the probable outcome when a finger reaches saturation and cannot accommodate the imposed flow.

IMPLICATIONS OF PREFERENTIAL FLOW FOR VADOSE ZONE MONITORING OF CONTAMINATION

One of the greatest uncertainties in vadose zone monitoring is the preferential flow of solutes resulting in an uneven distribution and with the possible effect of bypassing of samplers (Biggar and Nielsen, 1976). Traditional sampling schemes postulate that point measurements can be scaled up to field scale. Preferential flow clearly seems to invalidate this hypothesis (even in the case when geostatistical methods are employed). The concern about how preferential flow affects sampling in the vadose zone is not novel. It was identified by Beasley (1976) for forested watersheds, by Shaffer et al. (1979) and Shuford et al. (1977) for water sampling with suction cups in field soils, and by Wilson (1980) in a review on sampling methods in the vadose zone. However, no solutions were proposed. We will examine the effect of preferential flow on sampling for three sites in New York. The first is a structured soil in Willsboro, the second is a silt loam soil in Freeville, and the third is a coarse grained soil on Long Island.

Willsboro

Field experiments were performed on a tile-drained field at the Cornell University Experimental Farm in the town of Willsboro, NY. Prior to the experiments, the field had not been plowed for several decades. The soil is mapped as a Rhinebeck fine sandy loam. It contains a dense clay loam layer from 20 to 65 cm, which was penetrated by root and wormholes. The 20 cm thick topsoil consists of a fine sandy loam and the subsoil is a gravelly loam that becomes more sandy with depth.

An agricultural tile-drain system with parallel drains in the experimental plots was installed at a 10 m spacing in the fall of 1984 at a depth of 0.8 m. Piezometers at 20, 50, and 80 cm depths, 2, 7, and 12 m, respectively, from each of the tile lines, were installed in rows perpendicular to the tile line at 25, 65, and 105 m from the manhole in the native pasture of plots A and C. These wells sampled water which had passed through the vadose zone. In addition, piezometers at depths of 20 cm in each row were installed at distances of 5 and 10 m perpendicular to the tile line at distances of 45 and 85 m from the manhole. A total of 80 kg of chloride (as $CaCl_2.2H_2O$) was applied in a strip, 83 m long and 8 m wide, starting at 2 m uphill from the tile line and 35 m from the manhole in November 1984. Piezometer concentrations were determined 1, 3, 5, 7, 12, 14, and 16 days after the chloride application. During this period, 13 cm of water was applied by an irrigation system and 2 cm of precipitation was recorded. Soil samples were taken after 16 days at 20, 50, and 80 cm depths near the locations of the 50 and 80 cm deep piezometers, where tile outflow concentration was monitored continuously.

In Table 5.1, the arithmetic and geometric means of the soil solution concentra-

Table 5.1. Geometric and Arithmetic Mean of Chloride Concentration (mg/L) for Soil Samples and Piezometers During November 1984, at the Willsboro Site

	20 cm Depth				50 cm Depth		80 cm Depth	
	Day 2	Day 5	Day 16		Day 16		Day 16	
	Soil	Pzmt	Soil	Pzmt	Soil	Pzmt	Soil	Pzmt
PLOT A								
Arithmetic Mean	1879	1091	744	n.a.	456	1521	651	607
Geometric Mean					416	346	535	316
Standard Dev.	1975	2070	212	n.a.	190	3316	509	624
Coeff. of Variation	1.05	1.90	0.28		0.42	2.18	0.78	1.03
PLOT C								
Arithmetic Mean	1276	2097	450	n.a.	865	n.a.	771	1214
Geometric Mean					568		551	462
Standard Dev.	1302	3156	228	n.a.	857	n.a.	730	1809
Coeff. of Variation	1.02	1.51	0.51		0.99		0.95	1.49

tion for the soil cores and the piezometers at the 20, 50, and 80 cm depths are presented for the six-day period after the chloride application in which a total of 6 cm of water fell or was applied. For the piezometers, the geometric mean for the 50 and 80 cm depths, day 16, is much lower than the arithmetic mean. This is caused by very few high concentrations in one or two of the piezometers and much lower values in the remaining shallow wells. (See Figure 5.6, where the concentration for each piezometer in plot A at 20 cm is shown.) In contrast, the chloride concentration obtained by soil coring in Figure 5.7 is much more although here, also, large variation exists between individual sampling points. The reason for the large variation in the concentration, as shown by dye studies, is that all transport in the upper 60–70 cm of the profile occurs through the macropores. Thus, if the well screen is connected via a macropore with the surface, one will find a high chloride concentration after the well is bailed. Moreover, the distribution of chloride with depth indicates that there is significant bypass flow, as explained in more detail in the Freeville results. The low chloride concentrations in the piezometers 2 m from the tile line indicated that most of the chloride in the macropore system had already been leached to the tile line, while the soil matrix, even at day 16, contained a significant amount of chloride.

To examine if chloride concentrations in the soil core and piezometer were correlated, we plotted both concentrations at the 50 and 80 cm depths for day 16 in Figure 5.8. It is obvious that the correlation is poor. If we assume that the shallow wells are mostly influenced by the macropore water and the soil cores contain the total soil solution including the matrix water, it seems that matrix water and macropore flow are almost independent of each other. It also indicates that any statistical procedure using either method is highly suspect.

Chloride concentrations in the tile outflow during the November 1984 experimental period are presented in Figure 5.9. They are characterized by high peaks during the first rainfall event, followed by smaller peaks during subsequent rainfall events. The solute concentration peaks in the tile outflow correlate well with the 20 cm piezometers at the 2 m distance from the tile line, as postulated by Lawes et al. (1882). It was also expected that the tile lines could integrate preferential flow over a

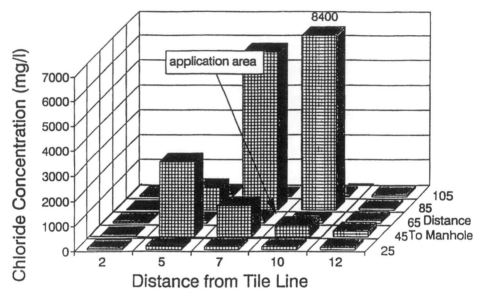

Figure 5.6.   Average chloride concentrations measured in shallow wells as a function of their location in plot A at the Willsboro site on day 3.

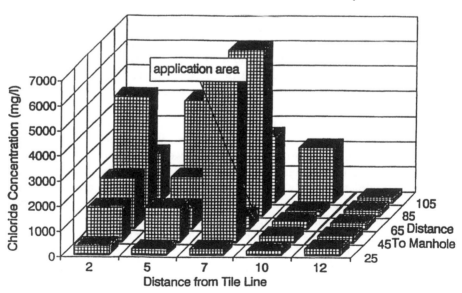

Figure 5.7.   Chloride concentrations of soil water obtained with soil coring as a function of their location in plot A at the Willsboro site on day 3.

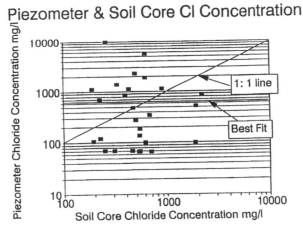

Figure 5.8. Comparison of chloride concentration obtained with soil coring and shallow well sampling for the 50 and 80 cm depths on day 16 at the Willsboro site.

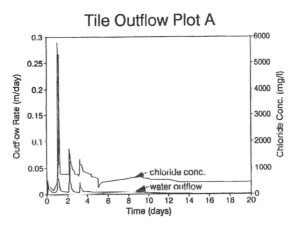

Figure 5.9. Tile concentration and outflow at the Willsboro site.

large area. Therefore, the experiment carried out during the 1985–1986 season looked particularly at the suitability of tile lines in measuring the ground-water loading. This was, indeed, the case as shown by Richard and Steenhuis (1988) on the same field, but for plots B and D.

Freeville

The Freeville site also shows a typical example of how samplers can be bypassed by solutes (Figures 5.10 and 5.11) (Boll et al., 1991). Four suction lysimeters were installed at a depth of 60 cm, and two each at 90, 120, 150, and 180 cm in a sandy loam with a 2 m depth to ground water. A 4 cm pulse of bromide, with a concentration of 8000 mg/L, was applied on day 0 followed by 4 cm of irrigation daily. The suction lysimeters at 60 cm were affected by preferential flow paths directly connected to the surface (some induced along the sampling tubes), giving rise to the widely varying breakthrough curves (Figure 5.10). The remaining sampling devices

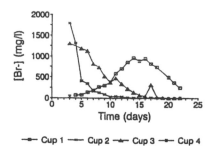

Figure 5.10.   Breakthrough curves for porous cup samplers at the 60 cm depth at the Freeville site.

did not appear to have this problem. Based on the convective-dispersive flow then, one would expect to see a high concentration, at first, at the shallower (90 cm) depth, and then, while concentrations were decreasing at the shallower depth, an increasing concentration at the lower depth. Also, one would expect that the peaks would be decreasing in concentration with depth. However, as shown in Figure 5.11, the peak concentration observed in the capillary fringe at 180 cm, for instance, occurred prior to the peak seen at 150 cm, which, in turn, arrived prior to the maximum concentration seen at the 120 cm depth. This pattern was caused by a preferential flow from the finer soils at the 90 cm depth through the coarse soil directly bypassing the soil suction lysimeter, to the ground-water table where it was stored in the capillary fringe and was picked up by the suction lysimeter at the 180 cm depth and then at 150 cm.

Long Island

The field experiment carried out on Long Island was to determine if a nematicide, Mocap, could be used to control the nematode population without contaminating the ground water. The field selected was located in the town of Jamesport, NY. The root zone was approximately 75 cm deep and consisted of sandy loam. The intermediate zone consisted of coarse sand with clay lenses at the 180 cm depth, gravel layers below the 2.3 m depth, and the water table at a depth of 2.7 m. The capillary fringe was estimated to be 30 cm in thickness. Porous cup lysimeters were installed at 60,

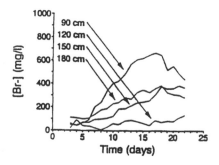

Figure 5.11.   Breakthrough curves for porous cups at 90, 120, 150, and 180 cm at the Freeville site. Each line is the average of two replicates.

75, 170, 180, 190, 230, and 260 cm depths. Suction was applied for one day each month. Figure 5.12 shows the concentration of Mocap obtained from those regions yielding samples. The samplers at 75, 170, 190, and 230 cm, which were all located in the intermediate zone, did not collect any water. Only samplers in the root zone and capillary fringe and one sampler at 180 cm (probably above a structural interface) yielded sufficient volume to be analyzed for Mocap. Porous cups in the capillary fringe yielded the samples of greatest volume.

Similar to the Freeville site, the low, but consistent, concentrations at all sampled depths (Figure 5.12) contradicted the classical convective-dispersive approach. Almost immediately after application (less than one month), there was a small, but elevated, concentration (0.2 ppb) at the 260 cm depth. Moreover, Figure 5.12 shows that the concentrations at all depths increased and decreased approximately at the same time. Both features are typical for soils in which preferential flow takes place and solutes are bypassing the samplers (Shaffer et al., 1979; Bailey, 1990).

This experiment clearly shows that the samplers between the root zone and the capillary fringe were bypassed by the water flowing in preferential flow paths, except those near a structural interface that intercepted some of the flow. Porous cup samplers in the capillary fringe gave the most consistent results. The results are, in many respects, similar to the findings of Starr et al. (1986) in Connecticut, in which only porous cup samplers in the root zone and capillary fringe gave consistent data.

Currently, there are two general approaches—claimed to be independent of the sampling device—that can be used to negate the effect of preferential flow. The first is Bouma's (1990) morphological approach in which a Representative Elementary Volume (REV) is used. These soil volumes are representative in the sense that they are large enough so that individual flow path differences are being averaged. According to Lauren et al. (1988), sampling volumes of 30 peds or more result in "representative" samples. Thus, REVs characterize "an average flow path" in which the soil heterogeneities can be studied as a stochastic or statistical phenomena (Parlange et al., 1988). In situations where concentrations on the order of parts per billion are harmful, any statistical description of heterogeneities is inadequate. Rather, individual preferential pathways that may be responsible for the transport of chemicals must be sampled. In this case, the second approach by Barcelona and

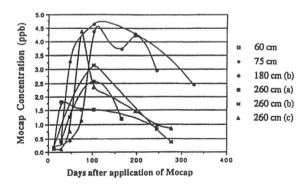

Figure 5.12.   Mocap concentration measured with porous suction cups at various depths below the surface at the Long Island site. The tensiometers located at 260 cm are within the capillary fringe.

Morrison (1988) to locate the sampling devices in the likely pathways of water and contaminants is feasible. Another approach where the ground water is at a shallow depth is to use artificial tile lines for sampling. The tile lines integrate samples over a field scale (Richard and Steenhuis, 1988; Everts and Kanwar, 1988).

SUMMARY AND CONCLUSIONS

Preferential flow represents the movement of water and solutes through preferred pathways in a soil. Preferential flow is much the same whether it occurs in structured soils or homogeneous sandy soils. In fact, preferential flow seems to be most common in soils at both ends of the spectrum. In dense clay soils, cracks provide the only channels through which flow can occur and, in most coarse soils (when rainfall intensity is low), instabilities resulting in preferential flow are a near certainty. Most current studies direct their attention on preferential flow due to cracks and macropores. Much less attention is placed on fingered preferential flow that occurs in sandy textures within the vadose zone.

Preferential flow is especially important for those chemicals that are toxic in parts per billion or trillion. For these chemicals, using an average approach in which each molecule samples all pore spaces is inadequate. It needs to be replaced by a method where individual flow paths are considered separately.

More research findings on preferential flow are required before we can significantly improve models of vadose water and solute flow currently being used. Also, for monitoring, we need to better understand the role of preferential flow, as illustrated in many previous studies: there is no quick and easy solution to determine the method, size, and number of samples required to achieve representative samples.

REFERENCES

Aburime, S.A., Water and Chemical Movement in Non-Homogeneous Soils, unpublished PhD thesis, Cornell University, Ithaca, NY, 1986, p. 197.

Ajuha, L.R., RZWQM Components Dealing with Water and Chemical Transport in Soil Matrix and Macropores, Internal Report, U.S. Dept. of Agriculture-Agricultural Research Service, National Agr. Water Quality Lab., Durant, OK, 1991.

Bailey, N.O., Solute Flux into Porous Cup Samplers Near the Water Table, unpublished PhD thesis, Cornell University, Ithaca, NY, 1990.

Barcelona, M.J., and R.D. Morrison, "Sample Collection, Handling and Storage: Water, Soils and Aquifer Solids," in *Methods for Ground Water Quality Studies, Proc. National Workshop*, D.W. Nelson and R.H. Dowdy, eds., Agricultural Research Division, University of Nebraska, Lincoln, NE, 1988, p. 49.

Beasley, R.S., "Contribution of Subsurface Flow from the Upper Slopes of Forested Watersheds to Channel Flow," *Soil Sci. Soc. Am. J.*, 40:955 (1976).

Beven, K., "Modelling Preferential Flow," in *Preferential Flow, Proc. National Symposium*, T.J. Gish and A. Shirmohammadi, eds., American Society of Agricultural Engineers, St. Joseph, MI, 1991, p. 1.

Biggar, J.W., and D.R. Nielsen, "Spatial Variability of the Leaching Characteristics of a Field Soil," *Water Resources Res.*, 12:78 (1976).

Boll, J., J.S. Selker, B.M. Nijssen, T.S. Steenhuis, J. Van Winkle, and E. Jolles, "Water

Quality Sampling Under Preferential Flow Conditions," in *Lysimeters for Evapotranspiration and Environmental Measures, Proc. International Lysimetry Symposium*, R.G. Allen, T.A. Howell, W.O. Pruitt, I.A. Walter, and M.E. Jensen, eds., American Society of Civil Engineers, New York, NY, 1991, p. 290.

Bouma, J., "Using Morphometric Expressions for Macropores to Improve Soil Physical Analysis of Field Soil," *Geoderma*, 46:3 (1990).

Corwin, D.L., B.L. Waggoner, and J.D. Rhoades, "A Functional Model of Solute Transport That Accounts for Bypass," *J. Env. Qual.*, 20:647 (1991).

Everts, C.J., and R.S. Kanwar, Quantifying Preferential Flow to a Tile Line with Tracers, Paper No. 88-2635, American Society of Agricultural Engineers, St. Joseph, MI, 1988, p. 15.

Glass, R.J., T.S. Steenhuis, and J.-Y. Parlange, "Wetting Front Instability as a Rapid and Far-Reaching Hydrologic Process in the Vadose Zone," *J. Contam. Hydrol.*, 3:207 (1988).

Glass, R.J., J.-Y. Parlange, and T.S. Steenhuis, "Wetting Front Instability, 1. Theoretical Discussion and Dimensional Analysis," *Water Resources Res.*, 25:1187 (1989a).

Glass, R.J., T.S. Steenhuis, and J.-Y. Parlange, "Wetting Front Instability, 2. Experimental Determination of Relationships Between System Parameters and Two-Dimensional Unstable Flow Field Behavior in Initially Dry Porous Media," *Water Resources Res.*, 25:1195 (1989b).

Glass, R.J., T.S. Steenhuis, and J.-Y. Parlange, "Mechanism for Finger Persistence in Homogeneous, Unsaturated, Porous Media: Theory and Verification," *Soil Sci.*, 148:60 (1989c).

Glass, R.J., J.-Y. Parlange, and T.S. Steenhuis, "Immiscible Displacement in Porous Media: Stability Analysis of Three-Dimensional, Axisymmetric Disturbances with Application to Gravity-Driven Wetting Front Instability," *Water Resources Res.*, 27:1947 (1991).

Hagerman, J., N.B. Pickering, W.F. Ritter, and T.S. Steenhuis, In Situ Measurement of Preferential Flow, American Society of Civil Engineers National Water Conference and Symposium, Newark, DE, 1989, p. 10.

Hegert, G.W., D.R. Bouldin, S.D. Klausner, and P.J. Zwerman, "Phosphorus Concentration-Water Flow Interaction in the Effluent from Manured Land," *J. Env. Qual.*, 10:338 (1981).

Helling, C.S., and T.J. Gish, "Physical and Chemical Processes Affecting Preferential Flow," in *Preferential Flow, Proc. National Symposium*, T.J. Gish and A. Shirmohammadi, eds., American Society of Agricultural Engineers, St. Joseph, MI, 1991, p. 77.

Hendrickx, J.M.H., L.W. Dekker, E.J. van Zuilen, and O.H. Boersma, "Water and Solute Movement Through a Water Repellent Sand Soil with Grass Cover," in *Validation of Flow and Transport Models for the Unsaturated Zone, International Conference and Workshop Proc.*, P.J. Wierenga and D. Bachelet, eds., College of Agriculture and Home Economics, New Mexico State University, Las Cruces, NM, 1988, p. 131.

Hill, D.E., and J.-Y. Parlange, "Wetting Front Instability in Layered Soils," *Soil Sci. Soc. Am. J.*, 36:697 (1972).

Hillel, D., and R.S. Baker, "A Descriptive Theory of Fingering During Infiltration into Layered Soils," *Soil Sci.*, 146:51 (1988).

Hoa, N.T., "A New Method Allowing the Measurement of Rapid Variations on the Water Content in Sandy Porous Media," *Water Resources Res.*, 17:41 (1981).

Kung, K.-J.S., "Preferential Flow in a Sandy Vadose Soil: 1. Field Observations," *Geoderma*, 46:51 (1990a).

Kung, K.-J.S., "Preferential Flow in a Sandy Vadose Soil: 2. Mechanism and Implications," *Geoderma*, 46:59 (1990b).

Kung, K.-J.S., J. Boll, J.S. Selker, W.F. Ritter, and T.S. Steenhuis, "Use of Ground Penetrating Radar to Improve Water Quality Monitoring in the Vadose Zone," in *Preferential Flow, Proc. National Symposium*, T.J. Gish and A. Shirmohammadi, eds., American Society of Agricultural Engineers, St. Joseph, MI, 1991, p. 142.

Lauren, J.G., R.J. Wagenet, J. Bouma, and J.H.M. Wösten, "Variability of Saturated Hydraulic Conductivity in a Glossaquic Hapludalf with Macropores," *Soil Sci.*, 145:20 (1988).

Lawes, J.B., J.H. Gilbert, and R. Warington, *On the Amount and Composition of the Rain and Drainage Water Collected at Rothamstead*, Williams Clowes and Sons, Ltd., London, 1882, p. 167. Originally published in *J. Royal Agr. Soc. of England XVII*, 241, 1881; *XVIII*, 1, 1882.

Liu, Y., T.S. Steenhuis, and J.-Y. Parlange, "Closed Form Solution for Finger Width in Sandy Soils at Different Water Contents," *Water Resources Res.*, 30(4):949–952 (1994).

Parlange, J.-Y., and D.E. Hill, "Theoretical Analysis of Wetting Front Instability in Soils," *Soil Sci.*, 122:236 (1976).

Parlange, J.-Y., T.S. Steenhuis, R.J. Glass, T.L. Richard, N.B. Pickering, W.J. Waltman, N.O. Bailey, M.S. Andreini, and J.A. Throop, "The Flow of Pesticides Through Preferential Paths in Soils," *New York's Food & Life Science Quarterly*, 18:20 (1988).

Philip, J.R., "Stability Analysis of Infiltration," *Soil Sci. Soc. Am. Proc.*, 39:1042 (1975).

Raats, P.A.C., "Unstable Wetting Fronts in Uniform and Nonuniform Soils," *Soil Sci. Soc. Am. Proc.*, 37:681 (1973).

Richard, T.L., and T.S. Steenhuis, "Tile Drain Sampling of Preferential Flow on a Field Scale," in *Rapid and Far Reaching Hydrologic Processes in the Vadose Zone*, P.F. Germann, ed., *J. Contam. Hydrol.*, 3:307 (1988).

Richards, L.A., "Capillary Conduction of Liquids in Porous Mediums," *Physics*, 1:318 (1931).

Selker, J.S., P. Leclerq, J.-Y. Parlange, and T.S. Steenhuis, "Fingered Flow in Two Dimensions, Part 1, Measurement of Matric Potential," *Water Resources Res.*, 28:2513 (1992a).

Selker, J.S., J.-Y. Parlange, and T.S. Steenhuis, "Fingered Flow in Two Dimensions, Part 2, Predicting Finger Moisture Profile and Measure of Conductivity," *Water Resources Res.*, 28:2523 (1992b).

Selker, J.S., T.S. Steenhuis, and J.-Y. Parlange, "Wetting Front Instability in Homogeneous Sandy Soils Under Continuous Infiltration," *Soil Sci. Soc. Am. J.*, 56:1346 (1992c).

Shaffer, K.A., D.D. Fritton, and D.E. Baker, "Drainage Water Sampling in a Wet Dual Pore System," *J. Env. Qual.*, 8:241 (1979).

Shalit, G., T.S. Steenhuis, J. Boll, L.D. Geohring, H.A.M. Hakvoort, and H.M. van Es, "Agricultural Tile Lines for Sampling Tillage Practices in Soils with Preferential Flow Paths," XVII General Assembly of the European Geophysical Society, Edinburgh, Scotland, 1992.

Shuford, J.W., D.D. Fritton, and D.E. Baker, "Nitrate Nitrogen and Chloride Movement Through Undisturbed Field Soil," *J. Env. Qual.*, 6:255 (1977).

Simpson, T.W., and R.L. Cunningham, "The Occurrence of Flow Channels in Soils," *J. Env. Qual.*, 11:29 (1982).

Stagnitti, F., T.S. Steenhuis, J.-Y. Parlange, B.M. Nijssen, and M.B. Parlange, "Preferential Solute and Moisture Transport in Hillslopes," in *Challenges for Sustainable Development, Proc. International Hydrology and Water Resources Symposium*, Institution of Engineers, Barton, Australia, 1991, p. 919.

Starr, J.L., J.-Y. Parlange, and C.R. Frink, "Water and Chloride Movement Through Layered Field Soil," *Soil Sci. Soc. Am. J.*, 50:1384 (1986).

Steenhuis, T.S., and W.H. van der Molen, "The Thornthwaite-Mather Procedure as a Simple Engineering Method to Predict Recharge," *J. Hydrol.*, 84:221 (1986).

Steenhuis, T.S., S. Pacenka, and K.S. Porter, "MOUSE: A Management Model for Evaluating Groundwater Contamination from Diffuse Surface Sources Aided by Computer Graphics," *Applied Agr. Res.*, 2:277 (1987).

Steenhuis, T.S., and R.E. Muck, "Preferred Movement of Non-Adsorbed Chemicals on Wet, Shallow, Sloping Soils," *J. Env. Qual.*, 17:376 (1988).

Steenhuis, T.S., and M.F. Walter, "A Nitrogen Model for Field Management," in *Proc. National Symposium Hydrologic Transport Modeling: Modeling Applications to Hydrology in the 1980's*, American Society of Agricultural Engineers, St. Joseph, MI, 1980, p. 246.

Steenhuis, T.S., B.M. Nijssen, F. Stagnitti, and J.-Y. Parlange, "Preferential Solute Movement in Structured Soils: Theory and Application," in *Challenges for Sustainable Development, Proc. International Hydrology and Water Resources Symposium*, Institution of Engineers, Barton, Australia, 1991, p. 925.

Tsjang, Y.W., and C.-F. Tsjang, "Flow Channeling in a Single Fracture as a Two-Dimensional Strongly Heterogeneous Permeable Medium," *Water Resources Res.*, 23:467 (1989).

van Dam, J.C., J.M.H. Hendrickx, H.C. van Ommen, M.H. Bannink, M.T. van Genuchten, and L.W. Dekker, "Water and Solute Movement in a Coarse Textured Water Repellent Field Soil," *J. Hydrol.*, 120:359 (1990).

van der Molen, W.H., "Desalinization of Saline Soils as a Column Process," *Soil Sci.*, 81:19 (1956).

van Es, H.M., T.S. Steenhuis, L.D. Geohring, J. Vermeulen, and J. Boll, "Movement of Surface-Applied and Soil-Embodied Chemicals to Drainage Lines in a Well-Structured Soil," in *Preferential Flow, Proc. National Symposium*, T.J. Gish and A. Shirmohammadi, eds., American Society of Agricultural Engineers, St. Joseph, MI, 1991, p. 59.

Wilson, L.G., Monitoring in the Vadose Zone: A Review of Technical Elements and Methods, Environmental Monitoring Systems Laboratory, Office of Research and Technology, U.S. Environmental Protection Agency, Las Vegas, NV, EPA-600/7-80-134, 1980, p. 168.

6

Basic Contaminant Fate and Transport Processes in the Vadose Zone—Inorganics

Donald D. Runnells

INORGANIC SUBSTANCES

Introduction

The physical-chemical processes that prevail in the vadose zone differ in many respects from processes that predominate in surface waters and in phreatic ground waters. Factors that uniquely determine the important processes in the vadose zone include the presence of a gas phase, a generally oxidizing environment, active biological and biochemical processes, and a large ratio of surface area to volume of water. Based on these characteristics, at least seven chemical processes can be identified as being potentially important in the vadose zone, as follows (see also an earlier publication by the author [Runnells, 1976], from which portions of the following discussion have been taken):

- precipitation/dissolution
- acid/base reactions
- substitution/hydrolysis
- oxidation/reduction
- gaseous transfer
- biological transformations
- sorption.

Revision of paper published in *Ground Water* in 1976. Used with permission of Ground Water Publishing Company.

Table 6.1.   Chemical Components Present in Documented Cases of Contamination of Ground Water

chromium	cyanide
cadmium	copper
zinc	selenium
lead	chloride (and hydrochloric acid)
fluoride	sulfate (and sulfuric acid)
iron	nitrate
barium	detergents
manganese	wood preservatives
nickel	organic solvents
silver	vanadium
molybdenum	gasoline and other fuels
boron	landfill leachates
radionuclides	pesticides
mercury	herbicides
aluminum	phosphates
lithium	tanning wastes
arsenic	brines

For many years it was felt that chemical processes in the vadose zone would protect underlying ground water from contamination. However, as shown in Table 6.1, numerous chemical contaminants have now been found in ground water throughout the world (summarized from several sources) (Fuhriman and Barton, 1971; Scalf et al., 1973; Cole, 1974; Miller et al., 1974; National Research Council, 1984; van der Leeden et al., 1990). For most of the cases represented in Table 6.1, the contaminants were introduced at the surface and simply overwhelmed the limited protective capacity of the vadose zone. Many years ago Miller et al. (1974) made the important observation that most instances of ground-water contamination become known because of obvious leaks or spills at the surface or because of complaints from subsequent users of ground water, not because of systematic programs involving the testing or monitoring of ground water.

PHYSICOCHEMICAL PROCESSES IN THE VADOSE ZONE

Introduction

In this section the possible role and importance of each of the seven processes listed above will be discussed.

Precipitation/Dissolution

It is self-evident that ground water dissolves natural rock materials. Rainwater and snowmelt that act as recharge to aquifers generally evolve into higher salinities and different chemical compositions as the water moves through soils and aquifer materials. The processes of dissolution and precipitation can be summarized by the following general reaction:

initial-water composition + initial solids →

final-water composition + final solids

The final composition of the vadose water will depend on the mineralogical composition of the soil or aquifer, the texture and permeability, and the time of contact (Colman and Dethier, 1986; Sparks, 1989). The longer the time of contact between water and solids, the greater the possible degree of chemical evolution.

The thermodynamic drive for dissolution or precipitation of a solid phase can be expressed as the computed saturation index (SI) (Hem, 1985):

$$SI = log [IAP/K_{sp}]$$

where IAP is the empirical ion activity product for a given mineral in the water of interest, and K_{sp} is the equilibrium solubility product constant for the same mineral at the temperature and pressure. Net dissolution should occur if the vadose water is undersaturated with respect to the solid, as indicated by a negative value of SI. Net precipitation can occur only if the SI has a positive value. An SI value of zero means that there is no thermodynamic tendency for net dissolution or net precipitation of a given mineral.

Matthes (1982) provides a summary of the composition of soil solutions, demonstrating how soil solutions differ from rainfall and snowmelt. The chemical changes result in part from the dissolution of minerals and gases. Carbon dioxide is added from atmospheric air, organic decay, dissolution of carbonate minerals, and biochemical processes. Sulfate and NO_3 are added to vadose water by biochemical processes, infiltration, and oxidation of organic matter. Calcium, Mg, Na, and SiO_2 are introduced from the weathering of the minerals and ion-exchange from mineral surfaces. As a rule, K is not significantly enriched in soil solutions because of strong adsorption on clays and soil colloids.

The extent of chemical modification of meteoric recharge water in the vadose zone will also depend on the balance between infiltration and evapotranspiration. In arid regions, evapotranspiration will result in vadose waters that are saline and saturated with rock-forming minerals. In contrast, in humid regions the recharging waters may not be greatly modified during rapid passage through the vadose zone.

Evidence that precipitation reactions can control concentrations of aqueous species, including contaminants, has been given by many researchers. For example, Lehman (1968) found that the concentrations of Fe and Zn in the ground waters of the Tucson Basin are apparently controlled by dissolved silica and pH. Calcium is usually controlled by equilibrium with calcite, or by ion-exchange on clays. The concentrations of Fe and Mn are commonly controlled by solid oxyhydroxides in the oxygenated vadose zone. The precipitation of fluorite (CaF_2) is an important controlling factor for dissolved F^- ion.

Recent work has shown that colloidal transport of components in ground water in the saturated zone may be much more important than previously thought, but the extent to which colloidal transport is important in the vadose zone has not been adequately studied. For example, Degueldre et al. (1989) found a concentration of approximately 1010 colloidal particles (40 to 1000 nm in diameter) per liter in water from a fracture in granite at a depth of 450 m below the land surface in the Swiss Alps. The particles were made up of a mixture of organic (C, N, O, S) and inorganic (Si, Ca, Sr, Mg, Sr, and Ba) elements. Ryan and Gschwend (1990) studied colloidal mobilization in ground water of the Pine Barrens of New Jersey. Using careful methods of collection to avoid mechanical disturbance of the well, Ryan and Gschwend (1990) found total concentrations of colloids up to 60 mg/L in the ground water, primarily as particles of clay; colloidal organic carbon was also present in all samples. These examples from the phreatic zone suggest that similar studies may be appropriate for vadose conditions.

Acid/Base Reactions

The pH is a critical factor in many natural reactions, including processes that affect the stability of solid minerals and precipitates in the vadose zone.

Natural ground waters generally exhibit pH values between about 6 and 9. The upper limit is usually established by the reaction between carbon dioxide gas and either limestone or caliche in the soil; this reaction is of major importance in arid and semiarid regions. Lower values of pH are usually found in shallow ground water from nonreactive rocks, such as granite or sandstone. Aqueous buffering can be provided by conjugate acid/base pairs, such as HCO_3^-/CO_3^{2-}, HSO_4^-/SO_4^{2-}, protonated/deprotonated organic matter, etc. Buffering of the pH of the water by solid silicates, carbonates, and organic matter is also important in some geochemical environments. Carbon dioxide gas can significantly influence the pH.

Substitution/Hydrolysis

These types of reactions (sometimes called "metathetical reactions") involve the transformation of dissolved inorganic or organic contaminants in ground water, according to the generalized reaction:

$$RX + A \leftrightarrow RA + X$$

in which R represents the main portion of the dissolved molecule, X is a replaceable element or group, and A is the dissolved species that replaces X. The product RA may be aqueous or solid and either more or less reactive than the original species.

Hydrolysis is a specific and important type of replacement reaction, described by:

$$RX + H_2O \leftrightarrow ROH + HX$$

or

$$RX + H_2O \leftrightarrow RH + XOH.$$

The resulting ROH or XOH may be either an aqueous or solid product, depending on the solubility of the substance and the pH of the water. Hydrolysis reactions can obviously be considered to be a type of acid/base reaction. The solubilities of such metals as Al, Fe, and Mn are strongly dependent on hydrolysis reactions and the pH of the water (the reader is referred especially to Hem [1985] for examples of the role of hydrolysis and pH in controlling the solubility of Al, Fe, and Mn).

Common examples of hydrolysis involve the precipitation of dissolved ferric (Fe^{3+}) iron, Al, and Mn by reaction with water to form solid oxyhydroxides plus aqueous H_3O^+. In an aerated environment the dissolved concentrations of these metals are held to low values by hydrolysis and precipitation.

In addition to ferric (Fe^{3+}) iron, several toxic metals form highly insoluble hydroxides in the range of pH of 6 to 9. These include Cu (above a pH of about 6.5), Cr (the $3+$ form only, above a pH of about 6), Ni (above a pH of 9), and Zn (above a pH of about 9). In some cases, coprecipitation of two or more contaminants may occur; for example, LeGendre and Runnells (1975) and Kaback and Runnells (1980) show that Mo is strongly coprecipitated with iron oxyhydroxide under moderately acidic conditions.

The importance of hydrolysis in both the vadose and phreatic zones is the result of the simple requirement that water be present for these reactions to occur.

Oxidation/Reduction Reactions

Oxidation and reduction (redox) reactions involve the transfer of one or more electrons between chemical elements. In the vadose zone, oxidation can produce relatively insoluble oxides that may control the solubility of many metals, such as Cu, Fe, Mn, Hg, Cr, and Ni. In the phreatic zone, or in swampy or flooded vadose environments, anaerobic conditions may prevail and lead to the mobilization of some species. Lehman (1968) found that rapid infiltration of domestic sewage in the Tucson Basin caused the loss of oxidizing conditions in the vadose zone and led to increased movement of Cu, Zn, Mn, Ni, and Pb through the soil. Masscheleyn et al. (1991) show the importance of redox reactions in controlling the forms and solubility of As in natural environments, and similar reactions are important in controlling the solubility of Se (Goering, et al., 1968).

Oxidation and reduction reactions can also have important effects on the speciation of dissolved elements and on their solubility. The reader is referred to Hem (1985) for a more detailed discussion of the role of oxidation and reduction reactions in aqueous speciation and solubility. A typical reaction would be the oxidation of S(-II) (probably as aqueous HS^-) to S($+VI$) (probably in the form of the aqueous species SO_4^{2-}), as follows:

$$HS^-(aq) + 4 H_2O(liq) \leftrightarrow SO_4^{2-}(aq) + 9 H^+(aq) + 8 e^-.$$

Conceivably the SO_4^{2-} produced by the oxidation of HS^- could then precipitate as a solid phase, such as gypsum, in the vadose zone in an arid climate.

Gaseous Transfer

Gases are involved in many chemical reactions in the vadose zone. The solubility of a gas in water is controlled by Henry's constant, K_h, according to the reaction:

$$[gas(aqueous)] = K_h [gas]$$

in which [ ] represents the activity of the gas and its dissolved counterpart. Henry's constant is a function of temperature and pressure. The molal solubility (moles of gas per kilogram of solvent) of the aqueous gas is also a function of the salinity of the water because of the effect on the molal activity coefficient of the dissolved gas. At temperatures near 298°K and total pressures near 1 atmosphere, the activity of the gas phase is closely approximated by the partial pressure, and the concentration is approximately equal to activity. Therefore, for many natural ground waters, the following simple relationship holds:

$$(\text{aqueous gas}) = K_h \, P_{gas}$$

where ( ) represents molality and P_{gas} represents the partial pressure of the gas (Hem, 1985), with Henry's constant in units of moles kg^{-1} atm^{-1}. For representative gases that may be important in vadose-zone chemistry, the values of Henry's constant in pure water at near-surface conditions range from approximately $10^{+0.15}$ mol kg^{-1} atm^{-1} for SO_2 to $10^{-3.2}$ mol kg^{-1} atm^{-1} for N_2 gas (Wilhelm et al., 1977).

The gas of most obvious geochemical significance is CO_2, which exerts a major control on the solubility of carbonate minerals and rocks. Over the range of partial pressures of CO_2 gas encountered near the surface, the solubility of carbonate minerals increases with increasing partial pressure of CO_2 gas. At near-surface conditions, the value (Wilhelm et al., 1977) of Henry's constant for CO_2 gas is $10^{-1.5}$ mol kg^{-1} atm^{-1}.

Other gases may also be of importance in the vadose zone. For example, in an oxygen-poor environment, the dissolution and exsolution of gaseous H_2S may control the solubility of sulfide minerals, such as pyrite (FeS_2). The presence of CO, H_2, and CH_4 may be indicative of biological activity.

Some inorganic species can be transformed into volatile species and thus move through the vadose zone. Perhaps the best known example is that of the bacterial reduction of dissolved SO_4 to H_2S gas, with loss of H_2S to the atmosphere (Kellogg et al., 1972). Mercury in solution can also be volatilized in anaerobic environments (Lagerwerff, 1972) or by reaction with dissolved humic acids (Alberts et al., 1974). Several organic compounds of As are volatile, and escape of As as a gas has been demonstrated for both aerobic and anaerobic soils (Woolson et al., 1971). Selenium is also subject to transport through volatilization (Lakin, 1973). And of course, the microbial reduction of NO_3 to NH_3 or N_2 gas is well documented (Smith et al., 1991), although the failure of this mechanism to protect the quality of ground water is demonstrated by numerous examples of nitrate pollution (Minear and Patterson, 1973).

Gaseous transfer can be important for the transport and fate of a wide variety of volatile and semivolatile anthropogenic contaminants commonly found in both the saturated and vadose zones, such as trichloroethylene, benzene, xylene, etc. Concentration gradients may exist in the vadose zone for these contaminants, allowing migration from one site to another.

Biological Transformations

Biological transformations, the sixth process, are important in degradation of organic and biologic contaminants, especially under oxidizing conditions in the vadose zone. However, biological transformation of inorganic contaminants can also be a significant process. The biological involvement of SO_4 and NO_3 has already been mentioned. In addition, As, CN, Hg, and Se are likely candidates for biologic fixation or volatilization, and Mo is strongly assimilated and concentrated by plants that are nitrogen-fixers (Allaway, 1977). Much research is currently being conducted on the utilization and enhancement of in situ biological transformations and removal of organic contaminants in the subsurface (Abramowicz, 1990; Sayler, 1990). Biological processes could be of great value in managing discharges to the vadose zone, but at present we know so little of the principles involved that each case must be studied and evaluated as a special situation.

Sorption

The composition of ground water may be strongly influenced by various processes of sorption. Ground water, especially in the vadose zone, is exposed to the surfaces of the rocks, minerals, and amorphous solids that comprise the framework of an aquifer, and ions and neutral molecules can be attracted to the surfaces of the solids. The process by which the aqueous species are attracted to the surfaces of the solids is termed "sorption," and the release of the sorbed species to the liquid phase is termed "desorption." The terms sorption and desorption are used here in a broad sense, to include any process involving the attraction and release of aqueous species to or from the surfaces of a solid particles. Within this broad definition, we include such specific processes as ion-exchange, adsorption or release of protons and hydroxyl ions from the surface of a solid, and interactions between neutral molecules and the surfaces of solids. Sorption and desorption are important for both inorganic and organic species.

The details of the reactions that take place between aqueous species and the surfaces of solids in an aquifer have been the subject of intensive study by numerous investigators for many years. For interesting examples of detailed investigations of sorption/desorption reactions, the reader is referred to the several books on various aspects of the subject (Kavanaugh and Leckie, 1980; Davis and Hayes, 1986; Stumm, 1987; Melchior and Bassett, 1990).

The chemical principle which underlies sorption and desorption reactions is that of attraction between aqueous species and receptive sites on the surfaces of solids. The attraction results from chemical bonding (electrostatic, covalent, van der Waals, hydrogen, etc.) between the dissolved species and the components of the solid surface. If the bonding is mainly electrostatic, and the reaction involves the equivalent exchange of one charged aqueous species for another that is already present on the surface site, the reaction is termed "ion-exchange." An example of an ion-exchange reaction of importance in ground water is:

$$Na_2X_{(s)} + Ca^{2+}(aq) \leftrightarrow CaX_{(s)} + 2\,Na^+(aq)$$

where X may represent any solid substrate. In natural ground-water systems, X may be clays, solid organics, or metallic (Fe, Mn, Al) oxyhydroxides. Ion-exchange reactions, such as the one above, can be described by an apparent equilibrium constant, K_{ex}, where:

$$K_{ex} = [(CaX)[Na^+]^2]/[(Na_2X)[Ca^{2+}]]$$

in which ( ) represents concentrations for the solids and [ ] represents activities for the aqueous ions. In general, K_{ex} is not a constant; it varies with the salinity of the water, the type and concentration of counterions in solution, the types of ions already in exchange sites on the solid, the competing ions in solution, and the proportion of exchange sites already filled on the surface of the solid. Values of K_{ex} must be determined experimentally for each reaction and each ground water of interest.

Sorption/desorption reactions are generally fast, reaching completion over time scales of a few minutes to a few hours. However, if the adsorbed species gradually become incorporated into the crystal structure of the solid phase, the reaction can become very slow, occurring over periods of weeks to months. To distinguish the mechanisms involved, reactions in which a species is incorporated into the structure of the solid are called "*ab*sorption" rather than "*ad*sorption."

Most of the theoretical and predictive modeling of sorption/desorption reactions has focused on bonding reactions on the surfaces of metal oxides and hydroxides, such as SiO_2, Al_2O_3, $Al(OH)_3$, Fe_2O_3, and $FeOOH$ (Parks, 1967; Davis et al., 1978; Davis and Leckie, 1978). The model reactions are represented as replacements of protons or hydroxyl ions from bonding sites on the surfaces, such as:

$$S\text{-}O^- + H^+ \leftrightarrow S\text{-}O\text{-}H$$
$$S\text{-}O\text{-}H \leftrightarrow S+ + OH^-$$
$$S\text{-}O\text{-}H + M^+ \leftrightarrow S\text{-}O\text{-}M + H^+$$
$$S\text{-}O\text{-} + M + \leftrightarrow S\text{-}O\text{-}M$$
$$S\text{-}O\text{-}H + M^{2+} \leftrightarrow S\text{-}O\text{-}M^+ + H^+$$

and so on, where S represents the main internal structure of the substrate. In this approach, an intrinsic constant of reaction, similar to a thermodynamic equilibrium constant, is assigned to the reactions involving the surface sites. The number of sites and the strength of the electrical field adjacent to the charged surface are the remaining parameters that enter into the predictive model. One recent example of such calculations, including a discussion of the parameters required, can be found in the study by Rea and Parks (1990) of sorption of organic molecules on quartz. Models of surface reactions and example calculations for solids in soils, applicable to soils and rock substrates, are presented in a particularly clear fashion by Gast (1977).

Clays, metallic oxyhydroxides, and solid organics can all be good substrates for sorption of various dissolved species. In arid regions, iron oxyhydroxide and various smectitic clays are probably the most important sorbents. With the exception of fractured shale or siltstone, consolidated bedrock will generally not be very effective as a sorbent. Virtually every ionic species and many nonelectrolytes will be sorbed and removed to some extent as ground water moves through an aquifer; of the common inorganic species, only Cl^- seems to pass through soils and alluvium without significant sorption.

A clear contrast exists between sorption of inorganic species in the vadose zone and partitioning of organics. Inorganic sorption is driven by either electrostatic bonding or surface complexation reactions on the surfaces of minerals and is dependent on the surface area available for reaction. In contrast, many organics are hydrophobic and partition favorably into solid organic matter, with the controlling factors being the octanol/water partitioning coefficient and the percentage of organic matter in the porous medium.

It is clear that processes of sorption depend on the aqueous species and on the physical and chemical properties of the aqueous solution and porous medium. It is useful to recognize that sorption is generally temporary, representing a kind of storage in the vadose zone, not a permanent removal of the contaminants.

Summary

Seven geochemical processes have been discussed, each of which may be important in controlling the chemistry of water in the vadose zone. The chemistry of the vadose system is complex, and simultaneous consideration of all possible chemical and physical reactions in ground water is an almost overwhelming task. However, the use of increasingly sophisticated geochemical computer codes (Melchior and Bassett, 1990; National Research Council, 1990; Mangold and Tsang, 1991; Wierenga, 1991) greatly facilitates consideration of many of the important reactions and processes.

STORAGE OF CONTAMINANTS IN THE VADOSE ZONE

Introduction

The vadose zone has significant potential value for the long-term storage of a variety of wastes, especially in arid regions. In fact, in the early stages of the present debate and dilemma about the safe storage of nuclear wastes, Winograd (1974) presented a detailed analysis of a plan to utilize the vadose zone of arid regions for such storage. As water scientists, we can envision similar useful applications in the discharge of other wastes. Bouwer (1974) has given an example of proper management of disposal of sewage effluent to the vadose zone in Arizona in order to maximize removal of dissolved NO_3.

Potential Capacity of the Vadose Zone for Storage of Metals

The capacity of the vadose zone can be large for removal and storage of contaminants. As an example of this large theoretical capacity, let us consider a hypothetical discharge of dissolved Cd from a metal-plating plant. The assumptions involved in this example are summarized in Table 6.2, involving an area of one acre and a depth of alluvium of 100 feet. The concentration of 3.7 mg/L dissolved Cd is that actually reported by Lieber et al. (1964) for the well-known example of ground-water contamination on Long Island. Let us assume that only about 25% of the estimated total sorptive capacity of the unconsolidated material is utilized. Based on the data listed in Table 6.2, approximately 190,000,000 pounds (about 86,500 metric tons) of cadmium would be removed and stored in the vadose zone. To give some idea of the great magnitude of this number, the total industrial consumption of cadmium in the United States in 1989 was only about 4,000 metric tons (Llewellyn, 1991), meaning that the hypothetical sorptive capacity described above would correspond to more than 20 years of the entire industrial consumption of cadmium in the U.S. At a concentration of 3.7 mg/L of dissolved cadmium, this amount would correspond to

Table 6.2. Assumptions Used in Hypothetical Example of the Capacity for Sorptive Removal of Dissolved Cd from Discharge to the Vadose Zone in an Arid Region

Assumptions:
1. Unconsolidated mantle with total of 5 weight percent adsorbing substrates (clay, plus iron, manganese, and aluminum oxyhydroxides).
2. Cation exchange capacity of 50 milliequivalents per 100 grams of sorbing substrates.
3. Bulk density of 2 grams/cc (124.9 lb/cubic foot).
4. Thickness of mantle = 100 feet (30.5 meters).
5. Area of disposal = 1 acre (0.405 hectare).
6. Efficiency of sorption is 25%.

about 19,000,000 acre-feet (2.35E13 liters) of wastewater. These enormous figures simply emphasize the fact that the vadose zone, especially in an arid region, offers a potentially important reservoir that can be managed for the long-term storage of contaminants.

Of course, the preceding example ignores competition or enhancement of sorption by other ions, competing chemical reactions, possible channelized flow of the waste-water through the vadose zone, and a great many other complicating factors. The example simply points out that a significant degree of purification of wastewater and storage of dissolved contaminants is possible under the vadose conditions that exist in an arid region.

Necessity for Site-Specific Study

Although significant processes of purification may occur in the vadose zone, it is important to remember that the capacity for purification is finite, and that there is always a danger of leakage and contamination of ground water. As discussed earlier, the thousands of examples of ground-water contamination show that it is possible to overwhelm the protective mechanisms in the subsurface with virtually any dissolved contaminant. This knowledge warns us of the necessity for thorough and competent studies of the hydrologic and chemical characteristics of a proposed site before discharge begins.

Measurements and Testing

Our knowledge of the specific physical-chemical processes that exist in the subsurface is primitive. We cannot predict from theory what will happen when a particular contaminant in a particular fluid matrix is discharged into a specific type of soil. We do have some general knowledge of the principles involved, as discussed earlier, but we remain ignorant of many of the specifics. For example, one must have a fairly complete knowledge of the hydrogeologic conditions and homogeneity of the earth materials at the proposed site of disposal. In order to obtain such information it will certainly be necessary to conduct fairly extensive field studies, possibly including drilling, to determine the rate and paths of movement of the fluid discharge. An example of one aspect of the work that may be necessary can be found in the study of seepage of effluent from septic tanks, published by Bouma et al. (1972).

Hajek (1969) long ago gave a good summary of some of the technical aspects of the tests and calculations that should be done to understand and predict the chemical interactions of wastewaters with soils, including consideration of the concepts of the distribution coefficient and retardation factor. Experiments must be run, not only to determine the extent of removal of contaminants from the discharge, but also to determine the possibility of later remobilization and flushing to ground water. With regard to storage, Winograd (1974) discussed the selection of a site for disposal to best avoid exhumation or mobilization of nuclear wastes by erosion or climatic change, respectively, over periods of hundreds of thousands of years.

In an earlier publication (Runnells, 1976) this author presented an example of the type of site-specific testing that is necessary for understanding and predicting the behavior of metals as a contaminated water passes through a soil. The soil chosen was from a mountainous region in northern New Mexico. The testing was designed to determine if the soil would remove Cu and Mo from a synthetic solution that simulated decant-water from a base-metal tailings pond. The results showed that the soil removed a very large amount of Cu from the water, but only a moderate proportion of the Mo. Although the experiment involved saturated conditions, it could have been reconfigured for the nonsaturated conditions of the vadose zone. The experiments yielded a retardation factor (R_f) of approximately 27 for Mo, corresponding to a distribution coefficient (Kd) of about 14 mL/gm. Use of these values (Hajek, 1969; Tamura, 1972), and assuming reasonable rates of flow of water in the vadose zone of an arid region resulted in a predicted travel time of thousands of years for dissolved Mo to reach a hypothetical water table at a depth of 30 meters (98 feet) in the region of interest. Because of the greater removal, Cu would require a much greater period of time to reach the ground water. Wierenga et al. (1975) present details of such calculations for B in the vadose zone potentially impacted by leachate from a fly-ash pond in southeastern Utah.

SUMMARY

Although the chemistry of the vadose zone is complicated, it is possible to conceptualize the chemistry in terms of just the seven following processes: precipitation/dissolution, acid/base reactions, substitution/hydrolysis, oxidation/reduction, gaseous transfer, biological transformations, and sorption. These processes control the solubility, transport, and fate of natural and anthropogenic compounds in the vadose zone. Modern geochemical computer codes give us the capability to make meaningful predictions of the behavior of metals and other components in the water of the vadose zone, including the rate and distance of migration of contaminants. In the context of disposal of contaminants, the vadose zone in arid regions can offer a very large capacity for long-term storage of contaminants.

REFERENCES

Abramowicz, D.A., "Biodegradation of PCBs," in *Proceedings: Environmental Research Conference on Groundwater Quality and Waste Disposal*, I.P. Murarka and S. Cordle, eds., EPRI EN-6749, Electric Power Research Institute, Palo Alto, CA 1990, p. 27–1.

Alberts, J.J., J.E. Schindler, R.W. Miller, and D.E. Nutter, Jr., "Elemental Mercury Evolution Mediated by Humic Acid," *Science*, 184:895 (1974).

Allaway, W.H., "Perspectives on Molybdenum in Soils and Plants," in *Molybdenum in the Environment*, Vol. 2, W.R. Chappell and K.K. Petersen, eds., (New York: Marcel Dekker, Inc., 1977), Chap. 1.

Bouma, J., W.A. Ziebell, W.G. Walker, P.G. Olcott, E. McCoy, and F.D. Hole, Soil Absorption of Septic Tank Effluent, University of Wisconsin Extension, Inform. Circ. 20, Madison, WI, 1972, p. 235.

Bouwer, H., "Design and Operation of Land Treatment Systems for Minimum Contamination of Ground Water," *Ground Water*, 12:140 (1974).

Cole, J.A., ed., Groundwater Pollution in Europe, Water Information Center, Inc., Port Washington, NY, 1974, p. 546.

Colman, S.M., and D.P. Dethier, eds., *Rates of Chemical Weathering of Rocks and Minerals*, (Orlando, FL: Academic Press, Inc., 1986), p. 603.

Davis, J.A., R.O. James, and J.O. Leckie, "Surface Ionization and Complexation at the

Oxide/Water Interface: 1. Computation of Electrical Double Layer Properties in Simple Electrolytes," *J. Colloid and Interfacial Sci.*, 63:480 (1978).

Davis, J.A., and J.O. Leckie, "Surface Ionization and Complexation at the Oxide/Water Interface. II. Surface Properties of Amorphous Iron Oxyhydroxide and Adsorption of Metal Ions," *J. Colloid and Interfacial Sci.*, 67:90 (1978).

Davis, J.A., and K.F. Hayes, eds., Geochemical Processes at Mineral Surfaces, ACS Symposium Series 323, American Chemical Society, Washington, DC, 1986.

Degueldre, C., B. Baeyens, W. Goerlich, J. Riga, J. Verbist, and P. Stadelmann, "Colloids in Water from a Subsurface Fracture in Granitic Rock, Grimsel Test Site, Switzerland," *Geochimica et Cosmochimica Acta*, 53:603 (1989).

Fuhriman, D.K., and J.R. Barton, Ground-Water Pollution in Arizona, California, Nevada, and Utah, U.S. Environmental Protection Agency, Water Poll. Control Res. Series 16060 ERU, 1971.

Gast, R.G., "Surface and Colloid Chemistry," in *Minerals in Soil Environments*, R.C. Dinauer, Ed., Soil Science Society of America, Madison, WI, 1977, p. 27.

Goering, H.R., E.E. Cary, L.H.P. Jones, and W.H. Allaway, "Solubility and Redox Criteria for the Possible Forms of Selenium in Soils," *Soil Sci. Soc. Amer. Proc.*, 32:35 (1968).

Hajek, B.F., "Chemical Interactions of Wastewater in a Soil Environment," *J. Water Pollut. Control Fed.*, 41:1775 (1969).

Hem, J.D., Study and Interpretation of the Chemical Characteristics of Natural Water, Third Edition, U.S. Geological Survey Water-Supply Paper 2254, 1985, p. 263.

LeGendre, G.R., and D.D. Runnells, "Removal of Dissolved Molybdenum from Wastewaters by Precipitates of Ferric Iron," *Environ. Sci. Technol.*, 9:744 (1975).

Kaback, D.S., and D.D. Runnells, "Geochemistry of Molybdenum in Some Stream Sediments and Waters," *Geochimimica et Cosmochimica Acta*, 44:447 (1980).

Kavanaugh, M.C., and J.O. Leckie, eds., Particulates in Water, Advances in Chemistry Series 189, American Chemical Society, Washington, DC, 1980.

Kellogg, W.W., R.D. Cadle, E.R. Allen, A.L. Lazarus, and E.C. Martell, "The Sulfur Cycle," *Science*, 175:589 (1972).

Lagerwerff, J.R. "Lead, Mercury, and Cadmium as Environmental Contaminants," in *Micronutrients in Agriculture*, Soil Sci. Soc. Amer., Madison, WI, 1972, p. 93.

Lakin, H.W., "Selenium in Our Environment," in *Advances in Chemistry Series No. 123*, E.L. Kothny, ed., American Chemical Society, Washington, DC, 1973, Chap. 6.

Lehman, G.S., Soil and Grass Filtration of Domestic Sewage Effluent for the Removal of Trace Elements, unpublished M.S. thesis, University of Arizona, 1968, p. 108.

Lieber, M., N.M. Perlmutter, and H.L. Frauenthal, "Cadmium and Hexavalent Chromium in Nassau County Ground Water," *J. Am. Water Works Assoc.*, 56:734 (1964).

Llewellyn, T.O., "Cadmium," in *Minerals Yearbook*, 1989, U.S. Bureau of Mines, Department of the Interior, Washington, DC, 1991, p. 1186.

Mangold, D.C., and C. Tsang, "A Summary of Subsurface Hydrological and Hydrochemical Models," *Rev. Geophysics*, 29:1 (1991).

Masscheleyn, P.H., R.D. Delaune, and W.H. Patrick, Jr., "Effect of Redox Potential and pH on Arsenic Speciation and Solubility in Contaminated Soil," *Environ. Sci. Technol.*, 25:1414 (1991).

Matthes, G., *The Properties of Groundwater*, J.C. Harvey, Translator, (New York: Wiley-Interscience, 1982), p. 406.

Melchior, D.C., and R.L. Bassett, eds., *Chemical Modeling of Aqueous Systems II*, ACS Symposium Series No. 416, American Chemical Society, Washington, DC, 1990.

Miller, D.W., F.A. DeLuca, and T.L. Tessier, Groundwater Contamination in the Northeast States, Environmental Protection Agency, Environ. Protect. Technology Series EPA-660/2-74-056, 1974, p. 325.

Minear, R.A., and J.W. Patterson, "Septic Tanks and Ground-Water Pollution," in *Ground-Water Pollution*, Proceedings of Groundwater Pollution Conference, Underwater Research Institute, St. Louis, MO, 1973, p. 53.

National Research Council, *Groundwater Contamination*, (Washington, DC: National Academy Press, 1984), p. 179.

National Research Council, *Ground Water Models—Scientific and Regulatory Applications*, (Washington, DC: National Academy Press, 1990).

Parks, G.A., "Aqueous Surface Chemistry of Oxides and Complex Oxide Minerals: Isoelectric Point and Zero Point of Charge," in *Equilibrium Concepts in Natural Water Systems*, W. Stumm, ed., Advances in Chemistry Series 67, American Chemical Society, Washington, DC, 1967, p. 121.

Rea, R.L., and G.A. Parks, "Numerical Simulation of Coadsorption of Ionic Surfactants with Inorganic Ions on Quartz," in *Chemical Modeling of Aqueous Systems II*, D.C. Melchior and R.L. Bassett, eds., ACS Symposium Series No. 416, American Chemical Society, Washington, DC, 1990, p. 260.

Runnells, D.D., "Wastewater in the Vadose Zone of Arid Regions: Geochemical Interactions," *Ground Water*, 14:374 (1976).

Ryan, J.N., and P.M. Gschwend, "Colloid Mobilization in Two Atlantic Coastal Plain Aquifers: Field Studies," *Water Resour. Res.*, 26:307 (1990).

Sayler, G.S., "Molecular Approaches for the Analysis and Recovery of Biodegradative Microbial Populations," in *Proceedings: Environmental Research Conference on Groundwater Quality and Waste Disposal*, I.P. Murarka and S. Cordle, eds., EPRI EN-6749, Electric Power Research Institute, Palo Alto, Ca, 1990, p. 32-1.

Scalf, M.R., J.W. Keeley, and C.J. LaFevers, Ground-Water Pollution in the South Central States, U.S. Environmental Protection Agency, Environ. Prot. Technology Series EPA-R2-73-268, 1973, p. 181.

Smith, R.L., B.L. Howes, and J.D. Huff, "Denitrification in Nitrate-Contaminated Groundwater: Occurrence in Steep Vertical Geochemical Gradients," *Geochimica et Cosmochimica Acta*, 55:1815 (1991).

Sparks, D.L., *Kinetics of Soil Chemical Processes*, (San Diego, CA: Academic Press, Inc., 1989), p. 210.

Stumm, W., ed., *Aquatic Surface Chemistry*, (New York: Wiley-Interscience, 1987).

Tamura, T., "Sorption Phenomena Significant in Radioactive Waste Disposal," in *Underground Waste Management and Environmental Implications*, T.D. Cook, ed., Amer. Assoc. Petrol. Geologists Memoir 18, Tulsa, OK, 1972, p. 318.

van der Leeden, F., F.L. Troise, and D.K. Todd, *The Water Encyclopedia*, Second Edition, (Chelsea, MI: Lewis Publishers, 1990), p. 808.

Wierenga, P.J., M.T. van Genuchten, and F.W. Boyle, "Transfer of Boron and Tritiated Water Through Sandstone," *J. Environ. Quality*, 4:83 (1975).

Wierenga, P.J., ed., "Validation of Flow and Transport Models for the Unsaturated Zone," *J. Contaminant Hydrol.*, 7:175 (1991).

Wilhelm, E., R. Battino, and R.J. Wilcock, "Low-Pressure Solubility of Gases in Liquid Water," *Chemical Rev.*, 77:219 (1977).

Winograd, I.J., "Radioactive Waste Storage in the Arid Zone," *Trans. Amer. Geophys. Union*, 55:10 (1974).

Woolson, E.A., J.A. Axley, and P.C. Kearney, "The Chemistry and Phytotoxicity of Arsenic in Soils: I. Contaminated Field Soils," *Soil Sci. Soc. Amer. Proc.*, 35:938 (1971).

7

Sorption and Transport of Organic Chemicals

Mark L. Brusseau

INTRODUCTION

There are several factors and processes that can affect the transport and fate of organic contaminants in the vadose zone, including advection by water flow, dispersion, gas-phase transport, and immiscible-liquid processes. Other processes of importance are hydrolysis, oxidation and reduction, volatilization, sorption, and biological transformation. Sorption and biological transformations are often the dominant reactions governing the fate and transport of chemicals in the subsurface. Elsewhere in this book, Runnells reviews hydrolysis and oxidation/reduction reactions. The focus of this section is volatilization, sorption, and transport of organic solutes.

VOLATILIZATION

Volatilization in the vadose zone is the transfer of a chemical from a liquid or solid phase to the vapor or gas phase; this gas phase may be the soil pore space or the atmosphere. Volatilization is governed by properties of the chemical and the soil, and by climatic factors. Chemical properties governing volatilization include vapor pressure, boiling point, and solubility. The distribution of a substance between vapor and liquid phases is commonly described by Henry's law as follows:

$$K_H = \frac{V_p}{C} \tag{1}$$

0–87371–610–8/95/$0.00 + $.50

where K_H = Henry's law constant
           V_p = vapor pressure of the chemical
           C = concentration of the chemical in water

Henry's law is valid for dilute solutions (e.g., mole fraction less than 0.001). Because many organic contaminants of interest are poorly soluble in water, the Henry's law relationship often holds from trace concentrations to near water-solubility levels. Accordingly, K_H may be estimated as the ratio of the saturated vapor density to water solubility. Alternatively, Henry's law constant may be estimated from the following equation (Dragun, 1988):

$$K_H = (V_p)(MW)/760(S) \tag{2}$$

where K_H = Henry's law constant (atm.m^3/mole)
           V_p = vapor pressure of the chemical (mm Hg)
           MW = molecular weight of the chemical (g/mol)
           S = solubility in water (mg/L)

The soil properties affecting volatilization include soil water content, bulk density or porosity, and sorption capacity (clay and organic matter content). Larger soil water contents result in smaller cross-sectional areas for mass transfer, which affects the effective gas-liquid mass transfer coefficients, and hence volatilization. Soil organic matter and clay content can retard emissions of volatile organic compounds by sorption effects; i.e., the presence of a third phase (sorbed phase) reduces the fraction of contaminant transferring to the gas phase.

EQUILIBRIUM SORPTION AND THE SORPTION ISOTHERM

In the context of contaminant transport in the vadose zone, an insight into sorption phenomena is important because chemicals strongly sorbed to a solid matrix are relatively immobile and, hence, not readily transported. Conversely, weakly sorbed chemicals are readily transported, serving as potential sources of ground-water pollution.

Elsewhere in this book, Runnells reviewed the basic fundamentals of sorption of inorganic compounds. Hence, the present discussion will focus on facets that are particular to organic compounds. In addition, most of the discussion will emphasize non-ionic organic compounds, since the sorption of ionic organic compounds is similar in a general manner to that of inorganic compounds (e.g., sorption by ion exchange).

The sorption isotherms generated for non-ionic organic solutes are typically described with the Freundlich isotherm:

$$S = K_f C^n \tag{3}$$

where S is the concentration of solute associated with the sorbent, C is the liquid-phase concentration, K_f is the Freundlich equilibrium sorption coefficient and n is the power function. The value for n usually ranges from 0.8 to 1. Several investigators have shown that the linear isotherm is a valid representation of sorption in many cases (cf., Brusseau and Rao, 1989). However, there is still some discussion of the occurrence, cause, and prevalence of nonlinear isotherms. For the case of n = 1, the Freundlich isotherm is equivalent to a linear isotherm.

The existence of nonlinear isotherms can have an impact on vadose zone monitoring activities. For example, estimates of solute distribution and total contaminant mass calculated with an assumption of a linear isotherm will be erroneous if the isotherm is actually nonlinear. A major potential impact of nonlinear isotherms is on solute transport. This will be discussed in a following section.

The Relationship Between Soil Organic Carbon, Moisture Content and Sorption

It has been well documented that the sorption of non-ionic organic compounds is controlled primarily by the organic matter components of soil. The contributions of inorganic components, such as clay minerals, to the sorption of non-ionic organic compounds has been shown to be, in most cases, relatively small. The positive linear correlation observed between sorption of non-ionic organic chemicals and soil organic matter is expressed in terms of the organic carbon partition coefficient, K_{oc}:

$$K_{oc} = \frac{\text{mass of sorbed chemical/mass of organic carbon}}{\text{mass of chemical in solution/volume of solution}} \qquad (4)$$

Several researchers have shown that this relationship may be valid in soils with an organic carbon content as low as 0.1%. The organic carbon partition coefficient is related to the distribution coefficient, K_d, as follows:

$$K_{oc} = K_d/f_{oc} \qquad (5)$$

where f_{oc} is the fraction of organic carbon.

There are conditions where organic matter may not predominate the sorption of organic compounds. The first is the case of ionic or highly polar compounds. These compounds can have significant interactions with inorganic soil components. A second case is where even relatively low polarity, non-ionic compounds interact significantly with inorganic components under conditions of extremely low moisture. When water is not present to solvate the soil surfaces, the sorption of organic compounds can increase many fold through vapor-sorbent interactions. However, soil in the vadose zone will rarely be under conditions of low moisture content sufficient to observe these large increases in sorption. Therefore, the assumption of no vapor-phase sorption will be valid for most vadose-zone systems (cf., Rao et al., 1989; Smith et al., 1990; Ong and Lion, 1991).

Rate-Limited Sorption

A large quantity of data exists that demonstrates the sorption of organic solutes by soils is not instantaneous. This "nonequilibrium" or rate-limited sorption phenomenon has been under investigation for several years. Early attempts to simulate the sorption kinetics of organic compounds involved simple "one-site" first-order models in which the sorption rate is taken as a function of the concentration difference between the sorbent and solution phases (cf., Oddson et al. 1970). This approach was based on that taken by researchers in chemical engineering (e.g., Lapidus and Amundson, 1952). This model has often failed to predict experimental data (e.g., Schwarzenbach and Westall, 1981; Wu and Gschwend, 1986; Boesten and van der Pas, 1988). Sorption data from batch experiments have been found to exhibit a two-stage approach to equilibrium: a short initial phase of fast uptake, followed by an extended period of much slower uptake. This pattern is followed by most, if not all, sorption reactions (cf., Brusseau and Rao, 1989). Generally, roughly 50% of the sorption occurs within the first few minutes to hours, with the remainder occurring over periods of days or months (e.g., Karickhoff and Morris, 1985; McCall and Agin, 1985; Brusseau and Rao, 1989; Ball and Roberts, 1991). This type of behavior lends itself readily to approximation with a bicontinuum approach. Models formulated with this conceptualization have been variously called "two-site," "two-compartment," or "two-box" models (cf., Brusseau and Rao, 1989).

The mechanism or mechanisms responsible for rate-limited sorption has been under investigation by several researchers. Several processes have been proposed as the causative mechanism for rate-limited sorption of organic solutes. They can be grouped into three categories (Brusseau and Rao, 1989): (1) transport-related nonequilibrium, which results from macroscopic-scale heterogeneous soil/aquifer properties (e.g., aggregates, macropores), (2) chemical nonequilibrium, which is caused by rate-limited sorbate-sorbent interactions, and (3) intrasorbent diffusion, i.e., rate-limited mass transfer within the sorbent matrix. The elucidation of causative mechanisms has been clouded somewhat by failure to differentiate between these three groups of mechanisms.

The sorption of non-ionic, low-polarity, hydrophobic organic chemicals is generally considered to be predominated by solvophobic interactions and, as such, is unlikely to be constrained by a rate-limited chemical reaction (cf., Brusseau and Rao, 1989). That such is the case was recently demonstrated by Brusseau et al. (1991a). Intrasorbent diffusion would, therefore, appear to be the most likely cause of the rate-limited sorption observed for many organic solutes. Diffusion-limited sorption has been proposed by several researchers as the probable cause for rate-limited sorption of organic solutes. The exact domain involved in the diffusion process has, however, been in question. It is likely that the domain will be a function of the type of organic compound and the nature of the soil. For non-ionic organic compounds, it appears that diffusion within sorbent organic matter may be the controlling process (Brusseau and Rao, 1989; Brusseau et al., 1991a).

TRANSPORT OF ORGANIC SOLUTES

The conventional equation used to describe one-dimensional, advective-dispersive transport of a sorbing solute under conditions of steady-state water flow is (Lapidus and Amundson, 1952):

$$\frac{\partial C}{\partial t} + \frac{\rho \partial S}{\theta \partial t} = D \frac{\partial^2 C}{dx^2} - v \frac{\partial C}{\partial x} \tag{6}$$

where C is solution-phase solute concentration (M/L^3), S is sorbed-phase solute concentration (M/M), D is dispersion coefficient (L^2/T), v is average pore-water velocity (L/T), ρ is bulk density (M/L^3), θ is soil-water content (L^3/L^3), x is distance (L), and t is time (T). Describing the $\partial S/\partial t$ term is the critical step in formulating transport models for sorbing solute. The typical approach has been to assume that sorption isotherms are linear, that desorption isotherms coincide with adsorption isotherms (singular isotherms), and that sorption-desorption is instantaneous. These assumptions greatly simplify the resultant expression for $\partial S/\partial t$, where $\partial S/\partial t = K_d$ $\partial C/\partial t$ (K_d = distribution coefficient). Upon substitution, the transport equation becomes

$$R \frac{\partial C}{\partial t} = D \frac{\partial^2 C}{\partial X} - v \frac{\partial C}{\partial X} \qquad (7)$$

where $R = (1 + (\rho/\theta) K_d)$ is the retardation factor. This equation will hereafter be referred to as the advective-dispersive (A-D) transport equation.

Equation 3 may be written in the following nondimensional form

$$R \frac{\partial C^*}{\partial T} = \frac{1}{P} \frac{\partial^2 C^*}{\partial X} - \frac{\partial C^*}{\partial X} \qquad (8)$$

by introducing the following dimensionless parameters

$$T = \frac{vt}{L} \qquad X = \frac{x}{L} \qquad (9)$$

Where L is system length and T is dimensionless time in terms of pore volumes.

The Peclet number (P) describes the shape of the solute breakthrough curve (i.e., magnitude of dispersion) and R describes its position (relative to the mean velocity of water). The solution to Equation 4 under flux-type boundary conditions is given by Brenner (1962) for various P values and with $R = 1$.

Inspection of model output shows that, as P increases, the breakthrough curve (BTC) approaches a symmetrical, sigmoidal shape (see Figure 7.1). However, for P less than approximately 10, the BTC is noticeably asymmetrical. This asymmetry is the result of a very large dispersive-transport component (i.e., large D).

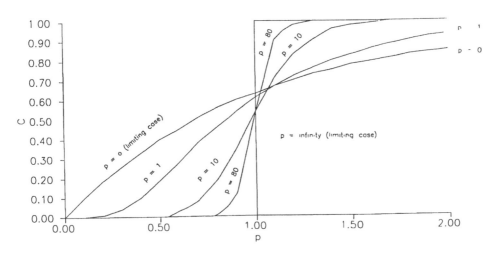

Figure 7.1. Brenner's solution to the nondimensional form of the advective-dispersion transport equation.

When one or more of the conditions that were assumed to derive Equation 4 are not met, the A-D model will not fully describe solute-transport data. In these situations, the data may be characterized as exhibiting nonideality. Many data sets have been reported that exhibit behavior deviating from that predicted by the simple model. This nonideal behavior has been attributed to several factors, including structured soil, rate-limited sorption reactions, diffusive mass transfer resistances, isotherm nonlinearity, sorption-desorption nonsingularity, and facilitated transport. As a majority of the past and present subsurface contamination modeling efforts have employed the A-D model and its attendant assumptions, it is important to analyze the validity of this approach. For cases where the A-D model is not valid, the effects of nonideality on contaminant transport, modeling, and remediation need to be assessed. The remaining sections deal with solute mass transfer and transport in structured soils, facilitated transport, solute transport in unsaturated soils, transport of rate-limited solute, and transport of solute having nonlinear sorption isotherms.

Solute Transport in Structured Soils

This discussion on transport in structured soils is taken largely from a recent review presented by Brusseau and Rao (1990). The literature is replete with reports on the influence of soil structure on solute transport. Such reports date back to over 100 years ago (Schumacher, 1864; Lawes et al., 1882), as noted by White (1985). The explanation that the observed nonideal transport (e.g., "preferential flow," "subsurface storm flow") is the result of nonunimodal pore-size distributions (i.e., nonuniform velocity field) is generally accepted. To model such systems, a two-domain or dual-porosity approach is often employed. With this approach, the porous medium is considered to be comprised of two domains: a "mobile" domain, where solute transport occurs by advection and dispersion, and an "immobile" domain (stagnant water), in which there is minimal advective flow.

Rapid transport in the mobile domain is accompanied by diffusive mass transfer of solutes between the mobile and immobile domains, which results in the latter behaving as sink/source components. Solute transport in such systems, as described by breakthrough curves, is characterized by early initial breakthrough and by "tailing" or delayed approach to relative concentration values of 0 or 1. Because access to some portion of the porous medium is constrained by diffusive mass transfer, solute in the system may be considered to be in a state of nonequilibrium. This phenomenon has been termed physical or transport nonequilibrium (TNE). An example of this phenomenon is shown in Figure 7.2, where a breakthrough curve for transport of a nonreactive tracer in an aggregated soil is shown. Compare the tailing exhibited in this case to the symmetrical, nontailed breakthrough curve obtained for transport of the same tracer through a nonaggregated soil (Figure 7.2).

Behavior attributable to TNE has been observed in experiments performed with aggregated, macroscopically heterogeneous (with respect to hydraulic conductivity), and fractured porous media as well as macroporous media. Several field studies of solute transport under controlled conditions, where observed nonideality was attributed to TNE, have been performed for inorganics (cf., Balasubramanian et al., 1973; Wild and Babiker, 1976). An early study of TNE in field-scale transport of an organic solute (herbicide picloram) was performed by Rao et al. (1974). Other studies for organics (Bowman and Rice, 1986; Jury et al., 1986) have since been reported.

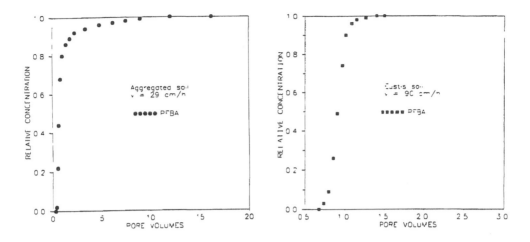

Figure 7.2. (*left*) Nonideal versus ideal transport of pentafluorobenzoate (PFBA) through an aggregated soil (Adapted from Brusseau, 1993). (*right*) Nonideal versus ideal transport of pentafluorobenzoate (PFBA) through a nonaggregated soil (Adapted from Brusseau, 1993).

Solute Transport Under Unsaturated Conditions

The fact that pore spaces in the vadose zone are rarely saturated with water leads to additional solute transport complexities. For example, Biggar and Nielsen (1960) performed miscible displacement studies under (water) unsaturated conditions and observed "progressive shifting of the BTC to the left" as soil-water content decreased. This behavior was attributed to the creation of "static sinks" upon desaturation. Similar results have since been reported by several researchers. The influence of unsaturated conditions upon nonideal solute transport (e.g., increased dispersion) is dependent upon the pore-size distribution of the porous medium. Two general cases can be described as follows (Brusseau and Rao, 1990).

Uniform Distribution

Nonideality will generally increase when soil-water content is reduced from saturation for a medium that has a relatively narrow, uniform distribution of pore sizes. Breakthrough curves obtained under saturated conditions will exhibit a relatively small amount of dispersion because of the uniform pore-size distribution. Upon desaturation, some fraction of the pores no longer readily transmit water. These pores thus become "immobile" domains that act as distributed diffusional sink/source components for solute. Breakthrough curves obtained under unsaturated conditions will exhibit a greater degree of dispersion and may be shifted leftward in comparison to the BTC obtained under saturation because of the impact of the immobile domains. Such behavior, where nonideality increases as soil-water content is reduced from saturation, has been reported by several researchers (cf., Biggar and Nielsen, 1960; Krupp and Elrick, 1968).

Further reduction in soil-water content can result in the effective (i.e., transmitting) pore-size distribution once again becoming narrow and uniform. Solute transport will thus become less nonideal, in comparison to that observed for the larger soil-water contents. Solute-transport nonideality then, will first increase as soil water content is reduced from saturation, reach a maximum at some critical value of soil-water content that is a function of the medium, and then decrease as soil-water content is further reduced. Such behavior was reported by Krupp and Elrick (1968).

Nonuniform Distribution

Solute transport under saturated conditions in porous media that are characterized by a wide pore-size distribution (e.g., presence of macropores), will exhibit nonideal behavior, as discussed in a previous chapter. The macropores drain upon desaturation, which results in a narrowing of the effective pore-size distribution. Solute transport nonideality will thus be reduced when soil-water content is reduced from saturation. Such behavior has been observed, for example, by Seyfried and Rao (1987).

Once the largest pores are drained, it is likely that the porous medium will thereafter exhibit behavior similar to that discussed in case (1) for uniform distribution. Further reduction in soil-water content will create immobile regions and result in increased nonideality. After the critical soil-water content is reached, further reductions will eliminate the immobile regions, resulting in a reduction in nonideality.

Transport of Rate-Limited Sorbing Solute

The occurrence of transport-related nonequilibrium during solute transport in structured soils is well documented, as discussed above. There exists, however, a significant amount of published data where nonideal behavior is exhibited in the absence of TNE; for example, nonsorbing solutes exhibiting symmetrical breakthrough curves, whereas sorbing solutes do not (cf., references cited in Brusseau and Rao, 1989). Such behavior suggests that the nonequilibrium is a result of a sorption-related mechanism (Brusseau and Rao, 1989). The influence of heterogeneous soil properties on water flow and thus solute transport can result in behavior that is similar to that caused by rate-limited sorption. However, it is important to note that the phenomena attributed to transport-related nonequilibrium are exhibited by nonsorbing, as well as sorbing, solutes. This highlights the point that the mechanism responsible for this physical "nonequilibrium" is a transport component, and that a sorption reaction itself is not involved. It is best to label this case as nonequilibrium or nonideal transport. In cases where nonequilibrium behavior is exhibited in the absence of heterogeneous or structured soil, it is probable that the mechanism is related to the sorption process.

Transport of Solute Having Nonlinear Sorption Isotherms

The isotherms for many organic compounds may be nonlinear, as discussed above. Such nonlinearity affects the transport of a solute by causing the retardation factor to be a concentration-dependent parameter. This nonconstant retardation factor results in the position of a breakthrough curve being dependent on the influent concentration. This is demonstrated for the case where n is less than one by the data presented by Rao and Davidson (1979), which is shown in Figure 7.3. Isotherm nonlinearity also influences the shape of the breakthrough curve. For the case of $n < 1$, the front portion of the wave is sharpened and the distal portion is extended (see Figure 7.3).

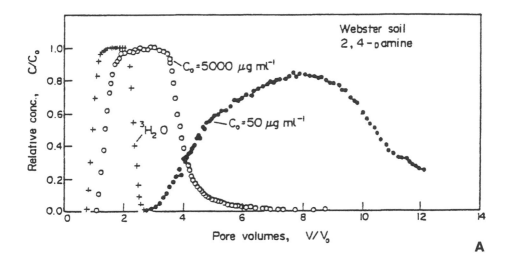

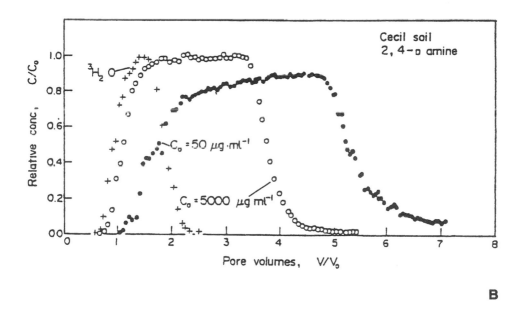

Figure 7.3. Effect of nonlinearity on solute transport, effluent breakthrough curves for 2,4-D-amine (C_0 = 50 and 5000 μg/mL) and tritiated water displacement through (A) Webster soil and (B) Cecil soil (from Rao and Davidson, 1979).

Facilitated Solute Transport

Facilitated transport is any process enhancing the transport of a solute at rates in excess of those expected from idealized Darcian flow and equilibrium sorption by a solid phase (Huling, 1989). As a consequence of facilitated transport, solutes normally considered to be relatively immobile [e.g., polyaromatic hydrocarbons (PAHs)] may be transported quickly through the vadose zone. Processes associated with facilitated transport include cosolvent effects and interactions with colloids.

Cosolvency involves the effect of organic solvents (such as methanol) on the solubility of nonpolar organic compounds. The presence of a cosolvent generally increases the solubility of a non-ionic organic chemical, resulting in decreased sorption and, hence, an increased mobility of these compounds. For nonpolar organics, as the concentration of a cosolvent increases in a mixture containing the organic compound of interest, the sorption coefficient decreases logarithmically (cf. Rao et al., 1985; Fu and Luthy, 1986; Wood et al., 1990). Thus, the retardation coefficient also decreases. Interestingly, cosolvents appear to affect the rate as well as the magnitude of sorption, with sorption-desorption rate constants increasing as cosolvent concentration increases (Brusseau et al., 1991b).

By definition, colloids are particles with diameters less than 10 microns (McCarthy and Zachara, 1989). Common colloids include clays, humic substances, viruses, and certain bacteria. Low-solubility contaminants may be mobilized by their affinity for easily translocatable colloidal substances. Two classes of colloids are associated with the cotransport of organic compounds; namely, organic and inorganic colloids. According to Huling (1989), organic colloids include biocolloids, such as bacteria, spores, and viruses; macromolecules such as humic substances; and nonaqueous phase liquids. Examples of inorganic colloids are clays, FeOOH, and precipitated $CaCO_3$ (McCarthy and Zachara, 1989). Little information is presently available on the association between inorganic colloids and organic pollutants and the effect of such an association on mobility enhancement (Huling, 1989).

Transport in the Field

Several factors that can cause nonideal solute transport have been briefly discussed in the preceding paragraphs. Although they have been discussed individually, it is possible, and even probable, that the transport of a particular solute will be affected simultaneously by several of them (cf., Brusseau, et al., 1989; Brusseau, 1992). Therefore, efforts to develop monitoring systems should consider all potential factors that may affect solute transport and fate in the vadose zone.

REFERENCES

Balasubramanian, V., Y. Kanehiro, P.S.C. Rao, and R.E. Green, "Field Study of Solute Movement in a Highly Aggregated Oxisol with Intermittent Flooding: 1. Nitrate," *J. Environ. Quality*, 2:359 (1973).

Ball, W.P., and P.V. Roberts, "Long-Term Sorption of Halogenated Organic Chemical by Aquifer Material-2. Intraparticle Diffusion," *Environ. Sci. Technol.*, 25:1237 (1991).

Biggar, J.W., and D.R. Nielsen, "Diffusion Effects in Saturated and Unsaturated Porous Material," *J. Geophys. Res.*, 65:2887 (1960).

Boesten, J.J.T.I., and L.J.T. Van Der Pas, "Modeling Adsorption/Desorption Kinetics of Pesticides in a Solid Suspension," *Soil Sci.*, 146:221 (1988).

Bowman, R.S., and R.C. Rice, "Accelerated Herbicide Leaching Resulting from Preferential Flow Phenomena and Its Implications for Ground Water Contamination," in *Proc. Conf.*

on Southwestern Ground Water Issues, AZ, 20–22 Oct., 1986, pp. 413–425, Natl. Water Well Assoc., Dublin, OH.

Brenner, H., "The Diffusion Model of Longitudinal Mixing in Beds of Finite Length. Numerical Values," *Chem. Eng. Sci.*, 17:229–243 (1962).

Brusseau, M.L., "Transport of Rate-Limited Sorbing Solutes in Heterogeneous Porous Media: Application of a One-Dimensional Multifactor Nonideality Model to Field Data," *Water Resour. Res.*, 28(9):2485–2497 (1992).

Brusseau, M.L., "The Influence of Solute Size, Pore-Water Velocity, and Intraparticle Porosity on Solute Dispersion and Transport in Soil," *Water Resour. Res.*, 29(4):1071–1080 (1993).

Brusseau, M.L., and P.S.C. Rao, "Sorption Nonideality During Organic Contaminant Transport in Porous Media," *CRC Critical Reviews in Environ. Control*, 19:33 (1989).

Brusseau, M.L., and P.S.C. Rao, "Modeling Solute Transport in Structured Soils: A Review," *Geoderma*, 46:169 (1990).

Brusseau, M.L., R.E. Jessup, and P.S.C. Rao, "Modeling the Transport of Solutes Influenced by Multi-Process Nonequilibrium," *Water Resour. Res.*, 25, 1989.

Brusseau, M. L., R.E. Jessup, and P. S. C. Rao, "Nonequilibrium Sorption of Organic Chemicals: Elucidation of Rate-Limiting Processes," *Environ. Sci. Technol.*, 25(1):134–142 (1991).

Brusseau, M.L., A.L. Wood, and P.S.C. Rao, "The Influence of Organic Cosolvents on the Sorption Kinetics of Hydrophobic Organic Chemicals," *Environ. Sci. Technol.*, 25:903–910 (1991b).

Dragun, J., The Soil Chemistry of Hazardous Materials, Hazardous Materials Control Research Institute, Silver Spring, MD, 1988.

Fu, I., and R.G. Luthy, "Effect of Organic Solvent on Sorption of Aromatic Solutes onto Soils," *J. Environ. Eng.*, 112:346–357 (1986).

Huling, S. G., Facilitated Transport, Superfund Groundwater Issue Paper, EPA/540/4–89/003, Robert S. Kerr Environmental Research Laboratory, Ada, OK, 1989.

Jury, W.A., H. Elabd, and M. Resketo, "Field Study of Napropamide Movement Through Unsaturated Soil," *Water Resour. Res.*, 22(5):749–755 (1986).

Karickhoff, S.W.. and K.R. Morris, "Sorption Dynamics of Hydrophobic Pollutants in Sediment Suspensions," *Env. Tox. Chem.*, 4:469 (1985).

Krupp, H.K., and D.E. Elrick, "Miscible Displacement in an Unsaturated Glass Bead Medium," *Water Resour. Res.*, 4:809 (1968).

Lapidus, L., and N.R. Amundson, "Mathematics of Adsorption in Beds. VI. The Effects of Lingitudinal Diffusion in Ion Exchange and Chromatographic Columns," *J. Phys. Chem.*, 56:984 (1952).

McCall, P.J., and G.L. Agin, "Desorption Kinetics of Picloram as Affected by Residence Time in the Soil," *Env. Tox. Chem.* 4:37 (1985).

McCarthy, J.F., and J.M. Zachara, "Subsurface Transport of Contaminants," *Environ. Sci. Technol.*, 23(5):496–502 (1989).

Oddson, J.K., J. Letey, and L.V. Weeks, "Predicted Distribution of Organic Chemicals in Solution and Adsorbed as a Function of Position and Time for Various Chemical and Soil Properties," *Soil Sci. Soc. Am. Proc.*, 34:412 (1970).

Ong, S.K., and L.W. Lion, "Effects of Solid Properties and Moisture on the Sorption of Trichloroethylene Vapor," *Water Res.*, 25(1):29–36 (1991).

Rao, P.S.C., R.E. Green, V. Balasubramanian, and Y. Kanehiro, "Field Study of Solute Movement in a Highly Aggregated Oxisol with Intermittent Flooding. II. Picloram," *J. Environ. Qual.*, 3:197 (1974).

Rao, P.S.C., and J.M. Davidson, "Adsorption and Movement of Selected Pesticides at High Concentrations in Soils," *Water Res.*, 13:375–342 (1979).

Rao, P.S.C., A.G. Hornsby, D.P. Kilcrease, and P. Nkedi-kizza, "Sorption and Transport of Hydrophobic Organic Chemicals in Aqueous and Mixed Solvent Systems: Model Development and Preliminary Evaluation," *J. Environ. Qual.*, 14:376–384 (1985).

Rao, P.S.C., R.A. Ogwada, and R.D. Rhue, "Adsorption of Volatile Organic Compounds on Anhydrous and Hydrated Sorbents: Equilibrium Adsorption and Energetics," *Chemosphere*, 18(11/12):2177–2191 (1989).

Schwarzenbach, R.P., and J. Westall, "Transport of Nonpolar Organic Compounds from Surface Water to Groundwater. Laboratory Sorption Studies," *Environ. Sci. Technol.*, 15:1360 (1981).

Seyfried, M.S., and P.S.C. Rao, "Solute Transport in Undisturbed Columns of an Aggregated Tropical Soil: Preferential Flow Effects," *Soil Sci. Soc. Am. J.*, 51:1434 (1987).

Smith, J.A., C.T. Chiou, J.A. Kammer, J.A., and D.E. Kile, "Effect of Soil Moisture on the Sorption of Trichloroethane Vapor to Vadose-Zone Soil at Picatinny Arsenal, New Jersey," *Environ. Sci. Technol.*, 24(5):676–683 (1990).

White, R.E., "The Influence of Macropores on the Transport of Dissolved and Suspended Matter Through Soil," *Soil Sci.*, 3:95, B.A. Stewart, ed., (New York: Springer-Verlag, 1985).

Wild, A., and I.A. Babiker, "The Asymmetric Leaching Pattern of Nitrate and Chloride in a Loamy Sand under Field Conditions," *J. Soil Sci.*, 27:460 (1976).

Wood, A.L., D.C. Bouchard, M.L. Brusseau, and P.S.C. Rao, "Cosolvent Effects on Sorption and Mobility of Organic Contaminants in Soils," *Chemosphere*, 21:575–585 (1990).

Wu, S., and P.M. Gschwend, "Sorption Kinetics of Hydrophobic Organic Compounds to Natural Sediments and Soils," *Environ. Sci. Technol.*, 20:725 (1986).

8

Biotransformation of Organic Compounds

R.M. Miller

INTRODUCTION

Organic compounds in the vadose zone can undergo a variety of biological and chemical transformations. Biological transformations range from partial to complete (Atlas and Bartha, 1993). The complete breakdown of an organic compound to carbon dioxide and water is called mineralization, which is a process that provides energy for growth and reproduction of cells. Incomplete transformations can also provide energy to support cellular growth, but can result in accumulation of dead-end intermediate products which are not further degraded. Cometabolism is a second type of incomplete transformation in which a partial oxidation of the substrate occurs but the degrading cell is not able to utilize the energy from the oxidation for growth. Finally, in some instances microbial transformations can result in polymerization or coupling of organic compounds leading to formation of products which are more complex and stable than the parent compound (Bollag, 1992).

In comparing transformation reactions which may occur in the vadose zone to those which occur in the ground-water region or surface soils, several generalizations can be made. Similar to surface soils but differing from the ground-water region, the primarily aerobic conditions found in the vadose zone will favor aerobic degradation and thus, growth of aerobic organisms (Colwell, 1989). Pockets of anaerobic activity may be found due to localized conditions which lead to reduced oxygen levels, e.g., high biodegradative activity, saturated microsites, or seasonally saturated regions of the vadose zone. Similar to the ground-water region (Fredrickson et al., 1991), the number of microorganisms found in the vadose zone is often small in comparison to surface soils (Colwell, 1989; Konopka and Turco, 1991). This is particularly true when describing microbial populations in terms of viable counts. Viable counts tend to neglect organisms that exist in a state of semistarvation, a condition common to many organisms in the vadose zone, an environment characterized by low amounts of available organic carbon. Also similar to the ground-water region, the vadose zone has limited exposure to the wide variety of organic compounds found in most surface soils. Therefore, long acclimation or adaptation times may be necessary before biodegradation occurs. It should be noted that nutrient concentrations (including oxygen) in the saturated region can vary greatly, depending upon rates of recharge and discharge. In saturated systems which have high flow rates, numbers and activities of microorganisms can be similar to that found in surface soils (Chapelle, 1993; Konopka and Turco, 1991).

To summarize conditions in the vadose zone, it might be expected that transformation reactions would be primarily aerobic. The development of localized anaerobic regions may result from unusually high microbial activity or saturated microsites. Due to the low numbers of microorganisms and poor nutrient conditions in the vadose zone, significant acclimation times may be necessary for selection and growth of specific microorganisms able to degrade a target contaminant (Konopka and Turco, 1991).

Even though differences in environmental conditions exist between the vadose zone and surface soils, transformation reactions that occur in the vadose zone will be governed by many of the same principles found in surface soils which are also primarily aerobic environments. In both the vadose zone and surface soils, there are many parameters which affect the rate and occurrence of biodegradation of organic compounds. Most important are the structure of the organic compound and environmental conditions, both of which are discussed in more detail in the following sections.

RELATIONSHIP BETWEEN STRUCTURE AND BIODEGRADABILITY

The rate at which an organic contaminant is degraded in the environment depends on several factors that are directly dependent on contaminant structure (Leahy and Colwell, 1990; Pitter and Chudoba, 1990). These factors include:

1. Genetic potential, or the presence and expression of appropriate degrading genes.
2. Bioavailability, or the effect of limited water solubility and sorption on the rate a substrate is taken up by a microbial cell.
3. Steric effects, or the extent to which substituent groups sterically hinder contact between the active site of the degrading enzyme and the organic molecule.
4. Electronic effects, or the extent to which substituent groups electronically interfere with the interaction between the active site of the enzyme and the substrate, or alter the energy required to break critical bonds in the molecule.

Genetic Potential

Onset of contaminant biodegradation generally follows a period of adaptation or acclimation, the length of which depends on the contaminant structure. Many organic contaminants have structures similar to those found in naturally-occurring organic matter. In some instances there may have been previous exposure to the contaminant through repeated pesticide applications or through accidental spills. For these contaminants, biodegradation pathways may already exist. Adaptation of microbial populations to this type of contaminant most commonly occurs by induction of enzymes necessary for biodegradation followed by an increase in the population of biodegrading organisms (Button et al., 1992; Estrella et al., 1993; Leahy and Colwell, 1990; Song and Bartha, 1990).

For other contaminants, naturally-occurring analogs may not exist and previous exposure may not have occurred. For these contaminants a second type of adaptation is required which involves a genetic change such as a mutation or a gene transfer (Boyle, 1992; Fulthorpe and Wyndham, 1992; van der Meer et al., 1992). This results in the development of new metabolic capabilities. The length of time needed for an adaptation requiring a genetic change is not yet predictable.

Bioavailability

Biodegradation is usually thought to occur if the appropriate microbial enzymes are present to catalyze the transformation of the compound of interest. As a result, most research to date on the actual biodegradative process has been concerned with isolation and characterization of biodegradative enzymes and genes. There are, however, two steps in the biodegradative process. The first is the uptake of the substrate by the cell, and the second is the metabolism or degradation of the substrate. In some cases this first step can be the limiting step in biodegradation, e.g., for organic compounds with limited water solubility (Bossert and Bartha, 1986; Miller and Bartha, 1989), and for organic compounds that are sorbed onto soils or sediments (Cork and Krueger, 1991).

Growth on an organic compound with limited water solubility poses a unique problem for microorganisms because the compound is not freely available in the aqueous phase. Most microorganisms require high water activity (>0.96) for active metabolism (Atlas and Bartha, 1993), and thus the contact between the degrading organism and an organic compound with low water solubility is limited. A direct consequence of limited contact is a slow rate of uptake (and biodegradation) of the compound. The compound may be present in a liquid or solid state, both of which can form a two-phase system with water. Liquid hydrocarbons can be more or less dense than water forming a separate phase above or below the water surface. For example, polychlorinated biphenyls (PCBs) and chlorinated solvents such as trichloroethylene (TCE) are denser than water forming a separate phase below the water surface. Solvents such as benzene and other petroleum constituents are less dense than water forming a separate phase above the water surface. There are three possible modes for microbial uptake of a liquid organic (Nakahara et al., 1977):

1. Direct contact of cells with the organic compound. This can be mediated by cell modifications such as fimbriae (Rosenberg et al., 1982) or cell surface hydrophobicity (Singer and Finnerty, 1984b; Zhang and Miller, 1994) which increase attachment of the cell to the organic compound, or by cell secretions such as surfactants which increase water solubility of the organic compound (Miller, 1994; Zajic and Panchel, 1976; Zhang and Miller, 1992);
2. Direct contact with fine or submicron size substrate droplets dispersed in the aqueous phase; and
3. Utilization of the solubilized organic compound.

Which mode predominates depends largely on the water solubility of the organic compound. In general, direct contact with the organic compound plays a more important role (modes 1 and 2), as water solubility decreases. For organic compounds in the solid phase, e.g., waxes, plastics or polyaromatic hydrocarbons (PAHs), there are only two modes by which a cell can take up the substrate: (a) direct contact with the substrate; and (b) utilization of solubilized substrate. Available evidence suggests that for solid phase organic compounds, utilization of solubilized substrate is most important (Chakravarty et al., 1972; Miller and Bartha, 1989; Wodzinski and Bertolini, 1972; Wodzinski and Coyle, 1974; Wodzinski and Larocca, 1977).

The second factor which affects bioavailability of an organic compound is sorption of the compound by soil or sediment. Details concerning sorption mechanisms are discussed elsewhere in this book, in the chapter by M.L. Brusseau. Depending on the sorption mechanism, organic compounds can be weakly (hydrogen-bonding, van der Waals forces, hydrophobic interactions) or strongly (covalent binding) bound to soil. Recent work has shown that weakly-bound or labile residues are available for biodegradation, while strongly bound residues are not (Novak et al., 1994; Steinberg et al., 1987; Scow, 1993). Sorption of labile residues is reversible; however, there is evidence to suggest that the proportion of labile residues decreases with length of exposure of the residues to soil (Calderbank, 1989). Thus, as contaminants age in the vadose zone, bioavailability, and therefore biodegradation, can be expected to decrease.

Steric Effects

The site at which the substrate and enzyme come into contact during a transformation step is the reaction site. The presence of branching or functional groups on a substrate can effectively block or sterically hinder enzymatic attack on the substrate carbon skeleton at the reaction site. This is illustrated in Figure 8.1 which compares two structures with C-8 carbon skeletons. Structure A is readily degradable. In contrast, the four methyl substituents in structure B inhibit degradation at both ends of the molecule. Branching or functional groups can also affect transport of the substrate across the cell membrane, especially if the transport is enzyme-assisted. Steric effects usually increase with the size of the functional group (Pitter and Chudoba, 1990).

Electronic Effects

In addition to steric effects, functional groups may contribute electronic effects which hinder biodegradation, by affecting the interaction between the substrate and

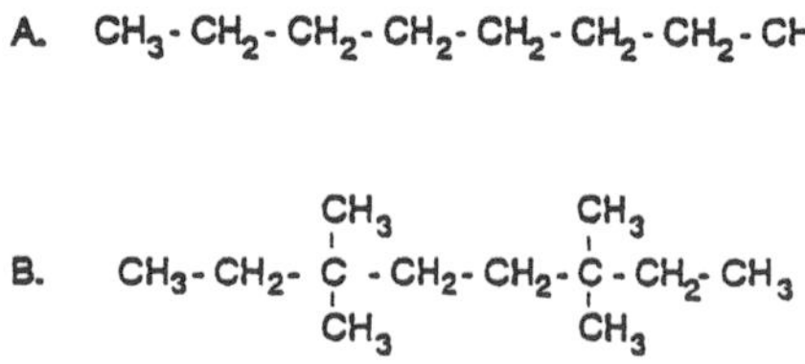

A. $CH_3 - CH_2 - CH_2 - CH_2 - CH_2 - CH_2 - CH_2 - CH_3$

B. $CH_3 - CH_2 - \overset{\overset{\displaystyle CH_3}{|}}{\underset{\underset{\displaystyle CH_3}{|}}{C}} - CH_2 - CH_2 - \overset{\overset{\displaystyle CH_3}{|}}{\underset{\underset{\displaystyle CH_3}{|}}{C}} - CH_2 - CH_3$

Figure 8.1. Effect of methyl substituents on biodegradability of octane.

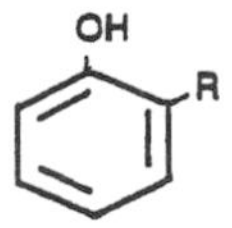

Figure 8.2. An ortho-substituted phenol where R = NO_2, Cl, OH, CH_3 or COO^-.

enzyme. Functional groups can be electron donating (CH_3) or electron withdrawing (Cl) and therefore can change the electron density of the substrate reaction center. In general, functional groups that add to the electron density of the reaction center increase biodegradation rates, while functional groups that decrease the electron density of the reaction center decrease biodegradation rates. To illustrate the relationship between functional group electronegativity and rate of biodegradation, Pitter and Chudoba (1990) compared the electronegativity of a series of ortho-substituted phenols with their biodegradation rates (Figure 8.2). Electronegativity followed the order: $NO_2 > Cl > OH > CH_3 > COO^-$, and relative rates of biodegradation were determined to be 13.9, 25.0, 55.5, 55.0, 94.8 mg g^{-1} h^{-1}, respectively.

ENVIRONMENTAL EFFECTS

There are a number of parameters that influence the survivability and activity of microorganisms in any environment. Since nutritional requirements vary among microorganisms, some microorganisms can grow in many different ecosystems while others are limited to a few specific ecosystems. Organic matter is the primary source of carbon for most heterotrophic microorganisms in the environment. The vadose zone is generally characterized by a low organic carbon content, one reason that microbial numbers in the vadose zone are low in comparison to surface soils. Occurrence and abundance of microorganisms in an environment are determined not only by nutrients but also by various physical and chemical factors. The most important in the vadose zone are: water activity, redox potential (oxygen availability), pH, and for some areas, salinity. These have been extensively reviewed by Atlas and Bartha (1993), and Paul and Clark (1989).

TRANSFORMATION REACTIONS

Aerobic Conditions

In the presence of oxygen, aerobic microorganisms oxidize organic carbon completely to carbon dioxide using oxygen as the terminal electron acceptor (oxygen is reduced to water) in a series of oxidation-reduction reactions used to produce energy for cell maintenance and growth. The vast majority of the organic carbon available to microorganisms in the vadose zone is material which has been photosynthetically fixed (plant material). However, in some instances anthropogenic activity results in addition of organic carbon in the form of industrial or agricultural chemicals such as petroleum products, organic solvents, or pesticides. Many of these chemicals are readily degraded in the environment because of their structural similarity to naturally-occurring organic carbon. However, some chemical structures may require long periods of adaptation, or have low bioavailability, or steric or electronic characteristics that result in slow to nonexistent biodegradation rates. The following sections discuss biotransformation reactions of the various types of organic compounds which are added to the environment through anthropogenic activity. These compounds can be divided into three structural classes: aliphatic, alicyclic, and aromatic. Initial degradation steps differ for each structural class. Some compounds, particularly pesticides, can have elements from more than one structural group.

Aliphatic Hydrocarbons

Aliphatic hydrocarbons include straight-chain and branched-chain structures. Industrial solvent wastes and petroleum industry by-products and spills are the primary sources of aliphatic hydrocarbon contaminants introduced into the environment (USEPA, 1984; Plumb, 1985). Many microorganisms in the environment can utilize aliphatic hydrocarbons as carbon sources (Britton, 1984; Leahy and Colwell, 1990; Pitter and Chudoba, 1990; Prince, 1993; Singer and Finnerty, 1984b). The following generalizations can be made about biodegradation of aliphatics by microorganisms.

1. Mid-size straight-chain aliphatics (n-alkanes C_{10} to C_{18} in length) are utilized more readily than n-alkanes with either shorter or longer chains. Long-chain n-alkanes are utilized more slowly, due to low bioavailability resulting from extremely low water solubilities (Miller and Bartha, 1989). For example, the reported water solubility of decane is 0.052 mg/L, while the solubility of octadecane (C_{18}) is 10-fold less (0.006 mg/L) (Singer and Finnerty, 1984b). Solubility continues to decrease with increasing chain length. In contrast, short-chain n-alkanes have higher aqueous solubility, e.g., the water solubility of butane (C_4) is 61.4 mg/L, but they are toxic to cells. Toxicity is caused by disruption of the cell membrane through interaction with membrane-bound proteins that function in transport and oxidation of aliphatics (Britton, 1984). In some cases it has been shown that the toxicity of short-chain n-alkanes can be reduced by the presence of a long-chain n-alkane. The protective effect is attributed to partitioning of the toxic hydrocarbon from the aqueous phase into the long-chain alkane, thereby reducing the concentration (Britton, 1984). Therefore, degradation rates may differ depending on whether the substrate is a pure compound or a mixture of compounds.
2. Saturated aliphatics are hydrocarbons with a carbon skeleton which is "saturated" with hydrogen, or contain only single carbon-carbon bonds. Unsaturated hydrocarbons contain one or more double (alkenes) or triple (alkynes) carbon-carbon bonds. In general, saturated aliphatics are degraded more readily than unsaturated ones (Britton, 1984).
3. Biodegradability of aliphatics is negatively influenced by branching in the hydrocarbon chain (Pitter and Chudoba, 1990). The degree of resistance to biodegradation depends on both the number of branches and on the positions of methyl groups in the molecule. Compounds with a quaternary carbon atom (4 carbon-carbon bonds) are extremely stable due to steric effects. Terminal quaternary carbons particularly inhibit biodegradation. This is illustrated in Figure 8.3, which shows that 2,2-dimethylheptane can

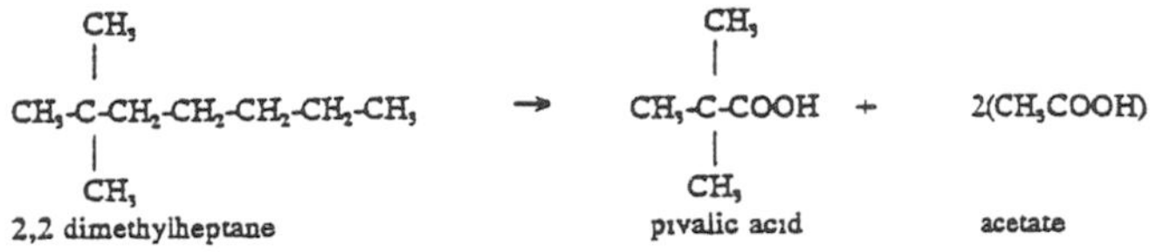

$$
\begin{array}{c}
CH_3 \\
| \\
CH_3\text{-}C\text{-}CH_2\text{-}CH_2\text{-}CH_2\text{-}CH_2\text{-}CH_3 \\
| \\
CH_3 \\
\text{2,2 dimethylheptane}
\end{array}
\rightarrow
\begin{array}{c}
CH_3 \\
| \\
CH_3\text{-}C\text{-}COOH \\
| \\
CH_3 \\
\text{pivalic acid}
\end{array}
+ \;\; 2(CH_3COOH) \atop \text{acetate}
$$

Figure 8.3. Branching inhibits further biodegradation of pivalic acid.

be oxidized to pivalic acid, which is then very resistant to further biodegradation (Catelane et al., 1977).

4. Biodegradability of aliphatics is negatively influenced by halogen substitution of the hydrocarbon chain. Studies of chlorinated aliphatics have shown that while monochloro-n-alkanes are degraded, the presence of two or three chlorines on the same carbon slows degradation significantly (Janssen et al., 1990).

Biodegradation pathways have been elucidated for many aliphatic compounds. These pathways are summarized in the following sections for alkanes, alkenes, and chlorinated aliphatics.

Alkanes.   Alkanes are usually considered to be the most readily biodegradable type of hydrocarbon and many microorganisms can degrade them (Britton, 1984). Since they are highly reduced molecules, alkane mineralization requires higher amounts of oxygen than is required for degradation of sugars or organic matter. This is reflected by the high biological oxygen demand (BOD) values reported for alkane biodegradation (Pitter and Chudoba, 1990). Aerobic biodegradation of alkanes can occur by the two pathways shown below. The most common is the direct incorporation of one atom of O_2 by a mixed function oxidase or monooxygenase enzyme (see pathway I). Alternatively, both oxygen atoms can be incorporated into the alkane (pathway II). Both pathways result in the formation of a primary fatty acid (monoterminal oxidation) which is the most common type of oxidation. There are also examples in the literature of diterminal oxidation where both ends of the alkane are oxidized, and of subterminal oxidation where an interior carbon is oxidized (Britton, 1984).

I. $R\text{-}CH_2\text{-}CH_3 \rightarrow R\text{-}CH_2\text{-}CH_2OH \rightarrow R\text{-}CH_2\text{-}CHO \rightarrow R\text{-}CH_2\text{-}COOH$

II. $R\text{-}CH_2\text{-}CH_3 \rightarrow R\text{-}CH_2\text{-}CH_2\text{-}OOH \rightarrow R\text{-}CH_2\text{-}CH_2OH \rightarrow R\text{-}CH_2\text{-}CHO \rightarrow R\text{-}CH_2\text{-}COOH$

The primary fatty acid formed in degradation of an alkane is subject to β-oxidation which cleaves off consecutive two-carbon fragments. The two-carbon fragment is removed by coenzyme A as acetyl-CoA which then enters the tricarboxylic acid (TCA) cycle for complete degradation to CO_2 and H_2O (Brock and Madigan, 1991). If the alkane has an even number of carbons, acetyl-CoA is the last residue. If the alkane has an odd number of carbons, propionyl-CoA is the last residue. Propionyl-CoA is then converted to succinyl-CoA which can then also enter the TCA cycle.

Alkenes.   Alkenes are hydrocarbons which contain one or more double bonds. The majority of alkene biodegradability studies have used 1-alkenes as model compounds (Britton, 1984). These studies have shown that alkenes and alkanes have comparable biodegradation rates. As illustrated below, the initial step in 1-alkene degradation can involve attack at the terminal (1) or a subterminal (2) carbon as described for alkanes, or attack at the double bond which can yield a secondary (3) or primary alcohol (4) or an epoxide (5) (Britton, 1984). Each of these initial degradation products will ultimately be oxidized into a primary fatty acid which is degraded as described for alkanes.

$$\rightarrow HOCH_2\text{-}CH_2\text{-}R\text{-}CH = CH_2 \qquad (1)$$

$$\rightarrow CH_3\text{-}CHOH\text{-}R\text{-}CH = CH_2 \qquad (2)$$

$$CH_3\text{-}CH_2\text{-}R\text{-}CH = CH_2 \quad \rightarrow CH_3\text{-}CH_2\text{-}R\text{-}CHOH\text{-}CH_3 \qquad (3)$$

$$\rightarrow CH_3\text{-}CH_2\text{-}R\text{-}CH_2\text{-}CH_2OH \qquad (4)$$

$$\rightarrow CH_3\text{-}CH_2\text{-}R\text{-}\overset{\displaystyle O}{\overset{\displaystyle \triangle}{CH\text{-}CH_2}} \qquad (5)$$

Halogenated Aliphatics. Chlorinated solvents such as trichloroethylene (TCE) have been extensively used as industrial solvents. As a result of improper use and disposal, these solvents are one of the most frequently detected types of organic contaminants in ground water (USEPA, 1984; Plumb, 1985). The need for efficient and cost-effective remediation of solvent contaminated sites has recently stimulated interest in the degradation of these C_1 and C_2 halogenated aliphatics.

Compared to aliphatics without halogen substitution, halogenated aliphatics generally degrade slowly. Janssen et al. (1990) report that 1-chloroalkanes ranging from C_1 to C_{12} were degraded as sole carbon sources in pure culture. Okey and Bogan (1965) demonstrated that degradation rates of 1-chloroalkanes, ranging from C_3 to C_{12}, increased with increasing alkyl chain length. These results can be explained by the decreasing electronic effects of the chlorine atom on the enzyme-carbon reaction center as the alkane chain length increased.

The presence of two or three chlorines bound to the same carbon atom seems to inhibit aerobic degradation (Janssen et al., 1990). More generally, aerobic conditions favor the biodegradation of compounds with fewer halogen substituents, while anaerobic conditions favor the biodegradation of compounds with a high number of halogen substituents (Vogel et al., 1987). However, for highly halogenated aliphatics, complete biodegradation under anaerobic conditions often does not take place; therefore, in some instances sequential anaerobic and aerobic treatment may be most useful (Fathepure and Vogel, 1991). Initial incubation under anaerobic conditions would decrease the halogen content, and subsequent addition of oxygen would create aerobic conditions to allow complete degradation.

Biodegradation of halogenated aliphatics occurs by three basic types of reactions:

1. **Substitution** is a nucleophilic reaction (the reacting species brings an electron pair) in which the halogens on a mono- or dihalogenated compound are substituted by a hydroxy group.

$$CH_3\text{-}CH_2Cl + H_2O \rightarrow CH_3\text{-}CH_2OH + H^+ + Cl^-$$

2. **Oxidation** reactions have been reported for a select group of monooxygenase and dioxygenase enzymes which act nonspecifically against a wide variety of highly chlorinated C_1 and C_2 compounds, e.g., trichloroethylene (TCE). This type of degradation is a cometabolic reaction requiring large substrate to chlorinated hydrocarbon ratios (Janssen et al., 1990). Reported examples of oxygenase/substrate systems which cometabolically degrade chlorinated hydrocarbons include: methane monooxygenase (MMO) produced by methanotrophic bacteria during growth on carbon sources such as methane or formate (Little et al., 1988; Oldenhuis et al., 1989); toluene dioxygenase (TDD) produced by several bacterial isolates during growth on toluene (Nelson et al., 1987; Wackett and Gibson, 1988); ammonia monooxygenase (AMO) produced by *Nitrosomonas europaea* during growth on ammonia (Vannelli et al., 1990); and propane monooxygenase produced by *Mycobacterium vaccae* JOB5 during growth on propane (Wackett et al., 1989). Shown below is an example of cometabolic oxidation of a C_1 compound, chloroform (see pathway I) and a C_2 alkene, TCE (see pathway II).

I. $CHCl_3 + O_2 \rightarrow C(OH)Cl_3 + H_2O \rightarrow CO_2 + 3Cl^- + 3H^+$

II. $ClHC = CCl_2 + O_2 \rightarrow ClHC\overset{\overset{O}{\|}}{-}CCl_2 \rightarrow ClHC(OH)\text{-}(OH)CCl_2 \rightarrow HCOOH + CO$
$$\rightarrow CHO\text{-}COOH$$

3. **Reductive dehalogenation** is mediated by reduced transition metal/metal complexes. Reductive dehalogenation generally occurs in an anaerobic environment, although the second of the reactions shown in Figure 8.4, formation of alkenes, can occur aerobically for a limited number of chlorinated compounds which have a higher reducing potential than O_2, e.g.,

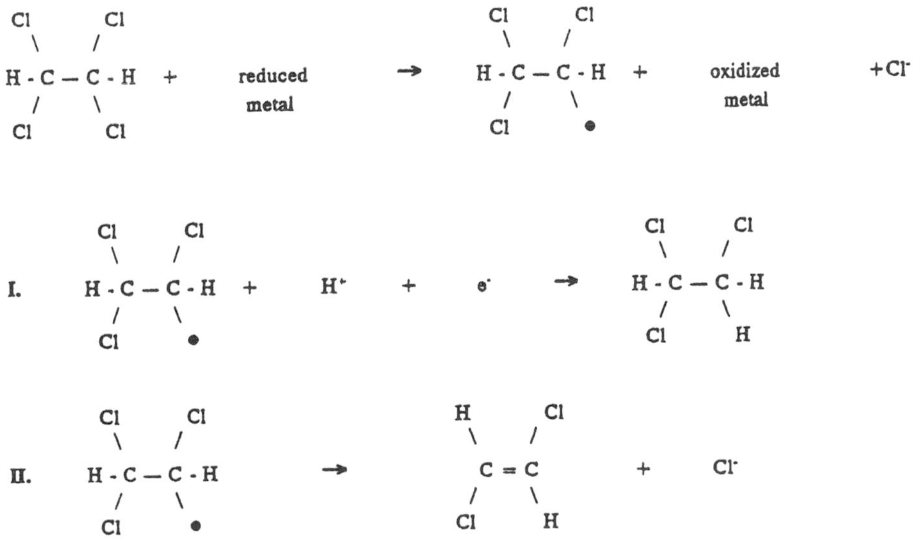

Figure 8.4. Reductive dehalogenation.

hexachloroethane and dibromoethane (Vogel et al., 1987). In the first step of reductive dehalogenation, electrons are transferred from the reduced metal to the halogenated aliphatic, resulting in an alkyl radical and free halogen. The alkyl radical can either scavenge a hydrogen atom (I), or lose a second halogen to form an alkene (II).

Alicyclic Hydrocarbons

Alicyclic hydrocarbons are saturated carbon chains which form a ring structure. There is a great variety of naturally occurring alicyclic hydrocarbons. For example, alicyclic hydrocarbons are a major component of crude oil, comprising 20% to 67% by volume. The various components range from simple, such as cyclopentane and cylcohexane, to complex, such as trimethylcyclopentane and various cycloparaffins (Perry, 1984). Other examples of complex naturally occurring alicyclic hydrocarbons include camphor, which is a plant oil, cyclohexyl fatty acids, which are components of microbial lipids, or the paraffins from leaf waxes (Trudgill, 1984). Anthropogenic additions of alicyclic hydrocarbons to the environment come from fossil fuel processing and spills, and the use of agrochemicals such as the pyrethrin insecticides.

Compared to aliphatic hydrocarbons, it is difficult to isolate pure cultures which degrade alicyclic hydrocarbons using enrichment techniques, so it is not surprising that there is no correlation between the capability to utilize n-alkanes and to fully oxidize cycloalkanes. Trower et al. (1985) isolated a *Xanthobacter* sp. from soil which was able to degrade cyclohexane completely. However, the majority of alicyclic hydrocarbon degradation is thought to occur in large part from commensalistic and cometabolic reactions where one organism converts cyclohexane to cyclohexanone via a cyclohexanol (step 1 and step 2) but is unable to lactonize and open the ring (Figure 8.5). A second organism which is unable to oxidize cyclohexane to cyclohexanone can perform the lactonization, ring opening, and mineralization of the remaining aliphatic compound (Perry, 1984; Trudgill, 1984).

Cyclopentane and cyclohexane derivatives that contain one or two OH, C = O, or COOH groups are readily metabolized. The importance of initial ring oxidation for degradation of cycloalkanes is indicated by two observations. First, degradation of unsubstituted alicyclic compounds is often cometabolic, and second, organisms capable of the degradation of cycloalkanols and cycloalkanones are so ubiquitous in their distribution that they are easily isolated from water, mud, and soil samples. In contrast, degradation of alicyclic derivatives is negatively affected by the presence of one or more CH₃ groups. Pitter and Chudoba (1990) examined a series of alkyl

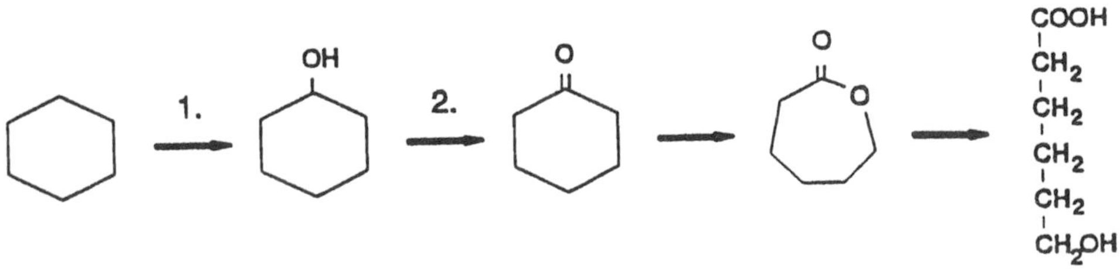

Figure 8.5. Degradation of cyclohexane.

derivatives of cyclohexanol and found that the rate of biodegradation decreased in the order: cyclohexanol > methylcyclohexanol > dimethylcyclohexanol.

Aromatic Hydrocarbons

Aromatic compounds contain at least one unsaturated ring system with the general structure C_6R_6, where R is any functional group. Benzene (C_6H_6) is the parent hydrocarbon of this family of unsaturated cyclic compounds and is unique in that it does not exhibit the high reactivity of typical polyenes. It is remarkably inert to many oxidizing reagents, stable to air, and tolerates many free radical initiators. This stability is due to the resonance energy which comes from the delocalization of electrons around the benzene ring. As a group, the benzene-like compounds were given the name "aromatic" because many of them have characteristic aromas.

Polyaromatic hydrocarbons (PAHs) are hydrocarbons containing two or more fused benzene rings. PAHs are synthesized naturally by plants, some bacteria, and fungi, and are also released by natural processes such as forest and grass fires (Cerniglia and Heitkamp, 1989). Anthropogenic activities such as fossil fuel processing and utilization are major sources of PAHs added to the environment. The biochemistry, genetics, and regulation of degradation of PAHs has been extensively studied (Cerniglia and Heitkamp, 1989; Gibson and Subramanian, 1984; Zeyer, et al., 1990). The results of this work can be summarized to aid in prediction of degradation in the vadose zone.

A wide variety of bacteria and fungi can carry out PAH transformations under a variety of environmental conditions. Under aerobic conditions, the most common initial transformation is a hydroxylation that involves the incorporation of molecular oxygen. The enzymes involved in these initial transformations fall into two groups, monooxygenases that incorporate one atom of molecular oxygen, and dioxygenases that incorporate both atoms of molecular oxygen into the PAH. In general, prokaryotic microorganisms (bacteria) transform PAH by an initial dioxygenase attack to cis-dihydrodiols. The cis-dihydrodiol is rearomatized to form a dihydroxylated intermediate, catechol. The catechol ring is cleaved by a second dioxygenase either between the two hydroxyl groups (ortho pathway) or next to one of the hydroxyl groups (meta pathway) as shown in Figure 8.6, and further degraded.

Eukaryotic microorganisms (fungi) initially attack PAH with a cytochrome P-450 monooxygenase incorporating one atom of molecular oxygen into the PAH and reducing the second to water, resulting in the formation of an arene oxide. This is

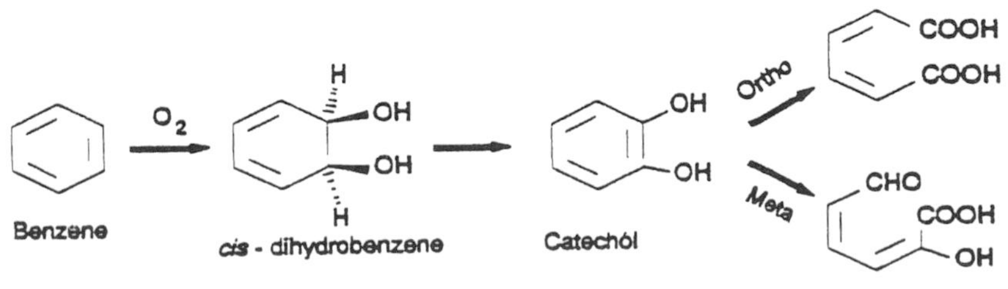

Figure 8.6. Incorporation of oxygen into the aromatic ring by the dioxygenase enzyme, followed by meta or ortho ring cleavage.

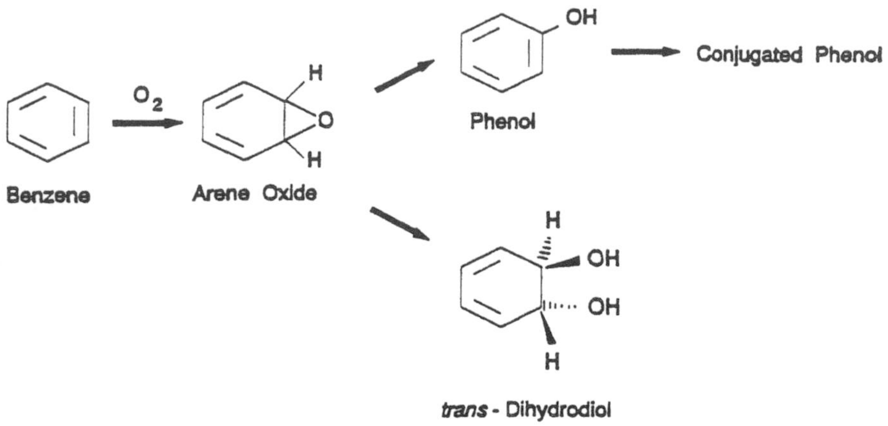

Figure 8.7. Fungal monooxygenase incorporation of oxygen into the aromatic ring.

followed by the enzymatic addition of water to yield a *trans*-dihydrodiol (Figure 8.7). Alternatively, the arene oxide can be isomerized to form phenols that can be conjugated with sulfate, glucuronic acid, glucose, and glutathione. These conjugates are similar to those formed in higher organisms, and seem to aid in detoxification and elimination of PAH.

The capability for degradation of PAH, especially chlorinated PAH, is often mediated by plasmids which are genetic elements separate from the chromosome within bacterial cells (Chakrabarty, 1976; Ghosal et al., 1985; Haas, 1983; Singer and Finnerty, 1984a). Plasmids can carry both individual genes and operons encoding for partial or complete biodegradation of PAH, and it may be possible to genetically manipulate bacterial plasmids and/or chromosomes to allow construction of bacterial strains with a broader biodegradation potential (Chaudhry and Chapalamadugu, 1991).

In general, PAHs of two or three condensed rings are transformed rapidly and often completely mineralized, whereas PAHs of four or more condensed rings are transformed much more slowly, often as a result of cometabolic attack (Cerniglia and Heitkamp, 1989; Gibson and Subramanian, 1984). Due to the hydrophobicity of PAH, transformation rates are often greatest at soil/water interfaces, and degradation rates can be influenced by environmental factors. Frequently, chronic exposure to PAH will result in increased transformation rates due to microbial adaptation.

Anaerobic Conditions

Even in well-aerated soils there are microenvironments with little or no oxygen. This can occur when the rate of oxygen consumption by microorganisms which oxidize organic materials in the microenvironment is greater than the rate of oxygen diffusion. Other more stable and generally larger anaerobic environments can develop in saturated regions of the vadose zone, e.g., perched layers. These saturated environments tend to become anaerobic due to the combined effects of the low solubility of oxygen in water (8 to 10 mg/L), the relatively slow diffusion of oxygen through water, and utilization of the available oxygen supply by active degradation of organic carbon that is present (Zehnder and Stumm, 1988).

In the absence of oxygen, organic material can alternatively be mineralized to carbon dioxide by anaerobic respiration, although this is a less efficient process than aerobic respiration. Anaerobic respiration uses alternate electron acceptors following a sequence which depends on the electron affinity of the electron acceptors. Nitrate has the highest electron affinity and is reduced first, followed by iron, manganese, sulfate, and then carbonate.

In comparison to aerobic biodegradation, far less is known about anaerobic degradation of organic compounds (Grbić-Galić, 1990). Interestingly, some compounds which are easily degraded aerobically are difficult to degrade anaerobically; for instance, saturated aliphatic compounds. On the other hand, highly halogenated compounds that are not degraded aerobically can be partially or completely degraded anaerobically. The following sections briefly discuss anaerobic degradation of aliphatic and aromatic organic compounds.

Aliphatic Hydrocarbons

Several generalizations can be made about the anaerobic biodegradation of aliphatics (Britton, 1984; Schink, 1988).

1. Although early reports suggested that saturated aliphatic hydrocarbons were degraded anaerobically, there is no conclusive evidence that saturated aliphatics are degraded in the absence of oxygen.
2. Unsaturated aliphatic hydrocarbons can be degraded anaerobically (Schink, 1985). The suggested pathway of biodegradation for unsaturated hydrocarbons is the hydration of the double bond to an alcohol, with further oxidation to a ketone or aldehyde, and finally formation of a fatty acid.
3. Shorter hydrocarbons seem to be degraded less effectively, and ethene (C_2H_4) was not degraded in enrichment cultures.
4. Branched hydrocarbons with subterminal double bonds are degraded less rapidly than hydrocarbons with terminal double bonds.
5. Aliphatics containing oxygen such as aliphatic alcohols and ketones are readily degraded by anaerobic enrichment consortia.
6. Many halogenated aliphatics are partially or completely degraded under anaerobic conditions. A highly halogenated compound will most likely undergo reductive dehalogenation (Figure 8.4) resulting in a less halogenated product which may then be amenable to oxidation (Vogel et al., 1987).

Aromatic Hydrocarbons

Oxygenated aromatic compounds such as benzoate are completely degraded under anaerobic conditions by microbial consortia working under different redox potentials. This process was described by Young and Häggblom (1989) as an anaerobic food chain. For example, a benzoate utilizer can be successfully isolated only by growing it in coculture with a methanogen or sulfate reducer. In this consortium, benzoate is transformed by one or more anaerobes to yield aromatic acids that in turn are transformed to methanogenic precursors such as acetate, carbon dioxide, or formate, which finally can be utilized by methanogens.

Until recently, nonoxygenated aromatic compounds were thought to be resistant

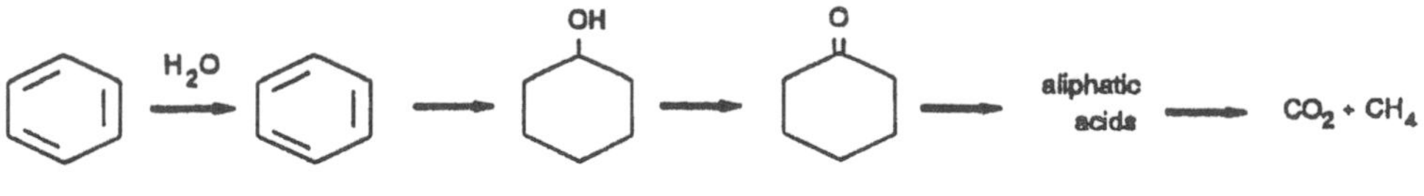

H_2O → OH → O → aliphatic acids → $CO_2 + CH_4$

Figure 8.8.

to anaerobic degradation. However, recent work has shown that a variety of non-oxygenated aromatic compounds can be degraded as shown in Figure 8.8 for benzene (Grbić-Galić, 1990).

As shown in Figure 8.8, complete degradation under anaerobic conditions results in a disproportionation of carbon to methane (CH_4) and CO_2. Disproportionation to CO_2 and CH_4 occurs under methanogenic conditions when CO_2 acts as a terminal electron acceptor and is also the final product of degradation.

REFERENCES

Atlas, R.M., and R. Bartha. *Microbial Ecology*, (Menlo Park, CA: Benjamin/Cummings Publishing Co., 1993).

Bollag, J-M. "Decontaminating Soil with Enzymes," *Environ. Sci. Technol.* 26:1876–1881 (1992).

Bossert I.D., and R. Bartha. "Structure-Biodegradation Relationships of Polycyclic Aromatic Hydrocarbons in Soil," *Bull. Environ. Contam. Toxicol.* 37:490–495 (1986).

Boyle, M. "The Importance of Genetic Exchange in Degradation of Xenobiotic Chemicals," in R. Mitchell, ed., *Environmental Microbiology*, (New York: Wiley-Liss, Inc., 1992), pp. 319–333.

Britton, L.N. "Microbial Degradation of Aliphatic Hydrocarbons, in D.T. Gibson, ed., *Microbial Degradation of Organic Compounds*, (New York: Marcel Dekker Inc., 1984), pp. 89–129.

Brock, T.D., and M.T. Madigan. *Biology of Microorganisms*, Sixth edition. (Englewood Cliffs, NJ: Prentice Hall, 1991).

Button, D.K., B.R. Robertson, D. McIntosh, and F. Juttner. "Interactions Between Marine Bacteria and Dissolved-Phase and Beached Hydrocarbons After the Exxon Valdez Oil Spill," *Appl. Environ. Micro.* 58:243–251 (1992).

Calderbank, A. "The Occurrence and Significance of Bound Pesticide Residues in Soil," *Rev. Environ. Contam. Tox.* 108:71–103 (1989).

Catelane, D., A. Colombi, C. Sorline, and V. Treccani. "Metabolism of Quaternary Carbon Compounds: 2,2-Dimethylheptane and *tert*-Butylbenzene," *Appl. Environ. Microbiol.* 34:351–357 (1977).

Cerniglia, C.E., and M.A. Heitkamp. "Microbial Degradation of Polyaromatic Hydrocarbons (PAH) in the Aquatic Environment," in U. Varanasi, ed., *Metabolism of Polycyclic Aromatic Hydrocarbons in the Aquatic Environment*, (Boca Raton, FL: CRC Press, 1989), pp. 41–68.

Chakrabarty, A.M. "Plasmids in *Pseudomonas*" *Ann. Rev. Genet.* 10:7–30 (1976).

Chakravarty, M., P.M. Amin, H.D. Singh, J.N. Baruah, and M.S. Iyengar. "A Kinetic Model for Microbial Growth on Solid Hydrocarbons," *Biotech. Bioeng.* 14:61-63 (1972).

Chapelle, F.H. *Ground-Water Microbiology and Geochemistry.* (New York: John Wiley and Sons Inc., 1993).

Chaudhry, G.R., and S. Chapalamadugu. "Biodegradation of Halogenated Organic Compounds," *Microbiol. Rev.* 55:59-79 (1991).

Colwell, F.S. 1989. "Microbiological Comparison of Surface Soil and Unsaturated Subsurface Soil from a Semiarid High Desert," *Appl. Environ. Micro.* 55:2420-2423 (1989).

Cork, D.J., and J.P. Krueger. "Microbial Transformations of Herbicides and Pesticides," in S.L. Neidleman and A.I. Laskin, eds., *Advances in Microbiology,* Vol. 36, (San Diego, CA: Academic Press Inc., 1991).

Estrella, M.R., M.L. Brusseau, R.S. Maier, I.L. Pepper, P.J. Wierenga, and R.M. Miller. "Biodegradation, Sorption, and Transport of 2,4-Dichlorophenoxyacetic Acid in Saturated and Unsaturated Soils," *Appl. Environ. Micro.* 59:4266-4273 (1993).

Fathepure, B.Z., and R.M. Vogel. "Complete Degradation of Polychlorinated Hydrocarbons by a Two-Stage Biofilm Reactor," *Appl. Environ. Micro.* 57:3418-3422 (1991).

Fredrickson, J.K., D.L. Balkwill, J.M. Zachara, S-M.W. Li, F.J. Brockman, and M.A. Simmons. "Physiological Diversity and Distributions of Heterotrophic Bacteria in Deep Cretaceous Sediments of the Atlantic Coastal Plain," *Appl. Environ. Micro.* 57:420-411 (1991).

Fulthorpe, R.S., and R.C. Wyndham. "Involvement of a Chlorobenzoate-Catabolic Transposon, Tn5271, in Community Adaptation to Chlorobiphenyl, Chloroaniline, and 2,4-Dichlorophenoxyacetic Acid in a Freshwater Exosystem," *Appl. Environ. Micro.* 58:314-325 (1992).

Ghosal, D., I.S.-You, D.K. Chatterjee, and A.M. Chakrabarty, "Plasmids in the Degradation of Chlorinated Aromatic Compounds," in D.R. Helinski, S.N. Cohen, D.B. Clewell, D.A. Jackson, A. Hollaender, eds., *Plasmids in Bacteria*, (New York: Plenum Press, 1985), pp. 667-686.

Gibson, D.T., and V. Subramanian. "Microbial Degradation of Aromatic Hydrocarbons," in D.T. Gibson, ed., *Microbial Degradation of Organic Compounds*, (New York: Marcel Dekker, Inc., 1984), pp. 181-252.

Grbić-Galić, D. "Anaerobic Microbial Transformation of Nonoxygenated Aromatic and Alicyclic Compounds in Soil, Subsurface, and Freshwater Sediments," in J-M. Bollag, G. Stotzky, ed., *Soil Biochemistry*, Vol. 6, (New York: Marcel Dekker, 1990), pp. 117-189.

Haas, D. "Genetic Aspects of Biodegradation by *Pseudomonads*," *Experientia* 39:1199-1213 (1983).

Janssen D.B., R. Oldenhuis, A.J. van den Wijngarrd. "Hydrolytic and Oxidative Degradation of Chlorinated Aliphatic Compounds by Aerobic Microorganisms," in D. Kamely, A. Chakrabarty, G.S. Omenn, eds., *Biotechnology and Biodegradation*, (Houston, TX: Gulf Publishing Company, 1990), pp. 105-125.

Konopka, A., and R. Turco. "Biodegradation of Organic Compounds in Vadose Zone and Aquifer Sediments," *Appl. Environ. Micro.* 57:2260-2268 (1991).

Little, C.D. A.V. Palumbo, S.E. Herbes, M.E. Lidstrom, R.L. Tyndall, and P.J. Gilmer. "Trichloroethylene Biodegradation by a Methane-Oxidizing Bacterium," *Appl. Environ. Micro.* 54:951-956 (1988).

Leahy, J.G., and R.R. Colwell. "Microbial Degradation of Hydrocarbons in the Environment," *Microbiol. Rev.* 54:305–325 (1990).

Miller, R.M. "Surfactant-Enhanced Bioavailability of Slightly Soluble Organic Compounds," in H. Skipper, ed., *Bioremediation — Science & Applications*, in press. Soil Science Society of America special publication, Madison, WI, 1994.

Miller, R., and R. Bartha. "Evidence from Liposome Encapsulation for Transport-Limited Microbial Metabolism of Solid Alkanes," *Appl. Environ. Micro.* 55:269–274 (1989).

Nelson, M.J.K., S.O. Montgomery, W.R. Mahaffey, and P.H. Pritchard. "Biodegradation of Trichloroethylene and Involvement of an Aromatic Biodegradative Pathway," *Appl. Environ. Micro.* 53:949–954 (1987).

Nakahara, T., L.E. Erickson, and J.R. Gutierrez. "Characteristics of Hydrocarbon Uptake in Cultures with Two Liquid Phases," *Biotech. Bioeng.* 19:9–25 (1977).

Novak, J.M., K. Jayachandran, T.B. Moorman, and J.B. Weber. "Sorption and Binding of Organic Compounds in Soils and Their Relation to Bioavailability," in H. Skipper, ed., *Bioremediation — Science & Applications*, in press. Soil Science Society of America special publication, Madison, WI, 1994.

Okey, R.W., and R.H. Bogan. "Apparent Involvement of Electronic Mechanisms in Limiting the Microbial Metabolism of Pesticides," *J. Water Pollution Control Fed.*, 37:692–698 (1965).

Oldenhuis, R., R.L.M. Vink, D.B. Jannsen and B. Witholt. "Degradation of Chlorinated Aliphatic Hydrocarbons by *Methylosinus trichosporium* OB3b Expressing Soluble Methane Monooxygenase," *Appl. Environ. Micro.* 55:2819–2826 (1989).

Paul E.A., and F.E. Clark. *Soil Microbiology and Biochemistry.* (New York: Academic Press Inc., 1989).

Perry, J.J. "Microbial Metabolism of Cyclic Alkanes," in R. Atlas, ed., *Petroleum Microbiology*, (New York: Macmillan, 1984), pp. 61–97.

Pitter, P., and J. Chudoba. "Biodegradability of Organic Substrates in the Aquatic Environment," (Boca Raton, FL: CRC Press, 1990).

Plumb Jr., R.H. "Disposal Site Monitoring Data: Observations and Strategy Implications," in B. Hitchon, and M. Trudell, eds., *Proceedings Second Canadian/American Conference on Hydrology.* (Dublin, OH: National Water Well Association, 1985).

Prince, R.C. "Petroleum Spill Bioremediation in Marine Environments," *Crit. Rev. Microbio.* 19:217–242 (1993).

Rosenberg, E., E.A. Bayer, J. Delarea, and E. Rosenberg. "Role of Thin Fimbriae in Adherence and Growth of *Acinetobacter calcoaceticus* RAG-1 on Hexadecane," *Appl. Environ. Micro.* 44:929–937 (1982).

Schink, B. "Degradation of Unsaturated Hydrocarbons by Methanogenic Enrichment Cultures," *FEMS Microbiol. Ecol.* 31:69–77 (1985).

Schink, B. "Principles and Limits of Anaerobic Degradation, Environmental and Technological Aspects," in A.J.B. Zehnder, ed., *Biology of Anaerobic Microorganisms*, (New York: John Wiley and Sons, 1988).

Scow, K.M. 1993. "Effect of Sorption-Desorption and Diffusion Processes on the Kinetics of Biodegradation of Organic Chemicals in Soil," in D.M. Linn, T.H. Carski, M.L. Brusseau, F-H. Chang, eds., *Sorption and Degradation of Pesticides and Organic Chemicals in Soil*, Soil Science Society of America special publication, Madison, WI, pp. 73–114.

Singer, J.T., and W.R. Finnerty. "Genetics of Hydrocarbon-Utilizing Microorganisms," in R. Atlas, ed., *Petroleum Microbiology*, (New York: Macmillan, 1984a), pp. 299–354.

Singer, M.E., and W.R. Finnerty. "Microbial Metabolism of Straight-Chain and Branched Alkanes," in R. Atlas, ed., *Petroleum Microbiology*, (New York: Macmillan, 1984b), pp. 1–59.

Song H-G., and R. Bartha. "Effects of Jet Fuel Spills on the Microbial Community of Soils," *Appl. Environ. Micro.* 56:646–651 (1990).

Steinberg, S.M., J.J. Pignatello, and B.L. Sawheny. "Persistence of 1,2-Dibromoethane in Soils: Entrapment in Intraparticle Micropores," *Environ. Sci. Technol.* 21:1201–1208 (1987).

Trower, M.K., R.M. Buckland, R. Higgins, and M. Griffin. "Isolation and Characterization of a Cyclohexane-Metabolizing *Xanthobacter* sp.," *Appl. Environ. Microbiol.* 49:1282–1289 (1985).

Trudgill, P.W. "Microbial Degradation of the Alicyclic Ring," in D.T. Gibson, ed., *Microbial Degradation of Organic Compounds*, (New York: Marcel Dekker, Inc., 1984), pp. 131–180.

U.S. Environmental Protection Agency, Summary Report: Remedial Response at Hazardous Waste Sites, EPA-540/2-84-002a, March 1984.

van der Meer, J.R., W.M. de Vos, S. Harayama, and A.J.B. Zehnder. "Molecular Mechanisms of Genetic Adaptation to Xenobiotic Compounds," *Microbiol. Rev.* 56:677–694 (1992).

Vannelli, T., M. Logan, D.M. Arciero, and A.B. Hooper. "Degradation of Halogenated Aliphatic Compounds by the Ammonia-Oxidizing Bacterium *Nitrosomonas europaea*," *Appl. Environ. Micro.* 56:1169–1171 (1990).

Vogel, T.M. C.S. Criddle, and P.L. McCarty. "Transformations of Halogenated Aliphatic Compounds," *Environ. Sci. Tech.* 21:722–732 (1987).

Wackett, L.P., G.A. Brusseau, S.R. Householder, and R.S. Hanson. "Survey of Microbial Oxygenases: Trichloroethylene Degradation by Propane-Oxidizing Bacteria," *Appl. Environ. Micro.* 55:2960–2964 (1989).

Wackett, L.P., and D.T. Gibson. "Degradation of Trichloroethylene by Toluene Dioxygenase in Whole-Cell Studies with *Pseudomonas putida* F1," *Appl. Environ. Micro.* 54:1703–1708 (1988).

Wodzinski, R.S., and D. Bertolini. "Physical State in Which Naphthalene and Bibenzyl are Utilized by Bacteria," *Appl. Micro.* 23:1077–1081 (1972).

Wodzinski, R.S., and J.E. Coyle. "Physical State of Phenanthrene for Utilization by Bacteria," *Appl. Micro.* 27:1081–1084 (1974).

Wodzinski, R.S., and D. Larocca, D. "Bacterial Growth Kinetics on Diphenylmethane and Naphthalene-Heptamethylnonane Mixtures," *Appl. Environ. Microbiol.* 33:660–665 (1977).

Young, L.Y., and M.M. Häggblom, "The Anaerobic Microbiology and Biodegradation of Aromatic Compounds," in D. Kamely, A. Chakrabarty, G.S. Omenn, Eds., *Biotechnology and Biodegradation*, (Houston, TX: Gulf Publishing Company, 1989), pp. 3–19.

Zajic, J.E., and C.J. Panchel. "Bio-Emulsifiers," *Crit. Rev. Microbiol.* 5:39–66 (1976).

Zehnder, A.J.B., and W. Stumm. "Geochemistry and Biogeochemistry of Anaerobic Habitats," in A.J.B. Zehnder, ed., *Biology of Anaerobic Microorganisms*, (John Wiley and Sons, 1988).

Zeyer, J., P. Eicher, J. Dolfing, R.P. Schwarzenbach. "Anaerobic Degradation of Aromatic Hydrocarbons," in D. Kamely, A. Chakrabarty, G.S. Omenn, eds., *Biotechnology and Biodegradation*, (Houston, TX: Gulf Publishing Co., 1990), pp. 33–40.

Zhang, Y., and R.M. Miller. "Enhanced Octadecane Dispersion and Biodegradation by a *Pseudomonas* Rhamnolipid Surfactant (Biosurfactant)," *Appl. Environ. Microbiol.* 58:3276–3282 (1992).

Zhang, Y., and R.M. Miller. "Effect of a *Pseudomonas* Rhamnolipid Biosurfactant on Cell Hydrophobicity and Biodegradation of Octadecane," *Appl. Environ. Micro.* 60(6):2101–2106 (1994).

9

Fate and Transport of Microorganisms in the Vadose Zone

David K. Powelson and Charles P. Gerba

INTRODUCTION

There are three groups of microorganisms that may travel through porous geologic media: protozoa, bacteria, and viruses. These organisms differ greatly in size and survival in the environment. Each of these groups of organisms have members that are capable of causing disease (pathogenic) in humans and other mammals (Table 9.1). Although transport of bacteria is also important in bioremediation, this section will focus on transport of pathogens. In particular we will emphasize virus transport, since their small size makes them more likely to be transported through the vadose zone.

Pathogenic organisms present in fecal wastes (sewage) are of major concern since they are almost always present in domestic wastes. Sources of enteric pathogens include sewage, septic tanks, landfills, and sewage sludge. If these organisms survive passage through the vadose zone they will contaminate ground water and may emerge in springs or from wells used as drinking water supplies. The use of contaminated ground water was responsible for 51% of all waterborne disease outbreaks in the United States during 1971–1982 (Craun 1985).

The risks of infection, disease, and mortality from water contaminated with virus has been calculated. For example, drinking water with a concentration of one hepatitis A virus particle in 10^4 L over a lifetime can result in a significant risk ($>1{:}100$) of contracting the disease (Gerba and Haas, 1988). In some cases even a single virus particle may cause infection (Ward and Akin, 1984). More than 10^6 infectious virus particles may be excreted per gram of feces by infected persons, and concentrations as high as 10^5 infectious virus particles per liter have been detected in raw sewage (WHO, 1979).

0–87371–610–8/95/$0.00 + $.50

Table 9.1.   Some Important Human Pathogens That May Be Transported Through the Vadose Zone

Microorganism and Size	Disease
Viruses (0.02–0.10 μm)	
Adenoviruses	Respiratory disease
Polioviruses	Poliomyelitis
Echoviruses	Meningitis
Coxsackie viruses	Myocarditis, meningitis
Hepatitis A virus	Infectious hepatitis
Rotaviruses	Diarrhea
Bacteria (0.5–1.0 μm)	
Campylobacter fetus	Diarrhea
Escherichia coli (pathogenic)	Diarrhea
Salmonella typhi	Typhoid fever
Shigella spp.	Bacillary dysentery
Vibrio cholerae	Cholera
Protozoans (1.0-12 μm)	
Giardia lamblia	Diarrhea
Cryptosporidium spp.	Diarrhea

The presence of pathogenic virus in wells and drinking water in most parts of the world has been well documented (Gerba 1984; Gerba and Rose, 1990). In a survey of 99 wells in Israel, Marzouk et al. (1979) found enteroviruses in 20% of the ground-water samples. In Minnesota, Goyal et al. (1989) detected human pathogenic viruses in 10 of 26 ground-water sources, and on two occasions coliphage were isolated from samples in which coliform bacteria were absent.

Microbial movement in the field under unsaturated flow conditions has received only limited, often anecdotal, study to date. Virus and coliform bacteria were found in ground water 27.5 m beneath a field of hay on clay loam irrigated with sewage effluent in Texas (Goyal et al., 1984). Gerba et al. (1984) isolated viruses from as deep as 30 m and from as far as 100 m from sewage treatment basins in Texas, Arizona, Michigan, and Israel. Septic tanks are the most frequently reported cause of ground-water contamination in the U.S. (Yates, 1985). Vaughn et al. (1983) detected virus originating from septic tanks that had passed through 3.6 m of vadose zone and 67 m of saturated zone.

Once viruses contaminate an aquifer, water treatment cannot be depended on to remove all virus. Keswick et al. (1985) found rotavirus in 3 out of 26 samples of finished drinking water from a full-scale treatment plant that met acceptable limits for turbidity, total coliform bacteria, and residual chlorine. The presence of coliphage (bacterial virus) in 82 out of 147 potable water samples led El-Abagy et al. (1988) to suggest that pathogenic viruses may also survive normal treatment and disinfection processes.

Adsorption Mechanisms

This section begins by discussing adsorption of microorganisms in general and is followed by analysis of adsorption of viruses in particular, since there are more quantitative data on viruses. Many of the mechanisms discussed apply to transport in saturated as well as in unsaturated conditions. The vadose zone exhibits an extra level of complexity due to the addition of the gas phase.

Microorganisms, like many organic chemicals, have groups on their surface that can gain or lose a proton depending on the pH of the suspending solution. The pH at which electrical charges balance is called the isoelectric pH (pH_{iep}). If a microbe gains protons (pH < pH_{iep}) it has a net positive charge; if it loses protons (pH > pH_{iep}) it has a net negative charge. The pH_{iep} of some microbes and other constituents of soil are listed in Table 9.2. Most microbes and soil materials are negatively charged in the neutral pH range, and so tend to repel each other.

Microbes approach soil solids due to convection or Brownian motion, but are usually electrostatically repelled. This process, called anion exclusion, may contribute to enhanced transport of microorganisms relative to water. In anion exclusion, the negatively charged microbial particles are pushed to the center of the pore where water velocity is higher than average due to Newton's law of viscosity (Jury et al., 1991).

The negative charge of soil solids is neutralized by cations held in close proximity to the surface of the solid. This is often referred to as the diffuse double layer. If the salt concentration of the water is great enough, the diffuse layer will compact, allowing microbes to contact the solid surface. Once close approach or contact with the solid surface occurs, van der Waals forces tend to hold the microbe to the solid.

Microbes have hydrophobic groups on their surface, and so tend to be forced out of solution and onto hydrophobic areas of the soil. This is not a simple relationship since humic material (generally the most important hydrophobic component of soil solids) is in equilibrium with the solution phase. Dissolved organic molecules may

Table 9.2. Isoelectric Points of Microorganisms and Soil Components[a]

Particle	pH_{iep}	Reference
Viruses		
Polio 1	4.5, 7.0[b]	Mandel (1971)
Polio 2	4.5, 6.5[b]	Murray and Parks (1980)
Echo 1, 5 strains	5.0–6.4	Zerda (1982)
Coxsackie A21	4.8, 6.1[b]	Murray and Parks (1980)
MS2 bacteriophage	3.9	Zerda (1982)
Bacteria		
Aerobacter aerogenes	< 1.5	Stumm and Morgan (1981)
Humic Material		
Wisconsin soil	< 2.5	Bohn et al. (1985)
Organic matter		
Minerals		
Quartz (SiO_2)	2.0	Stumm and Morgan (1981)
Kaolinite clay	4.6	Stumm and Morgan (1981)
Montmorillonite clay	2.5	Stumm and Morgan (1981)
Calcite ($CaCO_3$)	8.2	Stumm and Morgan (1981)
Corundum (Al_2O_3)	9.1	Stumm and Morgan (1981)
Hematite (Fe_2O_3)	6.7	Stumm and Morgan (1981)

[a]Adapted from Gerba (1984) and Stumm and Morgan (1981).
[b]Some viruses have two conformational states, each of which has a different isoelectric point (Mandel, 1971).

compete with or complex with microbes, and interfere with hydrophobic adsorption (Lance and Gerba, 1984a; Powelson et al., 1991).

Gas-liquid interfaces in unsaturated soil may also provide sites for adsorption of particles. Wan and Wilson (1994) and Wan et al. (1994) observed, using glass micromodels, colloidal polystyrene beads, clay particles, and bacteria concentrating at gas-liquid interfaces. Sorption increased with particle hydrophobicity, solution ionic strength, and positive electric charge of the particles. They suggested that initial adsorption was due to van del Waals and electrostatic interactions, followed by essentially irreversible adsorption due to capillary force. Wan and Wilson (1994) predict, "For a relatively hydrophobic strain of bacteria, even a small amount of residual gas can dramatically reduce . . . transport."

Generally, adsorption is quantified by the use of equilibrium adsorption isotherms which relate the number of microorganisms adsorbed per gram of solid (C_s) to the liquid concentration of the microorganism (C_l) under equilibrium conditions (Gerba et al., 1991). Virus adsorption to soils can be characterized by a Freundlich isotherm:

$$C_s = K_A C_l^{1/n}$$

where K_A and n are empirical constants.

For $n < 1$, multilayer adsorption is the case; $n > 1$ results in saturation adsorption; and $n = 1$ indicates single-layer adsorption, resulting in a linear isotherm. Multilayer adsorption occurs when more than one virus particle adsorbs to a site. Saturation adsorption occurs when all sites are occupied, resulting in reduced adsorption. Experimentally determined values of $1/n$ for viruses were summarized by Vilker and Burge (1980) and were found to range from 0.87 to 1.24. Since reported values for $1/n$ are statistically close to 1, $1/n$ can often be replaced by 1 in the Freundlich equation, yielding $C_s = K_A C_l$, the linear isotherm.

The value K_A, the linear adsorption constant, has been shown to be dependent on many factors. These factors include type of soil, type of microorganism, pH, chemical quality of the ground water, and presence of soluble organic matter. A summary of K_A values which have been determined experimentally for poliovirus type 1 for various soils and water types is shown in Table 9.3. Adsorption constants range from 0.72 to 1000 mL/g, depending on the soil and water type. K_A values may also vary greatly, depending upon the type of virus. As shown in Table 9.4, some viruses exhibit little or no adsorption to a loamy sand soil.

In modeling microbial transport, K_A can be used to estimate retardation of breakthrough. The retardation factor (R) measures the slowing of microbial transport relative to the bulk of the water due to reversible adsorption onto solids:

$$R = t_{bm}/t_{bw} \qquad (1)$$

where t_{bm} is microbe breakthrough time and t_{bw} is water breakthrough time. Estimated retardation (R_e) is related to K_A by:

$$R_e = 1 + (\rho_b K_A/\theta_v) \qquad (2)$$

where ρ_b is soil bulk density and θ_v is volumetric water content. Equation 2 may overestimate R for microbes due to size exclusion of colloids from small pores in soil, discussed below. An overestimate of R would result in an overestimate of the

Table 9.3. Experimentally Determined Adsorption Constants (K_A) for Poliovirus Type 1 Adsorption to Various Soils[a]

Soil Type	K_A (mL/g)	Water Type	Reference
Sand	5.8	Deep ground water	Yeager and O'Brien, 1979
Sand	1.4	Shallow ground water	Yeager and O'Brien, 1979
Sandy loam	499.	Deep ground water	Yeager and O'Brien, 1979
Sandy loam	66.	Shallow ground water	Yeager and O'Brien, 1979
Loamy sand	1,000.	Distilled water	Goyal and Gerba, 1979
Loamy sand	142.	Secondarily treated sewage effluent	Goyal and Gerba, 1979
Sand	0.72	Deionized water	Goyal and Gerba, 1979
Sandy loam	4.6	Deionized water	Goyal and Gerba, 1979
Clay	1,000.	Deionized water	Goyal and Gerba, 1979
Sandy clay loam	99.	Deionized water	Goyal and Gerba, 1979
Sand and gravel	3.	Tertiary-treated sewage effluent	Landry et al., 1979
Sand	2.5	Phosphate-buffered saline	Vilker et al., 1983

[a]Modified from Grosser, 1985.

microbial breakthrough time by Equation 1. Hornberger et al. (1992) found strong retention of bacteria in sand columns, but little retardation of breakthrough. They suggested that the use of equilibrium adsorption models for bacterial transport may not be adequate. The relationship between retardation, volumetric water content, and microbial adsorption needs more research.

Size Effects

Pathogenic microbes that may move through the vadose zone range in size from 0.020 to 8 μm. The maximum size of water-filled pores can be estimated from the capillary rise equation:

$$r = -2\sigma\cos\gamma/(\rho_w\, g\, h_m)$$

where r is the capillary-equivalent radius, σ is water-air surface tension, γ is the water-solid contact angle [see Letey et al. (1962) for some typical values], ρ_w is density of water, g is gravitational acceleration, and h_m is water potential head due

Table 9.4. Adsorption (K_A) Constants for Various Viruses to a Loamy Sand Soil[a]

Virus Type	K_A(mL/g)
f2	0.
MS2	0.20
Coxsackievirus type B4	0.
Echovirus type 1	1.5
Poliovirus type 1	1,000.
Rotavirus type SA-11	1.1

[a]Calculated from data of Goyal and Gerba (1979).

to matrix forces (Jury et al., 1991). For example, in a moist soil with $h_m = -250$ cm [$\sigma = 72.7$ g s^{-2}, $\gamma = 47°$, $\rho_w = 1.0$ g cm^{-3}, g = 980 cm s^{-2}], the maximum water-filled pore radius is calculated to be 4 μm. This is just large enough to permit some protozoa to be transported. As soil drys, h_m becomes more negative, and the maximum water-conducting pore size decreases. Microbes that have diameters greater than the maximum pore size will be strained out. Generally, organisms larger than 8 μm are unlikely to move through water-filled pores in unsaturated media, and viruses are the most likely to be transported.

The range of pore sizes in soil may enhance straining, but those microbes not removed by this mechanism may be transported at a rate faster than the average water flow, as predicted by Poiseuille's law:

$$J_l = r^2 \, \Delta \, h_T \, \rho_w \, g / (8 \, L \, v)$$

where J_l is the liquid flux or velocity through a circular tube of radius r, Δh_T is the difference in total water potential head, L is the length of the tube, and v is the liquid viscosity (Jury et al., 1991). Note that flux increases as the square of the pore radius. Since microbes can only be transported in pores larger than their diameter, nonadsorbing microbes arrive at a given point before the water center-of-mass. This is called the size-exclusion effect. Fontes et al. (1991) found that larger bacteria had broader breakthrough peaks which they attributed to size exclusion. The effects of size- or anion-exclusion may result in an R value less than 1.

In many situations the vadose zone is not homogeneous, but composed of alternating layers of different grain sizes. This enhances preferential flow paths and reduces treatment. Fingers of saturated flow may rapidly traverse coarse material (Glass et al., 1988). A perched water table may form above a fine-textured layer. Under these saturated conditions water may funnel down animal burrows, plant-root channels, or well bores, bypassing most of the vadose zone.

Depth-Dependent Removal

In addition to straining, bacteria and protozoa may be retained in porous media by various capture mechanisms such as interception, sedimentation, impaction, and hydrodynamic action (Corapcioglu et al., 1987). The combined effect of these processes can be simply defined as removal of the microbes from the percolating water over a certain flow length (Mathess et al., 1988):

$$C = C_0 \exp(-\lambda X) \tag{3}$$

where C_0 is the initial concentration of organisms, X is distance, C is the observed concentration, and λ is the filtration coefficient.

Filtration mechanisms depend on hydraulic conditions (flow velocity and flow direction). When these parameters change, the bacteria may be remobilized. The filtration coefficient (λ) may also change with time; as the bacteria or other particles accumulate, a filter layer develops which further reduces the diameter of pores available for microorganism movement (Krone et al., 1958). Also, very high microbial concentrations will induce flocculation and aggregation of microorganisms, which will further enhance clogging.

Mathess et al. (1988) reported filtration coefficients of 10 to 44.6 m^{-1} for enteric bacteria in a sandy soil. Jang et al. (1983) reported filter coefficients of 40 to 93 m^{-1}

for sandstone cores, depending on the type of bacteria. In a comparison of saturated and unsaturated conditions on poliovirus removal, Lance and Gerba (1984b) found that infiltration rates in the range of 0.325 to 1.00 m/d did not affect penetration depth of poliovirus, supporting the concept of depth-dependent removal. Analysis of their data using Equation 3 resulted in saturated $\lambda = 4.60$ m^{-1} and unsaturated $\lambda = 23.7$ m^{-1}.

Time-Dependent Removal

Microbial mortality (die-off, inactivation, noninfectiousness) is potentially a result of numerous factors. In the case of enteric bacteria, starvation, competition from endogenous microflora, or predation are the major factors. Viruses do not require a food source, and inactivation results from natural thermal denaturation or attack by microbial enzymes of the native microflora. Extremes in pH or the presence of toxic substances (e.g., heavy metals) may also affect microbial inactivation.

The inactivation or decay of microorganisms in water and soil has usually been described as a first-order reaction (Reddy et al., 1981; Crane and Moore, 1986; Yates and Yates, 1988):

$$C = C_o \exp(-\mu\, t) \tag{4}$$

where t is time and μ is the first-order decay constant. Here μ is an expression of the sum total of all factors which influence microorganism survival.

Note the similarity of Equation 3 and Equation 4. They differ in the independent variable that determines microbe concentration. The use of Equation 3 is appropriate when distance is the controlling factor, and variation in residence time does not affect concentration. Equation 4 is appropriate when residence time determines concentration, and distance traveled is not important. The predictive equations below (Equations 5 and 6) use the time-dependent Equation 4.

The decay constants determined for several enteric bacteria and viruses in soils with various moisture contents are listed in Table 9.5. In every case, μ increased with temperature and decreasing water content. The temperature effect may be due to increased predation, chemical degradation, or, for bacteria, starvation.

The mechanism for the water-content effect on μ is less clear. Competition between introduced and indigenous microbes is likely to be enhanced in the aerobic conditions of the vadose zone as compared to less aerobic conditions below the water table. Hurst et al. (1980) found that MS2 virus survival under aerobic nonsterile conditions was less than that occurring under aerobic sterile, anaerobic nonsterile, or anaerobic sterile conditions.

The physical presence of air-water interfaces in the vadose zone may also reduce virus survival. Powelson et al. (1990) found that MS2 bacteriophage was not removed during passage through one meter of saturated soil, but was 95% removed by unsaturated soil. The authors hypothesized that virus particles adsorbed to air-water interfaces in the unsaturated soil and were degraded by surface tension effects on the viral capsid, as described by Trouwborst et al. (1974). This hypothesis is supported by earlier studies (Adams, 1948) in which 7 bacteriophages were exposed to the air-water interface by shaking in test tubes. In all cases where phage suspensions were shaken, inactivation could be described as a first-order reaction (Equation 4). Adams (1948) found that presence of 1 μg/mL of gelatin protected the phages from inactivation at the air-water interface, and suggested that gelatin formed a monomo-

Table 9.5. First-Order Decay Coefficients (μ) for Some Sewage Microorganisms in Soils with a Range of Water Contents

Microorganism	T	Water Status	Decay Coeff.	Remarks
	°C		Day^{-1}	
	Boyd et al. (1969)			
Escherichia coli		$\theta_g{}^a$ (g/g)		Batch studies.
	20	0.50	0.26	Weld fine sandy loam
	20	0.40	0.33	(110 ppm Ca).
	20	0.20	0.33	
	20	0.10	0.54	
	20	0.50	0.25	Greeley fine sandy loam
	20	0.40	0.34	(473 ppm Ca).
	20	0.20	0.34	
	20	0.10	0.45	
	Kibbey et al. (1978)			
Streptococcus faecalis		ψ^b (bars)		Batch studies.
	4	0.0	0.032	Averages of 4–5 loamy
	4	0.3	0.050	soils.
	4	7.5	0.086	
	4	30.0	0.13	
	10	0.0	0.037	
	10	0.3	0.070	
	10	7.5	0.10	
	10	30.0	0.17	
	25	0.0	0.057	
	25	0.3	0.079	
	25	7.5	0.14	
	25	30.0	0.33	
	37	0.0	0.10	
	37	0.3	0.19	
	37	7.5	0.37	
	37	30.0	0.60	
	Powelson et al. (1990)			
Bacteriophage MS2	4	Saturated	0.0	Column transport study.
	4	Unsaturated	4.6	Loamy fine sand.
	Powelson et al. (1993)			
Bacteriophage MS2	18–30	Unsaturated	5.52	Field transport study.
Bacteriophage PRD1	18–30	Unsaturated	15.6	Coarse alluvium.
	Powelson and Gerba (1994)			
Viruses	22	Saturated	2.28	Column transport studies.
	22	Unsaturated	7.44	Averages for poliovirus 1 and phages MS2 and PRD1.

^{a}Gravimetric water content.
^{b}Soil water suction.

lecular film at the surface which prevented adsorption of virus capsid protein. Powelson et al. (1991) found that dissolved organic matter protected bacteriophage MS2 during unsaturated flow through soil columns.

Removal and Transport Models

Reddy et al. (1981) conducted an extensive review of bacterial decay rates and developed empirical equations to adjust μ for changes in the most important variables: temperature, moisture content, pH, and method of application. For moisture content changes:

$$\mu_2 = \mu_1 (F_{m2}/F_{m1}) \tag{5}$$

where μ_2 and μ_1 are decay coefficients at moisture content 2 and 1; and F_{m2} and F_{m1} are factors, defined below, at water contents 2 and 1. Using data of Boyd et al. (1969), Reddy et al. (1981) derived the following factor to adjust *Escherichia coli* decay constants for moisture changes:

$$F_m = 1.00 - (0.9\ \Theta_g)$$

where Θ_g is the gravimetric moisture content of the soil (g/g). Alternatively, if soil moisture tension is measured, as in Kibbey et al. (1978), Reddy et al. (1981) derived:

$$F_m = 0.303 + (0.0241\ \Psi)$$

where Ψ is the soil-water suction (bars).

Yates and Ouyang (1992) developed a mathematical model to predict virus transport in the unsaturated zone. The model modifies the convective-dispersive transport equation to allow for different decay coefficients for virus adsorbed on soil particles (μ_s) or suspended in solution (μ_l):

$$\frac{\partial}{\partial t} (\rho_b C_s + \theta_v C_l) = \frac{\partial}{\partial z} \left(\theta_v D \frac{\partial C_l}{\partial z} \right) - V_l \theta_v \frac{\partial C_l}{\partial z} - (\theta_v \mu_l C_l + \rho_b \mu_s C_s) \tag{6}$$

where C_l is the liquid concentration of the microorganism, C_s is calculated from K_A ($C_s = K_A C_l$), D is the hydrodynamic dispersion coefficient, and V_l is the average linear velocity of water ($V_l = J_l/\theta_v$). Furthermore, the decay coefficients do not have to remain constant in the model, but may vary depending on the temperature at a given depth. More work needs to be done to calibrate and test this model in the field.

The model of Yates and Ouyang (1992) allows for the effect of temperature because this is one of the most important factors controlling virus survival in the subsurface. Yates et al. (1985) studied the effects of numerous physical-chemical and microbial properties of ground water on virus survival in saturated conditions. They found that ground-water temperature was the most important predictor of virus inactivation. This was found to be true for two enteroviruses and the bacteriophage MS2. The linear regression equation for the inactivation coefficient of MS2 as a function of temperature was:

$$\mu_l = (-0.174 + 0.0342\ T)/\text{day}$$

where T is temperature (°C). Note that this equation indicates that at ground-water temperatures below 5°C, the rate of MS2 inactivation is negligible.

CONCLUSIONS

Exposure of microbial contaminants to vadose zone conditions should generally enhance removal of these microbes compared to saturated flow through the same media.

In unsaturated conditions the high-velocity, low interaction pores are drained. The smaller pore sequences enhance straining of pathogenic protozoa and bacteria. Microbes are forced closer to solid surfaces, increasing the effectiveness of adsorption processes. Adsorption aids removal by increasing the residence time in which the decay process may operate, and may result in irreversible adsorption. If adsorption is minimal, breakthrough times for microbes may be earlier than breakthrough for nonadsorptive chemicals. This is because microbes are excluded from smaller pores, and are transported through the larger, high-velocity pores that remain filled with water.

Microbe concentrations have been found to decline exponentially with distance or residence time. Increased temperature and presence of air-water interfaces in unsaturated soil increases the decay rates of introduced microorganisms.

The vadose zone's effect on microbial contaminants is complex, and it is difficult to predict removal efficiency. Clearly temperature, pore-size distribution, water velocity, and degree of unsaturation are important parameters. When possible, we recommend that the vadose zone be tested for microbe removal using nonpathogenic, easily-transported bacteriophages such as MS2. After model predictions and bacteriophage testing, when wastewater is actually percolating through the vadose zone, ground water should be monitored for the presence of enteric viruses and bacteria.

REFERENCES

Adams, M.H. "Surface Inactivation of Bacterial Viruses and of Proteins," *J. Gen. Physiol.* 31:417–431 (1948).

Bohn, H.L., B.L. McNeal, and G.A. O'Connor. *Soil Chemistry.* 2nd Ed. (New York: John Wiley & Sons, 1985).

Boyd, J.W., T. Yoshida, L.E. Vereen, R.L. Cada, and S.M. Morrison. Bacterial Response to the Soil Environment. Sanitary Engineering Papers, No. 5, Colorado State University, Fort Collins, CO, 1969.

Corapcioglu, M.Y., N.M. Abboud, and A. Haridas. "Governing Equations for Particle Transport in Porous Media," in J. Bear and M.Y. Corapcioglu, Eds., *Advances in Transport Phenomena in Porous Media*, (Dordrecht, The Netherlands: Martinus Nijhoff Pub., 1987), pp. 296–302.

Crane, S.R., and J.A. Moore. "Modeling Enteric Bacterial Die-Off: A Review," *Water Air Soil Pollut.* 27:411–439 (1986).

Craun, G.F. "A Summary of Waterborne Illness Transmitted Through Contaminated Groundwater," *J. Environ. Hlth.* 48:122–127 (1985).

El-Abagy, M.M., B.J. Dutka, and M. Damel. "Incidence of Coliphage in Potable Water Supplies," *Appl. Environ. Microbiol.* 54:1632–1633 (1988).

Fontes, D.E., A.L. Mills, G.M. Hornberger, and J.S. Herman. "Physical and Chemical Factors Influencing Transport of Microorganisms Through Porous Media," *Appl. Environ. Microbiol.* 57:2473–2481 (1991).

Gerba, C.P. "Applied and Theoretical Aspects of Virus Adsorption to Surfaces," *Adv. Appl. Microbiol.* 30:133–168 (1984).

Gerba, C.P., and C.N. Haas. "Assessment of Risks Associated with Enteric Viruses in Contaminated Drinking Water," in J.J. Lichtenberg et al., Eds., *Chemical and Biological Characterization of Sludges, Sediments, Dredge Spoils, and Drilling Muds*, Am. Soc. for Testing and Materials, Philadelphia, PA, 1988, pp. 489–494.

Gerba, C.P., Y. Marzouk, Y. Manor, E. Idelovitch, and J.M. Vaughn. "Virus Removal During Land Application of Wastewater: A Comparison of Three Projects," in *Future of Water Reuse*, Vol. 3, (Denver, CO: American Water Works Association Research Foundation, 1984), pp. 1518–1529.

Gerba, C.P. and J.B. Rose. "Viruses in Source and Drinking Water," in G.A. McFeters, Ed., *Drinking Water Microbiology*, (New York: Springer-Verlag, 1990), pp. 380–396.

Gerba, C.P., M.V. Yates, and S.R. Yates. "Quantitation of Factors Controlling Viral and Bacterial Transport in the Subsurface," in C.J. Hurst, Ed., *Modeling the Environmental Fate of Microorganisms*, (Washington, DC: American Society for Microbiology, 1991), pp. 77–88.

Glass, R.J., T.S. Steenhuis, and J.-Y. Parlange. "Wetting Front Instability as a Rapid and Far-Reaching Hydrologic Process in the Vadose Zone," *J. Contaminant Hydrology* 3:207–226 (1988).

Goyal, S.M., D. Amundson, R.A. Robinson, and C.P. Gerba. "Viruses and Drug Resistant Bacteria in Groundwater of Southeastern Minnesota," *J. Minn. Acad. Sci.* 55:58–62 (1989).

Goyal, S.M., and C.P. Gerba. "Comparative Adsorption of Human Enteroviruses, Simian Rotavirus, and Selected Bacteriophages to Soils," *Appl. Environ. Microbiol.* 38:241–247 (1979).

Goyal, S.M., B.H. Keswick, and C.P. Gerba. "Viruses in Groundwater Beneath Sewage Irrigated Cropland," *Water Research* 18:299–302 (1984).

Grosser, P.W. "A One-Dimensional Mathematical Model of Virus Transport," in N.N. Durham and A.E. Redelfs, Eds., *Proceedings of the Second International Conference on Ground Water Quality Research*, (Stillwater, OK: University Center for Water Research, Oklahoma State University, 1985), pp. 105–107.

Hornberger, G.M., A.L. Mills, and J.S. Herman. "Bacterial Transport in Porous Media: Evaluation of a Model Using Laboratory Observations," *Water Resour. Res.* 28:915–938 (1992).

Hurst, C.J., C.P. Gerba, and I. Cech. "Effects of Environmental Variables and Soil Characteristics on Virus Survival in Soil," *Appl. Environ. Microbiol.* 40:1067–1079 (1980).

Jang, L.K., P.W. Chang, J.E. Findley, and T.F. Yen. "Selection of Bacteria with Favorable Transport Properties through Porous Rock for the Application of Microbial-Enhanced Oil Recovery," *Appl. Environ. Microbiol.* 46:1066–1072 (1983).

Jury, W.A., W.R. Gardner, and W.H. Gardner. *Soil Physics*, 5th Ed., (New York: John Wiley & Sons, 1991).

Keswick, B.H., C.P. Gerba, J.B. Rose, and G.A. Toranzos. "Detection of Rotavirus in Treated Drinking Water," *Water Sci. Tech.* 17:1–6 (1985).

Kibbey, H.J., C. Hagedorn, and E.L. McCoy. "Use of Fecal Streptococci as Indicators of Pollution in the Soil," *Appl. Environ. Microbiol.* 35:711–717 (1978).

Krone, R.B., G.T. Orlob, and C. Hodgkinson. "Movement of Coliform Bacteria Through Porous Media," *Sewage Ind. Wastes* 30:1–13 (1958).

Lance, J.C., and C.P. Gerba. "Effect of Ionic Composition of Suspending Solution on Virus Adsorption by a Soil Column," *Appl. Environ. Microbiol.* 47:484–488 (1984a).

Lance, J.C., and C.P. Gerba. "Virus Movement in Soil During Saturated and Unsaturated Flow," *Appl. Environ. Microbiol.* 47:335–337 (1984b).

Landry, E.F., J.M. Vaughn, M.Z. Thomas, and C.A. Beckwith. "Adsorption of Enteroviruses to Soil Cores and Their Subsequent Elution by Artificial Rainwater," *Appl. Environ. Microbiol.* 38:680–687 (1979).

Letey, J., J. Osborn, and R.E. Pelishek. "Measurement of Liquid-Solid Contact Angles in Soil and Sand," *Soil Sci.* 93: 149–153 (1962).

Mandel, B. "Characterization of Type 1 Poliovirus by Electrophoretic Analysis," *Virology* 44:554–568 (1971).

Marzouk, Y., S.M. Goyal, and C.P. Gerba. 1979. "Prevalence of Enteroviruses in Ground Water of Israel," *Ground Water* 17:487–491 (1979).

Matthess, G., A. Pekdeger, and J. Schroeter. "Persistence and Transport of Bacteria and Viruses in Groundwater—A Conceptual Evaluation," *J. Contam. Hydrol.*, 2:171–188 (1988).

Murray, J.P., and G.A. Parks. "Poliovirus Adsorption on Oxide Surfaces," in M.C. Kavenaugh and J.O. Leckie, Eds., *Particulates in Water*, Adv. in Chem. Series. 189. American Chemical Society, Washington, DC, 1980, pp. 97–133.

Powelson, D.K., and C.P. Gerba. "Virus Removal from Sewage Effluents During Saturated and Unsaturated Flow through Soil Columns," *Wat. Res.*, 28:2175–2181 (1994).

Powelson, D.K., C.P. Gerba, and M.T. Yahya. "Virus Transport and Removal in Wastewater during Aquifer Recharge," *Wat. Res.* 27:583–590 (1993).

Powelson, D.K., J.R. Simpson, and C.P. Gerba. 1990. "Virus Transport and Survival in Saturated and Unsaturated Flow Through Soil Columns," *J. Environ. Qual.* 19:396–401 (1990).

Powelson, D.K., J.R. Simpson, and C.P. Gerba. 1991. "Effects of Organic Matter on Virus Transport in Unsaturated Flow," *Appl. Environ. Microbiol.* 57:2192–2196 (1991).

Reddy, K.R., R. Khaleel, and M.R. Overcash. "Behavior and Transport of Microbial Pathogens and Indicator Organisms in Soils Treated with Organic Wastes," *J. Environ. Qual.* 10:255–266 (1981).

Stumm, W., and J.J. Morgan. *Aquatic Chemistry.* 2nd ed. (New York: John Wiley & Sons, 1981).

Trouwborst, T., S. Kuyper, J.C. DeJong, and A.D. Plantinga. "Inactivation of Some Bacterial and Animal Viruses by Exposure to Liquid-Air Interfaces," *J. Gen. Virol.* 24:155–165 (1974).

Vaughn, J.M., E.F. Landry, and M.Z. Thomas. "Entrainment of Viruses from Septic Tank

Leach Fields Through a Shallow, Sandy Soil Aquifer," *Appl. Environ. Microbiol.* 45: 1474–1480 (1983).

Vilker, V.L., and W.D. Burge. "Adsorption Mass Transfer Model for Virus Transport in Soils," *Water Res.* 14:783–790 (1980).

Vilker, V.L., J.C. Fong, and M. Sayyed-Hoseyni. "Poliovirus Adsorption to Narrow Particle Size Fractions of Sand and Montmorillonite Clay," *J. Colloid Interface Sci.* 92:422–435 (1983).

Wan, J., and J.L. Wilson. "Visualization of the Role of the Gas-Water Interface on the Fate and Transport of Colloids in Porous Media," *Water Resour. Res.* 30:11–23 (1994).

Wan, J., J.L. Wilson, and T.L. Kieft. "Influence of the Gas-Water Interface on Transport of Microorganisms through Porous Media," *Appl. and Environ. Microbiol.* 60:509–516 (1994).

Ward, R.L., and E.W. Akin. "Minimum Infective Dose of Animal Viruses," *CRC Critical Reviews in Environmental Control* 14:297–310 (1984).

WHO. Human Viruses in Water, Wastewater, and Soil. Technical Report 639, World Health Organization, Geneva, 1979.

Yates, M.V., C.P. Gerba, and L.M. Kelley. "Virus Persistence in Groundwater," *Appl. Environ. Microbiol.* 49:778–781 (1985).

Yates, M.V., and Y. Ouyang. "VIRTUS, a Model of Virus Transport in Unsaturated Soils," *Appl. Environ. Microbiol.* 58:1609–1616 (1992).

Yates, M.V. "Septic Tank Density and Ground Water Contamination," *Ground Water* 23:586–591 (1985).

Yates, M.V., and S.R. Yates. "Modeling Microbial Fate in the Subsurface Environment," *Crit. Rev. Environ. Control* 17:307–343 (1988).

Yeager, J.E., and R.T. O'Brien. "Enterovirus Inactivation in Soil," *Appl. Environ. Microbiol.* 38:694–701 (1979).

Zerda, K.S., Ph.D. Dissertation, Baylor College of Medicine, Houston, TX, 1982, reported in C.P. Gerba, "Applied and Theoretical Aspects of Virus Transport," *Adv. Appl. Microbiol.* 30:133–168 (1984).

10

Understanding the Geologic Framework of the Vadose Zone and Its Effect on Storage and Transmission of Fluids

John H. Kramer and Barry Keller

INTRODUCTION

This chapter is designed to introduce readers to geologic issues concerning vadose zone investigations. The literature on vadose zone geology is voluminous, and it would be impossible to list all pertinent references; a partial list of authoritative texts and references is provided here for readers seeking more in-depth treatments than we present. Fetter (1988), Freeze and Cherry (1979), and Bouwer (1978) are traditional introductory texts which explain in more detail the effects of particle size distribution and lithology on contaminant transport. Nielsen and Johnson (1990) compile technical articles on specific issues of vadose zone sampling and monitoring. Devinney et al. (1990) provide detailed information on vadose zone sampling and biogeochemical processes pertinent to contaminant cleanups. General information on soils can be obtained from Brady (1990) and specific information on soil-contaminant processes from Dragun (1990). Reineck and Singh (1980) are an authoritative reference for interpreting depositional environments.

PURPOSE AND PARAMETERS

Vadose zone investigations and monitoring are commonly motivated by very practical concerns over threats of contamination to water resources. The purpose of geologic characterization of a site involving potential vadose zone transport of liquids and contaminants is to determine the type of transport mechanism or mechanisms which may exist between potential contaminant sources and a water table

0–87371–610–8/95/$0.00 + $.50
137

through mainly unsaturated geologic materials. Geologic materials may include rock, unconsolidated soils, or artificial fill. Vadose zone monitoring instruments should be placed so as to intercept flow of contaminants according to such mechanisms. In many cases this would be liquid flow; however, gas phase flow may also be important.

Contaminated sites, such as industrial facilities, are generally not located with regard to underlying features, so a site could be situated in a variety of geologic settings. It is important to have a general understanding of the genesis and depositional history of surface and near-surface material in order to predict the location of potential contaminant flow pathways, monitoring targets. Much of the instrumentation and transport theory traditionally employed in vadose zone monitoring has been developed as an extension of agricultural soil science, in which unconsolidated material has little, if any, lateral or vertical heterogeneity and is unaffected by adjacent rock types. Effective vadose zone monitoring networks require the synthesis of geologic knowledge and appropriate application of instrumentation and transport theory.

The parameters of interest for geologic characterization include spatial parameters such as the size of the region of concern, area of source or sources, distribution of depths to water table, types of geologic materials, and three-dimensional changes of materials. An important aspect is the expected flow regime. Many geologic materials may be expected to have a flow regime which can be macroscopically characterized as a continuum, such as a typical agricultural soil. However, other situations involving more indurated rocks are characterized by very heterogeneous flow, such as flow through fractures in crystalline rocks or along relatively permeable beds in sedimentary rocks. Continuum flow regimes are suitable for the kinds of monitoring devices developed for agricultural use and are also amenable to a number of computer modeling schemes, since mathematical representation is fairly straightforward. Heterogeneous flow regimes present much greater problems, both for monitoring and for modeling, but are often encountered in real situations, particularly in tectonically active areas.

TYPES OF GEOLOGIC MATERIALS

Since potentially contaminated sites may occur in any geologic setting, vadose zone investigations and monitoring require a general understanding of all types of surface and near-surface conditions, whereas other geologic applications, such as petroleum or mining geology focus on particular geologic settings (e.g., marine sedimentary rocks or mineralized alteration zones). Types of geologic materials are categorized here by the predominance of either "primary" or "secondary" liquid flow characteristics because the type of flow regime (the volume in which a particular flow type predominates) is of paramount importance. "Primary" flow refers to a continuum regime, which is amenable to mathematical modeling by some modification of Darcy's law. "Secondary" flow is the heterogeneous condition, which may be very difficult to express mathematically.

Materials Hosting Gas Phase Transport

Gas migration can occur in either the primary or secondary modes mentioned above. The driving forces can be diffusive chemical gradients, isothermal and thermally-induced density gradients, and advective pressure gradients. Geologic controls on gas phase transport concern available open contiguous pore space. As a general rule, when water content increases, gas permeability decreases.

Geologic material conducive to gas-phase transport has relatively high air-filled porosity. Examples are dry gravels and dry, well-sorted sands (or in the parlance of geotechnical professionals, poorly-graded sands). Backfill materials along subsurface pipelines, which are generally well-sorted coarse-sands, have high gas permeability. Materials with high bulk density, high clay content, broad particle size distributions and high residual water contents, or sparse fracture density are referred to as "tight" and are resistive to gas migration. Any material which is saturated will be impermeable to gas flow. Thus, the saturation percentage (volume of water per volume of porosity) and total porosity of geologic materials are very important field parameters to measure when dealing with gas-phase transport of contaminants.

Materials Hosting Primary Flow

Unconsolidated materials host primary flow regimes. Types of unconsolidated geologic materials include alluvial and windblown soils, deeply weathered soils, and some types of volcanic deposits, such as airfall and ash deposits. Such materials commonly exhibit uniform physical properties, due to the depositional or weathering process. Examples are shown in Figure 10.1. On published geologic maps, which may be used as an initial reference in the investigation of a site, most material exhibiting primary flow would be shown as Quaternary surficial deposits (usually yellow and/or designated by a "Q" prefix on map symbols), or recent volcanic deposits. Map units prefixed with "Qv" or "Tv" usually indicate recent volcanic deposits. An example of a portion of a geologic map is shown in Figure 10.2. For the sake of brevity, only the explanation describing Quaternary deposits is reproduced. Map symbols depicting the topography, culture, location, and attitude of bedding and faults, and the age and type of older map units also appear in the figure.

Flow velocities in primary flow are relatively slow, on the order of the saturated hydraulic conductivity, which is 10e-3 cm/sec or less for many materials. In the vadose zone, primary flow can be divided into two types, one driven by gravitational forces and one driven by capillary forces. Characterization and monitoring activities may differ, depending on the expected flow type.

Gravitational flow dominates when moisture content is high enough for gravity forces to overcome capillary forces; flow from a source will move downward with minor spreading. In many instances the flow may be driven by a continuously-existing saturated condition at the source, such as a leaking tank or pond. In such cases, a bulb-shaped saturated zone develops in which downward-flow velocity is determined by the hydraulic head, porosity, and saturated hydraulic conductivity of the material. The saturated bulb may not contact the water table if the vadose zone is deep relative to the leak flux, or if the leak is relatively young, and there exists unsaturated material beneath it. The magnitude and plunge of gravity-driven flow velocity in unsaturated material depends on moisture content, tension gradients, and unsaturated hydraulic conductivity, which in turn is a function of moisture content. In gravity-dominated flow, the flow direction would be expected to be primarily vertically downward for both saturated and unsaturated conditions. The implication for vadose zone monitoring is that instruments should be placed as closely as possible to directly below a potential source. A notable exception occurs when horizontal flow barriers impede downward flow and divert gravity-driven flow laterally

Figure 10.1. Photographs of geologic material which host primary flow; (a) alluvial material (Sierra Foothills, CA), (b) windblown sand (Death Valley, CA), (c) fluvial sands (Grand Canyon, AZ), (d) volcanic clastic rocks with ash fall causing undercut cliff, (West Elk Mountains, CO). Photos by John H. Kramer.

through more permeable material (e.g., sand layers overlying clay-rich layers). In this case lateral flow can evade monitoring instruments placed too deeply below the source.

Capillary-driven flow only occurs when materials are below the water content at which they drain freely, ("field capacity" as defined by Bear, 1979). Materials below field capacity generally occur near the surface or in climates where evapotranspiration exceeds precipitation. Capillary-driven flow from a source may move in any direction (laterally or possibly upward) because the moisture content is low enough that capillary forces overcome the pull of gravity. Flow velocities can be orders of magnitude slower than those for gravity-driven flow in the same materials. Flow occurs preferentially in fine-grained material (clays and silts) over coarse-grained material (sands and gravels) because fine-grained materials have greater particle-surface areas and consequently greater capillary attraction for water. The implication for vadose zone monitoring is that lateral spreading from a contaminant source may exceed downward percolation. Perimeter-monitoring strategies, in which instruments are placed around the area of a potential source rather than beneath it, may be adequate. Unruh et al. (1990) justified perimeter monitoring at a facility in a desert environment underlain by interbedded lake clays and silts because of a strong horizontal flow component. In seasonally arid climates, moisture, including contaminants, which spread laterally by capillary flow during the dry seasons can be mobilized downward by gravity flow during the wet seasons.

The field capacity represents the amount of aqueous phase liquid which may remain relatively immobile in the vadose zone, since it may move only in response to capillary forces or lose mass to vapor transport, but may not move downward in response to gravity. For unconsolidated materials, this amount of available volume is somewhat less than the total porosity, which has typical values in the range of 25% to 40%. Strategies to prevent further liquid, hence, contaminant, migration under primary flow conditions could involve maintaining the moisture content below field capacity of the host material.

Consolidated materials may also host primary flow. Massive sandstones with relatively uniform porosity are examples of geologic materials which can transmit primary flow; however, in most near-surface geologic settings consolidated materials are fractured and are better characterized by secondary flow.

Materials Hosting Secondary Flow

Lithified geologic materials which are exposed at the surface or are present in the subsurface at depths of interest for vadose zone monitoring are characterized by flow in heterogeneous zones such as fractures, solution cavities, or relatively permeable beds. This category includes rocks of all types which have achieved their state of induration at some greater depth, and have been transported to their present position by tectonic motions. Secondary flow may also occur in surface volcanic deposits, sedimentary deposits, and anthropogenic fill which have been compacted or partially cemented and subsequently fractured. Examples are shown in Figure 10.3. On published geologic maps, most pre-Quaternary units in the vadose zone exhibit secondary-flow characteristics; however, primary flow does occur in consolidated materials (e.g., bedrock aquifers such as found beneath the Great Plains).

Secondary flow may be very rapid in comparison with primary flow. At the extreme, flow may occur through open channels, with flow velocities similar to surface-stream flow, on the order of meters per second. Such channels may result

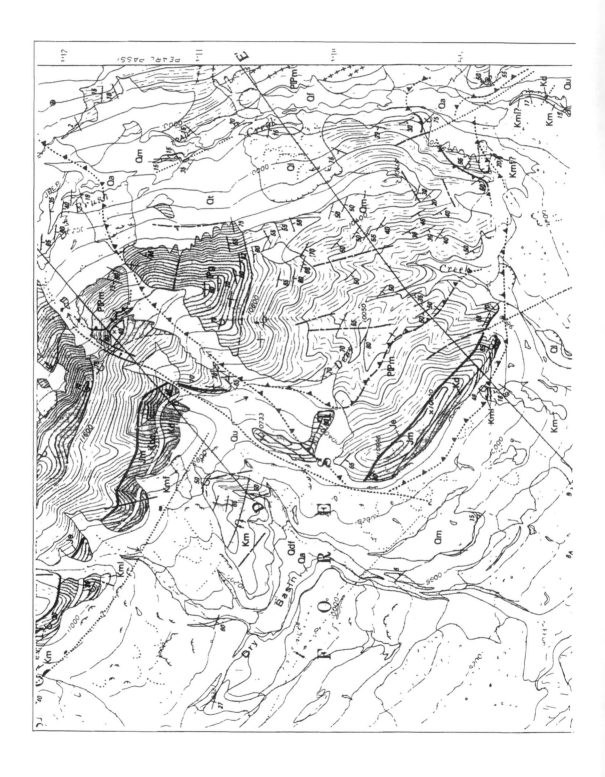

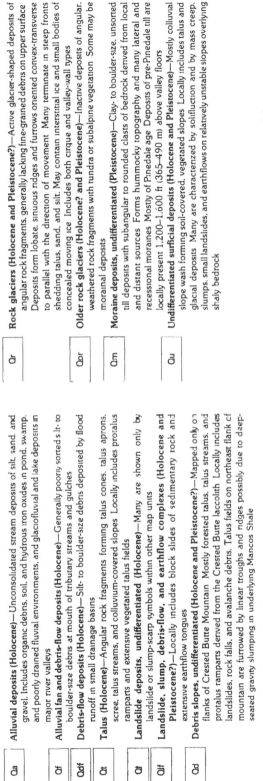

Figure 10.2. Portion of a geologic map (Gaskill et al., 1991) showing the map symbols and descriptions of various types of Quaternary surficial deposits.

a

b

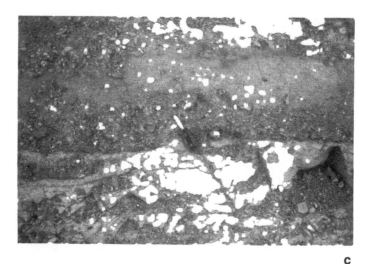

c

Figure 10.3. Photographs of geologic material which host secondary flow; (a) fractured intrusive rock (Cascade Range, WA), (b) deformed sedimentary rocks (Split Mountain, UT), (c) textural contrasts between depositional units (Sierra Foothills, CA), (d) cavernous limestone with flowing springs (Grand Canyon, AZ), (e) curving columnar joints in lavas (Grand Canyon, AZ),

(f) permeability contrasts at tilted geologic contacts, shown by different susceptibilities to erosion. (Devil's Slide, UT), (g) seepage from fractures in limestone (Santa Barbara, CA). Photos a-f by John H. Kramer; g by Barry Keller.

from tectonic fracturing, landsliding, land subsidence (e.g., from ll settlement or water and oil extraction), chemical erosion (e.g., cavernous limestone), clay dessication (e.g., invertisol soil types), cooling cracks or flow unit contacts in volcanic rocks (e.g., columnar joints and lava tubes), and other causes. Secondary flow occurs along discrete paths. For monitoring purposes, instruments would have to be correctly located to successfully detect contaminant transport. An instrument that was not in the flow path would fail to detect the flow.

A variety of spatially heterogeneous flow mechanisms may be operative in "secondary flow" situations. The flow mechanism of greatest concern for contaminant transport is gravity flow of liquid in open channels because flow events can be rapid and voluminous. Such events are typically short-lived, making them hard to monitor. Other mechanisms include combinations of channel flow and localized regions of primary flow with varying permeability. For instance, a deformed, poorly-consolidated sedimentary deposit might exhibit open channel flow in broken lithified beds, gravity-driven flow in sandy beds, and capillary-driven flow in clayey beds, leading to a very complex flow pattern.

CONTACTS BETWEEN TYPES OF GEOLOGIC MATERIALS

Vadose zone flow regimes may change abruptly at contacts between different types of geologic materials. Expected flow paths may change, and monitoring strategies that are appropriate in one regime may be ineffective in an adjacent regime. A number of types of contacts that could affect vadose zone flow with examples are discussed below.

Bedding Contact

Changes in the depositional environment of sedimentary rocks may produce permeability contrasts that affect flow, with flow preferentially directed along sandier units. If the bedding is flat-lying, this may lead to predominance of horizontal flow over vertical flow. This effect has been observed in a number of different types of sedimentary deposits, including alluvium, marine sedimentary rocks, continental fanglomerates, and deltaic clays. Proper identification of preferred flow paths allows placement of monitoring equipment in appropriate locations to detect potential contaminant flow. Figure 10.4 shows arrival times of a wetting front interpreted from neutron-probe readings in three monitoring tubes installed near an injection well at a pilot study for a soil flushing remediation project. The neutron probe is a useful instrument for detecting and measuring subsurface liquid flow (Kramer et al., 1992). The geologic material is flat-lying fine-grained alluvial material, predominantly silty clays and clayey silts interbedded with silty sands. The data clearly show preferential flow through sandy layers overlying clay-rich layers. Monitoring stations designed to intersect fugitive liquids escaping this site should be placed in sandy layers immediately above clay-rich layers.

Figure 10.5 shows a cross section from a constant-head leak test which compared the detection capabilities of several vadose zone instruments in tilted consolidated marine sedimentary rock at a hazardous waste landfill. A horizontal neutron probe access tube extended through tilted sandstone and claystone-siltstone units beneath a trench fitted with a line source controlled leak apparatus. Background neutron logs were collected, the line source was intentionally filled with water, and the access tube

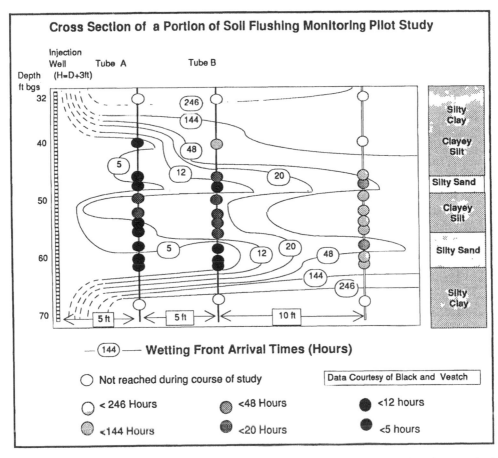

Figure 10.4. Preferential flow in sand-rich units is indicated from wetting front arrival times interpreted from neutron monitoring of an induced wetting front, (after Kramer et al., 1992).

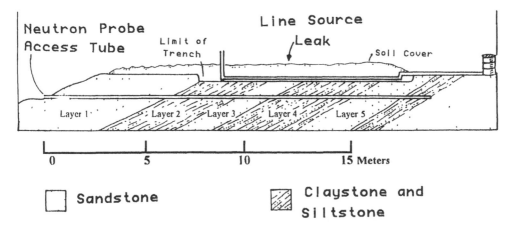

Figure 10.5. Cross section of tilted sandstone and claystone sedimentary rock showing layout of a simulated-leak detection test, (Kaman Sciences, Inc., 1987).

was periodically monitored. The predicted flow pathways were the lower bedding contacts for the sandier layers 1, 3, and 5. Figure 10.6 shows logs of neutron-probe readings taken in the horizontal neutron-probe access tube at four times during the test, at 30 hours, 4 days, 12 days and 16 days after the leak was started. Each graph in Figure 10.6 shows the background log (heavy line) for comparison with logs from each time. The background log exhibits low counts in sandy target horizons, as would be expected from the lower moisture content and clay mineral fractions present in this type of material. Figure 10.6 documents first breakthrough at the target pathway horizons at layer 1 after 30 hours, at layer 3 after 4 days, and at layer 5 after 12 days. This corresponds to actual flow velocities in the range of 10 e-3 to 10 e-4 cm/s, numerically similar to the saturated hydraulic conductivity for such material. Monitoring strategies applied to such identified preferred pathways have been implemented at hazardous waste and municipal landfills.

Unconformity

An unconformity is a surface of erosion or nondeposition — usually the former — that separates younger strata from older rocks (Billings, 1954). An example is illustrated in Figure 10.7, a photograph of colluvial deposits superimposed on the step-shaped sides of an ancient river canyon cut into horizontal strata. Within sequences of sedimentary rocks, or at depositional contacts of sedimentary rocks over crystalline rocks, an unconformity may correspond to a significant change in physical properties. In the vadose zone, such a contact may represent a change from primary flow in the overlying material to secondary flow below. This may be accompanied by a permeability contrast associated with the contact itself. In practical terms, an

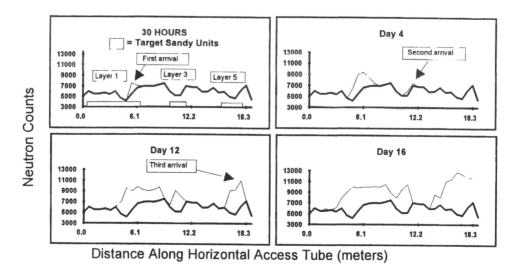

Figure 10.6.   Neutron probe logs of a horizontal access tube beneath a simulated leak from four times. Pre-leak background log is indicated by bold line in all graphs. The location of the predicted target flow pathways is shown by the labels Layer 1, Layer 3, and Layer 5 in the 30-hour log. Arrival and progress of wetting front in target sandy units is documented by above-background neutron counts during leak infiltration. (Kaman Sciences, Inc., 1987).

Figure 10.7.   Photograph of an angular unconformity. Colluvium (upper right) fills an ancient canyon in horizontal strata (lower left). The unconformity appears as jagged line from upper left to lower right, (Grand Canyon, AZ). Photo by John H. Kramer.

unconformity involving unconsolidated alluvial deposits overlying consolidated rocks is often referred to as "soil" over "bedrock." Vadose zone monitoring may be performed in a straightforward manner in the unconsolidated material, but may be very difficult in the consolidated rocks below. Figure 10.8 presents a hypothetical cross section showing several possible unconformity-related changes of flow pathways in the vadose zone. The geologist must be alert to the potential for such complications.

Fault Contact

A fault may juxtapose different parts of a single geologic unit or different units and may have any orientation. A fault surface may have either higher or lower permeability than the adjacent material, although higher permeability seems to be more common. Ground water flowing along faults supports the vegetation changes or patches of phreatophyte plants often observed along fault traces. Fault offset may cause a vadose zone on one side of a fault to abut a water table on the other.

Major tectonic faults, such as the San Andreas fault in California, may contain zones up to tens of meters in width of pulverized rock (fault gouge) that may locally have very low permeability. Smaller faults may be characterized by a "knife edge" contact between adjacent materials, as depicted in the photo of the Kern Bluff fault appearing in Figure 10.9. In this example, which is at a now-closed municipal landfill where both ground-water and vadose monitoring systems were installed,

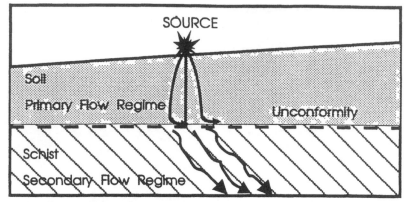

A) Secondary flow regime below an unconformity.

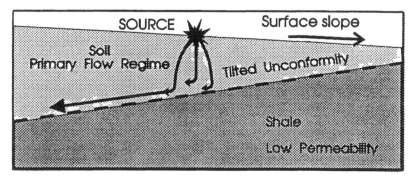

B) Tilted unconformity as an impermeable barrier diverts flow.

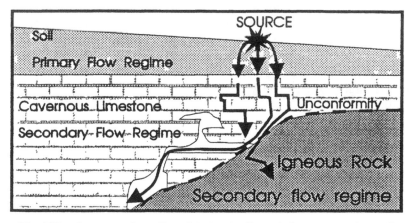

C) Tilted unconformity channels flow into secondary flow regime.   Resulting pathways can be in the opposite direction from surface slope.

Figure 10.8.   Diagram showing the potential effects unconformities have on vadose zone flow pathways.

Figure 10.9. Photograph of Wade Allmon pointing to a "knife-edge" outcropping of the Kern Bluff fault which juxtaposes sandy units against clay-rich units of contrasting permeability. Photo by Steve Cullen.

permeable, poorly-lithified sandstone is juxtaposed across the Kern Bluff fault against less permeable claystone. Based on well-log interpretations, the total vertical offset is about 36 meters (120 feet). The interpretation of the water-table geometry on the basis of monitoring-well water levels was important in determining the thickness of the vadose zone. As shown in Figure 10.10, two different interpretations of the water table, one with a hydraulic barrier at the fault dividing the water table into two laterally separated aquifers (Figure 10.10a) and the other with a continuous gradient across the fault (Figure 10.10b), would result in a difference of the total thickness of the vadose zone of 15 meters (50 feet) or more at locations near the fault. In this case, further characterization revealed the interbedded thin sands and clays were not sealed by the fault and the continuous-gradient interpretation (Figure 10.10b) was correct. At this site, evidence for stratigraphic controls on contaminant migration similar to that depicted in Figure 10.4 was obtained from vadose zone monitoring. Collection of soil-pore liquid samples in suction lysimeters at about the depth of the bottom of waste fill (which was about 91 meters [300 feet] above the water table) and at greater depth showed that contaminant migration took place along the top of individual horizontal claystone beds, with apparently very little vertically downward migration (Metcalf and Eddy, Inc. 1989).

An example of an inferred fault zone which corresponds to a ground-water blockage is presented in Figure 10.11, which shows water table depths across the north branch of the Modoc fault at a closed landfill site in Santa Barbara, CA. At this site an overdrafted ground-water basin has created a steep gradient, apparently across a relatively impermeable fault. The trace of the fault is poorly constrained by surface

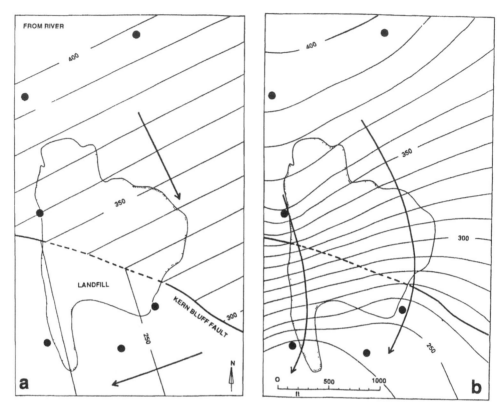

Figure 10.10.   Maps of two possible ground-water table contours across the Kern Bluff Fault (after Metcalf and Eddy, 1989).

geomorphology and outcrop, and its presence is primarily inferred from well data (Upson, 1951). This fault does not appear on conventional geologic maps (Dibblee, 1987), but is shown on maps in the water resources literature (Freckleton, 1989). In this discussion it will be assumed that the feature actually is a fault, instead of some other type of lateral permeability variation, such as a facies change or fold.

Piezometric heads differ by over 46 meters (150 feet) on either side of the fault, and one well (MW-6, Figure 10.11) drilled in the fault zone exhibits an intermediate head level. Interestingly, during the drilling of well MW-6, no water was encountered until below the current water elevation, indicating confined aquifer conditions within the fault zone. The fault zone acts like a leaky confining boundary to impede flow into the vadose zone southwest of the fault. The implication for vadose zone monitoring is that there may be distinct vadose zone thicknesses and permeabilities associated with each side of a fault, and within the fault zone itself.

SOURCES OF GEOLOGIC INFORMATION

Cost is often an important factor during investigation and initial characterization of a new site. Therefore, it is desirable to obtain as much information as possible by low-cost means prior to deploying costly equipment. Possible sources are presented here, approximately in order of increasing cost.

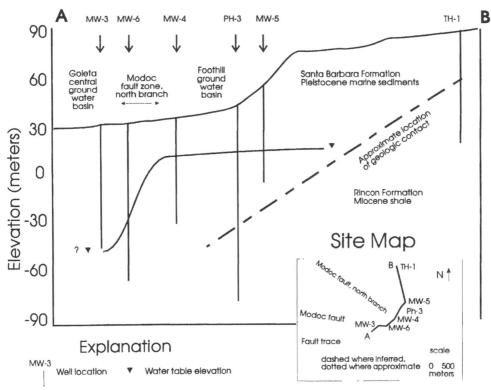

Figure 10.11. Cross section showing piezometric pressure drop across a flow-barrier fault in Santa Barbara, CA (after Staal, Gardner and Dunne, Inc., 1990).

Published Material

The amount of published material available for specific locations is extremely variable, depending largely on the extent of previous geologic investigations that have been carried out for other purposes. Geologic investigations for purposes other than contaminant investigations typically are done on a scale covering a much larger area. They therefore typically lack in specific detail, but can provide a starting point. Published geologic maps and professional articles provide a basis for determining the general type of rocks and unconsolidated geologic material, and the presence of faults and some types of geologic contacts. As mentioned above, primary flow regimes may be expected in areas mapped as alluvium or in some volcanic deposits, whereas most other map units would probably be characterized by secondary flow. An example of a portion of a geologic map is shown in Figure 10.2.

In areas where ground water is utilized for developed supplies, published articles may have information about depth to water tables; hence, the thickness of the vadose zone. In the United States, this information is often contained in United States Geological Survey Water Supply Papers.

Site-Specific Mapping

Surface geologic mapping of the site of interest and the surrounding areas may provide additional information, particularly where exposures of rocks or unconsoli-

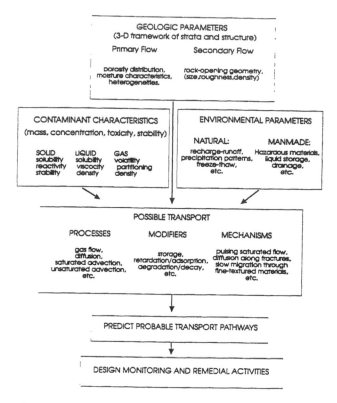

Figure 10.12. Schematic diagram showing the relationship between geologic parameters and steps in a vadose zone investigation.

dated material exist. However, in alluvial areas mapping may provide little information, and in industrialized areas there may be no vadose zone material visible at the surface.

Drilling

It is often necessary to drill test holes to obtain information about the thickness and types of rocks and other materials in the vadose zone, and about the distribution of moisture. Since drilling is relatively costly, all possible information available through the previously described methods should be obtained first. Several different drilling techniques are available. An appropriate technique for a particular site should be chosen based on the depth to be drilled, type of material to be drilled through, and the type of sampling to be performed. A drill hole provides information along a vertical line at one location. A number of drill holes may be necessary to obtain information over an area of interest, and some interpolation is required to interpret between holes.

Geophysical Exploration

Information on the lateral and vertical subsurface distribution of physical properties, which may be interpreted in terms of vadose zone hydrogeologic characteristics may be obtained by a variety of geophysical exploration techniques. The effectiveness of a particular technique depends on the physical properties of the vadose zone material at a particular site. Lateral or vertical variations in these properties must exist in order for the technique to provide useful information. Some techniques which may be effective in particular situations are presented here.

- Seismic refraction—measures one-way seismic wave travel times that are interpreted in terms of seismic velocity contrasts that may indicate different rock types.
- Seismic reflection—measures two-way seismic wave travel times that are interpreted in terms of seismic impedance contrasts that may indicate contacts between different rock units.
- Gravity—measures acceleration values at various locations that are interpreted in terms of density contrasts which can indicate the presence of different rock types.
- Magnetic—measures spatial changes in the Earth's magnetic field, indicating presence of iron as metal (generally buried man-made articles) or within minerals.
- Resistivity—measures electrical resistance of subsurface materials, that may indicate different rock types, moisture content, and some contaminant types or naturally occurring dissolved constituents.
- Electromagnetic—measures electromagnetic impedance contrasts, that may indicate different rock types or historic disturbances, such as filled excavations and water-bearing fractures.
- Ground penetrating radar—measures electromagnetic impedance at radar frequency (very high frequency), that may indicate shallow hard rocks, buried objects, filled trenches, and, in some cases, faults.
- Downhole methods—measures subsurface properties of rocks or unconsolidated material surrounding drill holes by several geophysical techniques. Very precise and specific stratigraphic information may be obtained in this way. Many geophysical tests require a fluid-filled borehole. These cannot be used in the vadose zone portion of a borehole unless it is artificially filled with fluid, which in many cases would be undesirable since such an action could mobilize contaminants or severely distort important features of the preexisting moisture distribution. Downhole methods which do not require a fluid-filled hole include nuclear methods such as natural gamma ray detection and spectroscopy, induction resistivity, and video logging. As noted above, the neutron probe instrument, which is easily deployed in the field, provides a useful downhole technique to determine moisture levels and liquid movement.

PROCEDURE TO CHARACTERIZE AND REMEDIATE A SITE

In approaching a new site for geologic characterization of the vadose zone, the following series of steps are appropriate, with modification as required for specific situations. In most cases, the investigation will be oriented toward some specific, previously identified problem, typically involving contaminants, rather than simply generating a body of knowledge, so the ultimate goal of contaminant identification and remediation should be borne in mind. Such problems might include escapes of nonaqueous phase liquid from tanks or surface impoundments, mobilization of soil contaminants by infiltration of precipitation or artificially applied water, migration of aqueous fluids with dissolved contaminants, and vapor phase transport.

Site Visit and Background Information

A site visit is generally very useful in visualizing the scale of investigation and becoming familiar with the local geology. Review of published background information on the regional and local geologic conditions is useful both before and after the site visit. An important question in any initial vadose zone investigation is the thickness of the vadose zone, so any published information on subsurface hydrology is especially pertinent.

Detailed Geologic Mapping

As noted above, a great deal of information may be gained at some sites by detailed surface mapping, while at others the surface material, either natural or artificial, may not be representative of materials at the depths of interest. At a site with rock outcrops, permeability distributions and the existence of faults and other contacts may be inferred from surface observations. At sites with appreciable topographic relief, material which forms the subsurface vadose zone in one location may be exposed in cliffs or hillsides nearby.

Subsurface Investigation

A method or combination of methods may be selected which is most likely to yield information at a particular site. In many cases, chemical contamination of vadose zone fluids is of primary interest, so some drilling will be necessary to obtain samples for analysis. Permanent monitoring equipment may also be installed at the time of drilling for samples, or installed later based on the information obtained.

Maps and Cross Sections

Interpretation of the information obtained may be most usefully displayed on maps and cross sections. Particular items which should be included on individual or composite displays include rock types and flow regimes, distribution of moisture content, including the water table and other saturated zones, and contaminant distribution. In the case of contamination with volatile organic compounds in the vadose zone, one particularly useful display involves plotting gas chromatograms obtained with a portable field instrument on a cross section showing drill hole locations.

Predict Future Migration

If contaminants are detected in the vadose zone, it is of interest to predict migration rates and future flow paths, in order to evaluate potential impacts. Geologic knowledge is important in identifying contaminant pathways. The information obtained and interpreted by the above steps allows such predictions to be made. A useful guide for such prediction is provided in the flow sheet shown in Figure 10.12. Inputs include geologic parameters discussed in this chapter as well as contaminant characteristics and environmental controls presented elsewhere in this book. The geologic input is the geometric framework which controls the location of potential contaminant transport pathways. When coupled with contaminant characteristics (e.g., volatility, solubility, etc.) and environmental factors, operative transport processes, modifiers, and mechanisms are identified, and the location of probable transport pathways can be predicted. The predicted pathways are the basis for design of monitoring and remediation efforts.

Monitoring Network Design or Remediation

If the vadose zone is uncontaminated, appropriate monitoring networks can be designed to provide early warning of contaminant migration. If the vadose zone is contaminated, the ultimate goal is to remediate and avoid further degradation of the subsurface, particularly of any underlying water table which has not yet been contaminated. The information from the geologic characterization contributes to the necessary background data upon which a monitoring or remediation scheme may be designed.

ACKNOWLEDGMENT

This work was partially funded through the Vadose Zone Monitoring Laboratory at the Institute for Crustal Studies, University of California Santa Barbara, by EPA cooperative agreement CR 816969-01-0 with the Environmental Systems Laboratory-Las Vegas, Lawrence Eccles, Project Officer, contribution #0154-31HW.

REFERENCES

Bear, J., *Hydraulics of Groundwater*, (New York: McGraw-Hill, 1979), p. 203.

Billings, M.P., *Structural Geology*, Second Edition, (Englewood Cliffs, NJ: Prentice-Hall, 1954).

Bouwer, H., *Groundwater Hydrology*. (New York: McGraw-Hill, 1978).

Brady, N.C., *The Nature and Properties of Soils*, (New York: MacMillan Publishing Company, 1990).

Devinney, J.S., L.G. Everett, J.C.S. Lu, and R. Stollar, *Subsurface Migration of Hazardous Wastes*, (New York: Van Nostrand Reinhold, 1990).

Dibblee, T.W., Jr., *Geologic Map of the Goleta Quadrangle*, Dibblee Foundation Map DF-07, Santa Barbara, CA, 1987.

Dragun, J., *The Soil Chemistry of Hazardous Materials*, Hazardous Materials Control Institute, Silver Spring, MD, 1988.

Fetter, C.W., *Applied Hydrogeology*, (Columbus, OH: Merrill Publishing Company, 1988).

Freckleton, J.R., Geohydrology of the Foothill Ground-Water Basin Near Santa Barbara, California. United States Geological Survey Water-Resources Investigation Report 89-4017, 1989.

Freeze, R.A., and J.A . Cherry, *Groundwater*. (Englewood Cliffs, NJ: Prentice-Hall, 1979).

Gaskill, D.L., F.E. Mutschler, J.H. Kramer, J.A. Thomas, and S.G. Zahony, Geologic Map of the Gothic Quadrangle, Gunnison County, CO, Map GQ-1689, U.S. Geological Survey, 1991.

Hillel, D., *Fundamentals of Soil Physics*, (New York: Academic Press, 1980).

Kettleman Hills Vadose Zone Demonstration, Unpublished Report, Kaman Sciences Inc., Santa Barbara, CA., 1987.

Kramer, J.H., S.J. Cullen, and L.G. Everett, "Vadose Zone Monitoring with the Neutron Moisture Probe," *Ground Water Monitoring Review*, 13:177 (1992).

Metcalf and Eddy, Inc., Report to Kern County California Department of Public Works, 1989.

Nielsen, D.M., and A.I. Johnson, Eds., Ground Water and Vadose Zone Monitoring, STP 1053, American Society for Testing and Materials, Philadelphia, PA, 1990.

Reineck, H.E., and I.B. Singh, *Depositional Sedimentary Environments*, (New York: Springer-Verlag, 1980).

Staal, Gardner and Dunne, Inc. Final Report, Solid Waste Assessment Test Water Quality Assessment Transfer Station Landfill, Santa Barbara, CA, 1990.

Unruh, M.E., C. Corey, and J.H. Robertson, "Vadose Zone Monitoring by Fast Neutron Thermalization (Neutron Probe): A 2-Year Case Study," in *Ground Water Management*, (Dublin, OH: NWWA, 1990), p. 431.

Upson, J. E., Geology and Ground-Water Resources of the South-Coast Basins of Santa Barbara County, California, U.S. Geological Survey Water-Supply Paper 1108, 1951, p. 27.

11

Estimating the Storage Capacity of the Vadose Zone

Stephen J. Cullen and Lorne G. Everett

INTRODUCTION

Characterization of parameters governing vadose zone storage capacity is critical to developing a baseline set of data which characterizes the static subsurface hydrology. To adequately characterize the storage capacity of the vadose zone, one must determine: (1) the depth and areal extent of the vadose zone, (2) the capacity for storage of liquids in the vadose zone, (3) the portion of pore space occupied by aqueous liquids, and (4) the portion of pore space occupied by soil gases.

Everett et al. (1984) listed physical properties of the vadose zone associated with storage of water including: (1) thickness, (2) bulk density, (3) porosity, (4) water content, (5) soil-water characteristics, (6) field capacity (also known as specific retention), (7) specific yield, and (8) fillable porosity. Other useful parameters related to water storage in unsaturated regions of the vadose zone include permanent wilting point water content and residual water saturation.

VADOSE ZONE THICKNESS

The terms vadose zone and unsaturated zone are not synonymous. The vadose zone can include both saturated (perched water) and unsaturated regions. By definition, saturated regions have reached their maximum storage capacity for water, whereas unsaturated zones have not. The total storage capacity for liquids in the vadose zone is equal to the porosity of the vadose zone. The thickness of the vadose zone is not constant, but fluctuates with changes in the level of the water table in response to ground-water depletion or recharge. Everett et al. (1984) present an expression to calculate the equivalent depth of water that is stored in a homogeneous vadose zone at a given mass water content:

$$d_w = \left(\frac{\theta_m}{100}\right)\left(\frac{\rho_b}{\rho_w}\right) \cdot d_{vz}$$

where

d_w = depth of water

ρ_b = bulk density of soil (defined in the following section)

ρ_w = density of water

d_{vz} = depth of the vadose zone

θ_m = percent mass water content (based on oven drying at 105°C).

This expression is useful when detailed information regarding the vadose zone is not available. When boring and sampling information is available, differences in the storage capacity of the strata that make up the vadose zone should be considered. Average depth of an individual strata (d_i) can be interpreted from borehole logs. The bulk density (ρ_{bi}) and mass water content (θ_{wi}) of the individual strata can be determined by physical analyses of soil core samples taken during drilling. The modified expression for determining the vadose zone storage then becomes:

$$d_w = \Sigma \left(\frac{\theta_{mi}}{100}\right)\left(\frac{\rho_{bi}}{\rho_w}\right) \cdot d_i$$

wherein the stored water content is calculated for each individual strata and the values are then summed to determine the total vadose zone storage.

The maximum storage capacity is equivalent to, and can be determined from, calculation of the porosity. The difference between the maximum storage capacity and storage represents the air-filled porosity and the amount of additional water that would be required to bring the water content in the vadose zone to saturation.

With the notable exceptions of perched and capillary water, the vadose zone, particularly in unsaturated regions, rarely reaches the maximum water-holding capacity. Even under situations of continuous water application for days at a time, subsurface materials in the unsaturated zone will retain entrapped air and undergo continuous drainage to layers beneath and lateral to the zone being wetted. Field-saturated water content is the term used to describe the water content of soils when completely saturated with the exception of entrapped air. Bouwer (1986) indicated that hydraulic conductivity under field-saturated conditions is approximately a factor of 2 less than that of truly saturated subsurface materials. Figure 11.1 illustrates the effect of entrapped air on water flow through an individual soil pore. Christiansen (1944) found the reduction to be 60% to 85% in laboratory experiments. Klute (1986) uses the terms natural saturation and satiated water content and indicates that it is 80% to 90% of true saturation.

True saturation of sediments really only occurs over extended periods of time when entrapped air can dissolve and disperse in pore-liquids. Truly saturated sediments and soils are usually associated with bogs, estuary areas, and perennially perched zones.

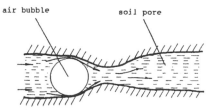

Figure 11.1.   Effect of entrapped air on water flow through a soil pore (adapted from Hillel, 1982).

BULK DENSITY AND POROSITY

Bulk density (usually denoted by the symbol ρ_b) of a soil is the mass of a given bulk volume of soil, where that volume includes the air, liquid, and solid phases. Restated, bulk density is the ratio of the oven-dry mass of a soil to its total volume (air + liquid + solid) as sampled. Bulk density reported on an oven-dry mass basis is also referred to in the literature as "dry bulk density." Bulk density is contrasted with the particle density (ρ_p), which represents the weight of a unit volume of the solid phase only. Both are typically expressed in units of grams per cubic centimeter (g/cc) in hydrologic work.

Some of the conclusions and inferences that can be drawn about the subsurface based on knowledge of the bulk density and particle density include:

1. Detection of pan layers in geologic profiles,
2. Determination and quantification of changes in lithology,
3. Correlation and verification of other physical measurements (e.g., hydraulic conductivity),
4. Determination of porosity and void ratio,
5. Evaluation of volume changes due to compaction and subsidence,
6. Conversion of gravimetric water content to volumetric water content,
7. Estimation of the weight of large soil volumes.

Regardless of the method employed to determine bulk density, the ultimate value used to represent the bulk density is dependent upon the moisture content at which the sample or measurement is made. This factor is by far most significant in clayey soils. Soils, particularly those of montmorillonitic and illitic clay mineralogy, are subject to shrinking and swelling. Because the volume of a shrinking-swelling soil changes with a change in its water content, values for bulk density should be reported along with the water content at which the sample or measurement was made. Typically, bulk density values are reported at existing field water conditions and a separate determination is made to quantify existing field water contents.

Methods for determining bulk density include clod methods, core methods, excavation methods, and radiation methods (Blake and Hartge, 1986). Clod methods involve collecting samples of natural clods or aggregations of soil. The clods are tied with fine wire or placed in fine-mesh nets and momentarily immersed in a solution of liquid plastic or paraffin. Volume of the clod is determined by using the Archimedes Principle of weighing the clod in air and in water (USDA, 1984). The clod method is typically used for relatively near-surface applications where excavation is possible. Clod samples are useful when the soil is too hard or brittle to sample with a coring device. Caution should be exercised when interpreting soil clod bulk density data. The method has a natural bias against inclusion of interaggregate pores, and will tend to produce relatively high measured values.

Core methods have been widely used in general subsurface hydrologic and hazardous waste contaminant investigations and may represent the only viable method for collecting soil bulk density samples from deep portions of the intermediate vadose zone. Soil core samples can be used for other physical and chemical determinations after the determination for bulk density has been made.

Excavation methods for determining bulk density require a reasonably level surface and are typically performed at the surface or at a surface created at depth by digging or trenching down to the level of interest. Bulk density is determined by weighing an amount of soil retrieved from a small cylinder-shaped hole excavated into the prepared surface. Once out of the excavation, the weight and moisture content of the excavated soil are determined. The volume of the excavation is determined by backfilling the excavated hole with material of a known bulk density. Blake and Hartge (1986) and ASTM (1991a, 1991b) discuss detailed methods for using excavation methods which are backfilled with No. 3 sand and a water-filled balloon. The disadvantage of the excavation methods is that they are restricted to relatively near-surface usage.

Radiation methods take advantage of the fact that bulk density influences soil gamma radiation transmission. Backscattering techniques are used extensively in surface geotechnical applications such as road building and dam construction (ASTM, 1991c), and less frequently to make downhole determinations as part of a contaminant investigation. Photons emitted from a gamma radiation source penetrate the surrounding material (air, casing, well packing, and geologic formation) and the resulting scattered photons are measured by the detectors. Shielding between the source and detectors reduces the number of photons traveling directly from source to detector. The number of detected photons is directly related to the overall density of the material being measured.

Downhole probes utilize an instrument configuration in which the source and detector are placed on the same axis (Figure 11.2), a fixed distance apart. The gamma photon source is typically Cesium-137 or Radium-226, while the detector consists of either gas-filled Geiger-Muller tubes or scintillation crystals of sodium-iodide. Since the density of the composite material is measured, a correction for water content is necessary to determine dry density. Paired water content determinations are required for maximum accuracy of the technique. Neutron moderation can be used to obtain paired water content measurements and is discussed in detail in Chapter 18. Instruments are commercially available which utilize both gamma radiation backscattering and neutron moderation to simultaneously determine bulk density and soil water content. Density determination by radiation methods is indirect and requires the use of a calibration curve. While calibration curves are typically supplied with commercial equipment, the curves do not necessarily hold for all soils because of differences in chemical composition (ASTM, 1991c).

Together, bulk density and particle density can be used to calculate porosity (e) from the expression:

$$e = \left[1 - \left(\frac{\rho_b}{\rho_p} \right) \right] \cdot 100$$

where,

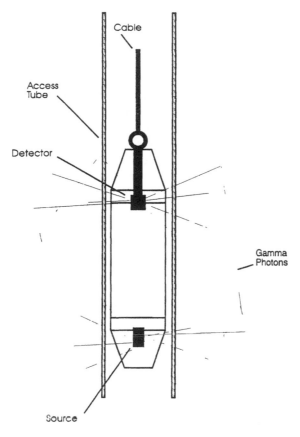

Figure 11.2. In-line configuration of gamma radiation source and detector used in downhole gamma radiation density measurements.

e = porosity expressed as a volumetric percent,
ρ_b = soil bulk density,
ρ_p = particle density of the solid matrix.

The particle density can be accurately measured using the air pycnometer technique (Blake and Hartge, 1986). Often, a particle density of 2.65 g/cm³ is assumed for many soils based on the density of quartz which is commonly prevalent in sandy soils. The particle density of some clays may, however, be less and the density of soils with appreciable amounts of ferromagnesian minerals, iron oxides, or heavy metals may be greater (Hausenbuiller, 1972). Organic-rich soils and subsurface strata may have a particle density less than 2.65 g/cm³. Brady (1984) states that organic matter has a typical particle density in the range of 1.1 to 1.4 g/cm³. Cullen (1981) measured mineral particle densities as low as 2.4 to 2.5 g/cm³ on volcanic ash soils in western Montana. As indicated in the equation above, accurate knowledge of the particle density is critical to calculation of the porosity based on the bulk density. Assumptions of particle density values should be validated by use of correlative data or by representative sampling and testing.

SOIL WATER CHARACTERISTICS

While the quantity of water (and dissolved contaminants) stored in unsaturated regions of the vadose zone varies with porosity and saturation percentage, the mechanism by which water is held in soils and sediments is related to the total water potential. The matric potential, or soil suction, with which water is held in the soil determines the direction and magnitude of water movement in soils and sediments, the ability to collect pore-liquid samples with lysimeters, and the unsaturated hydraulic conductivity of unsaturated regions of the vadose zone. A soil water characteristic curve (Figure 11.3) describes the relationship between soil water content and soil water matric potential and is unique for each given soil or sediment type.

Methods for characterizing the soil hydrologic properties include field methods and laboratory methods used in conjunction with soil core samples. Soil water characteristic curves can be determined in the field using side-by-side installations of instruments which measure matric potential and soil water content. Figure 11.4 illustrates a cross section of an experimental setup in which a neutron probe can be used to measure water content in combination with tensiometers to measure matric potential. Measurements can be made with the kind of instrumentation shown in the illustration during the use of a double-ring infiltrometer (Green et al., 1986). As the wetting front progresses beneath the infiltrometer, successive measurements are made with each instrument. Data collection can be facilitated by use of electronic data loggers. Paired measurements of matric potential (with the tensiometers) and water content (with the neutron probe, time domain reflectometer, or frequency domain capacitance probe) at increasing levels of wetness are used to construct wetting curves of the general shape shown in Figure 11.3. The approach is limited to near-surface applications because of the depth limitations associated with use of tensiometers. Hillel (1982) points out that this approach has also been subject to problems associated with soil heterogeneity and uncertainties over hysteretic phe-

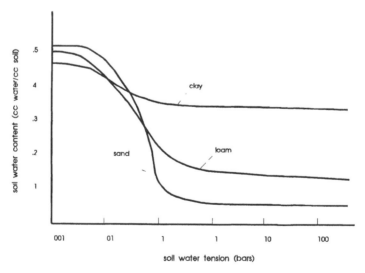

Figure 11.3.   Soil water characteristic (draining) curves for three representative mineral soil types.

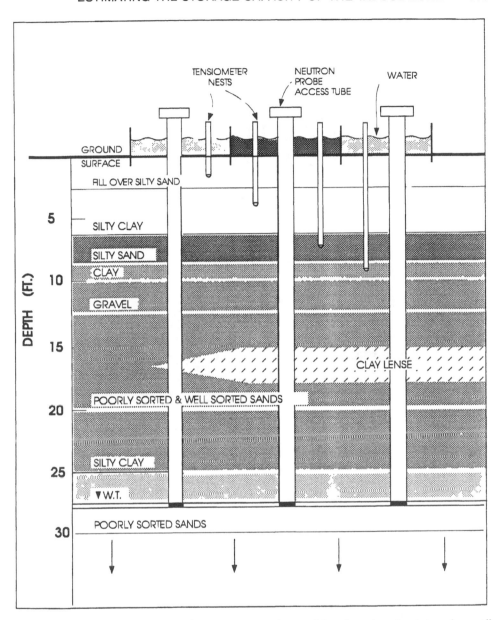

Figure 11.4. Field setup for using neutron probe and tensiometers to determine soil water characteristic curve.

nomena as they occur in the field. Bruce and Luxmoore (1986) present a detailed discussion on field procedures for developing soil water retention relationships.

Laboratory measurement of soil water characteristics is by far the more common (ASTM, 1991d; ASTM, 1991e; Klute, 1986) methodology. Either disturbed or undisturbed samples are collected in the field and transported to the laboratory. Undisturbed samples should be used whenever possible because of the effect that microstructural features and aggregation can have on the soil water relations, particularly in the low (less than 1 bar) soil suction range. The samples are loaded into the pressure extraction apparatus (Figure 11.5) and a series of increasingly positive air

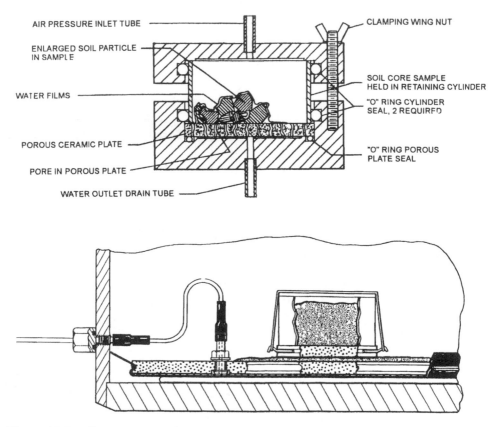

Figure 11.5.   Pressure extraction apparatus for soil water characteristic determinations: (a) Tempe-style pressure cell accepts undisturbed core sample; (b) partial cross section of pressure plate system which can be used for disturbed or undisturbed samples (Courtesy Soilmoisture Equipment Corp.).

pressures is applied to samples contained in a pressure vessel. The samples are saturated and placed in intimate hydraulic contact with a porous material. The porous material is configured in such a way that it is exposed to the applied pneumatic pressure on one side and to ambient air pressure on the other. Water in the samples which is held with a soil suction of absolute value less than the applied air pressure will move out of the soil, through the porous ceramic, toward the lower air pressure on the outside of the pressure vessel. By sequentially establishing a series of pressure equilibria and determining the associated water content at each pressure step, soil water characteristic curves, as shown in Figure 11.3, can be constructed.

At increasingly low soil water tensions, the larger pores become filled with water. The very largest pores (greater than about 60 microns) are termed "macropores." When filled, they hold a substantial amount of water and conduct water at appreciable flow rates under a hydraulic gradient. The natural macropore structure should be preserved when making determination of soil hydrologic properties at low soil water tensions. It is recommended that undisturbed core samples be used for determinations in the 0 to 1 bar range. Use of undisturbed samples is less critical for determinations above 1 bar. Water held at high soil water tensions is held in interstitial spaces and is much less vulnerable to disturbance by sampling techniques.

The amount of water stored at any given water potential is affected by the antecedent water conditions; in other words, whether the unsaturated zone is wetting or
draining (Figure 11.6). This concept, called hysteresis, is also illustrated graphically
in Figure 11.7. For much of the range of water potentials encountered in the field,
this is not an issue. However, in the range of soil suctions up to as high as one bar,
inadequate consideration of hysteretic effects can cause significant errors in representing field conditions. Instrumentation, similar in function to the equipment mentioned above, is available to characterize the imbibition (wetting) curve of the soil
water characteristic (Soilmoisture, 1992).

Soil water characteristics are unique to each soil/sediment type and representative
characterization data should be acquired for each significant hydrogeologic unit.
Soil water content and matric potential data are required to determine the mobility
of dissolved contaminants and the potential for collecting pore-liquid samples using
suction lysimeters.

FIELD CAPACITY

Field capacity is generally referred to as the water retained in a soil or sediment
against the force of gravity. By analogy, a sponge submersed in a tub of water will
become saturated, ignoring entrapped air. Once the sponge is lifted into the air,
water will at first quickly drain. Later, the sponge will virtually cease drainage. At
the point where drainage due to the force of gravity essentially ceases, the sponge is
said to be at its field capacity. In field soils and sediments, drainage does not cease at
field capacity but can continue at slow rates over extended periods of time as the
water redistributes in the profile due to potential gradients. Strictly speaking, the
term "field capacity" is vague and does not represent a completely definable measurement of the status of water in soils.

Taylor and Ashcroft (1972), however, demonstrated the concept for agricultural
irrigation (Figure 11.8). Cassel and Nielsen (1986) present a useful definition of field
capacity by Gardner (1960): " . . . that water content below which the hydraulic
conductivity is sufficiently small that redistribution of moisture in the soil profile
due to a hydraulic head gradient can usually be neglected." Application of the
concept requires a more exact, albeit somewhat arbitrary, definition such as the
water content at .1 bar soil suction. "Specific retention" is an equivalent term used
by hydrogeologists to describe the amount of water remaining in a formation after
dewatering. Use of these terms in vadose zone monitoring is important because they
represent a straightforward measure of the ability of soils and sediments to retain
water (and therefore dissolved contaminants) immediately after recharge or a contamination leak event.

Values for field capacity in soils and sediments vary widely. Hillel (1982) indicates
that field capacity may vary from about 4% by weight in sands to about 45% in
heavy clays. Certain organic soils may have a field capacity in excess of 100% water
content by weight. The drainage time required to reach field capacity after flooding
may also vary. Freely drained, sandy materials may reach field capacity within one
day after flooding (Cassel and Sweeney, 1974). Clay materials, on the other end of
the textural spectrum, may take up to several weeks to drain to field capacity
(Davidson et al., 1969).

Measurement and characterization of field capacity is best accomplished in situ.
In general, excess water is added to a field soil until it is thoroughly wetted to the

(a)   (b)

wetting   draining

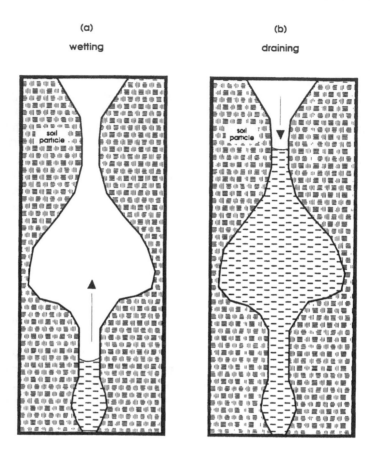

Figure 11.6.   "Ink bottle" explanation of the effect of antecedent moisture conditions on soil water content (hysteresis) at low soil matric potentials. Effective diameter of the smallest pore controls water entry or exit at any given soil water tension.

desired depth. In situ sensors (e.g., time domain reflectometry [TDR], capacitance probe, or neutron probe) can be used to monitor the strata of interest until the water content is essentially stable over time. This stabilized water content is taken as the field capacity of the strata. Alternatively, tensiometers or electrical resistance can be used as the in-situ sensors, and the field capacity expressed in terms of soil suction units. Cassel and Nielsen (1986) present a discussion of methodologies used to measure field capacity in the field and on laboratory samples.

In characterizing unsaturated regions of the vadose zone, all of these methods are limited by the ability to place sensors at depth or by the ability to move water deep into the geologic profile by infiltration. In situ characterization of field capacity should be conducted with caution when employed in a contaminant investigation, as introduction of large amounts of water into the unsaturated zone may mobilize potential contaminants which may exist as immobile compounds in otherwise dry strata. Subsequent monitoring efforts may also be confounded by the uncharacteristically high water content induced by the initial flooding.

While there is really no good substitute for the in-situ measurement of field capacity, estimates of field capacity are needed when in-situ characterization is not

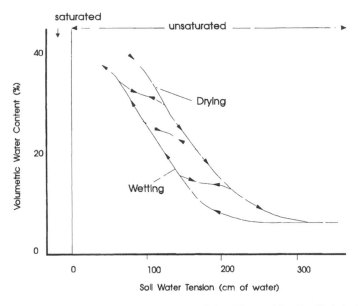

Figure 11.7. Hysteresis of water content, explained in part by the "ink bottle" effect, in a drying and wetting sand.

possible. Laboratory methods are described here but should be considered as approximations and clearly identified as such when reporting results. Laboratory methods will not necessarily yield the same results as field methods conducted on the same materials.

Laboratory measurement of field capacity should be made on undisturbed soil core samples taken from a borehole. The sample is then saturated, placed on a porous membrane, and brought to equilibrium with an applied pneumatic pressure in a pressure vessel identical to that described earlier and illustrated in Figure 11.5. Alternatively, suction can be applied using the Haines apparatus (Figure 11.9), as described in Day et al. (1967). The key to this approximation method is that a field capacity soil suction value for each soil must be identified. Small field trials can be initially performed to measure the in-situ field capacity soil suction and later applied to soil core samples submitted to the laboratory. More often, the value is assumed.

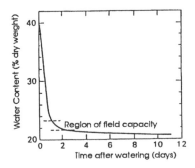

Figure 11.8. Determination of field capacity based on change in gravimetric water content over time (adapted from Taylor and Ashcroft, 1972).

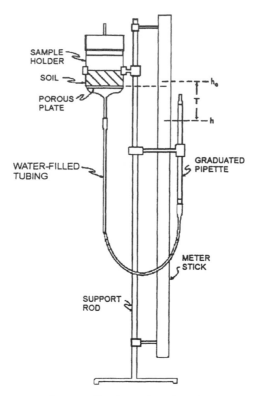

Figure 11.9. Haines apparatus: suction-based method for measuring field capacity and soil water characteristics (adapted from Day et al., 1967).

Typically, lower values (5 and 10 cb) are assumed for coarse-textured soils such as sands, 33 cb is assumed for medium-textured soils like loams, and 50 cb is used for finer-textured clayey soils (Cassel and Nielsen, 1986). For vadose zone hydrological work, the appropriate equilibrium suction is best identified by relating the selected suction value to the hydraulic conductivity function of the soil or sediment (Miller and Klute, 1967).

SPECIFIC YIELD

"Specific yield" is a term employed by hydrogeologists to characterize storage in an unconfined aquifer; that is, specific yield is " . . . the volume of water that an unconfined aquifer releases from storage per unit surface area of aquifer per unit decline in the water table" (Freeze and Cherry, 1979). Figure 11.10 shows the relationship between specific yield and specific retention for various geologic materials. As shown in Figure 11.11, the specific yield for a medium equals the porosity value minus the value of specific retention. Representative specific yield values for valley sediments in California are listed in Table 11.1.

A number of techniques are available for estimating the specific yield of the ground-water zone. If the storage properties of material above and below the water table are similar, the specific yield value determined for a ground-water zone can be assumed to approximate the specific yield in the vadose zone. This assumption may

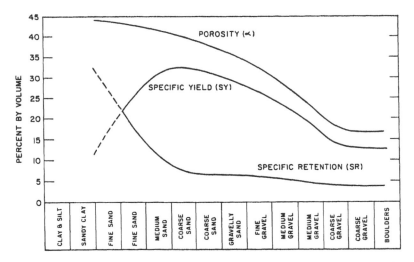

Figure 11.10. Relationship of porosity, specific yield and specific retention for various grain size distributions (Scott and Scalmanini, 1978).

be inappropriate at many locations because of variations in lithology. Techniques for estimating specific yield of the ground-water zone are reviewed in Everett et al. (1984).

Stallman (1967) suggests that a realistic estimate of specific yield can be obtained by observing variations in water content profiles during the decline of a water table. In particular, the specific yield can be calculated by using a sequence of water content profiles to determine the volume of water drained near the water table during a decline in the water table (Bouwer, 1978). Meyer (1963) used this technique to estimate temporal variations of the apparent specific yield near pumping wells.

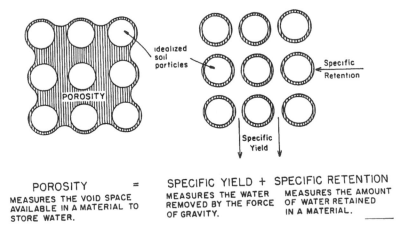

Figure 11.11. Conceptual relationship between porosity, specific yield, and specific retention (Scott and Scalmanini, 1978).

Table 11.1.   Representative Specific Yield Values

Material	Average Specific Yield (percent)
Clay	2
Silt	8
Sandy Clay	7
Fine Sand	21
Medium Sand	26
Coarse Sand	27
Gravelly Sand	25
Fine Gravel	25
Medium Gravel	23
Coarse Gravel	22

Source: Cooley, Harsh, and Lewis, 1972.

The use of neutron moisture logs appears to be the most suitable approach for estimating the specific yield of vadose zone sediments near the water table. Such logs can also be used to estimate the specific yield of deposits in other regions of the vadose zone where perched ground-water bodies are generated during cyclic recharge events. Allmon and Wells (1992) also employed the neutron probe technique to use water content profiling for identification of the capillary fringe zone (Figure 11.12).

FILLABLE POROSITY

The volume of water that an unconfined aquifer stores during a unit rise in water table per unit surface area is called the fillable porosity (Bouwer, 1978). As shown in Figure 11.13, the amount of water placed into storage (fillable porosity) during the rise of a water table is less than the amount released during drainage (specific yield). The difference reflects hysteresis caused by air entrapped in pore sequences during the rise of the water table.

Neutron moisture logs also allow estimates of the fillable porosity of sediments near the water table, i.e., the volume of water placed into storage per unit rise in water level. Because of entrapment of air during the rise of a water table, the fillable porosity will initially be less than the specific yield.

PERMANENT WILTING POINT AND RESIDUAL SATURATION

The term permanent wilting point (PWP) has been used by agriculturalists to identify the water content or soil suction dryer than that at which plants will irreversibly wilt and die. The value generally assigned for PWP suction is 15 bars, although the actual value will vary with the plant life under investigation, and can be as high as 80 bars. PWP water content is of value to hydrogeologic work in that data are readily available in most soil survey interpretations published by the Soil Conservation Service and it has been used as an approximation of the residual water content. When data are not readily available, laboratory determinations (ASTM 1991d and

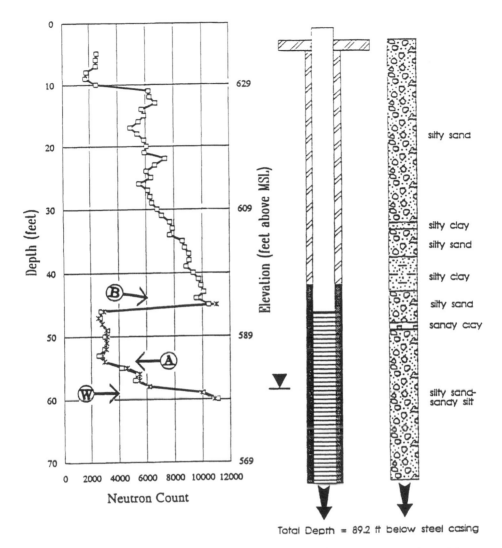

Figure 11.12.   Use of neutron logging technique to identify the top of the capillary fringe. A = approximate top of the capillary fringe; B = point at which well grout affects neutron probe counts; w = approximate level of the water table (Allmon and Wells, 1992).

1991e) of the 15-bar water content can be made from soil samples taken in the field. Undisturbed cores are not required for this analysis.

"Residual water content" has been defined as the water content at which the slope of the soil water characteristic curve becomes zero and at which the hydraulic conductivity approaches zero (van Genuchten, 1978). As is identified in the following chapter on characterizing water transmission, characterizing the hydraulic conductivity function in the intermediate vadose zone is a complex problem. Since the required data may be inadequate or absent altogether, models are often used to calculate the conductivity function from the more easily acquired soil water characteristic data (Mualem, 1976). Residual water content can usually be approximated by the PWP or 15-bar water content.

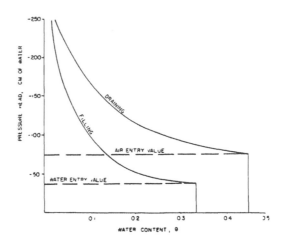

Figure 11.13.   Hysteretic effect of rising (filling) and falling (draining) water table on the satiated water content, and thus, fillable porosity (Wilson, 1980).

REFERENCES

Allmon, W.F., and J. Wells, Identification and Determination of the Capillary Fringe Zone Utilizing the Neutron Probe at Amboy, California. Metcalf and Eddy, Report to Waste Management of North America, 1992.

ASTM. D-1556 Test Method for Density of Soil in Place by the Sand-Cone Method, in *ASTM Annual Book of Standards*, Vol. 4.08, pp. 214–219. American Society of Testing and Materials, Philadelphia, PA, 1991a.

ASTM. D-2167 Test Method for Density and Unit Weight of Soil in Place by the Rubber-Balloon Method, in *ASTM Annual Book of Standards*, Vol. 4.08, pp. 269–272. American Society of Testing and Materials, Philadelphia, PA, 1991b.

ASTM. D-2922 Test Method for Density of Soil and Soil-Aggregate in Place by Nuclear Methods (Shallow Depth), in *ASTM Annual Book of Standards*, Vol. 4.08, pp. 359–365. American Society of Testing and Materials, Philadelphia, PA, 1991c.

ASTM. D-2325 Test Method for Capillary-Moisture Relationships for Coarse- and Medium-Textured Soils by Porous-Plate Apparatus, in *ASTM Annual Book of Standards*, Vol. 4.08, pp. 285–291. American Society of Testing and Materials, Philadelphia, PA, 1991d.

ASTM. D-3152 Test Method for Capillary-Moisture Relationships for Fine-Textured Soils by Pressure-Membrane Apparatus, in *ASTM Annual Book of Standards*, Vol. 4.08, pp. 285–291. American Society of Testing and Materials, Philadelphia, PA, 1991e.

Blake, G.R., and K.H. Hartge, "Bulk Density," in A. Klute, Ed., *Methods of Soil Analysis, Part 1, Physical and Mineralogical Methods*, 2nd edition, 1986, pp. 363–382.

Bouwer, H., "Intake Rate: Cylinder Infiltrometer," In A. Klute, Ed., *Methods of Soil Analysis, Part 1, Physical and Mineralogical Methods*, 2nd edition, 1986, pp. 619–633.

Bouwer, H., *Groundwater Hydrology*. (New York: McGraw-Hill, 1978).

Brady, N.C., *The Nature and Properties of Soil*, (New York: Macmillan Publishing Co., 1984).

Bruce, R.R., and R.J. Luxmoore, "Water Retention: Field Methods," in A. Klute, Ed.,

Methods of Soil Analysis, Part 1, Physical and Mineralogical Methods, 2nd edition, 1986, pp. 663–686.

Cassel, D.K., and D.R. Nielsen, "Field Capacity and Available Water Capacity," in A. Klute, Ed., *Methods of Soil Analysis, Part 1, Physical and Mineralogical Methods*, 2nd edition, 1986, pp. 901–926.

Cassel, D.K., and M.D. Sweeney, In Situ Soil Water Holding Capacities of Selected North Dakota Soils. North Dakota Agric. Exp. Stn. Bull. 495, 1974.

Christiansen, J.E., "Effects of Entrapped Air Upon the Permeability of Soils," *Soil Science*, 58:355–365 (1944).

Cooley, R.L., J.F. Harsh, and D.C. Lewis, *Principles of Ground-Water Hydrology, Hydrologic Engineering Methods for Water Resources Development*, Vol. 10, Hydrologic Engineering Center, U.S. Army Corps of Engineers, Davis, CA, 1972.

Cullen, S.J., The Characterization and Compaction of Forest Soils Forming in Three Parent Materials in Western Montana, MS thesis, Department of Plant and Soil Science, Montana State University, Bozeman, MT, 1981.

Davidson, J.M., L.R. Stone, D.R. Nielsen, and M.E. LaRue, "Field Measurement and Use of Soil-Water Properties," *Water Resour. Res.*, 5:1312–1321 (1969).

Day, P.R., G.H. Bolt, and D.M. Anderson, "Nature of Soil Water," in R.M. Hagan, et al., eds., *Irrigation of Agricultural Lands*, (Madison, WI: American Society of Agronomy, 1967), pp. 193–208.

Everett, L.G., L.G. Wilson, and E.W. Hoylman. *Vadose Zone Monitoring for Hazardous Waste Sites*, (Park Ridge, NJ: Noyes Data Corporation, 1984).

Freeze, R.A., and J.A. Cherry, *Groundwater*, (Englewood Cliffs, NJ: Prentice-Hall, 1979).

Gardner, W.R., "Dynamic Aspects of Water Availability to Plants," *Soil Sci.*, 89:63–73 (1960).

Green, R.E., L.R. Ahuja, and S.K. Chong, "Hydraulic Conductivity, Diffusivity, and Sorptivity of Unsaturated Soils: Field Methods," in A. Klute, Ed., *Methods of Soil Analysis, Part 1, Physical and Mineralogical Methods*, 2nd edition, 1986, pp. 771–798.

Hausenbuiller, R.L., *Soil Science, Principles and Practice*, (Dubuque, IA: William C. Brown Co., 1972), pp. 80–81.

Hillel, D., 1982. *Introduction to Soil Physics*, (New York: Academic Press, 1982).

Klute, A., "Water Retention: Laboratory Methods," in A. Klute, Ed., *Methods of Soil Analysis, Part 1, Physical and Mineralogical Methods*, 2nd edition, 1986, pp. 635–662.

Meyer, W.R., Use of a Neutron Moisture Probe to Determine the Storage Coefficient of an Unconfined Aquifer, Prof. Paper 450-E, U.S. Geological Survey, 1963, pp. 174–176.

Miller, E.E., and A. Klute, "The Dynamics of Soil Water: Part 1—Mechanical Forces," in R.B. Hagan et al., Eds., *Irrigation of Agricultural Lands*, (Madison, WI: American Society of Agronomy, 1967), pp. 209–244.

Mualem, Y., "Hysteretical Models for Prediction of the Hydraulic Conductivity of Unsaturated Porous Media," *Water Resour. Res.* 12:1248–1254 (1976).

Scott, V.H., and J.C. Scalmanini, Water Wells and Pumps: Their Design, Construction, Operation and Maintenance. Bull. 1889, Div. Agric. Sci., U. of California, 1978.

Soilmoisture Equipment Corp., Operating Instructions, Model 1250 Volumetric Pressure Plate Extractor, Santa Barbara, CA, 1992.

Stallman, R.W., "Flow in the Zone of Aeration," in Ven Te Chow, Ed., *Advances in Hydro-science*, Vol. 4, (New York: Academic Press, 1967).

Taylor, S.A., and G.L. Ashcroft, *Physical Edaphology: The Physics of Irrigated and Nonirri-gated Soils*, (San Francisco, CA: W.H. Freeman and Co., 1972).

USDA. Procedures for Collecting Soil Samples and Methods of Analysis for Soil Survey. Soil Conservation Service, Soil Survey Investigations, Report No. 1., Lincoln, NE, 1984.

van Genuchten, R., Calculating the Unsaturated Hydraulic Conductivity with a New Closed-Form Analytical Model. Research Report 18-WR-08, Department of Civil Engineering, Princeton University, Princeton, NJ, 1978.

Wilson, L.G., Monitoring in the Vadose Zone: A Review of Technical Elements and Meth-ods, EPA-600/7-80-134, U.S. Environmental Protection Agency, Las Vegas, NV, 1980.

12

Estimating the Ability of the Vadose Zone to Transmit Liquids

Herman Bouwer

PROCESSES AND EQUATIONS

This section reviews methods for characterizing the liquid-transmission potential of the vadose zone at waste disposal facilities *prior* to selecting and installing monitoring devices.

The simplest situation of flow from the surface to underlying ground water is a completely uniform, homogeneous vadose zone which, of course, is also completely fictitious. For that case, the infiltration and downward flow process can be described with Darcy's equation, assuming piston flow in the wetted zone (uniform water content) and constant, (negative) pressure or suction at the wetting front, and vertical flow (infinite or large extent of infiltrating area). This yields the familiar Green and Ampt equation which was developed in 1911 and has been well discussed in the literature (Bouwer, 1978, and references therein). The equation can be written as

$$v_i = K \frac{H_w + L_f - h_{cr}}{L_f} \qquad (1)$$

where

v_i = infiltration rate (length/time)
K = hydraulic conductivity of wetted zone
H_w = depth of water above soil
h_{cr} = critical pressure head of soil for wetting
L_f = depth of wetting front

0–87371–610–8/95/$0.00 + $.50

The wetting front is considered as an abrupt interface between wetted and nonwetted soil materials. The term v_i in Equation 1 is the Darcy velocity which expresses the infiltration rate as the rate of fall of the water surface above the ground if no water were added to the ponded water. The value of K normally is less than K at saturation because entrapped air is present in the wetted zone. Thus, K in Equation 1 refers to the "resaturated" K, which may be about one-half of K at saturation (Bouwer, 1966) for light-textured soils. For heavier soils, the resaturated K may be on the order of about one-fourth the saturated K. The suction at the wet front can be estimated as the water entry value of the soil (water displacing air), which is about one-half the air entry value (air displacing water). The suction at the wet front can be taken as the center of the suction range where most of the hydraulic conductivity reduction occurs in the S-shaped curves of plots of hydraulic conductivity versus soil moisture suction. The Green and Ampt equation also shows that increasing the water depth H_w will increase infiltration at the beginning of the infiltration process when L_f is small. However, if L_f is already large, increases in H_w will only yield small increases in the infiltration rate, and no increases at all if L_f is "infinity" (very large).

Real vadose zones are not uniform but stratified or otherwise heterogeneous. If hydraulic conductivities decrease with depth, the entire wetted zone will still be at resaturated K. For the stratified case, the relation between v_i and L_f is then calculated as a step-by-step process for the various layers and the average vertical K of the wetted zone and, hence, the final infiltration rate, is the harmonic mean of the K values of the various layers (Bouwer, 1969). If there is a definite perching layer, the height of the perched ground water can be calculated with Darcy's equation (see Bouwer, 1978, page 54). If this mound rises so high that the top of the capillary fringe gets close to the soil surface, reductions in infiltration rate will occur until, with continuing rise of the perched water table, infiltration can stop altogether if the perched water table level coincides with the water surface above the soil. The height of the capillary fringe in this context refers to the dynamic height of the capillary fringe under vertically downward flow. This height will be larger than the height of the capillary fringe under static condition without any vertical flow, and can be calculated with Darcy's equation. Where K increases with depth, the top part of the wetted zone will still be resaturated, but when the wet front reaches a layer that has a larger resaturated K than the infiltration rate, that layer will not completely "resaturate" and unsaturated flow will develop (Bouwer, 1976). This unsaturated flow eventually will be at unit gradient (gravity flow only) and the water content in the unsaturated layers will establish themselves at a value whereby the corresponding unsaturated K is equal to the infiltration rate (Bouwer, 1976).

An extreme situation of K decreasing with depth is obtained when a clogging layer develops on the soil surface, due to biological activity and/or deposition of fine sediment. This commonly occurs where there is a long-term inundation and infiltration, such as in ground-water recharge basins, wetlands, ponds for clean water or wastewater, lagoons, etc. The clogging layer then will be saturated, but the underlying material will be unsaturated. The infiltration rate v_i can then be calculated by applying Darcy's equation to the clogging layer (Bouwer, 1982), yielding

$$v_i = K_c \frac{H_w + L_c - h_i}{L_c} \qquad (2)$$

where

v_i = seepage or infiltration rate (dimension length/time),

H_w = water depth above clogging layer,

K_c = saturated hydraulic conductivity of clogging layer (length/time),

L_c = thickness of clogging layer,

h_i = pressure head of water at bottom of clogging layer.

If the clogging layer is too thin to determine its actual thickness and hydraulic conductivity, these two parameters are combined into one as L_c/K_c, which is called the hydraulic impedance and has the dimension of time (i.e., the time required for a unit amount of water to move through the clogging layer per unit gradient). Where infiltration is controlled by a clogging layer on the soil, Equation 2 shows that increasing the water depth essentially will give a linear increase in infiltration rates. However, increasing H_w also compresses the clogging layer, which then reduces K_c and may actually produce lower infiltration rates (Bouwer and Rice, 1989). The value h_i in Equation 2 is the pressure head in the unsaturated zone beneath the clogging layer. This is the pressure head or suction whereby the unsaturated hydraulic conductivity in the unsaturated zone is numerically equal to the infiltration rate. Procedures for solving Equation 2 and estimating h_i are discussed in a previous publication (Bouwer, 1982).

Another physically based infiltration equation was developed by Phillips, who treated the infiltration system as a diffusion process (Bouwer, 1978, and references therein). Since, in practice, infiltration is affected by many physical, chemical, and biological factors that change with time (crusting, clogging, dissolution and formation of gases in the wetted zone, changes in the status of the clay, soil compaction, temperature, etc.), physically based equations have their limitations and empirical equations, such as the Kostiakov equation (Bouwer, 1978, and references therein) that use coefficients calculated from actual infiltration measurements, may be more practical.

The infiltration equations discussed above estimate infiltration rates when the soil is flooded. These are the maximum infiltration rates that can occur, as controlled by water depths and hydraulic properties of the underlying soils. When the water arrives at the surface at a rate less than the maximum potential infiltration rate, as with natural rainfall or sprinkler irrigation, the infiltration rate will be equal to the application rate, and the Darcy flow in the soil will be equal to the application rate.

APPROXIMATIONS/CATALOGS

Solution of Darcy's equation for estimating rates of water movement in the vadose zone requires estimates or measurements of the saturated hydraulic conductivity, K. Knowing the texture of granular vadose-zone regions of interest (e.g., based on drill cuttings from test holes), order of magnitude estimates of K can be selected from the following (Bouwer, 1978):

Description	K (m/day)
Clay soils (surface)	0.01–0.2
Deep clay beds	10^{-8}–10^{-2}
Loam soils (surface)	0.1–1
Fine sand	1–5
Medium sand	5–20
Coarse sand	20–100
Gravel	100–1000
Sand and gravel mixes	5–100
Clay, sand, and gravel mixes (till)	0.001–0.1

Corresponding K values for consolidated materials are as follows:

Description	K (m/day)
Sandstone	0.001–1
Carbonate rock with secondary porosity	0.01–1
Shale	10^{-7}
Dense, solid rock	$< 10^{-5}$
Fractured or weathered rock (aquifers)	0.001–10
Fractured or weathered rock (core samples)	almost 0–300
Volcanic rock	0–1000

Mualem (1976) prepared a catalog of the unsaturated hydraulic properties based on data published in the literature. Additional catalogs of unsaturated flow properties for a large number of soils have been prepared to update and supplement Mualem's original catalog (Case et al., 1983, and Panian, 1987). Included among the properties are saturated K, values of matric potential as a function of degree of saturation or volumetric saturation, and unsaturated hydraulic conductivity as a function of matric potential or saturation (Case et al., 1983). The hydraulic properties of vadose zone sediments at a site of interest may be estimated by searching these catalogs for soils of similar textures.

MEASUREMENT

ASTM Standard Guide D 5126-90 (ASTM, 1990) reviews alternative field methods for available techniques for measuring K in the vadose zone. Table 12.1, extracted from this Standard Guide, summarizes these methods together with their advantages and disadvantages.* Following is a review of infiltrometers, the air-entry permeameter, and the infiltration gradient method used to measure K and/or infiltration rates.

---

*Standard Guide D 5126-90 ("Standard Guide for Comparison of Field Methods for Determining Hydraulic Conductivity in the Vadose Zone") may be purchased from ASTM, 1916 Race Street, Philadelphia, PA 19103.

Table 12.1.   Review and Comparison of Test Methods for Measuring Hydraulic Conductivity in the Vadose Zone

Characteristics	Single Ring Infiltrometer[a]	Double Ring Infiltrometer[a]	Double-Tube Test Method[b]	Air-Entry Permeameter[c]	Borehole Permeameter Methods — Free Surface[d]	Capillarity Fixed[e]	Capillarity Predicted[f]	2-Head Simultaneous Solution[e]	Instantaneous Profile[g]	Crust[g]	Empirical[h,i]
Relative accuracy	Low	Fair	Fair	Good	Poor	Good	Good	Variable	Good	Good	Low
Relative cost	Low	Low-Moderate	Moderate	Moderate	Low to Moderate				High	High	Low
Time required (at $K_{fs} = 10^{-5}$ cm/s)	< 4 h	4 h to 1 day	< 4 h	< 4 h		< 4 h			1 day	4 h to 1 day	4 h
Depth of Testing Possible	Surface	Surface	0 to 1 ft	0 to 1 ft		Any			0 to 5 ft	0 to 1 ft	Any
Range of K_{fs} (cm/s) for which test is suited	10^{-2} to 10^{-6}	10^{-2} to 10^{-6}, 10^{-6} to 10^{-8} (with flexible bag for inner reservoir)	< 10^{-6}	< 10^{-8}	< 10^{-8} (with precautions for temperature effects on reservoir volume)	< 10^{-6}				—	—
Advantages	Simple apparatus, rapid, can estimate K_{fs} from infiltration data, can increase diameter to reduce scale effects and edge effect	Similar to single ring	—	Measures vertical K_{fs} only, accounts for capillary effects	Simple numerical solution, good approximation for sands	Accounts for capillary effects	—	Simple solution, accounts for capillarity	Excellent method for deriving K (unsat) curve	Good for values of K near zero	Simple, rapid
Drawbacks	Lateral flow affects accuracy, measures infiltration not K_{fs}; surface crust reduces infiltration, measured on surface of soil only	Similar to single ring	Cumbersome apparatus, time-consuming numerical solution	Sometimes difficult to drive tube, difficult to identify wetting front in wet soil	Does not account for capillary effects, high error for medium to fine unstructured soil	Must assume ratio of capillary to flux effects, difficult to predict	Requires description data	Occasionally gives negative values	Time-consuming, affected by barometric changes	Time-economical, only one value of K (unsat) and water potential per test	Low accuracy

Source: ASTM D 5126–90. Standard Guide for Comparison of Field Methods for Determining Hydraulic Conductivity in the Vadose Zone. Copyright ASTM. Reprinted with permission.

[a] Bouwer (1986)
[b] Bouwer (1964)
[c] Amoozegar and Warrick (1986)
[d] Glover (1953)
[e] Reynolds and Elrick (1986)
[f] Mualem (1976)
[g] USBR (1978)
[h] Bouma and Denning (1972)
[i] Van Genuchten (1980)

Infiltrometers

Cylindrical devices, pushed 5 to 10 cm into the soil and covered with water inside, are commonly used to measure how fast water will infiltrate into a soil. Since the parameter of interest usually is the infiltration rate for vertically downward flow (for example, for the design of irrigation or recharge systems, or to predict seepage from ponds or reservoirs, etc.), the cylinder or infiltrometer should be large enough so that the effects of lateral divergence of the flow in the soil below and around the cylinder are negligible. This requires the use of a fairly large diameter infiltrometer (for example, 2 m or more). So-called "buffered" infiltrometers where a small inner cylinder of about 20 cm diameter is placed in the center of a larger cylinder of about 30 cm diameter with the idea that the outer ring or buffer absorbs all the divergence and the inner ring gives the true vertical infiltration flow as if the inundated area is very large, are not the answer. Sand-tank and electrical-analog studies more than 30 years ago by Schwartzendruber and Bouwer, (Bouwer, 1986, and references therein) have shown that while the flow below the inner ring is then mostly vertically downward, it is still at much higher gradient than it would be below a large inundated area, and thus still grossly overestimates the vertical infiltration rate for flooded areas of large extent. Despite these early investigations, the buffered cylinder infiltrometer myth is hard to die, and the device is still used and recommended today. A much better approach would be to use large single-ring infiltrometers or bermed flooded areas of at least about 2×2 m to minimize the effect of divergence and to get essentially vertical flow at unit hydraulic gradient when final infiltration is reached. The major source of error then may be deeper restricting layers, which can reduce infiltration rates when a large area is flooded or infiltrating. However, such reductions usually will not occur in a small infiltrometer test area because the perched mound can then spread out and will not rise high enough to reduce infiltration rates. Thus, measurement of the resaturated vertical hydraulic conductivity of the soil at the surface and deeper in the vadose zone can also be useful to estimate the infiltration capacity of a given site, especially if there is concern about adequate K deeper down.

Large Diameter Infiltrometers

A large diameter infiltrometer test is performed at the bottom of a large diameter hole, such as the "dry wells" with diameters of about 1.2 to 1.5 m that are used for "disposal" of urban runoff. For the test, the dry-well bottom is cleaned as much as possible to remove loose material. It is then covered with a few centimeters of clean sand that is much coarser than the vadose zone material at the hole bottom, and water is poured into the hole through a pipe down to the bottom, using a plate or other energy breaker on the bottom to prevent puddling or erosion. A constant, shallow water depth is then maintained just above the sand. Infiltration rates are calculated from the rate of inflow of water into the pipe or by periodically stopping the inflow rate and measuring the rate of fall of the water level in the hole. Periodic measurements are continued until the infiltration rate approaches a constant value, which then is a reasonable estimate of the resaturated K of the soil material below the hole bottom, because the hole diameter is large enough to minimize divergence and edge effects. Constant K may be approached in less than an hour for sandy materials and in several hours or more for finer textured materials below the hole bottom. Again, the test can be performed at different depths as the hole is drilled, to get a profile of K.

If the soils in the vadose zone are stony or gravelly, only the dry-well test can be used for in situ measurement of K, because rocks and stones interfere with the installation of the cylindrical devices required for the air permeameter and infiltration gradient techniques, discussed below. If the dry-well test cannot be used, about the only other approach to measure K of stony soils would be to measure the volume fraction of the rock and to measure K of the finer material between the rocks in the laboratory on disturbed samples. Knowing these two parameters, the bulk hydraulic conductivity of K_b of the soil-rock mixture in the vadose zone can then be calculated as (Bouwer and Rice, 1984)

$$K_b = K_s \frac{e_b}{e_s} \tag{3}$$

where

K_b = bulk hydraulic conductivity of sand and gravel mixture,
K_s = hydraulic conductivity of the sand fraction alone,
e_b = bulk void ratio (volume of voids divided by volume of solids) for sand and gravel mixture, and
e_s = void ratio of sand.

Air-Entry Permeameter for In Situ Determination of K

Since water pressures in the vadose zone are negative, traditional pumped well or auger hole tests are not possible, and K must be determined on artificially wetted zones in the vadose zone. For surface soils, the air-entry permeameter can be used. As indicated by Bouwer (1978), this is the simplest and quickest technique for measuring K in the absence of a water table. The air-entry permeameter consists of a metal cylinder, about 30 cm in diameter that is driven 10 to 20 cm into the soil (Figure 12.1). The soil is covered with a layer of coarse sand, and a disk is placed in the center to avoid soil erosion when water is applied. A lid assembly with standpipe and reservoir is clamped to the cylinder, water is poured into the reservoir, and the supply valve at the base of the standpipe is opened while water is continually added to the reservoir. After the cylinder is filled with water, the air-escape valve is closed and the reservoir is kept full while water infiltrates into the soil. When the wet front is expected to have reached a depth of about 10 cm, water is no longer added to the reservoir, a few time and water-level readings are taken to determine the rate of water-level fall in the reservoir, and the supply valve is closed (a few trial tests may be needed to determine how long infiltration should continue before the wet front has reached the desired depth). After closing the valve, the wet front no longer advances into the soil. Also, the pressure of the aboveground water inside the cylinder will decrease to negative values and reach a minimum when the air-entry value of the wetted zone is reached. At this point, air will start moving up through the wetted zone and eventually bubble up through the soil surface, increasing the pressure in the aboveground water inside the cylinder. As soon as minimum pressure is observed on the vacuum gauge, it is measured, the cylinder is immediately removed, and the height L_f of the wetted zone is measured. This can be done visually, using a shovel, or by pushing a rod down and observing the depth where the penetration resistance increases.

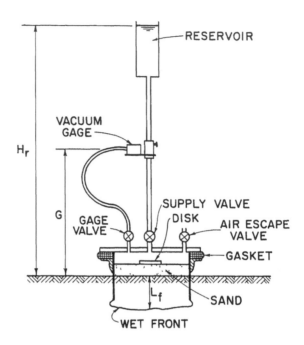

Figure 12.1. Air-entry permeameter.

Referring the minimum pressure in the aboveground water to the elevation of the wet front yields the air entry P_a of the wetted zone [$P_a = P_{min} + G + L_f$, where P_{min} is the minimum pressure head measured with the vacuum gauge and $G + L_f$ is the height of the gauge above the wet front (see Figure 12.1)]. The water-entry value, which is the pressure head at the wet front while it is moving down, is then taken as $1/2P_a$. Since the pressure head at the top and bottom of the wetted zone during infiltration, the height of the wetted zone, and the infiltration rate prior to closing the supply valve are known, K of the wetted zone can be calculated with Darcy's equation as

$$K = \frac{L_f(dH_r/dt)\ (r_r/r_c)^2}{H_t + L_f - 0.5\ P_a} \qquad (4)$$

where

$$L_f = \text{height of wetted zone at end of test}$$
$$dH_r/dt = \text{rate of fall of water in reservoir just before closing supply valve}$$
$$H_r = \text{height of water level in reservoir above soil just before closing supply valve}$$
$$r_r = \text{radius of reservoir}$$
$$r_c = \text{radius of cylinder}$$
$$P_a = \text{air-entry value of wetted zone (expressed as a negative water-pressure head)}$$

Since very precise data often are not possible or required because of nonuniformity and spatial variability, the air-entry permeameter can be simplified by estimating rather than measuring the water entry value for the soil, especially if the soils are sandy because their water entry values are small compared to the height of the standpipe above the soil surface. Also, water-entry values may be estimated for the calculation of K. Estimated values may be something like –5 to –10 cm for coarse sands, –10 to –30 cm for fine and loamy sands, –50 cm for loams, and –100 cm for structureless clays. While the air-entry permeameter is basically a surface-soil device, it can also be used to measure K of deeper layers by placing the device in pits or trenches excavated to these layers. Sloping trenches are useful because they allow measurement of K from the surface to a depth of about 2.5 m (depending on maximum depth of the trenching machine) to get a K profile. Advantages of the air-entry permeameter are simplicity of the equipment, low water use per test, and rapidity of tests.

Infiltration Gradient Method

Deeper in situ measurements of K can be made with the infiltration gradient technique (Bouwer, 1978, and references therein) which consists of two concentric tubes in a borehole that penetrate the hole bottom a small distance (for example, 10 cm for the outer tube and 3 cm for the inner tube). Both tubes are filled with water to equal levels to create a wetted zone with positive pressures below the hole bottom. After some time, the infiltration rate is measured in the inner tube, and small piezometers are pushed into the hole bottom at small increments at the time to measure the vertical hydraulic gradient. K is then calculated from the measured infiltration rate and hydraulic gradient. The pressures of the piezometers are measured with strain gauges for fast response (minimum displacement of water in the piezometer system). The method can be used as the hole is drilled, stopping at different depths to measure K.

MACROSCOPIC VELOCITIES

Infiltration rates as measured in actual systems or with infiltrometers or calculated with infiltration equations or from measurements of K all are expressed as a Darcy flux, which is the volume rate of flow per unit gross area (soil materials as well as pores) perpendicular to the flow. In actuality, however, water moves between the soil particles, through the pores. The velocity of water in the pores, called pore velocity, actual velocity, molecular velocity, or macroscopic velocity, thus is greater than the Darcy flux. The pore velocity V_p usually is calculated as the Darcy velocity, V_d, divided by the water content of the soil or

$$V_p = \frac{V_d}{\theta} \tag{5}$$

This equation is used for saturated flow as well as for unsaturated flow. For saturated flow, θ would be the porosity if all the water in the medium is mobile. Often, however, not all the pore water is moving like stagnant water in dead-end or blocked pores. In that case, θ should be taken as only the volumetric water content of the water that is actually moving (effective water content). Pore velocities are extremely important to know because they determine how fast pollutants travel through the vadose zone and aquifer. Pollutants that are adsorbed to the solid matrix travel slower than the pore velocity, according to the retardation factor (Bouwer, 1991 and references therein). Some pollutants that are repulsed by the solid matrix (anions, for example) or otherwise travel more in the center of the pores (viruses, for example) where velocities are higher, travel faster than indicated by the average pore velocity.

The above discussions all assume full-matrix flow, i.e., the water flows uniformly throughout the entire vadose zone or other porous medium. In reality, however, the flow more likely tends to be concentrated along "preferential" flow paths. These can be cracks, fissures, rootholes, wormholes, or other macropores, and in structured clay soils (Amoozegar-Fard et al., 1982; Beven and Germann, 1982; Dao et al., 1979; Germann and Beven, 1985; Kanchanasut et al., 1978; Scotter, 1978; Thomas and Phillips, 1979; Tyler and Thomas, 1977; White, 1985a; 1985b). However, preferential flow also can occur if fine or clogging layers overlie coarser materials (Bowman et al., 1987; Bowman and Rice, 1986a, 1986b; Jaynes et al., 1988; and Rice et al., 1986). If there is downward flow through such a profile, the underlying coarser material then does not wet up entirely, even when completely uniform, and the flow is concentrated in "fingers." Preferential flow may also be a matter of micro spatial variability. Needless to say, that preferential flow accelerates downward movement of water and contaminants in the vadose zone, and should be taken into account in modeling and monitoring underground transport of contaminants.

EPILOGUE

While ground-water and vadose-zone hydrology are based on well-defined physical principles, the complexity of the underground materials and all the other physical, chemical, and biological processes that are going on in the vadose zone and aquifer make accurate prediction of flow rates and contaminant transport impossible. For this reason alone, physical properties and transport rates of contaminants should never be expressed in more than two significant figures, and maybe only one!

REFERENCES

Amoozegar, A., and A.W. Warrick, "Hydraulic Conductivity of Saturated Soils, Field Methods," in *Methods of Soil Analyses, Part 1: Physical and Mineralogical Methods*, A. Klute, Ed., Agronomy Monograph No.9, American Society of Agronomy, Madison, WI, 1986, pp. 735–770.

Amoozegar-Fard, A., D.R. Nielsen, and A.W. Warrick, "Soil Solute Concentration Distributions for Spatially Varying Pore Water Velocities and Apparent Diffusion Coefficients," *Soil Sci. Soc. Am. J.*, 46:3–9 (1982).

American Society of Testing Materials, ASTM D 5126–90, Standard Guide for Comparison of Field Methods for Determining Hydraulic Conductivity in the Vadose Zone, ASTM Standards on Ground Water and Vadose Zone Investigations, ASTM Publication Code Number (PCN):03–418192–38, ASTM, Philadelphia, PA, 1992, pp. 134–143.

Beven, K., and P. Germann, "Macropores and Water Flow in Soils," *Water Resour. Res.*, 18:311–325 (1982).

Bouma, J., and J.C. Denning, "Field Measurement of Hydraulic Conductivity by Infiltration Through Artificial Crusts," *Soil Sci. Soc. Amer. Proc.*, 36:846–847 (1972).

Bouwer, H., "Measuring Horizontal and Vertical Hydraulic Conductivity of Soil with the Double-Tube Method," *Soil Sci. Soc. Amer. Proc.*, 28:19–23 (1964).

Bouwer, H., "Rapid Field Measurement of Air Entry Value and Hydraulic Conductivity of Soil as Significant Parameters in Flow System Analysis," *Water Resour. Res.*, 2:729–738 (1966).

Bouwer, H., "Infiltration of Water into Nonuniform Soil," *J. Irrig. and Drain. Div. ASCE*, 95(IR 4):451–462 (1969).

Bouwer, H., "Infiltration into Increasingly Permeable Soils," *J. Irrig. and Drain. Div. ASCE*, 102(IR 1):127–136 (1976).

Bouwer, H., *Groundwater Hydrology*, (New York: McGraw-Hill Book Company, 1978).

Bouwer, H., "Design Considerations for Earth Linings for Seepage Control," *Ground Water*, 20:531–537 (1982).

Bouwer, H., "Intake Rate, Cylinder Infiltrometer," in *Methods of Soil Analyses, Part 1, Physical and Mineralogical Methods*, A. Klute, Ed., Agronomy Monograph No.9, American Society of Agronomy, Madison, WI, 1986, pp. 825–844.

Bouwer, H., and R.C. Rice, "Effect of Water Depth in Groundwater Recharge Basins on Infiltration Rate," *J. Irrig. and Drain. Engr.*, 115:556–568 (1989).

Bouwer, H., "Simple Derivation of the Retardation Equation and Application to Preferential Flow and Macrodispersion," *Ground Water*, 29(1):41–46 (1991).

Bouwer, H., and R.C. Rice, "Hydraulic Properties of Stony Vadose Zones," *Ground Water*, 22:696–705 (1984).

Bowman, R. S., H. Bouwer, and R.C. Rice, "The Role of Preferential Flow Phenomena in Unsaturated Transport," in *Proc. Specialty Conf. Environ. Eng. Am. Soc. Civil Eng.*, New York, 1987, pp. 477–482.

Bowman, R.S., and R.C. Rice, "Accelerated Herbicide Leaching Resulting from Preferential Flow Phenomena and Its Implications for Groundwater Contamination," in *Proc. Conf. Focus on Southwestern Ground Water Issues*, Nat. Water Well Assoc., Scottsdale, AZ, 1986a, pp. 413–425.

Bowman, R.S., and R.C. Rice, "Transport of Conservative Tracers in the Field under Intermittent Flood Irrigation," *Water Resour. Res.*, 22(11):1531–1536 (1986b).

Case, C., M. Kautsky, P. Pearl, R. Goldfarb, S. Leatham, and L. Metcalf, Unsaturated Flow Properties, Data Catalog, Publication 45033, Water Resources Center, Desert Research Institute, University of Nevada, 1983.

Dao, T. H., T.L. Lavy, and R.C. Sorensen, "Atrazine Degradation and Residue Distribution in Soil," *Soil Sci. Soc. Am. J.*, 43:129–134 (1979).

Germann, P.F., and K. Beven, "Kinematic Wave Approximation to Infiltration into Soils with Sorbing Macropores," *Water Resour. Res.*, 21:990–996 (1985).

Glover, R.E., "Flow from a Test-Hole Located Above Ground-Water Level," *U.S. Bureau Rec. Eng. Meng.*, 8:69–71 (1953).

Jaynes, D. B., R.S. Bowman, and R.C. Rice, "Transport of a Conservative Tracer in the Field Under Continuous Flood Irrigation," *Soil Sci. Soc. Am. J.*, 52:618–624 (1988).

Kanchanasut, P., D.R. Scotter, and R.W. Tillman, "Preferential Solute Movement Through Larger Soil Voids. II. Experiments with Saturated Soil," *Aust. J. Soil Res.*, 16:269–276 (1978).

Mualem, Y., "A New Model for Predicting the Hydraulic Conductivity of Unsaturated Porous Media," *Water Resources Research*, 12:513–522 (1976).

Panian, T.F., *Unsaturated Flow Properties Data Catalog*, Volume 2, Water Resources Center, Desert Research Institute, University of Nevada, Publication No. 45061, NTIS, U.S. Department of Commerce, Springfield, VA, 1987.

Reynolds, D., and D.E. Elrick, "A Method for Simultaneous In-Situ Measurement in the Vadose Zone of Field Saturated Hydraulic Conductivity, Sorptivity and the Conductivity-Pressure Head Relationship," *Ground Water Monitoring Review*, 6:84 (1986).

Rice, R. C., R.S. Bowman, and D.B. Jaynes, "Percolation of Water Below an Irrigated Field," *Soil Sci. Soc. Am. J.*, 50:855–859 (1986).

Scotter, D. R., "Preferential Solute Movement Through Larger Soil Voids. I. Some Computations Using Simple Theory," *Aust. J. Soil Res.*, 16:257–267 (1978).

Thomas, G.W., and R.E. Phillips, "Consequences of Water Movement in Macropores," *J. Environ. Qual.*, 8:149–152 (1979).

Tyler, D.D., and G.W. Thomas, "Lysimeter Measurements of Nitrate and Chloride Losses from Soil under Conventional and No-Tillage Corn," *J. Environ. Qual.*, 6:63–66 (1977).

J.S. Bureau of Reclamation, Drainage Manual, U.S. Govt. Printing Office, 1978, pp. 74–97.

Van Genuchten, M.T., "A Closed Form Equation for Predicting the Hydraulic Conductivity of Unsaturated Soil," *Soil Sci. Soc. Amer. J.*, 44:892–898 (1980).

White, R.E., "The Analysis of Solute Break-Through Curves to Predict Water Redistribution During Unsteady Flow Through Undisturbed Structured Clay Soil," *J. Hydrol.*, 79:21–35 (1985).

White, R.E., "A Model for Nitrate Leaching in Undisturbed Structured Clay Soil During Unsteady Flow," *J. Hydrol.*, 79:37–51 (1985).

13

Tension Infiltrometers for the Measurement of Vadose Zone Hydraulic Properties

A. A. Hussen and A. W. Warrick

INTRODUCTION

An essential requirement in understanding and studying contaminated water movement in the vadose zone is the characterization of hydraulic properties, especially the saturated and unsaturated hydraulic conductivity. This includes hydraulic conductivity as a function of matric potential in a partially saturated porous material. Many methods have been developed to evaluate the hydraulic properties of porous materials by in situ and laboratory procedures. However, because of the sensitivity of hydraulic properties to soil structure, in situ methods are potentially more accurate. Recently, the disc infiltrometer has become a popular device for determining in situ hydraulic properties [unsaturated hydraulic conductivity (K_{wet}), saturated hydraulic conductivity (K_s), sorptivity (S) and the macropore capillary length (λ_c)] of soil in the wet region (White and Sully, 1987; Ankeny et al. 1988; Smettem and Clothier, 1989; Hussen, 1991; White et al., 1992; Hussen and Warrick, 1993a,b) where the majority of contaminated movement occurs. Water intake is measured from a source which is carefully controlled at a constant tension within a circular interface at the soil surface. In this chapter we will discuss the use of this device for measuring hydraulic properties in the vadose zone.

THEORY

Water flow from a tension infiltrometer is three-dimensional and is the result of gravity acting downward, capillary forces acting in all directions, and the particular geometry of the source. For early times, flow is nearly one-dimensional (Philip, 1969):

$$I = St^{0.5} \tag{1}$$

where I is cumulative intake per unit area (L) and t the time. The sorptivity S ($LT^{-0.5}$) is equal to the slope of I versus $t^{0.5}$. For large times, Wooding (1968) presented a solution for steady-state flow per unit area, Q (LT^{-1}) in the form

$$Q = K_{wet} \left[1 + \frac{4\lambda_c}{\pi r_o} \right] \tag{2}$$

where r_o is the radius of the disc infiltrometer (L), K_{wet} is the hydraulic conductivity (LT^{-1}) corresponding to the water supply potential (h_{wet}) and λ_c is the macropore capillary length (L). Wooding assumed that the soil is uniform, homogenous, non-swelling and has an exponential hydraulic conductivity function (Gardner, 1958):

$$K = K_s \exp \left(\frac{h}{\lambda_c} \right) \tag{3}$$

where K is the unsaturated hydraulic conductivity at matric potential h (L) and K_s the saturated hydraulic conductivity.

On the right side of Equation 2, the first term represents the contribution to the flow due to gravity and the second term represents the contribution of capillarity and source geometry. White and Sully (1987) showed that λ_c is a function of sorptivity and hydraulic conductivity,

$$\lambda_c = \frac{bS^2}{(\theta_{wet} - \theta_{dry})(K_{wet} - K_{dry})} \tag{4}$$

where θ_{wet} ($L^3 L^{-3}$) is the volumetric water content corresponding to the supply potential h_{wet}, θ_{dry} is the initial volumetric water content, K_{wet} and K_{dry} are the hydraulic conductivity values at θ_{wet} and θ_{dry}, respectively, and b is a constant (unitless). The value of b can be taken as 0.55 (White and Sully, 1987; Warrick and Broadbridge, 1992). For $K_{wet} >> K_{dry}$, from Equations 2 and 4 we get

$$Q = K_{wet} + \frac{2.2 S^2}{\pi r_o (\theta_{wet} - \theta_{dry})} \tag{5}$$

If S is known from Equation 1 based on early time measurements and Q is found for large times, then K_{wet} follows from Equation 5.

When steady-state flow values are known for two or more r_o values, K_{wet} can be found by methods described by Scotter et al. (1982), Yitayew and Watson (1986), Smettem and Clothier (1989) and Hussen (1991). For steady flow with two different disc radii values, Equation 2 is solved for two unknowns (K_{wet} and λ_c), and λ_c is calculated in the form

$$\lambda_c = \frac{\pi}{4} \left[\frac{Q_1 - Q_2}{\frac{Q_2}{r_1} - \frac{Q_1}{r_2}} \right] \tag{6}$$

where Q_1 and Q_2 (LT^{-1}) are the steady-state flow rates corresponding to disc radii r_1 and r_2, respectively. The value of K_{wet} follows from (2). If the steady-state flow (Q) is known for more than two radii, non-linear fitting procedures can be used along with Equation 2 to find a "best-fitting" λ_c and K_{wet}.

The same principle can be applied for a single disc radius with multiple tensions (Lien, 1989; Ankeny et al. 1991; Hussen, 1991; Reynolds and Elrick, 1991). When Gardner's relationship (Equation 3) is used, then Wooding's Equation 2 can be written as

$$Q = K_s \exp\left(\frac{h}{\lambda_c}\right) \left[1 + \frac{4\lambda}{\pi r_o} \right] \tag{7}$$

Measurement of steady-state flow for n tensions at the same site, yields n equations, each with two unknowns of K_s and λ. If n = 2, the value of λ_c is obtained by dividing two equations from Equation 7 with different tensions, which results in

$$\lambda_c = \frac{|h_2 - h_1|}{|\ell n(Q_2/Q_1)|} \tag{8}$$

where Q_1 and Q_2 are now steady-state flow rates at two different supply tensions h_1 and h_2. The value of K_s is found by substituting λ_c into Equation 7. If steady-state values are measured at three or more supply tensions, then K_s and λ_c can be evaluated using a best-fitting procedure (Hussen, 1991; Hussen and Warrick, 1993b).

COMPONENTS OF THE DISC TENSION INFILTROMETERS

Several scientists have described the design and use of disc tension infiltrometers (Perroux and White, 1988; Lien, 1989; Ankeny et al. 1988; Hussen, 1991) with only slight differences in operational components. In general, the disc tension infiltrometer consists of three major parts, which are summarized in the following.

Bubble Tower

This is essentially a marriotte-type device consisting of a short tube with one or more small air entry tubes, each with a pinch clamp at the top (see Figure 13.1). The tubes extend through a rubber stopper into the water at different depths, each corresponding to the tension to be imposed at the disc-soil interface. For Figure 13.1, the pressure P_s at the membrane (at the soil surface) is

$$P_s = P_a - \rho g (h_1 - h_2) \tag{9}$$

where P_a is the atmospheric pressure, h_1 the length of inlet tube below the water and h_2 the height of the connection tube from the soil surface, ρ is the density of water, and g the gravitational acceleration constant. If h_1 is greater than h_2, the water

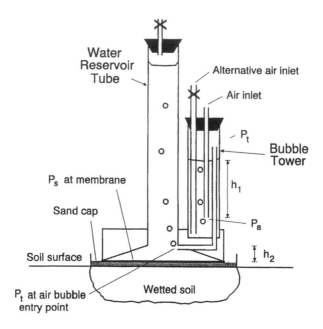

Figure 13.1. Diagram of the disc tension infiltrometer with major components. The supply water pressure is $h_{wet} = h_2 - h_1 < 0$.

pressure will be less than atmospheric by an amount corresponding to $P_s - P_a$ or the pressure head

$$h_{wet} = -h_1 + h_2 < 0 \qquad (10)$$

Reservoir

The reservoir supplies water to the disc. The water level can be monitored by either reading the water elevation (such as from a centimeter scale on the tube) or with pressure transducers.

Porous Baseplate

This is a circular plate covered on one side with a porous membrane (usually nylon) which serves to establish hydraulic continuity between the soil or other porous media and the reservoir. The air entry value of the membrane is dependent upon its mesh size. The normal air entry range for these membranes is between 25 to 50 cm; however, some membranes may approach values of 100 cm (White et al. 1992).

Accessories

In addition to the disc tension infiltrometer, there are some accessories required: a stop watch with at least 30 time laps to collect infiltration time, two water buckets to soak and fill the disc tension infiltrometers, and coarse sand as contact material between the soil surface and the disc membrane.

METHODS OF MEASURING HYDRAULIC PROPERTIES USING DISC TENSION INFILTROMETERS

The following are the basic principles and operational procedures of the three main methods used to measure hydraulic properties using disc tension infiltrometers:

Method 1. Conventional Method

One disc tension infiltrometer is used to measure flow rates at a single tension. The sorptivity is calculated from Equation 1 as the slope of cumulative infiltration versus the square root of time for short times (on the order of 10 to 200 s), as shown in Figure 13.2. The steady-state flow (Q) can be found by plotting cumulative infiltration versus time for large times (on the order of 1 to 3 h), as shown in Figure 13.3. Once the steady-state flow rate and sorptivity are calculated, and the initial and final volumetric water content are measured, the value of K_{wet} follows from Equation 5.

Method 2. Multiple Disc Method

Steady-state flow rates are found for two or more disc tension infiltrometer radii (r_o) at the same tension. The only measurement required is the steady-state flow at the same tension for two (or more) different disc radii in order to solve for λ_c from Equation 6 and K_{wet} from Equation 2. When steady-state flow for more than 2 disc sizes are used, a best fitting scheme is used based on Equation 2, to fit K_{wet} and λ_c.

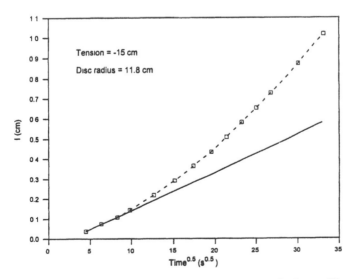

Figure 13.2. Cumulative infiltration versus square root at early times. The slope of the solid line is the sorptivity and dashed line is field data.

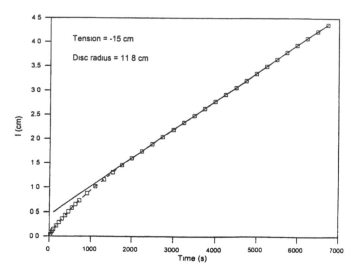

Figure 13.3. Cumulative infiltration vs. time. The solid line is the steady-state flow rate and the dashed line is field data.

Method 3. Multiple Tension Method

In this method, one disc tension infiltrometer is used to find steady-state flow at two or more supply tensions. The highest tension is used first, followed by the next lower tension and so on, to the lowest tension. If only two tensions are used, then K_{wet} and λ_c are found from Equations 8, 7, and 3. If more than two tensions are used, the K_{wet} and λ_c are best fit based on Equations 3 and 7. For example, nonlinear optimization fitting procedures can minimize the mean square error between measured field data and predicted data from Equation 7 (Hussen and Warrick, 1993b).

OPERATIONAL PROCEDURES

Method 1. Conventional Method

1. With minimum disturbance of the soil surface, prepare a smooth, level area about 30×30 cm for placing the disc tension infiltrometer.
2. Place a steel ring (0.2 cm thick) on the center of the prepared area by pressing it into the soil. The soil surface should be level so as to ensure complete contact between the ring and the soil.
3. Insert tared metal cans approximately 10 to 15 cm on either side of the ring. These will be removed later for determining the initial moisture content and bulk density.
4. Fill the inside area of the steel ring with moist sand contact material to the level of the top edge of the ring, using a ruler as a straight edge. Remove any excess sand material.
5. Remove the disc tension infiltrometer from the water storage bucket. Gently blow air into the air inlet tube of the bubble tower and clamp shut while it is bubbling to eliminate water inside the air inlet tube. The supply pressure head is equal to $-h_1 + h_2$ in Figure 13.1.

6. Place the disc tension infiltrometer back in the storage bucket and fill the reservoir tube with water using a suction pump to remove air from the reservoir tube.
7. Read and record the initial water height in the water reservoir tube on a data sheet. Enter predetermined depth marks to be read on the data sheet.
8. Unclamp the air inlet tube of the bubble tower. Gently place the disc tension infiltrometer on the center of the sand-filled ring.
9. Start the stop watch as soon as bubbles start to form in the bubble tower. Record the time readings whenever the bottom of the meniscus reaches the predetermined depth mark. Press the "stop" button when the meniscus reaches the last depth mark or when the one-hour reading has been taken.
10. Wash away the sand material attached to the bottom of the membrane and return to water bucket.
11. Use a spatula to remove sand material from the ring and collect a sample from the top 0 to 0.4 cm of soil into a tared metal can for final moisture content determination.
12. Recall and record the time readings stored in the stop watch on the data sheet. Remove the two soil sample cans from the soil for initial moisture contents and bulk density determinations.

Sample Calculation

Assume a disc tension infiltrometer was used to collect data on the soil surface with a supply pressure $h_{wet} = -h_1 + h_2 = -15$ cm. The collected data are listed in Table 13.1 as cumulative infiltration versus time. The radius (r) of the disc infiltrometer was 11.8 cm, initial volumetric water content was 0.018 (cm^3 cm^{3}), and final water was 0.274 (cm^3 cm^{-3}). We want the hydraulic conductivity at the supply potential ($h_{wet} = -15$ cm).

The first step is to calculate sorptivity (S) by plotting cumulative infiltration versus square root of early times as shown in Figure 13.2. The slope from Equation 1 gives $S = 0.0230$ cm s$^{-0.5}$.

The second step is to calculate steady-state flow (Q) which is equal to the slope of cumulative infiltration at large times (Figure 13.3) with the result $Q = 0.000581$ cm s^{-1}. Finally, from Equation 5, K_{wet} is

$$K_{wet} = 0.000581 - \frac{(2.2)(0.0230)^2}{(3.14)(11.8)(0.274 - 0.018)}$$

or $K_{wet} = 0.000458$ cm s^{-1} = 1.65 cm h^{-1}

Method 2. Multiple Disc Method

Follow the same steps as in Method 1 using two or three different disc radii, and calculate steady-state flow rates for the different disc radii.

Sample Calculation

Assume data is collected using two disc tension infiltrometers with $h_{wet} = -15$ cm tension and radii of $r_1 = 11.8$ and $r_2 = 5.2$ cm. First, calculate steady-state flow Q_1

Table 13.1. Cumulative Infiltration Data Collected from Disc Tension Infiltrometer at h_{wet} −15 cm

Time (sec)	I (cm)	Time (sec)	I (cm)
20.0	0.036	3740.0	2.618
40.2	0.073	3989.0	2.763
68.3	0.108	4238.0	2.909
96.2	0.145	4487.0	3.054
160.2	0.218	4736.0	3.200
228.7	0.291	4985.0	3.345
302.0	0.364	5234.0	3.490
381.0	0.436	5483.0	3.636
457.3	0.509	5732.0	3.781
539.5	0.582	5981.0	3.927
626.6	0.654	6230.0	4.072
714.8	0.727	6479.0	4.218
903.6	0.873	6728.0	4.363
1103.1	1.018	6977.0	4.508
1318.1	1.163	7226.0	4.654
1533.9	1.309	7475.0	4.799
1758.6	1.454	7724.0	4.945
1992.3	1.600	7973.0	5.090
2238.4	1.745	8222.0	5.236
2493.0	1.891	8471.0	5.381
2744.0	2.036	8720.0	5.526
2993.0	2.181	8969.0	5.672
3242.0	2.327	9218.0	5.817
3491.0	2.472	9467.0	5.963

and Q_2 for both discs by plotting cumulative infiltration versus large times (Figure 13.3). Suppose the results are

$$Q_1 = 0.000581 \text{ cm s}^{-1}$$

$$Q_2 = 0.000655 \text{ cm s}^{-1}$$

From Equation 6, λ_c is

$$\lambda_c = \frac{3.14}{4}\left[\frac{0.000581 - 0.000655}{\dfrac{0.000655}{11.8} - \dfrac{0.000581}{5.2}}\right]$$

or $\lambda_c = 1.033$ cm.

The K_{wet} is from Equation 2

$$K_{wet} = \frac{0.000581}{\left[1 + \dfrac{(4)(1.033)}{(3.14)(11.8)}\right]}$$

or $K_{wet} = 0.000523$ cm s^{-1} = 1.88 cm h^{-1}.

Method 3. Multiple Tension Method

Follow steps 1, 2, 4, 5, 6, and 7 as before. Then add the following:

1. Unclamp the air inlet tube for the largest tension of the bubble tower. Gently place the disc tension infiltrometer on the center of the sand cap.
2. Start the stop watch (or define $t = 0$) when the bubbles start. Record the time when the meniscus reaches the subsequent predetermined depth until reaching a steady-state flow, i.e., when the time interval between two predetermined depths becomes equal. Use another stop watch to take 7 to 8 readings for steady-state flow.
3. Without moving the disc tension infiltrometer, unclamp the air inlet tube for the second highest tension of the tension tower. Start the stop watch when the bubbles slow down (which happens very quickly compared to the first time). Record the times until steady-state is reached. Use the same watch as before to take 7 to 8 readings for steady-state flow.
4. Repeat the last step for the other tensions, and record the steady-state flows with time for each tension.
5. Recall and record the time for steady-state readings stored in the stop watch for each tension.

Sample Calculation

Assume data are collected of cumulative infiltration with time using a single disc tension infiltrometer at three tensions –15, –10, and –5 cm with a disc radius of 11.8 cm. Assume the calculated steady-state flows are 0.000745, 0.00104, and 0.001365 cm s^{-1} for h_{wet} = –15, –10, and –5 cm respectively.

There are two ways to calculate the unsaturated hydraulic conductivity, depending on whether two values of Q are considered at a time or all values are considered together. If two values are taken for h_{wet} = –15, and –10 cm, then apply (8) to give

$$\lambda_c = \frac{|(-10) - (-15)|}{\ell n[0.00104/0.000745]}$$

or λ_c = 14.98 cm.

Substitute the value of λ_c into Equation 7:

$$K_s = \frac{0.000745}{\exp\left(\frac{-15}{14.98}\right)\left[1 + \frac{(4)(14.98)}{(3.14)(11.8)}\right]}$$

or K_s = 0.000775 cm s^{-1} = 2.79 cm h^{-1}. Then values of K_s and λ_c are substituted in Equation 3 to calculate K_{wet} at –15, –10, and –5 cm tensions.

$$K_{wet(15)} = 2.79 \exp(-15/14.98)$$

or $K_{wet(15)}$ = 1.03 cm h^{-1}

Solving different combinations of tensions to solve for K_s generally gives similar, but slightly different, values. Alternatively, the best fitting method (Equation 7), is used for steady-state flow at different tensions (at least three) as input to find K_s and

λ_c. Any nonlinear optimization fitting program can be used to fit the data to Equation 7, in which the mean square error (MSE) is used to minimize the difference between field data and fitting data:

$$\text{MSE} = \sum_{i=1}^{n} [y(x_i) - \hat{y}(x_i)]^2 \qquad (11)$$

where $y(x_i)$ are the field and $\hat{y}(x_i)$ are the fitted data.

We fit the Q = 0.000745, 0.00104, and 0.001365 cm s^{-1} (Figure 13.4), resulting in

$$K_s = 0.000655 \text{ cm s}^{-1} = 2.36 \text{ cm h}^{-1}$$

$$\lambda_c = 16.86 \text{ cm.}$$

Then use Equation 3 to calculate K_{wet} at h_{wet} = -5, -10, -15 cm.

ADVANTAGES AND LIMITATIONS

The flow of both water (Perroux and White, 1988) and solutes (Clothier et al., 1992; Quadri et al., 1994) from disc tension infiltrometers can be interpreted in various ways to allow in situ measurement of vadose zone hydraulic properties, as well as parameters relating to the transport of chemicals. Although the disc tension infiltrometer functions near the wet end of the water potential range, this is the range where most transport is expected to occur. Hydraulic conductivity at tensions greater than 150 cm are often negligible in the terms of overall contaminant transport.

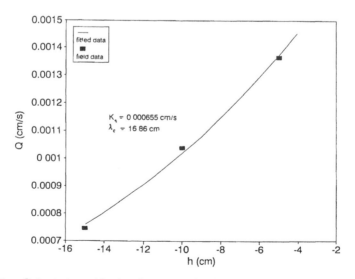

Figure 13.4. Calculation of hydraulic properties by fitting Equation 7 to field data.

The Advantage of Using Disc Tension Infiltrometers

- In situ measurement of vadose zone hydraulic (K_s, K_{wet}, S, and λ_c) and transport parameters (diffusion-dispersion coefficient, D) are possible.
- Results are generally fast, reliable, and repeatable.
- Devices are easy to use and inexpensive.

As mentioned in the previous section, there is more than one method to measure hydraulic properties using disc tension infiltrometers. Some methods are easier, faster and more stable than others. The conventional method (Method 1) requires relatively dry initial conditions. It requires measurements of θ_{dry}, θ_{wet}, sorptivity, and steady-state flow. The multiple disc radii method (Method 2) is easier to perform, and can be used in either dry or moist soil based on steady-state flow for two or more different radii and does not require θ_{dry}, θ_{wet} or sorptivity measurements. The multiple tension method (Method 3) with more than two tensions has an advantage over two tensions because the solutions tend to be more stable. Unlike Methods 1 and 2, this method tends not to give any false negative hydraulic conductivity values. The results are known immediately because all of the parameters are measured directly. Also, multiple tension measurements obtained without moving the disc infiltrometer minimize spatial variation differences between the tensions. The result of the multiple tension method with fitting procedures is averaged over other measurements, and is therefore more stable than solving two equations for two unknowns.

Hussen and Warrick (1993b) results indicated that there were no differences in the measurement values of hydraulic conductivity using large and small disc infiltrometers. For practical purposes, large discs are preferred because they are more stable in the field under windy conditions, in addition to averaging over a larger surface area.

Limitations and Sources of Error Using Disc Tension Infiltrometers

1. One of the common limitations, as mentioned previously, is the narrow tension range in which the disc tension infiltrometer operates. Most studies show that disc tension infiltrometers work well from 0 to 25 cm tension, but White et al. (1992) mentioned that the tension range could go up to 100 cm.
2. Disc tension infiltrometers are used to measure hydraulic and transport properties at the soil surface. They could be used at shallow depths if the soil is excavated to the depth of interest and the base of the excavation is leveled carefully.
3. The principal limitations of the disc tension infiltrometer are those associated with the simplifying assumptions of the analysis, which assume that the soil is uniform, homogeneous, and nonswelling. In practice, gradients in water content, soil-bulk density, soil layering, and changes in soil texture all occur near the soil surface, which can result in negative calculations of hydraulic conductivity (White et al. 1992), especially for Method 1.
4. Other difficulties arise using supply potentials close to zero in freshly cultivated soils. The strength of soil at zero or near zero potentials may not be sufficient to support the weight of the disc tension infiltrometer which causes a collapse of soil macropores and subsequently lead to negative values for unsaturated hydraulic conductivity.
5. The hydraulic conductivity of the sand cap which provides the contact

between the disc membrane and the soil surface or porous media should be higher than the soil or porous material, otherwise the sand cap will limit the movement of water into the soil.

6. Difficulties could arise generally when measurements are made in heavy-textured soils, but automated disc tension infiltrometers, such as described by Ankeny et al. (1988) overcome these difficulties.

7. Another error is that the time to approach steady-state flow can be considerably longer than what is commonly reported in the literature (10 min to 1 hour) (Warrick, 1992; Hussen and Warrick, 1993a). Underestimation of the steady-state flow could lead to high hydraulic conductivity values. Hussen and Warrick (1993a) reported that the steady-state flow results for a sandy loam soil were sensitive to the time of evaluation, up to 1.3 hours.

8. A disadvantage of Method 2 is the potential for yielding negative hydraulic conductivity values, due to spatial or experimental variations between the two different disc radii.

REFERENCES

Ankeny, M.D., M. Ahmed, T.C. Kaspar, and R. Horton, "Simple Method for Determining Unsaturated Hydraulic Conductivity," *Soil Sci. Soc. Am. J.*, 55:467–470 (1991).

Ankeny, M.D., T.C. Kasper, and R. Horton, "Design for an Automated Tension Infiltrometer," *Soil Sci. Soc. Am. J.*, 52:893–896 (1988).

Clothier, B.E., M.B. Kirkham, and J.E. McLean, "In Situ Measurement of the Effective Transport Volume for Solute Moving Through Soil," *Soil Sci. Soc. Am. J.* 56:733–736 (1992).

Gardner, W.R. "Some Steady-State Solutions of the Unsaturated Moisture Flow Equation with Application to Evaporation from a Water Table," *Soil Sci.*, 85:228–232 (1958).

Hussen, A.A., "Measurement of Unsaturated Hydraulic Conductivity in the Field," PhD dissertation, University of Arizona, Tucson, 1991.

Hussen, A.A., and A.W. Warrick, "Algebraic Models for Disc Tension Permeameters," *Water Resour. Res.*, 29:2779–2786 (1993a).

Hussen, A.A.,and A.W. Warrick, "Alternative Analyses of Hydraulic Data from Disc Tension Infiltrometers," *Water Resour. Res.*, 29:4103–4108 (1993b).

Lien, B., "Field Measurement of Soil Sorptivity and Hydraulic Conductivity," MS thesis, University of Arizona, Tucson, 1989.

Perroux, K.M.,and I. White, "Designs for Disc Permeameters," *Soil Sci. Soc. Am. J.*, 52:1205–1215 (1988).

Philip, J.R. "Theory of Infiltration," *Adv. Hydrosci.*, 5:215–305 (1969).

Quadri, M.B., B.E. Clothier, R. Angulo-Jaramilo, M. Vauclin and S.R. Green, "Axisymmetric Transport of Water Underneath a Disk Permeameter: Experiments and Numerical Model," *Soil Sci.Soc. Amer. J.* 58:696–703 (1994).

Reynolds, W.D., and D.E. Elrick, "Determination of Hydraulic Conductivity Using a Tension Infiltrometer," *Soil Sci. Soc. Am. J.*, 55:633–639 (1991).

Scotter, D.R., B.E. Clothier, and E.R. Harper, "Measuring Saturated Hydraulic Conductivity and Sorptivity Using Twin Rings," *Aust. J. Soil Res.*, 20:295–304 (1982).

Smettem, K.R.J., and B.E. Clothier, "Measuring Unsaturated Sorptivity and Hydraulic Conductivity Using Multiple Disk Permeameters," *J. Soil Sci.*, 40:563–568 (1989).

Warrick, A.W., "Models or Disc Infiltrometers," *Water Resour. Res.*, 28:1319–1327 (1992).

Warrick, A.W., and P. Broadbridge, "Sorptivity and Macroscopic Capillary Length Relationships," *Water Resour. Res.*, 28:427–431 (1992).

White, I., and M. J. Sully, "Macroscopic and Microscopic Capillary Length and Time Scales from Field Infiltration," *Water Resour. Res.*, 23:1514–1522 (1987).

White, I., M.J. Sully, and K.M. Perroux, "Measurement of Surface-Soil Hydraulic Properties: Disc Permeameters, Tension Infiltrometers and Other Techniques," pp. 69–104 in G.C. Topp et al., Ed., *Advances in Measurement of Soil Physical Properties: Bringing Theory into Practice*, SSSA Spec. Publ. 30. SSSA, Madison, WI, 1992.

Wooding, R.A., "Steady Infiltration from a Shallow Circular Pond," *Water Resour. Res.*, 4:1259–1273 (1968).

Yitayew, M., and J.E. Watson, "Field Methods for Determining Unsaturated Hydraulic Conductivity," Am. Soc. Agric. Eng. Paper 86–2570, 1986.

14

Estimating the Transport and Fate of Contaminants in the Vadose Zone Based on Physical and Chemical Properties of the Vadose Zone and Chemicals of Interest

Mark L. Brusseau and L.G. Wilson

INTRODUCTION

Estimates of the transport and fate of contaminants of interest at a waste-disposal facility provide guidelines for selecting vadose zone monitoring devices. For example, if such contaminants are nonpolar organic compounds, they will probably be strongly sorbed by the near-surface soils, and transport may therefore be limited. Accordingly, shallow monitoring units may suffice if other factors such as preferential flow are not important. However, highly soluble and mobile contaminants may require a full-scale vadose zone monitoring system. Estimates of the physical/chemical properties of chemicals and of the vadose zone are also required when modeling fate and transport in the subsurface.

The transport and fate of contaminants in the vadose zone can be affected by many factors. Thus, a complete understanding of contaminant transport and fate would require knowledge of a multitude of physical, chemical, and biological parameters. However, knowledge of a few parameters can provide a first-order level of understanding that would be useful for planning vadose zone monitoring systems, which can then be used to obtain more refined estimates. Properties of the soil and chemical compounds that are most relevant to transport and fate are briefly reviewed below. References to methods for estimating transport and fate based on these properties are included.

SOIL PROPERTIES

Soil properties of importance when estimating contaminant fate and transport are as follows:

- permeability
- texture
- structure
- pH
- water content
- soil-moisture release curve
- bulk density
- porosity
- exchange capacity
- organic carbon content
- oxides of iron and aluminum

Soil property data, primarily for agricultural areas throughout the U.S., are available from university extension services and the U.S. Soil Conservation Service. Panian (1987) prepared a catalog of unsaturated flow properties, including porosity, bulk density, and degrees of saturation versus matric potentials for many soil types. These data are available for depths extending to 60 inches below land surface. Properties of soils at greater depths in the vadose zone will require the use of drilling and sampling techniques described by Dorrance et al. in Chapter 24 of this book. Laboratory techniques for characterizing collected samples for the above-listed parameters are available in standard references, listed below after a discussion of specific properties.

Texture

Texture refers to the grain-size distribution of a soil, i.e., the relative distribution of the basic soil particles, sand, silt, and clay. Finer-textured soils generally have a greater sorptive capacity than coarser-textured soils. Additionally, unless they are structured (see below), finer-textured soils are less permeable, enhancing the contact time for sorption. References for conducting particle size analyses include Gee and Bauder (1986) and ASTM Standard Method 422 (ASTM, 1993a).

Structure

Structure refers to the degree of aggregation of the primary soil particles into structural units or peds. The presence of aggregated soil enhances the potential for preferential flow in the macropores between the aggregates, reducing the contact time for sorption. Kemper and Rosenau (1986) present methods for characterizing aggregate stability and size.

pH

Soil pH affects the mobility of some contaminants. For example, except for selenium, chromium, and arsenic at some valence states, the mobility of trace elements and heavy metals increases with decreasing pH (Fuller, 1977). In addition, many organic contaminants are ionizable (e.g., phenolics, amines) and their sorption is dependent on pH. Methods for characterizing soil pH are included in ASTM D4972-89 (ASTM, 1993b).

Bulk Density

As defined by Blake and Hartge (1986), bulk density is the ratio of the mass of dry solids to the bulk volume of soil. Bulk volume includes the volume of solids and pore space. Field methods for measuring bulk density are included in Blake and Hartge (1986) and ASTM Standard Method D1556-90 (ASTM, 1993c).

Porosity

Porosity (n) is the volume fraction of pore-space in a unit volume of soil. Bulk density and porosity are related through the equation:

$$n = \left[1 - \frac{\rho_b}{\rho_p} \right] \tag{1}$$

where ρ_b is the bulk density
ρ_p is the particle density (usually assumed to be 2.65 g/cm^3)

For saturated soil, porosity is used in the retardation equation:

$$R = 1 + \frac{\rho K_d}{n} \tag{2}$$

where R is the retardation factor, K_d is the distribution coefficient representing sorption of the solute by soil, and ρ is the bulk density of the soil.

Danielson and Sutherland (1986) review methods for determining total porosity and pore-size distribution. ASTM D4404-84 is a standard test method for determining the pore volume and pore-volume distribution of soil and rock by the mercury-intrusion method (ASTM, 1993d)

Water Content

The water content affects contaminant transport in several ways, such as controlling the portion of the porous medium participating in water flow. In addition, the water content or degree of saturation influences the degree of tortuosity affecting water and gas flow. During flow under unsaturated (water) conditions, the water content, θ(vol/vol), is used in place of the porosity, n, in the retardation equation; i.e.,

$$R = 1 + \frac{\rho K_d}{\theta} \tag{3}$$

Soil water also competes with contaminants for sorption sites on the soil, thus affecting the degree of sorption and volatilization.

Gardner (1986) presents direct and indirect methods for measuring water content of soils. ASTM Standard Test Method 2216 (ASTM 1993e) presents laboratory methods for determination of soil, rock, or saturated aggregate mixtures.

Ion Exchange Capacity

The sorption of inorganic as well as ionic organic compounds often occurs by ion exchange, as discussed by Runnells (1995). Hence, the exchange capacity, EC, of the soil is a major parameter of interest.

Organic Carbon Content

As discussed by Brusseau in Chapter 7 of this volume, the sorption of nonionic organic compounds is generally controlled by the organic fraction of the soil. Therefore, the organic matter or organic carbon content (f_{oc}) of the soil is another major parameter of interest.

Calculation of contaminant distribution and retardation factors requires knowledge of the moisture content (θ) and bulk density (ρ_b) of the soil. Determining values for these four parameters (EC, f_{oc}, θ, ρ_b) should be the focus of preliminary soil characterization efforts.

Composition of the Oxide Surface

The most significant factor affecting the mobility of negatively charged species in soils is the composition of the oxide surface (Bassett, 1992). For example, in the weathering process ferric iron precipitates as amorphous material, which often contains several hundred square meters per gram of available surface. Even though the amount of iron is only a fraction of a percent by weight in a soil, it is often the dominant surface affecting the sorption of metal contaminants because of this extremely high surface area.

Jackson, Lim, and Zelazny (1986) describe methods for characterizing oxides, hydroxides, and aluminosilicates.

PHYSICAL AND CHEMICAL PROPERTIES OF CONTAMINANTS

Physical and chemical properties of contaminants have a pronounced effect on transport and fate in the vadose zone. Predominant among these factors are the following:

- solubility
- vapor pressure
- diffusion
- biodegradability
- sorption potential
- structure

Solubility

Water solubility is one of the most important factors affecting the transport and fate of inorganic and organic contaminants in the subsurface. Highly soluble contaminants have relatively low sorption coefficients for soils, tend to volatilize from soils, and are more readily biodegraded (Lyman, 1982a, Montgomery and Welkom, 1990). Solubility of organic compounds in solvents (e.g., octanol) is also important for estimating transport and fate in soils (c.f., Lyman, 1982b). Contaminants that are soluble in water tend to be readily transported through the vadose zone. Conversely, compounds that are soluble in organic solvents and poorly soluble in water tend to exhibit high retardation and low mobility. The *Handbook of Chemistry and Physics* (Lide, 1992), *The Merck Index* (Budavari et al., 1989), and the *Hazardous Chemicals Information Annual* (Sax, 1986) are among the references containing data on the solubility of organic compounds in water and other solvents. Montgomery and Welkom (1990) summarize the solubilities of all of the Environmental Protection Agency's (EPA's) Priority Pollutants. Lyman (1820a,b) summarizes techniques for estimating solubilities of organic compounds.

Vapor Pressure

The vapor pressure of a contaminant affects its escaping tendency from soils. In Chapter 7 of this volume, Brusseau discusses the factors affecting volatilization from soils. As indicated in that chapter, the vapor pressure of a contaminant is estimated from the following relationship:

$$V_p = CK_H \qquad (4)$$

where

K_H = Henry's law constant
V_p = vapor pressure of the chemical
C = concentration of the chemical in water

Montgomery and Welkom (1990) include Henry's law constants and vapor-pressure values for EPA's priority pollutants. Alternatively, Henry's law constant may be estimated from the following equation:

$$K_H = (V_p)(MW)/760(S) \qquad (5)$$

where

K_H = Henry's law constant (atm.m^3/mole)
V_p = vapor pressure of the chemical (mm Hg)
MW = molecular weight of the chemical (g/mol)
S = solubility in water (mg/L)

Grain (1982) includes techniques for estimating vapor pressure of organic compounds, while Thomas (1982) reviews techniques for estimating volatilization from soils.

Biodegradability

Biodegradation is an important mechanism governing the fate of contaminants in the vadose zone and is dependent on several factors. Commonly, the biodegradation or disappearance rate of a chemical is expressed in terms of its half life, $t_{1/2}$, defined as the time needed for half of the concentration to react. Data on in situ biodegradation rates are sparse and existing values for given compounds differ because of differences in experimental techniques between researchers. Biodegradation or disappearance rate data for many organic compounds are listed by Howard et al., 1991. Scow (1982) provides aids for evaluating the biodegradation potential for organic compounds, including standard test procedures, chemical rules of thumb for biodegradability, and attempts by various investigators to estimate rates. She avoids presenting estimation techniques, because of the complexity of the biodegradation process.

Sorption of Organic Compounds

The distribution or sorption coefficient, K_d, is the primary chemical-associated parameter of interest for sorption. Various techniques have been developed for measuring these sorption coefficients; these will be discussed in the next chapter. While measurement is to be preferred, estimation techniques can be used to obtain an idea of approximate values for the sorption coefficient. Lyman (1982c) reviews methods for estimating sorption coefficients for organic compounds. Most of the estimation equations assume that sorption is controlled by organic matter and use an organic-carbon normalized sorption coefficient (K_{oc}):

$$K_{oc} = \frac{K_d}{f_{oc}} \tag{6}$$

where f_{oc} is the mass fraction of organic carbon associated with the soil. Regression equations relating K_{oc} to a specified property of the chemical compound, such as aqueous solubility (S) or octanol-water partition coefficient (K_{ow}) have been reported by many researchers. With knowledge of S or K_{ow}, which are available for most compounds, one can estimate K_{oc} for a particular compound. A value of K_d can then be estimated by use of the estimated K_{oc} and a value for f_{oc} (e.g., Equation 6). Several regression equations relating K_{oc} to water solubility and K_{ow} are listed in Table 14.1 and plotted on Figures 14.1 and 14.2, respectively.

Table 14.2 relates values for K_d and K_{oc} with mobility classes for a variety of pesticides in a soil with 2.5% organic matter. These relationships were based on thin-layer chromatography (TLR) experiments.

The sorption of organic compounds may not be instantaneous. In such cases, the time scale of sorption would be of interest. An inverse relationship between the first-order desorption rate coefficient and the equilibrium sorption coefficient has been reported for neutral organic compounds (Karickhoff and Morris, 1985; Brusseau and Rao, 1989). As an example, Brusseau and Rao (1989) presented the following relationship:

$$\log k_2 = -0.67 \log K_p + 0.3 \tag{7}$$

Table 14.1.   Regression Equations for the Estimation of K_{oc} [a]

Eq. No.	Equation	No.[b]	r^2[c]	Chemical Classes Represented	Reference
7	$\log K_{oc} = -0.55 \log S + 3.64$ (S in mg/L)	106	0.71	Wide variety, mostly pesticides	Kenaga and Goring, 1980
8	$\log K_{oc} = -0.54 \log S + 0.44$ (S in mole fraction)	10	0.94	Mostly aromatic or polynuclear aromatics, two chlorinated	Karickhoff et al., 1979
9	$\log K_{oc} = -0.557 \log S + 4.277$ (S in μ moles/L)	15	0.99	Chlorinated hydrocarbons	Chiou et al., 1979
10	$\log K_{oc} = 0.544 \log K_{ow} + 1.377$	45	0.74	Wide variety, mostly pesticides	Kenaga and Goring. 1980
11	$\log K_{oc} = 0.937 \log K_{ow} - 0.006$	19	0.95	Aromatics, polynuclear aromatics, triazenes and dinitroanaline herbicide	Brown et al.
12	$\log K_{oc} = 1.00 \log K_{ow} - 0.21$	10	1.00	Mostly aromatic or polynuclear aromatics; two chlorinated	Karickhoff et al., 1979
13	$\log K_{oc} = 0.94 \log K_{ow} + 0.02$	9	N/A[e]	s-Triazines and dinitroanaline herbicides	Brown, 1979
14	$\log K_{oc} = 1.029 \log K_{ow} - 0.18$	13	0.91	Variety of insecticides, herbicides and fungicides	Rao and Davidson. 1980
15[d]	$\log K_{oc} = 0.524 \log K_{ow} + 0.855$	30	0.84	Substituted phenylureas and alkyl-N-phenylcarbamates	Briggs, 1973
16[d,f]	$\log K_{oc} = 0.0067(P-45N) + 0.237$	29	0.69	Aromatic compounds, ureas, 1.3.5-triazines, carbamates. and uracils	Hance. 1969
17	$\log K_{oc} = 0.681 \log BCF(f) + 1.963$	13	0.76	Wide variety, mostly pesticides	Kenaga and Goring, 1980
18	$\log K_{oc} = 0.681 \log BCF(f) + 1.886$	22	0.83	Wide variety, mostly pesticides	Kenaga and Goring, 1980

Source: Lyman 1982c.

[a]$\log K_{oc}$ = soil (or sediment) adsorption coefficient; S = water solubility. K_{ow} = octanol-water partition coefficient; BCF(f) = bioconcentration factor from flowing water tests; BCF(t) = bioconcentration factor from model ecosystems; P = parachlor; N = number of sites in molecule which can participate in the formation of a hydrogen bond.

[b]No. = number of chemicals used to obtain regression equation.

[c]r^2 = correlation coefficient for regression equation.

[d]Equation originally given in terms of K_{om}. The relationship K_{om} = $\log K_{oc}/1.724$ was used to rewrite the equation in terms of K_{oc}.

[e]N/A = not available.

[f]Specific chemicals used to obtain the regression equation not specified.

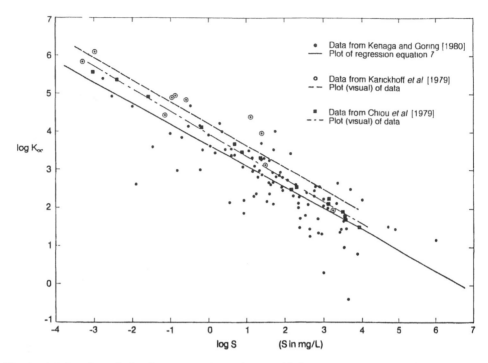

Figure 14.1. Correlation between adsorption coefficient and water solubility (Adapted from Lyman, 1982c).

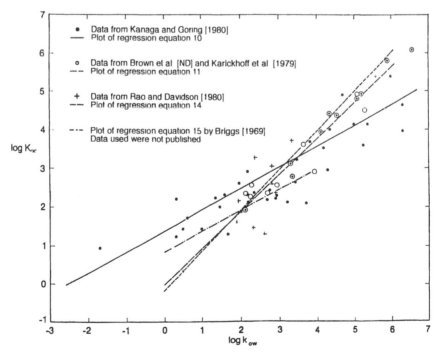

Figure 14.2. Correlation between adsorption coefficient and octanol-water partition coefficient (Adapted from Lyman, 1982c).

Table 14.2. Relative Relationships Among K_d, K_{oc}, and Mobility Classes for a Variety of Pesticides in a Soil with 2.5% Organic Matter

K_d	K_{oc}	Mobility Class	
>10	>2000	I.	Immobile
2–10	500–2000	II.	Low mobility
0.5–2	150–200	III.	Intermediate mobility
0.1–0.5	150–500	IV.	Mobile
<0.1	<50	V.	Very mobile

Source: After Dragun, 1988.

where k_2 is the first-order reverse desorption rate coefficient. The estimated or measured values of k_2 can be compared to the hydraulic residence time to evaluate the potential importance of rate-limited sorption (Brusseau and Rao, 1989).

It is important to recognize the assumptions employed in estimation techniques and to be cognizant of the constraints imposed by the limitations associated with these assumptions. For example, K_d values estimated by use of estimated K_{oc} values will be erroneous if the assumption of organic-matter-controlled sorption is not valid. For another example, the first-order rate coefficients presented in Equation 2 can be dependent on the time scale of measurement (Brusseau et al., 1991; Brusseau, 1992).

Sorption of Metals

Adsorption of metals by soils is highly sensitive to environmental parameters, particularly pH. Accordingly, estimating the movement of metals should be approached with great caution. Bassett (1992) used the chemical speciation model MINTEQ to calculate K_d values for the EPA's Priority metals based on soil chemical composition. Soil properties of particular importance were weight percent iron hydroxide, temperature, pH, types of aqueous ion pairs and complexes, and types of surface adsorption sites.

Structure of the Chemical Compounds

The structure of an organic compound often controls the solubility, sorption, and transformations characterizing the compound. For example, nonionic, low-polarity compounds are poorly soluble in water. This low solubility results in high sorption/retardation and low rates of water-phase transport. Conversely, an anionic organic compound usually has a much larger solubility and, therefore, lower sorption/retardation and greater transport potential. A cationic organic compound, while perhaps being fairly soluble in water, may have a high degree of sorption/retardation because of cation exchange. Clearly, the transport of organic compounds is dependent on their structural properties.

FINAL COMMENTS

A number of computer models are currently available for estimating the transport of chemicals in the vadose zone. In Chapter 17 of this book, Kramer and Cullen review the currently available, state-of-the-art models, many of which can be run on personal computers. The physical and chemical properties discussed in this and other chapters of this book are required to run such models. In using flow and transport models to aid in designing vadose-zone monitoring systems, the reader should be aware that computer modeling of transport in the vadose zone is still an imperfect art. Thus, according to Nielsen et al. (1990), ". . . the efficacy of accurately predicting the attenuation and eventual location of solutes or constituents in the vadose zone remains undeveloped." These authors listed the following constraints and limitations of predicting transport in the vadose zone:

- a lack of quantitative knowledge of local and regional sorption and exchange phenomena,
- spatial and temporal variations in pore-water velocity distributions, and
- lack of a unified approach to include all of the interacting physical, chemical, and biological approaches.

ACKNOWLEDGMENT

Figure modifications by Helen Wilson.

REFERENCES

American Society for Testing and Materials, Standard Test for Particle-Size Analyses of Soils, ASTM D422–63, *ASTM 1993 Annual Book of ASTM Standards*, Vol. 04.08 Soil and Rock; Dimensions Stone: Geosynthetics, American Society for Testing and Materials, Philadelphia, PA, 1993a, pp. 93–99.

American Society for Testing and Materials, Standard Test for pH of Soils, ASTM D4972–89, *ASTM 1993 Annual Book of ASTM Standards*, Vol. 04.08 Soil and Rock; Dimensions Stone: Geosynthetics, American Society for Testing and Materials, Philadelphia, PA, 1993b, pp. 1163–1165.

American Society for Testing and Materials, Standard Test for Density and Unit Weight of Soil in Place by the Cone Method, ASTM D1556–90, *ASTM 1993 Annual Book of ASTM Standards*, Vol. 04.08 Soil and Rock; Dimensions Stone: Geosynthetics, American Society for Testing and Materials, Philadelphia, PA, 1993c, pp. 221–226.

American Society for Testing and Materials, Standard Test for Determination of Pore Volume and Pore Volume Distribution of Soil and Rock by the Mercury Intrusion Porosimetry, ASTM D4404–84, *ASTM 1993 Annual Book of ASTM Standards*, Vol. 04.08 Soil and Rock; Dimensions Stone: Geosynthetics, American Society for Testing and Materials, Philadelphia, PA, 1993d, pp. 751–755.

American Society for Testing and Materials, Standard Test for Laboratory Determination of Water (Moisture) Content of Soil and Rock, ASTM D2216–92, *ASTM 1993 Annual Book of ASTM Standards*, Vol. 04.08 Soil and Rock; Dimensions Stone: Geosynthetics, American Society for Testing and Materials, Philadelphia, PA, 1993e, pp. 294–297.

Bassett, R.L., Determination of the Distribution Coefficients for the Sorption of Metal

Aqueous Ions onto Iron Rich Sediments in the Vadose Zone Using the Surface Complexation Adsorption Model. Research Report to the U.S. Environmental Protection Agency Under the Shallow Injection Well Initiative Program, 1992.

Blake, G.R., and K.H. Hartge, "Particle Density," in *Methods of Soil Analysis, Part 1, Physical and Mineralogical Methods*, A. Klute, Ed., Number 9 (Part 1) in Agronomy, American Society of Agronomy, Soil Science Society of America, Madison, WI, 1986, pp. 377–382.

Briggs, G.G., "A Simple Relationship Between Soil Adsorption of Organic Chemicals and Their Octanol/Water Partition Coefficients," *Proceedings 7th British Insecticide and Fungicide Conference*, The Boots Co., Ltd., Nottingham, Great Britain, 1973, pp. 83–88.

Brown, D.S., U.S. Environmental Protection Agency, Athens, GA, Personal communication to W.J. Lyman (cited in Lyman, 1982c), September 20, 1979.

Brown, D.S., S.W. Karickhoff, and E.W. Flagg, "Empirical Prediction of Organic Pollutant Sorption in Natural Sediments," no date, (cited in Lyman, 1982c).

Brusseau, M. L., "Sorption and Transport of Organic Chemicals," in *Handbook of Vadose Zone Characterization & Monitoring*, L.G. Wilson, L.G. Everett, and S.J. Cullen, Eds., (Chelsea, MI: Lewis Publishers, 1995).

Brusseau, M.L., and P.S.C. Rao, "Sorption Nonideality During Organic Contaminant Transport in Porous Media," *CRC Critical Reviews in Environmental Control*, 19:33 (1989).

Brusseau, M.L., and P.S.C. Rao, "The Influence of Sorbate-Organic Matter Interactions on Sorption Nonequilibrium," *Chemosphere*, 18:1691 (1989).

Brusseau, M.L., T. Larsen, and T.H. Christensen, "Rate-Limited Sorption and Nonequilibrium Transport of Organic Chemicals in Low Organic Carbon Aquifers," *Water Resour. Res.*, 27:1137 (1991).

Brusseau, M.L., "Nonequilibrium Transport of Organic Chemicals: The Impact of Pore-Water Velocity," *J. Contam. Hydrology*, 9:353 (1992).

Budavari, S., M.J. O'Neil, A. Smith, and P.E. Heckleman, *The Merck Index*, (Rahway, NJ: Merck and Co., 1989).

Chiou, C.T., L.J. Peters, and V.H. Freed, "A Physical Concept of Soil-Water Equilibria for Nonionic Organic Compounds," *Science*, 206:831–832 (1979).

Danielson, R.E., and P.L. Sutherland, "Porosity," in *Methods of Soil Analysis, Part 1, Physical and Mineralogical Methods*, A. Klute, Ed., Number 9 (Part 1) in Agronomy, American Society of Agronomy, Soil Science Society of America, Madison, WI, 1986, pp. 443–461.

Dorrance, D.W., L.G. Wilson, L.G. Everett, and S.J. Cullen, "A Compendium of Soil Samplers for the Vadose Zone," Chapter 24 in *Handbook of Vadose Zone Characterization & Monitoring*, L.G. Wilson, L.G. Everett, and S.J. Cullen, Eds. (Chelsea, MI: Lewis Publishers, 1995).

Dragun, J., The Soil Chemistry of Hazardous Materials, Hazardous Materials Control Research Institute, Silver Spring, MD, 1988.

Fuller, W.H., Movement of Selected Metals, Asbestos, and Cyanide in Soil: Application to Waste Disposal Problems, U.S. Environmental Protection Agency, EPA/2-77-020, 1977.

Gardner, W.H., "Water Content," in *Methods of Soil Analysis, Part 1, Physical and Mineralogical Methods*, A. Klute, Ed., Number 9 (Part 1) in Agronomy, American Society of Agronomy, Soil Science Society of America, Madison, WI, 1986, pp. 493–544.

Gee, G.W., and J.W. Bauder, "Particle-Size Analysis," in *Methods of Soil Analysis, Part 1, Physical and Mineralogical Methods*, A. Klute, Ed., Number 9 (Part 1) in Agronomy, American Society of Agronomy, Soil Science Society of America, Madison, WI, 1986, pp. 383–412.

Grain, C.F. "Vapor Pressure," in *Handbook of Chemical Property Estimation Methods, Environmental Behavior of Organic Compounds*, W.J. Lyman, W.F. Reehl, and D.H. Rosenblatt, Eds. (New York: McGraw-Hill Book Co., 1982), Chap. 14.

Hance, R.J., "An Empirical Relationship Between Chemical Structure and the Sorption of Some Herbicides in Soils," *Jour. Agric. Food Chem.*, 17:667–668 (1969).

Howard, P., R.S. Boethling, W.F. Jarvis, W.M. Meylan, and E.M. Michalenko, *Handbook of Environmental Degradation Rates*, (Chelsea, MI: Lewis Publishers, 1991).

Jackson, M.L., C.H. Lim, and L. W. Zelazny, "Oxides, Hydroxides, and Aluminosilicates," in *Methods of Soil Analysis, Part 1, Physical and Mineralogical Methods*, A. Klute, Ed., Number 9 (Part 1) in Agronomy, American Society of Agronomy, Soil Science Society of America, Madison, WI, 1986, pp. 101–150.

Jury, W.A., W.F. Spencer, and W.J. Farmer, "Behavior Assessment Model for Trace Organics in Soil: I. Model Description," *Journal of Environmental Quality*, 12:558–564 (1983).

Jury, W.A., W.J. Farmer, and W.F. Spencer, "Behavior Assessment Model for Trace Organics in Soil: II. Chemical Classification and Parameter Sensitivity," *Journal of Environmental Quality*, 13:567–572 (1984a).

Jury, W.A., W.F. Spencer, and W.J. Farmer, "Behavior Assessment Model for Trace Organics in Soil: III. Application of Screening Model," *Journal of Environmental Quality*, 13:573–579 (1984b).

Jury, W.A., W.F. Spencer, and W.J. Farmer, "Behavior Assessment Model for Trace Organics in Soil: IV. Review of Experimental Evidence," *Journal of Environmental Quality*, 13:580–586 (1984c).

Karickhoff, S.W., D.S. Brown, and T.A. Scott, "Sorption of Hydrophobic Pollutants on Natural Sediments," *Journal of Environmental Quality*, 13:241–248 (1979).

Karickhoff, S.W. and K.R. Morris, "Sorption Dynamics of Hydrophobic Pollutants in Sediment Suspensions," *Env. Tox. Chem.*, 4:469 (1985).

Kemper, W.D., and R.C. Rosenau, "Aggregate Suitability and Size Distribution," in *Methods of Soil Analysis, Part 1, Physical and Mineralogical Methods*, A. Klute, Ed., No. 9 in series Agronomy, American Society of Agronomy, Soil Science Society of America, Madison, WI, 1986, pp. 425–442.

Kenaga, E.E. and C.A.I. Goring, "Relationship Between Water Solubility, Soil Sorption, Octanol-Water Partitioning, and Concentration of Chemicals in Biota," in *Aquatic Toxicology*, J.C. Eaton et al., Eds., American Society for Testing and Materials, Philadelphia, PA, 1980, pp. 78–115.

Lide, D.R., Ed., *Handbook of Chemistry and Physics*, 73rd ed. (Boca Raton, FL: CRC Press, 1992).

Lyman, W.J., "Solubility in Water," in *Handbook of Chemical Property Estimation Methods in Environmental Behavior of Organic Compounds*, W.J. Lyman, W.F. Reehl, and D.H. Rosenblatt, Eds., (New York: McGraw-Hill Book Co., 1982a), Chap. 2.

yman, W.J., "Solubility in Various Solvents," in *Handbook of Chemical Property Estimation Methods in Environmental Behavior of Organic Compounds*, W.J. Lyman, W.F. Reehl, and D.H. Rosenblatt, Eds., (New York: McGraw-Hill Book Co., 1982b), Chap. 3.

Lyman, W.J., "Adsorption Coefficient for Soils and Sediments," in *Handbook of Chemical Property Estimation Methods in Environmental Behavior of Organic Compounds*, W.J. Lyman, W.F. Reehl, and D.H. Rosenblatt, Eds., (New York: McGraw-Hill Book Co., 1982c), Chap. 4.

Miller, R.M., "Biotransformations of Organic Compounds," in *Handbook of Vadose Zone Characterization & Monitoring*, L.G. Wilson, L.G. Everett, and S.J. Cullen, Eds., (Chelsea, MI: Lewis Publishers, 1995).

Montgomery, J.H., and L.M. Welkom, *Groundwater Chemicals Desk Reference*, (Chelsea, MI: Lewis Publishers, 1990).

Nielsen, D.R., D. Shibberu, G. E. Fogg, and D. R. Rolston, A Review of State of the Art: Predicting Contaminant Transport in the Vadose Zone, Water Resources Control Board, State of California, 1990.

Nofziger, D.L., and A.G. Hornsby, Chemical Movement in Layered Soils: User's Manual, Department of Agronomy, Oklahoma State University and Soil Science Department, University of Florida, 1987.

Panian, T.F., *Unsaturated Flow Properties Data Catalog*, Volume 2, Water Resources Center, Desert Research Institute, University of Nevada, Publication No. 54061, NTIS. U.S. Department of Commerce, Springfield, VA, 1987.

Rao, P.S.C., and J.M. Davidson, "Estimation of Pesticide Retention and Transformation Parameters Required in Nonpoint Source Pollution Models," in *Environmental Impact of Nonpoint Source Pollution*, M.R. Overcash and J.M. Davidson, Eds., (Ann Arbor, MI: Ann Arbor Science Publishers, Inc., 1980), pp. 23–67.

Runnells, D.D., "Basic Contaminant Fate and Transport Processes in the Vadose Zone," in *Handbook of Vadose Zone Characterization & Monitoring*, L.G. Wilson, L.G. Everett, and S.J. Cullen, Eds., (Chelsea, MI: Lewis Publishers, 1995).

Sax, N.I., *Hazardous Chemicals Information Annual*, No. 1., (New York: Van Nostrand Reinhold, 1986).

Scow, K.M., "Rate of Biodegradation," in *Handbook of Chemical Property Estimation Methods, Environmental Behavior of Organic Compounds*, W.J. Lyman, W.F. Reehl, and D.H. Rosenblatt, Eds. (New York: McGraw-Hill Book Co., 1982), Chap. 9.

Thomas, R.G. "Volatilization from Soil," in *Handbook of Chemical Property Estimation Methods, Environmental Behavior of Organic Compounds*, W.J. Lyman, W.F. Reehl, and D.H. Rosenblatt, Eds. (New York: McGraw-Hill Book Co., 1982), Chap. 16.

Tucker, W.A., and L.H. Nelken, "Diffusion Coefficients in Air and Water," in *Handbook of Chemical Property Estimation Methods, Environmental Behavior of Organic Compounds*, W.J. Lyman, W.F. Reehl, and D.H. Rosenblatt, Eds. (New York: McGraw-Hill Book Co., 1982), Chap. 17.

15

Laboratory Studies on Air Permeability

David S. Springer, Stephen J. Cullen, and Lorne G. Everett

OVERVIEW AND INTRODUCTION

The concept of air permeability of earth materials has been an important parameter to many branches of the earth sciences. Air permeability has been of particular interest to soil and agricultural scientists for decades as it relates to the aeration and gas exchange between soils and the atmosphere, to soil structure, and to the movement of surface and subsurface waters and its relation to irrigation problems. Additionally, chemical and petroleum engineers have exhaustively investigated the air permeability parameter as it relates to petroleum gas production and migration in oil field reservoirs. The more recent interest of soil gas movement in soils as it relates to the migration potential of hazardous, volatile vapors in the subsurface has focused renewed attention on the relationships between soil water content and the rate of gas movement through soils. Venting of gases from unsaturated soils has been used for the removal of methane from landfills, and noxious or explosive vapors that collect under and threaten buildings or utilities. A more recent offshoot of utilizing the permeable nature of soils to air flow is the evolution of a remedial strategy called soil vapor extraction, which involves inducing clean air flow through soils contaminated with highly volatile chemicals in an effort to strip them of contamination. The soil venting remedial technique is directly influenced by the air permeability parameter, where the design and engineering aspects and often the success hinge on its accurate quantification.

Earth materials occurring beneath the water table and in the overlying capillary fringe (within the lower vadose zone) are saturated with water, and thus do not have the capacity to freely transmit air. Above the capillary fringe, soil pore-water coexists with a gaseous phase. At lower water contents, materials can become sufficiently drained so that a semicontinuous air gap exists within the pores to transmit air or other gas mixtures. In simplest terms, air permeability of a porous material emerges at the point where an interconnected network of air-filled pores can transmit air when exposed to a driving force at opposing ends. The emergence of air permeability in a material (i.e., the "air-entry pressure") often occurs at water contents of 80% to 90% of the maximum water content value (Corey, 1986). From the point of air permeability emergence, as the water content, θ, continues to decrease within a porous material, a larger volume of the pores becomes filled with air, enabling a greater cross sectional area within the material to transport air. Since the total porosity of a material is the sum of the water content and air content, the air permeability parameter, k_a, is directly a function of a material's water content, θ.

The following sections present a brief historical review of air permeability determination in porous media, a discussion of the linearity of Darcy's law, circumstances which may lead to nondarcian flow behavior, including gas slippage and inertio-viscous effects, mathematical techniques appropriate toward characterizing the $k_a(\theta)$ function in earth materials, followed by an introduction and discussion of selected permeameter designs, laboratory apparatus, and experimental methodologies used in characterizing the $k_a(\theta)$ function in porous media.

REVIEW OF HISTORIC WORK

From both the theoretical as well as the experimental standpoint, the flow of fluids through capillaries has been fairly well understood for some time. Since the individual pores within a material are essentially interconnected, minute capillaries, the development of the flow of fluids through porous media has been treated along these same lines. Perhaps the most famous early work was completed by the French sanitary engineer Henry Darcy on beds of unconsolidated sands. He determined also that the quantity of flow is proportional to a coefficient, K, which is dependent upon the nature of the porous medium. Darcy found that the rate at which water is transmitted through a column of sand is directly proportional to the hydraulic head and cross sectional area of the sands, and inversely proportional to the length of flow through the sand. Darcy's law generally applies toward describing viscous, incompressible, laminar, fluid flow through porous materials. His empirical findings have been the subject of extensive research and have been repeatedly validated by experimenters, except in certain cases where tight materials are tested under low pressure fields, or in the presence of excessive velocities of fluid flow during measurement, where nonlinear flow behavior is manifested, which deviates from the linearity prescribed by Darcy's law.

Air Permeability of Rock and Synthetic Cores

The fundamental assumption that the permeability of a material is independent of the fluid used in its determination has since been addressed by numerous experimenters. The occurrence of gases flowing through a medium at rates greater than predicted by Darcy's law, a condition called slip flow, was first described in the literature by Kundt and Warburg in 1875. They showed that a layer of gas adjacent to a pore wall, in a stream flowing along that surface has a finite velocity with respect to the pore wall. In the specific case of flow through capillaries, this condition would produce flow rates greater than would be predicted by viscous flow

(Darcy's law). Thus, in effect, the gas slips past the pore wall, giving rise to the term slip flow. The phenomena of slip flow was later demonstrated under rigorous conditions using sintered glass material and sandstone core samples by L. J. Klinkenberg in 1941. Klinkenberg showed that the gas permeability of his test materials was dependent upon the mean free path of the permeant gas (the distance gas molecules travel between successive collisions), and therefore dependent upon the specific factors that influence the mean free path, which include the nature of the gas used in the measurement, and the pressure and temperature. He further demonstrated that the degree of slippage increases as the mean free path of the gas increases with respect to the pore radius.

Following Klinkenberg's classic work, the measurement of the air permeability parameter within porous materials was of widespread interest to the petroleum industry, where many studies were funded through the American Petroleum Institute (API) toward characterizing three-phase fluid flow and petroleum gas production and fluid migration in oil field reservoirs. Much of this historic work focused on measuring relative permeabilities of reservoir rock cores (very low permeability) using heterogeneous fluids including air or other gases in the presence of a coexisting water and/or petroleum liquid. Discrepancies between measured steady state gas versus liquid permeabilities in identical samples have been reported to exist in numerous studies; most values for water permeability being lower than air permeability as predicted by the phenomena of slip flow (Fancher, Lewis, and Barnes, 1933; Leverett and Lewis, 1941; Leas et al., 1950; Gates and Lietz, 1950; Estes and Fulton, 1956; and others).

The first physical experiments linking the dependence of the magnitude of the air permeability as a function of water content, $k_a(\theta_v)$, were reported by Wyckoff and Botset (1936). By introducing mixtures of gas and liquid through porous media at several controlled water contents (θ_v), they were able to demonstrate that the water permeability (k_w) increases and the air permeability (k_a) decreases with a corresponding increase in water content, (θ_v). The empirical results developed by this study were later formulated into a series of mathematical differential equations of flow by Muskat (1934).

Calhoun and Yuster (1946) determined permeabilities of Pyrex-glass and silica membranes using five different inert gases and various liquids. Their findings indicated that, in the absence of reactions of the measuring liquid with the porous body, permeabilities using various liquids were essentially equivalent whereas gas permeabilities were higher, the difference being proportional to the gas slippage factor calculated from the experiments.

Heid et al. (1950) completed air permeability determinations on 11 synthetic core and 164 natural rock core samples of low permeability from across the United States. In addition to the use of air in the permeability measurements, water, petroleum fractions, and brine water were used on selected cores, and the results evaluated. Experimental results demonstrated that air permeabilities measured at low differential pressures were everywhere greater than corresponding liquid permeabilities measured at the same pressures. As pressures were increased during experimentation, the air permeability was demonstrated to approach the liquid permeability, similar to Klinkenberg's earlier work.

Air Permeability of Native and Repacked Sediments

Adapting from and improving upon the methodologies and results for determining permeabilities on rock cores, more recent air permeability experiments have been completed on in situ (field) soils (Kirkham, 1946; Evans and Kirkham, 1949; Grover, 1955; and Weeks, 1978), in situ surface exposed rock fractures (Kilbury et al., 1986), and natural and artificial repacked materials (Buehrer, 1932; Corey, 1957; Stonestrom, 1987; Springer, 1993; Detty, 1992, and others).

Corey (1957) worked with two loam materials and employed the stationary liquid method at steady-state soil water contents during a drainage cycle at low pressures. He found that air permeability was initiated in these materials at a water saturation percentage of approximately 85%, with an extrapolated gas slippage factor as high as 10.5%. He concluded that the permeability to water of the saturated soils was about half the permeability to air of the dried soils ($k_w = 0.5\ k_a$), even when gas slippage was corrected for.

Kirkham (1946), Evans and Kirkham (1949), and Grover (1955) investigated field air permeabilities by introducing a steadily declining pressure field through a tube or cylinder inserted into shallow soils. The pressure decline measured from an air chamber was related to a measure of the in situ air permeability. King (1968) developed a similar approach by injecting air into an auger hole, and Boardman and Skrove (1966) into a packed-off section of drill hole.

Weeks (1978) conducted a study of in situ vertical air permeabilities in soils at five field sites by simultaneously comparing measurements of the change in barometric pressures at ground surface to measurements collected from nested vapor piezometers located at different depths below ground surface. Weeks used the field-derived air permeability values to calculate equivalent hydraulic conductivity values. With one exception, Weeks found that field-derived air permeabilities were greater than laboratory-determined results by several times. This he attributed to the inevitable compaction of laboratory core samples during sampling, structural changes of soils with depth, and to the small sample size of cores which may miss secondary permeability features (e.g., soil structure, fractures, etc.). Weeks established empirically that the air permeabilities of soils at field capacity are generally 0.6 to 0.8 times their value when air dry.

Stonestrom (1987) completed a thorough study focusing on the co-determination of a family of parametric functions which characterize unsaturated flow; namely, liquid permeability, air (gas) permeability, retentivity (matric potential), and air trapping, all as functions of a medium's water content. The materials tested included repacked Aiken loam, Oakley sand, and glass beads. Test runs were completed over complete wetting and drying cycles to establish hysteresis effects, and all reported air permeability values were corrected for gas slippage effects using the method of Klinkenberg (1941).

Springer et al. (1991) and Springer (1993) completed a series of air permeability experiments on two repacked glass bead materials, and two repacked alluvial sediments. Using a specially constructed air permeameter, the $k_a(\theta_v)$ function was characterized from the approximate air entry value to near air dry soil water conditions during a monotonic drainage cycle. The magnitude of air permeability was determined to vary most significantly in the intermediate range of soil water content, where, as the materials approached air dry conditions, the magnitude of change in the air permeability parameter was negligible.

Detty (1992) conducted a series of column experiments on unconsolidated, repacked Berino fine loamy sand and 30 mesh silica sand to investigate, among other objectives, factors affecting the air permeability measurement, including the effects of slip flow and visco-inertial flow. Detty reported a negligible contribution to gas flow via gas slippage effects, with substantial visco-inertial effects corresponding to elevated pressure gradients and increasing water contents within the test samples.

CALCULATION OF AIR PERMEABILITY

Introduction

The following section presents a rigorous mathematical derivation of the governing equation of flow for air permeability calculations using conventional laboratory apparatus from Darcy's law. Additional mathematical solutions for alternative flow geometries and boundary conditions are presented in Muskat, 1934, and Bear, 1972. If the reader is not concerned with this mathematical derivation, skip to the end of this section and refer to Equation 10 and the definition of variables.

Theory and Mathematics of One-Dimensional, Steady-State Gas Flow

In developing the theory of flow of compressible fluids through porous materials, the established principles from earlier works (Muskat, 1934; Kirkham, 1946) describing the processes of gas transport using principles of classic hydrodynamics have been adopted for this discussion. The basic physical conditions include a Darcian flow regime, steady-state, one-directional, isothermal, compressible gas flow through porous media. In the case of incompressible, vertical water flow through a porous medium, the simplified Darcian equation applies:

$$v = -k/\mu \, (dP/dz + \rho g) \qquad (1)$$

where: v = volume flux of fluid
k = intrinsic permeability
μ = viscosity of fluid
dP/dz = pressure gradient along the z direction
ρ = density of fluid
g = acceleration due to gravity

The product ρg is the gravitational force acting on a unit volume of fluid in a vertically downward direction opposing the driving gradient (dP/dz) on a unit volume in the upward direction. For the case of air flow, a compressible fluid, assuming isothermal conditions, neglecting the gravitational term is an appropriate assumption providing that the gravitational component is negligible in comparison to the force of the applied pressure gradient (Muskat and Botset, 1931). At typical experimental delivery pressures, when vertical flow regimes are used, the ρg term appearing in Equation 1 for gases is less than 1% of the pressure term, and can thus be neglected.

In order to address the compressibility of air flow through porous materials, it must be assumed that Darcy's equation is valid for describing liquid flow as well as gas flow through porous materials. Direct integration of Equation 1 for gas flow cannot be carried out since the volume flux (v) of gas is not a constant but is a function of z, increasing as the gas approaches the low pressure end of the sample.

Equation 1 can be integrated if v can be expressed in terms of mass where at steady state, mass flow will be a constant.

Assuming that the flowing gas behaves as an ideal gas according to the Ideal Gas Law:

$$PV = nRT \qquad (2)$$

where: P = pressure
                V = volume
                n = number of moles of gas
                R = ideal gas constant
                T = absolute temperature

Under isothermal conditions, Equation 2 can be rewritten according to:

$$PV = (m/MW) * RT \qquad (3)$$

where: n = (m/MW), and
                m = mass of gas
          MW = molar weight of gas

Equation 3 can be rewritten as:

$$P = (m/V) * (RT/MW) = \rho * (RT/MW) \qquad (4)$$

where: ρ = density of the gas

Assuming isothermal conditions, let (MW/RT) equal a constant, C, according to:

$$\rho = P * (MW/RT) = P * C = \rho \qquad (5)$$

We express the volume flux term in Equation 1 in terms of mass by multiplying both sides of the equation by ρ, where $\rho * v$ (density * volume = mass, and mass is conserved). Substituting Equation 5 into Equation 1:

$$\rho * v = \rho * -(k_a/\mu * (dP/dz)) \qquad (6)$$

where: k_a = air permeability (potentially slip enhanced)

This can be rewritten as:

$$(C * P) * v = (C * P) * -(k_a/\mu * (dP/dz))$$

The constant, C, cancels out, where after rearranging terms:

$$(P * v) \, dz = -k_a/\mu \, PdP \qquad (7)$$

Under steady state conditions, the product Pv is a constant, as is $-k_a/\mu$, therefore through integration of dz along the length of the permeameter, and PdP from the inlet to the outlet of the sample we arrive at:

$$(P_o v_o) \int dz = -(k_a/\mu) \int PdP$$

which is solved as follows:

$$P_o \, v_o \, L = -k_a/\mu * (P_i^2 - P_o^2)/2 \qquad (8)$$

where L is the length of the permeameter, the "$_o$" subscript denotes that the measurement is taken at the outlet of the permeameter, and the "$_i$" subscript denotes the measurement is taken at the sample inlet.

Rearranging Equation 8 and recognizing that Q_o, volume outlet flow, is equal to the product $v_o A$, where A is the cross-sectional area of the permeameter:

$$v_o * A = Q_o = -(k_a A/2\mu L) * ((P_i^2 - P_o^2)/P_o) \qquad (9)$$

Solving for k_a yields the governing equation:

$$k_a = Q_o * [(2 \, \mu L/A) * ((P_o)/(P_i^2 - P_o^2))] \qquad (10)$$

Equation 10, the extended Darcy equation for gas flow, describes the steady-state, isothermal, volumetric flow rate, Q_o, (cm^3/sec) of a compressible gas, with viscosity μ (g/cm sec), through a sample of length L (cm), and cross-sectional area A (cm^2), with inlet pressure (absolute) of P_i and outlet pressure of P_o (g/cm sec^2, or Pascals), where the resultant air permeability is calculated by solving for the k_a (cm^2) parameter. The parameter k_a is a measure of the air permeability (potentially slip enhanced) of a porous material at a given, steady-state volumetric water content, (θ_v). A series of measurements of the air permeability parameter over a range of volumetric water contents provides information as to the behavior of the $k_a(\theta_v)$ function.

Calculating Air Permeability Results from the Darcian Equation for Gas Flow

Direct k_a Determination

The validity of Equation 10 for describing air flow during experimentation is evaluated by directly determining k_a for test soils following each experimental run. At specific measured water contents, pressure fields, and flow rates during each test run, individual calculations of k_a are completed. However, since individually measured k_a values may be affected by experimental error and/or nonlinear flow conditions, the graphical approaches described in the following sections are used to evaluate these potential conditions.

Graphical k_a Determination

The validity of the Darcian gas flow equation (Equation 10) can also be evaluated using graphical methods (after Stonestrom, 1987). Separating Equation 10 into two components, which the authors designate as term "Q" equivalent to the volumetric outflow rate, and a term "P'" which groups the pressure, viscosity, and geometric components according to:

$$k_a = Q_o * [(2\mu L/A) * ((Po)/(P_i^2 - P_o^2))]$$

we can rewrite this expression as:

$$k_a = Q_o / [(A/2\mu L) * ((P_i^2 - P_o^2)/P_o)] \tag{11}$$

$$k_a = [\text{``Q'' term}] / [\text{``P'' term}]$$

where we define:

$$Q = Q_o$$

$$P' = 1 / [(2\mu L/A) \times (P_o)/(P_i^2 - P_o^2)].$$

Since the P' term is defined as the reciprocal of the group of terms presented in Equation 10, dividing Q by P' in Equation 11 results in an equivalent expression of the Darcian equation for gas flow (Equation 10). The units of the Q term are in cm³/sec, while the units of the P' term are in cm/sec, yielding units of cm² for the air permeability parameter which is standard notation. By plotting the steady-state Q term on the y-axis versus the P' term on the x-axis, k_a is thus defined as the slope of the line of best fit of the data points for a sample at a particular water content. Subsequent graphing using this approach must plot linearly and intersect the origin as prescribed by Darcy's law; otherwise, the test for the Darcian gas flow equation as a descriptive equation for compressible air flow through porous media is not strictly valid. As indicated by Equation 11, the units for slope of the line of best fit are in cm², which is consistent with the standard notation for air permeability. Figure 15.1 presents a graphical example of this type of analysis for experimental data.

NONDARCIAN GAS FLOW THROUGH POROUS MEDIA

Darcy's law effectively describes laminar, viscous fluid flow through porous media. In the laboratory, using conventional apparatus with typical imposed pressure gradients, liquid permeability measurements are characterized by Darcy's law. How-

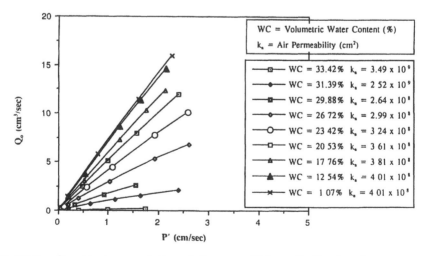

Figure 15.1. Graphical presentation of Q versus P' to verify Darcian flow.

ever, for gas permeability measurements, nondarcian flow effects may be manifested from turbulence, molecular motion, and pressure dependence on the compressibility of gas. Two nondarcian flow effects which will be further discussed include gas slippage and visco-inertial flow.

Gas Slippage

Gas slippage or slip flow arises from successive collisions between gas molecules at distances equal to the mean free path of the gas at the pore walls, which impart a nonzero velocity to the molecules in the direction of flow. This is contrary to laminar flow theory, which assumes zero velocity at the pore wall. Klinkenberg (1941) explored the mechanisms causing slip flow in porous media. Employing a conceptual model of porous media represented by a bundle of randomly oriented capillary tubes, Klinkenberg defined the following relationship between the "apparent" and "true" permeability of an idealized porous system to gas:

$$k_a = k' [1 + (4cl/r)] = k'[1 + (b/P_m)] \qquad (12)$$

where: k_a = "apparent" or slip enhanced permeability
k' = "true" or slip corrected permeability
c = constant
l = mean free path of gas
r = pore (or capillary) radius
b = slip parameter
P_m = mean pressure

From this relationship, it follows that the gas slippage is at a maximum when the mean free path, l, of the gas is equal to the pore radius, and under low mean pressures, P_m. Conversely, the effects of slip flow are expected to be at a minimum in coarse materials under higher mean pressures. As the mean pressure is progressively increased, the gas will approach a liquid in property as far as the mean free path is concerned, and therefore the gas permeability should approach the liquid permeability. By plotting the "apparent" gas permeability, k_a, versus the reciprocal of the mean pressure $(1/P_m)$, and extrapolating the data to an infinite pressure, "true," or slip corrected values of air permeability will be determined. The limiting situation in which the mean free path, l, of the gas molecules contributes to slip flow is where l is greater than either the diameter of the capillary or its length. Under these circumstances, only collisions between gas molecules and the pore walls occur which gives rise to "molecular streaming" or Knudsen flow, which takes place by a diffusive, as opposed to a viscous mechanism (after Dullien, 1979).

The current ASTM D 4525 (ASTM, 1990) standard test method entitled Permeability of Rocks by Flowing Air and API RP 27: Recommended Practice for Determining Permeability of Porous Media (API, 1952) utilize the identical equations and methods developed by Klinkenberg for determining the apparent permeability of materials as affected by gas slippage.

Weeks (1978) indicated that gas slippage is only significant for materials having an intrinsic permeability of less than about 1×10^{-10} cm^2. Massmann (1989) related the relative importance of slip flow and viscous flow for low pressure systems, where materials with a pore radii greater than 1×10^{-3} mm exhibit minimal effects of slip flow relative to Darcian flow. Silt and clay materials often demonstrate pore radii of this magnitude, whereas sands typically have pore radii on the order of 1×10^{-1} mm and greater. Based on these findings, it is prudent to determine and apply the gas slippage correction to test materials predominantly containing silt and clay materials, although gas slippage has been reported for coarser materials (Stonestrom and Rubin, 1989).

As gas slippage is a linear function of the reciprocal of the mean pressure, calculate the reciprocal of each mean pressure as follows (after ASTM D 4525, 1990):

$$2/(P_i + P_o) \qquad (13)$$

Plot the apparent air permeability parameter (k_a) determined during the experiments along the y-axis versus the reciprocal mean pressure along the x-axis for each test (see Figure 15.2). Draw a straight line through at least three plotted points to intersect the ordinate line at zero reciprocal mean pressure. The value of k' at the intersection of the y-axis is the slip corrected value of the test sample, which is theoretically equivalent to the liquid permeability, k_l, of the test sample (Klinkenberg, 1941; ASTM, 1990).

Visco-Inertial Effects

Nondarcian flow conditions also arise within porous media owing to excessive fluid velocities, giving rise to turbulent flow. Historically, the upper limit of laminar flow has been associated with the aid of the Reynold's number, a dimensionless number that expresses the ratio of inertial to viscous forces during flow (Freeze and Cherry, 1979). The contention that "Darcy's Law is valid as long as the Reynold's number based on average grain diameter does not exceed some value between 1 and 10" has been widely adopted as a rule of thumb (Bear, 1972, and others). An alternate approach toward analyzing linear visco-inertial gas flow was introduced by Rawlins and Schellhardt (1936) as follows:

$$Q_{sc} = C \, (dP^2)^n \qquad (14)$$

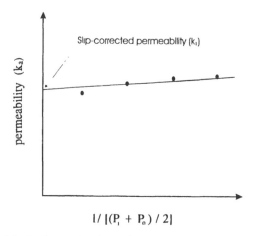

where: Q_{sc} = mass flux of gas
        C = constant depending on porous media
      dP = pressure gradient
       n = constant indicating the state or region of the visco-inertial flow regime.

Rawlins and Schellhardt had proposed that within an entirely viscous flow regime, the value of n was equal to one, corresponding to an expression of Darcy's law. As the flow regime progresses from viscous flow toward inertial flow, the value of n decreases until entirely inertial flow is achieved, whereby the value of n corresponds to 0.5. Experimentally, the value of n can be estimated by constructing a deliverability plot from experimental data (after Detty, 1992). Using Equation 14, the deliverability plot is prepared by plotting the log (dP^2) versus log Q_{sc}, where the value of n is estimated from the slope of the plot according to:

$$\log Q_{sc} = \log C + n \log (dP^2) \qquad (15)$$

Deliverability plots can be constructed to determine the nature of the visco-inertial flow regime and evaluate the representativeness of measured air permeability data (Detty, 1992).

METHODS OF DETERMINATION

General

The scope of methods for measuring k_a in porous materials encompasses scales varying from laboratory scale systems on singular soil cores to full scale field installations using multiple wells and complex flow systems. Although the scale of measurement can significantly affect the measured result and interpretations, discussion of scale effects is not treated herein. Also, both unsteady-state and steady-state methods of determination exist (Corey, 1986). As this review focuses on air permeability determination in the laboratory environment under steady-state flow conditions, for a discussion of field air permeability determination methods, the reader is directed to the following references for field scale systems: Weeks, 1978; Corey, 1986; Massmann, 1989; Johnson, Kemblowski, and Colthart, 1990; and U.S. EPA, 1991; and to Weldge (1952) for unsteady-state air permeability determination methods.

Steady-State Methods

In simplest terms, steady-state methods involve measuring the air permeability while maintaining a constant or uniform water content within a test sample. An important requirement of steady-state $k_a(\theta_v)$ characterization is to maintain a uniform water content, θ, along the entire axis of flow (Corey, 1986). Richards (1931) pioneered conventional steady-state hydraulic conductivity measurement using 100 kPa (1 bar) rated ceramic plates as capillary barriers which were regulated to control the liquid head across a sample. A drawback of using this method is that the water content profile is typically not uniform. As a means of achieving this end, a method called the stationary liquid method has been developed which involves holding the wetting phase stationary via capillary forces, permitting only the nonwetting phase to flow through the sample (Osoba et al., 1951; Leas et al., 1950). This method also avoids boundary effects which arise from discontinuities in the capillary properties of a porous media.

PERMEAMETER DESIGN AND CONSTRUCTION

Design Considerations

Several unique and innovative permeameter designs have been advanced in the scientific literature as a means of addressing the various demands of steady-state air permeability testing. The basic design features that all permeameters must exhibit include a functional sample-holding device capable of retaining a sample for measurement while not altering the geometry or physical properties of the sample. Integral to the sample-holding device is the inclusion of sample support end caps. End cap construction should allow for a tight seal against the ends of a sample, while easily facilitating air flow. Typically, end caps are constructed with a network of grooves (Corey, 1986; Stonestrom, 1987) to allow passage of air and water phases, and using a material of sufficient strength with little resistance to air flow (Springer et al., 1991). If significant resistance to air flow is realized in a particular design, a correction factor must be calculated and applied to the results of experimental runs to account for impedance to air flow. Because a pressure gradient must be developed for induced air flow to pass through a sample, the sample-holding device should adequately seal the edges of the sample so that air flow is only permitted to flow through the sample and not "short circuit" due to macropore development or wall effects.

As an alternative to ensuring a tight seal between the walls of the sample-holding device through physical design, samples can be sealed within a latex, wax, plastic, or other membrane, leaving only the ends exposed. Flexible membranes such as latex offer the added benefit where potential shrinkage or swelling of the sample arising from clay content during testing can be compensated for through applying pressure across the membrane or "sleeve" to ensure good contact (Daniel et al., 1984).

Permeameters designed to measure air permeabilities over a range of water contents must also provide a provision for manipulation and control of the water content. For steady-state flow measurements, this requirement has proved to be rather elusive. Centrifugal force techniques have been used to establish nearly uniform moisture distributions in soil (Nimmo et al., 1987). Porous plastics have also been successfully used in permeameter designs (Roseberg and McCoy, 1990); however, due to limitations in manufacturing, plastics exhibit limited abilities toward remaining conductive to fluids at higher pressures. However, ceramics, owing to their very small and uniform pore size, once saturated, will remain wet and serve as a fluid conduit for effective manipulation and control of water content in soil samples over a broad range of pressures. The range of pressures in which a given ceramic material will remain saturated, once presaturated, is controlled by the pore diameter and uniformity of pore size. The drawback of using very small pore size ceramics to accommodate wide pressure ranges during water desaturation is the reduction in the rate at which liquids are transmitted.

Permeameter Design Types

The American Petroleum Institute, Recommended Practice for Core-Analysis Procedure RP-40 (API, 1960), commonly known as "RP-40," presents several permeameter designs, each with specific capabilities in mind. The Hassler-type permeability cell is designed for cylindrical and cross-sectionally uniform cores (Hassler, 1944). The core is introduced within the cell and capped off by concentrically grooved end caps. A rubber diaphragm and high pressure liquid or air directed into the annulus in conjunction with heated plastic or pitch is used to ensure a good seal between the sample and the permeameter body (Figure 15.3).

The Fancher-type core holder presented in RP-40 accepts specimens placed within a metal compression yoke. A rubber compression ring is secured around the sample, which is supported by a lower metal base and an upper metal compression yoke, sealed by mechanical compression and rubber gaskets (Figure 15.4). Because these permeameters were developed several decades ago designed primarily for rock core analyses, the unfortunate drawback of these "compression cylinder type" designs is that test samples must be sufficiently hard and consolidated to resist any form of cracking or other deformation, as the air permeability measurement is extremely sensitive to the slightest crack.

A standard compression cell holder may be used as an alternative for core samples contained within metal rings (Figure 15.5). The ends of the core must be clean and open to air flow, and the core must be sealed along the sides of the mounting medium. Although this design dates back to the earlier compression type permeameters, it is entirely feasible to further adapt the standard compression holder to accept standard unconsolidated samples collected in metal rings. Typically, sample rings or sleeves are used to line core sampling devices driven into earth materials during drilling operations. Provided that adequate measures are included to prevent sample deformation and preferential flow along the solid ring inner wall, the standard compression holder is a viable and adaptable permeameter design.

Perhaps the most common type of air permeameter in use today is a modified permeameter originally designed for saturated or unsaturated hydraulic conductivity determinations. Sample holders equipped with ceramic end plates, such as a Tempe cell (Soilmoisture Equipment Corp.) offer several features readily adaptable to a

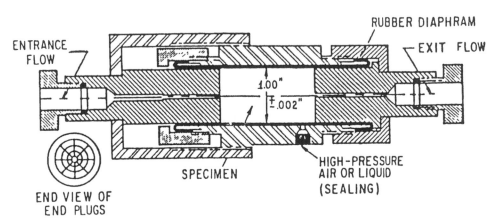

Figure 15.3. Hassler-type permeameter. Copyright ASTM. Reprinted with permission.

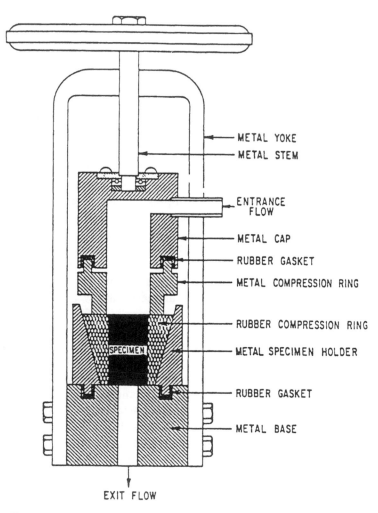

Figure 15.4. Fancher-type permeameter. Copyright ASTM. Reprinted with permission

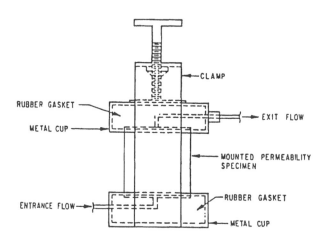

Figure 15.5. Compression cell permeameter for ring-mounted samples.

permeameter. Alternatively, more expensive and elaborate triaxial cells have been modified to accommodate samples mounted within membranes. Triaxial cell chambers have the added capability of introducing pressurized gas directly into the chamber between the sample and the chamber wall. This pressure serves as a radial confining pressure which can be used to mimic in situ soil conditions at depth.

Corey (1986) developed a unique air permeameter using the stationary water method for determining $k_a(\theta)$. The sample holder portion of this apparatus retains the sample in the annulus between a water-wetted ceramic cylinder and a rubber sleeve lining the inner wall of the sleeve cylinder. Short circuiting of air is prevented along sample-wall boundaries by applying a pneumatic pressure across the rubber sleeve of between 50 to 100 kilopascals (kPa). This pneumatic pressure also maintains contact between the sample and the inner ceramic cylinder; an important requirement to ensure control over water content, θ, and to accommodate potential shrinkage of the sample during desaturation.

Stonestrom (1987), and Stonestrom and Rubin (1989), adapted and modified the Corey (1986) permeameter design by adding peripheral features for co-determination of trapped-air, retentivity, and air and water permeability functions. The principal design modification was to use concentrically grooved ceramic end plates termed tessellate-plate flow cells to dually permit water desaturation and air flow through the sample.

Springer et al. (1991), and Springer (1993) developed a permeameter cell which was also adapted and improved upon from the Corey (1986) and Stonestrom (1987) designs. The permeameter cell, termed the "LEAP Cell," which stands for Liquid Extracting Air Permeameter, also uses an integral ceramic element. The sample holder is composed of a 100 kPa (1 bar) rated ceramic cylinder which dually acts as a sample holder and as a porous membrane permitting liquid extraction (Figure 15.6). A ceramic glaze applied at the ends of the ceramic cylinder constrains liquid movement perpendicular to the cylinder walls (radially outward from the sample core). This feature permits rapid dewatering of a sample in response to a uniform pressure gradient along the length of the core sample and thus the theoretical establishment of a uniform water content along the axis of flow, a capability that has not been adequately addressed by past researchers (Corey, 1986). At successive equilibrium stages, the sample's water content is gravimetrically determined without removing the sample from the cylinder, allowing a series of tests to be performed at various water contents without physically disturbing the test sample.

PERIPHERAL EQUIPMENT CONSIDERATIONS

Introduction

Owing to the wide variety of material types, sample geometries, and testing equipment, an equally wide variety of peripheral measurement devices are employed toward air permeability determination. In general, the following equipment items are essential toward measuring the variables for calculating the air permeability parameter: temperature sensing device, pressure sensing device, pressurized gas supply with regulation, flow sensing device, and an appropriate source of water or alternate test fluid. A few additional equipment items that prove invaluable include: a drying oven, humidifying device, relative humidity sensor, analytical balance, barometer, temperature control (either room temperature control or insulating jack-

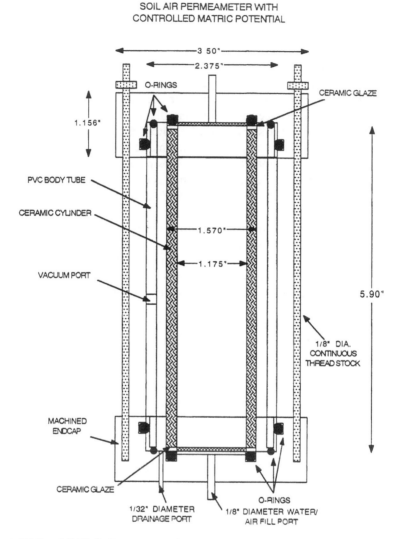

Figure 15.6. LEAP Cell permeameter.

ets), and graduated scales and calipers. The following section provides a limited discussion toward the appropriate selection of the above-listed equipment.

Temperature Sensing

Changes in the temperature of fluids affect both the fluid density and viscosity, both of which will affect the outcome of an air permeability test. The density of fluids is further a function of pressure, which is influenced by temperature. In light of these interrelationships, it is essential that if the temperature cannot be adequately controlled within the working space, it must at least be measured so corrections can be applied to appropriate parameters, and results may be expressed in terms of standard conditions. Thermocouple sensors offer a compact, accurate, and potentially inexpensive method of measuring temperature. In most applications, temperature sensing elements are positioned at opposing ends of the sample, out of the way of the airflow path.

Pressure Sensing

Pressure sensing devices may include gauges mounted on the pressure delivery system, sensors to measure the pressure drop across the sample and any appurtenances, such as restrictive end caps, etc., and tensiometric devices for measurement of matric potential or soil suction within the test sample. In general, for most medium-coarse grain, unconsolidated materials, moderate pressures are required to induce air flow through a sample. This requires the pressure regulator and gauge assembly to accurately regulate pressures in the range of ± a few millimeters of water.

For pressure measurement at the base of, across, and at the top of the sample, either sensitive pressure transducers or manometers can be appropriate. A very reliable, inexpensive, and relatively sensitive pressure sensing device is the U-tube manometer. Fluids of different viscosities can be introduced into the U-tube manometer to modify sensitivities as needed. Additionally, use of inclined manometers will further increase pressure sensitivity. As a minimum, pressure measurement is required at the base of the sample and at the effluent of the sample. This can be achieved using separate sensors at each location. If gas slippage corrections are not desired, a differential manometer may be positioned to sense pressures across the sample. It is suggested that both types be used.

Flow Sensing Device

Flow sensors are utilized primarily at the sample outlet in order to determine the air flow rate through the sample for the air permeability parameter calculation. ASTM D-4525 (ASTM, 1990) procedure for determining the air permeability of porous materials as well as RP-40 (API, 1960) both specify utilizing dry gas as the sample permeant. When adhering to these methods, which are most appropriate for oven-dry samples, either calibrated rotameters or highly accurate mass flow controller devices may be used as the volumetric flow sensor. However, when the desired air permeability test is based on a constant water content, and when using prehumidified air as the gas permeant, the increased relative humidity of the effluent gas can create additional difficulties. Water vapor can readily condense within a rotameter orifice as it exits the sample, reducing the effective diameter of the indicator tube, resulting in spurious readings. Similarly, water vapor can condense on the reference temperature sensor of a mass flow controller resulting in an artificially high flow rate, requiring further calibration considerations. In these cases, it is recommended that soap film bubble meters be used as flow sensing devices. These devices can be obtained through commercial sources offering a variety of flow ranges and options such as photo-actuated timers. Alternatively, very accurate soap film flowmeters can be fabricated from readily available parts with the assistance of a skilled glassblower (Corey, 1986). The primary advantages of using soap film flowmeters is that the resultant readings are independent of the relative humidity of the outlet airstream, and once these instruments are calibrated, periodic cleaning ensures that future calibrations are not necessary.

Water Supply

Water is typically used for filling manometers and humidifying devices. In these instances, an ultra pure deionized water source is appropriate to relax assumptions of fluid composition, density, and viscosity, and to avoid mineral buildup in flow lines. Additionally, the water should be deaired to avoid the potential of dissolved gases coming out of solution.

Ideally, the best fluids to add to soil samples for sample saturation and water content manipulation are the ambient fluids contained within the pores themselves. However, as it is often impractical to match fluid compositions for all test samples, a 0.005 M $CaSO_4$ solution, saturated with thymol has been successfully used (Klute and Dirksen, 1986). The calcium ion is designed to reduce the dispersion of clays in the samples, and the thymol will inhibit biological activity. If possible, the electrolyte concentration within the saturating solution should closely approximate that within the soil solution (after Corey, 1986).

Air Humidifier and Relative Humidity Probe

At equilibrium, the air-filled pores within most native earth materials, even when rather dry, exhibit relative humidity levels between 98% and 100%. As such, use of dry air (or other gas) as the permeant gas can effectively carry away moisture through advective flow during the measurement process and alter the original moisture content. For this reason, it is essential to pre-humidify the permeant gas prior to introducing the gas into the test sample. To achieve this end, humidifying devices of many designs have been utilized. Stonestrom (1987) passed permeant air through a woven fiberglass wick suspended above a water reservoir. Springer (1993) forced delivery air first through a fine porous stone immersed in water contained within a pressure vessel. Similarly, moist sponges have been used to these same ends. Another possible design incorporates the use of a water-filled glass Ehrlenmeyer flask fitted with a two-hole rubber stopper where one line for gas inlet is lowered to below the static water surface, while the other line is for the humidified gas outlet mounted at the exit port.

Additional Peripheral Equipment Items

In addition to those items mentioned above, if the experimenter wishes to conduct air permeability measurements on test materials while controlling the water content, a drying oven and analytical balance for determining sample mass is essential. Analytical balances should be acquired with a sensitivity of at least 0.1 grams. Conventional soil drying convection ovens require a minimum of 110°C ± 5°C temperature capability.

For accurately determining sample geometry, calibrated engineering scales and calipers are appropriate. Additionally, laboratory temperature control, with a continuously recording barometer is advised to provide for absolute pressure measurements and to account for requisite corrections, if significant, on viscosity and density parameters, and to express the pressure in terms of absolute pressure.

LABORATORY AIR PERMEABILITY DETERMINATION PROCEDURES

Characterization of the Test Sample

Prior to initiating air permeability testing, selected diagnostic soil tests are typically completed in order to characterize selected physical parameters of each experimental material. This often requires that sufficient sample amounts are collected at the onset of testing. A definitive list of appropriate soil tests and corresponding methods would be of limited use as the purpose and scope of characterization will certainly vary from one individual to another. However, as a practical guide, as a minimum the test soil type, texture, description of pertinent features, sample mass, sample dimensions, bulk density, and porosity should be measured, calculated, and/or estimated prior to experimentation. The experimenter should try to adhere to standardized experimental protocols and use standard units such as those established by the American Society for Testing and Materials (ASTM), United Soil Classification System (USCS), etc., whenever appropriate and possible in order to attain comparability and reproducibility of experimental protocol.

Bulk Density

Bulk density determination should be completed prior to and immediately following air permeability testing to establish whether sample deformation has taken place. For remolded or recompacted sample testing, the dry bulk density measurement should be used, while for native state testing, native state bulk density should be determined. As air permeability testing is very sensitive to even slight changes in the bulk density value (native or dry), when working with recompacted materials, if a correlation with native materials is desired, artificially created bulk density values should nearly match native state values.

Porosity

The porosity of the test material can be either estimated using a variety of sources (Freeze and Cherry, 1979; Fetter, 1980; Driscoll, 1986), measured using a variety of accepted test methods (Danielson and Sutherland, 1986), or calculated using the following relationship, for example:

$$n = (1 - \rho_b/\rho_p) \tag{16}$$

where:
n = porosity (dimensionless)
ρ_b = bulk dry density (g/cm^3)
ρ_p = particle density (g/cm^3)

Particle densities for test materials can be determined using standard laboratory devices such as air pycnometers, or estimated based on mineral constituency. For most mineral soils, a particle density of 2.65 g/cm^3 is a valid approximation.

Hydraulic Conductivity

Hydraulic conductivity is perhaps the most widely known and utilized parameter to hydrologists and soil scientists today. Knowledge of the hydraulic conductivity parameter of a test material serves both to further characterize specific soil properties affected by air permeability testing, as well as to provide a quality control check to compare measured air permeability values through the use of selected empirical equations. The hydraulic conductivity is related to the intrinsic permeability parameter (k) according to the following relationship (Driscoll, 1986):

$$K = k\rho g/\mu \qquad (17)$$

where: K = saturated hydraulic conductivity
          k = intrinsic permeability
          ρ = density of fluid
          g = acceleration due to gravity
          μ = fluid viscosity

Because the ρ and μ terms are properties of the fluid, and g is a constant, the intrinsic permeability, k, is dependent on properties of the medium only (Hubbert, 1940). However, as suggested by Childs (1969), this concept may be misleading when applied to natural earth materials because of physical changes in the medium that may occur upon exposure to different fluids, including waters of different chemical quality. The intrinsic permeability value calculated from the above (or similar) equation can be used to provide estimates of expected air permeabilities. It is cautioned however, that since the hydraulic conductivity value can vary over as many as 13 orders of magnitude in different earth materials (Freeze and Cherry, 1979), estimates of expected air permeability values using such equations should be viewed as order of magnitude approximations, at best.

Particle Size Analysis

A particle size analysis is a measurement of the size distribution of individual particles comprising a test material. Particle size analysis data can be presented and used in several ways, perhaps the most common of which is a particle size distribution curve. Particle size distribution is determined by passing a disaggregated sample through a series of sieves. The cumulative weight of particles caught on each sieve of an established particle size is typically plotted as a percent of the total sample weight against the logarithm of particle diameter. The resultant plot of data points establishes a particle size distribution curve, where average grain size and degree of sorting are indicated.

The importance of grain size in affecting permeability has given rise to the following expression called the Hazen Approximation Method (Hazen, 1911):

$$K = k\rho g/\mu = (CD_{10}^2)\rho g/\mu \qquad (18)$$

where: $k = CD_{10}^2$

In this equation, k is the intrinsic permeability, C is a dimensionless constant called the shape factor, usually held to include properties such as tortuosity, particle shape, sediment sorting, and porosity, and D_{10} is the effective grain size where 10% of the test material is finer, and 90% is coarser (Fetter, 1980).

Alternately, a second empirical equation can be utilized to estimate permeabilities based on a single grain size parameter which takes the following expression (Massmann, 1989):

$$k = 1,250 \times (D_{15})^2 \tag{19}$$

where: k = intrinsic permeability
              D_{15} = grain size for which 15% of particles are finer than by weight.

Other researchers have evaluated particle size distribution-permeability estimation methods such as the Hazen Method and have proposed different expressions including new parameters or modifications of existing parameters. Shepherd (1989) proposed a grain diameter exponent value of between 1.65 and 1.85 as opposed to "2" as shown in the Hazen Method. Krumbein and Monk (1942) found an empirical relationship for the permeability of unconsolidated sand as a function of mean grain diameter by weight and standard deviation. Berg (1970) added a porosity term to the equation developed by Krumbein and Monk. Gangi (1985) reanalyzed these data to relate the mean grain diameter by number. In short, many proposed estimation schemes have been proposed to relate the particle size distribution to the permeability of selected materials, yet no dependable correlation exists. Nevertheless, it is possible to roughly estimate reasonable permeability values based on particle size distribution curves, given practical experience and carefully applying predictive equations. These predictive equations can often serve as a check when attempting to diagnose out-of-control experimental results.

Sample Preparation and Introduction into the Sample Holder

Soil type factors largely in the selection of appropriate sample holding devices. As a minimum, the specimen holder should exhibit an internal diameter of at least ten times the diameter of the largest particle of the specimen, and a length of 1 to 2 times the diameter of the specimen (ASTM D 4525, 1990). Undisturbed native materials collected from boreholes at depth should be collected in cylindrical rings of equal or greater diameter to the diameter of the sample holder. Prior to testing, all samples should be visually inspected for cracks and overall sample tightness. If any macro-pores are noted to exist that are not desired or do not reflect actual conditions, the sample should be discarded or adequately trimmed to remove such features, and remeasured.

Remolded or artificially compacted test materials should be selected based on characteristics that render them suitable for air permeability testing. Materials low in clay content are amenable to artificial repacking where bulk densities and particle aggregation should remain stable in aqueous solutions (Stonestrom, 1987).

Sample Introduction-Undisturbed

For undisturbed, native state testing, after weighing and measuring the dimensions of the sample, the sample is loaded into the sample holder quickly in order to avoid evaporative moisture loss to the atmosphere. Use of a humidifying box can further minimize evaporative losses during sample introduction. Secure the end support caps over the sample and position the sample base onto the lower permeameter base. Position the top end plate over the sample holder, secure, and tighten to seal.

Sample Introduction Remolded and Recompacted

Initially, the test material should be dispersed and oven-dried at 110°C for at least 12 hours. Thoroughly homogenize the sample prior to packing within the sample holder. Several of a variety of packing procedures can be utilized (Reeve and Brooks, 1953; Stonestrom, 1987; and Springer et al., 1991) that offer reproducible results. With any methodology, however, the experimenter should strive to pack the test material as nearly homogeneous as is possible. Artificial packing of materials into sample holders is important because packing affects the pore-geometric factors which strongly influence fluid flow as revealed by theoretical analysis (Bear, 1972). After packing the sample, determine the mass of the sample plus holder and calculate the initial bulk dry density of the test material after subtracting the tare weight of the sample holder.

Saturating the Test Sample

If the experimenter desires to conduct air permeability testing at the native state water content only, skip to the next section. Otherwise, refer to the sequence described below.

Connect the sealed permeameter assembly to a deaired, water reservoir and slowly vacuum saturate from bottom to top by applying a slight vacuum to the top port of the permeameter to inhibit trapped air and the formation of air bubbles within the sample. This allows the water to slowly rise upward through the sample from bottom to top, pushing air upward and out of the sample during saturation.

Alternatively, an apparatus called a "hanging column" apparatus may be employed to saturate the test sample. Using this apparatus, a reservoir of deaired water is stored within a graduated beaker or similar reservoir attached to a ring stand. A tube runs from the outlet of the reservoir and is attached to the bottom port of the test sample holder. Initially, air is evacuated from the entire system. To initiate saturation, the elevation of the upper surface of the water reservoir is positioned even with the elevation of the lower surface of the test sample. Slowly, the elevation of the reservoir is incrementally raised by sliding the reservoir up the ring stand a measured distance, allowing water to flow into the sample to a height equal to the elevation of the reservoir. As some capillary wicking will usually occur during this procedure, incremental heights should remain small enough to avoid trapping air within the sample.

Permeameter Setup and Controlled Desaturation of Test Sample

Preliminary Setup

Following saturation, begin flowing air through the air humidifier, connect all flow lines, and calibrate and/or "zero" appropriate measurement devices. An example schematic of an air permeameter testing apparatus is shown in Figure 15.7.

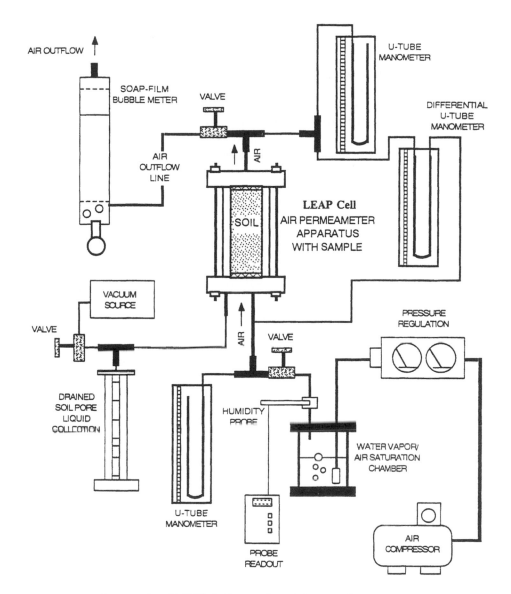

Figure 15.7. Schematic of LEAP Cell experimental apparatus.

If a flexible membrane is utilized in the permeameter, apply a positive pressure against the sleeve to seal the sample against the sample holder wall. If testing at in situ sample depths with a flexible wall permeameter, estimate the pressure exerted on the test sample due to overburden pressures and apply the corresponding air pressure against the sleeve. Allow the sample to consolidate overnight at the set delivery pressure.

Desaturation

This section applies toward permeameter designs that feature water conductive ceramics in their design, and specifically toward operation of the LEAP Cell. Apply air pressure at the base of the sample equal to the height of water within the sample (e.g., if a 5 cm length sample is being tested, apply a positive, vertically upward pressure of 5 cm) to counteract gravitational water flow downward through the sample. This approach, which is a modification of the Stationary Liquid Method (Brooks and Corey, 1964) allows the pressure difference between air and water to be equal at both ends of the sample. The air pressure at the top of the permeameter is atmospheric, as the outlet port is open to the atmosphere, while the air pressure at the bottom of the permeameter is maintained as constant through a sensitive pressure regulator. As pointed out by Corey (1986), it is critical to use a pressure regulation system capable of precise control as even small changes in the pressure of regulated air can be transmitted to the pressure sensing devices, and lead to spurious readings.

Apply a small (4–5 cm H_2O) vacuum to the port connected to the annulus of the inner ceramic and outer body tube on the LEAP Cell. This creates a gradient between the pressure within the sample and the pressure within the annulus, causing water to flow radially outward from the sample through the ceramic sample holder along the axis of flow. A fine vacuum regulator positioned between the permeameter and the vacuum source permits easy manipulation of vacuum gradients. Water should begin to visibly bead up on the outer surface of the ceramic sample holder. At this stage, note whether measurable flow is indicated by the flow sensing device. Where no flow is realized, repeat this procedure to incrementally desaturate the test sample. In practice, depending on soil texture, typically between 7% and 14% of the pore water by volume is removed before continuous flow is realized.

Taking a Measurement

At the stage where flow is first detected, measure the volumetric or mass flow rate, record the air pressure gradient applied across the sample, barometric pressure, and note the air temperature and relative humidity. After recording the first air permeability measurement, at the first air pressure gradient, increase the pressure into the sample at two progressively higher pressure gradients. Typically, for sandy materials, gradients of 5 cm H_2O, 15 cm H_2O, and 25 cm H_2O should be sufficient to generate the necessary data for air permeability calculation. Following completion of these measurements, decrease the pressure gradient back to the initial pressure gradient (5 cm H_2O in this instance) and repeat the measurement sequence. If the resultant air flow rate is off by more than 10% at the same differential pressure, it is likely that disturbance to the soil water equilibrium within the sample has occurred, and the test should be repeated.

It should be noted that at high water contents, particularly where air permeability is first measurable (i.e., the air entry pressure), desaturation tends to proceed very rapidly. As such, it is imperative that the operator maintain a constant air pressure to the base of the sample. Additionally, as the soil water content is very high at this stage, inlet air flow rates and differential pressures should be kept as low as possible, because higher air flow rates and differential pressures can give rise to physical liquid displacement. This occurs when the nonwetting fluid (air) is introduced into the test material at a high rate, while the wetting phase (water) exists as pore liquid occupying and blocking a pore throat. The pressurized air may build up and "push" the pore water to a new position, resulting in a temporary unsteady state. As a result of this condition, the menisci separating the two phases move to a new equilibrium position within the soil pores, i.e., the air phase penetrates smaller pores from which it displaces the water phase (Dullien, 1979).

Repeat this sequence at the next predetermined water content value. To assist in selecting appropriate desaturation increments, estimate the porosity of the test sample using Equation 16 presented earlier. A suggestion for calculating increments is as follows: divide the total porosity value by the total number of $k_a(\theta_v)$ test runs desired, (e.g., 40% porosity/ 5 runs = 8). Using this approximation, desaturating the volumetric water content by 8% each successive run (e.g., 32%, 24%, 16%, and 8%, and dry) should characterize the $k_a(\theta_v)$ function adequately. However, it is noted that as the water content is successively decreased within a test sample, the magnitude of k_a change with each additional desaturation increment diminishes. Therefore, since the behavior of the $k_a(\theta_v)$ function demonstrates the most pronounced changes at higher water contents, the experimenter may wish to weight the number of test runs toward the higher water content values as opposed to spacing the increments out linearly as outlined above.

Determining the Water Content

If only one $k_a(\theta_v)$ test is desired at the prevailing water content, remove the test sample from the sample holder and immediately weigh on an analytical balance. Determine the gravimetric water content according to the following equation:

$$\theta_g = [(W_1 - W_2)/(W_2 - W_c)] \times 100 = W_w/W_s \times 100 \qquad (20)$$

where: θ_g = water content, %.
           W_1 – mass of sample holder and sample, g.
           W_2 = mass of sample holder and oven-dried sample
           W_c = mass of sample holder, g.
           W_w = mass of water, g.
           W_s = mass of solid particles, g.

(after ASTM D 2216, 1990).

The gravimetric water content (θ_g) is in turn converted to the volumetric water content, (θ_v), using the following relationship:

$$(\theta_v) = (\rho_b)/(\rho_w) (\theta_g) \qquad (21)$$

where: (θ_v) = volumetric water content, %.
           (ρ_b) = bulk dry density of soil, g/cm^3.
           (ρ_w) = density of water (typically 1 g/cm^3).

Alternatively, if $k_a(\theta_v)$ testing over a range of water contents (θ_v), is desired, remove the ceramic sample holder/test sample assembly from the permeameter body and immediately weigh on an analytical balance. Repeat this procedure for each incremental water content immediately after testing. Replace the sample holder within the permeameter and repeat air permeability testing over the desired ranges of predetermined water contents. Upon completion of testing, remove the entire test sample from the sample holder, place on a weighed drying dish, and place in a laboratory drying oven set to 110°C ± 5°C. The oven dry weight of the sample (W_s) is determined upon completion of the final $k_a(\theta_v)$ test run, allowing the calculation of each incremental water content throughout the test. Since the ceramic sample-holder weight is involved in each water content calculation, it is imperative that potential evaporative water losses during removal and weighing for water content determination are minimized.

REPORTING

Experimental information associated with test results typically includes the soil type, laboratory number, and a description of the soil using a standard classification system. Additionally, associated physical test results and test methods (when available), and all raw experimental data and notes should be included as attachments to air permeability test results in a tabular format. The generated data should be expressed in standardized units, and at standard temperature and pressure conditions (STP) whenever possible. The standard notations for the k_a parameter is either in units of cm^2, or in darcies (1 darcy = 0.987^{-6} cm^2). Gas slippage corrected air permeabilities should be reported so that the slip correction is expressed as a percentage of the actual air permeability measurement. With the increased use of the k_a toward designing and engineering soil vapor extraction systems, the air conductivity value (cm/sec) is often reported alongside k_a. Equation 17 relates the air conductivity (cm/sec), to the air permeability, k_a, after inputting the appropriate values for the density, ρ, and the viscosity, μ, of air.

In the last decade, with the explosion of computer software specialty packages available to the user, data handling and graphical presentation software packages have been used increasingly to tabulate and present experimental data. Typically, graphical plots of test results with raw experimental data in tabular format are the most instructive vehicles for presenting the behavior of the $k_a(\theta_v)$ function in test materials. Reporting formats such as those presented in Figure 15.1 demonstrate the linearity of Darcy's law as a descriptive equation of gas flow. Similarly, analytical techniques including plotting experimental data using Equation 15 can be used to evaluate the nature of the flow regime, whether laminar or visco-inertial during experimentation (Detty, 1992). A more instructive method of presentation entails a simple plot of the experimentally determined k_a parameter as a function of the volumetric water content, θ_v for a given test sample (Figure 15.8). In this format, the behavior of the $k_a(\theta_v)$ is more easily conveyed, indicating a maximum value at the lowest θ_v values, and progressively decreasing with an increase in soil water content. Alternately, k_a may be plotted as a function of saturation percentage such that test results of different materials may be directly compared.

OVERVIEW

Despite the advances in permeameter design and experimental methodology that have been developed for characterizing the $k_a(\theta_v)$ function of porous materials, no singular method can effectively measure $k_a(\theta_v)$ for all soil types. Rigid wall permeameters such as those developed by API RP 40 (1960) can effectively measure air permeability in rock core samples but are limited to well indurated sediments due to the excessive compaction requirements. The rigid wall designs developed by Stonestrom (1987) and Springer et al. (1991) can effectively measure air permeability in silty and sandy materials for remolded sample types, and can be adapted to accept

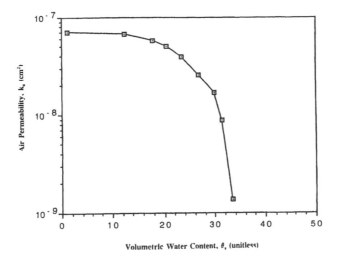

Figure 15.8. Graphical presentation of k_a versus θ.

undisturbed core samples of diameter equal to the sample holder, but have yet to be successfully tested in this configuration.

Methods to adequately seal off air "short circuiting" along sample boundaries during desaturation of clay bearing materials are limited to flexible wall designs such as proposed by Corey (1986), and triaxial cell designs. The design developed by Springer et al. (1991) permits $k_a(\theta_v)$ over a range of water contents from 0–100 kPa negative pressure without having to remove the sample from the sample holder. This allows a relaxation of the problems associated with sample disturbance during incremental desaturation, provided that evaporative losses to the ceramic sample holder are eliminated during weighing. Perhaps a hybrid design incorporating flexible wall capabilities for sample confinement with the water content manipulation capabilities of ceramics offers the most promise in future designs.

Additionally, the influence of hysteresis on affecting the outcome on air permeability test results has yet to be fully characterized. Stonestrom (1987) characterized the $k_a(\theta_v)$ function for Oakley sand over alternating wetting and drying cycles. Results plotted as k_a versus θ_v indicated the characteristic sigmoidal trace indicative of hysteresis. Within the broadest range of intermediate water contents, however, findings of alternating wetting and drying cycles demonstrated nearly linear behavior of the $k_a(\theta_v)$ function with an offset owing to hysteretic effects such as trapped air. Stonestrom (1987) also found that the $k_a(\theta_v)$ behavior of a single monotonic drying cycle from initial saturation was unique, suggesting that by standardizing testing protocol to evaluate a single drying cycle from initial saturation, test results of different materials may be compared directly without having to account for hysteretic effects.

Confidently extrapolating laboratory air permeability test results to characterize field conditions remains a difficult task. This is largely due to the inevitable disturbance imposed on soils during the sampling process potentially altering soil properties such as bulk density and structure, as well as the differences in field scale related features such as soil structure and macropore development that may not be adequately characterized during "point" sampling. To account for these differences in scale features, and owing in large part to the vast number of field sites undergoing remediation using the soil vapor extraction (SVE) technique, field air permeability testing is becoming relatively frequent and standardized. Typical field scale methods of air permeability determination entails measuring pressure fields from air piezometers imposed on subsurface soils from a centrally located soil vapor extraction well. Employing any of a variety of Darcian analog flow equations (Johnson et. al, 1988) to field-measured parameters collected during the test allows for calculation of air permeabilities. The characterization of the $k_a(\theta_v)$ function is particularly useful toward predicting and tracking changes in soil vapor extraction over time. This results because continuous operation of SVE systems tends to advectively remove soil water vapor from treatment areas over time. The net result is an overall increase in air permeabilities over time as described by the $k_a(\theta_v)$ function, providing that infiltration of precipitation and/or fluctuating ground-water levels do not significantly affect the existing soil moisture regime.

REFERENCES

ASTM. (American Society for Testing and Materials) D 4525, Standard Test Method for Permeability of Rocks by Flowing Air, Section 4 Construction, Vol. 04.08 Soil and Rock; Dimension Stone; Geosynthetics, Philadelphia, PA, 1990.

ASTM (American Society for Testing and Materials) D 2216, Standard Test Method for Laboratory Determination of Water (Moisture) Content of Soil, Rock, and Soil-Aggregate Mixtures, Section 4 Construction, Vol. 04.08 Soil and Rock; Dimension Stone; Geosynthetics, Philadelphia, PA, 1990.

API (American Petroleum Institute), Recommended Practice for Determining Permeability of Porous Media (RP 27), (Third Edition) API Production Department, Dallas, TX, 1952.

API (American Petroleum Institute), Recommended Practice for Core Analysis Procedure (RP 40), (First Edition), API Production Department, Dallas, TX, 1960.

Barrer, R.M., *Diffusion in and Through Solids.* (New York: Cambridge University Press, 1941).

Bear, J., *Dynamics of Fluids in Porous Media*, (New York: Elsevier Pub. Co., 1972).

Berg, R.R., "Method for Determining Permeability of Reservoir Rock Properties," *Trans. Gulf Coast Assoc. Geol. Soc.*, 20:303 (1970).

Boardman, C.R., and J.W. Skrove, "Distribution in Fracture Permeability of a Granitic Rock Mass Following a Contained Nuclear Explosion," *J. Petro. Tech.*, 18:5,619 (1966).

Brooks, R.H., and A.T. Corey, Hydraulic Properties of Porous Media, Hydrology Paper #3, Colorado State University, Fort Collins, CO, 1964.

Buehrer, T.F., "The Movement of Gases Through Soil as a Criterion of Soil Structure," *Ariz. Agri. Station Tech. Bul.*, 39:57 (1932).

Calhoun, J.C., and S.T. Yuster, "A Study of the Flow of Homogeneous Fluids Through Ideal Porous Media," *A.P.I. Drilling and Production Practice*, 335, 1946.

Childs, E.C., *An Introduction of the Physical Basis of Soil Water Phenomena*, (New York: John Wiley & Sons, 1969).

Corey, A.T., *Air Permeability, Methods of Soil Analysis, Part 1, Physical and Mineralogical Methods*, (Second Edition), Number 9 in the Series Agronomy, A. Klute, Ed., 1986, p. 1121.

Corey, A.T., "Measurement of Water and Air Permeability in Unsaturated Soil," in *Proc. Soil Sci. Soc. of Amer.*, 21:7 (1957).

Daniel, D.E., S.J. Trautwein, S.S. Boynton, and D.E. Foreman, "Permeability Testing with Flexible-Wall Permeameters," *J. Geotechnical Testing*, 7:3,113 (1984).

Danielson, R.E., and P.L. Sutherland, *Porosity, Methods of Soil Analysis, Part 1, Physical and Mineralogical Methods*, (Second Edition), Number 9 in the Series Agronomy, A. Klute, Ed., 1986, p. 443.

Detty, T.E., Determination of Air and Water Relative Permeability Relationships for Selected Unconsolidated Porous Materials, Master of Science thesis, University of Arizona, Tucson, AZ, 1992.

Driscoll, F.G., *Groundwater and Wells* (Second Edition), Johnson Division, St. Paul, MN, 1986.

Dullien, F.A.L., *Porous Media: Fluid Transport and Pore Structure,* (San Diego, CA: Academic Press, 1979).

Estes, R.K., and P.F. Fulton, "Gas Slippage and Permeability Measurements," *Trans. Amer. Inst. Metall. Eng.* 207:338 (1956).

Evans, D.D., and D. Kirkham, "Measurement of Air Permeability for Soil In-Situ," in *Proc. Soil Sci. Soc. of Amer.*, 14:65 (1949).

Fancher, G.H., Lewis, J.A., and K.B. Barnes, "Some Physical Characteristics of Oil Sands," Penn. State College Mineral Industries Experimental Station Bulletin 12, 65, 1933.

Fetter, C.W., *Applied Hydrogeology* (Second Edition) (Columbus, OH: Merrill Publishers, 1980).

Freeze, R.A., and J.A. Cherry, *Groundwater* (Englewood Cliffs, NJ: Prentice-Hall, Inc., 1979), p. 72.

Gangi, A.F., "Permeability of Unconsolidated Sands and Porous Rocks," *J. Geophysical Res.*, 90:B4,3099 (1985).

Gates, J.I., and T. Lietz, "Relative Permeabilities of California Cores by the Capillary-Pressure Method," *A.P.I. Drilling and Production Practice*, 1950, p. 285.

Grover, B.L., "Simplified Air Permeameters for Soils in Place," in *Proc. Soil Sci. Soc. of Amer.*, 19:414 (1955).

Hassler, G.L., U.S. Patent 2,345,935, 1944.

Hazen, A., "Discussion: Dams on Sand Foundations," *Trans. Amer. Soc. Civ. Eng.* 73:199 (1911).

Heid, J.G., J.J. McMahon, R.F. Nielsen, and S.T. Yuster, "Study of the Permeability of Rocks to Homogeneous Fluids," *A.P.I. Drilling and Production Practice*, 1950, p. 230.

Hubbert, M.K., "Theory of Ground Water Motion," *J. Geol.*, 48:785 (1940).

Johnson, P.C., M.W. Kemblowski, and J.D. Colthart, "Practical Screening Models for Soil Venting Applications," in *Proc. of the Conf. on Petrol. Hydro. and Org. Chem. in Grdwtr.: Prev., Detec., and Restor.*, J. Water Well Pub., Dublin, OH, 1990, p. 512.

Kilbury, R.K., T.C. Rasmussen, and D.D. Evans, "Water and Air Intake of Surface-Exposed Rock Fractures In Situ," *Water Resour. Res.*, 22:10,1431 (1986).

King, L.O., Mathematical Models for Underground Injection of Gaseous Wastes into the Vadose Zone. Battelle Northwest Laboratories Publ. BNWL-945, 1968.

Kirkham, D., "Field Methods for Determination of Air Permeability of Soil in Its Unsaturated State," in *Proc. Soil Sci. Soc. of Amer.*, 11:93 (1946).

Klinkenberg, L.J., "The Permeability of Porous Media to Liquids and Gases," *A.P.I. Drilling and Production Practice*, 1941, p. 200.

Klute, A., and C. Dirksen, *Hydraulic Conductivity and Diffusivity: Laboratory Methods, Methods of Soil Analysis, Part 1, Physical and Mineralogical Methods*, (Second Edition), Number 9 in the Series Agronomy, A. Klute, Ed., 1986, p.687.

Krumbein, W.C., and G.D. Monk, "Permeability as a Function of the Size Parameters of Unconsolidated Sands," *Trans. Amer. Inst. Min. Metall. Eng.*, 151:153 (1942).

Kundt, A., and E. Warburg, *Poggendorfs Ann. Physik*, 1875.

Leas, W.J., L.H. Jenks, and C.D. Russel, "Relative Permeability to Gas," *Trans. Amer. Inst. Min. Metall. Eng.* 189:65 (1950).

Leverett, M.C., and W.B. Lewis, "Steady Flow of Gas-Oil-Water Mixtures Through Unconsolidated Sands," *Trans. Amer. Inst. Min. Metall. Eng.* 142:107 (1941).

Massmann, J.W., "Applying Groundwater in Design," *J. Envir. Eng.* 115:1,129 (1989).

Muskat, M., "The Flow of Compressible Fluids Through Porous Media and Some Problems in Heat Conduction," *Physics*, 5:71 (1934).

Muskat, M, and H.G. Botset, "Flow of Gases Through Porous Media," *Physics*, 2:22 (1931).

Nimmo, J.R., J. Rubin, and D.P. Hammermeister, "Unsaturated Flow in a Centrifugal Field: Measurement of Hydraulic Conductivity and Testing of Darcy's Law," *Water Resour. Res.*, 23:1,124 (1987).

Osoba, J.S., G.G. Richardson, J.K. Kerver, J.A. Hanford, and P.M. Blair, "Laboratory Measurements of Relative Permeability," *Trans. Amer. Inst. Min. Metall. Eng.*, 192:47 (1951).

Rawlins, E.L., and M.A. Schellhardt, "Back-Pressure Data on Natural Gas Wells and Their Application to Production Practices," *U. S. Bureau of Mines Monograph*, 1936.

Reeve, R.C., and R.H. Brooks, "Equipment for Subsampling and Packing Fragmented Soil Samples for Air and Water Permeability Tests," in *Proc. Soil Sci. Soc. of Amer.*, 17:333 (1953).

Richards, L.A., "Capillary Conduction of Liquids Through Porous Mediums," *Physics*, 1:318 (1931).

Roseberg, R.J., and E.L. McCoy, "Measurement of Soil Macropore Air Permeability," in *Proc. Soil Sci. Soc. of America*, 54:969 (1990).

Shepherd, R.G., "Correlations of Permeability and Grain Size," *Ground Water*, 27(5): 633 (1989).

Springer, D.S., S.J. Cullen, and L.G. Everett, "Determining Air Permeability Under Controlled Soil Water Conditions," in *Proc. 5th Ann. Out. Act. Conf. Aquif. Rest., Gr. Wat. Monit., and Geophy. Meth.*, 5:119 (1991).

Springer, D.S., Determining the Air Permeability of Porous Materials as a Function of a Variable Water Content Under Controlled Laboratory Conditions, Master of Arts thesis, University of California, Santa Barbara, CA, 1993.

Stonestrom, D.A., Co-Determination and Comparisons of Hysteresis-Affected, Parametric Functions of Unsaturated Flow: Water Content Dependence on Matric Pressure, Air Trapping, and Fluid Permeabilities in a Non-Swelling Soil, Ph.D. dissertation, Stanford University, 1987.

Stonestrom, D.A., and J. Rubin, "Air Permeability and Trapped-Air Content in Two Soils," in *Water Resour. Res.*, 25(9):1959 (1989).

Tanner, C.B., and R.W. Wengel, "An Air Permeameter for Field and Laboratory Use," in *Proc. Soil Sci. Soc. Amer.*, 21:663 (1957).

U.S. EPA, Guide for Conducting Treatability Studies Under CERCLA: Soil Vapor Extraction (revised final draft), U.S. Environmental Risk Reduction Laboratory, Office R&D, Cincinnati, OH, 1991.

Van Bavel, C.H.M., and D. Kirkham, "Field Measurement of Soil Permeability Using Auger Holes," in *Proc. Soil Sci. Soc. of Amer.*, 13:90 (1949).

Weeks, E.P., Field Determination of Vertical Permeability to Air in the Unsaturated Zone, U.S.G.S. Paper 1051, U.S. Government Printing Office, Stock No. 024-001-03092-6, 1978.

Weldge, H.J., "A Simplified Method for Computing Oil Recovery by Gas or Water Drive," *Trans. Amer. Inst. Min., Metall. Eng.*, 195:91 (1952).

Wyckoff, R.D., and H.G. Botset, "The Flow of Gas-Liquid Mixtures Through Unconsolidated Sands," *Physics*, 7:325 (1936).

Yuster, S.T., "Homogeneous Permeability Determination," *A.P.I. Drilling and Production Practice*, 1950, p. 356.

16

Modeling Contaminant Transport in the Vadose Zone: Perspective on State of the Art

Graham E. Fogg, D.R. Nielsen, and D. Shibberu

INTRODUCTION

The increasing severity and geographic extent of subsurface contamination problems has resulted in growing demands for predictive models of transport through both the vadose and saturated zones. Consequently, the list of available approaches and computer codes for modeling transport has grown steadily (e.g., van der Heijde and Elnawawy, 1993; Pennell et al., 1990). The practitioner must make an informed decision of which code to use based on (1) the specific nature of the problem, (2) the capability of the algorithm and code, and (3) his or her experience. This chapter is intended to assist the practitioner in assessing the capability of available methods by providing a perspective on state-of-the-art of vadose zone modeling approaches. While we do not catalog and categorize available codes, as this was done recently by van der Heijde and Elnawawy (1993), we do review some of the key assumptions inherent to various modeling approaches to offer a framework for assessing their strengths and weaknesses. Such a framework would be less important if there existed one body of widely accepted theory that provided a foundation upon which to base transport modeling and prediction. On the contrary, considerable debate remains in the scientific literature as to the validity of various assumptions and theories underlying the modeling approaches for the unsaturated zone. This should not preclude

---

This chapter is an adaptation of the California State Water Resources Control Board (SWRCB) report, "A Review of the State of the Art: Predicting Contaminant Transport in the Vadose Zone," which was prepared for the SWRCB by Fogg et al., and is used with permission of the SWRCB.

prudent application of the models, but nevertheless strongly dictates the appropriate modes of application and interpretation.

A major recurrent theme in the literature is the need for more detailed field studies to better understand basic vadose zone processes that affect contaminant transport (e.g., Wierenga, 1989; Glass et al., 1989). The lack of such studies has meant that available theoretical transport models are "largely untested" (Butters and Jury, 1989), and considerable uncertainty over the validity of particular modeling approaches remains. Mantoglou and Gelhar (1987) state, "even the most basic behavior of large-scale unsaturated flow systems is presently not well understood."

In pointing out some of the shortcomings of vadose zone modeling techniques we are by no means advocating avoidance of these techniques. In fact, we believe that owing to the complexity of many field situations, application of sophisticated numerical or analytical models is requisite. Importantly though, the theoretical weaknesses and inadequate data in the typical vadose-zone transport model dictates that specific predictions by the model will be inaccurate. Indeed, the same can be said of most saturated-zone models. It follows logically that prediction or forecasting is not the chief strength of vadose-zone models; rather, the greatest advantage of vadose-zone models is the scientific framework they provide for obtaining a better understanding of a particular system. This better understanding often comes from model sensitivity analysis to determine consequences of fluctuations in model parameters and boundary conditions. Clearly, most vadose zone models should be viewed as quantitative conceptual models, and the actual predictions must still be made by the hydrologist, based on some combination of judgment, experience, and model results. In general, we use more sophisticated models in order to make fewer assumptions with regard to the flow or transport processes.

In view of the uncertainty of vadose-zone transport models, this chapter first reviews commonly employed model assumptions before considering the models themselves. Following the classification of Butters and Jury (1989), the models are organized into three groups: models based on the convection-dispersion equation (CDE) with constant coefficients (deterministic); models based on the CDE with random (spatially correlated), variable coefficients (stochastic); and transfer function models. Next, the problems of site characterization and the determination of input parameters are covered. Finally, the issue of model validation is discussed using relevant case studies.

MODEL ASSUMPTIONS

A comprehensive review of vadose zone processes is provided by Nielsen et al. (1986) in which the basic unsaturated flow equations are outlined and simplifying assumptions used to solve the equations for solute transport modeling are identified. The major assumptions are discussed below.

Steady-State Flow and Hysteresis

Most vadose zone models assume steady-state water flow even though this condition is rare near the soil surface in the field. Nearly 30 years ago in a comprehensive field experiment, Miller et al. (1965) showed that movement of solutes in the vicinity of the soil surface may indeed be manipulated or even controlled by the method of water application. The frequency and magnitude of each water application alters leaching efficiencies even for equal total amounts of water eventually applied. An understanding of such dynamic processes was clearly articulated nearly a century ago by Means and Holmes (1901). Their lucid description of leaching, based on a career of observing field conditions as they mapped virgin and cultivated soils in the western U.S., manifests an understanding of solute with dual porosity models (e.g., Gerke and van Genuchten, 1993). Nevertheless, Nielsen et al. (1986) felt that the general features of transport involving "long-term rates and amounts of solute leaving the upper part of the vadose zone" may be adequately described assuming steady-state water flow. Validity of the steady-state assumption for short-time processes, however, is questionable (Russo et al., 1989a). Experiments by Bowman and Rice (1986) analyzing solute transport under intermittent flood irrigation indicated water velocities much greater than predicted from water balance methods. Butters et al. (1989) also found large discrepancies between predicted and observed water velocities. Russo et al. (1989a) investigated the effects of hysteresis and vertical heterogeneity on transient solute transport. They concluded that under monotonic conditions, the steady-state water flow model could considerably overestimate the effective vertical solute velocity, while under nonmonotonic conditions, hysteresis became an important factor that caused retardation of solute movement near the surface zone.

Magnitude of the Anisotropy in the Hydraulic Conductivity

According to McCord et al. (1989), the possibility of variable anisotropy in the hydraulic conductivity has never been considered for predictive vadose zone modeling. Results of their field experiments using tracers in a large sand dune demonstrated variable anisotropy at the macroscopic scale, thereby confirming the results of other researchers (e.g., Yeh et al., 1985). Methods of accounting for this behavior were discussed; however, modeling efforts remain in their initial stages. The implications for significant lateral spreading during contaminant transport were noted by Stephens and Heermann (1988) and Yeh (1989) using results from field experiments.

One-Dimensional Flow and Transport

Most models of flow or transport in the unsaturated zone have been vertical and one-dimensional, ignoring velocity components in the horizontal plane. This arises from the fact that, due to gravity drainage, direction of flow in the vadose zone is predominantly downward. Prevalence of vertical drainage and the commonly accepted view that soil properties are structured in laterally extensive layers has led to a certain degree of complacency about appropriateness of one-dimensional models. In reality, soil or sediment layers are laterally discontinuous, and spatial variability in both recharge rates and hydraulic properties can lead to substantial horizontal flow in the vadose zone (Mantoglou and Gelhar, 1985). When field-monitoring data are interpreted solely within the confines of a one-dimensional model, a number of erroneous conclusions can result, including incorrect travel times and spurious assertions regarding the roles of other processes such as sorption and biodegradation. While such models may be "tweaked" to fit the data, they may serve as especially poor forecasters of future system behavior because they misrepresent multidimensional processes.

Fick's-Law Dispersion

In theory, the water moves with velocity v inside the pore of the porous medium; however, the v that we measure in the laboratory column or in the field represents an average of a number of local scale processes that includes molecular diffusion, tortuous flow around the solid portions of the porous medium, and tortuous flow due to heterogeneities larger than the pore scale. These local-scale deviations of fluid and solute velocities from the average v result in additional spreading and dilution of solutes, which is called dispersion. Convection-dispersion models attempt to account for this process by assuming that it is analogous to Fick's law of diffusion (concentration is proportional to concentration gradient times diffusion coefficient [D_0]), and substituting a hydrodynamic dispersion coefficient, D for D_0. D is intended to account for all deviations in local v from the average v due to physical processes of diffusion and heterogeneity in the porous medium. The Fick's law assumption is tantamount to assuming the deviations in local v from average v are characterized by a Gaussian (normal) distribution, and the rate of spreading is scale-independent.

The Fick's law assumption has been called into question by a number of investigators (e.g., Sposito et al., 1986; Matheron and de Marsily, 1980; Anderson, 1979) because: (1) the mechanical mixing term in D, dispersivity, has been observed to grow with the scale of measurement and with time of transport; (2) theoretical studies based on analytical stochastic approaches under fairly ideal conditions (Matheron and de Marsily, 1980; Gelhar and Axness, 1983; Dagan, 1984; Güven et al., 1984) have demonstrated that dispersivity only approaches a "Fickian" constant after the plume spreads over a significant distance that is equivalent to many characteristic mixing lengths of the heterogeneous porous medium, which could involve considerable distances and times (10s to 100s of years); and (3) knowledge of subsurface geometries of soil and geologic units suggests that deviations in v from the average would often be non-Gaussian (hence non-Fickian), and that correlation scales of the heterogeneities encountered by the plume typically increase as the plume encounters larger volumes of the porous medium (Fogg and Kreitler, 1981; Neuman, 1994). Non-Fickian behavior is especially easy to visualize when the porous medium contains preferred pathways such as interconnected macropores or high-permeability channels that facilitate relatively rapid transport.

Use of Adsorption Isotherms

The use of the distribution coefficient and adsorption isotherms to characterize reactive chemical transport is critically evaluated by Liu and Narasimhan (1989a). While this approach may be appropriate under certain conditions, simulation of contaminant movement that involves multi-species ion exchange reactions requires consideration of various other chemical reactions including acid-base, oxidation-reduction, and precipitation-dissolution that may significantly affect the concentrations of the sorbed species. Valocchi (1985) discusses conditions under which the local equilibrium assumption is justified. Models that can account for these reactions are usually limited to one dimension (e.g., Theis et al., 1981). Liu and Narasimhan (1989a,b) argue that a two-dimensional numerical grid is the minimum required for problems that involve migration of contaminants from a landfill. Their model is described as the first to be able to account for oxidation-reduction reactions in a three-dimensional field.

Even if models that could explicitly account for the significant chemical reactions were widely available, use of the models would be constrained by the lack of input data concerning attenuation rates for specific contaminants. According to a literature review by Rai and Zachara (1984), "quantitative predictions of chemical attenuation rates based upon mineralogy and groundwater composition cannot be made because only descriptive and qualitative information are available for adsorption/desorption mechanisms."

Over-reliance on contaminant adsorption isotherms for predictive purposes is criticized from a different perspective by Gruber (1990), who stressed the importance of considering the potential for contaminant accumulation and subsequent remobilization due to the subsequent intrusion by water with appropriate chemical characteristics.

Most models require certain combinations of the above assumptions since attempting to explicitly account for every significant flow and transport process would introduce unmanageable complexity without guaranteeing accuracy for any particular situation.

Assumption of Instantaneous Adsorption

Use of this assumption for transport modeling is being seriously questioned for a number of field conditions such as the presence of strongly adsorbing species or water flow under nonsteady conditions created by natural rainfall and man-induced infiltration events. A summary of the development of "physical" and "chemical" nonequilibrium models to account for deviations from observed behavior is given by Nielsen et al. (1986). Recent experimental evidence for the occurrence of mobile/immobile water partitioning is given by De Smedt (1989), who found such conditions during constant rate infiltration but not during flood infiltration. A model that explicitly accounts for both physical and chemical nonequilibrium was presented by Brusseau et al. (1989). The model described the behavior of herbicide and pesticide transport stemming from three different data sets.

VADOSE ZONE TRANSPORT MODELS

Transport models are classified herein as either deterministic, stochastic, or based on a transfer function concept. In a deterministic model the coefficients and boundary conditions are assumed to be known exactly throughout the model domain and, as such, the model produces only one set of outcomes per simulation. The user can nevertheless rerun the deterministic model many times to test effects of different inputs on model outputs.

In a stochastic model, the coefficients and/or boundary conditions are assumed to be uncertain in time or space based on some statistical construct. Stochastic models consequently produce not just one outcome but a statistical distribution or range of outcomes. Monte Carlo simulation is one type of stochastic simulation accomplished by systematically rerunning a deterministic model with many different (often 100s of) spatial distributions of parameters and/or boundary conditions in an attempt to fully characterize the probability of all possible outcomes. The stochastic approaches offer a way to deal rigorously with the uncertainty that is inherent to characterizations of the subsurface. Transfer function models will be discussed later.

Transport models can also be categorized as either analytical, numerical, or a combination thereof. The analytical models are based on a closed-form solution of the governing equation subject to relatively simple boundary conditions and spatially homogeneous or simple layered properties. These models can provide first-cut answers to more complex problems and can provide a great deal of insight into transport processes. Numerical models are typically based on finite difference or finite element techniques and can handle more complex variations in parameters and boundary conditions. Consequently, the numerical models are often better suited for modeling complex real-world scenarios than are analytical models. On the other hand, numerical models require more input data and may be more cumbersome to apply. Users of finite difference or finite element transport models based on solution of the convection-dispersion equation should be aware that the numerical approximations can lead to significant errors due to numerical (artificial) dispersion (Lantz, 1971). This problem can be avoided by using a different strategy based on particle-tracking methods (Konikow and Bredehoeft, 1978; Tompson, 1993).

Another type of model, often called the "dual-porosity model," has been designed to simulate those environments characterized by preferential flow through fractures, cracks, and macropores or interaggregate pores wherein advance of the water and solute can be highly nonuniform and non-Fickian. For recent research in this area and review of previous work, refer to Gerke and van Genuchten (1993a,b).

Convection-Dispersion Equation with Constant Coefficients (Deterministic)

The Committee on Ground Water Models Assessment (1990), in reference to the development of stochastically based methods to better characterize heterogeneity and dispersion, felt that subsurface hydrology is "on the threshold of a significant change in how the subsurface environment is interpreted." However, the classical deterministic theories are also being further developed and extended, allowing for greater sophistication in simulating and predicting contaminant transport.

The redox-controlled multiple species model by Liu and Narasimhan (1989a,b) serves to illustrate this point. Their DYNAMIX model is significantly advanced in its ability to efficiently model complex, reactive transport in two or three dimensions and make long-term predictions of contaminant attenuation. The model is applicable to both the saturated and unsaturated zones; however, only saturated cases were simulated.

With specific regard to the unsaturated zone, Selim et al. (1989) state that "mathematical models that describe transport of heavy metals in laboratory soil columns or in soil profiles under field conditions have only been recently introduced into the literature." They present a one-dimensional "nonlinear retention/release" model for chromium transport that assumes steady-state flow and equilibrium sorption. Tests of the model's predictive capabilities demonstrated its validity for some soils but not for other highly reactive soils. The model introduced by Brusseau et al. (1989) is less restrictive than the Selim model because it does not use the local equilibrium assumption. The model was apparently validated through independent predictions of published data sets for pesticide/herbicide transport; however, applicability of the model to inorganic contaminants subject to complex reactions is unclear. Lindstrom and Piver (1986) extended a two-dimensional ground-water model for application to problems involving transport of chemicals with low solubility at typical landfill sites. The model was intended for "realistic, long-time simulations" of transport in heterogeneous soils with variable rainfall.

Generally, the simulation and prediction of the fate of pesticides in the soil appear to have received more attention than other problems in the unsaturated zone. Jones et al. (1986) describe and apply the Environmental Protection Agency's widely used PRZM mass balance model for pesticides, finding it effective in identifying areas where pesticide application would negatively impact ground water. Wagenet and Hutson (1986) argue that the simplifying assumptions of the PRZM model have not been adequately tested. They present a research model intended to more accurately represent the processes involved in pesticide transport. The model was field tested for the case of aldicarb attenuation, and the approach was described as having "substantial promise" for predictive modeling of the fate of pesticides.

Another research area where the classical CDE is being extended is the area of multiphase flow. Abriola and Pinder (1985a,b) developed a one-dimensional model to simulate three-phase, air-water-nonaqueous phase liquid (NAPL) flow. A number of simplifying assumptions employed in developing the model (e.g., local equilibrium, neglect of hysteresis and gas phase convection) need to be investigated further. The authors (Pinder and Abriola, 1986) stated "there is still a great amount of data (with respect to quantifying input parameters) which must be collected before prediction of the multiphase migration of contaminants becomes feasible." Faust et al. (1989) presented a two-phase, three-dimensional flow model for NAPLs which was used to study contamination problems at two chemical waste landfills. Predictive simulations of NAPL movement were in general agreement with subsequent observations from bedrock drilling. Other results identified geohydrologic conditions that could act as barriers to the downward migration of NAPLs, and demonstrated the necessity of three-dimensional simulations for certain field problems.

Convection-Dispersion Equation with Random Variable Coefficients (Stochastic)

Despite the growing sophistication of models based on the classical CDE, the Committee on Ground Water Modeling Assessment (1990) stated that "there are very few documented cases for which deterministic solute transport models have been successfully applied to groundwater contamination problems involving complex chemical reactions." Success may depend strongly on inclusion of the spatial and temporal variability of large-scale flow and transport properties. In order to account for effects of this variability, stochastic concepts have been incorporated into the governing equations.

Van Genuchten and Jury (1987) classify stochastic approaches into three groups: scaling theories, Monte Carlo methods, and stochastic continuum models. In the scaling approach, soil water behavior in one soil is described using experimental or computed soil water variables from another soil that have been scaled appropriately (Clausnitzer et al., 1992). The Bresler-Dagan model (1983) is an often-cited example that uses the scaling approach. It was shown to be "quite accurate" in calculating statistical moments for conservative solute concentrations in spatially variable field soils, while it lacked accuracy in determining concentrations for particular values of the saturated hydraulic conductivity. The model is evaluated in detail by Sposito et al. (1986) due to its "completeness and demonstrated relevance to water quality management." Nevertheless, several of the critical assumptions used in the model are questioned. Mantoglou and Gelhar (1987) and Yeh et al. (1985) feel that the Bresler-Dagan model is inappropriate for waste leakage predictions, owing to its representation of the soil as a collection of one-dimensional, noninteracting columns.

Despite criticism to the contrary, Monte Carlo simulation often offers the best stochastic technique for addressing site-specific scenarios. Monte Carlo simulation has been criticized in comparison to stochastic continuum models (e.g., Dagan, 1989) as a predictive tool owing to (1) large demands on computer time and memory; (2) problems with achieving and verifying convergence, and (3) the difficulty of drawing general, fundamental conclusions from the results. Owing to exponential advances in computer hardware, the large demands on computer time and memory are now fairly tractable for many cases. "Problems with achieving and verifying convergence" refers to the issue of verifying that the statistical outcomes produced by the Monte Carlo model will converge on or adequately represent a particular stochastic partial differential equation. This problem can be of considerable concern when the purpose of the Monte Carlo modeling is to produce results that are fundamental to transport in a broad class of vadose-zone scenarios that are relatively simple with respect to style of heterogeneity and boundary conditions. However, when we wish to predict behavior of a real-world system having non-Gaussian distributions of properties, hysteresis, irregular geometries, spatially or temporally varying boundary conditions, etc., it is often necessary to characterize and represent the actual field scenario as closely as possible. Typically, the best way of accomplishing this is through Monte Carlo simulation implemented on a deterministic numerical model. The deterministic numerical model offers the capability of simulating complex field conditions, and various possible patterns in parameters or boundary condition for the Monte Carlo procedure can be generated externally by geostatistical or other methods (Fogg et al., 1990; Deutsch and Journel, 1992). The stochastic continuum models can provide invaluable information needed to intelligently design the Monte Carlo experiments.

Persaud et al. (1985) used Monte Carlo simulations to model one-dimensional, conservative solute transport in heterogeneous soils. The main goal was to use the simulations to generate probability density functions (pdf) of dispersivity and velocity which are not yet available, owing to a dearth of such observations in the field. Im (1990) used three Monte Carlo models to predict solute concentrations and arrival times (using confidence limits in the mg/L range) for hypothetical groundwater contaminant transport problems that included hydrolysis and biodegradation. Eleven input parameters were parameterized using 1 of 6 pdf's. A sensitivity analysis showed the importance of selecting representative pdf's and the need for more information on reaction rate constants.

Stochastic continuum models have been employed in the vadose zone by Parker and van Genuchten (1984), by Yeh et al. (1985), and Mantoglou and Gelhar (1987). It is important to remember that these models provide information on the mean behavior of the plume and provide little or no information on the details of flow. This further highlights the point that stochastic continuum models are primarily basic research tools and are not intended for detailed modeling of specific sites.

The Parker and van Genuchten model uses noninteracting, one-dimensional soil columns—an approach that the authors acknowledge is "likely not valid" for many situations. The most significant result of the approach of Yeh et al. (1985) studying unsaturated flow was the identification of moisture-dependent anisotropy as an important factor that could cause increased horizontal migration of pollutants. While Yeh's model was steady-state, Mantoglou and Gelhar (1987) produced a transient model for a stratified soil and gained insights into hysteresis, variable anisotropy, and effective unsaturated hydraulic conductivity.

Transfer Function Models

The transfer function model (TFM) approach was introduced into the vadose zone literature by Jury (1982) with the intention of avoiding the problems of measuring the complexly varying hydraulic and retention parameters of a field soil. This relatively simple stochastic approach relates inputs of solute mass into a transport volume of porous medium to losses of solute mass from a soil unit (Jury et al., 1986) without modeling directly the various physical and chemical processes operating within the transport volume. A philosophical premise of the TFM is that the spatially and temporally varying physical and chemical process operating within the transport volume is too complex to characterize and model accurately. The TFM is therefore posed as a generalized model that allows prediction of solute transport rates based on a calibrated relationship between observed solute inputs and outputs. Advantages of the technique are that it is conceptually simple and requires little characterization of the porous medium. A disadvantage is that the required calibration data on solute input and outflow can often only be measured over short space and time scales owing to slow migration of tracers in the subsurface. Consequently, representation of field-scale or regional-scale transport may require considerable extrapolation of local-scale results. Given that the TFM does not actually model the processes operating within the transfer volume, such extrapolation may lead to erroneous results (Mantoglou and Gelhar, 1985).

Jury's (1982) initial TFM was used in a field test to predict solute movement down to a depth of 3 m after calibration at the 30-cm depth. The model was further developed by Jury et al. (1986) to account for solutes that are subject to chemical and biological transformations in the soil. They derived an equation that related solute mass input rates to solute mass loss rates. In a companion paper, White et al. (1986) analyzed conservative solute transport through laboratory columns using TFM but did not attempt to predict breakthrough curves.

Predictive applications of the TFM were explored by White (1987) for the case of nitrate leaching. Data from observed leaching events were used to estimate the probability density function of solute travel times and to calibrate the TFM; however, methods of extending the TFM for predictive uses were only discussed in general terms.

Butters and Jury (1989) used experimental data from a large field experiment to compare the predictive performances of the 1982 Jury TFM and a one-dimensional, deterministic convection-dispersion model. Two cases were considered: predicting breakthrough curves down to a depth of 4.5 m after calibration at 30 cm, and predicting the entire solute travel depth profile at 300 days to a depth of 25 m after calibration at 4.5 m. In the first case, the TFM model provided "excellent representations" of the observed breakthrough curves to a depth of 1.8 m, while the CDE performed poorly at depths greater than 0.9 m. For the entire profile, the CDE better captured the shape of the solute distribution, but the maximum solute depth was underpredicted by at least 6 m. The reverse was true for the TFM, which accurately estimated the leading edge of the solute front but which failed to closely approximate the observed concentrations. These results were explained in terms of dispersion—the CDE better described the "far field regime" where the effective dispersion nears an asymptotic value, while the TFM better describes the "near field regime" where effective dispersion is growing linearly.

In an approach similar to the Jury TFM, Beven and Young (1988) presented an aggregated mixing zone (AMZ) model which allows greater flexibility in determining the form of the transfer function based on additional experimental data. In studies using lab columns, the AMZ model was especially successful in fitting breakthrough curves that are traditionally difficult to model. However, testing of the model for predictive uses was not performed.

SITE CHARACTERIZATION AND INPUT PARAMETERS

A major constraint on the application of any transport model is the difficulty in obtaining the data needed for accurate calibration. In a literature review of parameter estimation methods, Kool et al. (1987) state "difficulties in model calibration are nowhere more evident than in analyses of water and chemical transport in the vadose zone." They proceed to detail more promising parameter estimation methods, but state that application of these methods to the vadose zone has only begun. Mishra and Parker (1989) discuss two separate parameter estimation methods. Application of the first inversion technique to a hypothetical system yielded "excellent" predictions of parameters. The second technique, based on particle size distribution data for a field soil, was less successful in determining observed soil parameters.

Another basic problem in analysis of actual transport systems is characterization of the scale of the system in order to ensure representative data measurements and appropriate modeling approaches. Yeh et al. (1985) state that although unsaturated flow is adequately described by the classical governing equations for small scales such as laboratory columns, problems such as waste disposal evaluation "require quantification at much larger scales, e.g., hundreds of meters or more."

The issue of "how to extrapolate the classical small scale behavior to the pertinent field scale at which the soil parameters exhibit complex natural heterogeneity" remains a formidable obstacle. This same problem has been faced by ground-water hydrologists and petroleum investigators. They have implemented promising techniques involving use of geologic conceptual models in conjunction with geostatistical methods for modeling or simulating spatial variability in a Monte Carlo fashion (Lake and Carroll, 1986; Lake et al., 1991; Fogg et al., 1991; Koltermann and Gorelick, 1992). The underlying premise is that the unknown spatial patterns in transport processes cannot be adequately estimated unless one adequately understands what processes created the patterns. Below the root zone, these patterns are dictated by geologic processes (e.g., deposition of alluvial fans or fluvial materials); and because these processes are predictable based on observations of modern depositional systems, the geologic model can be used as a template for estimating spatial patterns in transport properties or for estimating statistical parameters of stochastic continuum or geostatistical models. Although this approach has been shown to yield more realistic simulations of displacement of oil by water (Fogg, et al., 1991; Hewett and Behrens, 1991), estimation of the geological heterogeneity at the pertinent scales remains a formidable task that will become gradually more tractable only through many more intensive characterizations of subsurface sites and of outcrop exposures. Furthermore, much additional research is needed on the problem of estimating transport properties based on geological textural attributes. Geostatistical methods for stochastically simulating spatially varying geological attributes or hydraulic properties based on "hard" (measured) and "soft" (inferred) data are already fairly well-developed (Deutsch and Journel, 1992).

Warrick and Yeh (1989) discussed the effects of scale on the determination of parameters such as hydraulic conductivity and dispersivity. The use of fractal geometry in calculating dispersivity was evaluated and though the approach was found "promising," the authors stress the need to verify the underlying assumption of self-similarity and reconcile the theory with experimental evidence that indicates asymptotic behavior for the dispersivity. Butters and Jury (1989) detailed the effects of scale by comparing average statistical moments at the local (sample) scale with the estimated field-scale values, finding significant differences especially for the dispersivity.

A conceptual framework for analysis of length scales is provided by, among others, Weber (1986) and Dagan (1986, 1989). Dagan recognized three fundamental length scales. The first (L) is divided into three basic representative scales: laboratory, local and regional, that are employed to "characterize the extent of the flow and transport domain." The second fundamental length scale is the correlation scale (I) that is associated with heterogeneity, while the third fundamental scale (D) is associated with the scale of measurement and sampling. Identification of the relevant length scales and their configuration is important in order to correctly employ geostatistical methods and apply simplifying assumptions to any transport model. Lake and Carroll (1986), Lake et al. (1991) and Fogg (1990) include examples of how geological approaches can be used to estimate correlation scales.

CASE STUDIES, MODEL VALIDATION

It is generally agreed that there is a great need for detailed field studies of vadose zone models, since, as Stein and McTigue (1989) state, "few data are available that show the evolution of a contaminant plume in the unsaturated zone." Ongoing, large-scale field experiments have only recently been undertaken to provide the necessary data for field validation of vadose zone models.

Difficulties in model validation are already evident in laboratory studies. Glass et al. (1989) examined the migration of conservative solutes in laboratory columns by calibrating a one-dimensional model with one column to predict solute transport in other columns. The predictions were especially poor when a small flow rate (such as encountered in the field) was used. Moreover, the model performed poorly even under settings only slightly different from those used for calibration. The authors concluded that extensive model validation and subsequent continued field monitoring are necessary to achieve accurate predictions of solute transport.

Focusing on the issue of actual field validation, Stein and McTigue (1989) presented results of their efforts to analyze and simulate chromium contamination in the unsaturated alluvium beneath a waste site. A one-dimensional, steady-state flow model coupled with a parameter estimation code was used. Modeling the observed concentrations (two data sets collected during 1983 and 1987) was seriously hindered by problems such as unknown surface boundary conditions, the presence of material heterogeneities at length scales comparable to the sample spacing, and the "extremely slow" transport processes, which meant that the sampling time scale was small compared to the "characteristic time for the evolution of the contaminant field." The authors felt that this last factor caused the greatest difficulty.

Trautwein and Daniel (1983) presented a case study concerning leakage from an evaporation pond. The focus of their investigation was on the likelihood of waste-water migrating through the 390-ft vadose zone and contaminating an underlying aquifer. Field and lab studies identified a contaminant front at a depth of 308 ft. To predict future contaminant migration, the worst-case scenario of no attenuation was assumed, and a one-dimensional, transient flow model was calibrated and used to predict the arrival time of the contaminant front. The general behavior and order of magnitude of arrival times were judged successful. On the other hand, criteria to judge the success of such predictions have generally not been established over speci-fied distances and times, and even if they were for specific cases, our ability to adequately monitor the vadose zone to guarantee the satisfaction of the criteria has not yet been proved.

CONCLUDING REMARKS

Despite the abundance of models that simulate flow in the vadose zone, there do not appear to be any multidimensional models that can accommodate a "multicon-stituent waste migrating through a heterogeneous vadose zone under uncontrolled hydraulic heads." The DYNAMIX model (Liu and Narasimhan, 1989a,b), which essentially represents the state-of-the-art for deterministic models, has only been applied to the saturated zone with fixed hydraulic head conditions. Additionally, the simulation did not consider heterogeneity. Contaminant attenuation was predicted for concentrations in the parts-per-million range; however, no confidence limits were provided.

The stochastic model developed by Im (1990) provided contaminant concentra-tion predictions in the ppm range along with 95% confidence limits. Although the model did allow for heterogeneity, it was limited to one dimension and was only applied to the saturated zone. Generally, application of stochastic models to con-taminant transport modeling in the vadose zone is still in the preliminary stages. For example, the Dagan stochastic model, despite its successful application at the Bor-den site, has not yet been applied to reactive solutes even in the saturated zone.

Transfer function models remain essentially untested for contaminant transport modeling. The ability of the Jury TFM to accurately predict the leading edge of a conservative solute front may not imply a similar capability for nonconservative solutes. Furthermore, the lack of representation of transport processes in the TFM leaves little scientific basis for extrapolating local-scale results to other space and time scales.

Because the reliability of models for contaminant transport has not been estab-lished even for site specific conditions, it appears that direct monitoring of the constituents in the vadose zone remains a necessity into the foreseeable future. Monitoring provides the opportunity to assess changes in concentrations of specific constituents at particular locations for better management alternatives, as well as to enhance the databases upon which our understanding rests. Although an exact measure of the water velocity or the position of a solute within the vadose zone is difficult to ascertain, monitoring allows observation of relative changes of concen-tration within prescribed limits of analytical and localized sampling procedures. Costs of sampling the vadose zone can be greater than those of sampling the satu-rated zone owing to the more complex behavior of fluids and solutes in the unsatu-rated subsurface environment. First, because water in the vadose zone generally exists at subatmospheric pressures, energy other than gravitational forces must be supplied to extract a sample of vadose water. Second, the subatmospheric condition requires the imposition of a porous medium between the extracted and resident water. This porous material has the possibility of reacting with the constituents to be monitored. Third, the time required to allow the resident fluid to move through the unsaturated zone to the porous material collector increases the cost of each sample. Small sample volumes require special, often expensive, handling and storage procedures and limit the number of different kinds of constituents to be quantified.

REFERENCES

Abriola, L.M., "Modeling Contaminant Transport in the Subsurface: An Interdisciplinary Challenge," *Reviews of Geophysics*, 25:125-134 (1987).

Abriola, L.M., and G.F. Pinder, "A Multiphase Approach to the Modeling of Porous Media Contamination by Organic Compounds. 1. Equation Development," *Water Resour. Res.*, 21:11-18 (1985a).

Abriola, L.M., and G.F. Pinder, "A Multiphase Approach to the Modeling of Porous Media Contamination by Organic Compounds. 2. Numerical Simulation," *Water Resour. Res.*, 21:19-26 (1985b).

Anderson, M.P. "Using Models to Simulate the Movement of Contaminant Through Groundwater Systems," *CRC Critical Reviews in Environmental Control*, 9:97-156 (1979).

Beven, K.J., and P.C. Young, "An Aggregated Mixing Zone Model of Solute Transport Through Porous Media," *Journal of Contaminant Hydrology*, 3:129-144 (1988).

Bowman, R., and R. Rice. "Transport of Conservative Tracers in the Field Under Intermittent Flood Irrigation," *Water Resour. Res.*, 22:1531-1536 (1986).

Bresler, E., and G. Dagan, "Unsaturated Flow in Spatially Variable Fields. Solute Transport Models and Their Application to Two Flow Fields," *Water Resour. Res.* 19:429-435 (1983).

Brusseau, M.L., R.E. Jessup and P.S.C. Rao, "Modeling the Transport of Solutes Influenced by Multiprocess Nonequilibrium," *Water Resour. Res.*, 25:1971-1988 (1989).

Butters, G.L., W.A. Jury and F.F. Ernst, "Field Scale Transport of Bromide in an Unsaturated Soil 1. Experimental Methodology and Results," *Water Resour. Res.*, 25:1575-1581 (1989).

Butters, G.L. and W.A. Jury, "Field Scale Transport of Bromide in an Unsaturated Soil, 2. Dispersion Modeling," *Water Resour. Res.*, 25:1575-1581 (1989).

Clausnitzer, V., J.W. Hopmans, and D.R. Nielsen, "Simultaneous Scaling of Soil Water Retention and Hydraulic Conductivity Curves," *Water Resour. Res.*, 28:19-31 (1992).

Committee on Ground Water Modeling Assessment, *Ground Water Models: Scientific and Regulatory Applications*. (Washington, DC: National Academy Press, 1990).

Dagan, G., "Solute Transport in Heterogeneous Porous Formations," *Journal of Fluid Mechanics*, 145:151-177 (1984).

Dagan, G., "Statistical Theory of Groundwater Flow and Transport: Pore to Laboratory, Laboratory to Formation, and Formation to Regional Scale," *Water Resour. Res.*, 22:120S-134S (1986).

Dagan, G., *Flow and Transport in Porous Formations*. (Heidelberg: Springer-Verlag, 1989).

De Smedt, F., "Transient Movement of Water and Solutes in an Unsaturated Porous Medium with Mobile and Immobile Water Phases," *Proceedings of the International Conference* and Workshop on the Validation of Flow and Transport Models for the Unsaturated Zone, P.J. Wierenga and D. Bachelet, Eds., New Mexico State University, 1989, pp. 64–70.

Deutsch, C.V., and A.G. Journel, GSLIB: Geostatistical Software Library and User's Guide, (New York: Oxford University Press, 1992).

Faust, C.R., J.H. Guswa, and J.W. Mercer, "Simulation of 3-Dimensional Flow of Immiscible Fluids Within and Below the Unsaturated Zone," Water Resour. Res., 25:2449–2464 (1989).

Fogg, G. E., "Emergence of Geologic and Stochastic Approaches for Characterization of Heterogeneous Aquifers," Proceedings of the Conference: New Field Techniques for Quantifying the Physical and Chemical Properties of Heterogeneous Aquifers, Dallas, TX, March 20-23, 1989, pp. 1-17.

Fogg, G. E., "Architecture and Interconnectedness of Geologic Media: Role of the Low-Permeability Facies in Flow and Transport," in Hydrogeology of Low Permeability Environments, S. P. Neuman and I. Neretrieks, Eds., (Hannover, Germany: Verlag Heinz Heise, 1990), pp. 19–41.

Fogg, G. E., and C. W. Kreitler, "Ground-Water Hydrology Around Salt Domes in the East Texas Basin: A Practical Approach to the Contaminant Transport Problem," Association of Engineering Geologists Bulletin, 18:387–411 (1981).

Fogg, G. E., F. J. Lucia, and R. K. Senger, "Stochastic Simulation of Interwell-Scale Heterogeneity for Improved Prediction of Sweep Efficiency in a Carbonate Reservoir," in Proceedings of the 2nd International Conference on Reservoir Characterization, L. W. Lake and H. B. Carroll, Jr., Eds., (San Diego, CA: Academic Press, 1991), pp. 355–381.

Gelhar, L.W., and C.L. Axness, "Three-Dimensional Stochastic Analysis of Macrodispersion in Aquifers," Water Resour. Res., 19:161-180 (1983).

Gerke, H.H., and M.T. van Genuchten, "A Dual-Porosity Model for Simulating the Preferential Movement of Water and Solutes in Structured Porous Media," Water Resour. Res. 29:305-319 (1993a).

Gerke, H.H., and M.T. van Genuchten, "Evaluation of a First-Order Water Transfer Term for Variably Saturated Dual-Porosity Flow Models," Water Resour. Res., 29:1225–1238 (1993b).

Glass, R.J., T.S. Steenhuis, G.H. Oosting, and J.Y. Parlange, "Uncertainty in Model Calibration and Validation for the Convection-Dispersion Process in the Layered Vadose Zone," Proceedings of the International Conference and Workshop on the Validation of Flow and Transport Models for the Unsaturated Zone, P.J. Wierenga and D. Bachelet, Eds., New Mexico State University, 1989, pp. 119–130.

Gruber, J., "Contaminant Accumulation During Transport Through Porous Media," Water Resour. Res., 26:99-107 (1990).

Güven, O., F.J. Molz, and J.G. Melville, "An Analysis of Dispersion in a Stratified Aquifer," Water Resour. Res., 20, 1337-1354 (1984).

Hewett, T.A., and R.A. Behrens, "Scaling Laws in Reservoir Simulation and Their Use in a Hybrid Finite Difference/Streamtube Approach to Simulating the Effects of Permeability Heterogeneity," in Proceedings of the 2nd International Conference on Reservoir Characterization, L.W. Lake and H.B. Carroll, Jr., Eds., (San Diego, CA: Academic Press, 1991), pp. 355–381.

Im, J.S., Application of Monte Carlo Simulation for the Analysis of Uncertainty in Ground-

water Contaminant Transport Modeling. National Council of the Paper Industry for Air and Stream Improvement (NCASI). Technical Bulletin No. 584 (1990).

Jones, R.L., G.W. Black, and T.L. Estes, "Comparison of Computer Model Predictions with Unsaturated Zone Field Data for Aldicarb and Aldoxycarb," *Environmental Toxicol. Chem.*, 5:1027-1037 (1986).

Jury, W.A., "Simulation of Solute Transport Using a Transfer Function Model, *Water Resour. Res.*, 18:363-368 (1982).

Jury, W.A., G. Sposito, and R.E. White, "A Transfer Function Model of Solute Transport Through Soil 1. Fundamental Concepts," *Water Resour. Res.*, 22:243-247 (1986).

Koltermann, C.E., and S.M. Gorelick, "Paleoclimatic Signature in Terrestrial Flood Deposits," *Science*, 26:1775-1782 (1992).

Konikow, L.F., and J.D. Bredehoeft, "Computer Model of Two-Dimensional Solute Transport and Dispersion in Groundwater," *U.S. Geol. Survey, Techniques of Water Resources Investigations*, Book 7, Chap. C2, 1978.

Kool, J.B., J.C. Parker, and M.T. van Genuchten, "Parameter Estimation for Unsaturated Flow and Transport Models—A Review," *Journal of Hydrology*, 91:255-293 (1987).

Lake L.W., and H.B. Carroll Jr., *Reservoir Characterization*, (Orlando, FL: Academic Press, 1986).

Lake L.W., H.B. Carroll Jr., and T.C. Wesson, *Reservoir Characterization II*, (San Diego, CA: Academic Press, 1991).

Lantz, R.B., "Quantitative Evaluation of Numerical Diffusion (Truncation Error)," *Society of Petroleum Engineering Journal*, 11:315-320 (1971).

Lindstrom, F.T., and W.T. Piver, "Vertical-Horizontal Transport and Fate of Low Water Solubility Chemicals in Unsaturated Soils," *Journal of Hydrology*, 86:93-131 (1986).

Liu, C.W., and T.N. Narasimhan, "Redox-Controlled Multiple-Species Reactive Chemical Transport 1. Model Development," *Water Resour. Res.*, 25:869-882 (1989a).

Liu, C.W., and T.N. Narasimhan, Redox-Controlled Multiple-Species Reactive Chemical Transport 2. Verification and Application," *Water Resour. Res.*, 25:869-882 (1989b).

Mantoglou, A., and L. W. Gelhar, "Large-Scale Models and Effective Parameters of Transient Unsaturated Flow and Contaminant Transport Using Stochastic Methods," Rpt. 299, Ralph M. Parsons Lab., Mass. Inst. of Technology, Cambridge, 1985.

Mantoglou, A., and L. W. Gelhar, "Stochastic Modeling of Large Scale Transient Unsaturated Flow Systems," *Water Resour. Res.*, 23:37-46 (1987).

Matheron, G., and G. de Marsily, "Is Transport in Porous Media Always Diffusive? A Counterexample," *Water Resour. Res.*, 16:901-917 (1980).

McCord, J.T., D.B. Stephens, and J. Wilson, "Field-Scale Unsaturated Flow and Transport in a Sloping Uniform Porous Medium: Field Experiments and Modeling Considerations," *Proceedings of the International Conference and Workshop on the Validation of Flow and Transport Models for the Unsaturated Zone*, P.J. Wierenga and D. Bachelet, Eds., New Mexico State University, 1989, pp. 256-267.

Means, T.H., and J.G. Holmes, "Soil Survey Around Fresno, Cal.," in *House of Representatives Document No. 526, USDA, Field Operations of the Division of Soils, 1900*, Government Printing Office, Washington, DC, 1901, pp. 333-384.

Miller, R.J., J.W. Biggar, and D.R. Nielsen, "Chloride Displacement in Panoche Clay Loam in Relation to Water Movement and Distribution," *Water Resour. Res.*, 1:63–73 (1965).

Mishra, S., and J.C. Parker, "Effects of Parameter Uncertainty on Predictions of Unsaturated Flow," *Journal of Hydrology*, 108:19–33 (1989).

Neuman, S. P., "Generalized Scaling of Permeabilities: Validation and Effect of Support Scale," *Geophysical Research Letters*, 21:349–352 (1994).

Nielsen, D.R., M.T. van Genuchten, and J.W. Biggar, "Water Flow and Solute Transport Processes in the Unsaturated Zone," *Water Resour. Res.*, 22:89–108S (1986).

Parker, J.C., and M.T. van Genuchten, "Flux-Averaged and Volume Averaged Concentrations in Continuum Approaches to Solute Transport," *Water Resour. Res.*, 20:866–872 (1984).

Pennell, K.D., A.G. Hornsby, R.E. Jessup, and P.S.C. Rao, "Evaluation of Five Simulation Models for Predicting Aldicarb and Bromide Behavior Under Field Conditions," *Water Resour. Res.*, 26:2679-2693 (1990).

Persaud, N., J.V. Giraldez, and A.C. Chang, "Monte-Carlo Simulation of Noninteracting Solute Transport in a Spatially Heterogeneous Soil," *Soil Science Society of America Journal*, 49:562–568 (1985).

Pinder, G.F., and L.M. Abriola, "On the Simulation of Nonaqueous Phase Organic Compounds in the Subsurface," *Water Resour. Res.*, 22:109S-119S (1986).

Rai, D., and J.M. Zachara, *Chemical Attenuation Rates, Coefficients and Constants in Leachate Migration, Volume 1: A Critical Review*. Electric Power Research Institute, Palo Alto, CA, 1984.

Russo, D., W.A. Jury, and G.L. Butters, "Numerical Analysis of Solute Transport During Transient Irrigation. 1: The Effect of Hysteresis and Profile Heterogeneity," *Water Resour. Res.*, 25:2109-2118 (1989a).

Russo, D., W.A. Jury, and G.L. Butters, "Numerical Analysis of Solute Transport During Irrigation. 2. The Effect of Immobile Water," *Water Resour. Res.*, 25:2119-2127 (1989b).

Selim, H.M., M.C. Amacher, and I.K. Iskandar, "Modeling the Transport of Chromium(VI) in Soil Columns," *Soil Science Society of America Journal*, 53:996-1004 (1989).

Sposito, G., W.A. Jury, and V.J. Gupta, "Fundamental Problems in the Stochastic Convection-Dispersion Model of Solute Transport in Aquifers and Field Soils," *Water Resour. Res.*, 22:77-88 (1986).

Stein, C.L., and D.F. McTigue, "Chromium Distribution Beneath a Contaminated Site: A Case for Model Validation," *Proceedings of the International Conference and Workshop on the Validation of Flow and Transport Models for the Unsaturated Zone*, P.J. Wierenga and D. Bachelet, Eds., New Mexico State University, 1989, pp. 392-400.

Stephens, D.B., and S. Heermann, "Dependence of Anisotropy in Saturation in a Stratified Sand," *Water Resour. Res.*, 24:770-778 (1988).

Theis, T.L., D.J. Kirkner, and A.A. Jennings, *Hydrodynamic and Chemical Modeling of Heavy Metals in Ash Pond Leachates*. Progress report, July 1, 1980 to August 31, 1981. U.S. Department of Energy (1981).

Tompson, A.F.B., "Numerical Simulation of Chemical Migration in Physically and Chemically Heterogeneous Porous Media," *Water Resour. Res.*, 29:3709-3726 (1993).

Trautwein, S.J., and D.E. Daniel, "Case History of Water Flow Through Unsaturated Soil," in *Role of the Unsaturated Zone in Radioactive and Hazardous Waste Disposal*. J.W. Mercer, P.S.C. Rao and I.W. Marine, Eds., 1983, pp. 229–253.

Valocchi, A.J., "Validity of the Local Equilibrium Assumption for Modeling Sorbing Solute Transport Through Homogenous Soils," *Water Resour. Res.*, 21(6):808–820 (1985).

van der Heijde, P.K.M., and O. A. Elnawawy, Compilation of Groundwater Models, U.S. Environmental Protection Agency, EPA/600/R-93/118 (1993).

van Genuchten, M.T., and W.A. Jury, "Progress in Unsaturated Flow and Transport Modeling," *Reviews of Geophysics*, 25:135–140 (1987).

Wagenet, R.J., and J.L. Hutson, "Predicting the Fate of Nonvolatile Pesticides in the Unsaturated Zone," *Journal of Environmental Quality*, 15:315–322 (1986).

Warrick, A.W., and T-C.J. Yeh, "Scale of Measurements, REV and Heterogeneities," *Proceedings of the Conference: New Field Techniques for Quantifying the Physical and Chemical Properties of Heterogeneous Aquifers*. F.J. Molz, J.G. Melville, and O. Güven, Eds., National Water Well Association, Dublin, OH, 1989, pp. 383–406.

Weber, K.J., "How Heterogeneity Affects Oil Recovery," in L.W. Lake and H.B. Carroll Jr., *Reservoir Characterization*, (Orlando, FL: Academic Press, 1986), pp. 487–544.

White, R.E., J.S. Dyson, R.A. Haigh, W.A. Jury, and G. Sposito, "A Transfer Function Model of Solute Transport Through Soil 2. Illustrative Applications," *Water Resour. Res.*, 22(2):248–254 (1986).

White, R.E., "A Transfer Function Model for the Prediction of Nitrate Leaching Under Field Conditions," *Journal of Hydrology*, 92:207–222 (1987).

Wierenga, P.J., and D. Bachelet, *Proceedings of the International Conference and Workshop on the Validation of Flow and Transport Models for the Unsaturated Zone*, New Mexico State University, 1989.

Yeh, T.C.J., "Analysis of One Dimensional Steady Infiltration in Heterogeneous Soils," *Proceedings of the International Conference and Workshop on the Validation of Flow and Transport Models for the Unsaturated Zone*, P.J. Wierenga and D. Bachelet, Eds., New Mexico State University, 1989, pp. 539–547.

Yeh, T.C.J., L.W. Gelhar, and A.L. Gutjahr, "Stochastic Analysis of Unsaturated Flow in Heterogeneous Soils. Observations and Applications," *Water Resour. Res.*, 21:457–464 (1985).

17

Review of Vadose Zone Flow and Transport Models

John H. Kramer and Stephen J. Cullen

INTRODUCTION

This chapter includes an overview of vadose-zone modeling with a summary of vadose-zone flow and transport processes, the approaches taken to model them, and a literature review of available models. It is intended to introduce readers with limited knowledge of vadose zone flow and transport processes to the problems confronted by modelers, and to provide a starting point for understanding how they are being attacked. The subject is a fast-evolving one with new approaches and codes being developed and amended continuously, so the table of models discussed and presented at the end of this summary is not complete. Most models currently in common usage are listed, and the table will assist readers in evaluating and accessing them.

OVERVIEW OF VADOSE ZONE MODELING

Usefulness of Vadose Zone Models

The vadose zone is a complex environment of unsaturated and saturated geologic material. The literature on the vadose zone is so extensive that any attempt to summarize it briefly will be incomplete. A partial list of works which presents the basics of vadose zone flow and transport includes Fetter, 1988; Everett et al., 1984; Hillel, 1982; Bear, 1979; Freeze and Cherry, 1979; and Bouwer, 1978. Vadose zone flow and transport are particularly difficult to model mathematically because of nonlinear interdependencies between chemical, physical, and hydraulic parameters (Fogg, 1995). In modeling flow processes only (without complications of chemical interactions), features such as hysteresis, nonlinear hydraulic-conductivity/pressure-head relationships, and the spatial variability inherent in geologic systems make any manageable mathematical representation an oversimplification of reality. An additional level of complications is introduced with attempts to model contaminant transport. Transport of single solutes, multi-species contaminants or immiscible contaminants involves numerous interdependent geochemical and biogeochemical reactions including degradation, chemical interferences and phase changes (partitioning of contaminants between adsorbed, liquid and gas phases). Therefore, one should not expect vadose zone modeling to be a foolproof, or even in many circumstances, a reliable predictor of future contaminant transport (Fogg et al., 1995). Nonetheless, despite the recognized difficulties of modeling contaminant flow and transport in the vadose zone, numerous bright and able minds apply mathematical models to this problem. The reason for this is that there exists no better way of:

- projecting into the future or the subsurface,
- predicting the effects of environmental changes (flood, drought, etc.),
- evaluating the effects of various potential remedial alternatives, or
- analyzing the importance of field data through computer model sensitivity analysis.

Modeling is commonly used for aspects of environmental investigations which cannot be measured or analyzed in more straightforward ways, including monitoring network design (Bumb et al., 1988) and long-term risk assessment. Vadose zone modeling is used in risk assessment to predict contaminant fate and transport. From a toxicological point of view, risk analysis is an inexact science (Lehr, 1990). The intelligent use of vadose zone modeling does not further degrade the certainty of conclusions regarding protection of human health, especially in the conservative worst-case scenarios commonly employed. In fact, there exists no other tool for predicting contaminant behavior in the subsurface over relevant time spans and distances.

Driving Forces and Controls

It is necessary for vadose-zone modelers to be familiar with unsaturated flow, because it can result in counter-intuitive transport pathways. Unsaturated flow can be divided into two types based on the predominant driving force: gravity-dominated flow, and capillary tension-dominated flow. Gravity-dominated flow, which occurs during drainage of wet soils, is directed downward and is intuitively simple to predict. By contrast, capillary tension-dominated flow, which occurs in dry to moist soils, proceeds in the direction of increasing pore water tension, which can be in any direction. Capillary water tension is caused by (1) adhesion of pore water molecules to soil particle surfaces, and (2) the attraction of water molecules for each other.

The pore size distributions of geologic materials create tension gradients which drive unsaturated flow. Small pore sizes generate large capillary forces which tend to pull water into the medium. Fine-grained and poorly-sorted materials generally have smaller pores and generate higher tensions than coarse-grained, well-sorted material of equal moisture content. For this reason, it is possible to have moist silt overlying dry sand. Water under tension will not flow from a fine pore to a larger one and unsaturated flow may occur more readily through fine-textured material than coarse-textured material. Coarse materials such as washed sand and gravel are unsaturated flow barriers. This commonly misunderstood fact is why subsurface drains will not evacuate water from unsaturated soils, and why pore water will not flow into a well screened in the unsaturated vadose zone.

The state of water saturation (partial to full) affects unsaturated-hydraulic conductivity. In dry material, pore water occurs in thin layers, adhered tightly to particle surfaces. Thus, flow is inhibited. With increasing saturation percentage, the pore-water layer thickens and pore-water tension decreases. Unsaturated-hydraulic conductivity increases in a nonlinear fashion. As the saturation percentage approaches 100%, tension approaches zero and hydraulic conductivity approaches the saturated-hydraulic conductivity. Saturation rarely reaches 100% because air is entrapped in small pore spaces. Entrapped air bubbles in pore spaces fill voids otherwise available for flow and account for why a rapidly-saturated soil has lower hydraulic conductivity than the same fully-saturated soil. Rapidly-saturated soils are said to be field-saturated. Field-saturated hydraulic conductivity can be measured in situ and is approximately 0.5 times the fully-saturated hydraulic conductivity. Bouwer (1978) uses entrapped air to explain, in part, why soil moisture characteristics and hydraulic conductivity functions of a soil can differ, depending on whether the soil is filling with water or draining. This phenomenon, called hysteresis, is a complication of soil physics that is only treated in the most advanced computer codes.

Transport Processes

Transport of soluble contaminants is affected by a host of chemical processes including decay, biotransformation, sorption and reaction. Nielsen et al. (1986) describe the nonlinear interaction between atmospheric, biologic, and geologic factors which combine to create an extremely complex flow and transport regime in the vadose zone. Freeze and Cherry (1979) present a discussion of contaminant transport in saturated media which is applicable to vadose zone transport with respect to accounting for retardation of solutes. Redox reactions greatly affect mobility of heavy metals which change solubility. Complex chemical flow models which account for eH/pH relations among several metals have been developed (Yeh, 1992; Liu and Narasimhan, 1989). Three-phase flow, treating immiscible liquids and gas phase transport are also treated by the most powerful codes.

Input Parameters

Input parameters are the Achilles' heel of vadose zone flow and transport model applications. They are critically important, because they are the foundation upon which mathematical models rest, and they are vulnerable because problems associated with scale and parameter heterogeneities make representative measurement elusive. Typical input parameters required for use in vadose zone flow and transport models and accepted methods for measuring them are listed in Table 17.1. Although not a comprehensive listing, Table 17.1 illustrates important characteristics in natural systems which are used in modeling.

Modelers must make practical input-parameter compromises concerning the mathematical rigor of the treatment, and/or the representativeness of input-parameter measurements to field conditions. The mathematical rigor is compro-

Table 17.1. Typical Input Parameters Required for Vadose Zone Flow and Transport Models

Input Parameters Typically Required for Vadose Zone Flow Models	Measurement Methodology or Reference	Additional Parameters Needed for Vadose Zone Transport Models	Measurement Methodology or Reference
Bulk Density	MOSA[a], 1986 Chap. 13, pp. 363–367	Specific Surface Area	MOSA, 1986 Chap. 16, pp. 413–423
Particle Density	ASTM[b], D 854–91	Cation Exchange Capacity	USEPA SW-846-9081
Porosity	MOSA, 1986 Chap. 18, pp. 443–461	Partitioning Coefficients	Sims et al., 1991[e]
Effective Porosity	Typically calculated from porosity and water content, but varies with model requirements	Diffusion Coefficient (liquid)	Sims et al., 1991
Mass Water Content	ASTM, D 2216–90	Diffusion Coefficient (gaseous)	Sims et al., 1991
Volumetric Water Content	MOSA, 1986 Chap. 21, pp. 493–495	Solubilities	Sims et al., 1991
Field Capacity Water Content	MOSA, 1986 Chap. 36, pp. 901–936	Biodegradation Rates	Sims et al., 1991
Field Capacity Soil Water Tension	MOSA, 1986	Chemical Degradation Rates	Sims et al., 1991
Residual Water Content	Estimated from 15-bar water content[c]	Radioactive Decay Rates	Sims et al., 1991
Infiltration Capacity	MOSA, 1986 Chap. 32, pp. 825–844	Organic Matter Content	ASTM, D 2974–87

Saturated Hydraulic Conductivity	ASTM D 2434–68(74) MOSA, 1986, Chap. 28, pp. 700–703
Soil Water Characteristic Curve	ASTM D2325–68(74). MOSA, 1986, Chap. 26 pp. 644–649
Conductivity/Pressure Head Relationship	MOSA, 1986 Chaps. 28, 29 and 30
Climatological Data (rainfall, ET, temperature, wind speed, etc.)	NOAA[d]

Source: Adapted from Cullen and Everett, 1993.

[a]MOSA = *Methods of Soil Analysis, Part 1.* 2nd ed. A. Klute, Ed., American Society of Agronomy, Madison, WI, 1986.

[b]ASTM = Annual Book of ASTM Standards.

[c]Residual Water Content (θ_r) specifies the maximum amount of water in a soil that will not contribute to liquid flow because of blockage from the flow paths or strong adsorption onto the solid phase. (van Genuchten et al., 1991). While it is a fitting parameter and without much physical meaning, it is often estimated by the 15-bar water content.

[d]NOAA = National Climatic Data Center, Federal Building, 37 Battery Park Avenue, Asheville, NC 28801–2733. Consult also EPA, 1993. *Subsurface Characterization and Monitoring Techniques: A Desk Reference Guide.* EPA/625/R–93/0036. CERI, Cincinnati, OH.

[e]Sims, R.C., J.L. Sims, and S.G. Hansen. *Soil Transport and Fate Database 2.0 and Model Management System.* R.S. Kerr Laboratory, Ada, OK, 1991. Other data bases of compound chemical properties are available and suitable for use. Determination of degradation rates often requires complex laboratory studies and these values are typically obtained from sources similar to the one cited above.

mised when (1) no attempt is made to compute the effects of known processes, (2) simplifying assumptions are made, or (3) empirical factors are introduced to simulate the results of complex hydraulic phenomena. An example of this is the "interconnectedness index," an input parameter used in the model SESOIL. Such factors are not measurable, and can be assigned based only on the performance of the model to different soil types. The index is adjusted so that the simulation results fit the known, anticipated, or desired field situation. The representativeness of input parameter measurements is compromised when more parameters are required by the mathematics than can be practically measured, or when the inherent variability of parameter values is not sufficiently characterized. The input parameter compromises depend on the power of the model.

Less powerful computer codes treat generalized volumes or simplified processes. Although they require fewer input parameters, making them easier to implement, they are not well-suited for applications involving detailed modeling of transient flow, complex geometries, irregular boundary conditions, or multiphase flow and transport. Powerful computer codes handle many conditions and processes but require hard-to-measure input parameters (e.g., species-specific diffusion coefficients, coupled relative hydraulic-conductivity phase-saturation functions, soil-specific adsorption coefficients, etc.). This problem is usually handled by estimating average or "effective" parameters for unavailable input requirements, or using default values from previous experience. It is generally impractical to collect enough site-specific data to characterize the mean and variance of the many input measurements required for powerful codes.

Model sensitivity to input parameters can be evaluated by comparing the results of several model simulations of the same scenario run at different parameter values (Nofziger et al., 1993; Ladwig and Hensel, 1993; Odencrantz et al., 1992). Sensitivity analysis is intended to provide an understanding of the range of reasonable results due to parameter heterogeneities for a specific modeled scenario. The most influential parameters can be identified for intensive and careful measurement in the field. Nofziger et al. (1993) point out that "there is abundant evidence that the sensitivity and uncertainty are highly dependent on the scenario being modeled and the parameters used." A specific sensitivity analysis is therefore appropriate for each modeled scenario.

The most credible modeling applications incorporate measured, site-specific hydraulic parameters. These parameters should be measured in situ where possible, in order to minimize errors attributable to sampling and analytic techniques. It is most desirable to measure a suite of the most important parameters throughout the modeled volume. A well-characterized suite is one to which a statistical model can be applied, and a measure of the mean, variance, and possible spatial correlation can be calculated. These characteristics aid in evaluating the uncertainty associated with model output through sensitivity analysis or stochastic treatment. Site-specific parameter measurements tie the model into the actual field setting, providing a link to reality.

A probabilistic argument can be advanced for generic modeling using databases of typical or estimated effective input parameters. The argument states that uncertainty in the model results will be no greater if generic input is used than measured site-specific input, because uncertainty in site-specific input parameters is high. This argument is used to justify modeling in situations where it is impossible to collect sufficient field data, and often extended to situations where it is inconvenient to do so. The danger in this approach is that the model result has no relation to site-specific conditions, and the inexperienced practitioner can be seduced to create a fabrication based on preconceived bias.

Pertinent Literature

Donigian and Rao (1987) give an excellent review of vadose-zone processes and modeling, and compare selected soil-leaching models circa 1984. Morel-Seytoux (1988) provides in-depth articles on modeling theory and practice for the advanced practitioners. A more current comparison of models is provided by van der Hiejde (1993) and Wallin and Wilson (1992). Hern and Melancon (1987) also contains well written articles on model validation, chemical transport, and other subjects of interest in vadose-zone modeling. Mangold and Tsang (1991) review hydrological and hydrochemical models, including vadose zone models. Two important sources of up-to-date information on computer modeling codes are: (1) The International Ground Water Modeling Center in Golden, Colorado; and (2) The Center for Subsurface Modeling Support (CSMoS) at the Robert S. Kerr Environmental Research Laboratory, Ada, Oklahoma.

Public domain software, such as VS2DT (Lappala et al., 1987; Healy, 1990) developed through public agencies tend to be well tested and have very usable documentation. VS2DT outperformed other codes in a benchmarking study that compared several codes against the same analytically solved problem, and against experimental data (Carovillano, 1993). VS2DT uses a finite difference approach, whereas other public-domain 2-D models, SUTRA (Voss, 1984) and SWMS-2D (Simunek et al., 1992) are finite-element solution models. Other public-domain software developed through government agencies includes many models which treat one-dimensional leaching. In addition to public domain codes, there are a number of computer models developed at research institutions or in the private sector which can be licensed or used through contracting services.

Modeling Solution Methods

Analytical Method

Analytical models use exact mathematical solutions of partial differential equations to represent flow and transport. Hydraulic properties are assumed to be homogeneous in the domain of interest. Approximate anisotropies can be treated through modification of axes scales. Moisture content at any point in space is correlated through pre-characterized soil moisture relations for the soils of interest. In the vadose zone, the nonlinear aspects of the variable hydraulic-conductivity relationships are estimated with continuous functions. Model outputs are exact solutions to the continuously variable pressure field through time and space. New developments in analytical solutions to saturated-flow problems employ the analytic element method (Strack, 1989), which facilitates application to irregular domains by linking analytic solutions across element boundaries. This approach will probably soon be applied to the vadose zone as well.

Advantages of analytic models include ease of use, time efficiencies, exact 3-D solutions, and others as discussed by McKee and Bumb (1988). Analytical solutions can be used to quickly generate graphics for reports. Disadvantages are the unrealistic treatment of soil systems as homogeneous, and the need for regular geometric boundaries amenable to exact mathematical representation.

Finite-Difference Method

This numerical technique is used in several well-known vadose zone programs (e.g., CHEMFLO, VLEACH, VS2DT). The solution technique is well described as applied to saturated problems (Bear, 1979; Freeze and Cherry, 1979; Bouwer, 1978). Its advantage is that it is conceptually simple, employing mass balance (accounting for changes in storage of liquids or solutes) in discrete spatial blocks progressing through time. A domain of specified dimension (1-D, 2-D, or 3-D) is divided (discretized) into N rectangular boxes, the centers of which are nodes of a grid. Each box can have uniquely variable hydraulic properties. Initial and boundary conditions are set, including soil water tension (negative pressure head), water pressures (in perched horizons), and concentrations of solutes. The mass balance of water (or solute) at each node is represented by writing the partial differential flow equations across each of the appropriate box boundaries (two for 1-D columns, four for 2-D grids, and six for 3-D grids). Since each node has as many neighboring nodes as boundaries (except at boundary nodes where conditions must be set), a system of N equations and N unknowns is generated. The pressure field at each node is found by solving the N "finite differences" for pressure simultaneously at pre-selected time steps. A steady-state starting pressure field can be found by using initial estimates of appropriate pressure heads at each node and employing iterative "relaxation" techniques to minimize residuals. The system can then be stressed by changing boundary conditions. Sources can be placed in the initial pressure field to simulate infiltration from point sources, stream recharge, etc., and sinks can be defined to simulate seepage faces or free drainage. Numerical solutions at different time steps represent snapshots in time of transient flow conditions. Discretization of the domain and selection of time steps are important for examining detail in the domain of interest, particularly in materials with distinct air-entry pressures subjected to rapid wetting where high contrast in hydraulic conductivity will create numerically unstable mathematical systems if grid sizes are too large. (Solutions tend to oscillate outside convergence criteria). Very small grid sizes, short time steps, adjustments for an appropriate effective hydraulic conductivity for the time step and cell spacing (upstream weighting), and adjustments to convergence criteria may be required to obtain a solution.

Advantages of the technique are that it is relatively efficient and versatile. Disadvantages are that node spacings are block-like. Diagonal boundaries are represented in stair-step fashion, introducing error at irregular boundaries. The approach is limited when attempting to model anisotropic flow regimes in which the favored flow direction does not coincide with the grid axes. This situation would occur when attempting to model flow through deformed rocks, folded or slumped soil horizons, or across tilted unconformities. In these cases, the simulated hydraulic conductivities across the grid-cell boundaries may be less than the actual hydraulic conductivity oblique to the boundaries. Geologic volumes with variable, nonorthagonal favored hydraulic conductivity directions may require multiple adjoining modeled domains.

Finite-Element Method

Finite-element codes are conceptually and mathematically more difficult to understand than finite difference models but are more versatile for modeling complex geometries and heterogeneous hydraulic conductivity fields (Bear, 1979). This is the technique most often applied in advanced vadose zone modeling of multiple processes and complex domains (e.g., Celia and Binning, 1992; Yeh et al., 1993; Simunek et al., 1992). The finite element method breaks down the problem of spatial domain into polygonal elements with nodes at the corners. The polygonal elements are treated as regions for which approximate integrated values of the parameter of concern (usually pore pressure) are solved locally using interpolating functions between the nodes. Flow equations are set up for each element, assuming parameter values at the nodes, and matrices of equations (both local and global) are solved simultaneously such that the changes in first derivatives (i.e., pressure gradients between nodes and between elements) are minimized. Finite-element solutions are standard practice in modern engineering design, and are used to model problems analogous to subsurface flow such as stress distribution or heat transfer.

Advantages of finite element models are that finite element techniques can treat irregular boundaries and boundary conditions naturally (Yeh et al., 1993). Finite element techniques can also treat changes in principal flow directions within anisotropic formations. Disadvantages are its mathematical complexity and processing time, which can be longer than finite difference models.

Stochastic Methods

Any of the previously described deterministic mathematical solution methods can be used in a stochastic treatment. Stochastic treatments consider input parameters to be random variables, and any particular solution to be only a single "realization" from a distribution of potential solutions. Any one realization is associated with a probability that it represents the average realization. The most commonly used stochastic treatment is called the Monte Carlo method. Numerous runs of a model are made, each with a different set of input parameters chosen randomly from a statistically characterized population of possible input parameter values (e.g., log-normal hydraulic conductivities). These numerous runs, in turn, build a population of solutions which can be characterized by a mean and standard deviation. The probability of any given outcome can be estimated from this sample of possible solutions as described by Massmann et al. (1991). An alternative to the Monte-Carlo procedure is the theoretical approach in which the variance of the inputs can be carried through calculations and used to estimate the distributions of the possible solutions. This approach yields interesting insights. For example, Yeh et al. (1985) predicted state-dependent anisotropy by correlating hydraulic-conductivity anisotropies (Kx/Kz) to saturation percentage, variance in particle-size distribution parameters, and layer thickness.

The advantage of a stochastic approach is that uncertainty in the input parameters is reflected in the solution, providing a measure of the uncertainty in the findings. A disadvantage is that the technique requires numerous modeling runs involving a well organized effort and extensive computer time.

Other Methods

Other approaches including nontheory based empirical models (e.g., transfer-function models) have been employed with some success on a site-specific basis (Jury, 1982). An empirical model is based on observations through time or space which are fitted, usually involving a regression, to a simplified mathematical expression. These can be scaled to extend past the domain of the original empirical observations. Empirical models have no theoretical basis and if transferred from one place to another may become invalid. These models can also not be used to interpret the importance of physically measurable properties. Projecting such models into the future is uncertain, particularly in transient flow regimes where input and through-flow may not bear any relation to the conditions for which the model was derived.

Model Classification

Table 17.2 summarizes specific vadose zone models that are currently well known or widely used. Further information on models in Table 17.2 can be obtained from the listed contacts. Models are generally classified according to their capabilities and solution approach (van der Heijde, 1993; Fogg et al., 1995; Donigian and Rao, 1987). The simplest and most commonly employed models are one-dimensional (1-D) leaching models, primarily used in risk assessment of ground-water contamination. Numerous one-dimensional leaching models have been developed by public and private sources to address soil-contaminant problems and transport to the water table. Their use is limited to downward transport through characterized soil columns. Some of these models were designed for risk analysis of pesticide applications to agricultural soils, but have been adapted to modeling the fate and transport of heavy metals and other contaminants by using appropriate contaminant-specific values of distribution coefficients; some are specific to landfarming of oily wastes, generation of leachate at landfills, and other problems.

Advantages of the 1-D leaching models are: (1) the relative simplicity of the modeled geometry and, hence, application of the model, and (2) their acceptance among regulators (CWRCB, 1989; US EPA, 1991). SESOIL and VLEACH are examples of 1-D leaching models that have gained acceptance from regulators. SESOIL is a seasonal compartment-type model which spills water from one compartment to the next. It can handle multiple layers and accounts for volatilization. For this reason it was used in California to develop the risk tables in the Leaking Underground Fuel Tank Program. VLEACH, another soil leaching model available through the U. S. Environmental Protection Agency (EPA), is cited in a record of decision. It should be pointed out that VLEACH is a transport model only, and does not simulate flow or calculate flux. Disadvantages of 1-D leaching models are that they use simplifications to approximate actual processes and cannot simulate lateral flow and transport, although lateral flow between columns can be approximated by adjusting input.

Two- and three-dimensional models employ a number of solution techniques to simulate flow and transport in complex domains for a variety of contaminants and conditions. Selection of an appropriate model for a given problem requires a familiarity with the solution techniques and the processes to be simulated. The models described in Table 17.2 are a sample of the variety of available two- and three-dimensional models.

Various computer codes have been developed to calculate parameters used in vadose zone flow and transport models. These are used to calculate soil hydraulic properties from limited experimental data for input into flow models, or to fit experimental data (e.g., tracer tests) and extract hydraulic properties. The most powerful and well accepted is the RETC code (van Genuchten, et al., 1991) which calculates soil hydraulic properties based on various theoretical models and assumptions from limited measured data.

Table 17.2. Vadose Zone Models and Their Applications

Model Name	Author(s)	Contact Address	Model Description	Features and Processes Modeled
		One-Dimensional Models		
CADIL[a]	Emerson, C.J. Thomas, B. Luxmoore, R.J. (1984)	Computer Sciences Oak Ridge National Laboratory Oak Ridge, TN 37831	A model to simulate chemical transport through soils and the effect of temperature on chemical degradation.	advection dispersion biodegradation immiscible flow
CHEMFLOW	Nofziger, D.L. Rajender, K. Nayudu, S.K. Su, P-Y (1989)	J. Williams CSMoS Robert S. Kerr Lab US EPA Ada, OK 74078	A finite difference solution of Richard's equation for flow, and the convection-dispersion equation for chemical transport	advection disperson adsorption
CHEMRANK[a]	Nofziger, D.L. Rao, P.S.C. Hornsby, A.G. (1988)	Institute of Food and Agriculture Sciences University of Florida Gainesville, FL 32611	A package which uses four schemes, based on rates of movement, and relative rates of mobility and degradation, for screening of organic chemicals for their potential to leach into groundwater.	leaching degradation mobility
CMLS[a]	Nofziger, D.L. Hornsby, A.G. (1988)	Department of Agronomy. Oklahoma State University Stillwater, OK 74078 Department of Soil Science University of Florida Gainesville, FL 32611	The model estimates location of the peak concentration of non-polar organic chemicals as they move through the soil in response to downward movement of water, and the relative amount of chemical remaining in the profile at any time.	leaching degradation stochastic treatment
HYDRUS	Kool, J.B. van Denuchten, M.T. (1991)	U.S. Salinity Laboratory 4500 Glenwood Drive Riverside, CA 92501	A model to simulate flow and solute transport which incorporates mathematic treatment of hysteresis.	hysteresis root water uptake ionic/molecular diffusion dispersion nonlinear adsorption decay
HYTEQ	Kool, J.B. (1990)	HydroGeoLogic. Inc. 503 Carlisle Drive Herndon, VA 22070	A code for modeling multicomponent chemical transport comprising HYDRUS and MINTEQ modules	Multispecies chemical reactions including complexation, sorption, ion exchange, chemical precipitation

Table 17.2. Vadose Zone Models and Their Applications (Continued)

Model Name	Author(s)	Contact Address	Model Description	Features and Processes Modeled
PESTAN (PESTicide ANalytical, Model)[a]	Enfield, C.G. Carsel, R.F. Cohen, S.Z. Phan, T. Walters. D.M. (1982)	CSMoS Robert S. Kerr Lab US EPA Ada, OK 74078	PESTAN, PESTicide ANalytical Model, for evaluating the transport of organic pollutants through the vadose zone. PESTAN is available with a user-friendly preprocessor. Output contains values for volumetric water content, pore water velocity, pollutant velocity, initial depth of pollutant slug and a soil column vs time profile.	linear adsorption first-order degradation hydrodynamic dispersion
PRZM-2 Pesticide Root Zone Model	Mullins, J.A.. Carsel, R.F. Scarborough, J.E. Ivery, A.M. (1992)	Environmental Research Lab. Office of Research and Development. US EPA, Athens. GA	A finite difference model used for estimating pesticide leaching which eliminates numerical dispersion by the method of characteristics.	soil temperature, volatilization vapor phase transport, irrigation, microbial transformation
RITZ: Regulatory Investigative Treatment Zone Model[a]	Nofziger, D.L. Williams, J.R. Short, T.E. (1988)	J. Williams CSMoS Robert S. Kerr Lab US EPA Ada, OK 74078	The model is an interactive analytical model requiring relatively few parameters for simulation of the fate of hazardous chemicals during land treatment of oily wastes.	volatilization degradation sorption leaching
RUSTIC[b]	Dean, J.D. Huyakorn, P.S. Donigian, A.S. Voos, K.A. Schanz, R.W. Meeks, Y.J. Carsel, R.F. (1988)	R.F. Carsel Environmental Research Lab. Office of Research and Development. US EPA Athens. GA	RUSTIC incorporates three models. root zone model (PRZM), vadose zone module (VADOFT), and saturated zone module (SAFTMOD) to predict pesticide rate and transport through the vadose zone and saturated zone.	leaching decay volatilization hydrodynamic dispersion advection sorption degradation

Model	Authors	Address	Description	Processes
SESOIL	Bonazountas. M. Wagner. J.M. Goodwin. B. (1984) Watson, D.B. Brown, S.M. (1985) Hetrick D. M. Travis, S.K. Leonard, S.K. Kinerson. R.S. (1985) Hetrick, D.M.	Annette Nold Office of Toxic Substances U.S. EPA Washington. D.C.	A model for long-term environmental pollutant fate simulations, a compartmental model designed to describe flow. sediment transport. pollutant transport and transformation, pollutant transport to ground water, and soil quality.	infiltration-runoff evapotranspiration leaching degradation volatilization
within risk assessment systems: RISKPRO	Scott, S.J. Barden, M.J. (1993)	John Thomas General Sciences Corporation 6100 Chevy Chase Dr. Laurel. MD 20707	A risk assessment system using SESOIL	user interface
DSS: Decision Support System	Spence, L. Salhotra, A. (in prep.)	American Petroleum Institute 1220 L Street, N.W. Washington, D.C. 20005	A risk assessment system using SESOIL	user interface
SOILCO2	Simunek. J. Suarez, D.L. (1992)	U.S. Salinity Laboratory 4500 Glenwood Drive Riverside, CA 92501	A model for simulating carbon dioxide production and transport.	liquid-gas transport of CO_2, heat transfer, plant production/uptake
TETrans: Trace Element Transport	Corwin, D. Waggoner. B. (1990)	U.S. Salinity Laboratory 4500 Glenwood Drive Riverside, CA 92501	A model to simulate the vertical movement of nonvolatile inorganic and organic chemicals through the vadose zone using simplified input parameters.	plant water uptake infiltration drainage partitioning adsorption exchange
VADSAT 2.0	Unlu, K. Kemblowski, M.W. Parker, J.C. Stevens, D. Chong, P.K. Kamil, I. (1992)	American Petroleum Institute (API) 1220 L Street, N.W. Washington, D.C. 20005	An interactive program to simulate the movement of compounds in land-disposed wastes to ground water.	convective-dispersive solute transport of oily wastes, adsorption, volatilization, leaching.

Table 17.2. Vadose Zone Models and Their Applications (Continued)

Model Name	Author(s)	Contact Address	Model Description	Features and Processes Modeled
VIP	Stevens, D.K. Grenney, W.J. Yan, Z. (1989)	Civil and Environmental Engineering Dept. Utah State University Logan, Utah 84322–4110	A finite difference model to simulate the land treatment of hazardous organic wastes using two soil layers, application and treatment zones.	degradation partitioning dispersion (water and air) advection (water and air) oxygen usage
VLEACH[a]	Turin, J. (1990)	CSMoS Robert S. Kerr Lab US EPA Ada, OK 74078	A finite difference model which does not simulate flow, but estimates contaminant leaching through the vadose zone.	advection gas diffusion sorption/desorption partitioning volatilization
Two-Dimensional Models				
DYNAMIX	Liu, C.W. Narasimhan, T.N. (1989)	Department of Materials Science and Mineral Engineering, University of California, Berkeley. CA	A model simulating dynamic mixing of multispecies chemical systems to simulate chemical transport.	diffusion-dispersion redox. acid-base reaction, complexation, precipitation-dissolution. kinetic dissolution
HELP–Hydrologic Evaluation of Landfill Performance Version 2 Version 3	Schroeder, P.R. Peyton, R.L. McEnroe, B.M. Sjostrom, J.W. (1988) Schoeder et al. (in press, 1994)	Robert Landeth Risk Reduction Engineering Laboratory Office of Research and Development, U.S. EPA Cincinnati, OH 45268	A quasi 2-D model to facilitate rapid, economical estimation of the amounts of surface runoff, subsurface drainage and leachate from a variety of landfill designs.	precipitation, surface storage, runoff, infiltration, percolation, evapotranspiration, soil moisture storage, and lateral drainage
HYDRO-GEOCHEM	Yeh, G.T. (1992)	Dr. George Yeh Dept. of Civil Engineering Penn State Univ. University Park, PA	A finite element model to simulate flow and transport.	multi-species redox-dependent reaction solutes

Model	Reference	Address	Description	Processes
MMOC2 from VSAFT	Yeh, J.T-C. Srivastava, R. Guzman, A. Harter, T. (1993)	Department of Hydrology and Water Resources. University of Arizona. Tucson, AZ 85721	A finite element model to simulate water flow and chemical transport through variably saturated porous media.	advection dispersion adsorption decay
PRINCETON UNSAT2D	Celia, M.A. (1992)	Princeton University Princeton, NJ	A model employing a lumped finite element procedure to simulate flow and transport of miscible solutes.	advection adsorption decay
UNSAT2[a]	Davis, L.A. Neuman. S.P. (1983)	Dept. of Hydrology and Water Resources University of Arizona. Tucson, AZ 85721	A finite element model to simulate flow and transport in variably-saturated, nonuniform anisotropic porous media.	Infiltration, evapotransporation advective transient flow
SUTRA	Voss, C.I. (1984)	U.S. Geological Survey Box 25046 M.S 413 Denver Federal Center Denver, CO 80225	Finite element model to simulate transient or steady state flow in variably saturated media with transport of energy or reactive single species solutes.	capillary convection dispersion diffusion, advection, adsorption reaction
SWMS-2D	Simunek, J. Vogel, T. van Genuchten. M.T. (1992)	U.S. Salinity Laboratory USDA. ARS 4500 Glenwood Drive Riverside, CA 92501	Finite element model to simulate flow and transport	advection. dispersion evapotranspiration volatilization degradation decay. chemical precipitation
TOUGH2 Successor to TOUGH	Pruess. K. (1991)	Pruess. K. Earth Science Div. Mailstop 50E Lawrence Berkeley Lab, University of California. Berkeley, CA 94720	A multi-dimensional integrated finite difference code written for use on computers capable of 64-bit arithmetic (i.e. Cray, or IBM RISC System/6000 Workstations).	nonisothermal flow. multicomponent fluids, multiphase fluids, in porous and fractured media
VADOSE	McKee, C.R. Bumb, A.C. (1988)	In Situ Inc. P.O. Box 1 Laramie, WY 82070	Analytical model using a Boltzman distribution to model soil moisture characteristics.	3-D flow in unsaturated soil

Table 17.2. Vadose Zone Models and Their Applications (Continued)

Model Name	Author(s)	Contact Address	Model Description	Features and Processes Modeled
VAM-2D (from SATURN)	HydroGeoLogic, Inc. (1993)	HydroGeoLogic, Inc. 503 Carlisle Drive Herndon, VA 22070	A finite element model to simulate flow and transport.	advection dispersion
VS2D VS2DT	Lappala, E.G. Healy, R.W. Weeks, E.P. (1987) Healy, R.W. (1990)	U.S. Geological Survey Box 25046 M.S. 413 Denver Federal Center Denver, CO 80225	Finite difference model for simulating flow and transport in variably saturated porous media.	advection dispersion degradation
FEHM	Zyvoloski, G.A.	Los Alamos National Lab Los Alamos, NM 87545	A finite element code to model air/water and heat flow, and multiple chemically reactive sorbing tracers.	dual porosity dual permeability advection sorption convection
MMOC3	Yeh, J.T-C. Srivastava, R. Guzman, A. Harter, T. (1993)	Department of Hydrology and Water Resources University of Arizona Tucson, AZ 85721	A finite element model to simulate flow and solute transport in variably saturated soils	advection dispersion adsorption decay
NUFT, Nonisothermal Unsaturated-saturated Flow and Transport	Nitao. J.J.	Lawrence Livermore National Laboratory L-206 P.O. Box 808 Livermore, CA 94550	A suite of multiphase. multicomponent models employing finite difference solution	advection adsorption degradation dispersion 3-phase flow gaseous diffusion
TRACR3D	Travis, B.J. Birdsell, K.H. (1991)	Los Alamos National Lab Los Alamos, NM 87545	A finite difference flow and transport model to simulate multiple chemically reactive, radioactive, and sorbing tracers, and microbial processes.	advection dispersion sorption decay air flow
VAM-3D	HydroGeologic, Inc. (1993)	HydroGeoLogic. Inc. 503 Carlisle Drive Herndon, VA 22070	A finite element model to simulate flow and transport.	advection dispersion adsorption decay

CXFIT	Parker, J.C. van Genuchten, M.T. (1984)	U.S. Salinity Laboratory 450C Glenwood Drive Riverside, CA 92501	A code using nonlinear least squares inversion methods to fit transport parameters from laboratory and field tracer experiments	dispersion coefficients retardation factors degradation constants
RETC	van Genuchten, M.T. Leij, F.J. Yates, S.R. (1991)	U.S. Salinity Laboratory 450C Glenwood Drive Riverside, CA 92501	A code which regresses experimental data to calculate soil moisture characteristics and hydraulic conductivity function.	Fitted parameters from this program are used in numerous flow models to calculate continuous hydraulic conductivity functions.
with GMI graphical interface, pre and post processors	Zhang, R.			user interface for RETC

[a]References from van der Heijde (1993) from the International Ground Water Modeling Center.
[b]Summary prepared by Wallis and Wilson (1992).

Application of Models

Application of models requires four crucial steps. First and second, as discussed above are (1) selection of the model most appropriate to the task, and (2) determining the most important input parameters for field measurement. The third step is calibrating the model to conform with site-specific conditions. This requires assigning values to the input parameters and defining boundary conditions that will drive the model to represent the particular target situation. The difficulty of this task increases with the complexity of the model and the number of observation points. The fourth and last step involves matching the modeling results against field observations to gauge how accurately the model simulated reality. This process is frequently referred to as "validation," a term criticized in the recent literature because it connotes veracity and promotes unfounded confidence in the mathematical simulation which, at best, is only a shadow of truth. "History matching" has been proposed as a more descriptive and less misleading term (Bredehoeft and Konikow, 1993). Others feel that "validation" has a very specific meaning within numerical mathematical modeling and should be retained (McCombie and McKinley, 1993). Without a history matching step, any model application is unreliable. Unfortunately many problems to which vadose zone modeling must be applied do not lend themselves to validation. First, the variability in the geologic environment is frequently too great. A cogent argument that vadose zone modeling in complex geology can never be "validated" is that it is impossible to sample enough vadose zone flow pathways. Second, effective history matching is frequently impossible due to the lack of historical data. Vadose zone processes can be so slow, and monitoring networks so young, that there is no meaningful time through which to match a history.

It should not be assumed that application of a computer model is straightforward. To get "up and running" could require some involved hardware and software decisions and a lot of personnel time. Most computer source codes for powerful fate and transport models are written in Fortran, which will not run on popular desktop machines unless compiled into DOS or Macintosh compatible formats. Some models are available in compiled versions, but compatibility problems are not uncommon. Although many model codes have preprocessors that build control and input files, some of the most powerful codes do not.

CONCLUSIONS

The state-of-the-art in computer modeling of vadose-zone flow and transport is that reliable prediction of contaminant flow in space and time has not been realized (Fogg et al., 1995). Therefore, computer-modeled simulations alone are not a reliable basis upon which to (1) justify that monitoring systems are unnecessary, (2) evaluate the efficacy of containment barriers, (3) develop monitoring schedules, or (4) define appropriate monitoring networks. Nonetheless, computer modeling can be a useful design tool when combined with expert opinion. Monitoring/modeling in an iterative combination can be beneficial to understanding site-specific processes that affect contaminant transport in the vadose zone. Furthermore, modeling represents the only available means of simulating potential contaminant transport into the future, between monitoring stations, or beyond monitoring networks. Despite predictive uncertainties, computer models are the best available tool for quantifying environmental risk.

Numerous computer models have been developed for vadose zone flow and transport problems. Well-tested models exist in the public domain. Models suitable for a variety of vadose zone applications have also been and are being developed in private and research settings. Computer codes are available to simulate complex processes such as multispecies, multiphase flow and transport in geologically complicated domains.

ACKNOWLEDGMENT

This work was partially funded through the Vadose Zone Monitoring Laboratory at the Institute for Crustal Studies, University of California/Santa Barbara, by EPA cooperative agreement CR 816969-01-0 with the Environmental Systems Laboratory-Las Vegas, Lawrence Eccles, Project Officer, contribution #0153-30HW.

REFERENCES

ASTM, American Society for Testing and Materials Annual Book of Standards, ASTM, Race Street, Philadelphia, PA, published annually.

Bear, J., *Hydraulics of Groundwater*, (New York: McGraw Hill, 1979), Chapter 6.

Bonazountas, M., J.M. Wagner, and B. Goodwin, *"SESOIL" A Seasonal Soil Compartment Model*. EPA Contract No. 68-01-6271, by Arthur D. Little, Cambridge, MA, for Office of Toxic Substances, US EPA, Washington, DC, 1984.

Bouwer, H., *Groundwater Hydrology*, (New York: McGraw-Hill, 1978), Chapter 2.

Bredehoeft, J.D., and L.F. Konikow, Editorial – "Ground Water Models: Validate or Invalidate," *Ground Water*, 31(2):178 (1993).

Bumb, A., C. McKee, R.B. Evans, and L.A. Eccles, "Design of Lysimeter Leak Detector Networks for Surface Impoundments and Landfills," *Ground Water Monitoring Review*, 8(2):148-160 (1988).

Carovillano, R.L., "A Comparison of Four Unsaturated Flow Models," in *Proceedings 1993 Ground Water Modeling Conference*, Colorado School of Mines, 1993, P-106.

Celia, M.A., and P. Binning, "A Mass Conservation Numerical Solution for Two-Phase Flow in Porous Media with Application to Unsaturated Flow," *Water Resour. Res.*, 28 (10):2819-2828 (1992).

Corwin, D., and Waggoner, B., *TETrans: Trace Element Transport*. Macintosh Version 1.6: Research Report No. 121, IBM PC-compatible Version 1.5, Research Report No.123, U.S. Salinity Laboratory, Agricultural Research Service, U. S. Department of Agriculture, Riverside, CA, 1990.

Cullen, S.J., and L.G. Everett, *Permit Writer's Guidance Document for Monitoring Unsaturated Regions of the Vadose Zone at RCRA, Subtitle C Facilities*. Report (2nd Draft) to U.S. Environmental Protection Agency, University of California, Santa Barbara, 1993.

CWRCB, California State Water Resources Control Board, *State of California Leaking Underground Fuel Tank (LUFT) Field Manual: Guidelines for Site Assessment, Cleanup, and Underground Storage Tank Closure*. Leaking Underground Fuel Tank Task Force, Sacramento, CA, 1989.

Davis, L.A., and S.P. Neuman, *Documentation and User's Guide: UNSAT2 – Variably Satu-

rated Flow Model. NUREG/CR-3390, U.S. Nuclear Regulatory Commission, Washington, DC, 1983.

Dean, J.D., P.S. Huyakorn, A.S. Donigian, Jr., K.A. Voos, R.W. Schanz, Y.J. Meeks, and R.F. Carsel, *Risk of Unsaturated/Saturated Transport and Transformation of Chemical Concentrations (RUSTIC); Volume 2: Users Guide*. EPA/600/3-89/048b, US EPA, ORD/ERL, Athens, GA, 1989.

Donigian, A.S., Jr., and P.S.C. Rao, "Overview of Terrestrial Processes and Modeling," in *Vadose Zone Modeling of Organic Pollutants*, S.C. Hern and S.M. Melancon, Eds., (Chelsea, MI: Lewis Publishers, 1987).

Emerson, C.J., B. Thomas, Jr., and R.J. Luxmoore. *CADIL: Model Documentation for Chemical Adsorption and Degradation in Land*. ORNL/TM-8972, Oak Ridge National Laboratory, Oak Ridge, TN, 1984.

Enfield, C.G., R.F. Carsel, S.Z. Cohen, T. Phan, and D.M. Walters, "Approximating Pollutant Transport to Ground Water," *Ground Water*, 20(6):711-722 (1982).

Everett, L.G., L.G. Wilson, and E.W. Hoylman, *Vadose Zone Monitoring for Hazardous Waste Sites*, (Park Ridge, NJ: Noyes Data Corporation, 1984).

Fetter, C.W., *Applied Hydrogeology*, (Columbus, OH: Merrill Publ. Co., 1988), pp. 87-101.

Fogg, G.E., D.R. Nielsen, and D. Shibberu, "Modeling Contaminant Transport in the Vadose Zone: Perspective on State of the Art," Chapter 16 in *Handbook of Vadose Zone Characterization & Monitoring*, L.G. Wilson, L.G. Everett, and S.J. Cullen, Eds. (Chelsea, MI: Lewis Publishers, Inc., 1995).

Freeze, R.A., and J.A. Cherry, *Groundwater*, (Englewood Cliffs, NJ: Prentice-Hall, 1979), pp. 39-46.

Healy, R.W., *Simulation of Solute Transport in Variably Saturated Porous Media with Supplemental Information on Modifications to the U.S. Geological Survey's Computer Program VS2D*. U. S. Geol. Survey Water-Resources Investigations Report 90-4025, Denver, CO, 1990.

Hern, S.C., and S.M. Melancon, *Vadose Zone Modeling of Organic Pollutants*. (Chelsea, MI: Lewis Publishers, 1987).

Hetrick, D.M., C.C. Travis, S.K. Leonard, and R.S. Kinerson, *Qualitative Validation of Pollutant Transport Components of an Unsaturated Soil Zone Model (SESOIL)*. ORNL/TM-10672, Oak Ridge National Laboratory, Oak Ridge, TN, 1988.

Hetrick, D.M., S.J. Scott, and M.J. Barden, *The New SESOIL User's Guide*, PUBL-SW-200-93, Wisconsin Department of Natural Resources, 1993.

Hillel, D., *Introduction to Soil Physics*, (New York: Academic Press, 1982).

HydroGeologic, Inc., *VAM2D*. HydroGeologic Software Sales, Herndon, VA, 1993.

Jury, W. A., "Simulation of Solute Transport Using a Transfer Function Model," *Water Resour. Res.*, 22(2):243-247 (1982).

Klute, A., Ed., *Methods of Soil Analysis, Part 1: Physical and Mineralogical Methods, 2nd Edition*. (Madison, WI: American Society of Agronomy, 1986).

Kool, J.B., *HYTEQ Computer Model for One-Dimensional Flow and Multicomponent Solute Transport with Chemical Equilibria Speciation Version 1.0*. HydroGeologic, Inc., Herndon, VA, 1990.

Kool, J.B., and M.T. van Genuchten, *HYDRUS. One-Dimensional Variably Saturated Flow and Transport Model Including Hysteresis and Root Water Uptake*. U.S. Salinity Laboratory, U.S. Department of Agriculture, Agricultural Research Service, Riverside, CA, 1991.

Lappala, E.G., R.W. Healy, and E.P. Weeks, *Documentation of Computer Program VS2D to Solve the Equations of Fluid Flow in Variably Saturated Porous Media*. U.S. Geol. Survey Water Resources Investigations Report 83–4099, 1987.

Lehr, J.H., Editorial—"Toxicological Risk Assessment Distortions: Part II—The Dose Makes the Poison," *Ground Water*, 28(2):170 (1990).

Liu, C.W., and T.N. Narasimhan, "Redox-Controlled Multiple-Species Reactive Chemical Transport 1. Model Development," *Water Resour. Res.*, 25(5):869–882 (1989).

Mangold, D.C., and C-F. Tsang, "A Summary of Subsurface Hydrological and Hydrochemical Models," *Reviews of Geophysics*, 29(1):51–79 (1991)

Massmann, J., A.R. Freeze, L. Smith, T. Sperling, and B. James, "Hydrogeological Decision Analysis: 2. Applications to Ground Water Contamination," *Ground Water*, 29(4):536–548 (1991).

McCombie, C., and I. McKinley, Guest Editorial—"Validation-Another Perspective," *Ground Water*, 31(4):530 (1993).

McKee, C.R., and A.C. Bumb, "A Three-Dimensional Analytical Model for Aid in Selecting Monitoring Locations in the Vadose Zone," *Ground Water Monitoring Review*, 8(2):124–136 (1988).

Morel-Seytoux, H.J., Ed., "Unsaturated Flow in Hydrologic Modeling Theory and Practice," *NATO ASI Series, Series C: Mathematical and Physical Series*, 275, (Norwell, MA: Kluwer Academic Publishers, 1988).

Mullins, J.A., R.F. Carsel, J.E. Scarborough, and A.M. Ivery. *PRZM-2 User's Manual, Version 1*. EPA/600/R-93/046, US EPA Environmental Research Laboratory, Athens, GA, 1992.

Nielsen, D.R., M.T. van Genuchten, and J.W. Biggar, "Water Flow and Solute Transport Processes in the Unsaturated Zone," *Water Resour. Res.*, 22(9):89–108 (1986).

Nofziger, D.L., J-S. Chen, and C.T. Haan, *Evaluation of Unsaturated/Vadose Zone Models for Superfund Sites*, EPA-600/ R-93/184, U.S. Environmental Protection Agency, R.S. Kerr Environmental Research Lab., Ada, OK, RSKERL-Ada-9331, 1993.

Nofziger, D.L., K. Rajender, S.K. Nayudu, and P-Y Su, *CHEMFLO: One-Dimensional Water and Chemical Movement in Unsaturated Soils*. EPA/600/8-89/076, U.S. Environmental Protection Agency, R.S. Kerr Environmental Research Laboratory, Ada, OK, 1989.

Nofziger, D.L., J. Williams, and T.E. Short, *Interactive Simulation of the Fate of Hazardous Chemicals During the Land Treatment of Oily Wastes: RITZ Users Guide*. EPA/600/ 8-88-001, US Environmental Protection Agency, 1988.

Nofziger, D.L., P.S.C. Rao, and A.G. Hornsby, *CHEMRANK: Interactive Software for Ranking the Potential of Organic Chemicals to Contaminate Groundwater*. Inst. of Food and Agricultural Sciences, University of Florida, Gainesville, 1988.

Nofziger, D.L., and A.G. Hornsby, *Chemical Movement in Layered Soils: Users Manual*. Circular 780, Inst. of Food and Agricultural Sciences, University of Florida, Gainesville, (also: Computer Software Series CCS-30, Agric. Experiment Station, Div. Agric., Oklahoma State University, Stillwater, OK), 1988.

Odencrantz, J.E., J.H. Farr, and C. Robinson, "Transport Model Parameter Sensitivity for Soil Cleanup Level Determinations Using SESOIL and AT123D in the Context of the

California Leaking Underground Fuel Tank Manual," *Journal of Soil Contamination*, 1(2):159–182 (1992).

Parker, J.C., and M.T. van Genuchten, *Determining Transport Parameters from Laboratory and Field Tracer Experiments.* Bulletin 84-3, ISSN 0096–6088, Virginia Agricultural Experiment Station, Virginia Polytechnic Institute and State University, Blacksburg, VA, 1984.

Pruess, K., *TOUGH2-A General-Purpose Numerical Simulator for Multiphase Fluid and Heat Flow.* LBL-29400, Earth Sciences Division, Lawrence Berkeley Laboratory, Berkeley, CA, 1991.

Schroeder, P.R., R.L. Peyton, B.M. McEnroe, and J.W. Sjostrom, *Hydrologic Evaluation of Landfill Performance, HELP Model, Volume 3, User's Manual for Version 2.* Internal Working Document EL 92-1 Rpt. 1, U.S. Army Corps of Engineers Waterways Experiment Station, Vicksburg, MS, 1988.

Sims, R.C., J.L. Sims, and S.G. Hansen, *Soil Transport and Fate Database 2.0 and Model Management System.* Department of Civil and Environmental Engineering, Utah State University, Logan, UT, 1991.

Simunek, J., and D.L. Suarez, *The SOILCO2 Code for Simulating One-Dimensional Carbon Dioxide Production and Transport in Variably Saturated Porous Media, Version 1.1.* Research Report No. 127, U. S. Salinity Laboratory, Agricultural Research Service, U.S. Department of Agriculture, Riverside, CA, 1992.

Simunek, J., T. Vogel, and M. T. van Genuchten, *The SWMS-2D Code For Simulating Water Flow and Solute Transport in Two-Dimensional Variably Saturated Media, Version 1.1,* Research Report No. 126, U.S. Salinity Laboratory, Agricultural Research Service, U. S. Department of Agriculture, Riverside, CA, 1992.

Stevens, D.K., W.J. Grenney, and Z. Yan, *VIP: A Model for the Evaluation of Hazardous Substances in Soil.* Civil and Environmental Engineering Department, Utah State University, Logan, UT, 1989.

Strack, O.D.L., *Groundwater Mechanics,* (Englewood Cliffs, NJ: Prentice Hall, 1989).

Travis, B.J., and K.H. Birdsell, *TRACR3D: A Model of Flow and Transport in Porous Media, Model Description and User's Manual.* LA-11798-M, Manual, UC-814, Los Alamos National Laboratory, Los Alamos, NM, Issued: April 1991.

Turin, J., *VLEACH: A One-Dimensional Finite Difference Vadose Zone Leaching Model.* Report prepared for US EPA Region 9., CH2M Hill, Redding, CA, 1990.

Unlu, K., M.W. Kemblowski, J.C. Parker, D. Stevens, P.K. Chong, and I. Kamil, "A Screening Model for Effects of Land-Disposed Wastes on Groundwater Quality," *Journal of Contaminant Hydrology*, 11:27–49 (1992).

US EPA, Environmental Protection Agency, Consent Decree, US EPA Phoenix Goodyear Airport Superfund Site, Appendix B, 1991.

van Genuchten, M. T., F.J. Leij, and S.R. Yates, *The RETC Code of Quantifying the Hydraulic Functions of Unsaturated Soils.* EPA/600/2-91/065, U.S. Environmental Protection Agency, R.S. Kerr Environmental Research Lab., Ada, OK, 1991.

van der Heijde, P.K.M., *Identification and Compilation of Unsaturated/Vadose Zone Models,* EPA-600/, Environmental Protection Agency, R.S. Kerr Environmental Research Lab., Ada, OK, RSKERL-Ada-9332, 1993.

Voss, C.I., *Saturated-unsaturated Transport (SUTRA),* U.S. Geological Survey Water Resources Investigation, 84-4369, 1984.

Wallin, R.W., and L.G. Wilson, *Development of Guidelines for Regulating Depths of Storm-Water Wells to Minimize Ground-Water Pollution*, Final Report to U.S. Environmental Protection Agency under the Shallow Well Initiative Program Phase 2 Final Report. Available through The University of Arizona, Department of Hydrology and Water Resources, 1992.

Watson, D.B., and S.M. Brown, *Testing and Evaluation of the SESOIL Model*. Anderson-Nichols and Co., Palo Alto, CA, 1985.

Yeh, G.T., *HYDROGEOCHEM: A Coupled Model of Hydrologic Transport and Geochemical Reaction in Saturated-Unsaturated Media*. Short Course on Modeling Reactive Multispecies-Multicomponent Transport Through Subsurface Media. Department of Civil Engineering, Pennsylvania State University, University Park, PA, 1992.

Yeh, J.T-C., R. Srivastava, A. Guzman, and T. Harter, "A Numerical Model for Water Flow and Chemical Transport in Variably Saturated Porous Media," *Ground Water*, 31(4):634 (1993).

Yeh, J. T-C., L.W. Gelhar, and A.L. Gutjahr, "Stochastic Analysis of Unsaturated Flow in Heterogeneous Soils. Observations and Applications," *Water Resour. Res.*, 21:465–471 (1985).

18

Vadose Zone Monitoring with the Neutron Moisture Probe

John H. Kramer, Stephen J. Cullen, and Lorne G. Everett

INTRODUCTION

The neutron moisture probe is increasingly being applied to vadose zone hydro-geologic problems such as site characterization, contaminant leak detection and monitoring, remediation, infiltration, and recharge. Ground-water scientists and engineers should be acquainted with the principles of operation, limitations, and advantages of neutron measurements. This chapter reviews the theory and application of neutron moisture probes and presents results from its use. The first part provides background necessary to understand the technique. The second part describes three monitoring projects which successfully employed the method. The intent of the authors is to provide a useful guide to the advantages of this monitoring approach.

BACKGROUND OF NEUTRON MOISTURE MEASUREMENT

Neutron soil moisture measurement was established in agriculture (van Bavel, 1963, Gardner, 1965) before environmental monitoring needs were identified. In 1972, the first major ground-water study funded by EPA (Everett, 1980) recognized that neutron moderation was a cost-effective technology. Everett et al. (1984) defined the application of vadose zone monitoring at solid and hazardous waste landfills and gave examples at hazardous waste impoundments. Six landfills in Los Angeles County were characterized with neutron measurements (Everett, 1985) to comply with California's Calderon Bill. The first horizontal tube application of neutron measurements was reported by Brose and Shatz (1986). Unruh et al. (1990)

Reprinted by permission of the Ground Water Publishing Company. Copyright 1992.

report precise, accurate, reliable, and cost-effective monitoring by nondestructive means at a landfill using neutron data. Neutron monitoring was used beneath a solid-waste landfill in conjunction with deep (greater than 250 feet) installation of high-pressure/vacuum lysimeters hung from a neutron access casing by Cullen et al. (Cullen et al., 1991). The concept of using neutron probes in perforated casing as part of a monitoring and passive remediation strategy was presented at the U.S.A./U.S.S.R. hydrology conference by Everett et al. (1990). The use of neutron monitoring to support passive remediation or hydrocarbon stability is receiving increased attention. In response to regulatory interpretations concerning waste facility siting, neutron data are now being used to estimate the effective height of the capillary fringe above the water table.

Neutron Moderation Theory

Neutron probes take advantage of the neutron moderation process in which fast neutrons emitted from a radioactive source are moderated, or slowed, by collisions with surrounding atoms. Slowed neutrons, also called thermalized neutrons, diffuse through interatomic space like gas particles until they are absorbed or captured by receptive atoms. Thermalized neutron capture emits a pulse of detectable energy, which is counted in a neutron probe detector.

The neutron moderation process is dominated by neutron-hydrogen collisions. Fast neutron collisions with relatively heavy soil atoms result in little loss of neutron velocity, whereas collisions with hydrogen atoms (of equal mass) result in appreciable neutron slowing. Thus, relatively high hydrogen density (moles/cc of soil) results in rapid (near source) neutron moderation.

Hydrogen in geologic material occurs as water, mineralogically bound $H+$, organic soil components, and organic liquids (e.g., petroleum contaminants). Water is nearly always the greatest source of hydrogen in the subsurface. Therefore, as a dry soil is wetted, the thermalized neutron density near a neutron source will increase. Elements vary in their capture cross sections, or the propensity to absorb thermalized neutrons. Of the elements present in soils and monitoring well systems, chlorine and boron are particularly high. Neutron capture by these elements may significantly lower neutron probe counts and should be considered in arid regions where salt buildup occurs in soils. The effects of chlorine in polyvinylchloride (PVC) casing on neutron moisture probe data do not significantly interfere with wetting front detection (Keller et al., 1990). Schneider and Greenhouse (1992) report significant count decreases resulting from chlorine present in liquid perchlorethylene (PCE). Additional discussion of theoretical concepts is presented by Greacen (1981), Keys and MacCary (1983), and Welex (1978).

Neutron Probe: Instrument Design

The term neutron probe describes two basic types of instruments: neutron-porosity loggers (type 1), used primarily in liquid-filled boreholes, and neutron moisture probes (type 2), used in dry access tubes. Both types use a fast neutron source and thermal neutron detector(s) but differ in source strength and detector geometry. The result is that neutron counts in type 1 instruments are inversely proportional to hydrogen density, whereas they are directly proportional in type 2 instruments. Ground-water investigators should be aware of the type of instrument being employed before interpreting results.

Type 1 neutron-porosity probes are well established in the oil industry, where they are used to infer porosity in saturated sediments. Type 1 probes employ strong sources (250 mCi to 5 Ci) to enhance sampling volume and dual detectors to isolate borehole effects. These probes, which are subject to stringent regulatory oversight, utilize long tool lengths (4 to 21 feet) and sophisticated, expensive logging vehicles. They have been successfully used to monitor the vadose zone (Hargis and Montgomery, 1983), and are available by subcontracting borehole logging specialists.

Type 2 neutron moisture probes are designed for precise measurement of soil moisture in small-diameter (2-inch) access tubes, but have also been employed in larger access tubes (Keller et al., 1990; Kramer et al., 1990a; Tyler, 1985, 1988; Hammermeister et al., 1985). These probes employ a low-strength source (10 to $50 \geq$ mCi) and a single, near-source detector. Similar to density gauges used by many geotechnical firms, they require minimal operator training (an eight-hour radiation safety course) and are subject to less regulatory oversight than the type 1 probes described. The moderate one-time acquisition cost ($4300 to $6900), low maintenance, one-person operation, light weight (less than 30 pounds), and small tool length (1 foot) keep the cost per sampling event low and make it the method of choice for diverse vadose zone monitoring tasks, especially in situations in which frequent sequential data are required. Hereafter, all references in this chapter to neutron probes refer to this type of instrument.

Instrument Capabilities

Neutron probes can be calibrated to measure soil moisture changes on the order of 1 to 5 volume percent. Although designed for measuring pore water, neutron probes are also sensitive to hydrocarbon liquids, as demonstrated by Kramer et al. (1990). Figure 18.1 shows data from five different positions in a test chamber which were

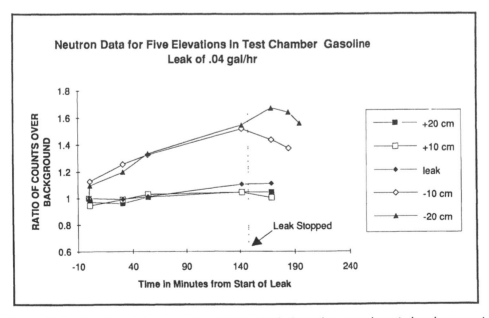

Figure 18.1.   Neutron data from slow gasoline leak detection experiment showing count increase in response to leak.

obtained before, during, and after the initiation of a slow gasoline leak halfway up the chamber. After initiating the leak, count increases were observed at elevations below the leak, but no count increases were observed at or above the leak elevation. The counts below the leak elevation decrease after the leak is cut off. The data indicate that the neutron moisture gauge is sensitive to slow liquid gasoline leaks, but not to increasing gasoline vapors which, though present, did not cause count increases at levels above the leak.

The radius of influence for neutron moisture probes varies with source strength, hydrogen density, solids density, and chemistry. Practical limits are from 6 to 24 inches from the point between the source and detector (Kramer et al., 1990a; Silvestri et al., 1991). The cloud of thermalized neutrons is compact in wet and/or dense soils, and expanded in dry and/or loose soils. Strong neutron absorbers, including boron and chlorine, will diminish the radius of influence.

Sensitivity of the neutron probe to wetting fronts depends on the magnitude of the possible moisture change and the masking effects of grout and casing. The neutron probe is most sensitive in ungrouted access tubes with minimal distance between the tube and soils (Teasdale and Johnson, 1970). Small-diameter (2-inch) driven metal casings are best. Steel, aluminum, stainless steel, and PVC casings at diameters up to 4 inches, installed in borings and backfilled with native material, have also been successfully used.

The neutron moisture probe can be applied in grouted wells. Well grout must cure to maturity before grout water will stabilize. For neat cement grout mixed at 5 gallons of water per 94-pound sack of cement, the stabilization period is on the order of seven days. Wetter grout will require much longer stabilization periods.

In laboratory tests at very low background moisture contents, grout thickness up to 3 radial inches did not obscure the detection of wetting fronts with increase of volumetric water content of 9 percent (Keller et al., 1990). Neutron data from a follow-up field test at more realistic background moisture levels are presented in Figure 18.2. A mock well (consisting of a grouted 10-inch diameter borehole with 4-

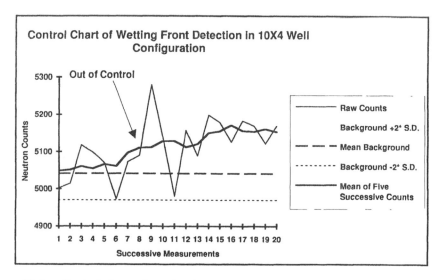

Figure 18.2.   Control chart of wetting front detection by neutron probe through 3 inches of grout and 4-inch diameter PVC casing in loamy soils.

inch PVC casing) from the aforementioned study was buried in loamy soils, allowed to stabilize for three months, and tested for wetting front detection. Figure 18.2 shows that under these field conditions, a wetting front could be detected in a 4-inch PVC casing through 3 radial inches of grout. Under wetter field conditions in clay-rich soils, however, a 3-inch grout thickness may fully mask neutron response to wetting. As a general rule, grout thickness greater than 3 radial inches is inappropriate for neutron moisture probes, although limited success has been achieved through thicker grout in arid settings.

Instrument Shortcomings

The neutron moisture probe cannot distinguish chemical species (e.g., leachate from ground water) and is sensitive only to changes in hydrogen density or neutron-absorbing solutes and liquids. Neutron data cannot be used to distinguish between gasoline and water, which have similar hydrogen densities (Kramer et al., 1990), nor to detect a steady-state flow situation that has no changing hydrogen content. In high background moisture environments, such as heavy clay-rich soils, under conditions of fully masking grout, or in the presence of migrating boron or chlorine, detection of wetting fronts with a neutron probe would be uncertain or impossible.

Counting Statistics

Neutron moisture probe data are amenable to statistical analysis. A statistical model, the continuous normal approximation of a Poisson distribution, can be used to approximate the population from which monitoring data are taken, and the significance of changes in monitoring data can be inferred.

The Poisson distribution, a special case of the binomial distribution appropriate to radioactive decay processes, offers a convenient estimate of the sample standard deviation as the square root of the mean:

$$\sigma = x^{(1/2)} \tag{1}$$

where σ = standard deviation and x = sample mean.

An estimate of the standard deviation is useful because it forms the basis for a variety of popular statistical testing techniques for continuous normal distributions. The continuous normal distribution is applicable to the Poisson distribution for sample sizes greater than 15. This is the reason that one probe manufacturer uses 16-second counts as a standard. The electronics of the probe actually count 16 one-second periods and report the mean of this subsample. Equation 1 can then be applied to obtain an estimate of the population standard deviation from any 16-second count.

To compare neutron probe counting statistics with the Poisson distribution model, 900 separate counts were collected at five different counting times in two materials, a clay and a dry sand, in barrels in the laboratory (100 counts for each material and time). The statistics are presented in Table 18.1. In all cases, Poisson-based calculations of the standard deviation (S.D.$_2$) were in close agreement with the standard deviations calculated for the whole data set (S.D.$_1$). Table 18.1 demonstrates the validity of using the square root of the mean as an estimate of the standard deviation.

Table 18.1.   Effect of Counting Time on Statistical Parameters Used to Interpret Neutron Probe Data

Parameter	Wet Clay Neutron Probe Counts				Dry Sand Neutron Probe Counts			
Counting Time (sec)	4	16	32	64	4	16	32	764	256
Minimum	2958	12405	24678	50028	141	612	1282	2560	10816
Maximum	3312	12879	25374	51212	200	726	1430	2840	11264
Range	354	474	696	1184	59	114	148	280	448
Mean	3143	12667	15045	50650	171	677	1362	2649	11232
Median	3149	12669	25042	50546	169	680	1362	2686	11024
Variance	3472	10698	25058	50547	171	637	1034	2682	11380
S.D.$_{-1}$	58.9	103.4	158.3	224.8	13.1	25.2	32.2	51.8	106.7
C.V.$_{-1}$	.019	.008	.006	.004	0.77	.037	.024	.019	.010
S.D.$_{-2}$	56.1	112.5	158.3	225.1	13.1	26.0	36.9	51.9	105.0

n = 100 measurements at each counting time.

S.D.$_{-1}$ = standard deviation calculation based on assumption of a normal distribution.

C.V.$_{-1}$ = coefficient of variation calculation based on assumption of a normal distribution.

S.D.$_{-2}$ = standard deviation calculation based on the assumption of the Poisson distribution (i.e., square root of the mean).

Instrument Precision

Precision, or the reproducibility of successive measurements, depends on instrument reliability, background stability, and counting time. The electronics in commercially available neutron probes are stable enough that, given an invariant background, 1% precision 68% of the time is routinely attained by adjusting counting times (CPN Corp., 1978). The precision improves with longer counting time, as shown by the decreasing coefficient of variation (C.V.: the ratio of one standard deviation to the mean) in Table 18.1. Assuming the normal approximation of the Poisson distribution model, neutron probe precision at the 68% confidence level can be expressed as:

$$P = 100 * x^{(1/2)}/x \qquad (2)$$

where P = precision in percent and x = sample mean count.

A useful rule of thumb is that a precision of ± 1% in 68 out of 100 samples can be expected by counting long enough to record 10,000 counts. For example, if a 16-second measurement recorded 5000 counts, then a 32-second reading would be required to attain 1% precision. If a four-second measurement of 2500≥ counts is obtained, the precision would be 2% ($100*2500^{(1/2)}/2500$). Note that on instruments that report counts normalized to a specific time equivalent (e.g., 16-second), the actual number of counts at the selected measurement time must be used (in the 32-second example, the reported measurement would be 5000 even though 10,000 counts were measured). Higher confidence in the precision is obtained by using multiples of the standard deviation in Equation 2. For the 32-second example, the precision is expected to be ± 2% in 95 out of 100 samples ($2*100*10,000^{(1/2)}/10,000$ = 2%).

An Example Validation of Short Counting Time Methodology

For a practical monitoring scheme, short counting times are economically advantageous in terms of field personnel time and safety. In a landfill monitoring study (Cullen et al., 1991) 16-second counting periods were undesirable. To support the use of shorter count times, an experiment was conducted to determine if significant differences existed between four-second counts and 16-second counts (both normalized to 16-second counts).

In the two-treatment experiment, the first treatment consisted of two sets of readings made in a borehole using 16-second counting periods. The second treatment was made separately and identical to the first, except that the readings were made using four-second count times. The means of the four-second readings from each depth were paired with the means of the 16-second readings from the same depths to minimize variability which could be attributed to stratigraphic differences.

Based on a paired t-test of the data as shown in Table 18.2, no difference was detected between the results of the four-second readings and those of the 16-second readings. This conclusion only applies to the site where these data were collected. The use of short counting-time intervals should be validated on a site-by-site basis.

Table 18.2.   Paired Samples T-Test on 4-Second vs. 16-Second Counts of Neutron Probe at Background Monitoring Access Tube of a California Solid Waste Landfill

Depth (Feet Below Ground Surface)	4-Second Counts	16-Second Counts
3	4094.0	4207.0
8	6142.0	6297.5
13	7660.0	7647.5
18	8878.0	8865.0
23	2578.0	2450.5
28	2598.0	2527.0
33	3860.0	3758.0
38	5032.0	4914.0

n = 8   sample t = .584
mean difference = 21.94   tabulated t = 3.499
sd difference = 106.27   alpha = .01

Count Ratios

Neutron probe data are often reported as count ratios, which are ratios of counts at sampling locations to counts on the same instrument in a stable standard material. Count ratios are defined as follows:

$$R = C/S \tag{3}$$

where R = Count Ratio, C = Measured counts, and S = Counts in standard material.

The ratio is used to offset instrument drift between sampling events due to component degradation or other factors. A disadvantage to reporting probe data as ratios is that random error in both the standard (ratio denominator) and the measurement (ratio numerator) can compound. To minimize error, the mean of two or more standard counts taken before and after the measurements should be used in ratios. Sinclair and Williams (1979) show that averaging several count ratios will reduce precision error to a very small amount of total variance. Averaging standard counts or count ratios over a large number of sampling events may obscure real moisture changes. In monitoring applications of long duration (i.e., years), ratios will eventually become desirable and standard counts before and after each monitoring event should be recorded and averaged.

The standard material may be any stable volume of neutron moderators. Manufacturers provide a shielded case which can be used as a standard but which may be subject to external influences. Greacen et al. (1981) recommend a large drum of water fitted with a centered access tube as a standard if portability is not a problem. Such standards will produce repeatable measurements to within 1% of the mean approximately 68% of the time.

Figure 18.3 compares three types of standard measurements for the same probe: an instrument case shield on a bench (or truck tailgate), a case shield on the ground (pavement, cement, or compacted earth), and an access tube in a 55-gallon drum of water. The case standards run on the bench include three extreme outliers that may

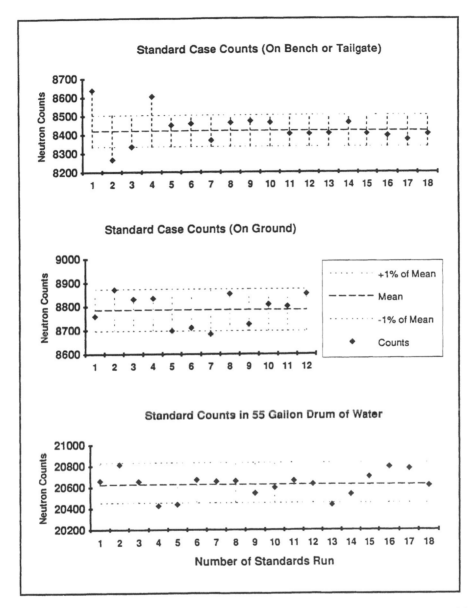

Figure 18.3. Three neutron standards: case shield on a bench or tailgate, case shield on ground surface, and drum of water.

have been caused by some external source of neutron moderators and absorbers near the case shield. It is easier to control external conditions for case standards run on the ground. Figure 18.3 shows that, for the instrument used, the precision in the water standard is no better than the precision of the case standard on a dense surface. As a general rule it is best to use a standard that approximates the count rate obtained in the target soils; this practice removes any effect of variance at different count rates.

Greacen et al. (1981) state that count ratios simplify comparisons of measurements from different experimental conditions, access tube geometries, or instruments. It should be cautioned, however, that ratios are not direct conversions between instruments. Each instrument has a unique combination of source strength and detector sensitivity which affect ratios. Figure 18.4 shows neutron ratios collected during an induced recharge event in an affected well and a control well. A change of instruments occurred between the sixth and seventh measuring events. Ratios in the control well show a downward bias associated with the instrument change which, if uncorrected, would mask the magnitude of moisture change in the affected well. Bias corrections will vary with the moisture contents measured because each instrument has a unique calibration coefficient (see next section). Comparisons between two instruments over a range of measured moisture levels require at least a two-point (wet and dry) calibration standard with each instrument.

Count ratios are also used to standardize measurements taken at different counting time intervals to develop a single soil calibration curve that will work for all counting time intervals.

Calibration

Neutron probe data can be reported as calibrated units of volumetric soil moisture content. This form of reporting is desirable if data are to be used to estimate soil water and soluble contaminant mobility in conjunction with a known soil water characteristic curve and associated knowledge of unsaturated hydraulic conductivity as a function of partial saturation. By measuring soil water content, one can assess the likelihood of liquid movement. If the soil water content is below a critical value (unique to each soil), then the likelihood of contaminant mobility is negligible.

Calibration curves traditionally are made in the laboratory (drums) or in the field by measuring the neutron counts from a given probe in a target soil at two or more known moisture contents and regressing these to a linear model. At least two of the soil moisture contents used in the calibration should span the moisture contents of interest in the natural system. Such a regression takes the form:

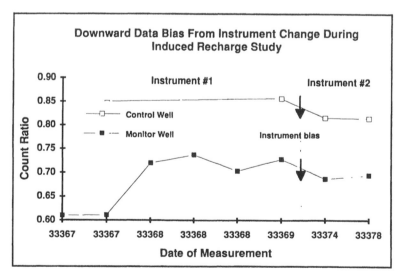

Figure 18.4. Neutron count ratios from two instruments in two wells documenting bias due to instrument change during a recharge event.

$$\theta = \theta_o + \beta\, n \qquad (4)$$

where θ = volumetric moisture content, θ_o = Y axis intercept, β = calibration coefficient, and n = neutron counts or count ratio.

Holland (1969) recommends regressing n on θ, where n is dependent for drum calibrations, and where error in t is small. In field calibrations, most workers chose t as the dependent variable, as in Equation 4, because it is desirable to know what moisture content corresponds to a given neutron count.

Laboratory drum calibrations can be performed in 55-gallon drums fitted with centered access tubes (identical to the field access tubes) and packed with target soils at different known water contents and densities (Silvestri et al., 1991). Water contents can be adjusted by mixing measured amounts of water with soils before packing the drums. Densities can be measured by weighing the tared drums or volumetrically sampling the packed material. Although this method results in even moisture distribution within the drum, it is difficult to accurately reproduce the density and pore-size distribution of the field soils. Another technique is to pack the drum with dry soil and saturate it from the bottom up (to exclude air entrapment). This technique is only good for two-point calibrations because moisture stratification can occur within the drum at less than saturation. To avoid this problem, drums may be fitted with candle extractors, tensiometers, TDR (time domain reflectometry), or capacitance probes to control tension and measure moisture content during calibrations.

Field calibrations can be performed by destructively volume sampling soils near an access tube (five samples are desirable) and measuring water contents gravimetrically before, during, and after saturation events, or by implanting a TDR or capacitance probe to measure changing soil moisture. These measurements can be correlated to neutron measurements taken during a wetting event.

In most hydrogeologic applications, it is impractical to obtain multiple gravimetric samples at a given position for calibration regressions (e.g., in deep borings in thin-bedded strata), or to monitor induced saturation events in situ (e.g., if contaminant mobilization is possible). It may also be impossible to obtain adequate representative soil samples for drum calibrations. In these cases, approximate calibrations can be generated using limited gravimetric data or by applying calibration curves from similar soils. Silvestri et al. (1991) showed that a factory calibration is generally applicable to sandy soils, but that probes should be calibrated to the soils in which they are to be used. They provide a calibration curve for a particular probe in wet clay soils and note that the calibration appears nonlinear at greater than 40% moisture content. This is most likely due to instrument design; loss of detector efficiency occurs as the cloud of thermalized neutrons collapses to within the geometry of the detector.

Calibrations are prone to error from uncertainty in the quality and representativeness of samples upon which they are based. In many geologic settings each monitoring position deserves a specific calibration curve based on a unique combination of field density and bound hydrogen. The number of calibrations can be impractically large, necessitating the grouping of different soils that actually have clearly distinct moisture retention properties. The resulting approximate calibration curve is an information filter, reducing sensitivity to changes at any single position. Consider data in Figure 18.5 from neutron moisture logs in a 2000-foot horizontal access tube. The variations in counts between positions are numerous, yet the pattern of

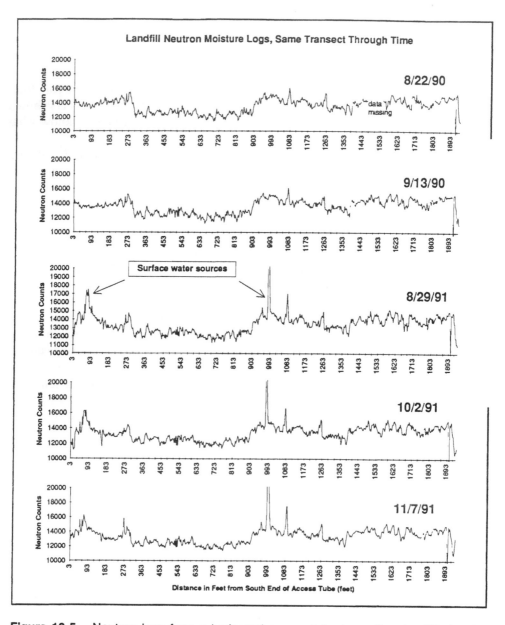

Figure 18.5. Neutron logs from a horizontal access tube beneath a landfill showing distinctly repeatable patterns, spatial variability, and pinpointed moisture anomalies.

peaks is distinct and repeatable from one log to the next (excluding obvious anomalous peaks to be discussed later). These variations indicate the presence of numerous distinct soil-pore distributions, each with a unique soil water characteristic curve and associated neutron calibration. Variograms calculated for these moisture data indicate that at distances greater than 6 feet the data are uncorrelated, which means that at least 330 distinct calibrations would be required to provide quantitative moisture contents along this transect. Because it is impossible to properly collect enough samples to perform these calibrations, some grouping and averaging schemes must be used if calibrated moisture contents are required. If quantitative moisture contents are not needed, background comparison of raw data should be used.

Background Comparison Techniques

When monitoring for changes in moisture content, the question of what represents a meaningful change has traditionally been addressed by attempting to calibrate neutron counts to site soils and to monitor for an arbitrarily chosen level of significant change [e.g., 5% volumetric moisture change (Unruh et al., 1990)]. Neutron data are better used by comparing neutron counts at each position through time to an established background. Analysis of raw data is more sensitive to subtle changes in hydrogen density than data processed through a calibration filter. Only after trends are noted in sequential count data are estimates of volumetric moisture content changes needed for risk assessment or predictive modeling efforts.

In control charting, one useful statistical approach to anomaly definition (U.S. EPA, 1989), a level of significance is assigned within which sequential samples may vary above or below established background. When a sample exceeds the significance level, the process is considered out of control and anomalies are investigated. In the example illustrated by Figure 18.2, a moving window of five sequential count measurements at a single monitoring position advances stepwise through the data set. The mean of each window is plotted at the central time position. If no trend exists in the data, the sequence of window means should be distributed normally about the background mean. If trends do exist, as is the case in Figure 18.2, deviation will be apparent. The plot of window means identifies when the increasing trend began. In the case of contaminant leaks, estimates of leak duration are useful to assessing contaminant volume. This approach is very appropriate to neutron data, which are real time, inexpensive, and repeatable. Out-of-control conditions can be verified by repeat sampling.

EXAMPLES OF NEUTRON MOISTURE MONITORING APPLICATIONS

This section includes descriptions of three field applications of neutron probe monitoring: a logging effort to identify potential monitoring horizons, evaluation of saturation fronts at a soil flushing project, and long-term monitoring beneath a waste facility.

Case 1: Detailed Neutron Logging to Identify Monitoring Horizons

Neutron moisture logs can be useful stratigraphic indicators, as shown by the example of reconnaissance neutron data from three grouted wells shown in Figure 18.6. Data were collected by one person in less than two hours onsite. Grout thickness was less than 2 radial inches and casing was aluminum. Potential flow pathways are indicated by arrows. These positions were chosen because they represent sandy material, which would be a rapid lateral transport pathway in the event of saturated

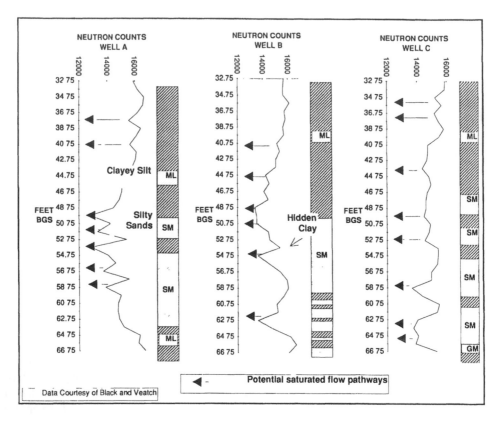

Figure 18.6. Descriptive and neutron logs of three monitor wells. Potential saturated flow pathways are identified based on neutron data.

infiltration events, over finer grained material (ML). Much slower unsaturated flow may occur in the ML (clay-rich) horizons.

Descriptive logs are shown in Figure 18.6 alongside neutron logs. Theory predicts that the sandy material would generate relatively low neutron counts and the clay-rich material would generate higher counts. The two logging techniques are in rough agreement; however, the descriptive log does not match the neutron log in detailed variations. For example, high neutron counts do not correlate well to sand-rich horizons in the physical logs at several positions (e.g., Well B, 53.75 feet). High neutron counts indicate that a clay layer was not detected in the lithologic logs. In fact, the hidden clay layer at 53.75 feet in Well B correlates well with the thin clay at 53 feet in Well A (also detected in the neutron log from Well A at 52.75 feet). A similar clay occurs in Well C at 53.75 feet. The combination of neutron and lithologic logs permits a case to be made for the existence of a continuous clay across the site, a potentially important control to subsurface flow.

Neutron logging is repeatable, whereas one-time boring logs are inherently less accurate unless the boreholes are continuously cored and logged at considerable expense. Therefore, neutron logging can be a cost-effective reconnaissance tool to locate specific, potential, or suspected flow pathways in existing wells or borings that cannot be continuously cored.

Case 2: Soil Flushing Project

A site underlain by fine-grained alluvial material, including interbedded silty sands, clayey silts, and gravels, contains chromium-contaminated soil from the surface to the water table at 80 feet. A ground-water contamination plume in excess of 1 mile long is being captured and treated onsite. The top 20 feet of soil is to be excavated and removed at a later date and will pose no long-term threat to ground water; however, the deeper nonexcavated soils between 20 to 80 feet could be a continuing future source of contamination. A pilot project to evaluate the potential for flushing the soils between 20 and 80 feet with treated ground water was initiated. Flushing is intended to transport contaminants adsorbed on soil particles to the ground-water pumping and treatment system.

The objective of neutron monitoring was to observe progressive saturation of strata near an injection well and measure arrival of wetting fronts. A vertical injection well screened between 20 and 80 feet was pressurized with a 3-foot head of potable water while monitoring tubes, installed 5, 15, and 30 feet from the injection well, were periodically logged with a neutron probe. Figure 18.7 shows example neutron data from two tubes (5 and 30 feet from the injection well) at six depths plotted with time of injection. Strata with high ratios are clay-rich horizons which show few, if any, meaningful changes, whereas sandy horizons (low ratios) show distinct wetting front arrivals. Strata at the 60.3-foot depth change in character from sandy at Tube A to clay-rich at Tube C.

Wetting front arrival times interpreted from these data are shown on a cross section of the site in Figure 18.8. Neutron monitoring positions are depicted as spheres to schematically represent the sampling volume of the neutron moisture probe, and patterned to indicate approximate wetting front arrival times. Contours of wetting front arrival times graphically illustrate vadose zone transport pathways at the site.

The monitoring revealed complex vadose zone transport processes including the presence of sharp fronts, gradual fronts, and hydrodynamic waves in which saturation builds up at the wetting front and dissipates after it passes. It is concluded that

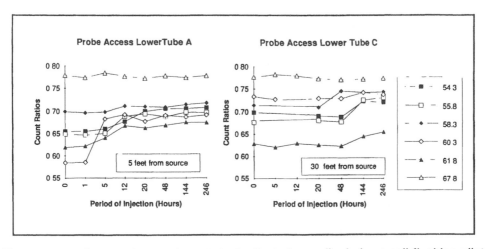

Figure 18.7. Count ratios vs. time at six depths in two wells during a soil flushing pilot study.

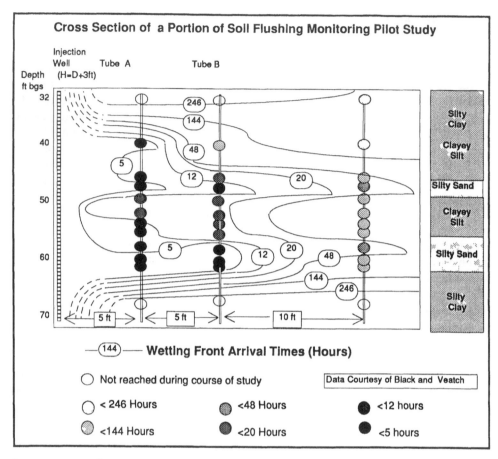

Figure 18.8.   Cross section showing contoured wetting front arrival times interpreted from neutron data.

vadose zone flow at the site is complex, characterized by rapid lateral transport in saturated pulses through discontinuous sandy horizons. Specific transport pathways have been identified using the neutron moisture probe data.

Case 3: Long-Term Landfill Monitoring Strategies

One of the most promising applications for neutron moisture monitoring is at waste facilities where monitoring of horizontal access tubes beneath the facilities can detect incipient liner failure before ground water is impacted. Advantages pointed out by Kramer and Everett (1990) and Kramer et al. (1991) are the low cost per sampling event, high precision, and high sampling density. Automation possibilities can systemize the monitoring task, eliminating human error and providing consistently high-quality data through post-closure phases.

Figure 18.5 shows four neutron logs from a 2000-foot long access tube beneath the leachate collection system of a municipal landfill. Several anomalous neutron data peaks indicate moisture anomalies developed between the second and third logs. The anomaly at 988 feet is a very sharp spike representing a moisture increase over less than 9 feet of the tube. An associated smaller anomaly occurs at 1083 feet.

A second major moisture anomaly at approximately 70 feet is also apparent. Both major anomalies are related to surface water drainage and operational practices at the landfill and probably do not represent leachate leaks. However, they serve to demonstrate the sensitivity of the system to possible leachate leaks, particularly the ability of moisture logs to pinpoint a potential problem area.

An interesting aspect of this set of data is the extraordinarily high count numbers which exceed the standard counts for the same probe access tube configuration in a 55-gallon drum of water. These high counts can only be recreated by partially submerging the neutron probe in a puddle of water; therefore, this discrepancy is tentatively explained by the presence of water in the pipe itself, which was perforated in order to collect gas samples. Proof will come with confirmatory liquid sampling by remote suction lysimeter. The lysimeter will only be able to collect a liquid sample if there is a puddle in the pipe which contacts the ceramic cup in the lysimeter. The sample will also serve to confirm the presence or absence of leachate at this location.

CONCLUSIONS

Innovative applications of the neutron moisture probe to vadose zone monitoring problems are on the rise. It is fast becoming an important component of the geoscientist's toolbox because it can provide high-quality data at relatively low cost. We have provided information on the theory of neutron moderation, and on issues of data collection that will aid potential users. We have demonstrated its application to characterize stratigraphic variability, to identify induced vadose zone flow at a soil flushing pilot project, and as a long-term monitoring strategy at waste facilities.

ACKNOWLEDGMENTS

This work was supported by EPA Cooperative Agreement CR 816969–01–0 between the Environmental Systems Monitoring Laboratory-Las Vegas, Project Officer Lawrence Eccles, and the Institute for Crustal Studies at the University of California Santa Barbara, contribution # 0090–23HW.

REFERENCES

Brose, R.J., and R.W. Shatz, "Neutron Monitoring in the Unsaturated Zone," in *Proceedings, First National Outdoor Action Conference on Aquifer Restoration, Ground Water Monitoring and Geophysical Methods*, NWWA, Dublin, OH, 1986, pp. 455–467.

CPN Corporation, Operator's Manual, 503 Hydroprobe Moisture Depth Gauge, Martinez, CA, 1978.

Cullen, S.J., W.F. Allmon, and B.K. Keller, China Grade Sanitary Landfill: Vadose Zone Monitoring Program, Report to County of Kern, Department of Public Works, Metcalf and Eddy, Inc., Santa Barbara, CA, 1991.

Everett, L.G., L.A. Eccles, and D. A. Blakely, "Neutron Moderation Applications to Hydrocarbon Site Risk Assessment, Monitoring and Remediation," presented at First USA/USSR Joint Conference on Environmental Hydrology and Hydrogeology, Leningrad, USSR, 1990.

Everett, L.G., Vadose Zone Monitoring Designs for Six Landfills in Los Angeles County, Los Angeles County Sanitation Districts, Kaman Tempo, 1985.

Everett, L. G., L.G. Wilson, and E.W. Hoylman, *Vadose Zone Monitoring for Hazardous Waste Sites*, (Park Ridge, NJ: Noyes Data Corp., 1984).

Everett, L.G., *Groundwater Monitoring*, (New York, NY: General Electric Company Publishing Corp., 1980).

Gardner, W.H., "Water Contents," in *Methods of Soil Analysis, Part I*, C.A. Black Ed., Agronomy Monograph No. 9, Am. Soc. Agron., Madison, WI, 1965, pp. 82–127.

Greacen, E.L., Ed. *Soil Water Assessment by the Neutron Method*, Division of Soils CSIRO, Adelaide, Australia, 1981.

Hammermeister, D.P., C.R. Kneiblher, and J. Klenke, "Borehole-Calibration Methods Used in Cased and Uncased Test Holes to Determine Moisture Profiles in the Unsaturated Zone, Yucca Mountain, NV," in *Proceedings of the NWWA Conference on Characterization and Monitoring of the Vadose (Unsaturated) Zone*, Denver, CO, 1985.

Hargis and Montgomery Inc., Results of Construction and Testing of Neutron Logging System Hughes Aircraft Co. Manufacturing Facility, Tucson, Arizona, Report to Hughes Aircraft Corp., Tucson, AZ, 1983.

Holland, D.A., "The Construction of Calibration Curves for Determining Soil Moisture Content from Radiation Counts," *J. Soil Sci.* 20:132–140 (1969).

Keller, B.R., L.G. Everett, and R.J. Marks, "Effects of Access Tube Material and Grout on Neutron Probe Measurements in the Vadose Zone," *Groundwater Monitoring Review*, 10(1):96–100 (1990).

Keys, W.S., and L.M MacCary, "Application of Borehole Geophysics to Water-Resources Investigations," Book 2 Chapter E1, in *Techniques of Water Resources Investigations*, U. S. Geological Survey, 1983.

Kramer, J.H., and L.G. Everett, "Proactive Post-Closure Vadose Zone Monitoring Strategy Using Neutron Logs," presented at Project Earth—Meeting the Demands GRCDA, Western Regional Conference, Ontario, Canada, 1990.

Kramer, J.H., L.G. Everett, L.A. Eccles, and D.A. Blakely, "Contamination Investigations Using Neutron Moderation in Grouted Holes: A Cost-Effective Technique," in *Minimizing Risk to the Hydrologic Environment*, A. Zporozec, Ed., (Dubuque, IA: Kendall/Hunt Publishing Co., 1990), pp. 234–242.

Kramer, J. H., L.G. Everett, and S.J Cullen, "Innovative Vadose Zone Monitoring at a Landfill Using the Neutron Probe," in *Proceedings of the Fifth National Outdoor Action Conference of Aquifer Restoration*, Ground Water Monitoring and Geophysical Methods. NWWA, Dublin, OH, 1991.

Kramer, J. H., L. G. Everett, and L. A. Eccles, "Effects of Well Construction Materials on Neutron Probe Readings with Implications for Vadose Zone Monitoring Strategies," in *Ground Water Management, Number 2*, NWWA, Dublin, OH, 1990a, pp.1303–1317.

Schneider, G.W., and J.P Greenhouse, "Geophysical Detection of Perchlorethylene in a Sandy Aquifer Using Resistivity and Nuclear Logging Techniques," in *Proceedings SA-GEEP*, Chicago, April 26–28, 1992.

Silvestri, V., G. Sarkis, N. Bekkouche, M. Soulie, and C. Tabib, "Laboratory and Field Calibration of a Neutron Depth Moisture Gauge for Use in High Water Content Soils," *Geotechnical Testing Journal*, ASTM, 1991, pp. 64–70.

Sinclair, D.F., and J. Williams, "Components of Variance Involved in Estimating Soil Water Content and Water Content Change Using a Neutron Moisture Meter," *Aust. J. Soil Res.* 17:237–70 (1979).

Teasdale, W.E., and A.I. Johnson, Evaluation of Installation Methods for Neutron-meter Access Tubes, in *U.S. Geol. Survey Prof. Paper 700-C* C237-C241, 1970.

Tyler, S., "Moisture Monitoring in Large Diameter Boreholes," in *Proceedings of Conference on Characterization and Monitoring of the (Unsaturated) Vadose Zone*, NWWA, Dublin, OH, 1985.

Tyler, S., "Neutron Moisture Meter Calibration in Large Diameter Boreholes," *Soil Sci. Soc. Am.*, 52:890 (1988).

U.S. EPA, Statistical Analysis of Ground Water Monitoring Data at RCRA (Resource Conservation and Recovery Act) Facilities: Interim Final Guidance Document. U.S. Dept. of Commerce, NTIS# PB89-151047, 1989.

Unruh, M.E., C. Corey, and J.M. Robertson, "Vadose Zone Monitoring by Fast Neutron Thermalization (Neutron Probe): A 2-Year Case Study," in *Ground Water Management, Number 2, NWWA*, Dublin, OH, 1990, pp. 431–444, 1303–1317.

van Bavel, C.H.M., "Neutron Scattering Measurements of Soil Moisture: Development and Current Status," in *Proc. Int. Symp. Humidity and Moisture*, Washington, DC, 1963, pp. 171–184.

Welex, *Neutron Logging*, A Haliburton Co., Houston, TX, 1978.

19

Discussion of "Vadose Zone Monitoring with the Neutron Moisture Probe"

Michael A. Williams

This chapter will clarify information incorrectly presented by Kramer et al., 1992, concerning the Soil Flushing Project (Case 2). In addition, the practical application of neutron monitoring and the monitoring tube design used at our client's facility will be discussed.

The article incorrectly states that the top 20 feet of soil will be excavated and removed. However, remedial alternatives concerning the top 20 feet have not yet been selected and will be addressed at a later time. The article also states that a vertical injection well was screened from 20 to 80 feet below land surface (bls). Actually, the pilot study injection wells were screened from 20 to 70 feet bls. Because the ground-water surface was at approximately 80 to 82 feet bls, the intention was to avoid direct interconnection between ground water and flushing water until the flushing water had moved laterally for some distance. The monitoring tube spacing is stated incorrectly in the text and on Figures 7 and 8. Monitoring tubes were spaced at 5, 10, and 20 feet from the injection well. Three additional monitoring tubes were located at approximately 26, 52, and 70 feet from the injection well.

OVERVIEW

The Soil Flushing Project has progressed from a pilot study to full-scale implementation. The purpose is to flush highly mobile and soluble hexavalent chromium from unsaturated soils to the ground water. Ground water is currently pumped and treated through an onsite treatment system and reinjected as flushing water or discharged in accordance with a NPDES permit.

Neutron probe monitoring was selected to verify the success of flushing unsaturated pore water from approximately 20 to 80 feet bls. The method is inexpensive, nondestructive, and repeatable.

DESIGN

The first step in designing a monitoring network capable of measuring moisture content changes in the deep unsaturated zone was to select a proper drilling technique. The most important factor in maximizing neutron penetration is to minimize borehole size. At the same time, it is often important to log the borehole and to collect soil physical and chemical samples. In some geologic environments, hole integrity may be a concern.

With these factors in mind, we chose the dual wall percussion hammer drilling technique. The percussion hammer uses a powerful diesel hammer to drive dual steel tubes into the ground. Compressed air is forced through the dual tubes, and borehole materials are removed through the center of the tubes and collected at a discharge "cyclone." The drilling can be extremely quick and efficient. For example, at one location where sampling was not required, we drilled to 90 feet bls in approximately 45 minutes.

The two most commonly used dual tube sizes are 9-inch and 6.6-inch outside diameter (OD). However, we selected the 5.5-inch OD by 3.25-inch inside diameter (ID) because it was optimal for minimizing borehole diameter and allowing sampling and monitoring tube construction through the dual tubes.

The importance of limiting borehole and well diameter has been shown by other researchers. Keller et al., 1990, demonstrated that although polyvinyl chloride (PVC) had a greater masking effect than stainless steel, the greatest inhibiting factor for neutron penetration was large diameter wells. In addition, Amoozegar et al., 1989, found that slurry backfills yielded neutron count rates as good as access tubes tightly fitted to the borehole; however, increasing access hole size resulted in decreased sensitivity.

Our monitoring tube design consisted of 2.5-inch nominal, Schedule 80 aluminum pipe with approximately 10 feet of PVC well screen placed at the bottom of each monitoring tube. The monitoring tube served as both a monitoring well and neutron probe access tube. The monitoring tube was constructed much like a typical monitoring well (Figure 19.1). However, the water-to-cement ratio used in the backfill was carefully controlled to adhere to previous laboratory results by Kramer et al., 1990; that is, each mix of grout contained 5 gallons of water per 94-pound sack of cement, with approximately 5% sodium bentonite powder. According to the results presented by Kramer et al., this mix should require approximately seven days for proper maturation.

Grout curing was verified approximately two weeks after monitoring tube construction by monitoring neutron count drift for three consecutive days. Five discrete depths were selected, and a mean count rate was determined by taking three consecutive 64-second measurements. Neutron count drift was minimal based on the extremely low standard deviations and coefficients of variation (Table 19.1). Therefore, grout maturation had stabilized.

SUCCESS

The monitoring tube design was successfully used to monitor moisture content changes in the unsaturated zone, including the development of preferential pathways. Approximate lateral migration rates were inferred from peak neutron counts at monitoring tubes located at discrete distances from the injection well. For in-

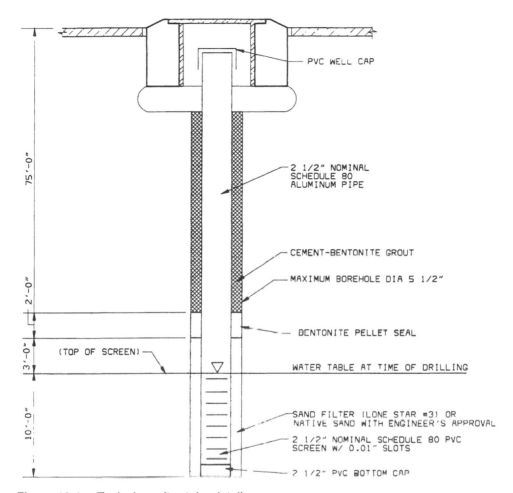

Figure 19.1. Typical monitor tube detail.

stance, in the Kramer et al. article (Figure 18.7), a dramatic peak is depicted at the 60.3-foot depth of the monitoring tube located 5 feet from the injection well. This peak occurred between 5 and 20 hours from the start of injection.

Coarse-grained lithologies showed the most dramatic changes in neutron counts. This is because pre-flushing moisture contents were low in the sandy horizons and preferential flow probably occurred through these highly permeable zones. Clay horizons showed limited success in flushing, as evidenced by gradual increases in neutron counts. Consequently, we assume that limited flushing is occurring in these zones.

Soil moisture changes observed through neutron monitoring techniques were used to aid in the design of a full-scale soil flushing system. Design of this system included calibration of an unsaturated flow model using the USGS numerical code SUTRA (Voss, 1990). Lateral moisture content changes were simulated, and a maximum radius of influence was determined for the placement of full-scale injection wells.

Although we had hoped that a saturated wedge or mound would develop near the injection well, water level elevations remained nearly the same. This may indicate

Table 19.1. Post Grout Stabilization Monitoring, Monitoring Tube MT-1

Date	Neutron Counts				
	18.3 ft.	27.8 ft.	48.3 ft	58.3 ft	67.8 ft
04/29/91	15758	16318	14774	15222	16006
	15792	16356	14797	15327	16126
	15980	16405	14756	15179	16016
Mean	15843	16360	14776	15243	16049
04/30/91	15923	16234	14730	15415	16196
	15871	16342	14825	15318	16109
	15930	16273	14781	15499	16086
Mean	15908	16283	14779	15411	16131
05/01/91	15780	16359	14584	15392	16080
	15934	16292	14674	15439	16093
	15849	16327	14756	15312	16067
Mean	15854	16326	14671	15381	16081
Cumulative Mean	15868.56	16322.89	14741.89	15344.78	16086.56
Standard Deviation	28.25	31.38	49.91	73.21	33.39
Coefficient of Variation (%)	0.18	0.19	0.34	0.48	0.21

that moisture fronts were advancing, but matric potentials (negative pressure heads) were still present, much like an induced capillary fringe. In addition, regional drawdown of the ground-water surface may have inhibited the development of a mound. A system of nested monitoring wells would have failed to observe the moisture increases because a positive pressure head was not established.

PROBLEMS

During drilling of the monitoring tubes, representative soil samples were collected and analyzed for dry bulk density (EM 1110–2-1906), particle density/specific gravity (ASTM D854), and gravimetric moisture content (ASTM D2216). Volumetric moisture content was calculated from the following relationship (Hillel, 1982):

$$\theta = w*Pb/Pw$$

where:
θ = volumetric moisture content (cm^3/cm^3),
Pb = dry bulk density (gm/cm^3),
Pw = density of water (gm/cm^3), and
w = gravimetric moisture content (gm/gm).

Soil porosity was then calculated as:

$$f = 1 - (Pb/Ps)$$

where:
f = porosity (cm^3/cm^3), and
Ps = particle density (gm/cm^3).

By knowing the volumetric moisture content and the porosity, we hoped to use the relationship of θ divided by f to determine the initial percentage of saturation and to prove complete saturation at the conclusion of the pilot study.

However, during the second round of soil sampling, several moisture content values were lower than initial moisture content values at the same depth. Neutron probe measurements indicated that the moisture contents should be higher.

Several factors may have caused the anomalous results. First, during the confirmation sampling, hollow stem auger drilling was used and ambient temperatures reached as high as 100° Fahrenheit (F). Visual evidence of steam rising from the split-spoon sampler indicated that moisture may have been lost.

In addition, the neutron probe gauge averages over an interval of approximately one foot. Soil samples were collected as discrete grab samples. Therefore, some inherent uncertainty existed concerning whether the neutron probe measurements and the soil moisture contents could be correlated. In addition, undisturbed samples collected and measured for dry bulk density are tenuous at best.

For the full-scale system, we have employed a strategy of qualitative monitoring in which neutron counts are taken at one-foot intervals and count ratios are determined to provide a nearly continuous neutron log. Zones of increased count ratios help show where flushing is occurring (Figure 19.2).

CONCLUSIONS/RECOMMENDATIONS

Neutron monitoring has been an invaluable part of our flushing design. This study is one of the few successful field applications of neutron monitoring in grouted boreholes. By closely monitoring the water-to-cement ratio and by minimizing borehole size, neutron monitoring can be an effective part of an investigative, remedial, or monitoring design.

We recommend using neutron logs as shown in Figure 19.2, rather than monitor-

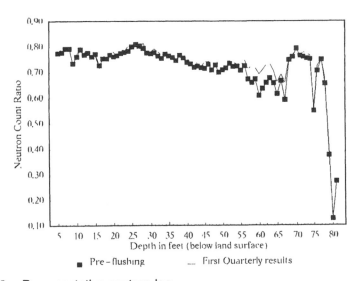

Figure 19.2. Representative neutron log.

ing at discrete depths to determine precise moisture contents. Qualitative monitoring will continue to be an important part of our full-scale monitoring.

REPLY TO THE PRECEDING DISCUSSION BY MICHAEL A. WILLIAMS OF "VADOSE ZONE MONITORING WITH THE NEUTRON MOISTURE PROBE." BY JOHN H. KRAMER, VADOSE ZONE MONITORING LABORATORY, INSTITUTE FOR CRUSTAL STUDIES, UCSB, CA 93106–1100

Inaccuracies corrected by Mr. Williams resulted from our misinterpretation of drawing scales and do not affect the conclusions drawn in the original paper concerning the style of vadose zone transport of the usefulness of the neutron monitoring technique, However, additional information presented by Mr. Williams sheds light on three important issues concerning vadose zone monitoring with the neutron moisture probe.

Access Tube Construction

Investigators at several sites have either reported problems or are seeking information on successful deep neutron access tube design and installation. Minimal distance from probe to soil is crucial to a usable access tube. Although driven small-diameter casings (2–4 inch) accomplish this, they may not be feasible in many instances for the following reasons: refusal, drive point collapse, concern about compaction resulting from the shouldering aside of material in front of a drive point, or concern over cross contamination along the outside of the access tube. Other potential approaches to deep access tube installation include sonic drilling or boring small (i.e., 2–4 inch diameter) pilot holes into which tight fitting casings are driven. These techniques are unproven, whereas Williams' method, which employs dual wall percussion hammer drilling techniques and stipulates minimum borehole size, minimal grout thickness, and carefully controlled grout water contents, resulted in successful installations. Even with this design, however, neutron monitoring may be of limited use in moist, clay-rich material where the neutron probe radius of measurement is small and the count measurements are near the maximum of the instrument range.

Calibrations

The difficulties of applying calibrations reported by Williams support conclusions in Kramer et al. that calibration attempts are exercises in data filtering which may reduce sensitivity by introducing unknown variability.

Continuous Log Monitoring

Williams' Figure 19.2 is an example of how detailed logging by neutron moisture probe is an effective monitoring tool, not only to document induced saturation at a soil flushing project, but to scan geologic strata for preferential flow pathways (e.g., at property boundaries and compliance points). As predicted in Figure 18.6 of Kramer et al., regions of relatively low background counts are demonstrated to be vadose zone flow pathways by Williams' results. Detailed characterization of the vadose zone using neutron logs can indicate potential contaminant flow pathways which might be missed by point sampling from lysimeters or discrete neutron monitoring positions.

REFERENCES

Amoozegar, A., K.C. Martin, and M.T. Hoover, "Effect of Access Hole Properties on Soil Water Content Determination by Neutron Thermalization," *Soil Sci. Soc. Am. J.*, 53:330–335 (1989).

Hillel, D., *Introduction to Soil Physics*, (London, England: Academic Press, 1982).

Keller, B.R., L.G. Everett, and R.J. Marks, "Effects of Access Tube Material and Grout on Neutron Probe Measurements in the Vadose Zone," *Ground Water Monitoring Review*, 10(5):96–100 (1990).

Kramer, J.H., S.J. Cullen, and L.G. Everett, *Ground Water Monitoring Review*, 12(3):177 (Summer 1992).

Kramer, J.H., L.G. Everett, and L.A. Eccles, "Effects of Well Construction Materials on Neutron Probe Readings with Implications for Vadose Zone Monitoring Strategies," in *Proceedings of the Fourth Annual Outdoor Action Conference on Aquifer Restoration, Ground Water Monitoring and Geophysical Methods*, presented by NWWA and the U.S. EPA EMSL, 1990.

Voss, C.I., A Finite-Element Simulation Model for Saturated-Unsaturated, Fluid-Density-Dependent Ground-Water Flow with Energy Transport or Chemically-Reactive Single Species Solute Transport, Version 06902D. U.S. Geological Survey National Center, Reston, VA, 1990.

20

Tensiometry

T.-C. Jim Yeh and Amado Guzman-Guzman

INTRODUCTION

Movement of water in the vadose zone (i.e., the variably saturated zone above the regional ground-water table) is of considerable interest in studies of landfill and hazardous waste sites, ground-water recharge investigations, irrigation management, and civil engineering projects. Many of these studies have relied on soil water content measurements which are useful for many purposes. However, evaluation of the energy status of the soil water in the vadose zone is necessary for defining the direction of water and contaminant movement, and studying the soil-water-plant relationship.

Water in unsaturated media moves from regions of high total energy to those of low total energy. The total energy status of water in unsaturated media is sometimes called the total soil water potential, which is commonly defined (Corey and Klute, 1985) as the sum of gravitational (ϕ_g), matric-capillary-(ϕ_m), and osmotic-solute-potential (ϕ_o).

$$\phi_T = \phi_g + \phi_m + \phi_o \tag{1}$$

Osmotic potential is related to differences in solute concentration. It may play an important role in some cases but does not, in general, significantly affect mass or water flow and can often be ignored (Hillel, 1971). The gravitational potential, resulting from the attraction by the earth's gravitational force on the fluid mass, is considered as the potential energy stored in a fluid due to its vertical position in reference to an elevation datum. As such, it is equal to the product of the specific weight of the fluid (γ) and the fluid's elevation (z) relative to the datum

$$\phi_g = \gamma z \tag{2}$$

0-87371-610-8/95/$0.00 + $.50
© 1995 by Lewis Publishers

Therefore, fluids at high elevations have larger gravitational potentials than those at low elevations. The matric potential, which attracts and binds water to the soil, is a result of the interaction between cohesive and adhesive forces of air, water, and the soil matrix. This potential can be related to the surface tension (σ) of the fluid-matrix interface and inversely proportional to the radius (r) of the water menisci

$$\phi_m = \frac{2\sigma \cos\alpha}{r} \tag{3}$$

where α is the interfacial angle between the water and soil, usually taken as zero. At full saturation, the radius of the menisci becomes infinite resulting in a capillary pressure equal to zero. The total soil water potential is commonly expressed in terms of Energy per unit Mass, Weight or Volume; in SI units, it has units of Joules/Kg, Meters, and N/m^2, respectively. In terms of energy per unit weight, total potential is written as (disregarding the osmotic potential)

$$\phi_T = z + \psi \tag{4}$$

and is expressed in units of length (usually in meters of water). ψ is called the pressure head and includes both pressure and matric components. Under saturated conditions ψ is positive (hydrostatic), while under partially saturated it is negative. The pressure head is called differently by different authors; names such as tension, matric potential, suction and capillary pressure are found in the literature. For partially saturated soils, pressure head and matric potential are assumed negative; and tension, suction and capillary pressure are conventionally taken as positive (i.e., tension = −pressure head). Also, it is customary to report pressure head as gauge pressure and thus soil water potential, or tension, for a fully saturated medium is said to be equal to zero (gauge), while the absolute pressure is still the local magnitude of the atmospheric pressure.

To illustrate the effect of matric potential on the movement of water, consider a partially saturated soil consisting of a coarse-textured- and a fine-textured-material. If these portions are situated at the same elevation (no difference in the gravitational potential) and have the same initial water content, water will gradually move from the coarse-textured to the fine-textured-material. This movement can be attributed to the fact that pores in the fine-textured material are smaller than those in the coarse-textured portion and induce a larger tension (a more negative pressure potential) to the interstitial water in the former. Water in this material, thus, has a lower total water potential than that in the coarser material, resulting in water movement from coarse-textured material to the fine.

A tensiometer is an instrument designed to measure the matric potential of a porous medium. The first tensiometer was reported by Gardner et al. in 1922. Several different devices are available for measuring the matric potential, but for the range between 0 m and about 8 m of tension, tensiometers are the most appropriate devices. The following sections provide a brief discussion on the principles of the tensiometer, some commonly used tensiometer systems, and field installation procedures. General comments on different tensiometer systems are given in the last section. More detailed information about tensiometers can be found in other reference books (e.g., Hillel, 1971 and Cassel and Klute, 1986).

PRINCIPLES OF TENSIOMETERS

Figure 20.1 shows the typical parts of a tensiometer for field use. The tensiometer consists of a porous cup, a connecting tube, an airtight cap, and a readout device to "measure" the soil tension. Tension measurements require, in most cases, the porous cup and the connecting tube to be filled with deaerated water. The purpose of the porous tip is to allow hydraulic contact between the water inside the tensiometer and the water in the soil, and to prevent air from entering the tensiometer. Two important characteristics of the porous cup are: its air-entry value, the pressure magnitude at which air starts displacing water from an initially saturated porous cup; and its hydraulic conductivity which controls the responsiveness of the tensiometer. The air-entry value is also known as the bubbling pressure of the porous cup.

When a tensiometer is placed in an unsaturated soil, the water inside the cup comes in hydraulic contact with water in the soil. If soil water potential is low, water will flow out through the cup due to the difference in potential until a state of equilibrium is reached between the soil and the cup. Since the tensiometer is sealed from the atmosphere, water outflow creates a pressure depression (a vacuum with respect to the initial atmospheric pressure) in the tensiometer. An ideal tensiometer maximizes the pressure change resulting from water flowing into or out of the instrument, and minimizes the equilibration time between the fluid (generally water) in the instrument and water in the surrounding soil. At equilibrium, the pressure depression (P) in the tensiometer equals the sum of the pressure due to the water column above the porous cup (γL), and the soil matric potential of the soil water (ψ). Therefore, to determine the true soil water tension, the height of the water column must be subtracted from the gauge reading on the tensiometer. Once equilibrium is established with soil water, the instrument responds to further changes of soil water tension; water flows in or out of the tensiometer responding to local changes in the soil saturation. Determining the soil matric potential is, then, a matter of measuring the pressure depression (vacuum) inside the tensiometer. This pressure is normally measured by a manometer, which may be a simple water- or mercury-filled U-tube, a vacuum gauge, or an electrical pressure transducer. Field units for soil

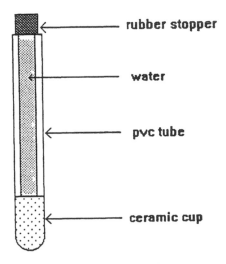

Figure 20.1. Schematic illustration of a field tensiometer.

water tension are Bars or Kilo-Pascals (1 Bar approximately equal to 10 m of H_2O, 101 KPa, 1 atm, and 100 J/Kg).

In practice, matric potential measurements by tensiometry are limited to tension values below 0.85 Bar at an atmospheric pressure near 1 Bar. This limiting pressure range is mainly due to possible vaporization of liquid water and varies with the elevation of the field site. Also, if the soil gas (air and water vapor) pressure exceeds the gas pressure inside the tensiometer, and their difference is larger than the bubbling pressure of the porous cup, air will penetrate the tensiometer curtailing the operation of the instrument. Porous cups with higher bubbling pressures may help to alleviate this problem. Therefore, the user must have some idea about the range of tensions that may be encountered in the field during project life. In theory, tensiometers as large as 26 feet are possible; however, tensiometer columns exceeding 6 feet in length are impractical.

TENSIOMETER SYSTEMS

Although many different types of tensiometer designs are available, their basic operation and principles are the same. The major difference stems from the way the pressure depression within the tensiometer is measured. A brief description of four commonly used field tensiometer systems is presented in the following section.

Mercury-Water Manometer System

This type of system relies on manometers to register the pressure depression in the tensiometer. It consists of a capped tensiometer filled with water, and a U-shaped water tubing which connects the tensiometer to a mercury reservoir (Figure 20.2). If the soil water potential surrounding the porous cup is smaller than that of the water in the tensiometer, water will be drawn out, creating a pressure depression in the system. This causes the mercury in the manometer to rise. The matric potential (ψ) of the soil is determined by subtracting the pressure of the water column between the tip of the ceramic cup and the water-mercury interface (γL) from the pressure as indicated by the mercury rise (P)

$$\psi = P - \gamma L \tag{5}$$

An "all-water" manometer system is possible; however, its practical application is limited to tension values under 6 feet.

Jet-Filled Type

The jet fill tensiometer (Figure 20.3) is a product of Soilmoisture Equipment Co. (Santa Barbara, CA). A similar system is produced by Irrometer Co. (Riverside, CA). The system depicted in Figure 20.3 is a gauge type tensiometer equipped with a detachable water reservoir at the top. The reservoir is covered with a flexible top and a spring valve that closes or opens the connection between the tensiometer and the reservoir. A push on the flexible cover injects water into the body of the tensiometer, removing accumulated air with a small disturbance to the soil tension. The pressure gauge provides a direct reading of the soil matric potential since the gauge is adjusted to account for the pressure due to the water column above the porous cup.

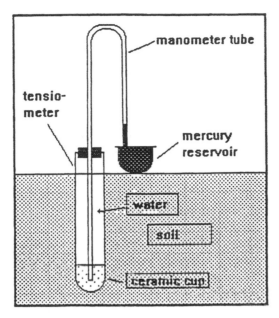

Figure 20.2. Schematic illustration of a mercury-water manometer type tensiometer system.

The accuracy of this kind of tensiometer is limited by the resolution of the pressure gauge. Typically, tension measurements from this instrument have an accuracy of ±10 cm of water. More accurate measurements (on the order of ±1 cm) of tension can be obtained with mercury-water manometers or pressure transducers systems such as those described in the following paragraphs.

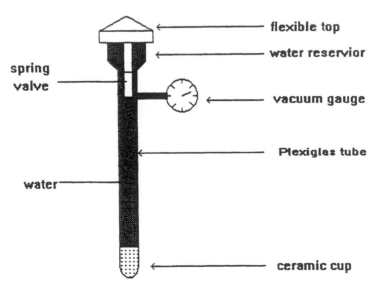

Figure 20.3. Jet fill type tensiometer system.

Quick-Draw Probe

The quick-draw probe is a product of Soilmoisture Equipment Co. It consists of a thermally isolated tube with a porous ceramic tip on the bottom, a vacuum gauge at the top, and a "Null Knob" cap (Figure 20.4). The Null Knob is used to adjust the pressure inside the probe so that a small amount of water can be injected into or sucked from the surrounding soil. This adjustment allows one to quickly bracket the range of soil-water pressure in the soil. Soil water tension is measured directly from the gauge as long as the pressure due to the water column above the porous cup is already taken into account.

Hand-Held Pressure Transducer Type

As illustrated in Figure 20.5, this tensiometer is a transducer-measured system. The tensiometer consists of a ceramic porous cup connected to a PVC tube (Soil Measurements Systems; Tucson, AZ), and a clear Plexiglas tube of 2–3 inches attached at the top of the PVC tube. In order to measure soil water tension, the system is filled with water (not necessarily deaerated) to a level a few centimeters below the top of the Plexiglas tube. The system is then sealed with a septum stopper filled with silicon sealant in such a way that a small amount of air is trapped in the system. A tensimeter consisting of a hypodermic needle connected to a pressure transducer, wired to a digital readout device, is used to measure the soil matric potential. When the hypodermic needle of the tensimeter is inserted into the tensiometer through the septum stopper, the readout device registers the vacuum inside the system. If the pressure due to the water column in the tensiometer is known, the soil water suction is computed as the difference between the tensimeter measurement and this pressure.

Insertion of the needle introduces a small disturbance into the closed tensiometer

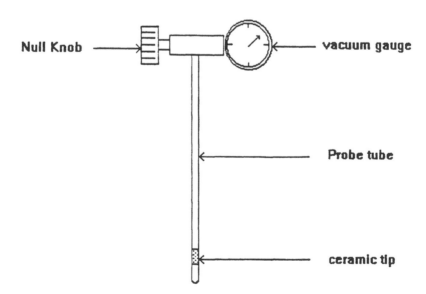

Figure 20.4.   Quick draw type tensiometer system.

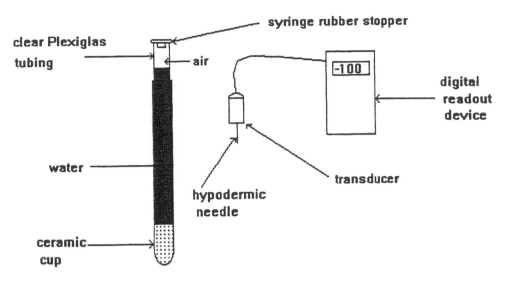

Figure 20.5. Transducer type tensiometer system.

system, but the air space left in the system acts as a buffer to compensate the disturbance. Theoretical analysis of the air buffer is given in Yeh (1982).

Pressure transducers have become increasingly popular in soil water potential measurements. Modern electronic transducers allow for rapid and more accurate determinations than do manometers and pressure gauges. Several authors (e.g., Strebel et al., 1973; and Long, 1982), have reported the use of pressure transducers in conjunction with tensiometers. A major advantage of these instruments is that their measurements can be collected with a data logger system, allowing for auto-mated measurements or detailed monitoring of the transient behavior of the soil water potential. Accuracy of pressure transducers is superior to those of pressure gauges or water-mercury manometers and allows for more precise determinations of soil tension gradients.

FIELD PROCEDURES

Any tensiometer should be tested for leaks prior to usage. The test may be performed by submerging the tensiometer into a pail of water and allowing the porous cup to soak overnight. Water should move through the porous cup into the tensiometer barrel. If the water does not move into the barrel, the walls of the porous cups are plugged and need to be reconditioned by gently brushing its surface with a wire-brush, or by sanding. Test for leaks by emptying the water from the tensiometer barrel, submersing the porous cup, and securely capping the tensiome-ter. Then apply air pressure steps up to a magnitude corresponding to the air entry value of the porous cup. If leaks occur at pressures below the specified bubbling pressure, the tensiometer is defective. The location of the leak should be easily found during this procedure. Leaks commonly occur at the contact between the ceramic cup and the barrel. Invisible cracks in the ceramic cup or poorly manufac-tured cups may be responsible for the leaks. In this latter case, the tensiometer should be discarded.

After testing, remove the cap from the tensiometer and stand the tensiometer upright in a pail of water and allow for the cup to resaturate overnight. Fill tensiometer with deaerated water, which can be prepared by boiling and cooling the water. Hypodermic-needle tensiometers may be filled with tap water. Then, apply a small vacuum to the tensiometer to remove air bubbles from the ceramic cup and those adsorbed to the walls of the tensiometer barrel. For tensiometers being used under freezing conditions, a methanol-water solution (30% methanol by volume) may be used (Wendt et al., 1978). The use of fluid other than water may introduce some bias into the tension measurements due to the contrast between the contact angles of tensiometer solution and soil water with the soil surface. A small concentration of algicide should be added to the water to prevent the growth of algae in the tensiometer.

A tensiometer may be installed in moist loose soil at shallow depths by pushing it into the soil to the desired depth. However, the installation is sometimes completed through an access hole which has the same diameter as the barrel of the tensiometer. An auger with a diameter slightly larger than that of the barrel of the tensiometer can also be used to prepare the access hole. After the hole is completed, a heavy slurry of water and cored soil is poured into the hole before installing the tensiometer to improve the hydraulic contact between the ceramic cup and the surrounding soil. Good hydraulic contact is necessary to obtain correct soil-water tension measurements. Once the installation is completed, the annulus between the tensiometer barrel and the access hole should be backfilled to avoid movement of water along the annular space.

Measurements may begin several days after installation to allow equilibration between the tensiometer's water, the slurry, and the surrounding soil water. The time required to reach equilibrium varies with the texture and wetness of the soil. In general, one or two days are needed in coarse-textured soils and one week for fine-textured soils. Daily monitoring of the tensiometer gauge should provide a clear indication of when the equilibrium state is achieved.

COMMENTS ON THE TENSIOMETER SYSTEMS

The design and construction of the mercury-water- or water-manometer type tensiometer system is relatively simple. The cost of this system is low. In fact, for some laboratory experiments this tensiometer type may provide accurate measurement of matric potential. However, for field applications, plumbing for the manometer is rather cumbersome when simultaneous measurements of matric potentials at several locations are desired. Air coming out of solution from water in the tensiometer is the major shortcoming of this type of instrument. Air is likely to be trapped in the manometer tube and may lead to inaccurate readings of the matric potential. Also, some health-related concerns with the used of mercury may limit the utilization of such instruments.

The jet-filled tensiometer system is a convenient device for monitoring matric potential in the field. The flexible reservoir cover (Soilmoisture Equipment Co.) allows for convenient filling, and it removes accumulated air while producing a small disturbance in the soil tension. The cost of each system is higher than that of the manometer-type tensiometer.

The quick draw probe is another convenient way to measure soil potential in soft soils and shallow depths. It can be easily installed into the soil without augering a large access hole. If necessary, a slurry mix can be used to improve the contact between the ceramic cup and the surrounding soil. This tensiometer type allows for rapid field collection of a large amount of matric potential data. A reading of the matric potential can be quickly made if proper "approximating" procedure is followed. This procedure introduces a disturbance which may require a long time to reach an equilibrium. In general, this procedure works well in the wet and coarse materials. For other soils, early readings obtained from the system may be rather uncertain because of the long re-equilibration times. The probe is commonly manufactured with a length of less than two feet, and thus the measurements of soil potential are limited to these depths.

The transducer type system is a very convenient apparatus for delineating spatial variability of matric potential in a large field (Yeh et al., 1986). Due to its higher precision, with respect to the manometer and gauge system it is quite adequate to delineate soil tension gradients. Measurement of the matric potential can be done easily by inserting the hypodermic needle into the tensiometer, and matric potential readings can be obtained in just a few seconds. A large number of measurements in a large field site can be completed in a short period of time using many tensiometers but only one transducer device. As such, the system allows collection of a nearly simultaneous set of potential measurements in a large field. The tensiometers can be easily assembled by the user, and the materials are inexpensive. The major cost for this kind of tensiometer system arises from purchase of a transducer/readout device assembly. This type of tensiometer may be a cost-effective alternative for collecting a large amount of matric potentials in a swift manner.

Care should be taken when this system is used in the field, however. The small amount of air left in the tensiometer may expand and contract if the tensiometer is not well-insulated. Thus, the soil tension reading may become misleading. Also, the user should verify that the hypodermic needle is not clogged and the needle does not touch the water in the tensiometer during each measurement.

In principle, the tensiometer types discussed above could be equipped with a pressure transducer to perform the tension measurements. A number of companies offer inexpensive pressure transducers which can be powered with hand-held batteries and read with the aid of a volt-meter. The widespread availability of electronic technology will allow pressure transducers to become more widely used in this type of application.

REFERENCES

Cassel, D.K., and A. Klute, "Water Potential: Tensiometry," in *Methods of Soil Analysis, Part 1-Physical and Mineralogical Methods*, Second Edition, A. Klute, Ed., American Society of Agronomy, Inc. and Soil Science Society of America, Inc., Madison, WI, 1986, Chap. 23.

Corey, A.T., and A. Klute, "Reviews of Research: Application of the Potential Concept to Soil Water Equilibrium and Transport," *Soil Sci. Soc. Am. J.*, 49:3-11 (1985).

Gardner, W., O.W. Israelsen, N.E. Edlesfsen, and D. Clyde, "The Capillary Potential Function and its Relation to Irrigation Practice," *Phys. Rev.* 20:196 (1922).

Hillel, D., *Soil and Water, Physical Principles and Processes*, (New York: Academic Press, 1971).

Long, F.L., "A New Solid State Device for Reading Tensiometers," *Soil Sci.* 93:204–207 (1982).

Strebel, O., M. Renger, and W. Geisel. "Soil Suction Measurements for Evaluation of Vertical Flow at Greater Depths with a Pressure Transducer Tensiometer," *J. Hydrol.* 18:367–370 (1973).

Wendt, C. W., O. C. Wilke, and L. L. New, "Use of Methanol-Water Solutions for Freeze Protection of Tensiometers," *Agron. J.* 70:890 (1978).

Yeh, T-C J., Stochastic Effects of Spatial Variability on Unsaturated Flow, PhD dissertation, New Mexico Tech, Socorro, 1982, Appendix A.

Yeh, T-C. J., L. W. Gelhar, and P. J. Wierenga, "Observations of Spatial Variability of Soil-Water Pressure in a Field Soil," *Soil Science*, 142:7 (1986).

21

Energy-Related Methods: Psychrometers

Todd C. Rasmussen and Shirlee C. Rhodes

INTRODUCTION

The measurement of the direction and magnitude of contaminant migration in the subsurface is often hampered by the heterogeneous and complex nature of most geologic media. An alternative to direct measurement of subsurface migration routes and rates in the unsaturated zone is to indirectly determine water flow using the direction and magnitude of the hydraulic gradient and the hydraulic conductivity of the medium. The direction and magnitude of fluid flow is governed by the tensorial form of Darcy's law:

$$q = -\underline{\underline{K}}(\psi)\,\underline{i} \tag{1}$$

where
$$\underline{i} = \nabla H$$
$$H = z - \psi + \phi$$

and

$$q = \text{fluid flux vector, } m \cdot s^{-1};$$
$$\underline{\underline{K}}(\psi) = \text{suction-dependent hydraulic conductivity tensor, } m \cdot s^{-1};$$
$$\underline{i} = \text{hydraulic gradient, dimensionless;}$$
$$H = \text{total hydraulic head, } m;$$
$$z = \text{elevation head, } m;$$
$$\psi = \text{matric suction head, } m; \text{ and}$$
$$\phi = \text{osmotic potential head, } m.$$

The estimation of the hydraulic gradient requires knowledge of the distribution of the matric suction and the osmotic potential. The hydraulic conductivity tensor function can be determined in the laboratory using core segments extracted from the site of interest, or from field-scale permeability tests using water or air (see, e.g.,

0-87371-610-8/95/$0.00 + $.50

Chapter 28 by Evans and Rasmussen in this book). Because the hydraulic conductivity tensor is a strongly nonlinear function of the matric suction, it is important to know the ambient matric suction so that the appropriate value of hydraulic conductivity can be determined for the site.

An important tool for estimating the fluid potential, and hence the flow gradient, is the psychrometer. This device measures the potential of the water vapor present in the subsurface atmosphere. If the potential of the water vapor is equal to the potential of the pore fluid, then the psychrometer provides a means for identifying the pore fluid potential. An additional use of the thermocouple psychrometer is the determination of the moisture characteristic curve for fluid potentials greater than those readily attained using porous plates, approximately 5 bars. The fluid potential of field-collected samples is estimated by allowing samples to equilibrate at specified water contents, thus extending the characteristic curve to conditions much drier than normally obtained from pressure extraction vessels.

THEORY OF PSYCHROMETRIC OPERATION

Psychrometers measure the vapor-phase water activity in the subsurface atmosphere surrounding the sensor. The water activity is related to the relative humidity and the vapor pressure using:

$$a = h / 100 = p / p_o \tag{2}$$

where
- a = water vapor activity, dimensionless;
- h = relative humidity, percent;
- p = ambient water vapor pressure, kPa; and
- p_o = maximum (saturated) water vapor pressure at ambient temperature, kPa.

The saturated vapor pressure, p_o, is a function of the ambient temperature, T (K), and can be approximated using:

$$p_o = \exp(19.017 - 5327/T) \tag{3}$$

For water vapor in equilibrium with the liquid phase, the water activity is related to the fluid pressure by:

$$\Psi = (\rho R T / M) \log_e(a) \tag{4}$$

where $\Psi = \phi - \psi$

and

- Ψ = fluid potential, $J \cdot m^{-3}$;
- ρ = density of water, 998.21 $kg \cdot m^{-3}$ at 20°C;
- R = ideal gas constant, 8.314510 $J \cdot K^{-1} \cdot mol^{-1}$;
- T = ambient temperature, K; and
- M = molecular weight of water, 18.05128 $g \cdot mol^{-1}$.

The fluid potential calculated using Equation 4 is generally used to estimate the matric suction of the liquid water present in pores. The fluid potential in the pores may also include, however, the osmotic potential associated with solutes dissolved in the pore fluid. It is important to note that osmotic potentials may be significant when elevated solute concentrations are present. The pore fluid chemistry can be used, if necessary, to estimate the magnitude of the osmotic potential.

The dew-point temperature, T_o, is the temperature at which the ambient atmosphere must be chilled in order for water to condense (i.e., for $p_o = p$). An approximate expression for the dew-point temperature is:

$$T_o = [T^{-1} - \log_e(a)/5327]^{-1} \qquad (5a)$$

The dew-point depression, ΔT_d, is the difference between the ambient and dew-point temperatures, or approximately:

$$\Delta T_d = (T - T_o) = T [1 - (1 - T \log_e(a)/5327)^{-1}] \qquad (5b)$$

Table 21.1 presents the relationship between relative humidity, dew-point depression temperatures, and matric suctions for an ambient temperature of 20°C (273 K).

Psychrometers are a generic class of instruments used to measure the relative humidity that employ a dry bulb thermometer to measure the ambient temperature, and a second, wet bulb thermometer to measure the lower temperature associated with an evaporating fluid. For most subsurface applications, the ambient water vapor pressure will be near the saturation vapor pressure, i.e., the relative humidity will generally be greater than 99% and the dew-point depression will be less than 0.2°C. Most liquid-filled thermometers lack sufficient precision to provide measurements within ± 0.1°C, much less the precision ±0.001°C required for matric suction measurements near 1 bar. Thermocouple psychrometers are used instead to measure the small temperature depressions associated with relative humidities in the subsurface.

Table 21.1. Relationship Between Relative Humidity, Dew-Point Depression, and Matric Suction

Relative Humidity (%)	Dew-Point Depression (°C)	Matric Suction (bars)
100	0.0000	0
99.95	0.008	0.7
99.9	0.016	1.3
99.5	0.081	6.8
99	0.16	13.5
98	0.33	27.2
95	0.82	69.1
90	1.7	142
80	3.6	301
50	11	934
30	18	1623
20	24	2169
10	33	3104
1	59	6207

TYPES OF THERMOCOUPLE PSYCHROMETERS

Thermocouple psychrometers employ a bimetallic junction (e.g., copper-constantan or chromel-constantan), known as a thermocouple. Figure 21.1 illustrates the primary components of the thermocouple junction. The junction produces an electric current that is related to the junction temperature. This coupled heat-current phenomenon is known as the Peltier effect. The junction temperature can be found by measuring the current generated by the thermocouple junction, or, conversely, the temperature of the junction can be changed by inducing a current through the junction.

Thermocouple psychrometers use either the "wet-bulb" or "dew-point" methods to determine the water activity of the atmosphere surrounding the junction. Both methods use a thermocouple to measure the ambient temperature. The wet-bulb psychrometer measures the temperature of a "wet" junction (i.e., a junction covered with water), while the dew-point psychrometer measures the temperature at which saturated vapor pressure equals the ambient vapor pressure.

Wet-Bulb Psychrometry

The wet-bulb temperature is determined either by placing a small drop of water on a copper-constantan thermocouple junction (Richards method), or by cooling the thermocouple using the Peltier effect until water condenses on it (Spanner method). Cooling of the thermocouple is monitored as the water evaporates. For both methods, the drop in temperature on the wet-bulb (i.e., the wet-bulb depression) varies with the rate of evaporation. The evaporation rate, in turn, increases as the water activity surrounding the thermocouple decreases.

The equilibrium temperature attained by the wet bulb depends on factors that influence heat flow toward the junction and vapor flow away from the junction. Heat fluxes to the wet junction result from conduction through the connecting wires, from radiation to the bulb from the walls of the sample chamber, and from conductive and convective heat gain from the surrounding air. These heat fluxes are controlled by the dimensions of the sample chamber and thermocouple components. Cooling at the wet junction is proportional to the product of the latent heat of vaporization and the evaporation rate, a function of the water activity within the chamber and the diffusivity of water in air. These factors generally increase with temperature. The diffusivity of water in air also decreases with atmospheric pressure.

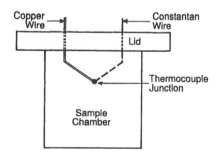

Figure 21.1. Schematic diagram of Peltier junction and psychrometer chamber.

The effects of temperature and pressure are incorporated using the conventional psychrometer equation (Campbell, 1979):

$$a = 1 - \Delta T \, (s + g) \, / \, p_o \qquad (6)$$

where s = slope of the saturation vapor pressure curve, $Pa \cdot K^{-1}$;
         g = psychrometer constant, $Pa \cdot K^{-1}$; and
         ΔT = wet-bulb depression, K.

The psychrometer constant, g, varies with atmospheric pressure and is relatively insensitive to temperature changes. The slope of the saturation vapor pressure curve, s, and the saturated vapor pressure of water, p_o, are significantly affected by temperature changes.

Dew-Point Psychrometry

An improved psychrometer technique uses a feedback loop to control the Peltier cooling rate, thereby maintaining the temperature of the measuring junction at the dew point (i.e., the temperature at which the ambient vapor pressure equals the saturated vapor pressure, $p = p_o$). When held at the dew-point temperature, a wet thermocouple junction neither loses water through evaporation nor gains water through condensation because heat flow to the wet junction from its surroundings is exactly offset by adjustments in the cooling current. The measurement is independent of the rate of heat flow to the wet junction and, thus, no water vapor diffusion occurs. This method eliminates most of the temperature and pressure dependence associated with wet-bulb measurements. The lowest matric suction that can be reliably measured in the laboratory using this device is approximately 1 bar under ideal conditions, with a lower limit of approximately 3 to 5 bars for field applications.

Chilled-Mirror Psychrometry

The chilled-mirror psychrometer improves the accuracy of the water activity measurement by employing a mirror that is chilled to near the dew-point temperature. Slight changes in the temperature of the mirror surface results in condensation on the surface of the mirror when the mirror surface drops to the dew-point temperature. The induced condensation causes a change in the light reflection properties of the surface. Monitoring of the change in light reflection from the mirror can then be used to determine the dew-point temperature in the laboratory from 1.5 to over 2500 bars.

USGS Four-Wire Current Psychrometer

While psychrometers are normally installed near the ground surface at depths of less than a few meters, psychrometers have been installed to a depth of over 350 m in deep unsaturated volcanic rocks (Montazer et al., 1988). A new method suitable for monitoring in deep unsaturated media (up to 600 m) has been developed by Merrill Instruments and J.P. Rousseau of the U.S. Geological Survey, Denver. The device uses a 4-wire thermocouple psychrometer in current mode rather than in the voltage mode to compensate for long line lengths. These psychrometers are able to measure matric suctions between 0.5 to 80 bars.

PSYCHROMETER CALIBRATION

Regular calibration of thermocouple psychrometers is necessary due to the possibility of corrosion of the metals used to create the thermocouple junction. Another reason for regular calibration is the possible accumulation of soluble salts on the junction which affects the measured wet-bulb and dew-point temperatures.

Psychrometers can be calibrated using a salt solution of known water activity. The salt solutions are used to create an osmotic potential which maintains a prescribed water vapor activity in the atmosphere above the salt solution. One possible design for the calibration chamber uses a closed, thermally-insulated chamber partially filled with a reference solution. The psychrometers to be calibrated are placed in the air space above the solution (Figure 21.2), assuring that the salt solution does not come into direct contact with the psychrometer or psychrometer cable. Sufficient time should be allowed for the contents of the chamber to come into thermal equilibrium, e.g., overnight for a 100-mL solution placed in a 500-mL chamber (Rhodes, 1993).

Two methods using osmotic potentials are used for establishing the known water activities. The first method employs salt solutions (e.g., NaCl or KCl) of variable concentrations to impose a range of calibration points. Table 21.2 presents a list of NaCl and KCl salt solutions and their associated relative humidities and matric suctions. The osmotic potential associated with various solutions can be calculated using Raoult's law and Kelvin's law (Rasmussen and Evans, 1986):

$$a = \frac{55.556}{55.556 + O_s} \tag{7a}$$

and

$$\phi = RT/V \log_e(a) \tag{7b}$$

where O_s is the solute osmolality, obtained from standard handbooks of chemistry and physics. Because the osmotic potential, ϕ, is a function of temperature, calibration values should be recalculated if the reference solution varies from 20°C.

A significant problem associated with the use of this technique lies in the inability

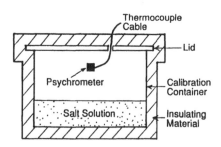

Figure 21.2. Chamber for psychrometer calibration using salt solutions.

Table 21.2.   Salt Solutions Useful for Psychrometric Calibration (at 20°C)

Salt Concentration (molality)	Relative Humidity (%)		Matric Suction (bars)	
	NaCl	KCl	NaCl	KCl
0.0	100	100	0	0
0.1	99.7	99.7	4.5	4.5
0.2	99.3	99.3	9.0	8.9
0.3	99.0	99.0	13.5	13.2
0.4	98.7	98.7	17.9	17.5
0.5	98.4	98.4	22.4	21.9
0.6	98.0	98.1	27.0	26.3
0.7	97.7	97.8	31.6	30.5
0.8	97.4	97.4	36.2	34.9
0.9	97.0	97.1	40.9	39.2
1.0	96.7	96.8	45.6	43.7

to maintain a constant salt solution concentration. Evaporation of the salt solution may cause an unacceptable increase in the salt concentration, and, hence, osmotic potential, while condensation may cause a decrease in the salt solution concentration.

Saturated salt solutions are used to avoid the problems associated with changing salt solution concentrations (Rhodes, 1993). In this method, a surplus of salt is added to a volume of water so that if evaporation or condensation were to occur, then salt would precipitate or dissolve to maintain a constant concentration. Table 21.3 presents a list of selected salts, with the relative humidities and water potentials given by their saturated solutions at 20°C.

For both methods, the psychrometer(s) are placed in at least two, and preferably four, different salt solutions that represent the range of matric suctions at the site of interest. After allowing for the temperature and vapor pressure within the chamber to equilibrate, readings of psychrometer output are obtained, and the psychrometers are moved to another chamber which contains a different fluid potential. The psychrometer reading and chamber temperature should be recorded for each solution once the reading becomes steady. Once the calibration data have been collected, the observed readings should be compared to the theoretical values specified by the manufacturer. If the psychrometer departs from the theoretical value beyond the

Table 21.3.   Saturated Salt Solutions Useful for Psychrometer Calibration (at 20°C)

Saturated Salt	Relative Humidity (%)	Matric Suction (bars)
$Pb(NO_3)_2$	98	27
$Na_2SO_3 \cdot 7H_2O$	95	69
K_2HPO_4	92	113
$ZnSO_4 \cdot 7H_2O$	90	142
K_2CrO_4	88	173
KBr	84	236
NH_4Cl	79	315
$NaClO_3$	75	389

tolerance specified by the manufacturer, then either the psychrometer should be discarded or a new calibration curve should be produced from the readings and used to convert measurements to water potential.

Because thermocouple psychrometers are usually calibrated with salt solutions of known water potential, the accuracy of a sample measurement depends upon reproducing conditions similar to those under which the calibration was performed. Because calibration conditions (e.g., identical barometric pressure) can rarely be duplicated exactly, Equation 6 can be employed to correct for temperature and pressure dependencies if the wet-bulb technique is employed.

FIELD INSTALLATION OF PSYCHROMETERS

It is extremely important that only the smallest possible installation hole surrounding the psychrometer be used. A small installation hole with minimal backfilling should be used to minimize disturbance to the geologic medium surrounding the probe. The hole should be backfilled with a silica flour or native geologic media to minimize void spaces. Clays of low permeability should not be used as backfill material for two reasons. First, clays require long times to equilibrate with the ambient matric suction. Second, clays may be of such low permeability that they prevent equilibration of water vapor between the psychrometer and the surrounding porous medium. For these reasons, sealing materials such as grouts and bentonite clays should be avoided during borehole completion to minimize absorption or release of water in the region surrounding the monitoring interval.

Figure 21.3A demonstrates one possible field installation strategy. A small diameter hole is placed vertically to the desired depth. A drilling method should be used that does not employ water or air (e.g., a percussion tool or auger may be suitable). Psychrometers should be emplaced in duplicate pairs for the purpose of evaluating the magnitude of instrument drift. A large weight can be attached to the thermocouple leads near the psychrometer pairs to help lower the probes to the desired depth, and to provide a large thermal mass to minimize thermal transients. The hole is then

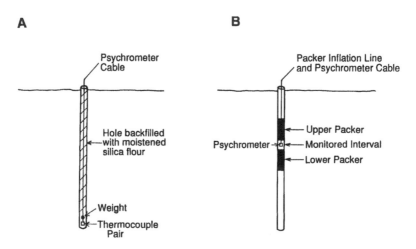

Figure 21.3.   Field installation of thermocouple psychrometer using (A) backfill and (B) inflatable packers.

backfilled with a silica flour moistened to reproduce the anticipated matric suction at the site of interest.

Another possible field installation strategy is to use inflatable packers. In this application, a monitoring interval is flanked by two packers to isolate the zone of interest (Figure 21.3B). The packers are expanded to seal the interval using compressed gas lines connected to a compressor or bottled gas on the surface. Because the expanding gas cools adiabatically when the packers are inflated, a large thermal perturbation will be observed and must be allowed to dissipate prior to obtaining a reliable reading. The advantage of using inflatable packers lies in the ability to reposition the psychrometers so that greater spatial coverage can be obtained using the same instruments.

INTERPRETATION OF PSYCHROMETER MEASUREMENTS

As described above, psychrometric measurements can be made using either: (a) the dew-point method, (b) the wet-bulb method, or (c) a combination of both. Figure 21.4 presents representative outputs for a sample at 25 bars and 32.8°C. It can be noted that the output reading in microvolts shows a more pronounced deflection using the dew-point method than for the wet-bulb method. The deflection to a stable microvolt reading is the value used to calculate the matric suction.

For some observations, however, no clear deflection is observed. Instead, a more gradual drift in reading is observed over time. This drift is especially apparent in the wet-bulb method, and when the dry-bulb technique is used near the upper or lower limit of its range. Under these conditions, utilization of both modes provides redundant estimates of matric suction, along with an indication of measurement accuracy. In general, better accuracy is indicated if both techniques provide consistent matric suction estimates.

Figure 21.5 presents results of laboratory psychrometric readings for 105 rock core segments obtained from the Apache Leap Tuff Site in central Arizona (Rasmussen et al., 1990). The cylindrical core segments (measuring 1 cm in length and 1

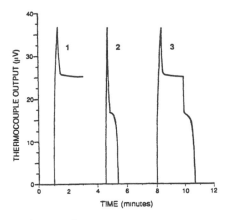

Figure 21.4. Thermocouple psychrometer response curves for (1) dew-point, (2) wet-bulb, and (3) combined modes. Osmotic potential is 25 bars suction and temperature is 32.8°C.

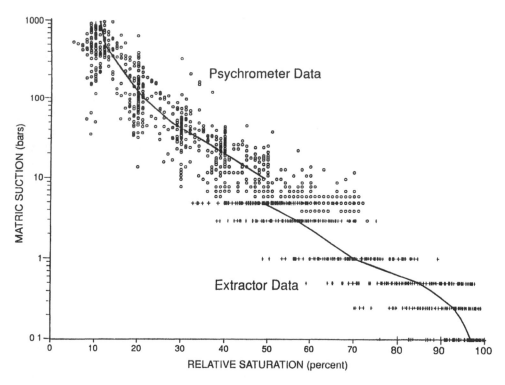

Figure 21.5.   Laboratory moisture characteristic (drying) curves for 105 core segments from the Apache Leap Tuff Site, central Arizona. Extraction and psychrometer curves show all observations (O,X) and mean (lines).

cm in diameter) were first saturated in a sterile, deionized solution containing 0.001 M CaSO$_4$. The segments were then allowed to dry on a precision balance to reach a target relative saturation (between 70 and 10% saturation in 10% increments). The segments were then transferred to a psychrometer chamber, allowed to equilibrate, and the matric suction was determined. Also shown for comparison are moisture characteristic curves (drying curves) obtained for 105 cores using an extractor vessel technique. A reasonably good agreement can be observed by extrapolating between the two sets of data.

An interesting field application of thermocouple psychrometers was conducted at Kartchner Caverns in southern Arizona. Relative humidities very near saturation within the cave were monitored to determine the magnitude and direction of water vapor movement resulting from the enlargement of the cave entrance. Figure 21.6 presents the interpreted matric suctions within the cave atmosphere. It is apparent from the figure that the two stations closest to the entrance are significantly drier than more distant stations. If it can be assumed that vapor flux within the more remote sections of the cave is small and that the cave atmosphere is in equilibrium with the subsurface media surrounding the cave, then the matric suction of the rock surrounding the cave can be estimated to be approximately 5 bars. No apparent trend can be discerned in the cave matric potential beyond the first two stations.

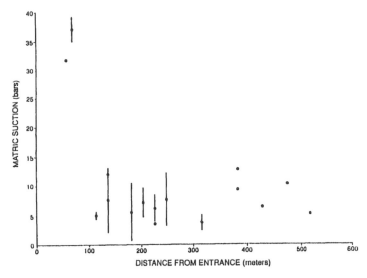

Figure 21.6. Observed matric suctions in Kartchner Caverns, southern Arizona. Circles indicate mean observed matric suction as a function of distance from the entrance. Vertical bars indicate one standard deviation errors about mean.

PSYCHROMETRY ERRORS

Accurate determinations of matric suctions in the subsurface require an understanding of the potential sources of transient and systematic errors. Errors from the following causes should be considered:

- temperature fluctuations with time,
- temperature gradients,
- vapor pressure gradients, and
- wet junction characteristics.

Laboratory psychrometers are designed to measure the water potential of soil samples by sealing the thermocouple just above or within the sample. One design allows the soil samples to be packed into several stainless steel cups, which are rotated to meet the thermocouple. These are fitted either with a Richards-type junction modified with the addition of a ceramic bead or with a Peltier-type wet junction. Measurements made with this equipment may be significantly affected by temperature fluctuation with time, particularly when the sample is considerably smaller than the dimensions of the chamber. The accuracy is greatest when the sample completely surrounds the junction. The effect of temperature fluctuations can be minimized by enclosing the device in a constant temperature bath or placing the device in contact with a large heat sink to reduce the rate of temperature changes.

Field psychrometers utilize thermocouples protected by cups or bulbs of various sizes, usually ceramic or fine mesh, to provide a protected space within the in situ environment. Field measurements are often subject to thermal gradients. Potential errors due to nonisothermal media can be eliminated if the psychrometer is oriented so that its axis of symmetry is perpendicular to the temperature gradient, or if a thermal mass (e.g., a weight) is placed near the device to minimize thermal transients.

Thermocouple psychrometer devices are currently available that measure both wet-bulb and dew-point temperatures in the field. Field units should be emplaced in duplicate to allow multiple devices to be used for assuring accuracy. Also, the devices should be regularly removed and recalibrated to determine whether corrosion or contamination of the bimetallic junction may have occurred. Psychrometers should be emplaced in such a manner as to assure minimal void space surrounding the device, and backfill material should be minimized by using an installation hole that closely matches the dimensions of the psychrometer. Also, psychrometers protected using ceramic bulbs may require prolonged equilibration times compared to fine mesh psychrometers.

While vapor pressure gradients may be caused by extraneous sources or sinks for water vapor, this problem is rare with Peltier psychrometers and can be reduced if samples are not significantly smaller than their sample chambers. Both laboratory and field psychrometer measurements may be affected by wet junction characteristics. Ideal psychrometer conditions assume free evaporation from a spherical wet junction surface. Yet actual size and shape of the wet area depends upon its wetting characteristics, among other factors, and can be contaminated with use. Such changes can be eliminated by cleaning or recalibrating.

Deviations from ideal psychrometer operating conditions may cause significant loss of psychrometer sensitivity, and substantial errors can be introduced if temperature differences between the reference junction and the liquid phase are not controlled to within $\pm 0.001\,°C$ (Rawlins and Campbell, 1986). Thus, there is a critical need to accurately measure temperatures; a small temperature difference error results in a large error in the calculated fluid potential.

Caution should be exercised when interpreting psychrometric readings collected near a site containing volatile liquids. The psychrometric equation assumes that water is the first liquid to condense from the soil atmosphere. For sites where other volatile liquids are present in sufficient quantity to achieve partial pressures near their vapor-phase saturation pressure, the first liquid to condense may be a liquid other than water. For example, if a benzene-saturated site is suspected, and the wet-bulb depression is smaller or the dew-point temperature is higher for the benzene gas than for water, then the benzene may condense prior to water. If this possibility is confirmed, then the vapor pressure of benzene can be determined, but no information about water saturation is possible.

A further complication results when two vapors form an azeotrope upon condensation. An azeotrope is a mixture of liquids whose components do not undergo dilution or concentration as the mixture is distilled. For the previous example using benzene, the condensate may consist of a mixture of benzene and water, with the benzene comprising approximately 90% of the mixture by weight. As the mixture evaporates during a test reading, the concentrations of both components remain constant, altering the psychometric relationship which assumes the presence of pure water. This binary azeotropic system is further complicated if more than two volatile components are present.

It should be emphasized that the caveats described here do not preclude the use of psychrometers at contaminated sites. While the methods may not provide quantitative evidence regarding fluid potentials, temporal changes in readings can provide crucial information regarding direction and magnitude of migration of water and contaminants.

ADDITIONAL INFORMATION

An excellent, although somewhat dated, summary of thermocouple psychrometry theory and applications can be found in Brown and Van Haveren (1972). Rawlins and Campbell (1986) also provide an excellent overview of thermocouple psychrometry.

REFERENCES

Brown, R.W., and B.P. Van Haveren, "Psychrometry in Water Relations Research," Proceedings of the Symposium on Thermocouple Psychrometers, Utah State University, March 17–19, 1971, Logan, UT, 1972.

Campbell, G.S., "Improved Thermocouple Psychrometers for Measurement of Soil Water Potential in a Temperature Gradient," *J. of Physics, E: Sci. Instruments*, 12:1–5 (1979).

Montazer, P., E.P. Weeks, F. Thamir, D. Hammermeister, S.N. Yard, and P.B. Hofrichter, "Monitoring the Vadose Zone in Fractured Tuff," *Ground Water Mon. Rev.*, 8:72–88 (1988).

Rasmussen, T.C., and D.D. Evans, Unsaturated Flow and Transport Through Fractured Rock — Related to High-Level Waste Repositories, Phase II, NUREG/CR-4655, U.S. Nuclear Regulatory Commission, Washington, DC, 1986.

Rasmussen, T.C., D.D. Evans, P.J. Sheets, and J.H. Blanford, Unsaturated Fractured Rock Characterization Methods and Data Sets at the Apache Leap Tuff Site, NUREG/CR-5596, 139 pp., U.S. Nuclear Regulatory Commission, Washington, DC, 1990.

Rawlins, S.L., and G.S. Campbell, "Water Potential: Thermocouple Psychrometry," in A. Klute, Ed., *Methods of Soil Analysis*, Part 1, 2nd Edition, 1986, pp. 597–618.

Rhodes, S.C., "Moisture Characteristic Curves for Apache Leap Tuff: Temperature Effects and Hysteresis, Superior, Arizona," Unpublished MS thesis, Department of Hydrology and Water Resources, The University of Arizona, Tucson, AZ, 1993.

22

Electric and Dielectric Methods for Monitoring Soil-Water Content

Ian White and S.J. Zegelin

INTRODUCTION

Information on the temporal and spatial variation of the water content of soil and other porous materials is central to understanding and managing a host of hydrologic, environmental, meteorological, agricultural, engineering, and industrial processes. It has long been recognized that reliable, robust, and automated techniques for the in situ monitoring of soil-water content in these varied processes can be extremely useful, if not essential. Over the past 70 years this recognition has fostered the investment of a considerable amount of ingenuity in developing such techniques.

It is, of course, naive to expect that a single technique will emerge which will satisfy all diverse needs for water content measurement. Rather, the selection of an appropriate technique or suite of techniques will depend on the specific application and the perceived strengths and weaknesses of the technique. It should also be appreciated that, while a knowledge of water content is adequate for many water balance studies, it may be insufficient for applications in which soil-water dynamics are required. There, it is the spatial gradient of the energy state of water in the material that determines flow, not the gradient of water content. To characterize soil-water flow the distribution of the soil-water potential is usually required as well.

The accepted standard technique for measuring the water content of porous materials involves oven drying a sample of the material at a prescribed temperature and sometimes for a prescribed time (Gardner, 1986). This inevitably involves destructive sampling. It is also time consuming and is therefore not well suited to in situ monitoring over extended periods or when "on-line" measurements are needed. For such instances indirect methods are needed.

Most physical properties of soils and porous materials vary systematically with changes in water content (Gardner, 1987). Many of these have been employed for indirect in situ water content determination. Ideally, any property selected for such use should depend uniquely on water content. In practice, this is seldom so. The electrical properties of porous materials, in particular, have long been considered to have substantial potential for water content measurement. This potential has been explored extensively.

To electrical engineers, soil is a lossy, dispersive dielectric medium whose electrical properties depend on water content and soil composition. Cashen in 1932 investigated the use of both capacitance and electrical conductivity for water content measurement. Soon after, Smith-Rose (1933) showed that the dielectric properties of electrically nonconducting soils were influenced strongly by water content. Following these pioneering investigations, interest in the electric and dielectric properties of soils as a means to monitor soil-water content has blossomed, particularly in the last 20 years. Advances in electronics and the arrival of microcomputers have enabled sophisticated techniques to be developed, introduced, and used on a routine basis in the field.

In this chapter we review electric and dielectric techniques for determining soil-water content. We describe the principles underpinning the technique and examine critically their strengths and weaknesses. Electrical conductance and dielectric techniques are discussed. We include electrical conductivity, electromagnetic induction, moisture blocks, capacitance, time domain reflectometry (TDR), and microwave and radar techniques in our examination.

ELECTRICAL CONDUCTANCE TECHNIQUES

The electrical resistance or conductivity of a soil or porous solid is not a unique function of its water content. It also depends on the amount of electrolyte present in the soil-water solution, on its composite minerals, particularly on the concentration and mobility of their surface charges, on the soil pore structure, and on the soil temperature. Three methods of using electrical conductivity have been employed. The first involves the direct measurement of the soil's electrical conductivity through the insertion of conductivity probes in the soil. The second uses transmitter coils placed on or above the soil surface to induce, electromagnetically, eddy currents in the soil that are detected by receiver coils. The third is an indirect method in which the electrical resistance of a block of porous material of known properties is monitored after the block has been inserted in the soil.

Soil Electrical Conductivity

Rhoades et al. (1976) suggested on theoretical grounds that the dependence of electrical conductivity of a soil, σ (S m^{-1}), on volumetric water content, θ [L$^3 \cdot$L^{-3}], may be approximated by

$$\sigma = \sigma_w(a_1\theta^2 + b_1\theta) + \sigma_s \qquad (1)$$

In Equation 1 σ_w is the electrical conductivity of the soil-water solution, σ_s is the contribution of surface charges to the bulk conductivity, and a_1 and b_1 are considered soil dependent empirical "constants." The contribution of σ_w to bulk soil conductivity arises from the presence of ionic species in the soil solution, while that of σ_s

is due to the presence of surface charges on the soil minerals and the countervailing ions in the double layers surrounding those surface charges. Clay minerals, in general, have large surface charge densities. The temperature dependence of σ in Equation 1 arises through the temperature dependence of σ_w and σ_s. Rhoades et al. (1976) listed values of a_1 and b_1 for a few, select soils. We see from Equation 1 that, provided σ_w and σ_s are independent of θ and remain unchanged by any other factors, θ is given by

$$\theta = \sqrt{a_2\sigma + b_2} - b_1 \qquad (2)$$

Here the "constants" a_2 and b_2 are given by $a_2 = 4a/\sigma_w$ and $b_2 = b_1{}^2 - \sigma_s/\sigma_w$. To use σ to determine θ, calibration curves must be constructed for each case under study.

Early attempts to measure water content through σ using an alternating current (ac) supplied to a two-electrode insertion probe were frustrated because of large variations in the contact resistance between the electrodes and the soil (Edlefsen and Anderson, 1941). The four-electrode probe, originally devised by Wenner (1916) for use in geophysical prospecting, eliminates the problem of contact resistance by measuring current and voltage between different pairs of electrodes (Wenner, 1916; McCorkle, 1931; Kirkham and Taylor, 1950).

A schematic diagram of the four-electrode probe, with electrode separation s at insertion depth z, is shown in Figure 22.1. The outer two electrodes of the four-electrode probe are connected to an ac power supply producing a known measured current I through the soil. By measuring the potential V_4 between the inner two electrodes while current I passes between the outer electrodes, the specific electrical conductivity σ of the soil may be determined from (Kirkham and Taylor, 1950)

$$\sigma = (n/4\pi s)(I/V_4) \qquad (3)$$

with the dimensionless probe geometry term n given by

$$n = 1 + 2[(1 + 4z^2/s^2)^{-1/2} - (4 + 4z^2/s^2)^{-1/2}] \qquad (4)$$

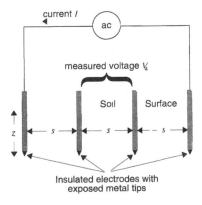

Figure 22.1. Schematic of four-electrode resistance probe.

In the limit of very large electrode spacing compared with insertion depth, Equation 4 gives $n = 2$, while for large insertion depths compared with electrode spacing $n = 1$.

For a fixed Wenner array geometry, Equation 3 simplifies to

$$\sigma = f_T/sR_T \qquad (5)$$

Here R_T is the measured electrical resistance at soil temperature T and f_T is a factor that includes both geometry and temperature which adjusts the reading to a reference temperature of 25°C.

The principal advantages of the Wenner array electrical conductivity technique for determining water content are its ease of use, the simplicity and low cost of the equipment, and the relatively large volume of soil sampled. Its most serious limitation arises from the fact that it is not water per se that conducts electricity, but rather the ions dissolved in the water or associated with the surface charges on the soil mineral particles. It requires, therefore, a great leap of faith to use σ to determine water content. We see from Equations 1 and 2 that this faith will often be misplaced because the assumed "constants" in these equations vary with minor, but quite common, changes in the soil-water solution composition and concentration and/or changes in the surface charge.

To use the technique to determine θ unequivocally, soil-water solution conductivity and surface conduction must be determined independently. This clearly destroys the advantage of simplicity. Except for highly specialized applications, the soil electrical conductivity technique on its own is not suited to routine water content determination (Gardner, 1987).

Electromagnetic induction offers the possibility of surveying large areas of soil rapidly. As we shall see, it suffers from similar problems to other electrical conductivity techniques (Rhoades, 1992).

Electromagnetic Induction

In the electromagnetic induction technique magnetic fields are imposed at the soil surface to induce electric currents in the underlying soil-water solution. The induced magnetic fields associated with these induced currents are then detected. A transmitter coil carrying an audio frequency ac current is placed on or near the soil surface (McNeill, 1992). The primary magnetic field, H_P, generated by the transmitter current induces eddy currents in the soil that in turn produce their own secondary magnetic fields, H_S. Both the secondary and primary fields are detected by a receiver coil at the soil surface as shown in Figure 22.2. The ratio of these two fields at the receiver coil provides an estimate of the apparent electrical conductivity of the soil, σ_{EM}. This method is also used for the aerial detection of ore bodies. It is related to ground-penetrating radar, which we will consider later.

The strength of the induced currents is a function of depth in the soil, the field strength at that depth, and both the local water content and soil electrical conductivity, together with the distribution of both between the depth in question and the soil surface. It is also dependent on the measurement frequency and the separation and configuration of the transmitter and receiver coils. The strength of the induced secondary magnetic fields is therefore a complicated function of soil electrical conductivity. In general, the task of separating out soil-water content at a particular depth from all the other factors that contribute to a single measurement represents a

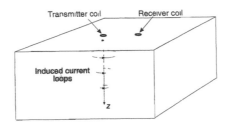

Figure 22.2. Schematic of operation of an electromagnetic induction soil conductivity meter.

mathematically ill-posed problem. This problem can be alleviated by ensuring that the induced secondary fields are small compared to the primary field, and by changing the spatial separation and orientation of the receiver and transmitter coils by varying the frequency of the imposed current or by following, in the time domain, the decay of induced fields when the transmitter is turned off (transient electromagnetic induction, TREM). It should be noted that measurements in the time domain are simply related, through Fourier transforms, to measurements in the frequency domain, FEM. Longer TREM decay times, for example, correspond to lower frequencies.

The ratio of the induced secondary magnetic field to the imposed primary field, H_S/H_P, when the transmit and receive coils are in the horizontal dipole configuration (plane of the coils perpendicular to the soil surface) is given by (McNeill, 1980)

$$(H_S/H_P)_H = 2[1 - 3/x^2 + (3 + 3x + x^2)e^{-x}/x^2] \qquad (6)$$

Here the subscript H refers to horizontal dipoles, $x = s\sqrt{(i\omega\mu_0\sigma)}$, with $i = \sqrt{-1}$, ω the angular frequency, μ_0 the permeability of free space, and s the separation between transmit and receive coils.

When the coils are in the vertical dipole configuration (both coils parallel to the soil surface) the field ratio is

$$(H_S/H_P)_V = (2/x^2)[9 - (9 + 9x + 4x^2 + x^3)e^{-x}] \qquad (7)$$

Here the subscript V refers to vertical dipoles. If $x << 1$, then the induction number, B, given by $B = x/(2i)^{1/2}$, is also small. For small B Equations 6 and 7 both reduce to (McNeill, 1980)

$$(H_S/H_P)_H = (H_S/H_P)_V = i\omega\mu_0\sigma s^2/4 \qquad (8)$$

So we see in Equation 8 that at low induction numbers, the field ratio detected at the receiver coil is directly proportional to the soil electrical conductivity and to the square of the separation between coils, and that the phase of the secondary field leads the primary by 90°. The low induction number criterion places bounds on the frequency which can be used for fixed coil spacing and soil conductivity. This condition is $\omega << 2/\mu_0\sigma s^2$. For frequencies above this limit, or for increasing σ, the linear relationship between field ratio and conductivity in Equation 8 no longer holds.

Equation 8 does not tell us how the distribution of conductivity within the soil influences the field ratio. As a rule of thumb, 70% of the measured response comes from soil to a depth of $3/4$ of the coil spacing when the coils are in the horizontal dipole configuration (plane of coils perpendicular to the soil surface). As the coil spacing is increased, secondary fields are induced at greater depth in the soil and the sampling depth of the instrument is increased (Rhoades and Halvorson, 1977).

The conductivity σ_{EM} determined by the induction method can be considered the weighted spatial average of the soil conductivity over the field penetration depth z_1,

$$\sigma_{EM} = (1/z_1) \int_0^{z_1} g(z)\sigma(z)dz \qquad (9)$$

We have assumed here, as is usual practice, that the conductivity is heterogeneous in only the vertically downward direction. Here $g(z)$ is the spatially dependent weighting function. With the coils in the horizontal dipole mode the response is most heavily weighted to the soil region closest to the surface, so that $g(z)$ is a monotonic function with depth. It is assumed that, at low induction numbers, the depth penetration is relatively insensitive to soil properties. This means that the relative contribution of soil at any depth in the soil profile is only a function of the ratio of that depth to inter-coil spacing, z/s.

McNeill (1980) gives the relative response of the measured secondary magnetic field, or apparent soil conductivity in the horizontal dipole, R_H from all soil below a depth z as

$$R_H(z/s) = \sqrt{4(z/s)^2 + 1} - 2(z/s) \qquad (10)$$

When the coils are turned through 90° into the vertical dipole mode the differential response is essentially not influenced by the near-surface soil. In this coil configuration, about 70% of the response comes from the soil to a depth of 1.5 times coil spacing. For vertical dipole configuration McNeill (1980) gives the reduced depth dependence of the relative response, R_V to all soil below a depth z as

$$R_V(z/s) = 1/\sqrt{4(z/s)^2 + 1} \qquad (11)$$

Rotation of the plane of the coils therefore provides a method for determining whether conductivity increases, decreases or is constant with depth (McNeill, 1992). Another way of extracting the conductivity at different depths from EM signals is to measure σ_{EM} with the transmit and receive coils at various heights above the soil surface and then to solve the resulting set of simultaneous equations (Rhoades and Corwin, 1981). Alternatively, lower frequencies are less attenuated and therefore penetrate deeper into the soil. Information from profiles of conductivity may be disentangled by either selecting several different frequencies, by sweeping through a range of frequencies, or by watching the decay of signal in the TREM mode.

In order to deconvolute the actual σ at any depth in the profile a model of the spatial distribution of σ is necessary. The simplest one usually employed is the layered model in which conductivity varies only in the vertical direction in discrete layers, each of which have fixed properties (McNeill, 1980). For example, in a three-layered soil model in which the top soil layer has thickness z_1 and the interface between the second and third soil layers occurs at depth z_2, and where the electrical conductivities of the top, middle and lowest soil layers are σ_1, σ_2, and σ_3 respectively, the overall soil electrical conductivity measured by the EM technique is given by (McNeill, 1980)

$$\sigma_{EM} = \sigma_1[1 - R_{H,V}(z_1)] + \sigma_2[R_{H,V}(z_1) - R_{H,V}(z_2)] + \sigma_3 R_{H,V}(z_2) \qquad (12)$$

where $R_{H,V}$ is given by Equation 10 or 11.

It can be seen from Equation 12 that, in practice, additional information about the depth dependence of soil properties, independent of the EM technique, is required.

McNeill (1980) has discussed case histories of the use of EM to survey soil conductivity over larger areas. These illustrate the spatial variations in soil conductivity that can be discovered with ease using the technique in various configurations. There are also good examples of the fact that the strength of the EM technique lies in its use as an interpolating tool between positions of known soil properties.

Others have examined the use of both TREM and the frequency domain, FEM, for estimating ground-water recharge in arid and semiarid regions (Page, 1969; Cook et al., 1989; Cook et al., 1992). In these climatic regions, with low surface relief, it has been assumed that recharge rate is a decreasing function of clay content which must be determined empirically for each location, and that recharge occurs one-dimensionally. These studies concluded that the EM response was predominantly due to clay content, and that recharge is generally influenced by the soil texture in the root zone. The conclusions appear to be highly location-dependent. Without independent point measurements of recharge, the estimation of recharge in any quantitative way from EM measurements seems to involve an even higher order ill-posed problem.

The major advantages of the electromagnetic induction method are that it does not need to be inserted in the ground, it is easy and quick to operate, it can provide estimates over large areas and substantial depths (of order 10 m) and it can detect small variations in soil conductivity. It suffers from the same disadvantage as the direct conductivity technique in the determination of soil-water content, because it also gives the bulk electrical conductivity of the soil, not water content. In addition, some assumed model of the distribution of electrical conductivity with depth is required to interpret data. Because of these, the electromagnetic induction technique must be recognized as a valuable technique for interpolating between sites where direct measurements of water content, soil-water conductivity, clay content and type have been made.

Electrical Resistance Blocks

The direct insertion probe method of measuring electrical conductivity suffers because of the differing contributions to bulk conductivity of surface charge and soil pore structure and their spatial distribution in field soils. Bouyoucos and Mick (1940) attempted to remove this uncertainty by embedding electrodes in porous gypsum blocks that were then placed in the soil. This indirect technique relies on the equilibration of soil-water potential in the block with that in the surrounding soil. Some of the gypsum in the block dissolves in the imbibed water to provide electrolyte. These changes are followed by monitoring the electrical resistance of the block, usually with an ac or pulsed direct current (dc) bridge. This resistance is, of course, a function of the temperature of the block.

Other porous materials such as fiberglass have also been used as the block (Coleman and Hendrix, 1949). These rely on the electrolytes in the soil-water solution itself to provide conduction. Because of this, inert porous blocks are sensitive to small changes in ambient soil-water electrolyte concentrations. Gypsum blocks are much less sensitive to such changes in slightly saline soils because they generate the electrolyte by self-dissolution.

The necessary equilibration process between soil and block water content after insertion, as well as when soil-water content changes naturally, means that resistance blocks respond to soil matrix potential, not soil-water content. Because of this, natural wetting and drying cycles give rise to hysteresis in the block response so that blocks must be calibrated under both wetting and drying regimes. In addition, because of the time scale for changes in the soil-water content, blocks seldom fully equilibrate with the soil. Also, for the wet range of soil-water content, the change in resistance of a block with change in water content is small and the precision is usually low (Gardner, 1986).

Gypsum blocks gradually dissolve, and this is accelerated in sodic soils and in the presence of a fluctuating water table. Their pores may also become plugged by deposited material. These changes alter the calibration curve between electrical resistance of the block and soil-water content with time. Inert blocks are more stable with time but suffer from drift, due to soil solution changes.

The advantages of resistance blocks lie in their cheapness, their ease of installation, their relative simplicity of operation, and in the fact that many blocks may be multiplexed from a single bridge. These strengths have made resistance blocks attractive for use in scheduling irrigation. However, changes in their calibration with time and ambient soil solution, their nonequilibration with soil-water and hysteresis problems mean that they are less useful for more accurate determination of water content.

It is clear in the above that all electrical conductance measurements, whether direct or indirect, suffer from similar difficulties. The principal problem is that both electrical conductivity and water content must be determined independently, preferably at several times, to use these techniques to monitor soil-water content with any precision. This problem and other considerations have led to investigations of dielectric techniques.

DIELECTRIC TECHNIQUES

The use of dielectric techniques to determine water content in the vadose zone has blossomed over the last decade. We will examine in some detail the basis of dielectric methods and the factors that complicate their use in the field.

Dielectric Constant

The potential between two electrically charged plates immersed in a nonconducting dielectric material is less by a factor $1/K$ than the potential between the same, identically charged and spaced plates in a vacuum. The macroscopic quantity K is known as the dielectric constant, relative permittivity or specific inductive capacity of the material. Its value depends on both the polarizability of the dielectric material induced by, and the angular frequency, ω of, the imposed electric field.

The induced polarization of the material between the charged plates arises from two effects. The first is due to the polarization of the distribution of electrons around the atoms of the dielectric due both to distortions of the electron cloud relative to the nucleus (electronic induction polarizability, γ_E), and to distortions of one partly charged atom relative to another (atomic induction polarizability, γ_A). The second is polarization of the material due to the reorientation of permanent molecular dipoles (dipole moment, μ) in the dielectric by the imposed field. The relation between dielectric constant and induced polarization and dipole reorientation for a dielectric in the gas phase or in dilute nonpolar solutions is (Le Fevre, 1953)

$$(K - 1)/(K + 2) = (4\pi\rho N/3M)(\gamma_E + \gamma_A + \mu^2/3kT) \tag{13}$$

In Equation 13 ρ is the density of the dielectric, N is Avogadro's number, M is the molecular weight, k is Boltzmann's constant, and T the absolute temperature. Here only μ depends on temperature. The second term in parentheses on the right hand side of Equation 13 is called the total molecular polarization. When γ_E, γ_A, and μ are zero, as they are for a vacuum, Equation 13 gives K equal to 1. We now examine typical values of dielectric constant for soil constituents.

Dielectric Constant of Water and Soil Constituents

The dipoles of water molecules in the liquid phase interact collaboratively to form a liquid with a large polarizability and therefore a large static, or zero frequency, dielectric constant $K(0)$. At 25°C the $K(0)$ of liquid water is 78.54 (Hasted, 1972). Air and fused silica at the same temperature have dielectric constants of 1.00054 and 3.78, respectively. Many soil minerals and organic matter have dielectric constants similar to quartz, usually in the range 3 to 5.

These large disparities between the dielectric constant of water, K_w, and other soil constituent solids, K_s, and air, K_{air}, have long been appreciated and suggest that the measurement of dielectric constants of porous materials may be useful for monitoring their water contents.

Effect of Measurement Frequency on Dielectric Constant

When the electric field imposed on a dielectric material is alternated at different angular frequencies, ω, it is found that K is not constant, as its name inappropriately suggests, but varies with imposed frequency. This is because the constituent molecules require some finite time, the relaxation time, τ, to adjust to the changing field strength. This relaxation process also gives rise to phase lags between the imposed field and the material's response to it. Debye (1929) showed that Equation 13 must be modified to account for this molecular relaxation,

$$(K - 1)/(K + 2) = (4\pi\rho N/3M)[\gamma_E + \gamma_A + \mu^2/(1 + i\omega\tau)3kT] \tag{14}$$

In Equation 14 $i = \sqrt{-1}$. We see here the frequency dependence of K and the fact that K is a complex quantity, composed of real, K', and imaginary, K'', components. The real part varies in phase with the changing imposed electric field. The imaginary part is the component which varies totally out of phase with the field. The complex dielectric constant may be therefore expressed as

$$K = K' - iK'' \tag{15}$$

For materials with a single relaxation time, Debye (1929) showed that K could be related to the static dielectric constant, $K(0)$, and the residual high frequency component (the constant component which is found at frequencies much higher than the characteristic frequency of dipole relaxation process), $K(\infty)$

$$[K - K(\infty)]/[K(0) - K(\infty)] = 1/(1 + i\omega\tau) \tag{16}$$

Equation 16 describes the semicircular Cole-Cole plot used to depict the dependence of K' and K'' on frequency. Dielectrics with a spectrum of relaxation times are described by a relationship similar to Equation 16 but are complicated by the inclusion of an exponential parameter which describes the spectrum width. It follows from Equations 15 and 16 that the real and imaginary parts of the dielectric constant for dielectrics with a single τ are given by

$$[K' - K(\infty)]/[K(0) - K(\infty)] = 1/[1 + (\omega\tau)^2] \tag{17}$$

and

$$K''/[K(0) - K(\infty)] = \omega\tau/[1 + (\omega\tau)^2] \tag{18}$$

For aqueous solutions $K(\infty)$ is 5.4 and appears to be independent of electrolyte composition (Pottel, 1973). In many applications it is the value of $K(0)$ that is required. When the dielectric material has appreciable electrical conductivity, ionic drift must be taken into account and the complex dielectric constant is (Kraus, 1984)

$$K(\omega) = K'(\omega) - i[K''(\omega) + \sigma_0/(\varepsilon_0\omega)] \tag{19}$$

Here, σ_0 is the dc or zero frequency electrical conductivity, and ε_0 is the permittivity of free space. Somewhat confusingly, the real component, K', is usually called the dielectric constant and the imaginary part, K'', is known as the dielectric loss factor. It is clear from Equation 19 that the material's electrical conductivity, σ, contributes to dielectric losses also. We shall return to this later.

The relaxation time of molecules in a liquid depends on temperature, T, viscosity, η, and the moment of inertia of the molecule. We can estimate this time approximately through application of Stokes' law as $\tau = 4\pi\eta r_m^3/kT$, where r_m is the molecular radius. Using typical values for water we find a relaxation time of about 10^{-10} seconds. At measurement frequencies greater than the relaxation frequency, $1/\tau$, or about 10 GHz, relaxation effects become increasingly important in water as its molecules can no longer follow the oscillations of the imposed field.

An added attraction of using dielectric constant to determine water content is that, at frequencies below 10 GHz, water's dielectric constant, K_w, is essentially independent of ω and its dielectric loss factor is small as well (Hasted, 1972). Above 10 GHz, K_w decreases because the relaxation time of water molecules is larger than their exposure time to the imposed field. This is shown in Figure 22.3a and the corresponding change in K'' for water is depicted in Figure 22.3b (Hoekstra and Delaney, 1974).

The frequency dependence of the soil minerals' dielectric constant between 0.45 to 35 GHz appears to be small (Campbell and Ulrichs, 1969). At low frequencies, soil type has a marked impact on dielectric constant, but at frequencies above 50 MHz this is less so. The electrical conductivity of the soil plays a complicating role here as

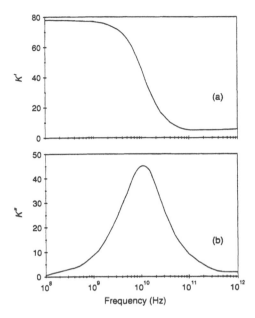

Figure 22.3. (a) Real component, and (b) imaginary component of dielectric constant as a function of frequency at 25°C.

well. Interfacial polarization at the electrode surface, due to the soil's electrical conductivity, may give rise to unacceptable extraneous capacitance effects. The relaxation frequency of the large dipoles associated with this polarization is around 27 MHz. Above this frequency interfacial polarization becomes less important, particularly above 50 MHz. For practical purposes then, dielectric determinations of water content in soils appear to be constrained to the frequency range 0.05 to 10 GHz.

In many determinations of dielectric constant of soils, particularly by the capacitance and TDR techniques, it has been tacitly assumed that the measured dielectric constant is the real part of the dielectric constant, irrespective of the measurement frequency. In essence, such measurements give an apparent dielectric constant, K_a, which is equal to the dielectric constant only when the contributions of K'' and σ to K are negligible (Topp et al., 1980). Both these contributions give rise to dielectric losses.

Dielectric Losses

Dielectric losses in porous materials containing water arise from two sources. The first is due to the relaxation time of molecule dipoles discussed above. The phase lag between imposed field and response of the dielectric is characterized by the loss angle δ. Dielectric losses are usually characterized by the tangent of the loss angle, which is known as the loss tangent or power factor. For molecular relaxation in liquids such as water, the Debye theory (1929) gives tanδ as

$$\tan\delta = \{4 \times 10^6 \rho N \ [(K + 2)\pi\mu]^2/(27KMkT)\} \{\omega\tau/[1 + (\omega\tau)^2]\} \qquad (20)$$

For soils, however, important losses also occur because of the electrical conductivity of the soil solution and the soil itself. Here the loss tangent is (Kraus, 1984)

$$\tan\delta = (K'' + \sigma_0/\omega\varepsilon_0)/K' \tag{21}$$

In many applications of the TDR and capacitance techniques in natural porous materials it has been assumed that $\tan\delta \ll 1$. Clearly at low frequencies in conducting soils, or at high frequencies when K'' is significant, or in highly conducting soils at all frequencies, $\tan\delta$ may not be negligible. Indeed, in some conducting soils it is often not possible to determine dielectric constant because of severe signal attenuation.

The dielectric constants of polar liquids are not only dependent on the frequency of the imposed field, they are also dependent on the strength of the imposed electromagnetic field (Le Fevre, 1953). This dependence is only important at high strength fields (2.5 MV·m^{-1}). It is negligible for all fields used in the dielectric determination of water content, particularly when compared with the influence of temperature on dielectric constant.

Influence of Temperature on Dielectric Constant

It is clear from the Equations 13 and 14 that, in materials with large dipole moments, dielectric constant depends on temperature. For liquid water, as the temperature is increased, thermal motions of water molecules increase and oppose their tendency to orient in an electromagnetic field. This lowers the dielectric constant. The temperature dependence of the static dielectric constant $K_w(0)$ is given by (Weast, 1974)

$$K_w(0) = 78.54[1 - 4.579 \times 10^{-3}(T - 25) + 1.19 \times 10^{-5}(T - 25)^2 - 2.8 \times 10^{-8}(T - 25)^3] \tag{22}$$

with T the temperature in degrees Celsius. The temperature dependence of the dielectric constant of the other soil constituents is much less than that of water (Campbell and Ulrichs, 1969). Nonetheless, for near-surface determinations of dielectric constant, accurate measurements require information on the surface soil temperature.

Effect of Dissolved Electrolytes on Dielectric Constant

Electrolytes dissolved in the soil-water decrease both the real and imaginary components of the dielectric constant because of less mobile hydration shells surrounding the ions (Pottel, 1973). For NaCl and KCl solutions, this decrease appears linear for concentrations below 0.5 mole l^{-1} and the decrease in the relative static dielectric constant is given approximately by

$$K(0)_\Gamma/K(0)_{\Gamma=0} = 1 - [10/K(0)_{\Gamma=0}]\Gamma \tag{23}$$

Here Γ is concentration in mole l^{-1}, $K(0)_\Gamma$ and $K(0)_{\Gamma=0}$ are the static dielectric constants of electrolyte solution of concentration Γ and pure water ($\Gamma = 0$), respectively. At low concentrations other salts show similar relationships. This implies that calibration curves, $K(\theta)$, even for soils which have small surface conductance, should show a dependence on the electrolyte composition of the soil solution. For the usual concentration of electrolytes found in soil we believe this will be a small, second-order effect. We now examine the relationship between K and water content.

Relationship Between Dielectric Constant and Volumetric Water Content

A relation between K and volumetric water content, θ, must be established if we are to use dielectric constant to determine soil-water content. The task of predicting the bulk K from the individual dielectric constants of mixtures of materials has been attempted by physicists for more than 100 years. Despite this, no simple relation exists between K and θ. Two approaches have been used; one employing mixing laws, the other empirical.

If soil is considered a mixture of three phases: water, soil and air, then the mixing law approach, e.g., Tinga et al., (1973) gives

$$K = [\theta K_w^{\beta} + (1 - \phi)K_s^{\beta} + (\phi - \theta)K_{air}^{\beta}]^{1/\beta} \tag{24}$$

In Equation 24 ϕ is the soil porosity (ratio of the volume of total pore space to the total volume of soil) and β is a geometric factor that depends on the spatial arrangement of the three phase mixture and its orientation in the imposed field. For isotropic two-phase mixtures $\beta = 1$ while for layered two-phase mixtures in which the layering is perpendicular to the field $\beta = -1$ (Birchak et al. 1974). Roth et al. (1990) found a "best fit" value of $\beta = 0.46$ for a range of soils. Because it is not possible, in general, to predict β or K_s a priori, Equation 24 is essentially an empirical equation, with parameters β and K_s. Its strength lies in the fact that it exhibits the correct behavior as $\theta \to 0$ and 1. We note that the derivation of Equation 24 assumes that the appropriate dielectric constant for water in soil is just that for pure water, K_w. To use Equation 24 to determine θ from measured K, values of porosity or bulk density are required in addition to K_s and β. The parameter β can also account for any contributions from the frequency-dependent complex dielectric constants of the soil constituents.

An alternate approach has been to use a convenient, empirical polynomial relation between K_a and θ. Topp et al. (1980) used a third-order polynomial,

$$\theta = a_3 + b_3 K_a + c_3 K_a^2 + d_3 K_a^3 \tag{25}$$

and evaluated the constants in Equation 25 experimentally using the TDR technique to determine K_a and oven drying at 105°C to find θ. They found that when the constants were given by $a_3 = -5.3 \times 10^{-2}$, $b_3 = 2.92 \times 10^{-2}$, $c_3 = -5.5 \times 10^{-4}$ and $d_3 = 4.3 \times 10^{-6}$, Equation 25 described the $\theta(K_a)$ relationships of a wide range of soil textures and porosities, provided $\theta < 0.6$.

Although Equation 25 does not give the correct limiting behavior as $K \to 1$ or $\theta \to 1$, it has proved useful for many soils, particularly those with coarser texture. For other materials, such as coal, different constants are required in Equation 25 which depend on the constituent materials of the medium (Zegelin et al., 1992). Use of Equation 25 does not require a priori estimation of K_s or ϕ. In addition it pragmatically makes no assumption concerning the dielectric constant of soil-water.

The relationships described by Equations 24 and 25 concentrate essentially on the way the real or apparent dielectric constant varies with water content. Reviews of the water content dependence of the imaginary part of the dielectric constant show a wider range of water content, frequency, and soil type dependence (Cihlar and Ulaby, 1974). Some measurements, in the frequency range 0.03 to 0.5 GHz, show that K'' has a greater water content dependence than K'. This seems to be an area that requires additional research.

Because of the wide range of materials which make up porous materials in nature, a truly "universal" relationship between θ and K, which does not require a priori information on composition, is phantasmagoric. If absolute water contents are required, then the applicability of Equations 24 or 25 must be tested for the particular application, or else a $\theta(K)$ calibration curve must be constructed using oven drying of the material. However, where only changes in stored soil-water are required, we have found Equation 25 gives acceptable values of those changes ($\pm 10\%$) for a wide range of soil types and situations (Zegelin et al., 1992). Equation 25 has an additional advantage in that it does not assume that the dielectric constant of water contained in porous materials is identical to that of free liquid water.

Effect of Bound Water on Dielectric Constant

Not all water in porous materials is as thermodynamically free as it is in liquid water. What we call "bound" and "free" water in porous material is, to some extent, a question of semantics. We will adopt here the working definition that water bound to mineral surfaces or to electrolyte ions is thermodynamically restricted because it is less able to reorient to an imposed electromagnetic field. It follows that bound water therefore possesses a smaller static dielectric constant than free liquid water. Evidence suggests that this bound water is restricted to about one to two molecular layers adjacent to the solid surface or ion. This water may not be lost on oven drying at 105°C. Commonly used values of K for water close to an interface is 6 for the first monomolecular layer of water and 32 for the second (Phillips, 1975). However, others claim that the bulk properties of pure water are applicable up to the interface (Hunter, 1966). Using our working definition, bound water may have a K anywhere between that for ice and pure liquid water.

Some mixing law models include bound water as a separate component (Dobson et al., 1985). In both the three-phase mixing law model of Equation 24 and the empirical approach of Equation 25 it appears that bound water would be incorporated into the solid phase dielectric constant. Dobson et al. (1985) used the Maxwell-de Loor equation to describe the effect of bound water on the soil's static dielectric constant,

$$K(0) = \frac{3K_s + 2(\theta - \theta_b)(K_w - K_s) + 2\theta_b(K_b - K_s) + 2(\phi - \theta)(K_{air} - K_s)}{3 + (\theta - \theta_b)(K_s/K_w - 1) + \theta_b(K_s/K_b - 1) + (\phi - \theta)(K_s/K_{air} - 1)} \quad (26)$$

Here θ_b and K_b are the volume fraction and dielectric constant of bound water. Dirksen and Dasberg (1994) have recently applied this four-component mixing law to soils with a range of specific surface areas, bulk densities, and mineral composition. They assumed, initially, that bound water was restricted to a single monomolecular layer and that K_b was identical to the high frequency limit of K for ice, 3.2. Good agreement between measured $K(\theta)$ and that predicted from Equation 26 could only be achieved by regarding the number of molecular layers of bound water and its dielectric constant to be variable parameters. In some cases the number of bound water layers had to be increased to three and K_b had to be enlarged to as much as 50. While Equation 26 may give a better description of the effects of bulk density and bound water on the soil's dielectric constant than the empirical relation, Equation 25, it requires that K_s, K_b, θ_b, and the soil's bulk density be known. We note here that Dirksen and Dasberg (1994) did not take into account the fact that they measured K_a, which includes a contribution from both the soil's bulk electrical conductivity and K'', rather than $K(0)$.

Brisco et al. (1992) have used a capacitance technique to examine the water content dependence of the real part of the dielectric constant in the measurement frequency range 0.45 to 9.3 GHz. They found that the empirical calibration of Topp et al. (1980) fitted their results remarkably well for the P (0.45 GHz) and L (1.25 GHz) microwave bands, and saw no reason to include bound water or soil textural effects in calibrations for any but low water contents for all frequencies studied. This is consistent with the conclusions of Rao et al. (1990), who found that the textural effect of bound water was limited to volumetric water contents below 0.1 to 0.15.

In many applications in environmental porous materials, bound water may be of secondary importance. Most of the soil-water that flows, transports nutrients and solutes, and is readily evaporated or taken up by plants, appears not to be as thermodynamically constrained as bound water. This labile water appears to be lost upon oven drying at 105°C. The mixing law models of Equations 24 and 26 assume that the water content of interest is that which has the dielectric constant of free liquid water. The empirical relation of Equation 25 makes no assumption about the dielectric constant of soil-water. It is calibrated by relating K_a to water lost on oven drying at 105°C.

We expect that the influence of bound water will only be significant at low water contents or in heavy clay soils. At low water contents, the individual dielectric constant of the soil mineral and organic constituents will also be important. It is therefore at low water contents or in heavy clay soils that we expect to find the greatest deviation between calculated and actual values when using Equations 24 and 25 with "universal" constants. As a corollary, we expect that the empirical relation Equation 25 will be most successful for higher water contents and for lighter textured soils.

We will now turn to common experimental techniques for finding the dielectric constant of soils in situ.

Techniques for Measuring the Dielectric Constant of Porous Materials

We examine here four techniques that have been used to measure the dielectric constant of soils and other porous materials; capacitance, time domain reflectometry, TDR, and microwave transmission and reflection.

Capacitance

When a potential is placed across the plates of a capacitor containing a dielectric, charges induced by polarization of the material act to counter the charges imposed on the plates. Ideally, the capacitance, C (in Farads), between two parallel plates of area A and separation s, is related to the dielectric constant K by

$$C = K\varepsilon_0 A/s \tag{27}$$

In the derivation of Equation 27 it is assumed that the lateral dimensions of the plate are much larger than the plate spacing s, and that all other sources of capacitance are insignificant. When the capacitor's geometry differs from that of the ideal parallel plate, and when inter-electrode and lead capacitance, C_e, are significant, Equation 27 becomes (Wobschall, 1980)

$$C = K\beta + C_e \tag{28}$$

Here β is a constant that depends on electrode spacing, orientation, and size and the invariant ε_0. The presence of electrolytes and mobile surface charges in soils tends, at low measurement frequencies, to produce interfacial polarization at the electrode surfaces, causing C_e to swamp the contribution by the soil's dielectric constant.

These problems plagued early attempts to use direct measurements of capacitance to determine soil-water content and for a long time discouraged interest in the technique (Gardner, 1987). The recognition that interfacial polarization could be overcome by using measurement frequencies above 50 MHz has renewed interest in the capacitance technique as an effective tool for monitoring in situ changes in soil-water content (Thomas, 1966). Advances in electronics have permitted the routine use of cheap high frequency circuits in the 50 to 150 MHz range, thus increasing the accessibility of the technique (Dean et al., 1987).

Measurement Principles.   In recent improvements to the capacitance technique, the capacitor containing the volume of soil to be measured forms part of the feedback loop of an inductance-capacitance, $L_i C$, resonance circuit of a Colpitts or Clapp high frequency oscillator (Wobschall, 1980; Dean et al., 1987). The resonance angular frequency of the oscillator, ω_r, is related to the capacitance of the soil probe C through

$$\omega_r = \left(1/\sqrt{Li}\right) \sqrt{(1/C + 1/C_i)} \tag{29}$$

In Equation 29 L_i is the inductance of the feedback loop and C_i the oscillator's capacitance. The relation between ω_r and the dielectric constant of the soil follows from Equations 28 and 29,

$$\omega_r = \left(1/\sqrt{Li}\right) \sqrt{[1/(K\beta + C_e) + 1/C_i]} \tag{30}$$

It is clear from Equation 30 that to detect changes in water content accurately, β, C_e, and C_i need to be minimized and kept constant. Fortunately, measurement of changes in ω_r to one part in 50,000 is relatively straightforward. The relationship between ω_r and volumetric water content should follow from combining Equation 30 with either Equation 24 or 25.

At 150 MHz the resonance frequency of the circuit may be altered by the capacitive effect of nearby bodies. Special precautions such as isolating the resonance circuit from the measurement circuit by using fiber optic cables, or else by placing all high frequency circuits within the capacitance probe, are necessary to prevent unwanted frequency shifts (Dean et al., 1987). Special attention to probe geometry is also necessary.

Probe Geometry. The geometry of the parallel plate capacitor is optimal since almost all the electric field is contained between the plates, and the contained field strength distribution varies as the reciprocal of distance from the plate, as shown in Equation 27. Such parallel plate probes have been widely used in laboratory determinations of water content of porous materials, particularly samples of stored grains, but their use in the field is less convenient because of plate insertion and soil disturbance problems.

More recently designed capacitance probes use split cylindrical electrodes that may be buried in the soil or positioned at different depths down plastic access tubes embedded in soil, as shown in Figure 22.4. The oscillator circuit and other electronics are placed within the cylindrical electrode probe (Dean et al., 1987).

It is clear from Figure 22.4 that not all the field between the cylindrical electrodes propagates into the soil. Some also flows through the plastic access tube and through the interior of the probe. The relative amounts of the field penetrating the probe, access tube, and soil compartments will depend on the radius of the cylindrical electrodes, the gap between the probes, and the relative dielectric constants of the compartments. As the radius and gap become smaller and as the soil becomes wetter, we expect less of the field will be proportioned to the soil compartment. The dielectric material between the cylindrical electrodes must have a low dielectric constant to ensure an adequate and accurate response to low soil dielectric constant, i.e., low soil-water content.

Zone of Influence. Two critical questions arise concerning any measurement probe placed in a porous material: over what region does the probe measure; and what is the spatial weighting of its response within that region? Dean et al. (1987) attempted to address those questions for the capacitance probe through an approximate experimental analysis of the region of influence of a probe similar to that in Figure 22.4. It is clear from Figure 22.4 that most of the field strength will be concentrated in the gap region between the plates. In normal use at least part of this region is occupied by the plastic access tube.

Dean et al. (1987) found that the region of influence is indeed restricted to a relatively narrow disc-shaped region surrounding the probe and centered on the gap between the electrodes. The probe is most sensitive to the region immediately adjacent to this gap. This means that the probe is very sensitive to any air gap between the probe, access tube and the soil, and that special care must be exercised in installation (Bell et al., 1987). A rigorous analysis of the effect of probe radius, plate gap width, plate width, and access tube thickness on the zone of influence and the spatial sensitivity of capacitance probes has yet to be undertaken.

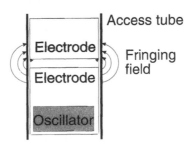

Figure 22.4. Capacitance probe cylindrical electrodes for use with plastic access tubes.

Response to Water Content Changes.   Equation 30 together with either Equation 24 or 25 should give a relationship between the circuit's resonance frequency and the volumetric water content of the soil. It is clear from the form of these equations that as θ increases, there is a nonlinear decrease in ω_r. Published data do show such a decline in resonance frequency with ω_r decreasing by 29% when the capacitance probe is moved from air to pure water (Bell et al., 1987). Unfortunately these results were not compared with the response predicted from Equation 30.

Extant calibration curves for different soils have used a very narrow water content range and have assumed that calibration is linear over that range. Somewhat disturbingly, these calibration curves show an almost ninefold variation in slope (Bell et al., 1987). This may indicate that the assumed constants in Equation 30 are in practice not constant, or it may be due to electrical conductivity of the soil, whose effect on the capacitance probe's performance appears not to have been explored systematically. Whatever the reason for the considerable disparity been calibration curves, these differences mean that calibration curves must be constructed for each site.

The stability, sensitivity to temperature change, and repeatability of measurements with the capacitance probe have been examined. It is found that measurement repeatability is better than 0.005 volumetric water content, and sensitivity to small changes in volumetric water content in dry materials is large. This repeatability and sensitivity are part of the strength of the capacitance probe technique.

Strengths and Weaknesses of Capacitance Techniques.   The strengths of modern capacitance probes are: their ability to be left in situ to log water content changes; the rapidity and ease of measurements; the portability of the apparatus; the repeatability of measurements; their extreme sensitivity to small changes in water content, particularly at the dry end of the water content range; their precise depth resolution due to their disc-like zone of influence; the fact that they do not involve ionizing radiation; and the relative cheapness of the probe.

The weaknesses of the capacitance technique include: the fact that a calibration curve must be constructed for each site; the large disparity between published calibration curves and expected behavior from Equation 30; the relatively small zone of influence of capacitance probes and their sensitivity to the region immediately adjacent to the probe; the effect of access tubes on its performance; the fact that the effect of electrical conduction on capacitance probe performance is not well documented; and its sensitivity to air gaps surrounding the probes.

There is still a considerable amount of development to be carried out on the capacitance probe. A rigorous analysis of the zone of influence and spatial weighting function of the probe needs to be undertaken. The reason for the mismatch between the expected behavior of calibration curves from Equation 30 and that found experimentally should be investigated, and the effect of soil electrical conductivity on capacitance probe response needs to be determined. Many of these factors have been considered in the time domain reflectometry technique.

Time Domain Reflectometry (TDR)

In the time domain reflectometry technique a very fast rise-time voltage pulse is propagated along a transmission line filled with or embedded in a material whose dielectric constant is required. This method is known also as cable radar, and is used widely for finding faults in cables. If the transmission line is open-circuited, then the pulse is reflected back from the end of the line.

The down and return travel time of the pulse in the transmission line depends on the length of the line and the dielectric constant of the material in which the line is embedded. Fellner-Feldegg (1969) appears to have been the first to have used TDR to measure the dielectric constant of liquids. Its use in the vadose zone to monitor soil-water content was pioneered by Topp et al. (1980).

Measurement Principles.   Provided the geometry of the transmission line probe is suitable, the TDR pulse propagates along the line, and the material filling it, in the transverse electric and magnetic (TEM) mode. When it reaches the end it is totally reflected. Impedance mismatches between the probe and the cable connecting the probe to the pulse generator and receiver cause partial reflections of the pulse. These reflections determine the signal amplitudes of the reflected pulses. Figure 22.5 shows the time course of a voltage pulse as it travels from the pulse generator to the end of the transmission line and back.

For a transmission line of length L, the total down and return travel time of a TEM wave is simply related to the wave's phase velocity, v, by

$$t = 2L/v \qquad\qquad (31)$$

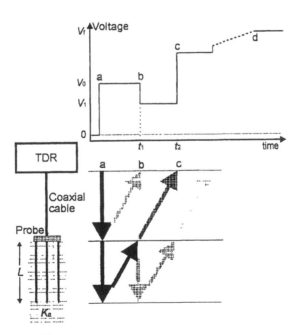

Figure 22.5.   Idealized TDR voltage vs. time trace and 3-wire TDR probe with wire length L. Travel time is $t = t_2 - t_1$ in uniform material with apparent dielectric constant K_a. Positions a, b, c, and d correspond to signal production, reflection from start of probe, reflection from end of probe, and final asymptotic voltage after all reflections, respectively. Voltage levels V_0, V_1, and V_f are output voltage level, voltage level after reflection from probe start, and final asymptotic voltage level respectively (from D.K. Cassel, R.G. Kachanoski, and G.C. Topp, personal communication).

The relation between phase velocity and the dielectric constant of the material filling the transmission line is given by (von Hippel, 1954)

$$v = c/\{K'[1 + (1 + \tan^2\delta)^{1/2}]/2\}^{1/2} \tag{32}$$

Here c is the velocity of TEM waves in free space, that is the speed of light ($\approx 3.0 \times 10^8$ m.s^{-1}), and $\tan\delta$ is the loss tangent defined by Equation 21. Topp et al. (1980) and Topp and Davis (1985) employed the now standard assumption that $\tan\delta \ll 1$ to derive the dependence of apparent dielectric constant on travel time,

$$K_a = (ct/2L)^2 \tag{33}$$

From Equation 33, which is the fundamental TDR equation, it can be seen that the larger the dielectric constant of the material in the transmission line, the longer the pulse travel time. Note that in heterogeneous materials, whose dielectric constant varies down the length of the probe, the measured travel time is summed over these variations so that the observed K_a is an average over the length L. If z is the coordinate along the length of the probe and $K_a(z)$ represents the variation of dielectric constant along the probe, the average $\overline{K_a}$ is

$$\overline{K_a} = \left[(1/L) \int_0^L \sqrt{K_a(z)}dz\right]^2 \tag{34}$$

The relation between the apparent dielectric constant and the real component of the dielectric constant can be found by comparing Equation 33 with Equations 31 and 32,

$$K_a = K'[1 + (1 + \tan^2\delta)^{1/2}]/2 \tag{35}$$

If the contributions of the soil's electrical conductivity and/or its dielectric loss factor are significant, then the apparent dielectric constant can exceed the real component of the dielectric constant. We might expect this to occur in heavy textured montmorillonitic soils with large surface charge densities, particularly those whose soil solutions are conductive as well.

Setting aside these complications, we see that the TDR technique involves determination of K of a sample through the measurement of transit time of a fast-rise TEM voltage pulse in a transmission line of known length placed in the sample.

The components of a TDR system consist of a fast-rise voltage pulse generator and an accurate signal analyzer connected by cable to a transmission line probe or through a multiplexer switch to an array of probes embedded in soil. Part of the art in the technique lies in its ability to measure accurately the extremely short travel times, usually in the 1 to 100 nanosecond range, associated with pulse propagation in probes that are usually less than 1 m in length. To achieve this, either 50 ohm output impedance commercial cable testers or specially made microprocessor-controlled TDR water content devices are employed (Skaling, 1992). The fast-rise voltage pulse of these devices essentially contains a window of all frequencies up to 3 GHz. When cable testers are used, signal analysis, data storage, and probe switching are carried out using a small, dedicated computer or data logger (Zegelin et al., 1989; Baker and Allmaras, 1990; Heimovaara and Bouten, 1990). These allow automatic analysis of water content in only a few seconds.

An additional part of the art of TDR lies in the minimization of extraneous partial reflections. The partial signal reflections shown in Figure 22.5 occur wherever there is a mismatch in impedance. These reflections cause information loss and interfere with the identification of the primary reflected pulse from the end of the probe. In any TDR system these impedance mismatches must be minimized.

The impedance of the dielectric filled transmission line probe, Z, is

$$Z = Z_0/\sqrt{K} \qquad (36)$$

where Z_0 is the characteristic impedance of the line. For certain transmission line geometries, such as coaxial lines, Z_0 can be calculated a priori. For other lines Z_0 may be measured by filling or immersing the line in a reference material of known dielectric constant, K_{REF} and measuring the reflection coefficient, ρ_r, of the line. Z_0 is then found from (Kraus, 1984)

$$Z_0 = Z_{TDR} \sqrt{K_{REF}} \; (1 + \rho_r)/(1 - \rho_r) \qquad (37)$$

where Z_{TDR} is the output impedance of the TDR device. The reflection coefficient is determined from the output and reflected voltages, V_0 and V_1, shown in Figure 22.5, and

$$\rho_r = (V_1/V_0) - 1 \qquad (38)$$

When the transmission line impedance matches the output impedance of the TDR device, $\rho_r = 0$. For a complete short circuit $\rho_r = -1$. Note that when a transmission line probe is immersed in a material whose water content changes abruptly along the length of the probe, partial reflection can also occur from that change. In designing TDR probes, special attention needs to be given to matching their characteristic impedance to the output impedance of the TDR system (Zegelin et al., 1989).

Insertion Probes. The coaxial transmission line probe may be considered the standard TDR probe. Its characteristic impedance can be matched exactly to the TDR system and can be calculated from the geometry of the probe which is filled or immersed in vacuum or air ($K = 1$), (Chipman, 1968).

$$Z_0 = 1/(2\pi\sqrt{K})\sqrt{\mu_0/\varepsilon_0} \; \ln (2s/d)$$
$$= 60 \ln(2s/d) \qquad (39)$$

Here $2s$ is the diameter of the outer conductor and d is the diameter of the inner, coaxial conductor. Equation 39 assumes that there are no end effects contributing to the probe's characteristic impedance. A typical coaxial probe is shown in Figure 22.6. While coaxial cells have been used widely in the laboratory, where they may be packed with soil or other porous materials, they are less easy to insert in the field. For field applications various insertion probes have been designed and used. These are also shown in Figure 22.6.

The two-wire probe was the first insertion probe used in field applications of the TDR technique (Topp et al., 1980; Topp and Davis, 1985). Its major limitation lies in the fact that the parallel line probe is mismatched to the coaxial TDR system, and this produces considerable signal and information loss. To overcome this mismatch a balancing transformer, a balun, is placed between the probe and the TDR coaxial

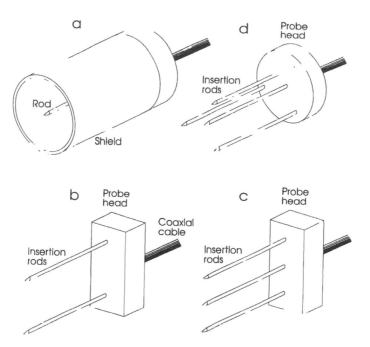

Figure 22.6. A selection of TDR probes, (a) coaxial cell, (b) parallel two-wire, (c) three-wire and (d) four-wire probes.

cable. This balun itself can be a source of noise and causes problems in analyzing signals from short probes or from probes immersed in conducting soils.

Coaxial emulating three- and four-wire probes, also illustrated in Figure 22.6, were introduced to overcome these problems (Zegelin et al., 1989). These probes are better matched to the coaxial TDR cable, their characteristic impedances approach that of a coaxial probe, they do not require a balun, and they produce much clearer and less noisy TDR signals. A typical TDR signal from a 2 m long probe inserted in a forested site is shown in Figure 22.7 (Zegelin et al., 1992). Here there was a gradation of water content with depth; however, the sharp rise of the reflected signal is still evident.

Irrespective of the number of wires used in the insertion probe, the design parameters that may be varied are the wire spacing, s, the wire diameter, d, and the length of the probe, L. The constraint for TDR operation is that signal propagation must be in the TEM. This limits the maximum wire spacing to the order of 0.1 m. We have found excellent TDR operation at this spacing (Zegelin et al., 1989). In essence this means that TDR probes are limited to sample diameters of about 0.2 m. Upper limitations on wire diameter arise because of problems with probe insertion and soil disturbance, and lower limitations on d arise because of measurement volume sensitivity, which we will discuss later (Knight, 1992).

There are also upper and lower limitations on the length of TDR probes that may be used. The lower limit is imposed by the accuracy of the time-of-travel measurement of the TDR device. This is currently of order 0.1 ns. This limits insertion probes to lengths of not less than 50 mm. In our experience probes of even 0.1 m have reduced accuracy because of this timing limit. The upper limit on probe length is decided partly by the strength of arm (or weakness of head) of the person who

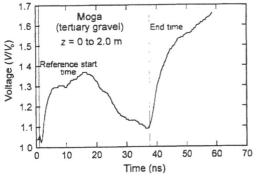

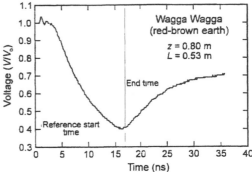

Figure 22.7. Time domain reflectometry trace of normalized voltage versus time for a Tertiary gravel with negligible electrical conductivity at Moga, New South Wales. Results for a three-wire probe (L = 2.0 m, d = 6.35 mm, s = 0.05 m). Measured water content θ = 0.140.

Figure 22.8. Time domain reflectometry trace of normalized voltage versus time for red-brown earth with massive clay subsoil at Wagga Wagga, New South Wales. Results for a three-wire probe (L = 0.53 m, d = 6.35 mm, s = 0.05 m). Measured water content θ = 0.340.

inserts the probe. We have inserted probes of length 2 m and obtained excellent results as seen in Figure 22.7. However, a more severe limit to probe length arises due to losses of the TDR signal in the soil. These losses can be severe in saline or heavy clay soils and can limit probe lengths to as little as 0.1 m. A typical example of the effect of such attenuation losses on the TDR signal is shown in Figure 22.8 for a soil with a heavy clay subsoil at Wagga Wagga, New South Wales.

The difficulty of insertion of TDR probes and the degree of soil disturbance clearly increases with the number of wires attached to the probe. The balance between these and the increased signal clarity generated seems to be reached with the three-wire probe. An alternate solution to the problem of insertion lies in the use of surface probes that do not need to be inserted in the soil.

Surface Probes. Surface probes were designed for use in situations in which probes could not be inserted, or to prevent soil disturbance or to measure water contents close to surfaces (White et al., 1991; White and Zegelin, 1992). Probes for use on surfaces, within product streams, and for use down unlined boreholes are shown in Figures 22.9 and 22.10. These surface probes are essentially either leaky strip-lines as in Figure 22.9, or straight strip-line probes as in Figure 22.10. The design parameters for the leaky strip line probes are central wire diameter, d, spacing to ground shield, s, length of probe, L, width of shield extension, w, and dielectric constant of the filler, K_f. Because part of the electric field is propagated through the filler, K_f must be as small as possible. A typical TDR trace for a leaky strip-line probe is shown in Figure 22.11. These probes have about half the sensitivity to water content changes of insertion or coaxial probes. For the strip-line probes in Figure 22.10, design parameters are just central wire diameter, d, distance to shield, s, and length of wire, L. The sensitivity of these probes to changes in water content is identical to coaxial or insertion probes.

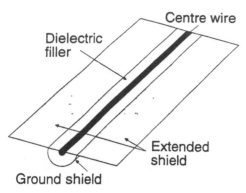

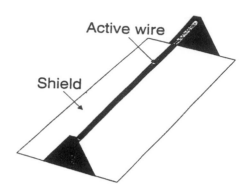

Figure 22.9. Leaky strip-line surface probe with center wire diameter *d*, center wire to ground shield spacing *s*, shield extension of width *w*, and probe length *L*.

Figure 22.10. Straight strip-line probe with center wire diameter *d*, center wire to shield distance *s*, and probe length *L*.

While the sensitivity of insertion and coaxial probes can be readily measured, the critical questions about the performance of TDR probes centers on their zone of influence and spatial weighting within that zone.

Zone of Influence and Spatial Weighting.   It is a relatively straightforward matter to calculate the approximate electric field distribution around TDR probes (Zegelin et al., 1989). Field distributions for two-, three- and four-wire insertion probes are shown in Figure 22.12. It can be seen that a great deal of the field is clustered relatively close to the transmission lines, and we expect that the probes will be most sensitive to the zones immediately adjacent to the wires. While the field distribution gives an approximate picture of the spatial sensitivity of TDR probes, more exact information can be found from the distribution of electromagnetic energy around the probe. Knight (1992) recently estimated the approximate two-dimensional spatial weighting function for coaxial and two-wire TDR probes by considering the effect of a small perturbation to the spatial distribution of the dielectric constant.

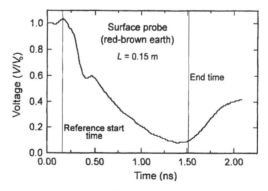

Figure 22.11.   Typical time domain reflectometry trace of normalized voltage versus time for a leaky strip-line probe (L = 0.14 m, d = 12.7 mm, s = 10.0 mm, w = 40.0 mm). Measured water content θ = 0.230.

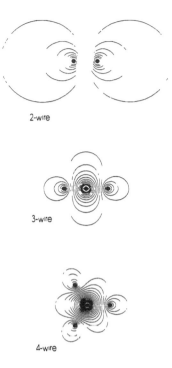

2-wire

3-wire

4-wire

Figure 22.12.   Dimensionless electric field distribution normal to the direction of probe insertion for a material of uniform dielectric constant for two-wire, three-wire, and four-wire probes with equal wire-to-wire spacing s and wire diameter d.

He showed that the weighting function was proportional to the distribution of electromagnetic energy around the probes.

The spatial weighting function $w_0(x,y,z)$ gives the apparent local dielectric constant at the point (x,y,z), $K_1(x,y,z)$, as a weighted average of the spatially distributed dielectric constant $K(x,y,z)$,

$$K_1(x,y,z) \; = \; \int_V K(x',y',z')w_0(x-x',y-y')dx'dy' \tag{40}$$

Here it is assumed that we are only interested in the effect of variations in K in the radial, x, y, directions and that there are no variations in K along the length of the probe, z. Knight (1992) found that the normalized w_0 for the coaxial probe is, to first order,

$$w_0(r) \; = \; [2\pi\ln(d/2s)]^{-1}r^{-2} \tag{41}$$

In Equation 41 r is the radial coordinate, d the diameter of the inner wire, and s the radius of the coaxial shield. We see that the spatial sensitivity for this probe geometry drops off rapidly with distance from the central wire. Similar results were also found for the two-wire probe. These results were in general accord with the experimental measurements of probe sensitivity by Baker and Lascano (1989). Similar results have been observed with three-wire probes (Zegelin et al., 1992). Because of the extreme sensitivity of the probe to the "skin" immediately surrounding the probe, Knight (1992) suggested that the ratio of d to $2s$ should be greater than or equal to 0.1, and that d should be as large as possible compared with the average pore size of the material.

The radial sensitivity of TDR probes means that they, like capacitance probes, are subject to measurement errors due to any air-gaps surrounding the probes (Annan, 1977a and b). They therefore have to be inserted with considerable care.

If the soil is uniform in the radial direction but varies in water content along the length of the probe, then the TDR technique gives the simple spatial average of the water content along the length of the probe. Vertically inserted TDR probes therefore provide a depth average of water content. Probes inserted horizontally from a pit provide an average at the plane of insertion. This then permits us to determine soil-water content profiles quite accurately. We will now discuss the response of TDR probes to changes in soil-water content.

Response to Water Content Changes. A typical calibration curve found using the TDR technique for a fine sand is shown in Figure 22.13. We compare this with the "universal" calibration curve, Equation 25 with appropriate constants found by Topp et al. (1980) to describe the calibration curves of a wide range of soils,

$$\theta = -5.3 \times 10^{-2} + 2.92 \times 10^{-2} K_a - 5.5 \times 10^{-4} K_a^2 + 4.3 \times 10^{-6} K_a^3 \qquad (42)$$

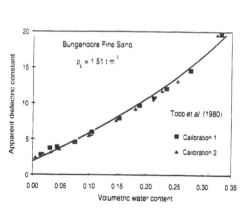

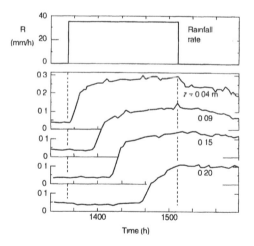

Figure 22.13. Calibration curve of apparent dielectric constant, K_a, against volumetric water content for Bungendore fine sand. Solid line is Equation 43.

Figure 22.14. Time dependence of soil-water contents measured using multiplexed horizontal three-wire probes at four depths, z, in a soil profile in a dry sclerophyll forest during a simulated rainfall event at Bungendore, New South Wales. Time is time of day; R is rainfall rate.

We see in Figure 22.13 that there is the expected discrepancy between Equation 42 and the measurements at the dry end of the calibration curve for $\theta < 0.05$. It is in this region that we expect contributions of the soil's particular mineral composition to be evident. In general the "universal" relation works well for many soils, particularly at the wet end and for lighter textured soils. Problems arise in peat soils and in heavy clay or conducting soils. In these materials individual calibration curves may be required.

The principal use of TDR lies in the monitoring of soil-water storage and water balance over extended periods. In Figure 22.14 we show the response of four multiplexed TDR probes buried horizontally to a simulated pulse of rain applied to the soil surface in the field. The arrival of the wetting front at each of the four depths can be clearly identified. So too can the subsequent drainage of the soil at the top two depths on the cessation of rainfall. Comparison between the amount of water added to the soil profile and that determined from the TDR measurements agreed within ±10%. The soil here was the fine sand used in the calibration in Figure 22.13. Comparisons can also be carried out between TDR determined changes in soil-water storage and changes in stored water measured by lysimeters over wetting and drying periods (Zegelin et al., 1992). These comparisons are given in Figures 22.15 and 22.16. Again they show agreement between TDR and the actual water lost or gained to within better than 10%. Two things are notable about the results in Figures 22.15 and 22.16. The first is that the TDR probes were located some 40 m from the lysimeter. The second is that the soil here has a heavy clay subsoil. If bound water is a problem in heavy clay soils, its effect on using TDR to determine changes in water balance is clearly second order.

The ability of TDR to detect small changes in water content will depend on the length of the probes used, the presence or absence of soil attenuation, and the amount of noise picked up by the TDR signal. In order to illustrate the small changes that can be observed we show, in Figure 22.17, cycling of water content in the surface layers of a soil detected by TDR due to the presence of a water table.

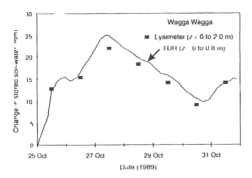

Figure 22.15. Comparison of TDR determined soil-water store during a wetting period with that measured with a weighing lysimeter over a 7 day period at Wagga Wagga, New South Wales.

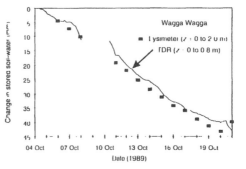

Figure 22.16. Comparison of TDR determined depletion in soil-water store due to evapotranspiration with that measured with a weighing lysimeter over a 16 day period at Wagga Wagga, New South Wales.

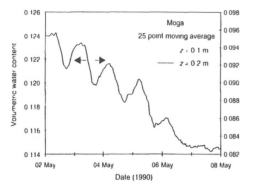

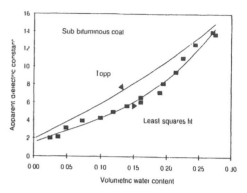

Figure 22.17. Example of cycling in soil-water content revealed through the use of 25 point moving average of 0.25 h TDR readings with horizontal probes (*L* = 0.5 m, *d* = 6.35 mm, *s* = 0.05 m) installed at soil depths of 0.1 and 0.2 m.

Figure 22.18. Calibration curve of apparent dielectric constant against volumetric water content for sub-bituminous coal. Solid curve is the empirical calibration for soils, Equation 42.

Theoretical considerations of simultaneous heat and mass transfer suggest that such cycling should take place (Yang and White, 1990).

Calibration curves for materials other than soil can be constructed as well. In Figure 22.18 we show the $\theta(K_a)$ calibration curve for sub-bituminous coal. It is clear that the universal calibration Equation 42 is not applicable here. For coal we find that the rank of the coal influences its calibration curve. When the coal is graphite-like, as it is in anthracite or semi-anthracite, bulk conductivity totally attenuates the return signal.

Electrical Conductivity and TDR Performance. Electrically conducting soils, whether conductive through surface charges on their clays or salinity in the soil solution, cause attenuation of the TDR signal. Occasionally the attenuation of the voltage pulse can be so severe that the TDR technique cannot be used.

Dalton and his colleagues proposed that the attenuation of the TDR signal could be used to estimate the bulk conductivity of the soil (Dalton et al., 1984; Dasberg and Dalton, 1985; Dalton and van Genuchten, 1986). This proposal has been examined in some detail by Topp et al. (1988), Yanuka et al. (1988), Zegelin et al. (1989). The latter authors used the thin sample analysis of Giese and Tiemann (1975) to provide an estimate of bulk conductivity, $\sigma_{G\text{-}T}$, based on easily measured voltages in the TDR trace. This estimate is given by

$$\sigma_{G\text{-}T} = [K_a^{1/2}/(120\pi L)](V_1/V_f)[(2V_0 - V_f)/(2V_0 - V_1)] \tag{43}$$

The voltages V_0, V_1, and V_f are shown in Figure 22.5. We have found that Equation 43 provides a useful estimate of the bulk soil electrical conductivity. A comparison between electrical conductivity estimated using Equation 43 and that directly measured with a conductivity bridge is shown in Figure 22.19 (Zegelin et al., 1989).

Despite the good agreement shown in Figure 22.19, alternate, supposedly superior

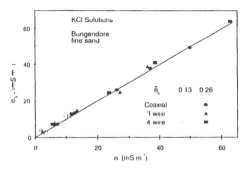

Figure 22.19. Comparison of bulk soil electrical conductivity, estimated from the TDR trace and Equation 43 for Bungendore sand which is wet at two average water contents with KCl solutions, with that measured using an ac conductivity bridge.

methods of estimating conductivity from TDR measurements have been proposed (Dalton, 1992). Part of the problem in determining σ from TDR seems to be confusion over what is measured in a TDR trace (see, e.g., Nadler et al., 1991). Another contribution to the problem may come from the continuing use of the assumption that K_a is identical to the real part of the dielectric constant, K', in Equation 19.

Several methods of overcoming the severe attenuation caused by electrically conducting soils are being investigated. One is the use of probes with wires that are coated with a thin insulating film. The effect of this film is unknown and the whole question of the impact of electrical conductivity on TDR response is an area still under active research.

Strengths and Weaknesses of the TDR Technique. The strengths of TDR technique include: its ability to be left in situ to log water content changes rapidly and almost simultaneously over many multiplexed sites; the rapidity and ease of measurements; the portability of the technique; the approximate "universal" calibration curve for light-textured soils, particularly at higher water contents; the repeatability of measurements; its precise depth resolution when horizontally inserted probes are used; the simple depth average of the water content profile provided by vertically inserted probes; its ability to estimate soil electrical conductivity; and the fact that it does not involve ionizing radiation.

Its weaknesses include: the relatively small zone of influence of TDR probes and their sensitivity to the region immediately adjacent to the probe wires; the sensitivity to air gaps surrounding the probes; attenuation of signal caused by salinity or highly conducting heavy clay soils; the failure of the "universal" calibration curve for heavy clay soils or at low water contents; the current price of measurement equipment; and the fact that TDR provides measurements over a relatively small soil sample.

In some applications measurements of water content over large areas are required. We will now examine techniques to provide estimates of water contents over such large areas.

Microwave Emission, Reflectance, and Transmission

The dielectric techniques described above are confined to sample volumes constrained between or near probes typically of the order of 0.1 to 1 m in length and 0.1 m in diameter. In many applications information on soil-water distribution is required over a much larger volume. The measurement of the transmission, reflectance, or emission of microwaves in or from soil has long been considered a key to unlock information about soil-water distribution on larger scales.

The microwave region of the spectrum lies in the wavelength band from 1 m to 1 mm with corresponding frequencies of 0.3 to 300 GHz. This bandwidth overlaps with the range of interest for water content determination, 0.05 to 10 GHz. Radiation in this energy band is associated with the rotational modes of the water molecule. The strength of absorption or emission of microwaves by soil depends on the rotational modes and dielectric constants of the medium and water, their relative concentrations, and their attenuation coefficients at the imposed frequency. Microwave absorption by water is, of course, the basis of microwave cooking. The potential for using microwave absorption to determine water content in porous materials has long been appreciated and is in routine use in industries such as bulk cement handling (Schmugge, 1980).

Several microwave techniques have been employed to determine water contents of porous materials. These include measurements of sample impedance at microwave frequencies (Equation 36), of transmission and reflection coefficients, of frequency shift, of polarization change, and of complex resonant frequency. These techniques are confined normally to fairly complex and fragile equipment in the laboratory (Stuchley and Hamid, 1972; Cronson, 1986). One exception is the portable dielectric probe, which was specifically designed for field use on small samples.

The Portable Dielectric Probe.   The portable dielectric probe (PDP) was introduced by Brunefeldt (1987). The PDP technique uses a microwave reflectometer to measure the reflection amplitude and phase of an open-ended coaxial cable probe. With this technique both the real and imaginary components of the dielectric constant are determined. This technique is operated in the P (0.45 GHz), L (1.25 GHz), C (5.3 GHz), and X (9.3 GHz) band microwave ranges. As with other techniques, measurements are critically dependent on probe geometry.

The tip of the portable dielectric probe consists of an open-ended coaxial line which is placed against the material. With this geometry the field penetrates a very short distance into the porous material. The diameter of the outer and inner coaxial conductors determine the penetration depth. In practice, the tip geometry and measurement circuitry determine the range of dielectric constants which can be measured with this probe.

The zone of influence of the portable dielectric probe is small, typically of the order 0 to 10 mm. The depth of this zone depends on the dielectric constant of the material under test and is larger for materials with small dielectric constant. The weighting function for PDPs is yet to be determined.

The response of the portable dielectric probe to changes in water content seems "better behaved" than for existing capacitance probes. This may be in part due to the fact that both the real and imaginary components of dielectric constant are determined with the PDP technique. Because K' is measured directly, the response of the PDP may be compared immediately with the expected responses from either Equations 24 or 25. Brisco et al. (1992) found at both 0.45 and 1.25 GHz that the dependence of K' on θ was identical to Equation 25 with the constants being those found by Topp et al. (1980) for the TDR technique. It seems, then, that the PDP technique, when operating in the P (0.45 GHz) and L (1.25 GHz) bands, does not require individual calibration for each site.

The strengths of the PDP are its portability, the fact that both real and imaginary components of the dielectric constant may be measured using it, and the finding that, like TDR, its response to water content change follows the empirical relation Equation 42, so that it may not require calibration for each site. Its weaknesses are that the as yet unanalyzed spatial weighting function of the probe is unknown, and the fact that the zone of influence is small.

Microwave and radar techniques have the potential to monitor large volumes of soil. We shall concentrate here on microwave emission and reflectance, and ground penetrating radar, as these are microwave techniques that are applicable in the field over large measurement scales.

Emission and Reflectance.   The emission of microwaves from the soil surface can be monitored by the passive microwave technique of radiometry. Measurement of the microwave reflectance of a surface involves the use of an active technique such as radar. Both passive and active techniques permit the sensor to be mounted on a mobile platform such as a truck, aircraft, or satellite. The depth of soil that contributes to the measurement in both passive and active systems depends on the wavelength of the microwaves, the soil-water content, soil type, and surface roughness (Jackson, 1988).

Measurement Principles.   The power reflection coefficient, r_f, of the soil surface can be approximated by (Cihlar and Ulaby, 1974)

$$r_h = [(p - \cos\psi)^2 + q^2]/[(p + \cos\psi)^2 + q^2]$$
$$r_v = [(K' \cos\psi - p)^2 + (K'' \cos\psi - q)^2]/[(K' \cos\psi + p)^2 + (K'' \cos\psi + q)^2] \quad (44)$$

Here r_h and r_v are the horizontal and vertical polarization, $\cos\psi$ is the cosine of the angle of incidence relative to the nadir and p and q are given by:

$$p = 2^{-1/2}\{[(K' + \cos^2\psi - 1)^2 + (K'')^2]^{1/2} + (K' + \cos^2\psi - 1)\}^{1/2}$$
$$q = 2^{-1/2}\{[(K' + \cos^2\psi - 1)^2 + (K'')^2]^{1/2} - (K' + \cos^2\psi - 1)\}^{1/2} \quad (45)$$

We see from Equation 44 that different responses will be observed, depending on the polarization of the detecting and emitting arrays and on the angle, ψ, of observation. For a smooth surface the emissivity in either vertical or horizontal polarization, $e_{v,h}$ is related simply to the reflectance by

$$e_{v,h} = 1 - r_{v,h} \quad (46)$$

The intensity of observed thermal emission from a surface by a radiometer in the microwave region is proportional to the brightness temperature of the surface, T_b. At some height above the surface, T_b is

$$T_b = \chi(r_f T_{sky} + T_e) + T_{atm} \quad (47)$$

In Equation 47, χ is the transmittance of the atmosphere between the radiometer and the surface, T_{sky}, T_e, and T_{atm} are the brightness temperatures of the sky, earth, and atmosphere, and r_f is the reflectance of the surface. At longer wavelengths atmospheric effects are minimal.

Thermal microwave emissions from soil are generated within the soil volume. The amount of energy leaving the soil is dependent on the distribution of dielectric constant in the soil and the surface reflectivity. Neglecting atmospheric effects, the brightness temperature of the surface can be written as the weighted average of the sub-surface temperature profile, $T(z)$ (Njoku and Kong, 1977)

$$T_b = \int_{-\infty}^{0} T(z)g[K(z)]dz \tag{48}$$

Here $g[K(z)]$ is a weighting function that includes the surface reflectivity. This weighting function will depend on the distributions of water, clay, and electrolyte in the soil, the wavelength of the microwaves, and the surface reflectance. Wet soil can exhibit a T_b 100°K cooler than dry soil.

From Equation 48 it is obvious that the attenuation of microwaves by soil plays a crucial role in either active or passive microwave techniques. The penetration of microwaves into a medium with relative magnetic permeability of one, which appears appropriate for soil, is characterized by the attenuation coefficient, α, which depends on the wavelength, λ, of the radiation, the dielectric constant, and the loss tangent of the medium:

$$\alpha = (2\pi/\lambda)\{(K'/2)[(1 + \tan^2\delta)^{1/2} - 1]\}^{1/2} \tag{49}$$

In general, because K' and $\tan\delta$ are functions of depth, so too is α. The power of the microwave radiation at any depth z, $P(z)$, is related to the power at the soil surface $P(0)$ through

$$P(z) = P(0)\exp(-2\alpha z) \tag{50}$$

We see in Equations 49 and 50 that shorter, higher frequency microwaves will have a larger attenuation coefficient than longer wavelengths, and will consequently "see" shallower depths of soil. For example, in dry soil radiation with $\lambda = 0.23$ m has an effective penetration depth of 100 m, whereas for $\lambda = 0.075$ m microwaves this depth is reduced to less than 0.5 m. The depth to which the radiation penetrates depends not only on λ but also on K' and $\tan\delta$, both of which are frequency dependent as well. It follows that the water, solute, clay, and soil mineral content distributions as well as the wavelength of the radiation all determine penetration depth in soils.

In addition to the properties identified in Equations 44 to 50, soil surface roughness, surface cover, and type and abundance of vegetative cover play important parts in determining the emission and reflectance or scattering coefficient of natural surfaces when sensed remotely.

Response to Water Content Changes.   A wide range of choice of operating parameters is offered in both passive and active microwave systems. These include geometry and configuration of the transmit and receive antennae, polarization, microwave frequency, angle of incidence of microwaves, height above ground surface, and "footprint" size of the surface region monitored. Variation of many of these parameters appears necessary because of the general, ill-posed nature of the problem of determining soil-water content from reflected or transmitted microwaves. This ill-posed nature is reflected in the conflicting results that have been produced in both emission and reflectance studies.

Detailed studies with both truck and aircraft mounted passive microwave detectors have shown that the L-band microwave region (1.4 GHz, $\lambda = 0.21$ m) appears to be the most useful for water content detection (Wang and Chowdrey, 1981; Njoku and O'Neill, 1982; Blanchard and O'Neill, 1983; Schmugge et al., 1980). However, it has been found that even in this region, both surface emissions and reflectance are sensitive to soil type as well as water content. In most studies brightness temperature or emissivity, reflectance or scattering coefficient have been correlated directly with soil-water content without any attempt to evaluate the dielectric properties of the soil.

It is possible that the observed soil texture dependence of both passive and active measurements may include a considerable contribution from the tangent of the loss angle arising from the imaginary part of the dielectric constant and the bulk conductivity of the soil. This possibility does not seem to have been explored in a systematic way. Schmugge (1980) demonstrated that some of the soil texture dependence could be removed if water content was expressed as a fraction of the "field capacity" of the soil. Field capacity is the presumed water content at which internal drainage ceases, (Veihmeyer and Hendrickson, 1955) and is an imprecise quantity which must be determined or estimated independently of any microwave measurement. Measuring field capacity does have the advantage of including implicitly some information on soil texture, since clay soils have a higher field capacity than sandy soils. It cannot, however, account for surface and soil solution conductivity.

In addition to soil texture dependence, surface cover and roughness have a significant impact on response, as does the in general unknown penetration depth of the radiation. A change in surface roughness from a smooth to a rough surface can affect surface brightness temperature as much as changing from a wet to a well drained soil. Vegetation in a plant canopy attenuates emitted or backscattered radiation from the soil as well as emitting or backscattering itself. Differences in scattering coefficients are observed between irrigated and nonirrigated crops (Dickey et al., 1974). However, we require more precision from these techniques for water balances than just a qualitative indication of soil wetness or dryness.

A final complication occurs when attempting to ground-truth these remote sensing techniques. This arises because of the size of the microwave "footprint" of the detection system at the soil surface. Often this footprint will encompass several soil types or even includes water bodies, varying soil moisture regimes, and disparate vegetation type and cover. Here the task of assessing independently the appropriate soil moisture for comparison is difficult if not impossible. This has led in some instances to the use of simulation models to provide ground-truth for comparison with microwave observations.

Strengths and Weaknesses.   The major strength of microwave emission and reflection techniques lies in their ability to rapidly cover large areas of the Earth's surface, particularly when aircraft or satellites are used. The main weaknesses of the techniques are that responses are dependent not only on soil moisture, but on soil type, surface cover and type of cover, and surface roughness. Also, the depth of soil "seen" by these techniques is a function of the soil-water content distribution. We believe the impact of soil electrical conductivity on microwave emission and reflection needs to be investigated systematically.

Ground Penetrating Radar

In ground penetrating radar the transmit and receive aerials may either be placed on the ground surface or inserted down boreholes. It differs slightly from the active radar reflectance techniques described previously in that the transmission path for microwaves is wholly within the soil. A considerable amount of research has been devoted to ground penetrating radar because of its potential for locating buried objects such as pipes, mines and underground tunnels (Peters and Young, 1986). Many of the principles underlying emission and reflectance techniques are directly applicable to ground probing radar.

Measurement Principles.   There are two frequency windows, at low frequency and high frequency, which permit the transmission of radar pulses in soil (Peters and Young, 1986). These frequency windows are determined by the dielectric constant and electrical conductivity of the soil, and by the desired penetration depth. The low frequency window is below 10 kHz and is within the frequency range used in electromagnetic induction measurements. Here we shall consider only the high frequency, microwave region.

The simplest case to interpret is where the transmit and receive aerials are placed within the soil, ensuring that the path length for transmission is known. For this case the attenuation due to the soil can be determined readily. Measurements in the 3 to 30 GHz range indicate that the attenuation is related linearly to water content (Akggarwal and Johnston, 1985). If it is assumed that the tangent of the loss angle is small then Equation 49 can be expanded to give

$$\alpha = (2\pi/\lambda)[(K'/4)(\tan^2\delta)]^{1/2} \tag{51}$$

Using the definition of $\tan\delta$ in Equation 21, we find

$$\alpha = (\pi/\lambda)(K'' + \sigma_0/\omega\varepsilon_0)/(K')^{1/2} \tag{52}$$

It is clear from Equation 52 that the attenuation coefficient is not only dependent on the real and imaginary parts of the dielectric constant, but also on the soil electrical conductivity. At the microwave frequencies normally selected for ground penetrating radar, around 10 GHz, the magnitudes of K' and K'' for water are approximately equal. So, in Equation 52 K', K'', and σ_0 are all significant functions of water content. Equations 1 and 42 indicate that all three quantities are, to a first approximation, quadratic functions of water content. Therefore the observed linear dependence of attenuation on θ must be considered fortuitous.

For separately buried transmit and receive antennae the travel time of the pulse may also be used to estimate the apparent dielectric constant using the fundamental TDR equation, Equation 33, in which $2L$ is the spacing between the antennae. The detection of the pulse may be complicated by scattered secondary pulses whose arrival time will depend on the distribution of soil properties. Here, as in all dielectric techniques, both real and imaginary components of the complex dielectric constant, and electrical conductivity contribute to K_a.

The application of buried transmit and receive antennae seems to offer some promise in monitoring water content changes at fixed locations over larger sample volumes than can be obtained with capacitance or TDR probes. The introduction of solid state microwave devices means that relatively low cost, robust devices are available for use in soil monitoring. Preliminary tests of these devices appear encouraging (Rasmussen and Campbell, 1987).

When the transmit and receive aerials are placed on the soil surface, the travel path length of microwaves through the soil is unknown, as is the distribution of soil-water and soil properties. The determination of water content from such measurements presents an ill-posed problem that is common to many geophysical techniques. Separation distance, antenna polarity, and radar frequency may all be varied in an attempt to solve this difficulty. It appears however that when antennae are placed on the soil surface, ground penetrating radar can only provide information on water content when coupled with an independent measurement technique such as TDR (Davis et al., 1977).

Strengths and Weaknesses.   As with microwave emission and reflectance techniques, the major strengths in ground penetrating radar lie in the large possible sample volumes and its ability to scan large areas when surface antennae are used. It is, however, for fixed, separately buried antennae that the technique appears to offer most promise, particularly when positioned down boreholes. Low-cost solid-state microwave sources offer considerable potential in this application. For antennae placed on the surface, the radar pulse path length is unknown. Independent local measurements of water content distribution appear the only way to break the ill-posed nature of this difficulty. In this technique, as in all other dielectric methods, soil electrical conductivity, through the presence of electrolytes and surface conduction, complicates measurements.

CONCLUDING COMMENTS

Soil is a lossy, dispersive, dielectric medium. Its ability to transmit electrical currents and electromagnetic radiation depends on the composition of the three-phase mixture which makes up soil. A vitally important component of that mixture is water, and it is no recent discovery that soil-water plays a major role in determining the electrical and dielectric properties of soil. Considerable effort has been spent in attempting to solve the inverse problem of determining the water content of soil from measured electrical or dielectric properties. There is, however, a nexus between water content, electrical conductivity, and dielectric properties which impinges on all measurement strategies, whether conductance or dielectric.

We have attempted here to give a critical overview of the use of electric and dielectric methods for determining water content in soils. We conclude, as others before have, that direct electrical conductivity measurements determine the soil's electrical conductivity, not water content. The indirect technique of measuring the conductivity of porous blocks inserted in the soil is a cheap but only qualitative tool for monitoring soil-water content.

Advances in electronics have enabled measurements to be made at frequency ranges which remove many of the problems encountered in early dielectric measurements. Capacitance and TDR techniques have been developed to measure the apparent dielectric constant of soils. Both require the insertion of probes into the soil. The measurement volume of these probes is now at least partly understood and their response to water content change is well documented.

In the TDR technique, an approximate "universal" calibration curve between dielectric constant and volumetric water content has been found for light textured soils, with low clay contents and low specific surface areas. As the clay content or electrical conductivity of the soil increases, the contributions of conductivity and the imaginary part of the dielectric constant to the apparent dielectric constant may become significant, requiring local calibration curves to be constructed. Further work is required on this problem. The TDR technique permits rapid multiplexed measurements of soil-water at a range of soil locations and depths. Without prior calibration, accuracies in water balance of ±10% seem attainable.

For the capacitance technique, quite large variations in calibration curves have been found. Here the relation between measured response and theoretical expectations and the response to soil electrical conductivity appear to have been unexplored. Additional work on both seems necessary. Capacitance and TDR techniques both suffer from a similar problem; measurements are most sensitive to a relatively small volume of soil close to probes.

The emission, reflectance, and transmission of microwaves through microwave radiometry or radar reflectance, scattering or ground penetration techniques offers the possibility of measurements over much larger scales. For these techniques soil type, or texture, and soil conductivity as well as water content contribute to observed responses. The problem, however, of determining water content is ill-posed because, in addition to the unknown soil properties, we do not know a priori the penetration depth of the microwaves. This problem may be solved by inserting independent transmit and receive antennae in the soil at a known separation; however, this restricts the technique to fixed locations. In this buried configuration, ground penetrating radar offers opportunities for monitoring larger soil volumes, although there are still significant problems to be addressed.

It is clear that the problem of monitoring soil-water continues to be both important, and a challenge. In the past 60 years considerable ingenuity has been used to devise useful and appropriate techniques. There seems little doubt that this ingenuity and inventiveness will continue.

LIST OF SYMBOLS

α	attenuation coefficient
β	geometric factor
Γ	concentration
γ_A	atomic induction polarizability
γ_E	electronic induction polarizability
δ	loss angle
ε	permittivity
ε_0	permittivity of free space ($\approx 8.85419 \times 10^{-12}$ F·m^{-1})
η	viscosity
θ, θ_v	volumetric water content
θ_b	volume fraction of bound water
$\varkappa$	lumped conductivity term for electromagnetic induction
λ	wavelength
μ	dipole moment

μ_0	permeability of free space ($4\pi \times 10^{-7}$ H·m^{-1})
π	ratio of circle circumference to diameter (≈ 3.14159265)
ρ	density
ρ_b	soil bulk density
ρ_r	reflection coefficient
σ	electrical conductivity
$\sigma(z)$	depth dependent electrical conductivity
σ_0	zero frequency electrical conductivity
σ_{EM}	electrical conductivity determined by electromagnetic induction
σ_{G-T}	Giese and Tiemann bulk electrical conductivity
σ_s	electrical conductivity due to surface charges
σ_w	electrical conductivity of soil-water solution
τ	relaxation time
ϕ	porosity
χ	transmittance
ψ	angle relative to nadir
ω	angular frequency
ω_r	resonance frequency
$a_1, a_2, \ldots, a_i$	constants
A	area
$b_1, b_2, \ldots, b_i$	constants
B	induction number
c	speed of light ($\approx 2.997925 \times 10^8$ m.s^{-1})
$c_1, c_2, \ldots, c_i$	constants
C	capacitance
C_e	inter-electrode and lead capacitance
C_i	oscillator capacitance
d	diameter
$d_1, d_2, \ldots, d_i$	constants
e	base of natural logarithms (≈ 2.71828)
$e_{v,h}$	emissivity in either vertical or horizontal polarization
f_T	factor that adjusts reading to reference temperature T
$g(z)$	spatial weighting function
H_P	imposed primary magnetic field
H_S	induced secondary magnetic field
$(H_S/H_P)_H$	ratio of secondary to primary magnetic fields for horizontal dipoles
$(H_S/H_P)_V$	ratio of secondary to primary magnetic fields for vertical dipoles
i	$\sqrt{-1}$
I	electric current
k	Boltzmann's constant ($\approx 1.38062 \times 10^{-23}$ J.K^{-1})
K	complex dielectric constant
K_a	apparent dielectric constant
K_{air}	dielectric constant of air
K_b	dielectric constant of bound water
K_f	dielectric constant of filler
K_l	apparent local dielectric constant
K_{REF}	dielectric constant of reference material
K_s	dielectric constant of constituent solids
K_w	dielectric constant of free water
K'	real component of dielectric constant

K''	imaginary component of dielectric constant
$K(\infty)$	residual high frequency component of dielectric constant
L	length
L_i	inductance
M	molecular weight
n	dimensionless probe geometry term
N	Avogadro's number
P	power of microwave radiation
r	radial coordinate
r_f	power reflection coefficient
r_h	horizontal power reflection coefficient
r_m	molecular radius
r_v	vertical power reflection coefficient
R_H	relative response of horizontal dipole
R_T	measured electrical resistance at temperature T
R_V	relative response of vertical dipole
s	separation
t	time
T	temperature
T_{atm}	brightness temperature of atmosphere
T_b	surface brightness temperature
T_e	brightness temperature of earth
T_{sky}	brightness temperature of sky
v	velocity
V	voltage
V_0	output voltage
V_1	reflected voltage
V_4	potential between inner two electrodes of four-electrode probe
V_f	final asymptotic voltage
w	width of surface probe shield
w_0	spatial weighting function
x	spatial co-ordinate in horizontal direction
y	spatial co-ordinate in horizontal direction
z	spatial co-ordinate in vertical direction
Z	impedance
Z_0	characteristic impedance of transmission line
Z_{TDR}	output impedance of TDR device

ACKNOWLEDGMENTS

The authors are grateful to G. Clarke Topp for his many helpful discussions, for his encouragement and for his constructive comments. We thank our colleagues John H. Knight for stimulating discussions and Rae Fry for expert editorial assistance. We acknowledge assistance for this work from the CSIRO Land and Water Care Program, and from NERDCC under Grant Number 1138, and from the Grains Research Council under Grant Number CSJ1W.

REFERENCES

Akggarwal, S.K., and R.H. Johnston, "The Effect of Temperature on the Accuracy of Microwave Moisture Measurements on Sandstone Cores," *IEEE Trans. Instrum. and Measur.*, IM-34: 21 (1985).

Annan, A.P., "Time Domain Reflectometry—Air Gap Problem for Parallel Wire Transmission Lines," *Report of Activities, Part B; Geol. Surv. Pap., Geol. Surv. Can.*, 77–1B: 59 (1977a).

Annan, A.P., "Time Domain Reflectometry—Air Gap Problem for Coaxial Line," *Report of Activities, Part B; Geol. Surv. Pap., Geol Surv. Can.*, 77–1B: 55 (1977b).

Baker, J.M., and R.R. Allmaras, "System for Automating and Multiplexing Soil Moisture Measurement by Time Domain Reflectometry," *Soil Sci. Soc. Am. J.*, 54: 1 (1990).

Baker, J.M., and R.J. Lascano, "The Spatial Sensitivity of Time Domain Reflectometry," *Soil Sci.*, 147: 378 (1989).

Bell, J.P., T.J. Dean, and M.G. Hodnett, "Soil Moisture Measurement by an Improved Capacitance Technique, Part II. Field Techniques, Evaluation and Calibration," *J. Hydrol.*, 93: 79 (1987).

Birchak, J.R., C.G. Gardner, J.E. Hipp, and J.M. Victor, "High Dielectric Constant Microwave Probes for Sensing Soil Moisture," *Proc. IEEE*, 62(1): 93 (1974).

Blanchard, B.J., and P.E. O'Neill, "Estimation of the Hydraulic Character of Soils with Passive Microwave Systems," in *Advances in Infiltration Proceedings National Conference on Advances in Infiltration*, December 12–23, Chicago, Am. Soc. Ag. Eng, St. Joseph, Michigan, 1983.

Bouyoucos, G.J., and A.H. Mick, "An Electrical Resistance Method for the Continuous Measurement of Soil Moisture Under Field Conditions," *Michigan Agric. Exp. Stn. Tech. Bull. 172*, East Lansing, MI, 1940.

Brisco, B., T.J. Pultz, R.J. Brown, G.C. Topp, M.A. Hares, and W.D. Zebchuk, "Soil Moisture Measurement Using Portable Dielectric Probes and Time Domain Reflectometry," *Water Resour. Res.*, 28: 1339 (1992).

Brunefeldt, D.R., "Theory and Design of a Field-Portable Dielectric Measurement System," in *IEEE International Geoscience and Remote Sensing Symposium Digest, Vol. 1, IEEE Cat. No. 87CH2434-9*, Institute of Electric and Electronic Engineers, New York, 1987, p. 559.

Campbell, M.J., and J. Ulrichs, "Electrical Properties of Rocks and Their Significance for Lunar Radar Observations," *J. Geophys. Res.*, 74: 5867 (1969).

Cashen, G.H., "Electrical Capacity and Conductivity of Soil Blocks," *J. Agr. Sc.*, 22: 146 (1932).

Chipman, R.A., *Theory and Problems of Transmission Lines*, Schaum's Outline Series, (New York: McGraw-Hill Book Co., 1968), p. 96.

Cihlar, J., and F.T. Ulaby, "Dielectric Properties of Soils as a Function of Water Content," *Remote Sensing Lab, RSL Tech. Rep., Univ. Kansas Center for Research Inc.*, 1974, p. 177.

Coleman, E.A., and T.M. Hendrix, "The Fibreglass Electrical Soil-Moisture Instrument," *Soil Sci.*, 67: 425 (1949).

Cook, P.G., M.W. Hughes, G.R. Walker, and G.B. Allison, "The Calibration of Frequency

Domain Electromagnetic Induction Meters and Their Possible Use in Recharge Studies," *J. Hydrol.*, 107: 251 (1989).

Cook, P.G., G.R. Walker, G. Buselli, I. Potts, and A.R. Dodds, "The Application of Electromagnetic Techniques to Groundwater Recharge Investigations," *J. Hydrol.*, 130: 201 (1992).

Cronson, H.M., "Time Domain Measurements of Components and Materials," in *Time Domain Measurements in Electromagnetics*, E.K. Miller, Ed., van Nostrand, Reinhold Elec. Comp. Sci. & Eng. Series, New York, 1986.

Dalton, F.N., "Development of Time Domain Reflectometry for Measuring Soil-Water Content and Bulk Soil Electrical Conductivity," in *Advances in Measurement of Soil Physical Properties: Bringing Theory into Practice*, G.C. Topp, W.D. Reynolds, and R.E. Green, Eds., Soil Science Society of America Inc., Madison, WI, 1992, p. 143.

Dalton, F.N., W.N. Herkelrath, D.S. Rawlins, and J.D. Rhoades, "Time Domain Reflectometry: Simultaneous Measurement of Soil-Water Content and Electrical Conductivity with a Single Probe," *Science (Washington, D.C.)*, 224: 989 (1984).

Dalton, F.N., and M.T. van Genuchten, "The Time Domain Reflectometry Method for Measuring Soil-Water Content and Salinity," *Geoderma*, 38: 237 (1986).

Dasberg, S., and F.N. Dalton, "Time Domain Reflectometry Field Measurements of Soil Water Content and Electrical Conductivity," *Soil Sci. Soc. Am. J.*, 49: 293 (1985).

Davis, J.L., G.C. Topp, and A.P. Annan, "Measuring Soil-Water Content in Situ Using Time Domain Reflectometry Techniques," *Geol. Survey Canada, Paper 77–1B*: 33 (1977).

Dean, T.J., J.P. Bell, and A.J.B. Baty, "Soil Moisture Measurement by an Improved Capacitance Technique, Part 1. Sensor Design and Performance," *J. Hydrol.*, 93: 67 (1987).

Debye, P., *Polar Molecules*, (Dover, New York), 1929, p. 77.

Dickey, F.M., J.C. King, J.C. Holtzman, and R.K. Moore, "Moisture Dependency of Radar Backscatter from Irrigated and Non-Irrigated Fields at 400 MHz and 13.3 GHz," *IEEE Trans. Geosci. Elect.*, GE-12: 19 (1974).

Dirksen, C.E., and S. Dasberg, "Four Component Mixing Model for Improved Calibration of TDR Soil Water Content Measurements," *Soil Sci. Soc. Am. J.*, 57:660 (1994).

Dobson, M.C., F.T. Hallikainen, and M.T. El-Rayes, "Microwave Dielectric Behaviour of Wet Soil. Part II. Dielectric Mixing Models," *IEEE Trans. Geosci. Remote Sensing*, GE-23: 35 (1985).

Edlefsen, N.E., and A.B.C. Anderson, "The Four-Electrode Resistance Method for Measuring Soil-Moisture Content Under Field Conditions," *Soil Sci.*, 51: 367 (1941).

Fellner-Feldegg, H., "The Measurement of Dielectrics in Time Domain," *J. Phys. Chem.*, 73: 616 (1969).

Gardner, W.H., "Water Content," in *Methods of Soil Analysis. Part I, Physical and Mineralogical Methods*, 2nd. edition, Klute, A., Ed., Am. Soc. Agron. & Soil Sci. Soc. Am., Madison, WI, 1986, p. 493.

Gardner, W.R., "Water Content: An Overview," in *International Conference on Measurement of Soil and Plant Water Status*, R.J. Hanks, and R.W. Brown, Eds., Utah State Univ., Logan, 1987, p. 7.

Giese, K., and R. Tiemann, "Determination of the Complex Permittivity from Thin-Sample Time Domain Reflectometry, Improved Analysis of the Step Response Waveform," *Adv. Mol. Relaxation Processes*, 7: 45 (1975).

Hasted, J.B., "Liquid Water: Dielectric Properties," in *Water, A Comprehensive Treatment Vol. 1, The Physics and Physical Chemistry of Water*, F. Franks, Ed., (New York: Plenum Press, 1972), p. 255.

Heimovaara, T.J., and W. Bouten, "A Computer-Controlled 36-Channel Time Domain Reflectometry System for Monitoring Soil-Water Content," *Water Resour. Res.*, 26: 2311 (1990).

Hoekstra, P., and A. Delaney, "Dielectric Properties of Soils at UHF and Microwave Frequencies," *J. Geophys. Res.*, 79: 1699 (1974).

Hunter, R.J., "The Interpretation of Electrokinetic Potentials," *J. Colloid Interfac. Sci.*, 22: 231 (1966).

Jackson, T.J., "Research Toward an Operational Passive Microwave Remote Sensing System for Soil Moisture," *J. Hydrol.*, 102: 95 (1988).

Kirkham, D., and G.S. Taylor, "Some Tests of a Four-Electrode Probe for Soil Moisture Measurement," *Soil Sci. Soc. Am. Proc.* 14: 42 (1950).

Knight, J.H., "The Sensitivity of Time Domain Reflectometry Measurements to Lateral Variations in Soil-Water Content," *Water Resour. Res.*, 28: 2345 (1992).

Kraus, J.D., *Electromagnetics*, 3rd ed., (New York: McGraw-Hill, 1984).

Le Fevre, R.J.W., *Dipole Moments: Their Measurement and Application in Chemistry*, (London: Methuen & Co., 1953).

McCorkle, W.H., "Determination of Soil Moisture by the Method of Multiple Electrodes," *Tex. Agr. Exp. Sta. Bul.*, 1931, p. 426.

McNeill, J.D., "Electromagnetic Terrain Conductivity Measurement at Low Induction Numbers," Tech. Note TN-6, Geonics Ltd., Ont., 1980.

McNeill, J.D., "Rapid, Accurate Mapping of Soil Salinity by Electromagnetic Ground Conductivity Meters," in *Advances in Measurement of Soil Physical Properties: Bringing Theory into Practice*, G.C. Topp, W.D. Reynolds, and R.E. Green, Eds., Soil Science Society of America Inc., Madison, WI, 1992, p. 209.

Nadler, A., S. Dasberg, and I. Lapid, "Time Domain Reflectometry Measurements of Water Content and Electrical Conductivity of Layered Soil Columns," *Soil Sci. Soc. Am. J.*, 55: 938 (1991).

Njoku, E.G., and J.A. Kong, "Theory for Passive Microwave Remote Sensing of Near-Surface Soil Moisture," *J. Geophys. Res.*, 82: 3108 (1977).

Njoku, E.G., and P.E. O'Neill, "Multi Frequency Microwave Radiometer Measurements of Soil Moisture," *IEEE Trans. Antennas Propag.*, AP-28: 680 (1982).

Page, L.M., "Use of Electrical Resistivity Methods for Investigating Geologic and Hydrologic Conditions in Santa Clara County, California," *J. Hydrol.*, 7: 167 (1969).

Peters, L., and J.D. Young, "Applications of Subsurface Transient Radar," in *Time Domain Measurements in Electromagnetics*, E.K. Miller, Ed., (New York: van Nostrand Reinhold, 1986).

Phillips, M.C., "Hydration and the Stability of Foams and Emulsions," in *Water, An Advanced Treatise, Vol. 5, Water in Disperse Systems*, F. Franks, Ed., (New York: Plenum, 1975), p. 133.

Pottel, R., "Dielectric Properties," in *Water, A Comprehensive Treatment Vol. 3, Aqueous Solutions of Simple Electrolytes*, F. Franks, Ed., (New York: Plenum Press, 1973), p. 401.

Rao, P.V.N., C.S. Raju, and K.S. Rao, "Microwave Remote Sensing of Soil Moisture: Elimination of Soil Textural Effects," *IEEE Trans. Geosci. Remote Sens.*, 28: 148 (1990).

Rasmussen, V.P., and R.H. Campbell, "A Simple Microwave Method for the Measurement of Soil Moisture," in *International Conference on Measurement of Soil and Plant Water Status*, Vol. 1, Utah State Univ., Logan, 1987, p. 275.

Rhoades, J.D., "Instrumental Field Methods of Salinity Appraisal," in *Advances in Measurement of Soil Physical Properties: Bringing Theory into Practice*, G.C. Topp, W.D. Reynolds, and R.E. Green, Eds., Soil Science Society of America Inc., Madison, WI, 1992, p. 231.

Rhoades, J.D., and D.L. Corwin, "Soil Electrical Conductivity—Depth Relations Using an Inductive Electromagnetic Soil Conductivity Meter," *Soil Sci. Soc. Am. J.*, 45: 255 (1981).

Rhoades, J.D., and A.D. Halvorson, "Electrical Conductivity Methods for Detecting and Delineating Saline Seeps and Measuring Salinity in Northern Great Plains Soils," ARS W-42 US Govt. Print. Off., Wash. D.C. 1977.

Rhoades, J.D., P.A.C. Raats, and R.J. Prather, "Effects of Liquid-Phase Conductivity, Water Content, and Surface Conductivity on Bulk Soil Electrical Conductivity," *Soil. Sci. Soc. Am. J.* 40: 651 (1976).

Roth, K., R. Schulin, H. Flühler, and W. Attinger, "Calibration of Time Domain Reflectometry for Water Content Measurement Using a Composite Dielectric Approach," *Water Resour. Res.*, 26: 2267 (1990).

Schmugge, T.J., "Effect of Texture on Microwave Emission from Soils," *IEEE Trans. Geosci. Remote Sensing*, GE-18: 353 (1980).

Schmugge, T.J., T.J. Jackson, and H.L. McKim, "Survey of Methods for Soil Moisture Determination," *Water Resour. Res.*, 16: 961 (1980).

Skaling, W., "TRASE: A Product History," in *Advances in Measurement of Soil Physical Properties: Bringing Theory into Practice*, G.C. Topp, W.D. Reynolds, and R.E. Green, Eds., Soil Science Society of America Inc., Madison, WI, 1992, p. 169.

Smith-Rose, R.L., "The Electrical Properties of Soils for Alternating Current at Radio Frequencies," *Proc. Roy. Soc. London*, 140: 359 (1933).

Stuchley, S.S., and M.A.K. Hamid, "State of the Art Microwave Sensors for Measuring Non-Electrical Quantities," *Int. J. Elec.*, 33(6): 617 (1972).

Thomas, A.M., "In-Situ Measurement of Moisture in Soil and Similar Substances by Fringe Capacitance," *J. Sci. Instrum.*, 43; 1966.

Tinga, W.R., W.A.G. Voss, and D.F. Blossey, "Generalised Approach to Multiphase Dielectric Mixture Theory," *J. Appl. Phys.*, 44: 3897 (1973).

Topp, G.C., and J.L. Davis, "Time Domain Reflectometry (TDR) and its Application to Irrigation Scheduling," in *Advances in Irrigation, Vol. 3*, D. Hillel, Ed., (Orlando, FL: Academic Press, 1985), p. 107.

Topp, G.C., J.L. Davis, and A.P. Annan, "Electromagnetic Determination of Soil-Water Content: Measurement in Coaxial Transmission Lines," *Water Resour. Res.*, 16: 574 (1980).

Topp, G.C., M. Yanuka, W.D. Zebchuk, and S.J. Zegelin, "Determination of Electrical Conductivity Using Time Domain Reflectometry: Soil and Water Experiments in Coaxial Lines," *Water Resour. Res.*, 24: 945 (1988).

Veihmeyer, F.J., and A.H. Hendrickson, "Does Transpiration Decrease as the Soil Moisture Decreases?" *Trans. Am. Geophys. Un.*, 36: 425 (1955).

von Hippel, A.R., Ed., *Dielectric Materials and Applications*, (Cambridge, MA: MIT Press, 1954).

Wang, J.R., and B.J. Chowdrey, "Remote Sensing of Soil Moisture over Bare Fields at 1.4 GHz Frequency," *J. Geophys. Res.*, 86: 5277 (1981).

Weast, R.C., Ed., *Handbook of Chemistry and Physics*, 55th edition, (Cleveland, OH: CRC Press, 1974), p. E-55.

Wenner, F., "A Method of Measuring Earth Resistivity," *U.S. Dept. Com. Bur. Standards Sci. Paper*, 1916, p. 258.

White, I., and S.J. Zegelin, "Measurement of Moisture Content and Electrical Conductivity," *US Patent 5,136,246*, 4 August, 1992.

White, I., S.J. Zegelin, and G.F. Russell, "Measurement of the Water Content of Coal Using Time Domain Reflectometry," NERDCC final report, CSIRO Centre for Env. Mech., Canberra, Tech. Rep. no. 44, 1991.

Wobschall, D., "A Frequency Shift Dielectric Soil Moisture Sensor," *IEEE Trans., Geoscience and Remote Sensing*, GE-18: 288 (1980).

Yang, J.Z., and I. White, "A Model of Coupled Water, Water Vapour and Heat Transport in Porous Media and a Simulation Analysis of Evaporation," *Math. Comput. Simulation*, 32: 161 (1990).

Yanuka, M., G.C. Topp, S.J. Zegelin, and W.D. Zebchuk, "Multiple Reflection and Attenuation of Time Domain Reflectometry Pulses: Theoretical Considerations for Applications to Soil and Water," *Water Resour. Res.*, 24: 939 (1988).

Zegelin, S.J., I. White, and D.R. Jenkins, "Improved Field Probes for Soil-Water Content and Electrical Conductivity Measurement Using Time Domain Reflectometry," *Water Resour. Res.*, 25: 2367 (1989).

Zegelin, S.J., I. White, and G.F. Russell, "A Critique of the Time Domain Reflectometry Technique for Determining Field Soil-Water Content," in *Advances in Measurement of Soil Physical Properties: Bringing Theory into Practice*, G.C. Topp, W.D. Reynolds, and R.E. Green, Eds., Soil Science Society of America Inc., Madison, WI, 1992, p. 187.

23

Applying Electrical Resistance Blocks for Unsaturated Zone Monitoring at Arid Sites*

Joseph P. Hayes and David C. Tight

INTRODUCTION

Electrical resistance blocks, made of gypsum or other porous materials, have been used to measure soil moisture in agricultural applications for decades. These blocks may be a useful and often overlooked tool for unsaturated zone monitoring applications at landfills or hazardous waste sites. The blocks are most accurate at soil-water tensions between 0.5 to 15 bars. Other common unsaturated zone monitoring devices, such as tensiometers, lysimeters, and manometers, will only operate in the wetter ranges of soil-water tension, below about 0.7 bar. Because they require damp soils, these other instruments are impractical in dry areas, such as the southwestern United States. Electric resistance blocks, however, will operate at conditions ranging from saturation to extreme aridity and, therefore, are well suited for unsaturated zone monitoring applications in the southwest.

Electrical resistance blocks (also known as soil-moisture blocks or gypsum blocks) typically consist of a pair or series of electrodes imbedded in a small block of suitable porous material. These blocks are buried and achieve moisture equilibrium with native earth materials. Changes in the water content of the porous material are reflected in changes in the electrical properties of the block. Changes in resistance are typically measured with a wheatstone bridge resistance meter and can be calibrated to reflect soil-moisture or soil-suction values.

These relatively simple, inexpensive instruments have been used extensively in agricultural research, and their operational properties are well established. However, these blocks are not widely used outside of agriculture, and many professionals working on ground-water contamination issues are not familiar with their operation. This chapter presents a brief review of the background information, installation procedures, and operation of electrical resistance blocks. In addition, a summary of a test installation evaluating the effectiveness of electrical resistance blocks at a southern California landfill site is presented. A wider understanding of the uses and limitations of these instruments is likely to generate new applications as the demand for monitoring the movement of water through the unsaturated zone continues to grow.

BACKGROUND AND OPERATIONAL INFORMATION

Electrical resistance blocks measure soil moisture as a function of the resistance between electrodes imbedded in porous material. The resistance is a measurement of energy potential and is determined by the soil-matric potential, or soil tension. Electrical resistance blocks were first developed over 40 years ago (Bouyoucos and Mick, 1948) and have been used primarily for applications in soil science and agricultural research. The basic elements of an electrical resistance block are a pair of electrodes, a porous matrix, which contains the electrodes and reaches moisture equilibrium with the native soil, and a recorder, which measures electrical resistance (see Figure 23.1). In reviewing irrigation research tools, Campbell and Campbell (1982) noted that electrical resistance blocks made of gypsum were simple and useful instruments; the main disadvantages were uncertainty in calibration accuracy and limited sensitivity at soil moisture potentials wetter than about 0.3 bar.

Blocks have been constructed with a wide variety of designs and from a variety of materials over the years. The porous matrix material should provide a stable and durable setting for the electrodes. Blocks have been constructed from gypsum, ceramic, nylon, fiberglass, and dental stone powder (Bouyoucos, 1952 and Morrison, 1983). Gypsum blocks have been successful because gypsum buffers against changes in salinity, which can strongly affect the electrical resistance of the block independent of moisture changes. Gypsum blocks are typically constructed with a 4 parts water to 5 parts gypsum solution, which maintains an electrolyte concentration of 2,200 to 2,400 parts per million (Bouyoucos, 1952). Buffering capacity would be important in landfill monitoring application where leachate migration could cause sudden change in salinity. However, gypsum blocks are subject to dissolution from moisture, which limits their lifespan in field installations. In wet or frequently saturated soils the blocks may only last 1 or 2 years. To reduce dissolution, gypsum is often combined with other materials, such as nylon, fiberglass, or resin. Combination of gypsum and other materials may also improve the sensitivity of the block in damp settings, since gypsum blocks are insensitive in soils with less than 0.3 bars tension (Bourget et al., 1958). A fiberglass-gypsum combination has been reported as the best combination for readings between 0 and 15 bars (Morrison, 1983).

Fiberglass blocks are more durable than gypsum in many soil environments, but they have demonstrated a chemical reaction and susceptibility to decomposition from soil salinity. Fiberglass does not provide any buffering capability, and is highly sensitive to changes in soil salinity. Nylon and nylon fabric "sheets" have been used to construct highly durable blocks that respond rapidly to moisture changes. Nylon fabric sheets imbedded with a flat mesh of electrodes have been used to provide a very thin matrix that minimizes variations in moisture between the inner and outer

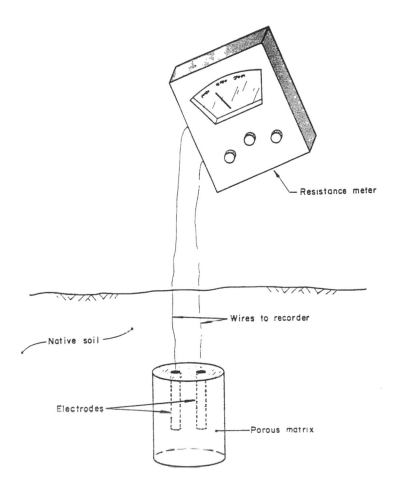

Figure 23.1. Cylindrical soil moisture block.

portions of the block. However, nylon has no buffering capacity and is usually combined with gypsum.

The electrodes for blocks have been made from a variety of metallic elements and steels. Screen type electrodes and multiple electrode sets have been built to maximize contact with the matrix material, although most blocks have a single pair of electrodes. To minimize stray electrical currents, both the electrodes and the blocks themselves are commonly cylindrical.

A wheatstone bridge resistance meter is used to obtain measurements from the blocks. Resistance increases with decreasing soil moisture. Some meters are constructed so that a high meter reading indicates high soil moisture and a low reading indicates low soil moisture. The electric resistance of the blocks has traditionally been made using AC current because a continuous direct current (DC) polarizes the blocks. However, modern datalogger systems that use a very brief DC pulse have been used successfully to automatically operate a network of installed soil-moisture blocks (Strangeways, 1983). The lower power and simple circuitry required to read these blocks make them ideal for automatic data acquisition.

Calibration

Calibration is required to convert the electric resistance measurements. Resistance readings should be calibrated as a function of soil tension rather than soil moisture. Standard calibration curves exist for commercially available soil-moisture blocks, but in situ or soil-specific calibration may be desired for improved accuracy. The relationship between resistance and soil tension varies with each soil and may also vary with each block under wetting versus drying conditions (hysteresis). The relationship between resistance and soil moisture varies even more widely because of differences in soil moisture characteristics. Calibration curves are generally constructed for either wetting or drying conditions because of variations caused by hysteresis. Calibration for wetting conditions would provide more accurate readings during recharge events and might be preferable for monitoring or leak detection installations. Calibration for drying conditions would provide more accurate readings during drying periods and would be preferable for irrigation or infiltration studies.

Calibration can be accomplished by placing a block in a representative sample of soil and measuring the soil tension with a pressure plate apparatus (Morrison, 1983), or by placing a block in soil in a wire basket and reading the gravimetric soil moisture as the sample is wetted or dried (Bouyoucos and Mick, 1948). In arid regions with highly saline soils, in situ calibration using gravimetric verification of soil samples is often required due to salinity effects on electrical resistance. In general, a precision of more than 5% to 10% should not be expected, and errors as great as 100% have been reported (Gardner, 1965). Moist soil-moisture blocks are not precise instruments and are better suited to measuring approximate or relative moisture changes.

Installation

Installation of the blocks is relatively simple, but must be done carefully to achieve intimate hydraulic contact with the native soil. This hydraulic contact is difficult to establish in consolidated materials or in rock debris or gravels. The electrical resistance blocks should be thoroughly saturated with distilled water prior to installation. Some researchers recommend that new blocks go through three cycles of wetting and drying before installation in order to remove air from the pore spaces of the blocks and ensure accurate readings (Everett et al., 1984).

Installation of the block in the sidewall of a trench or borehole is preferred for shallow depths due to the minimal disturbance of overlying soil required. Individual holes of the approximate size of the block can be made by hand, and cuttings from the hole can be used to fill any voids between the block and the soil. The borehole or trench is then backfilled and compacted either to the depth of the next installation or to the surface. Single or multiple sensors can also be placed in deeper boreholes and backfilled with a slurry of 200-mesh silica sand followed by the excavated soil. The soil should be recompacted in the borehole to its original density to prevent preferential infiltration. In multiple installations, this procedure is repeated at each targeted monitoring zone. In arid settings and low permeability materials, the amount of slurry added should be kept to the minimum amount necessary to cover the block in order to minimize the effects from water added to the slurry.

The borehole diameter should be only slightly larger than the diameter of the block to be installed in order to promote a good hydraulic contact between the block and the soil. Electrical resistance blocks may not perform well in coarse sandy soils or in soils with significant shrink-swell activity because of problems maintaining a good contact with the soil (Everett et al., 1984). Even with good soil contact, some lag time will occur between the arrival of a moisture front and an increase in the moisture content of an installed block.

RESULTS OF A FIELD STUDY

In the summer and fall of 1987, a field study was conducted on an unsaturated zone monitoring program at a landfill site in southern California to comply with requirements of Subchapter 15 of the California Administrative Code. Under this statute, unsaturated zone monitoring is required at solid waste disposal sites in California wherever feasible. Because this landfill site is located in a semi-arid climate and receives less than 15 inches of rainfall per year, it was believed that soil-water tensions would frequently be too high for lysimeters to obtain water samples from the unsaturated zone. The authors designed a program using electrical resistance blocks and tensiometers to measure soil-water tensions at the site and evaluate the feasibility of using suction lysimeters as an unsaturated zone sampling device. This investigation also allowed us to compare the accuracy and operational range of electrical resistance blocks to tensiometers and evaluate the ability of these instruments to detect wetting fronts.

Electrical resistance blocks and tensiometers were installed at five stations at the site in June 1987. Figure 23.2 shows a diagram of the tensiometers used in the study. Measurements were collected for approximately 5 months. Data for the two types of instruments were compiled and compared for each station. Moisture information from both types of instruments was also compared to moisture content analyses made on soil samples obtained during instrument installation. These comparisons indicate that the electrical resistance blocks can be reliable sensors of soil-moisture changes and can continue to operate at conditions too dry for conventional tensiometers. However, water added during installation influenced the electrical resistance blocks much more than expected and, in some cases, biased readings for several months. The installation and results of this investigation are described in greater detail below.

Monitoring Device Installation

Five instrument stations were installed around the perimeter of the landfill. Four stations are located near waste disposal areas, and one station is located several hundred yards upgradient to serve as a background station. Electrical resistance blocks constructed of stainless steel electrodes in a gypsum matrix were installed at several levels at each station. In addition, a tensiometer was installed at each station at a depth corresponding to a shallow gypsum block's setting. Table 23.1 lists the depths of the instruments and the native materials encountered at each installation.

Individual gypsum blocks were installed at the bottom of borings ranging from 3 to 23.5 feet deep. Shallow borings (3 to 6 feet) were hand augered, and deeper borings (7 to 23.5 feet) were drilled with air-rotary equipment. Hand-augered borings were advanced to 6 inches above the chosen total depth, and two 1.9-inch diameter, "undisturbed" soil samples were obtained with a U.S. Army Corps of Engineers soil sampler. These samples were preserved and analyzed for gravimetric

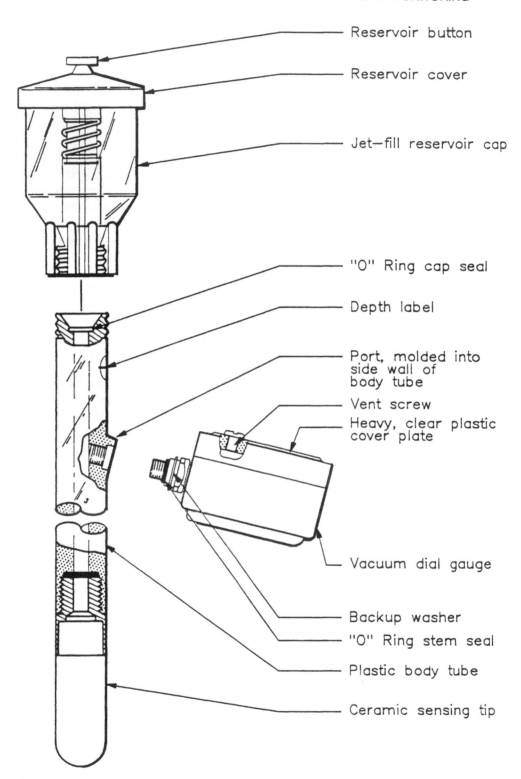

Figure 23.2.   Jet fill tensiometer.

Table 23.1. Unsaturated Zone Instrument Depths

Station Number	Depth		Native Material
	Instrument	(feet)	
1-A	G	23.5	Clayey Sandstone
1-B	G	10.5	Clayey Sandstone
1-C	G	3	Clayey Sandstone
1-T	T	3	Clayey Sandstone
2-A	G	5	Clayey Sandstone
2-T	T	5	Clayey Sandstone
3-A	G	5	Alluvium-(Silty Sand)
3-T	T	5	Alluvium-(Silty Sand)
4-A	G	3	Alluvium-(Silty Sand)
4-T	T	3	Alluvium-(Silty Sand)
5-A	G	4.5	Alluvium-(Silty Sand)
5-T	T	4.5	Alluvium-(Silty Sand)

G = Gypsum Block
T = Tensiometer

moisture content by gravimetric analyses. In deeper borings, an air-rotary core bit was advanced to 6 inches above the chosen total depth, and a core sample was obtained and analyzed for moisture content by gravimetric analysis.

Gypsum Block Installation

The gypsum blocks were installed in late June 1987 in the holes that were hand augered or drilled during soil sampling at the five vadose zone stations. The depth of each gypsum block is shown in Table 23.1.

All gypsum blocks were presoaked for approximately 30 minutes before installation, in accordance with the manufacturer's recommendations. The shallow gypsum blocks were installed in the following procedure.

After the soil samples were collected from the hand-augered hole, a thick slurry of silica flour and water was added to the hole to provide a backfill for the gypsum block. Sufficient slurry was added to the hole to fill about half the hole created by the soil sampler.

The gypsum block was lowered to the bottom of the hole, and additional silica slurry was added to cover the block. After waiting approximately 10 minutes for the silica to settle out and surround the block, the hole was backfilled with native cuttings. A shovel handle was used to tamp the backfill firmly in place. Finally, approximately the top 2 feet of the wire leads that connect to the gypsum block were encased in 1/2-inch polyvinyl chloride (PVC) conduit and completed, with about 1 inch of the PVC sticking up above the ground.

Deeper gypsum block installations were completed at locations 1-A, 1-B, 2-B, and 3-B (see Table 23.1) in holes drilled with an air-rotary rig. After soil samples were collected from the hole, the drilling rig was used to blow any cuttings out of the hole and to place a 1-1/2-inch steel tremmie pipe at the bottom of the hole. The tremmie pipe was forced 6 inches into the sandstone at the base of the hole to create a smaller diameter hole for the gypsum block placement. The gypsum block was placed inside the tremmie pipe and lowered to the bottom of the hole with 1/2-PVC conduit around the wire leads that connect to the block. Approximately 1 gallon of thick silica slurry was added to cover the gypsum block. After waiting approximately 10 minutes for the silica to settle, the hole was backfilled with cuttings to the surface. The wire leads from the gypsum block were then encased in PVC conduit as described for the shallow blocks.

Tensiometer Installation

Ceramic-tipped tensiometers were installed at the site to measure soil suction in the shallow zone. Tensiometers were installed at the same depths as shallow gypsum blocks to provide a correlation with gypsum block measurements (see Table 23.1).

Preparation of tensiometers prior to installation requires equipment assembly, testing for vacuum leaks, purging of air bubbles, and saturation of the ceramic cup. For installation, a borehole was advanced with a hand auger to 6 inches above the target depth. The hand auger was removed, and a tensiometer insertion tool was driven 6 inches to create a close-fitting hole for the ceramic tip of the tensiometer. The insertion tool was removed, and the tensiometer was placed into the hole, establishing a firm contact between the ceramic cup and the ground. A small amount of thick silica slurry was added, and the hole was backfilled and tamped to the surface with cuttings.

RESULTS OF FIELD INVESTIGATION

Field data are presented in Figures 23.3 through 23.8. The instruments, which were installed in June 1987, were monitored periodically for approximately 5 months through the late summer and early fall. A single rainfall event, a storm that deposited approximately an inch of rain, occurred 129 days after the instruments were installed. Gypsum block readings were converted from resistance to soil tension using a calibration curve prepared by the gypsum block manufacturer. Soil tensions were read from the tensiometers in centibars. While a characteristic curve relating soil tension to soil moisture was not prepared for the site, standard curves for sandy loam and clay loam soils were used to compare soil tension readings to results of soil moisture analyses performed on soil samples taken during installation of the devices. Soil tensions agreed reasonable well with these soil moisture results except where noted below. Data from both tensiometers and gypsum blocks have been plotted together for each instrument station to allow comparison of results.

At low to moderate soil tension (10 to 60 centibars), tensiometers and gypsum blocks gave very similar results. Figure 23.3 shows data from Station 2, where both types of instruments were installed at a depth of 5 feet. Data from the two instruments fluctuate slightly around an average soil tension of approximately 50 centibars, do not differ by more than 10 centibars, and often gave nearly identical results. This station remained at nearly constant soil-moisture conditions throughout the study.

Soil moisture decreased markedly during the study at other stations exposed to more direct sunlight and a greater degree of evapotranspiration. Figure 23.4 shows data from Station 4, where both types of instruments were installed at a depth of 3 feet in clayey fill. This clayey material developed desiccation cracks and high soil tensions as it dried. The tensiometer responded more rapidly to the rising soil

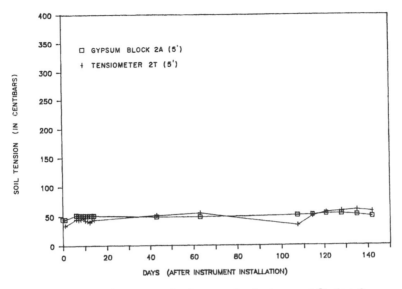

Figure 23.3.   Unsaturated zone monitoring results, Instrument Station 2.

tensions (Days 1 to 15 after installation), but was unable to register tensions greater than 50 centibars (see Figure 23.4). In contrast, the gypsum block indicated that soil tensions rose to 400 centibars by the end of the summer, eight times greater than the measurement limit of the tensiometer. During this period, the tensiometer continued to register 50 centibars or less as residual vacuum within the instrument dissipated. Numerous other investigations have indicated that ceramic-tipped tensiometers will not operate above 60 to 80 centibars (Morrison, 1983).

Other stations also registered this similarity of readings between the two types of instruments below approximately 50 centibars of tension and divergence of readings

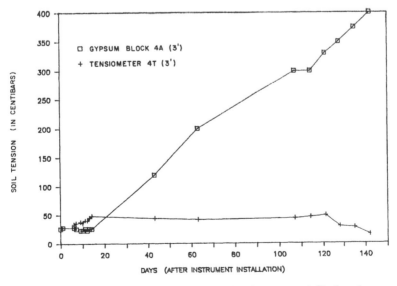

Figure 23.4.   Unsaturated zone monitoring results, Instrument Station 4.

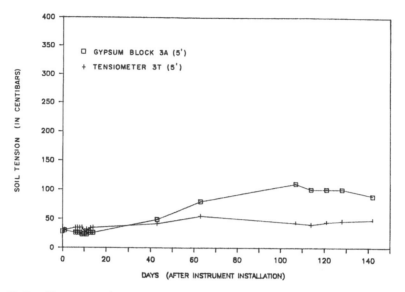

Figure 23.5.   Unsaturated zone monitoring results, Instrument Station 3.

at drier conditions. Figures 23.5 and 23.6 show results from Stations 3 and 4, respectively. Both of these stations showed consistent readings between gypsum blocks and tensiometers for several weeks after the instruments were installed at equal depths. However, tensiometer pressures level off around 50 centibars, while gypsum block readings increase to 105 centibars at Station 3 and 120 centibars at Station 4. These data indicate the advantage of using gypsum blocks to monitor soil tension in dry climates, where pressures above 50 centibars may be expected.

Not all gypsum block installations at the site were successful. Several of the block installations consistently indicated near saturated conditions, which did not agree

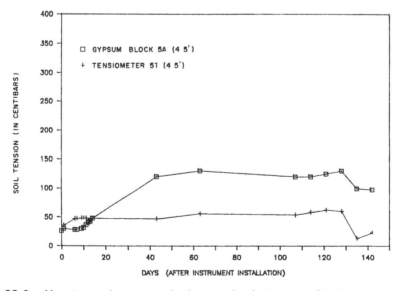

Figure 23.6.   Unsaturated zone monitoring results, Instrument Station 5.

with soil-moisture measurements made on soil samples. Figure 23.7 presents data from Station 1, where blocks were installed in clayey sandstone at 3, 10.5, and 23.5 feet. After initial fluctuations, all blocks registered low soil tensions at the bottom end of their effective range (20 centibars) for the entire study period, indicating fairly damp conditions. Volumetric moisture analysis of samples obtained during installation showed that native moisture contents at Station 1 were relatively dry (7.5% to 14.6%), well below saturation. Although a characteristic curve to relate soil tension to soil moisture was not prepared for the site because of lack of data, the characteristic curve for a Yolo clay loam soil indicates that a soil tension of 20

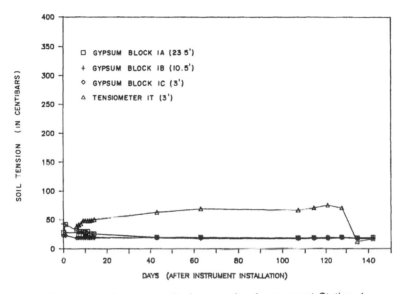

Figure 23.7. Unsaturated zone monitoring results, Instrument Station 1.

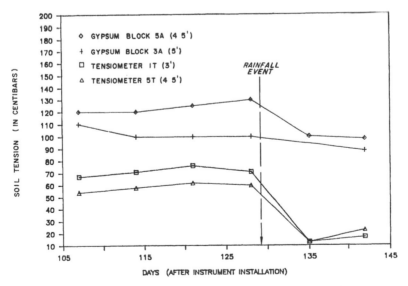

Figure 23.8. Unsaturated zone monitoring results, response to rainfall event.

centibars is equivalent to a soil moisture of approximately 30%. While the actual moisture content at the blocks is not known, it is clear that readings from these blocks indicated a higher moisture content than found in soil samples. It is believed that this apparent increase in moisture was caused by several gallons of water added during installation as a silica sand slurry. Because of the low permeability of the clayey materials, this water did not dissipate, and prevented the blocks from reaching equilibrium with the native materials. The tensiometer installed at the site recorded tensions up to 65 centibars.

Both tensiometers and gypsum blocks recorded wetting trends following a rainfall event at the site. Figure 23.8 shows the response to rainfall of the tensiometer and block at Station 5, as well as the block from Station 3. The magnitude of the response varied from a 40 centibar drop in tension at Station 5 to no response at Stations 2 and 4. It is not known if the stations that did not respond did not receive adequate moisture to cause a decrease, or were not in adequate hydraulic contact with the soil. However, the tensiometers at these stations also failed to register an increase in moisture, indicating that topographic or stratigraphic differences prevented any significant recharge from the rain at these locations.

CONCLUSIONS FROM FIELD INVESTIGATION

Several conclusions are evident from this simple field investigation. Given successful installation, gypsum blocks can provide approximately equivalent data to tensiometers at soil tensions from 20 to 50 centibars. At tensions above 50 centibars, tensiometers lose accuracy and cease registering tension increases, while gypsum blocks continue to function. For this reason, the blocks are clearly superior for unsaturated zone monitoring applications in dry settings where soil tensions are likely to exceed 50 centibars. However, several block stations were evidently biased for several months by water added during installation and effectively rendered useless. While some blocks clearly registered the arrival of a moisture front caused by rainfall, other blocks did not. Some degree of uncertainty and inaccuracy remains inherent in the use of gypsum blocks, suggesting that redundant installation should be part of any monitoring plan.

SUMMARY

In general, electrical resistance blocks can be an effective and economical tool for soil-moisture measurements. Their main limitations are uncertainty of calibration accuracy and potential problems from salinity changes or gypsum dissolution. Other potential problems include obtaining proper soil contact in deeper installations and pronounced hysteresis during calibration and field use. Electrical resistance blocks are best suited to applications measuring broad changes in moisture or measuring changes in moisture in dry soils where tensiometers and lysimeters will not operate. They have been used in the past for a variety of applications, including irrigation scheduling, infiltration studies, and plant drought stress research (Fischback, 1977). Potential application in characterizing or monitoring ground-water contamination sites are:

pipeline leak monitoring
impoundment leak monitoring

monitoring recharge basins
infiltration rate studies
unsaturated zone transport studies
scheduling lysimeter sampling
investigation feasibility of lysimeter operation
landfill liner leak monitoring
leachate sump leak monitoring

Other applications are certainly possible, as the demand for monitoring the movement of water through the unsaturated zone continues to grow.

REFERENCES

Bourget, S., D. Elrick, and C. Tanner, "Electrical Resistance Units for Moisture Measurements: Their Moisture Hysteresis, Uniformity, and Sensitivity," *Soil Science*, 86:298 (1958).

Bouyoucos, G., and A. Mick, "A Fabric Absorbtion Unit for Continuous Measurement of Soil Moisture in the Field," *Soil Science*, 66:217 (1948).

Bouyoucos, G., "Methods for Measuring the Moisture Content of Soils Under Field Conditions. Frost Action in Soils," National Research Council Publication 213, Highway Research Board, Special Report No. 2, 1952, p. 64.

Campbell, G., and M. Campbell, "Soil Moisture Measurements," *Advances in Irrigation*, D. Hilled, Ed., (New York: Academic Press, 1982).

Everett, L.G., L.G. Wilson, and E.W. Hoylman, *Vadose Zone Monitoring for Hazardous Waste Sites*, (Park Ridge, NJ: Noyes Data Corporation, 1984).

Fishback, P.E., *Scheduling Irrigations by Electrical Resistance Blocks*, NebGuide G 77-340, Institute of Agriculture and Natural Resources, University of Nebraska, 1977.

Gardner, W.H., "Water Content," *Methods of Soil Analyses*, C.A. Black, Ed., Agronomy No. 9, Amer. Soc. Agron., Madison, WI, 1965.

Morrison, R.D., *Ground-Water Monitoring Technology*, Timco Manufacturing, Inc, Prairie Du Sac, WI, 1983.

Strangeways, I.C., "Interfacing Soil Moisture Gypsum Blocks with a Modern Data-Logging System Using a Simple, Low Cost, DC Method," *Soil Science*, 136:322 (1983).

24

A Compendium of Soil Samplers for the Vadose Zone

D.W. Dorrance, L.G. Wilson, L.G. Everett, and S.J. Cullen

INTRODUCTION

Vadose zone investigations generally include soil sampling. In the context of this chapter, soil sampling is defined as the process of obtaining depth-wise, matrix samples from the entire vadose zone. Table 24.1 lists some of the reasons for soil sampling. Our discussion will focus on obtaining soil samples for pore liquid extraction and analyses. However, we recommend using collected samples to identify geologic layering in the vadose zone to aid in the subsequent installation of monitoring devices, such as suction samplers.

Elsewhere in this book, Wilson et al. discussed the problems in relating pore-liquid quality data from samples collected by core sampling versus suction samplers. Fleming (1992) referencing Kreft and Zuber (1978), described the fundamental difference between the two methods. Thus, core sample data represent resident concen-

Table 24.1. Reasons for Collecting Soil Samples

1. Lithologic description
2. Hydrogeologic testing
3. Geotechnical testing
4. Soil gas analyses
5. Extraction of microorganisms
6. Pore-liquid and soils chemical analyses

---

*This chapter is based on ASTM D 4700–91, Standard Guide for Soil Sampling from the Vadose Zone. Copies may be purchased from ASTM, 1916 Race Street, Philadelphia, PA 19103.

trations (the amount of solute per unit volume of fluid in the system at the time of sampling), yielding a "snapshot" of solute distribution at a particular time (Fleming, 1992). In contrast, concentration data from suction samplers reflect the mass of solute per unit volume of fluid passing through a given cross section during the sampling time interval, i.e., they represent flux concentration measurements (Fleming, 1992).

Soil sampling involves inserting a device into the ground to retain and recover a sample. In general the hand-operated samplers, developed by soil scientists and agronomists, are used in shallow soils (e.g., less than 6 m deep, depending on conditions). Mechanically-driven devices for deeper sampling have been developed by geotechnical engineers and modified by ground-water hydrologists. For the most part, samplers that are used below the water table may also be used in the vadose zone. Drilling methods which do not use fluids are preferred for advancing samplers in the vadose zone.

This chapter reviews a variety of available soil samplers and mechanical sample recovery methods. We also briefly review special approaches for sampling soils for volatile organic compounds (VOCs) and microorganisms. The applications and limitations of many of the samplers presented here are described by Hvorslev (1949), Shuter and Teasdale (1989), Aller et al. (1989), Boulding (1991), Brown et al. (1991), Davis et al. (1991), Leach and Ross (1991), Norris et al. (1991), and The Eastern Research Group, Inc. (1993).

SOIL SAMPLERS

We use the term "soil sampler" to describe any sampling device to obtain sediments from the vadose zone. Table 24.2 summarizes the sampler types that are discussed in this section, including screw-type augers, barrel augers, tube-type samplers, and bulk samplers. Table 24.2 also groups the samplers into hand-operated samplers and samplers used in conjunction with multipurpose or auger-drill rigs. Note that some of the samplers (e.g., thin-walled tube samplers) fit into both categories. Table 24.3 (derived from ASTM, 1991a, and Hern, Melancon, and Pollard, 1986) lists criteria to consider when selecting soil sampling devices for vadose zone monitoring. Sampling devices are evaluated for these criteria in Table 24.4 (after EPA, 1986 and Lewis et al., 1991).

Auger Samplers

Screw-Type Augers

Screw-type augers (Figure 24.1) are hand-operated samplers. They are essentially a small diameter wood auger with the cutting side flanges and tip removed (Soil Survey Staff, 1951). Variations on this design include the ship auger (Figure 24.1a), closed spiral auger (Figure 24.1b), and the Jamaica open spiral auger (Figure 24.1c) (Acker, 1974). The auger is fitted onto a length of tubular rod. The upper end of this rod is threaded onto extension rods. As many extensions are used as are required to reach the total drilling and sampling depth. A wooden or metal handle fits into a tee-type coupling, screwed into the uppermost extension rod.

During drilling, the handle is twisted manually to advance the auger into the soil.

Table 24.2.   Soil Samplers for the Vadose Zone

Hand Operated Samplers

1. Screw-type augers
        Ship augers
        Closed spiral augers
        Jamaica open spiral augers

2. Barrel augers
        Post-hole augers (also called Iwan-type augers)
        Dutch-type augers
        Regular or general purpose barrel augers
        Sand augers
        Mud augers

3. Tube-type samplers
        Soil sampling tubes (also called Lord samplers)
        Veihmeyer tubes (also called King tubes)
        Thin-walled tube samplers (also called Shelby tubes)
        Ring-lined barrel samplers
        Piston samplers

4. Hand-held power flight augers

5. Bulk samplers (shovels, trowels, etc.)

Samplers Used in Conjunction with Multipurpose or Auger Drill Rigs

1. Thin-walled tube samplers (also called Shelby tubes)

2. Split-barrel drive samplers (also called Split Spoons)

3. Ring-lined barrel samplers

4. Continuous-sample tube systems

5. Piston samplers

6. Solid-stem augers

7. Bucket augers

Table 24.3.   Criteria for Selecting Soil Samples

1. Ability to obtain an encased core sample, an uncased core sample, a depth-specific representative sample or a sample according to requirements of specific analyses.

2. Sample-size requirements.

3. Suitability for sampling various soil types.

4. Maximum sampling depth.

5. Suitability for sampling soils under various moisture conditions.

6. Ability to minimize cross contamination.

7. Accessibility to the sampling site and general site trafficability.

8. Personnel requirements and availability.

9. Safety requirements.

Table 24.4. Criteria for Selecting Soil Sampling Equipment

Type of Sampler	Obtains Core Samples	Most Suitable Soil Types	Operation in Stony Soils	Suitable Moisture Conditions	Relative Sample Size	Labor Requirements	Manual or Power
A. Mechanical Sample Recovery							
1. Hand-Held Power Augers	No	Coh/ Coh'less	Unfav	Interm.	Large	2+	Power
2. Solid Stem Flight Augers	No	Coh/ Coh'less	Favorable	Wet to Dry	Large	2+	Power
3. Hollow-Stem Augers	Yes	Coh/ Coh'less	Favorable	Wet to Dry	Large	2+	Power
4. Bucket Augers	No	Coh/ Coh'less	Favorable	Wet to Dry	Large	2+	Power
5. Backhoes	No	Coh/ Coh'less	Favorable	Wet to Dry	Large	2+	Power
B. Samplers							
1. Screw-Type Augers	No	Coh	Unfav	Interm.	Small	Single	Manual
2. Barrel Augers							
a. Post-Hole Augers	No	Coh	Unfav	Wet	Large	Single	Manual
b. Dutch Augers	No	Coh	Unfav	Wet	Large	Single	Manual
c. Regular Barrel Augers	No	Coh	Unfav	Interm	Large	Single	Manual
d. Sand Augers	No	Coh'less	Unfav	Interm	Large	Single	Manual
e. Mud Augers	No	Coh	Unfav	Wet	Large	Single	Manual
3. Tube-Samplers							
a. Soil Samplers	Yes	Coh	Unfav	Wet to Dry	Small	Single	Manual
b. Veihmeyer Tubes	Yes	Coh	Unfav	Interm	Large	Single	Manual
c. Shelby Tubes	Yes	Coh	Unfav	Interm	Large	2+	Both
d. Ring-Lined Samplers	Yes	Coh'less	Favorable	Wet to Interm	Large	2+	Both
e. Continuous Samplers	Yes	Coh	Unfav	Wet to Dry	Large	2+	Power
f. Piston Samplers	Yes	Coh	Unfav	Wet	Large	2+	Both
g. Zero-Contamination Samplers	Yes	Coh	Unfav	Wet to Interm	Small	2+	Both
h. Split-Spoon Samplers	Yes	Coh	Unfav	Interm	Large	2+	Both
4. Bulk Samplers	No	Coh	Favorable	Wet to dry	Large	Single	Manual

Source: From EPA/530–SW–86–040, 1986; Lewis et al., 1991.

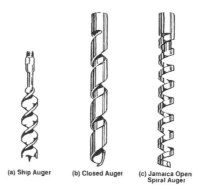

(a) Ship Auger   (b) Closed Auger   (c) Jamaica Open Spiral Auger

Figure 24.1. Spiral type augers (after Acker, 1974).

The auger is advanced to the full length of the barrel and then removed. Upon removal of the tool, soil from the deepest penetration is retained on the auger flights. A foot pump operated hydraulic system has been developed to advance augers up to 4.5 in. (11.43 cm) in diameter. This larger diameter allows insertion of other sampling devices if desired (Soiltest, 1983)

Screw-type augers operate better in wet, cohesive soils than in dry soils. Sampling in very dry (e.g., powdery) soils may not be possible with these augers, as soils will not be retained on the auger flights. In addition, if the soil contains gravel, cobbles, or rock fragments larger than about one-tenth of the hole diameter, drilling may not be possible (Bureau of Reclamation, 1974).

Barrel Augers

Barrel augers are also hand-operated units. They consist of a bit with cutting edges, a short tube or barrel within which the soil sample is retained, and two shanks. The shanks are welded to the barrel at one end and threaded at the other end. Extension rods are attached as required to reach the total sampling depth. The uppermost extension rod contains a tee-type coupling for a handle.

The sampler is turned to advance the barrel vertically into the ground. When the barrel is filled, the unit is withdrawn from the soil cavity and soil is removed from the barrel.

Barrel augers generally provide larger samples than screw-type augers. They can penetrate up to one foot per minute in shallow clays, silts, and fine grained sands (Art's Manufacturing, 1988). However, they do not work well in gravelly soils, caliche, or semi-lithified deposits.

Five common barrel augers are post-hole augers (Figure 24.2), Dutch-type augers (Figure 24.3), regular or general purpose barrel augers (Figure 24.4a, 24.4b), sand augers (Figure 24.4c), and mud augers (Figure 24.4d). The post-hole auger (Figure 24.2) is the most common and readily available barrel auger (Acker, 1974). There are two types of drilling systems for this design; one has a single rod and handle, and the other has two handles. In stable, cohesive soils, post-hole augers with single rods can be advanced up to 25 feet (7.62 m) (Acker, 1974). A.I. Johnson (personal communication, 1993) reports advancing these augers by hand to 65 ft using aluminum pipe for extensions.

The Dutch-type auger is a smaller variation of the post-hole auger design. As

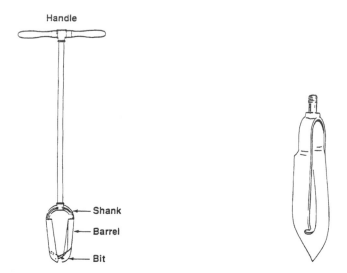

Figure 24.2. Post-hole type barrel auger.   **Figure 24.3.** Dutch-type auger (after Art's Manufacturing and Supply, 1988).

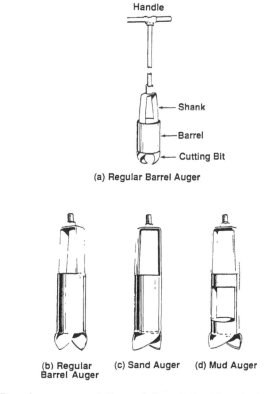

(a) Regular Barrel Auger

(b) Regular Barrel Auger   (c) Sand Auger   (d) Mud Auger

Figure 24.4. Barrel auger variations (after Art's Manufacturing and Soilmoisture Equipment Corp.).

shown in Figure 24.3, the pointed bit is attached to two narrow, curved body segments, welded onto the shanks. This tool is best suited for sampling wet, clayey soils and organic peats.

Figures 24.4a and 24.4b depict a version of the regular barrel auger commonly used by soil scientists and county agricultural agents. As shown, the barrel portion of this auger is completely enclosed. Barrel augers are available which include barrels and/or cutting bits made from high carbon steel and stainless steel. Finishes, such as paint, on the augers prevent rust but may contaminate samples destined for chemical analyses. As with the post-hole auger, the cutting blades are arranged so that soil is loosened and forced into the barrel as the unit is rotated and pushed into the ground. Plastic, stainless steel, PTFE (polytetrafluoroethylene or Teflon^R) or aluminum liners can also be used (Art's Manufacturing, 1988). The regular barrel auger is suitable for use in loamy soils.

In dry, sandy soils it may be necessary to use sand augers which include penetrating bits that are more overlapping to retain the sample in the barrel (see Figure 24.4c). The mud auger is a variation available for sampling wet, clayey soils. As shown in Figure 24.4d, the barrel is designed with open sides to facilitate sample extraction.

Advantages and Limitations of Auger Samplers

Eastern Research Group Inc. (1993) lists the following advantages and disadvantages of auger samplers:

Advantages

1. Relatively inexpensive,
2. Readily available, and
3. Most types require only a single operator.

Disadvantages

1. Cross contamination by wall sloughing during insertion and removal of sampler is a major limitation.
2. Samples obtained with augers and the bulk samplers are disturbed. Therefore, the samples are not suitable for tests requiring undisturbed samples, such as hydraulic conductivity measurements. In addition, soil structure is disrupted and small-scale lithologic features cannot be examined. When representative samples are desired, a smaller diameter tube sampler should be inserted to the bottom of the borehole without touching the sides of the borehole.
3. The churning action and exposure of samples to air promotes loss of VOCs of interest; loss may be accentuated by heat generated during sampling.
4. Sampling may be difficult for some samplers in very dry soils; also, sampling depths may be limited in stony or indurated soils. The Eijkelcamp stony soil auger is recommended for stony soils and asphalt (Eastern Research Group, Inc., 1993).
5. Sampling to great depths (e.g., greater than 10 ft) even under ideal conditions is cumbersome compared with power-drive augers.

Tube-Type Samplers

Tube-type samplers generally have smaller diameters and larger body lengths than barrel augers. Sampling with these units requires forcing the sampler in vertical increments into the soil. Unlike augers, tube samplers are not rotated during insertion into soil. When the sampler is filled at each depth, the assembly is pulled back to the surface. Commercial units are available with attachments which allow foot or hydraulic pressure application to force the sampler into the ground, or with drop-hammers to drive the sampler into the ground (Soilmoisture Equipment Corp., 1988). A vibratory head has also been developed to advance these samplers in noncohesive soils (VI-COR Technologies, 1988). A low-weight, mechanical hammer-drive unit is also available (Clements Associates, no date).

Samples recovered from tube-type samplers are often shorter than the distance pushed. This is due to compaction during sampling and due to friction between soil and the tube ID becoming greater than the shear strength of the soil in front of the tube. This causes soil in front of the advancing tube to be displaced laterally rather than enter the tube (Hvorslev, 1949). However, in general, shorter tubes provide less disturbed samples than longer tubes.

These units are not as suitable for sampling in compacted, gravelly soils as are the barrel augers. They are preferred if a relatively undisturbed sample is required. Commonly used varieties of the tube-type samplers include soil sampling tubes, Veihmeyer tubes (also called King tubes), thin-walled tube samplers (also called Shelby tubes), ring-lined barrel samplers, and piston samplers.

Open-Sided Soil-Sampling Tubes

As depicted in Figure 24.5, the soil-sampling tube consists of a hardened cutting tip, a cut-away barrel, and an uppermost threaded segment. The cut-away barrel is designed to facilitate lithologic examination and to allow for easy removal of soil samples. However, exposure of the sample to air promotes the loss of VOCs. Generally, the tube is constructed from high strength alloy steel (Clements Associates, no date) or chromed molybdenum steel (Soilmoisture Equipment Corp., 1988). Two modified versions of the tip are available for sampling in wet or in dry soils. The sampling tube is attached to extension rods (tubing) to attain the required sampling depth. A cross-handle is attached to the uppermost rod. The soil sampling tube works best in soft, clayey, cohesive soils. If the soil contains cobbles or rock fragments larger than the cutting tip diameter, sampling may not be possible. If the soil is cohesionless, it will not be retained in the tube. With time, the cutting tip will be damaged and worn dull. Most units are designed so that this part can be replaced.

Veihmeyer Tubes

In contrast to the soil sampling tube, the Veihmeyer tube is a long, solid tube. As shown in Figure 24.6, this sampler consists of a replaceable bevelled tip which is threaded into the lower end of the tube, and a drive head threaded into the upper end of the tube. Alternative tip designs are available for special conditions. The sampler is constructed of hardened steel. Prior to sampling, the inside of the tube is sometimes coated with a lubricant to facilitate extrusion. However, subsequent chemical and physical analyses should be performed to determine if the lubricant

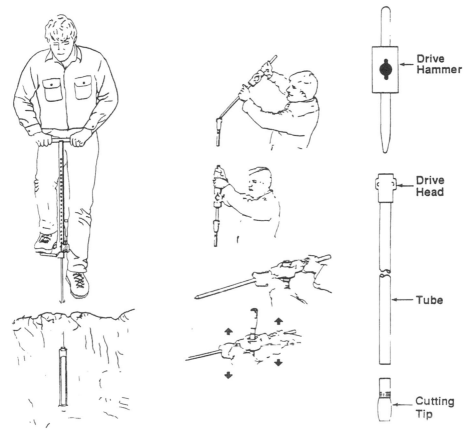

Figure 24.5. Soil sampling tube (after Clements Associates, Inc.).   **Figure 24.6.** Veihmeyer tube (after Soilmoisture Equipment Corp.).

will contaminate samples, biasing the results. Because the Veihmeyer sampler is a solid tube and is fitted with a drop hammer, it can generally be used in harder soils than the soil sampling tube.

Thin-Walled Tube Samplers (Shelby Tubes)

A thin-walled tube (Shelby tube) sampler is depicted in Figure 24.7. Shelby tubes are commonly available in carbon steel but can be manufactured from other metals. During manufacturing, the advancing end of the sampler is rolled inwardly and machined to a cutting edge that has a smaller diameter than the tube ID. The sampler tube is usually connected with set screws to a sampler head which in turn is threaded to connect with extension rods. The head contains a ball check valve (ASTM, 1991b) serving as an air vent as the sampler is pushed into soil.

The Shelby tube is pushed into soil by hand, with a jack-like system or with a hydraulic piston. Shelby tubes are best used in clays, silts, and fine grained sands. They can be advanced with the hydraulic system of most drill rigs in fine grained sands that are loose to moderately consolidated, or in clays and silts that are soft to firm. If the soils are cohesionless, they probably will not be retained in the tube. If

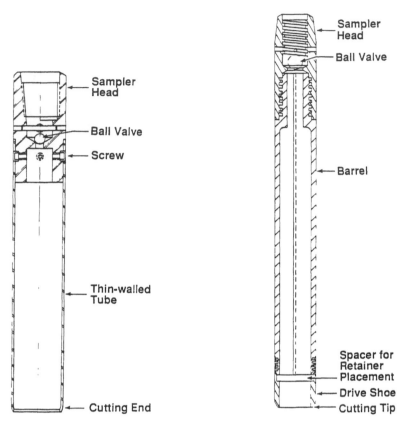

Figure 24.7. Thin-walled tube sampler (after Mobile Drill Co., Inc.).

Figure 24.8. Split barrel drive sampler (after Diedrich Drilling Equipment, Inc.).

consolidated soils, hard soils, or gravelly soils are encountered, driving the sampler may be required. However, tubes may buckle or cutting edges may be damaged under the drive stress (see ASTM, 1991b). A spring-loaded barrel has been developed to protect the Shelby tube from buckling when sampling these soils (Pitcher Drilling, 1986).

Split-Barrel Drive Samplers (Split-Spoon Samplers)

This sampler is used extensively in geotechnical exploration. The split-barrel drive sampler consists of two barrel halves, a drive shoe, and a sampler head containing a ball check valve, all of which are threaded together (see Figure 24.8). For some models, liners are installed for encasing the collected samples. A plastic or metal retainer basket, or a flap valve is often fitted into the drive shoe to prevent samples from falling out during retrieval.

The sampler is threaded onto drilling rods and is lowered to the bottom of the boring, most commonly through the flights of a hollow stem auger. The sampler is then driven into the soil with blows from a drop hammer attached to the drill rig. The hammer usually weighs 140 lb, and is operated by the driller. The sampler is usually driven out 1 to 2 feet ahead of the drill bit. The sampler is extracted from the soil by upward blows with the drop hammer (ASTM, 1991c). Once the sample is retrieved, the split spoon can be opened and the sample removed.

Split-barrel drive samplers can be used in all soil types if the larger grain sizes can enter through the opening of the drive shoe. Because the sampler can be fitted with a retainer basket, it is typically used in place of thin walled tubes when cohesionless soils are to be sampled.

Ring-Lined Barrel Samplers

The ring-lined barrel sampler consists of a one-piece barrel or two split-barrel halves, a drive shoe, rings, and a sampler head. The waste barrel provides a space above the rings into which disturbed soil, originally at the bottom of the hole, can move (ASTM, 1991d). Both hand-operated samplers (Figure 24.9) and mechanically driven samplers (Figure 24.10) are available. Mechanically-driven versions are often used in conjunction with hollow-stem augers. In this case, the sampler is mounted on the end of a drill string, and lowered to the bottom auger flight. The sampler is then driven into the soil with blows from a drop hammer attached to the drill rig. It is important to drive the sampler deep enough to allow all disturbed soil to move through the rings into the waste barrel. The rings, which are usually brass, fit snugly inside the barrel. Once filled with a sample, the rings can be removed as one unit and

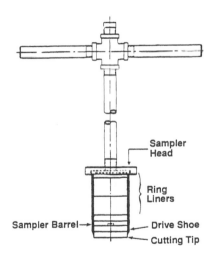

Figure 24.9. Hand-operated ring-lined barrel sampler (after Soilmoisture Equipment, Inc.).

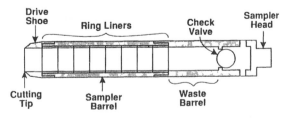

Figure 24.10. Ring-lined barrel sampler.

placed into a capped container. Alternately, the individual soil-filled rings can be capped with plastic or PTFE sheets and then sealed with wax and adhesive tape to retain moisture or chemicals. Inasmuch as barrel samplers are more rigid than thin walled tubes, they can be driven into hard soils and soils containing sands and gravels which might damage thin walled tubes. Barrel samplers are generally designed with replaceable cutting tips.

Continuous Sample Tube System

Continuous sample tube systems fit within a hollow-stem auger column. The assembly can be split- or solid-barrel and can be used with or without liners of various metallic and nonmetallic materials (Riggs, 1983) (see Figure 24.11). The sampler may also be fitted with a plastic or metal retainer basket, or a flap valve to prevent cohesionless soils from falling out of the sampler during retrieval (Riggs, 1983).

Typically, this sampler is 5 feet long, fitting within the lead flight of a hollow-stem auger. The sampler is locked in place inside the auger column with the basal end protruding slightly beyond the end of the column. As the column is advanced, soil enters the nonrotating sampling barrel. After a 5-foot advance, the sampler is withdrawn. The sampling tube is available either as a solid or a split barrel, accommodating a liner. If desired, the sample can be segmented and individual segments sealed and capped. Alternatively, the lined core sample is available for laboratory column studies (e.g., leaching studies).

The continuous-sample tube system replaces the pilot-head assembly in the hollow-stem auger column. Accordingly, the sampling speed is greatly increased.

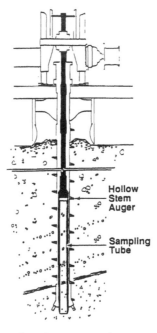

Figure 24.11.   Continuous sample tube system (courtesy of Central Mine Equipment Co.).

The continuous sample tube system is best used in clays, silts, and in fine-grained sands. It can be used to sample soils that are much more consolidated or harder than can be sampled with Shelby tubes.

Piston Samplers

Locally saturated, cohesionless soils and very soft soils or sludges may not be retained in most samplers, even when fitted with retainer baskets or flap valves. Piston samplers can be used in these situations (see Hvorslov, 1949). The sampler consists of a sampling tube, extension pipe attached to the tube, an internal piston, and rods connected to the piston and running through the extension pipe (see Figures 24.12 and 24.13). The piston fits snugly inside the tube. The particular design reported by Zapico et al. (1987) consists of a hardened steel drive shoe, an outer core barrel, and an inner liner (Zapico et al., 1987). These samplers are sometimes called "Waterloo samplers." The inner liners are often built, as needed, out of aluminum or plastic tubing. The length of the sampler is 5 ft (1.5m), corresponding to the length of an auger flight so that the sampling and drilling operation can be synchronized and rapidly advanced (Zapico et al., 1987). Leach and Ross (1991) described a modified Waterloo sampler which includes a ball valve in the

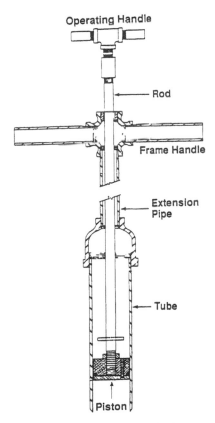

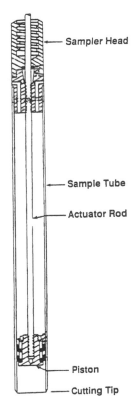

Figure 24.12. Hand-operated piston sampler (after Brakensiek et al., 1979).

Figure 24.13. Piston sampler (Diedrich Drilling Equip., Inc.).

sampler drive cap. The valve relieves internal pressure between the top of the piston and the sampler drive cap as the piston is drawn up the sample tube. Another variation designed for sampling peat has a cone-shaped piston (Acker, 1974). Sharma and De Datta (1985) described a cylindrical sampler for use in puddled soils and situations where the sample would flow out of most samplers. The design includes a basal shutter which retains the sample while the sampler is withdrawn from the soil.

Prior to sampling, the piston is placed at the base of the tube. For hand-operated units, the sampler is pushed into the ground with the handle. For mechanically-operated units, the sampler is attached to drill rods and lowered through the bore-hole or hollow stem auger column to the bottom of the hole (top of the sampling interval). As the tube is advanced, the piston is held stationary or pulled upward with the attached rods. At the bottom of the sampling interval, the sampler is twisted to break suction which might have developed at the tube-soil interface. The sampler is then pulled to the surface. The sample is retained because of suction which develops between the piston and the sample. This suction is stronger than the suction at the bottom of the sampler which would tend to extract soil from the sampler. Even so, it is often useful to twist the sampler with the drill rods prior to retrieval to ensure that the sample will not be pulled out of the sampler. Upon retrieval, the sample is extruded by using the piston to force the sample out of the tube. However, it should be noted that in some soils, this action may cause compaction of the sample (A.I. Johnson, personal communication, 1993).

Average recovery ratios greater than 0.9 can be attained with this sampling tool (Zapico, 1987). However, because the sampler depends on development of suction between the sample and the piston, it may not work in unsaturated, coarse-grained sands and gravels with the air-filled voids which provide pathways for suction to be relieved. Samples collected with piston samplers are relatively undisturbed. Zapico et al. (1987) described techniques for extracting fluid samples directly from liners, and for converting liners into permeameters.

Probe-Drive Samplers

This sampler, sometimes referred to as a Porter sampler, consists of a pointed, sealed tube which is driven to a sampling depth using either a hand-operated slide hammer or a power-driven hydraulic system. The pointed end is part of a piston which can be pulled back before the sampler is driven into undisturbed soil. The recovered sample is either removed by a piston extruding device or the sampler ends are capped and the sampler is sent to a laboratory. For difficult conditions, a pilot hole is created using a retractable drive point before inserting the sampler. Starr and Ingleton (1992) describe a probe-drive system [which they call a "drive point/piston sampler (DPPS)"] for collecting samples of sand, silt, and soft clay to depths up to 60 feet (18.3 m) below land surface. The method is not suitable for sampling stony materials.

"Zero Contamination" Tube Samplers

These tube samplers are lined with thin-walled acetate or stainless steel liners for recovering undisturbed samples (Figure 24.14). The sampler is segmented into upper and lower sections (Figure 24.14). The lower section of the tube contains a drive

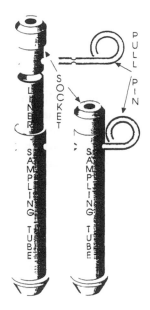

Figure 24.14. Zero Contamination sampler (courtesy of Clements and Associates).

point. The upper section (described as "socket" on the figure) accommodates a pin for locking the upper and lower sections together. The unit shown on Figure 24.14 is used for sampling at the surface. A drive out pin is used when sampling at depth. The sampler is forced into the soil either by a foot-operated jacking device or by a power-driven system. Upon removal of the sampler, the liner is extracted from the tube. The ends of the liner are sealed with caps to minimize loss of VOCs. Samples measure 0.8 inches in diameter; optional lengths are 12, 18, and 24 inches.

Advantages and Disadvantages of Tube Samplers

Eastern Research Group, Inc. (1993) lists the following advantages and disadvantages of tube samplers:

Advantages

1. Relatively available and inexpensive, and most manually-operated types require only a single operator.
2. Fairly undisturbed soil cores can be obtained permitting the characterization of soil lithology.
3. Less exposure to air (except open-sided soil samplers) enhances recovery of VOCs compared with auger samplers; recovery is particularly enhanced when liners are used.
4. When combined with an auger, manually-obtained tube samples can be extracted from depths up to 20 ft in unconsolidated sediments free of stones and gravel. Power-driven samplers facilitate tube sampling from even greater depths.

Disadvantages

1. May be difficult to extract sample.
2. Not suitable for consolidated sediments (especially true for thin-walled samplers), very dry or very wet soils, and loose soils.
3. Manually-obtained samplers may be difficult to pull out of the soil.
4. Sampling depths, using manual samplers, are usually limited to less than about 6 ft; this may be increased by the use of a pilot hole constructed by an auger. Sampling depths with power-driven equipment are limited only by the power train and local geological conditions.

Bulk Samplers

Samples may be obtained with simple bulk samplers which do not fit into the categories described above. They are often used for preliminary screening (Eastern Research Group, 1993). Such samplers include spades, shovels, trowels, scoops, and knives. For example, Bordner, Winter, and Scarpino (1978) recommend the following procedure for obtaining soil samples for microbial analysis:

> . . . scrape the top one inch of soil from a square foot area using a sterile scoop or spoon. If a subsurface sample is desired, use a sterile scoop or spatula to remove the top surface of one inch or more from a one-foot square area. Use a second scoop or spoon to take the sample. Place samplings in a sterile one-quart screw-cap bottle until it is full. Depending on the amount of moisture, a one-quart bottle holds 300–800 grams of soil. Label and tag the bottle carefully and store at 40°C until analyzed.

Advantages and Limitations of Bulk Samplers

According to Eastern Research Group (1993), the following are the advantages and limitations of bulk samplers:

Advantages

1. Inexpensive and readily available,
2. Require only a single operator,
3. Can be easily decontaminated, reducing sampling time,
4. Can be transported to remote areas,
5. Can be used to obtain large samples, and
6. Some samplers (e.g., soil moisture tins) may minimize loss of VOCs.

Disadvantages

1. Samples are disturbed,
2. Reproducibility of sample size may be poor, and
3. Sampling depth limited to the near surface.

MECHANICAL SAMPLE RECOVERY METHODS

Drill rigs used to obtain vadose zone samples are identical to those used to sample from beneath a water table. Commonly used drill rigs such as cable tool and rotary units are not recommended as they generally require drilling fluids. Air-rotary drilling is also undesirable for obtaining samples for pore liquid or gas extraction because they may alter the composition of the sample. The mechanical sampling devices reviewed include trench sampling using backhoes, hand-held power augers, solid-stem flight augers, bucket augers, and hollow stem augers. When sampling with these units, samples are brought to the surface and stored in suitable containers.

Trench Sampling Using Backhoes

Soils may be sampled from a trench or pit excavated for that purpose by a backhoe. Samples are collected with knives, trowels, or shovels. Occasionally, samples are collected from the sides or the bottom of the trench or pit with hand augers or tube-type samplers.

Sampling is performed only after the backhoe has moved away from the trench or pit. When the trench or pit is in unstable material or is more than five feet deep, the sampling technician should only enter the trench or pit after it has been shored up or the sidewalls have been cut back to within the angle of repose (see Occupational Safety and Health Administration regulations). In these situations, samples are more commonly collected at the surface from the bucket of the backhoe as excavation occurs.

The maximum sampling depth for the trench or pit method is dictated by the reach of the backhoe, the soil type, and the moisture content of the soil. Maximum depths of up to 20 ft (6.10 m) can be obtained in moist clays. Maximum depths of less than 10 ft (3.05 m) are common in dry sands. Samples can be collected from the side walls of the excavation or from the backhoe bucket. Samples should be taken from the center of the backhoe bucket to prevent cross contamination from the bucket surface, and to prevent inclusion of materials which may have fallen to the trench bottom. However, when this is done it is difficult to accurately estimate the depth from which the sample was obtained.

Advantages

1. Trenches are useful for obtaining lithologic information since cross sections of the vadose zone can be studied and photographed.
2. Trench or pit sampling is often used in areas with difficult access since backhoes are designed to travel on rough terrain. However, because the process involves excavating a much larger hole than drilling methods, chances of encountering underground utilities are increased.

Disadvantages

1. If soil cores are not taken from the sidewalls, the samples are disturbed.
2. Loss of VOCs occurs during construction of the trench.

3. Sampling depth is limited compared to drilling techniques (except the hand-held power auger samplers).

Hand-Held Power Augers

A very simple, commercially-available auger consists of a flight auger attached to and driven by a small air-cooled engine (see Figure 24.15) or electric motor powered by a portable generator. Two handles on the head assembly allow two operators to guide the auger into the soil. Throttle and clutch controls are integrated into grips on the handles.

As the auger rotates into soil, cuttings advance up the flights and are discharged at the surface. Soil samples can be collected from material augered to the surface, or from the auger flights after pulling the auger out of the ground. Alternatively, samples can be collected with other samplers (e.g., thin-walled tubes) from the borehole bottom after auger removal.

Samples collected from the surface or from the flights are disturbed and are not suitable for some uses. In addition, it is difficult to determine the depth from which the soil came, and cross-contamination concerns are amplified. The auger operates well in most soils. However, if the soil is cohesionless, it may not be retained on the flights and sampling in that fashion may not be possible. If the soil contains cobbles or boulders, drilling may not be possible. If the auger "hangs up" on an obstruction, the machine will start to rotate at the surface, causing operator safety concerns. An alternate design which transfers the torque to a separate engine prevents this problem (Little Beaver, 1988).

Other advantages and disadvantages of hand-held auger power samplers are the same as discussed for solid-stem auger samplers.

Bucket Auger Drilling and Sampling

The bucket auger is a large diameter cylindrical bucket with auger-type cutting blades on the bottom (see Figure 24.16). The bottom is hinged to allow cuttings to be emptied out. These units are commonly used in the southwest for the construction of "dry wells," vadose zone wells facilitating the drainage of urban runoff.

The bucket is rotated into the vadose zone until the bucket is full. Sampling

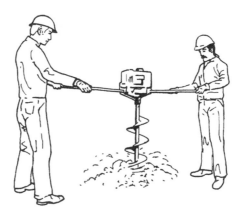

Figure 24.15. Hand-held power auger.

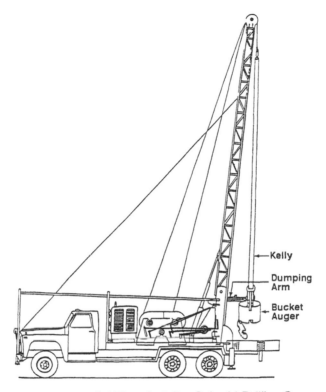

Figure 24.16.   Bucket auger and drilling rig (after Calweld Drilling Company).

consists of extracting auger or scooped samples from the interior of the bucket after it has been retrieved. The bucket auger is best suited for sampling from relatively stable clays as caving problems are amplified by the larger hole diameter. Generally, boulders up to 1/3 of the bucket diameter can be picked up by the bucket. Common sampling depths are less than 50 ft (15.24 m), but holes up to 250 ft (76.20 m) deep have been drilled (Driscoll, 1986; Scalf et al., 1981).

Advantages

1. Sampling can be used in conjunction with construction of monitor wells.
2. Large samples can be readily obtained.

Disadvantages

1. Cross contamination is a major concern with bucket augers, resulting from the introduction of soil into the bottom of the hole during the raising and lowering of the bucket (Norris et al., 1991). This problem may be mitigated by collecting core samples from the interior of the bucket.
2. Samples are disturbed.
3. Large boulders can impede drilling and may have to be individually removed from the hole before sampling can continue (Driscoll, 1986).
4. Sampling depth is limited to noncaving material.

5. Availability of rigs may be a problem (Eastern Research Group, Inc., 1993).

Solid-Stem Auger Drilling and Sampling

Figure 24.17 illustrates a solid-stem auger. The tools used for solid-stem auger drilling include: auger sections, the drive cap, and the cutter head. Head types include fish tail or drag bits for use in cohesionless materials, and clay or stinger bits for use in more consolidated material (Scalf et al., 1981).

As the auger column is rotated into soil, cuttings are retained on the flights. The augers are then removed from the hole and samples are taken from the retained soil (see ASTM, 1991e). Alternatively, samples are obtained by removing the auger flights, lowering a sampling tool in the open hole, and driving the sampler into undisturbed soil (Figure 24.17). Both methods are subject to cross contamination by caving and sloughing of the bore hole wall as the auger flights are removed and reinserted. This is a particular problem when the aim of soil sampling is to obtain depth-wise contaminant profiles.

Typical drilling depths with solid-stem augers range from 50 ft (15 m) to 120 ft (37 m). However, drilling depths up to 200 ft (61 m) have been reported (A.I. Johnson, personal communication, 1993). The greater drilling depths are attained in firm, silty, and clayey soils. However, the depth to which the hole will remain open for sampling once the auger column has been removed is usually less than the maximum drilling depth. If cascading water or cohesionless soils are encountered, it can be expected that the hole will cave at that depth. The sample depth measurement, as taken from its location on an auger, is not accurate. This is because soil moves up the flights in an uneven fashion as the auger column is advanced. As with hollow-stem augers, solid-stem augers are often painted by the driller or manufacturer. It is good practice to remove this paint before drilling. The majority of the paint can be removed by sand blasting. As with all sampling devices, decontamination (e.g., steam cleaning) should be performed between holes when chemical analy-

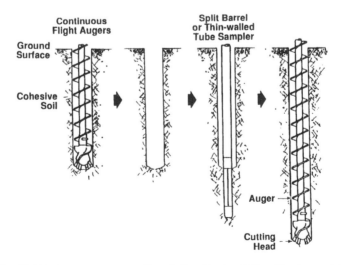

igure 24.17. Solid-stem auger sampling (after Central Mine Equipment Co.).

ses are to be performed on the samples. This is especially important with the solid-stem auger, as it doubles as the drilling and sampling tool.

Advantages

1. In unconsolidated material, solid-stem drilling rigs are fast; plus, they are highly mobile (Eastern Research Group, Inc., 1993).
2. Large samples are readily obtained.
3. Rigs are readily available.

Disadvantages

1. As indicated in a previous paragraph, the major problem with solid-stem augers is cross-contamination of samples, and sampling depths are uncertain. Thus, soil samples are unreliable unless tube samplers are used, slowing drilling speed, and these can only be taken in stable sediments (Eastern Research Group, Inc., 1993).
2. VOCs are lost by exposure to air and by heat generation.
3. Sampling depths are limited, especially when using under-powered rigs.

Hollow-Stem Augers

As the name implies, hollow-stem augers consist of a sequence of auger flights with a hollow interior, a pilot assembly, and center rod, and an auger head with replaceable carbide cutting teeth (Figure 24.18). Hollow-stem augers are specified by the inside diameter of the auger flights and not by the hole size that is drilled (Davis et al., 1991). Hollow-stem augers are available with inside diameters ranging from 2.5 inches (5 cm) to 10.25 (26 cm). The standard flight length is 5 ft (1.5 m).

As with solid-stem auger rigs, samples brought to the surface can be recovered directly from the flights. A better approach, particularly when desiring an undisturbed sample, is to insert tube-type samplers through the hollow core of the auger flights. In most cases, hollow-stem augers with some type of cylindrical sampler provide the greatest level of assurance that the soil sampled was not carried downward by the drilling or sampling process.

To use a sampler within a hollow-stem auger, the auger column and pilot assembly are rotated to the top of the desired sampling interval (Figure 24.19). The pilot assembly and center rod are then removed and the sampler is inserted through the hollow axis of the auger column. Alternatively, a nonrotating capped sample tube replaces the pilot assembly to facilitate quick removal when the desired sampling depth is reached (Leach and Ross, 1991). In some situations, the pilot assembly is removed completely to speed up insertion of sampling tubes. This is not recommended, however, in saturated cohesionless material, such as may occur in some perched ground-water regions. Under these conditions, hydrostatic pressure may force sand up into the annulus. To overcome this problem, some drillers install a wooden or metal knock-out plate within the auger head after removing the pilot assembly (Leach and Ross, 1991). At the desired sampling depth, the sampler is lowered into the annulus and the knock-out plate is driven out of the way. Two problems with this method are (1) it is undesirable to leave a foreign object in the hole, and (2) the block may restrict the entrance to the sampler (Leach and Ross, 1991).

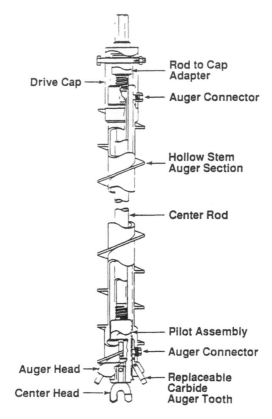

Figure 24.18.   Hollow-stem auger components (after Central Mine Equipment Co.).

The sampler is lowered to the sampling depth by attaching it to center rods or by using a wireline assembly. When the sampler is attached to center rods, a sample is collected by pushing or driving the sampler into undisturbed soil with the rig's hydraulic system or with a drop hammer. When a wireline is used, the sampler is locked into place ahead of the lower-most auger and advanced into the sampling

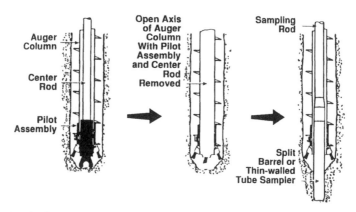

Figure 24.19.   Hollow-stem auger sampling (after Hackett, 1987).

interval by rotating the auger column. Commonly-available samplers used in hollow stem augers include split spoon samplers, ring samplers, and continuous samplers. This method minimizes cross contamination between sampling intervals.

Riggs and Hatheway (1988) reported that the maximum drilling depth, in feet, with 3.25 in. (8.26 cm) to 4.25 in. (10.80 cm) diameter augers, can be estimated by multiplying the available horsepower at the drill spindle by 1.5. This estimate does not take into account large cobbles and boulders which slow down or impede drilling. It is generally recognized that the maximum drilling depth of hollow-stem augers in unconsolidated materials is 150 ft (45.72 m) (Driscoll, 1986). However, recently, some units have been manufactured capable of drilling to depths of 300 ft (91.44 m) (Leach and Ross, 1991).

Advantages

1. These units are commonly available throughout the United States; they are highly mobile, and set-up time is minimal (Eastern Research Group, Inc., 1993).
2. Relatively undisturbed cores are obtained and, especially when using continuous-tube samplers, the lithology can be accurately determined.
3. Besides their role in obtaining core samples, hollow-stem augers facilitate installing vadose zone monitoring devices, such as suction samplers and monitor wells. In fact, perforated flights allow the flight train to serve as a monitor well (Hackett, 1987). They can also be used for some down-hole geophysical logging, e.g., natural gamma logging (Eastern Research Group, Inc., 1993).
4. When encountered during drilling, perched ground water is available for sampling, e.g, by bailing or advancing a screened well point (Eastern Research Group, Inc., 1993).

Disadvantages

1. Cannot be used in consolidated formations and drilling is inhibited in profiles with coarse boulders and cobbles (Eastern Research Group, Inc., 1993).
2. Sampling depth is limited when using underpowered rigs.
3. Sleeves and liners inserted into samplers limit the loss of VOCs by exposure to air. However, the heat generated when pounding samplers into place promotes some loss.

SPECIAL PROBLEMS WHEN SAMPLING FOR VOCs

Soil sampling for VOCs requires special care to avoid loss of sample through volatilization. Elsewhere in this book, Lewis and Wilson review in detail the procedures for sampling vadose zone sediments for VOCs. Techniques that provide disturbed samples easily exposed to the atmosphere are not recommended. In contrast, devices obtaining core samples with liners that can be quickly capped and sealed are generally suitable (Lewis et al., 1991). Lewis et al. (1991) describe a commercially-available subcoring device for extracting a 5-gram sample typically required for low-level purge and trap techniques. This design minimizes the exposure of the sample to the atmosphere and reduces the time it is exposed. The collected sample is extruded into a 40-mL U.S EPA vial with a cap which includes a special adapter. This adapter replaces the septum in the vial, permitting the vial to be heated for the purge and trap collection of soil VOCs in the field [Associated Design and Manufacturing Company (no date)].

SAMPLING FOR MICROORGANISMS

As pointed out by McNabb and Mallard (1984), relatively little attention has been paid to collecting undisturbed soil and rock samples for microbial analysis. The problems of sampling for chemical constituents and physical parameters are compounded when sampling for microorganisms by the need to maintain sterile conditions. For example, the outer surface of a core sample will be contaminated by microorganisms carried along by the wiping action of the cutting point and the inner walls of the tube (McNabb and Mallard, 1984). To avoid this problem, subcores are recommended. Wilson et al. (1983) developed a coring device consisting of a drive shoe (including a core retainer), a core barrel, and an adapter (Figure 24.20) for sampling at the bottom of an augered hole. As described by McNabb and Mallard (1984), immediately upon recovery of the sample, the drive shoe is removed. Next, the sample within the core barrel is extruded by a hydraulic ram through a sterile paring device (Figure 24.21). This action removes the outer skin of the core sample contaminated during sampling (Dunlap, no date). As the core is extruded, the first 5 to 8 cm are broken off using a sterile scalpel, and either used for other purposes or discarded. The subsample is then retained within a sterile container.

Leach and Ross (1991) described a technique for collecting aseptic core samples in the field using a specially designed anaerobic glove box. Dimensions of the box are $23.6 \times 35.4 \times 47.2$ inches ($60 \times 90 \times 120$ cm). Included in the glove box are presterilized sample containers and sterile stainless steel core paring devices. The box is purged with nitrogen gas to essentially remove internal oxygen. A sample tube containing a soil core is inserted into the box through an iris port and about 3.9 inches (10 cm) of sample is extruded through a 1.97 inch (5 cm) paring device and carefully broken away to expose an aseptic face. Paring removes the possibly contaminated outer region of the sample. Subsequently, 39.4 inch (100) cm long samples

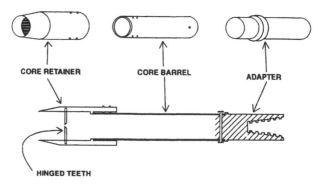

Figure 24.20. Core sampling device for microorganisms (after Wilson et al., 1983. Reprinted by permission of John Wiley & Sons, Inc.).

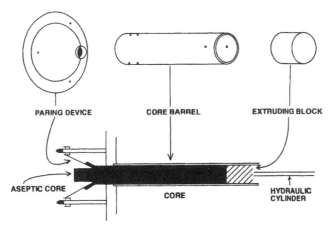

Figure 24.21. Extruding device for obtaining aseptic cores (after Wilson et al., 1983. Reprinted by permission of John Wiley & Sons, Inc.).

are extruded into labeled, sterile sample containers. The collected samples are packed in ice for transport to a laboratory (Leach and Ross, 1991).

CONCLUDING REMARKS

Soil sampling is an important procedure for assessing contaminant transport from surface sources through the vadose zone. This compendium presents an overview of currently available soil sampling devices and methodologies and is intended to assist the many new practitioners of vadose zone investigations in the selection of appropriate soil sampling tools for a variety of applications which may be encountered in the field.

REFERENCES

Acker, W.L. Basic Procedures for Soil Sampling and Core Drilling, Acker Drill Co. Inc., Scranton, PA, 1974.

Aller, L., T.W. Bennett, G. Hackett, R.J. Petty, J.H. Lehr, H. Sedoris, D.M. Nielsen, and J.E. Denne, *Handbook of Suggested Practices for the Design and Installation of Ground-Water Monitoring Wells*, Environmental Monitoring Systems Laboratory, Office of Research and Development, U.S. Environmental Protection Agency, Las Vegas NV, EPA 600/4-89/034 (Published by the National Water Well Association, Dublin, OH), 1989.

American Society for Testing and Materials. *ASTM D420, Standard Guide for Site Characterization for Engineering Purposes*, Annual Book of ASTM Standards, Section 4, Construction, Volume 04.08 Soil and Rock; Dimension Stone; Geosynthetics. American Society for Testing and Materials, 1916 Race Street, Philadelphia, PA, 1991a.

American Society for Testing and Materials. *ASTM D1587, Standard Practices for Thin-Walled Tube Sampling of Soils*, Annual Book of ASTM Standards, Section 4, Construction, Volume 04.08 Soil and Rock; Dimension Stone; Geosynthetics. American Society for Testing and Materials, 1916 Race Street, Philadelphia, PA, 1991b.

American Society for Testing and Materials. *ASTM D1586, Standard Method for Penetration Test and Split-Barrel Sampling of Soils*, Annual Book of ASTM Standards, Section 4,

Construction, Volume 04.08 Soil and Rock; Dimension Stone; Geosynthetics. American Society for Testing and Materials, 1916 Race Street, Philadelphia, PA, 1991c.

American Society for Testing and Materials. *ASTM D3550, Standard Practice for Ring-Lined Barrel Sampling of Soils*, Annual Book of ASTM Standards, Section 4, Construction, Volume 04.08 Soil and Rock; Dimension Stone; Geosynthetics. American Society for Testing and Materials, 1916 Race Street, Philadelphia, PA, 1991d.

American Society for Testing and Materials. *ASTM D1452, Standard Practice for Soil Investigations and Sampling by Auger Borings*, Annual Book of ASTM Standards, Section 4, Construction, Volume 04.08 Soil and Rock; Dimension Stone; Geosynthetics. American Society for Testing and Materials, 1916 Race Street, Philadelphia, PA, 1991e.

American Society for Testing and Materials. *ASTM D4700-91, Standard Guide for Soil Sampling from the Vadose Zone*, Annual Book of ASTM Standards, Section 4, Construction, Volume 04.08 Soil and Rock; Dimension Stone; Geosynthetics. American Society for Testing and Materials, 1916 Race Street, Philadelphia, PA, 1991f.

Art's Manufacturing and Supply, Sales Division, Catalog of Products, American Falls, ID, 1988.

Associated Design and Manufacturing Company, Catalog 105, Alexandria, VA, no date.

Bordner, R., J. Winter, and P. Scarpino, Microbiological Methods for Monitoring the Environment, U.S. Environmental Protection Agency, Monitoring and Support Laboratory, EPA-600/8-78-017, 1978.

Boulding, J.R., Description and Sampling of Contaminated Soils: A Field Pocket Guide, U.S. Environmental Protection Agency, EPA/625/12-91/002.

Brainard Kilman, Sales Division, Catalog of Products, Stone Mountain, GA, 1988.

Brakensiek, D.L., H.B. Osborn, and W.L. Rawls, *Field Manual for Research in Agricultural Hydrology, Agriculture Handbook No. 224*, Science and Education Administration, United States Dept. of Agriculture, Washington, DC, 1979 (revised), pp. 258–275.

Brown, K.W., R.P. Breckenridge, and R.C. Rope, "Soil Sampling Reference Field Methods," in *U.S. Fish and Wildlife Service Lands Contaminant Monitoring Operations Manual*, Appendix J. Prepared by Center for Environmental Monitoring and Assessment, Idaho National Engineering Laboratory, Idaho Falls, ID, 1991.

Bureau of Reclamation, *Earth Manual*, U.S. Dept. of the Interior, United States Government Printing Office, Washington, DC, 1974.

Clements Associates Inc. Sales Division, *J.M.C. Soil Investigation Equipment*, Catalog No.10, Newton, IA, no date.

Davis, H.E., J. Jehn, and S. Smith, "Monitoring Well Drilling, Soil Sampling, Rock Coring, and Borehole Logging," in *Practical Handbook of Ground-Water Monitoring*, D.L. Nielsen, Ed. (Chelsea, MI: Lewis Publishers, 1991), pp. 195–237.

Diedrich Drilling Equipment, Sales Division, Catalog of Products, LaPorte, IN, 1988.

Driscoll, F.G., *Groundwater and Wells*, 2nd ed., Johnson Division, St. Paul, MN, 1986.

Dunlap, W.J., "Some Concepts Pertaining to Investigative Methodology for Subsurface Process Research," in N.N. Durhamn and A.E. Redlefs, Eds., *Proceedings of the Second International Conference on Ground Water Quality Research*, University Center for Water Research, Oklahoma State University, Stillwater, OK, pp. 167–172, no date.

Eastern Research Group, Inc., *Subsurface Characterization and Monitoring Techniques: A Desk Reference Guide, Volume I: Solids and Ground Water*, Center for Environmental

Research Information, U.S. Environmental Protection Agency, Cincinnati, OH, and Environmental Systems Laboratory, U.S. Environmental Protection Agency, Las Vegas, NV, EPA/625/R-93/003a, 1993.

Fleming, J.B., Field Investigations of Solute Transport Using Alternative Sampling Methodologies, unpublished MS thesis, Colorado State University, Fort Collins, CO, 1992.

Gillham, R.W., and S.F. O'Hannesin, "Sorption of Aromatic Hydrocarbons by Materials Used in Construction of Ground-Water Sampling Wells," in *Ground Water and Vadose Zone Monitoring*, D.M. Nielsen, and A.I. Johnson, Eds., ASTM STP 1053, American Society for Testing and Materials, Philadelphia, PA, 1990, pp. 108–122.

Hackett, G., "Drilling and Constructing Monitoring Wells with Hollow-Stem Augers; Part 1: Drilling Considerations," *Ground Water Monitoring Review*, 7:51–62 (1987).

Hern, S.C., S.M. Melancon, and J.E. Pollard, "Generic Steps in the Field Validation of Vadose Zone Fate and Transport Models," in *Vadose Zone Modeling of Organic Pollutants*, S.C. Hern and S.M. Melancon, Eds., (Chelsea, MI: Lewis Publishers, 1986), pp. 61–80.

Hvorslev, M. J., *Subsurface Exploration and Sampling of Soils for Civil Engineering Purposes*, U.S. Army Corp of Engineers, Waterways Experiment Station, Vicksburg, MS, 1949.

Kreft, A., and A. Zuber, "On the Physical Meaning of the Dispersion Equation and Its Solution for Different Initial and Boundary Conditions," *Chemical Engineering Science*, 33:1471–1478 (1978).

Leach, L.E., and R.R. Ross, "Aseptic Sampling of Unconsolidated Heaving Soils in Saturated Zones," in *Groundwater Residue Sampling Design*, R.G. Nash and A.R. Leslie, Eds., ACS Symposium Series 465, American Chemical Society, Washington, DC, 1991, pp. 334–348.

Lewis, T.E., A.B. Crockett, R.L. Siegrist, and K. Zarrabi, Soil Sampling and Analysis for Volatile Organic Compounds, E.P.A. Ground-Water Forum Issue, EPA/540/4-91-001, 1991.

Little Beaver Inc., Sales Division, Catalog of Products, Livingston, TX, 1988.

McNabb, J.F., and G.E. Mallard, "Microbiological Sampling in the Assessment of Groundwater Pollution," in *Groundwater Pollution Microbiology*, G. Bitton and C.P. Gerba, Eds., (New York: John Wiley and Sons, 1984), pp. 235–260.

Mobile Drilling Co. Inc., Sales Division, Catalog of Products, Indianapolis, IN, 1988.

Norris, F.A., R.L. Jones, S.D. Kirkland, and T.E. Marquardt, "Techniques for Collecting Soil Samples in Field Research Studies," in *Groundwater Residue Sampling Design*, R.G. Nash and A.R. Leslie, Eds. ACS Symposium Series 465, American Chemical Society, Washington, DC, 1991, pp. 349–356.

Pitcher Drilling Co., Sales Division, Product Literature, Palo Alto, CA, 1986.

Riggs, C.O., "Soil Sampling in the Vadose Zone," in *Proceedings of the MWWA/U.S. EPA Conference on Characterization and Monitoring of the Vadose Zone*, National Water Well Association, Las Vegas, NV, 1983, pp. 611–622.

Riggs, C.O., and A.W. Hatheway, "Ground-Water Monitoring Field Practice: An Overview," in A.G. Collins and A.I. Johnson, Eds., *Ground-Water Contamination: Field Methods*, ASTM Special Technical Publication 963, ASTM, Philadelphia, PA, 1988, pp. 121–136.

Shuter, E., and W.E. Teasdale, *Application of Drilling, Coring, and Sampling Techniques to*

Test Holes and Wells, Techniques in Water Resources Investigations, United States Geological Survey, Book 2, Chapter F1, 1989.

Soiltest Inc., Materials Testing Division, Catalog of Products, Evanston, IL, 1983.

Soilmoisture Equipment Corp., Sales Division, Catalog of Products, Santa Barbara, CA, 1988.

Solinst, Sales Division, Instrumentation for Soil and Rocks, Catalog of Products, Burlington, Ontario, Canada, 1988.

Scalf, M.R., J.F. McNabb, W.J. Dunlap, R.L. Cosby, and J. Fryberger, *Manual of Groundwater Sampling Procedures*, National Water Well Association, Dublin, OH, 1981.

Sharma, P.K., and DeDatta, S.K., "A Core Sampler for Puddled Soils," *Soil Science Society of America Journal*, 49:1069–1070 (1985).

Soil Survey Staff, *Soil Survey Manual*, U.S. Dept. of Agriculture, Superintendent of Documents, Washington, DC, 1951.

Starr, R.C., and R.A. Ingleton, "A New Method for Collecting Core Samples Without a Drilling Rig," *Ground Water Monitoring Review Journal*, 12:91–95 (1992).

U.S. Environmental Protection Agency, Permit Guidance Manual on Unsaturated Zone Monitoring For Hazardous Waste Land Treatment Units, EPA/530-SW-86-040, 1986.

VI-COR Technologies Inc., Sales Division, Catalog of Products, Bellevue, WA, 1988.

Wilson, J.T., J.F. McNabb, B.H. Wilson, and M.J. Noonan, "Biotransformation of Selected Organic Pollutants in Ground Water," *Developments in Industrial Microbiology*, 24:225–233 (1983).

Wilson, J.T., and McNabb, J.F., "Biological Transformation of Organic Pollutants in Groundwater," EOS-Transactions, 64:505–507, no date.

Zapico, M.M., S. Vales, and Cherry, J.A., "A Wireline Core Barrel for Sampling Cohesionless Sand and Gravel Below the Water Table," *Ground Water Monitoring Review*, 7:75–82 (1987).

25

Soil Sampling for Volatile Organic Compounds

Timothy E. Lewis and L. Gray Wilson

INTRODUCTION

Volatile organic compounds (VOCs) are the most commonly encountered class of compounds at Superfund and other hazardous waste sites (McCoy, 1985; Plumb and Pitchford, 1985; Plumb, 1987; Arneth et al., 1988). Table 25.1 ranks the most commonly encountered compounds at Superfund sites. Many VOCs are considered hazardous because they are mutagenic, carcinogenic, or teratogenic and are commonly the targeted contaminants in site remediation projects. Volatile organic compounds are present at most sites, and indeed, are so ubiquitous that it has been suggested that they be used as indicators for the presence of other organic compounds (Plumb, 1987).

The intent of this chapter is to familiarize site investigators, field sampling personnel, engineers, chemometricians, decision makers, and all involved parties with the problems encountered in sampling soils for VOCs and the current progress toward identifying, quantifying, and reducing the measurement errors associated with the soil sampling process. A brief discussion of the factors that affect the retention and migration of VOCs through soil is presented. Before any sampling activity begins, a sound soil sampling and analysis plan should be formulated, which clearly states the data quality objectives (DQOs) necessary for the intended use of the data. Thus, the next part of the chapter discusses the development of a sampling and analysis plan and DQOs. The next section presents the decisions that must be made in selecting the proper sampling tools. The chapter concludes with examples to demonstrate the approaches that have been used by various site investigators to obtain the most representative samples possible for VOC measurements.

Table 25.1.   Most Frequently Occurring Organic Compounds at Superfund Sites

Frequency Rank	Compounds	Mean Concentration %	Frequency Detected
1	toluene	1.021	38
2	o-xylene	0.839	37
3	ethylbenzene	0.228	31
4	methylene chloride	0.078	17
5	bis(2-ethylhexyl)phthalate	0.021	15
6	naphthalene	0.027	13
7	perchloroethylene	0.135	12
8	2-butanone	0.690	11
8	phenol	0.241	11
9	2-methylphenol	0.383	9
9	trichloroethene	0.217	9
9	4-methyl-2-pentanone	0.117	9
10	benzene	0.058	8
10	acetone	0.665	8

Note: Samples collected from drums, tanks, pits, ponds, and trucks. Samples are expected to be concentrated. Samples taken from 221 sites in 41 states prior to 1984. A total of 133 organic compounds were identified.

FACTORS AFFECTING VOC RETENTION AND CONCENTRATION IN SOIL SYSTEMS

To understand why soil VOC measurement is such a challenge to even the most experienced and knowledgeable site investigation team, an understanding of the factors that influence VOC retention and concentration in soil systems is necessary. These factors are covered in much greater detail in other chapters of this book, but merit reiteration at this point in the context of soil sampling.

One of the characteristics of VOCs that makes their quantification in soils so difficult is that they may coexist in four phases: gaseous, liquid (dissolved), NAPL, and solid (sorbed to solids). [Note: "Sorbed" is used throughout this chapter to connote physicochemical adsorption as well as phase partitioning.] VOCs can exhibit extreme mobilities, especially in the vapor phase, where their gas diffusion coefficients can be four times greater than their liquid diffusion coefficients. A graphic illustration of the equilibrium relationships between these phases is shown in Figure 25.1. Note that sorption equilibrium relationships are assumed. However, nonequilibrium sorption behavior is often observed (Brusseau et al., 1991a; Brusseau et al., 1991b).

Several external and intrinsic chemical factors can affect the concentration and retention of VOCs in soils. These factors can be divided into five categories: VOC chemical properties, soil chemical properties, soil physical properties, environmental factors, and biological factors (Lewis et al., 1991). A brief summary of factors in four of the five categories is found in Table 25.2. The table is by no means exhaustive, but simply offers the reader an overview of the various factors which interact to control VOCs in the soil environment at the time a sample is collected. The cited references should be consulted for a more detailed discussion.

Soil-moisture content is a critical external factor in determining the phase parti-

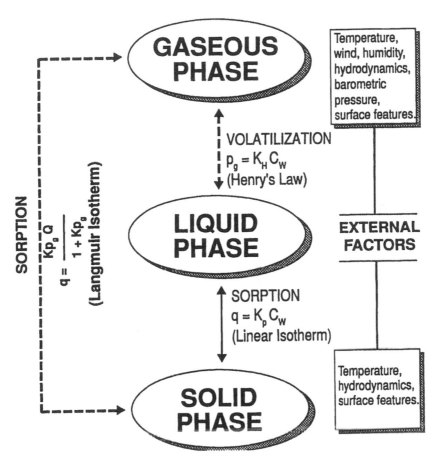

Figure 25.1. Equilibrium relationships for phase partitioning of VOC in soil systems (taken from Lewis et al., 1991).

tioning and sorption of VOCs in dry soils. At unsaturated moisture levels (i.e., less than monolayer coverage) the sorptive uptake of VOCs can be adequately described based upon knowledge of their sorptive behavior under saturated conditions, the air/water distribution coefficient (Henry's Constant) and the organic carbon partitioning coefficient (K_{oc}) (Lion et al., 1990). However, at soil moisture contents lower than those necessary for an average particle surface coverage of four to eight layers of water, sorption of vapors will be increased above that which is due to partitioning into the aqueous and organic carbon phases (Lion et al., 1990). The greater adsorption at extremely low moisture contents is attributed to adsorption onto clay mineral surfaces. Thus, the moisture content largely controls the relative contributions of mineral and organic soil fractions to the uptake of VOCs (Chiou and Shoup, 1985; Smith et al., 1990). Ong and Lion (1991) have shown that by normalizing TCE adsorption data for monolayer coverage at zero percent relative humidity, clay mineral surface area served as a good indicator of the adsorptive capacity of dry solids.

Temperature is also an important external environmental factor which may influence the phase partitioning of VOCs, especially in the near surface layer, where temperature fluctuations occur. Smith et al. (1990) found that there was a temporal

Table 25.2. Factors Affecting VOC Concentrations in Soils

Factor	Common Abbrev.	Units	Effects on VOC Concentrations in Soil	References
VOC Chemical Properties				
solubility	C_w	mg/L	Affects fate and transport in water, affects water/air partit., organic carbon partit.	Roy and Griffin (1985)
Henry's Constant	K_H	(atm-m^3)/mole	Constant of proportionality between water and gas phase concentrations; temperature and pressure dependent.	Shen and Sewell (1982) Spencer et al. (1988)
vapor pressure	v.p.	mm Hg	Affects rate of loss from soil.	Shen and Sewell (1982)
organic carbon part. coeff.	K_{oc}	mg VOC/g C	Adsorption coefficient normalized for soil organic content.	Farmer et al. (1980)
octanol/water part. coeff.	K_{ow}	mg VOC/ mg octanol	Equilibrium constant for distribution of VOC between water and an organic (octanol) phase. Used to estimate K_{oc}.	Voice and Weber (1983)
boiling point	b.p.	°C	Affects co-evaporation of VOC and water from soil surface.	Voice and Weber (1983)
soil/water distribution coefficient	K_d	[1]	Equilibrium constant for distribution of contaminant between solid and liquid phases.	Voice and Weber (1983)
Soil Chemical Properties				
cation exchange capacity	CEC	meq/100 g	Estimates the number of negatively charged sites on soil particles where polar or charged VOC may sorb; pH dependent.	
hydrogen ion concentration (activity)	pH	$-\log [H^+]$	Influences a number of soil processes that involve non-neutral organic partitioning; affects CEC and solubility of some VOCs.	

Property	Symbol	Units	Description	Reference
total organic carbon content	TOC	mg C/g soil	An important partitioning medium for non-polar, hydrophobic (high K_{oc}) VOCs; sorption of VOCs in this medium may be highly irreversible.	Chiou et al. (1988), Farmer et al. (1980)

Soil Physical Properties

Property	Symbol	Units	Description	Reference
particle size distribution or texture	A	% sand, silt, clay	Affects infiltration, penetration, retention, sorption, and mobility of VOCs. Influences hydraulics as well as surface area-to-volume ratio (s.a. $\propto K_d$).	Richardson and Epstein (1971)
specific surface area	s.a.	m^2/g	Affects adsorption of VOCs from vapor phase; affects soil porosity and other textural properties.	Karickhoff et al. (1979)
bulk density	p_b	g/cm^3	Used in estimating mobility and retention of VOCs in soils; will influence soil sampling device selection.	Spencer et al. (1988)
porosity	n	%	Void volume to total volume ratio. Affects volume, concentration, retention, and migration of VOCs in soil voids.	Farmer et al. (1980), Shen and Sewell (1982)
percent moisture	Θ	% (w/w)	Affects hydraulic conductivity of soil and sorption of VOCs. Determines the dissolution and mobility of VOCs in soil.	Farmer et al. (1980), Chiou and Shoup (1985)
water potential	pF	m	Relates to the rate, mobility, and concentration of VOCs in water or liquid chemicals.	
hydraulic conductivity	K	m/d	Affects viscous flow of VOCs in soil water depending on degree of saturation, gradients, and other physical factors.	

Table 25.2. Factors Affecting VOC Concentrations in Soils (Continued)

Factor	Common Abbrev.	Units	Effects on VOC Concentrations in Soil	References
Environmental Factors				
relative humidity	R.H.	%	Could affect the movement, diffusion, and concentration of VOCs; interrelated factors; could be site specific and dependent upon soil surface–air interface differentials.	Chiou and Shoup (1985)
temperature	T	°C		
barometric pressure		mm Hg		
wind speed		knots	Relevant to speed, movement, and concentration of VOCs exposed, removed, or diffusing from soil surface.	
ground cover		%	Intensity, nature, and kind, and distribution of cover could affect movement, diffusion rates, and concentration of VOCs.	

variation in TCE concentration in the soil-gas phase above a contaminated aquifer. In February, July, October, and December, the respective shallow ground-water temperatures were 8.9, 16.0, 13.2, and 9.3°C. The highest TCE soil-gas levels were found in July and October. The increased ground-water temperatures during these months caused an increase in the TCE vapor pressure and, therefore, an increase in the steady-state concentration profile of TCE in the vadose-zone soil gas. Kerfoot (1991) demonstrated the thermodynamic principles that account for the phase transfer at elevated temperatures. However, in a nonaqueous solvent, e.g., parathion in hexane, adsorption onto partially hydrated soils was shown to increase with increasing temperature (Yaron and Saltzman, 1972). In this case, parathion (a weaker adsorbate) competes with water (a more powerful adsorbate) for adsorption sites on soil mineral surfaces. An increase in temperature weakens the energetic interaction of minerals with water to a greater degree than with parathion, assisting the latter to compete more favorably for adsorption sites (Chiou, 1989).

The biological factors that affect VOC retention in soils can be divided into microbiological and macrobiological factors. At the microbiological scale, endemic microbial populations in the soil can alter VOC concentrations by various transformations. Although plants and animals metabolize a diversity of chemicals, the activities of the higher organisms are often minor compared to the transformations affected by heterotrophic bacteria and fungi residing in the same habitat (Lewis et al., 1991). The interaction between competing organisms and environmental factors, such as dissolved oxygen, oxidation-reduction potential (Eh), pH, temperature, availability of other nutrient substrates, and salinity, often control biodegradation. The physical and chemical characteristics of the VOC, such as solubility, volatility, and octanol/water partitioning coefficient (K_{ow}) also influence the ability of the compound to undergo biodegradation. Table 25.3 presents some examples of the microbial alterations of some commonly encountered soil VOCs. In general, the halogenated alkanes and alkenes are metabolized by soil microorganisms under anaerobic conditions (Kobayashi and Rittman, 1982; Bouwer, 1984), whereas the halogenated aromatics undergo aerobic decomposition.

On a macro scale, biological factors can influence the retention of VOCs in the saturated and vadose zones, and the soil surface. Table 25.4 presents some of the macrobiological influences on VOC retention in soil systems. Biofilms may accumulate in the saturated zone and participate in the biodegradation and bioaccumulation of VOCs from ground water. The biofilm, depending on its thickness, may impede ground-water flow (Kobayashi and Rittman, 1982). Plant roots with their associated mycorrhizal flora may enhance VOC removal via biodegradation and bioaccumulation. The root channels may act as conduits for increasing the air diffusion of VOCs through the soil. Similarly, animal burrows and holes may serve as paths of least resistance for VOC migration. These holes may range in size from capillary-size openings, created by worms and nematodes, to large-diameter tunnels excavated by burrowing animals. These openings may increase the depth to which surface spills penetrate the soil. A surface covering of vegetation acts as a significant barrier to volatilization of VOCs into the atmosphere. Some ground-water and vadose-zone models (e.g., RUSTIC) include subroutines to account for vegetative cover (Dean et al., 1989).

The above discussion illustrates the complexities of VOC behavior in the vadose zone. The scientist must decide which of these factors is playing a major role in VOC retention and partitioning at the site under investigation in order to knowledgeably

Table 25.3. Microbiological Factors Affecting VOCs in Soil Systems

Organism(s)	Compound(s)	Conditions	Remarks/Metabolite(s)
Various soil microbes	pentachlorophenol	aerobic	tetra-, tri-, di-, and m-chlorophenol (Kobayashi and Rittman, 1982)
	1,2,3- and 1,2,4-trichlorobenzene	aerobic	2,6-; 2,3-dichlorobenzene; 2,4- and 2,5-dichlorobenzene; CO_2 (Kobayashi and Rittman, 1982)
Various soil bacteria	trichloroethane, trichloromethane, methylchloride, chloroethane, dichloroethane, vinylidiene chloride, trichlorethene, tetrachloroethene, methylene chloride, dibromochloromethane, bromochloromethane	anaerobic	reductive dehalogenation under anoxic conditions (i.e., <0.35 V) (Kobayashi and Rittman, 1982)
Various soil microbes	tetrachloroethene	anaerobic	reductive dehalogenation to trichloroethene, dichloroethene, and vinyl chloride, and finally CO_2 (Vogel and McCarty, 1985)
Various soil microbes	^{13}C-labeled trichloroethene	anaerobic	dehalogenation to 1,2-dichloroethene and not 1,1-dichloroethene (Kleopfer et al., 1985)
Various soil bacteria	trichloroethene	aerobic	mineralized to CO_2 in the presence of a mixture of natural gas and air (Wilson and Wilson, 1985)
Actinomycetes	chlorinated and nonchlorinated aromatics	aerobic	various partial breakdown products mineralized by other microorganisms (Lechevalier and Lechevalier, 1976)
Fungi	DDT	aerobic	complete mineralization in 10–14 days (Johnson, 1976)
Pseudomonas sp. Acinetobactyer sp. Micrococcus sp.	aromatics	aerobic	organisms were capable of sustaining growth in these compounds with 100% biodegradation (Jamison et al., 1975)
Blue-green algae (Cyanobacteria)	oil wastes	aerobic	biodegradation of automobile oil wastes and crankcase oil; light required (Cameron, 1963)
Acetate-grown biofilm	chlorinated aliphatics	aerobic	no biodegradation observed (Bouwer, 1984)
	chlorinated and nonchlorinated aromatics	methanogenic	nearly 100% biodegradation observed (Bouwer, 1984)
		aerobic	nearly 100% biodegradation (Bouwer, 1984)
		methanogenic	no biodegradation observed (Bouwer, 1984)

Table 25.4. Macrobiological Factors Affecting VOCs in Soil Systems

Factor	Zone	Effects
Biofilms	saturated	biodegradation, bioaccumulation, formation of metabolites that are more or less toxic than parent compound, thick biofilm may retard saturated flow
Plant roots	capillary fringe to vadose	mycorrihizal fungi may biodegrade or bioaccumulate VOCs, root channels may serve as conduits for VOC migration
Animal burrows and holes	vadose	may act as entry point for downward migration of surface spills and serve as conduit for upward VOC migration
Vegetative cover	soil surface	serves as barrier to volatilization from soil surface and retards infiltration of surface spills

plan a sampling and analysis design. The following principles can be applied for writing and implementing a soil sampling and analysis plan.

SOIL SAMPLING AND ANALYSIS PLANNING ACTIVITIES

Up-front planning is the essential component for obtaining representative and meaningful data from soil VOC sampling activities. Figure 25.2 illustrates the sequence of events used to plan a VOC sampling and analysis activity. Note that the actual field implementation of the sampling activity occurs only after considerable time and effort is spent on planning. While it is the resolve of the authors to present a discussion on soil sampling, it would be remiss to lead anyone into believing that by following a step-by-step field method guide to the letter, meaningful data are always obtained. Therefore, the following sections are presented to stress the importance of presampling activities.

Define Goal

In any sampling excursion, whether it be in air, water, or soil, the goal should be determined before sample number one is taken. This is especially important in the case of soil VOCs. This is the major driving force in the selection of the appropriate sampling approach and devices used to obtain a representative sample in the investigation (Lewis et al., 1991). In many cases, unfortunately, how the data *can* be used is determined after the results are received. This is the cause for failure of many site investigations. The approach should not be to take the soil sample first, then after the data are in hand, decide what the sample tells you about the site. It must be the other way around. As F. Pitard has stated in his book (Pitard, 1989) on sampling theory and sampling practice:

> The total sampling process cannot be investigated backward or in retrospect because there is neither a scientific nor a legitimate way to do so. The only logical approach is a thorough, preventative examination of the selection, materialization, and preparation processes which is our only possible assur-

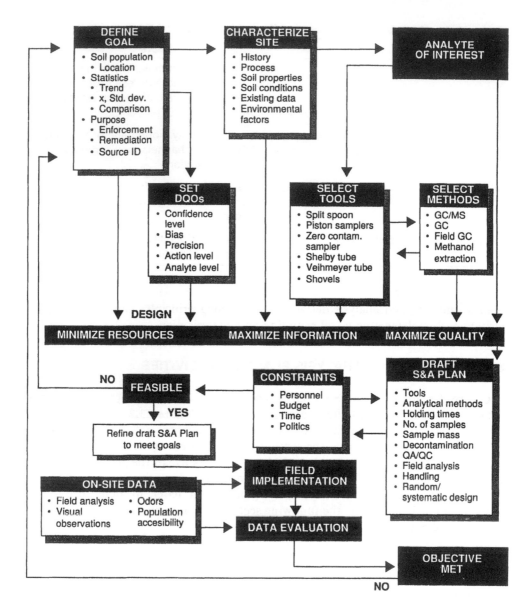

Figure 25.2.   Flowchart for planning and implementation of a soil sampling and analysis activity (taken from Lewis et al., 1991).

ance of obtaining a precise and accurate (i.e., representative) estimate of the true unknown critical content of a lot.

The goals of a sampling exercise and the intended data use may include but are not limited to:

- potential source of the contaminant
- extent of the contamination
- phase partitioning of the contaminant

- fate and transport
- populations at risk (human or ecological)
- compliance
- remedial recourse
- remedial effectiveness

A few brief examples will show how data use will determine the sampling strategy.

Risk Assessment—VOC Emissions from a Landfill Adjacent to a School Playground

The largest risk imposed to children at this site would be from VOC emissions from the surface of the landfill. A sampling strategy involving collection of numerous soil cores from the site would not provide any information as to emission rates from the landfill. Measuring emission rates from the surface of the soil with a surface flux chamber at various times under different meteorological conditions would provide useful data for risk assessment purposes coupled with air monitoring on the playground.

Remedial Alternatives—Delineation of Hydrocarbon Contamination in a Sandy Soil for Excavation

Soil-gas surveying would probably be used first to delineate the extent of the contamination. Soil samples would then be collected with a coring device for confirmatory analysis. Given the low VOC-adsorption capacity of sandy soils, undetectable levels of VOCs may be found. The remediation technique would change from excavation to vapor stripping. However, advances in soil sampling methodologies are becoming so efficient that they are able to maintain the integrity of VOCs in all four phases. Such an efficient sampling device may not be useful because the solid phase concentration would be overestimated. Large sums of money would be spent to excavate large areas of essentially uncontaminated soil. Knowledge of the phase partitioning is needed to determine the optimal remedial alternative. An integrated sampling approach employing soil-gas and solid sampling, and some preliminary phase partitioning calculations would provide useful estimates and allow for informative decision making.

The domains of interest must also be determined. In the planning stages the geostatistician should be made aware of the existence of varying soil types, horizons, lenses, etc. This allows the statistician to recommend a stratification in the sampling design that will greatly enhance precision (Homsher et al., 1992). The domains of stratification may include surface (two dimensions) or subsurface environments (three dimensions), hot spots, a concentration greater or less than an action limit, or the area above a leaking underground storage tank (Lewis et al., 1991). Statistical designs employed in soil sampling are beyond the scope of this chapter. These designs may be random, systematic, or judgmental. The type of statistics that may be generated from the target domain data must be considered prior to sample and analysis planning. Possible statistics of interest may include: average analyte concentration and the variance about the mean, comparative statistics as to whether the observed level is significantly above or below an action level, and temporal and spatial trends.

Set Data Quality Objectives

Data must be of sufficient quality to meet the goals of the sampling activity and be suitable for its intended use. Data quality objectives (DQOs) are used by the U.S. Environmental Protection Agency (EPA) to facilitate up-front planning (Homsher et al., 1992). The EPA requires that all monitoring and measurement programs establish DQOs based on the anticipated end uses of the data (Stanley and Verner, 1985). The data quality objective process may be considered to be the up-front interactive process between the data user (i.e., the decision maker) and the data generator (i.e., the supporting technical team) which defines the error level (uncertainty) that is acceptable in the decision/application process. The level of error that is acceptable is defined by the ability of the data to confirm, reject, or discriminate among hypotheses formulated by the data user. The quality of the data can be quantified with respect to the DQOs, thereby allowing the data user to evaluate the hypotheses with a known level of confidence.

The importance of the decision maker in the planning process is evidenced in the very definition of DQO. The decision maker must be able to relate to the site investigation team what the consequence is of making a wrong decision based on the data. The consequences of making a wrong decision are numerous; environmental, human health related, economic, or social. In many cases, either the decision makers are not involved until after the sampling activities are complete, or they unable or unwilling to provide this information. What results is a DQO formulating process that is largely driven by technical and analytical considerations alone. For the most part, the reason for making a wrong decision based on soil VOC data stems from negative bias and/or false negatives. However, false positives may occur.

A DQO may be defined as the minimum confidence level (MCL) required to make a decision from the data. DQOs can be classified into six elements: detectability, precision, accuracy, representativeness, comparability, and completeness. Measurements of VOCs in soils are prone to numerous sources of error. The problems inherent in sampling soils for VOCs manifest themselves as sources of error in all of the objectives for maintaining high data quality, most notably precision, accuracy (bias), representativeness, and comparability. These four elements will be discussed in the following sections.

Precision

Precision is the ability to obtain reproducible measurements from a soil system and is usually defined as percent relative standard deviation (%RSD). With variation in soil properties ranging from a few percent to several hundred percent over a few centimeters (Armson, 1977; Black, 1965; Mason, 1983), high precision (i.e., low %RSD) is difficult to attain for even the most well-behaved and conservative analytes of interest, let alone the evasive class of compounds known as VOCs. So-called "duplicate" soil samples collected within 10 cm of each other (e.g., with a split-spoon sampler) have exhibited one or more orders of magnitude differences in VOC concentrations.

Accuracy

Accuracy measures the bias in the measurement system. Sources of error are the sampling process, sample handling, field and laboratory introduced contamination, preservation and storage, sample preparation, and analysis. Since knowledge regarding the extent of contamination is critical to assessing risk, selecting remedial approaches, and determining the degree of cleanup achieved, it is essential that they be based on accurate measurements of the VOC concentrations present. Data generated from the analysis of VOC-contaminated soils have been used with the presumption that the measurements are indeed accurate. These accuracy estimates, however, have been based primarily on matrix spikes, matrix spike duplicates, and internal surrogates, all having analytically based measurement quality control functions. Unfortunately, these procedures do not reveal the true severity of the problems associated with sampling, sample handling, and preparation procedures. Since sampling bias for soil VOCs is so difficult, if not impossible to quantify, it is often necessary to assume that it is absent. This is obviously far from the truth.

It has been estimated that up to 90% of all environmental measurement variability can be attributed to the sampling process (Homsher et al., 1992). The recovery of VOCs from soils has been particularly variable. The source of variation in VOC analyte recovery can be associated with any or all of the single steps in the process, including sample collection, transfer from the sampling device to the sample container, sample shipment, sample preparation for analysis, and sample analysis. Sample collection and handling activities have large sources of random and systematic errors compared to the analysis itself (Barcelona, 1989). Negative bias (i.e., measured value less than true value) is perhaps the most significant and most difficult to delineate and control. Negative bias is primarily caused by volatilization losses during soil sample collection, storage, and handling. Up to 80% losses have been reported in the sampling process for certain VOCs (Siegrist and Jennsen, 1990). Other factors contributing to VOC negative bias are chemical and biochemical transformations prior to analysis. "Apparent" losses from freshly-spiked or "real-world" samples may be caused by irreversible adsorption and/or slow desorption kinetics in the analytical recovery procedure (e.g., purge-and-trap). It is typical for system measurement error bias to range from –100% to +25% (Siegrist, 1990; 1991).

Representativeness

Representativeness expresses the degree to which sample data accurately and precisely represent a characteristic of the population. Representativeness is the exclusive property that synthesizes both accuracy and precision. Representativeness not only encompasses the accuracy and precision attainable by the selected sampling methodology, but also the spatial and temporal components of measurement uncertainty. Representativeness does not end in the field. Once a field person has obtained a "representative" sample at the right time in the right place following the right sampling protocols, the laboratory analyst must then obtain a subsample from the containerized sample without jeopardizing accuracy and precision in the process. There are currently no proven procedures for compositing or homogenizing soil samples for volatile organic analysis (VOA) without loss of analytes.

Comparability

Comparability is a qualitative parameter expressing the confidence with which one data set can be compared with another. Data sets obtained from the same site using identical sampling methodologies may erroneously be compared if the environmental conditions extant at the time of sampling are not similar. Similarly, comparing two data sets, one at the start and one at some time after remedial treatment has been in progress, may falsely lead to the conclusion that the cleanup was successful. In the first data set, the laboratory extraction procedure was very rigorous, providing near 100% recovery of the total analyte concentration from the real-world matrix. In the second data set, a recovery method was used that was only 20% efficient. The "apparent" lower concentration in the post-remediation data set is actually due to the poorer extraction efficiency of the method. The two sets of data are nevertheless considered comparable because both laboratory methods exhibited good precision and comparable analyte recoveries (accuracy) for freshly spiked soil audit material and internal surrogates.

Implement a Quality Assurance Program

Once the DQOs have been defined, procedures must be implemented to ascertain whether the DQOs are being met. This is achieved by implementing a quality assurance (QA) program. Quality assurance is often neglected in field sampling. If the error associated with sample collection and handling is large, then the best laboratory QA program will be inadequate (van Ee et al., 1990). Quality assurance is a system of activities that provides the producer or user of the data with assurance that it meets the defined standards of quality (i.e., the DQOs). It consists of two separate but related components, quality control (QC) and quality assessment. Quality control is a system of activities used to control the quality of the measurement data so that they meet the needs of the user. Quality assessment provides an objective measure of the quality of data produced. Quality assurance is driven by the DQO process.

Sampling for soil VOCs is not readily amenable to the traditional methods for assessing and controlling field sampling measurement uncertainty. For most inorganic and nonvolatile organic contaminants, field sampling measurement uncertainty can be estimated by collection of collocated samples, field splits (duplicates), composites, and field audit samples. Collection of collocated samples may offer information on the precision of a soil VOC sampling method, provided the short-range spatial variability is not overwhelming, and the method does not permit losses of VOCs through volatilization. Field splits and composites require disruption of the soil matrix that will also result in significant losses through volatilization. A methanol immersion procedure is showing early signs as an effective means of compositing soil samples in the field for volatile organic analysis (VOA) (Siegrist and Jennsen, 1990; Urban et al., 1989). This procedure will be discussed in greater detail later in the chapter. Field audit samples or performance evaluation materials (PEMs) for soil VOCs are not commercially available in suitable quantities or formulations for use in the field. Attempts to generate PEMs for site-specific use have been reported in the literature (Lewis et al., 1990), but results suggest the need for further research.

When VOC concentrations are at or near an action or detection limit, the need for a rigorous QA/QC program is greater. At the present time, the only area where rigorous QA/QC protocols can be implemented is in the laboratory. Given the paucity of QA activities that are currently employable in the soil VOC sampling realm, the only means of ensuring that a representative sample is obtained is by selecting the most efficient sampling tool(s) and using sample handling procedures that reduce losses of VOCs through volatilization. Siegrist (1991) has suggested that more rigorous sample collection and sample handling practices (i.e., those practices that maintain the integrity of VOCs in the sample) be considered when accurate and precise measurements of VOCs having high volatility and low solubility (e.g., trichloroethene) are needed. Regardless of the relative volatility and solubility of a VOC or the proximity of its concentration to an action or detection limit, accuracy and precision are paramount and the most efficient sampling device and sample handling procedure should be used, all other factors being equal (e.g., cost, logistics). If less accurate or precise measurements are needed to make a decision based on the data, the rigor of laboratory QA/QC protocols can more easily be reduced. This reduction can be achieved by reducing the number and frequency of quality control samples with substantial analytical cost savings. However, if the laboratory receives samples that are collected in the field using methods with varying degrees of rigor, then adjustments in laboratory QA/QC may compound the measurement uncertainty. Given the elusiveness of VOCs in soils, it would be more appropriate to set accuracy and precision in the field to the highest levels possible and allow the laboratory QA manager to adjust the QA/QC requirements, based on the desired data end use.

For a more detailed discussion of QA/QC practices in soil sampling the reader is referred to a user's guide on quality assurance in soil sampling by Barth et al. (1989), and a document which discusses the errors associated with soil sampling by van Ee et al. (1990).

Characterize Site

The next stage in planning involves site characterization. Considerable information can be gleaned from some up-front detective work. As much as possible should be learned about the site prior to sample collection. Review of aerial photographs, historical records, and manifests can provide a great deal of information as to the nature and location of wastes on the site. Conversations with truckers who deliver wastes to the site can provide keen insights not available from historical records or manifests (U.S. EPA, 1992).

A walk-over inspection of the site should be performed to gather information on a smaller scale. This approach is semiquantitative. Photographs and sketches can be later inspected offsite, thus greatly reducing the amount of time field crews are on the site. Subsequent sampling operations can focus on areas with possible sources of hazardous materials using evidence gathered during the walk-over.

If the site is inactive or abandoned, both historical and newly-acquired aerial photographs can be valuable in inventorying and evaluating inactive waste disposal sites. However, since many abandoned sites have been covered over for years and are no longer recognizable as such on recent aerial photographs, historical aerial photographs are often more useful in inactive waste site investigations. Historical aerial photographs are not only useful in compiling land use/cover, environmental and other physical site-specific data, but are also useful in directing ground investigation teams to exact locations for conducting sampling operations (U.S. EPA, 1992). Changes in vegetative cover on the site noted on aerial photographs may offer additional evidence of the distribution of contaminants. Areas may be devoid of plants due to soil disturbance, direct toxicity of the organic contaminants, or displacement of oxygen from the soil by organic vapors (U.S. EPA, 1992).

Various state and federal agencies have photographed most of the United States in recent years. As discussed by Erb et al. (1981), at least one photograph exists for any land area within the continental United States. Most of the federal photographic data have been catalogued by the U.S. Geological Survey's National Cartographic Information Center.

The remoteness of a site and the physical setting may restrict access and, therefore, equipment selection. Such factors as vegetation, steep slopes, rugged or rocky terrain, overhead power lines or other overhead restrictions, or lack of roads, can contribute to access problems. Availability can also be a limiting factor in sampling equipment selection. In some cases, the desired equipment may be dedicated to major long-term projects and thus be unavailable (Lewis et al., 1991).

The presence of underground utilities, pipes, electrical lines, tanks, and leach fields, can also affect selection of sampling equipment. If the location or absence of these hazards cannot be established, it would be desirable to conduct a nonintrusive survey of the area and select a sampling approach that minimizes hazards. For example, hand tools and a backhoe are more practical under such circumstances than a large hollow stem auger. Under some circumstances, it may not be practical to collect deep soil samples. The presence of ordnance, drums, concrete, voids, pyrophoric materials, and high hazard radioactive materials, may preclude some sampling and require reconsideration of objectives or development of alternate sampling designs (Lewis et al., 1991).

The selection of a sampling device may be influenced by other contaminants of interest such as pesticides, metals, semivolatile organic compounds, radionuclides, and explosives. Where the site history indicates that the matrix is other than soil, special consideration should be given to device selection. Concrete, reinforcement bars, scrap metal, and lumber need to be sampled with tools other than those used for soils.

As mentioned previously, soil characteristics can vary orders of magnitude over a given site. Inasmuch as certain soil characteristics may directly or indirectly influence VOC retention, the necessity to characterize their distribution on the site is evident. Breckenridge et al. (1991) have covered soil characterization for hazardous waste site assessment in greater detail.

In general, the physical and chemical properties of soils vary in a stochastic manner which tends to follow a classic statistical distribution. For example, properties such as bulk density and effective porosity tend to be normally distributed (Campbell, 1985), whereas saturated hydraulic conductivity is often log-normal in distribution. On a coarse scale, soil maps are useful for determining the variation in soil type on the site and predicting pollutant retention (Fuller and Warrick, 1985). On this basis, similar soils should have similar VOC retention. However, many environmental factors (e.g., rain, temperature) may cause departures from the predictions. The USDA Soil Conservation Service recognizes classes of mineral soils based upon pollutant attenuation which are considered to significantly affect the movement of organic chemicals. These classes are:

- particle-size classes (sand, silt, clay)
- soil consistency classes
- mineral classes (primary and secondary clay minerals)
- calcareous and reactions (pH) classes (soil acidity, alkalinity)
- soil temperature classes
- soil depth classes
- soil slope classes
- classes of coatings (on sands)
- classes of cracks

At a minimum, organic carbon content, particle size distribution, bulk density, and percent moisture content should be determined at different locations and depths to adequately characterize a site and provide information to the statistician for devising a sampling design. These factors affect VOC and water retention, soil consistency, the capacity to shrink and swell, and many other properties that directly influence the phase partitioning and retention of the VOCs and the ultimate choice of sampling device. Computer models generally are used only at the final stages of a remedial investigation/feasibility study (RI/FS). However, modeling techniques can be used throughout the entire RI/FS process to assist in sampling device selection by estimating the phase partitioning of the VOCs (Lewis et al., 1991).

Select Tools

The overall efficiency of the entire VOC measurement process will only be as good as the weakest link. Soil sampling, defined here as the collection and containerization of a representative sample from the site, is perhaps the weakest link.

The characteristics of the soil material being sampled, therefore, have a marked effect upon the sampling devices used. A site investigator must evaluate this information in concert with the type of VOCs to be sampled and the depth at which a sample is to be collected before a proper sampling strategy can be selected. Specific characteristics that must be considered are:

1. Is the soil compacted, rocky, or rubble filled? (If the answer is yes, then either hollow stem augers or pit sampling should be used.)
2. Is the soil fine grained? (If yes, use split spoons, Shelby tubes, liners, or hollow stem augers.)
3. Are there flowing sands or saturated soils? (If yes, use samplers such as piston samplers that can retain these materials.)

In addition to soil characteristics, other factors must be considered in the selection of a sampling device and sampling procedures. These factors include the number of samples to be collected, available funds, site limitations, ability to sample the target domain, whether or not field screening procedures are to be used, the size of sample needed, and the required precision and accuracy as given in the DQO. The number of samples to be collected can greatly affect sampling costs and the time required to complete a site characterization. If many subsurface samples are required, it may be possible to use soil-gas sampling coupled with onsite analysis as integrated screening techniques to reduce the area of interest, and thus the number of samples required. Such an integrated sampling approach may be applicable for cases of near surface contamination (Lewis et al., 1991).

Some sampling equipment is difficult to decontaminate in the field and requires the use of decontamination pads with impervious liners, wash and rinse troughs, and careful handling of large equipment. These procedures take time, require extra equipment and ultimately affect site characterization costs. Decontamination is thus an important factor to be considered in device selection (Lewis et al., 1991).

Table 25.5 lists selection criteria for various types of commercially available soil sampling devices based on soil type, moisture status, and power requirements. The sampling device needed to achieve a certain sampling and analysis goal can be located in the table. Once a potentially suitable device has been selected, the supplier of such a device can be identified in Table 25.6. Table 25.6 is a partial list of

Table 25.5. Criteria for Selecting Soil Sampling Equipment

Type of Sampler	Obtains Core Samples		Most Suitable Soil Types		Operation in Stony Soils		Suitable Soil Moisture Conditions			Relative Sample Size		Labor Req'mnts		Power Req'mnts	
	Yes	No	Coh	Coh'less	Fav	Unfav	Wet	Dry	Interm	Small	Large	Single	2+	Manual	Power
A. Mechanical Sample Recovery															
1. Hand-held power augers		X	X	X		X	X		X		X		X		X
2. Solid stem flight augers		X	X	X	X			X	X		X		X		X
3. Hollow-stem augers	X		X	X	X	X	X	X	X		X		X		X
4. Bucket augers		X	X	X	X		X	X	X		X		X		X
5. Backhoes		X	X	X	X		X	X	X		X		X		X
B. Samplers															
1. Screw-type augers		X	X			X			X	X		X		X	
2. Barrel augers															
a. Post-hole augers		X	X			X	X				X	X		X	
b. Dutch augers		X	X			X	X				X	X		X	
c. Regular barrel augers		X	X			X		X			X	X		X	
d. Sand augers		X		X		X		X			X	X		X	
e. Mud augers		X	X			X	X				X	X		X	
3. Tube type samplers															
a. Soil samplers	X		X			X	X	X	X	X		X		X	
b. Veihmeyer tubes	X		X			X			X		X	X		X	
c. Shelby tubes	X		X			X			X		X	a	X	X	X
d. Ring-lined samplers	X		X		X		X	X	X		X	a	X	X	X
e. Continuous samplers	X		X			X	X	X	X		X	a	X		X
f. Probe-drive samplers	X		X			X	X		X		X	a	X		X
g. Piston samplers	X		X			X	X				X	a	X	X	X
h. Zero-contamination samplers	X		X			X	X		X	X		a	X	X	
i. Split spoon samplers	X		X			X	X		X		X	a	X	X	X
4. Bulk samplers		X	X	X		X	X	X	X		X	X		X	

Source: Adapted from EPA/530–SW–86–040, 1986

[a] All hand-operated versions of tube samplers, except for continuous samplers. can be worked by one person.

commercially available soil sampling devices that are currently in use for sampling soils for VOC analysis. The list is by no means exhaustive, and inclusion in the list should not be construed as an endorsement for their use.

SAMPLING DEVICE APPLICATIONS

There are currently no standardized procedures for sampling soils for VOC analyses. Several types of samplers are available for collecting intact and disturbed samples. Samples are usually removed from the sampler, which often disturbs intact samples, and placed in glass jars or vials and sealed with Teflon-lined caps. However, practical experience and recent field and laboratory research has suggested that simple procedures such as these may lead to significant loss of VOC.

Proper use of the appropriate soil sampling device is critical for maintaining the integrity of VOCs in the soil sample. Some devices are used to obtain undisturbed soil samples using either hand-held or mechanical devices. Other devices obtain disturbed samples. Devices are available for sampling either the soil solution (interstitial pore water), flow in saturated regions (e.g., perched ground water), or soil gas. The devices utilized in these techniques may be permanently installed in the vadose zone, allowing time-sequenced sampling at depth-wise intervals. Fiber optic sensors may provide real-time monitoring in saturated zones.

The following sections discuss the types of soil sampling devices commonly used and how efficient they are in maintaining the integrity of VOCs in the sample. For additional details see Chapter 24 in this book.

Mechanical Sample Recovery Methods

Drill rigs used for vadose zone soil sampling are identical to those used to sample ground water. Commonly used drill rigs such as cable tool and rotary units are not recommended because they generally require drilling fluids that may contaminate the sample. Air rotary drilling is also undesirable for obtaining samples for VOC because the forced air may cause excessive loss of VOCs. Mechanical sampling devices include two-man operated screw augers, solid-stem flight augers, bucket augers, and hollow-stem augers. Further discussions of these devices follows.

Hand-Held Power Augers

A simple power-driven auger consists of a solid flight auger attached to and driven by a small air-cooled engine. Operators guide the auger flight into the ground using handles. Throttle and clutch controls are integrated into grips on the handles. As the auger rotates into soil, cuttings are brought to the surface. Soil samples can be collected from material augured to the surface, or from the auger flights after pulling the auger out of the ground. The drilling procedure results in churning of the sample, with possible loss of VOC. A better approach is to collect samples from the base of the hole created using other samplers (e.g., thin-walled samplers). Care must be exercised to ensure that gasoline or oil from the engine does not drip into the hole during sampling.

Solid-Stem Flight Auger

This device is a larger scale version of the simple hand-held flight auger. This sampler is commonly mounted on a vehicle and driven by an independent power supply. Again, the churning action may result in a loss of VOC, and additionally there is a problem of cross contamination between sampling intervals.

Hollow-Stem Augers

As the name implies, hollow-stem augers consist of a sequence of auger flights with hollow interiors. To use a sampler within a hollow-stem auger, the auger column and pilot assembly are rotated to the top of the desired sampling interval. The pilot assembly and center rod are then removed and the sampler is inserted through the hollow axis of the auger column. The sampler may be lowered to the sampling depth by attaching it to center rods or by using a wireline assembly. When the sampler is attached to center rods, a sample is collected by pushing or driving the sampler into undisturbed soil with the rig hydraulic system or with a drop hammer. When a wireline is used, the sampler is locked into place ahead of the lower-most auger and advanced into the sampling interval by rotating the auger column. Common samplers used inside hollow-stem augers include split spoon samplers, ring samplers, and continuous samplers. Liners may be used with samplers to aid in sample removal and decontamination, and reduce cross contamination. This method minimizes cross contamination between sampling intervals, and, depending on the sampler, minimizes the loss of VOC by allowing the collection of an undisturbed core.

Bucket Auger Drilling and Sampling

The bucket auger drilling rig consists of a large diameter bucket with cutting blades mounted on the bottom. The bottom is hinged to allow cuttings to be dumped out. The churning action of this sampler severely disrupts the soil matrix and promotes the loss of VOCs. Alternatively, the bucket is brought to the surface and, before dumping, a tube sampler is inserted into the interior of the collected soil mass.

Backhoes

Backhoes are used to create a trench to a desired depth in the vadose zone. Samples may be obtained from the backhoe bucket, but this is not recommended for VOC sampling because disturbance of the soil in the bucket is likely to have caused significant release VOCs. A better approach is to sample from the trench walls using a tube sampler, such as those discussed below.

Soil Samplers

Screw-Type Augers

The screw or ship auger is essentially a small diameter wood auger from which the cutting side flanges and tip have been removed. Variations on this design include the closed spiral auger and the Jamaica open spiral auger (Acker, 1974). During drilling, the handle is twisted manually to drive the auger screws into the soil. Upon removal of the tool, soil from the deepest penetration is retained on the auger flights. The sample is scraped off the flights, using a spatula, into a sample container. The entire sample is exposed to the air during sampling, recovery, and insertion into a container, permitting potentially significant loss of VOCs.

Barrel Augers

Barrel augers include a bit with cutting edges, a short tube or barrel for retaining the soil sample, and two shanks. The shanks are welded to the barrel at one end and onto extension rods. The uppermost extension rod allows the insertion of a handle which is threaded at the other end. The sampler is turned to advance the barrel vertically into the ground. When the barrel is filled, the unit is withdrawn from the soil cavity and soil is removed from the barrel. As with the screw-type sampler, VOCs can be lost during the sampling process. Five common barrel augers are posthole augers (also called Iwan-type augers), Dutch-type augers, regular or general purpose barrel augers, sand augers, and mud augers.

Tube-Type Samplers

Sampling with these units involves forcibly pushing or driving the sampler into the soil. When the sampler is filled at a sampling depth, the assembly is pulled back to the surface. Tube samplers are particularly suited when a relatively undisturbed sample is required. Tube-type samplers include soil sampling tubes with cut-away barrels, Veihmeyer tubes (also called King tubes), thin-walled tube samplers (also called Shelby tubes), ring-lined barrel samplers, split-spoon samplers, piston samplers, probe drive samplers, and zero contamination samplers. These samplers are driven or otherwise forced into the soil. Liners are available for most of these types of samplers. A problem is that heat generated during this process may drive off VOCs. For additional information see Acker (1974).

Soil Sampling Tubes with Cut-Away Barrels. These manually-operated samplers consist of a hardened cutting tip, a cut-away barrel, and an uppermost section which threads onto extension rods. The section rods are connected with a handle. The cut-away barrel is designed to facilitate lithologic examination and to allow for easy removal of soil samples. Given that the sample is exposed to the atmosphere in the cut-away section, there will be considerable loss of VOC using these samplers.

Veihmeyer Tubes. In contrast to the soil sampling tube, the Veihmeyer tube is a long, solid tube. This is a manually-operated device. Occasionally, the inside of the tube is coated with a lubricant to facilitate sample removal. The chemical composition of the lubricant should be determined to ensure that the lubricant will not contaminate the sample. The Veihmeyer sampler is fitted with a drop hammer for driving the sampler into the soil profile. The sample is often extruded from the sampler by inverting and tapping on the upper end. Alternatively, a small piston, with an outside diameter slightly smaller than the ID of the tube, can be used to force out the sample. Volatile organics will be lost during the extrusion process. These tubes can be equipped with liners to minimize VOC loss during recovery.

Shelby Tubes.   Shelby tubes are thin-walled tube samplers. The sampler tube is connected to a sampler head which is threaded to connect with extension rods. The Shelby tube is pushed into the soil by hand using extension rods and a handle, with a jack-like system, or with a hydraulic piston. Tubes are also adapted to fit onto a drill rod for insertion into hollow-stem augers. Samples are extruded from Shelby tubes using a hydraulic ram. Alternatively, for VOC sampling, the extremities of the sampler are sealed with plastic and PTFE caps to minimize VOC loss, and the entire sampler is sent to the laboratory for analysis. However, plastic and PTFE are known to be permeable to VOCs. Some researchers put aluminum foil inside the cap, then wrap the cap with tape, hoping to reduce the loss of VOCs through the plastic or PTFE cap. However, a completely airtight seal cannot always be achieved, and the solvents on the tape may contaminate the encased sample.

Ring-Lined Barrel Samplers.   The ring-lined barrel sampler consists of a one-piece barrel or two split-barrel halves, a drive shoe, rings, and a sampler head. Barrel samplers are more rigid than thin-walled tube samplers. The ring liners are made from various materials, such as stainless steel, acetate, and brass. The rings fit snugly inside the barrel. Depending on sampler size, one to three rings fit within the sampler. Once filled with a sample, the rings are removed from the sampler. For the one-piece barrels, the rings are removed using an extruder. For split-barrel units, the ends are unscrewed, the barrel is split apart and the rings are removed individually. In either case, the rings are capped with a small PTFE sheet and a plastic or PTFE cap which is sealed with wax or adhesive tape. This technique minimizes the loss of VOCs, although some VOC dissipates from the base of the sampler as it is withdrawn from the soil and when the ring segments are initially separated. Furthermore, loss of VOCs may also occur by permeation through the caps. There may be an advantage to using devices which can accommodate three ring samplers in that the middle sampler may be least disturbed during removal.

Continuous Sample Tube System.   Typically, this sampler is 5 feet long, fitting within the lead flight of a hollow-stem auger. The sampler is locked in place inside the auger column with the basal end protruding slightly beyond the end of the column. As the column is advanced, soil enters the nonrotating sampling barrel. After a 5-foot advance, the sampler is withdrawn. The sampling tube is available either as a solid or split barrel, accommodating a liner. The ends of the sampler are capped and sealed to prevent loss of VOC. Additionally, if desired, the sample can be segmented and individual sections sealed and capped. Permeation of VOCs through the cap may also lead to losses.

Probe-Drive Samplers.   This sampler consists of a pointed, sealed tube which is driven to a sampling depth using either a slide hammer or a power-driven hydraulic system. The pointed end is part of a piston which can be pulled back before the sampler is driven into undisturbed soil. The unit is then extracted. The sample is either removed by a piston-extruding device or the sampler ends are capped and the sampler is sent to a laboratory. This approach minimizes the loss of VOC in the field. For difficult site conditions, a pilot hole is created using a retractable drive point before inserting the sampler.

Piston Samplers.   Components of a piston sampler include a sampling tube, an extension pipe attached to the tube, an internal piston, and rods connected to the piston and running through the extension pipe. The sampler can be pushed into the ground with the handle or mechanically driven into the ground with a drop hammer. The piston is held stationary or pulled upward with the attached rods as the sampler is inserted in the soil. At the sampling depth, the sampler is twisted to break suction at the tube-soil interface. The sample is retained within the sampler because of the suction which develops between the piston and the sample. Upon retrieval, the piston forces the sample out of the tube. Volatile organic compounds are likely to be lost as the sample is extruded and placed within a sample container. The suction may also cause migration of VOCs, and complicate vertical concentration profile determinations.

Zero Contamination Tube Samplers.   These tube samplers are lined with thin-walled acetate or stainless steel liners for recovering undisturbed samples. The lower end of the tube contains a drive point. The upper end accommodates a socket which is locked in place with a pin. The sampler is forced into the soil either by a foot-operated jacking device or by a power-driven system. Upon removal of the sampler, the liner is extracted from the tube. The ends of the liner are sealed with caps. As with ring samplers, this sampler probably minimizes the loss of VOCs. However, some VOC loss will occur out of the base of the unit as it is withdrawn from the soil, and by permeation through the caps.

Split-Barrel Drive Samplers (Split-Spoon Samplers).   These samplers consist of two split-barrel halves, a drive shoe, and a sampler head containing a ball check valve, all of which are threaded together. Liners are installed for encasing the collected samples. The sampler is threaded onto drilling rods and is lowered to the bottom of the boring, most commonly through the flights of a hollow stem auger. The sampler is then driven into the soil with blows from a drop hammer attached to a drill rig. The sampler is withdrawn by upward blows from the drop hammer. The sampler is usually driven out 1 to 2 feet ahead of the drill bit. Once the sample is retrieved, the split spoon can be opened and the sample removed. The ends of the liner are then capped and sealed to minimize loss of VOCs.

Bulk Samplers

Commonly, soil samples are obtained from the near surface using shovels, scoops, trowels, and even spatulas. These devices can be used to extract soil samples from trenches and pits excavated by back hoes. Samples can be collected from the sides or the bottom of the trench or pit with hand augers or tube-type samplers. Bulk samplers such as shovels and trowels expose the sample to the atmosphere, enhancing loss of VOCs.

Criteria for the selection of the soil sampling devices discussed above are summarized in Table 25.5. A partial list of the suppliers of various soil sampling devices can be found in Table 25.6. Devices most suitable for sampling soils for VOCs are shown in Table 25.7. Devices where the soil samples can be easily and quickly removed and containerized with the least amount of disturbance and exposure to the atmosphere are highly recommended (Lewis et al., 1991). Liners are available for most of the samplers recommended in Table 25.7.

HANDBOOK OF VADOSE ZONE CHARACTERIZATION & MONITORING

Table 25.6. Examples of Commercially Available Soil Sampling Devices

Manufacturers	Sampling Device	Specifications Length (inches) I.D. (inches) Sampler Material	Liners	Features	Users
Associates Design & Manufacturing Co. 814 North Henry Street Alexandria, VA 22314 703–549–5999	Purge and Trap Soil Sampler	3 0.5 stainless steel		Will rapidly sample soils for screening by "Low Level" Purge and Trap methods.	ORNL Oak Ridge, TN 615–576–6690
Acker Drill Co. P.O. Box 830 Scranton, PA 717–586–2061	Heavy Duty "Lynac" Split Tube Sampler	18 & 24 1½ to 4½ steel	brass, stainless	Split tube allows for easy sample removal.	Empire Soils Edison, NJ 201–287–2224
	Dennison Core Barrel	24 & 60 1⅞ to 6⁵⁄₁₆	brass	Will remove undisturbed sample from cohesive soils.	
AMS Harrison at Oregon Trail American Falls, ID 83211 800–635–7330	Core Soil Sampler	2 to 12 1½ to 3 alloy, stainless	stainless, plastic aluminum, bronze Teflon	Good in all types of soils.	IT Corp. Knoxville, TN 615–690–3211
	Dual Purpose Soil Recovery Probe	12, 18 & 24 ¾ and 1 4130 alloy, stainless	butyrate, Teflon stainless	Adapts to AMS "up & down" hammer attachment. Use with or without liners.	
	Soil Recovery Auger	8 to 12 2 & 3 stainless	plastic, stainless Teflon, aluminum	Adaptable to AMS extension and cross-handles	
Concord, Inc. 2800 7th Ave. N. Fargo, ND 58102 701–280–1260	Speedy Soil Sampler	48 & 72 3/16 to 3½ stainless	acetate	Automated system allows retrieval of 24 in. soil sample in 12 sec.	

Manufacturer	Product	Specifications	Liner Material	Comments	Additional Source
CME Central Mine Equip. Co. 6200 North Broadway St. Louis, MO 63147	Continuous Sampler	60, 2½ to 5⅜, steel, stainess	butyrate	May not be suitable in stony soils. Adapts to CMS auger.	Datum Exploration Long Beach, CA 213-595-6551
	Bearing Head Continuous Sample Tube System	60, 2½, steel, stainess	butyrate	Versatile system. Adapts to all brands of augers.	
Diedrich Drilling Equip. P.O. Box 1670 Laporte, IN 46350 800-348-8809	Heavy Duty Split Tube Sampler	18 & 24, 2, 2½, 3, steel	brass, plastic, stainless, Teflon	Full line of accessories are available.	Exploration Tech. Madison, WI 608-258-9550
	Continuous Sampler	60, 3, 3½, steel, stainless	brass, plastic stainless, Teflon	Switch-out of device easily done.	
Geoprobe Systems 607 Barney St. Salina, KS 913-825-1842	Probe Drive Soil Sampler	24, 1, alloy steel		Remains completely sealed while pushed to depth in soil.	Roy F. Weston 1001 Galaxy Wy Concord, CA 94520
Giddings Machine Co. P.O. Drawer 2024 Fort Collins, CO 80522 313-482-5586	Coring Tubes	48 & 60, ⅞ to 2⅜, 4130 molychrome	butyrate	A series of optional ⅝ in. slots permit observation of the sample.	Mobay Chemical, Inc. Kansas City, MO 816-242-2000
JMC Clements and Associates R.R. 1 Box 186 Newton, IA 50208 515-792-8285	Environmentalist's Sub-soil Probe	36 & 48, 0.9, nickel plated	PETG plastic, stainless	Adapts to drop-hammer to penetrate hardest of soils.	EES Dallas, TX 214-239-0773
	Zero Contamination Sampler™	12, 18 & 24, 0.9, nickel plate	PETG plastic, stainless	Adapts to power probe.	

Table 25.6. Examples of Commercially Available Soil Sampling Devices (Continued)

Manufacturers	Sampling Device	Specifications			Features	Users
		Length (inches) I.D. (inches) Sampler Material	Liners			
Mobile Drilling Co. 3807 Madison Ave. Indianapolis, IN 46227 800–428–4475	"Lynac" Split Barrel Sampler	18 & 24 1½	brass, plastic		Adapts to mobile wireline sampling system.	Datum Exploration Long Beach, CA 213–595–6551
Soiltest, Inc. 66 Albrecht Drive Lake Bluff, IL 800–323–1242	Zero Contamination Sampler™	12, 18 & 24 0.9 chrome plated	stainless, acetate		Hand sampler good for chemical residue	Lockheed-ESC Las Vegas, NV 702–798–2676
	Thin Wall Tube Sampler (Shelby)	30 2½, 3, 3½ steel			Will take undisturbed samples in cohesive soils and clays.	
	Split Tube Sampler	24 1½ tp 3 steel	brass		Forced into soil by jacking, hydraulic pressure or driving. Very popular type of sampler.	
	Veihmeyer Soil Sampling Tube	48 & 72 ¾ steel			Adapts to drop hammer for sampling in all sorts of soils.	
Sprague & Henwood, Inc. Scranton, PA 18501 800–344–8506	S & H Split Barrel Sampler	18 & 24 2 to 3½	brass, plastic		A general all-purpose sampling device designed for driving into material to be sampled.	Harding Lawson Reno, NV 702–329–6123

Source: Taken from Lewis et al., 1991.
Note: This list is not exhaustive. Inclusion in this list should not be construed as endorsement for use.

Table 25.7. Soil Samplers for VOC Analysis

Recommended	Not Recommended
Split spoon w/liners	Solid flight auger
Shelby tube (thin wall tubes)	Drilling mud auger
Hollow-stem augers	Air drilling auger
Veihmeyer or King tubes w/liners	Cable tool
Piston samplers[a]	Cut-away barrels
Zero contamination samplers[a]	Hand augers
Probe-drive samplers	Barrel augers
Free-drainage samplers	Scoop samplers
Perched ground-water samplers	Excavating tools;
	e.g., shovels, backhoes,
	Suction lysimeters
	Filter-tip samplers

Source: Taken from Lewis et al., 1991.
[a]May sustain VOC losses if not used with care.

Liners for Soil Sampling Devices

When sampling for VOC it is advisable to avoid interactions between the sample and the liner and/or sampler. Such interactions may include either adsorption of VOCs from the sample or release of VOCs to the sample. Gillman and O'Hannesin (1990) studied the sorption of six monoaromatic hydrocarbons onto seven materials. The hydrocarbons included benzene, toluene, ethylbenzene, and o-, m-, and p-xylene. The materials examined were stainless steel, rigid PVC, flexible PVC, PTFE, polyvinylidene fluoride, fiberglass, and polyethylene. Stainless steel showed no significant sorption during an eight week period. All polymer materials sorbed all compounds to some extent. The order of sorption was as follows: rigid PVC < fiberglass < polyvinylidene fluoride < PTFE < polyethylene < flexible PVC. Stainless steel or brass liners should be used since they exhibit the least adsorption of VOC. Other materials such as PVC or acetate may be used provided that contact time between the soil and the liner material is kept to a minimum.

Liners also serve to make soil removal from the coring device much easier and quicker. The liner can run the entire length of the core or can be pre-cut into any desired length. Stainless steel and brass liners have been capped and shipped to the laboratory for sectioning and analysis. It is unlikely that capping will prevent losses of VOCs. Acetate liners are available, but samples should not be held in these for any extended period of time due to adsorption onto and permeation through the material. Alternatively, the entire contents of the liner section can be emptied into a wide-mouth glass jar. This method causes significant losses of VOCs and is not recommended. Smaller aliquots can be taken from the center of the pre-cut liner using subcoring devices and the soil plug extruded into VOA vials. Sample transfer methods are perhaps the most critical and least understood step in the sampling and analysis procedure and are discussed in greater detail in the next section.

TRANSFER OF SOIL SAMPLES FROM DEVICE TO CONTAINER

The key point in sample transfer, whether in the field or in the laboratory, is to minimize disturbance and the amount of time the sample is exposed to the atmosphere. It is more prudent to transfer the sample rapidly to the container than to accurately weigh an aliquot, spend considerable time reducing headspace, or homogenizing the sample in the laboratory. A device that can obtain the approximate aliquot size required by the analytical laboratory and can efficiently transfer that aliquot directly into the container is recommended. Several designs are available for obtaining various aliquot sizes. Most subcoring devices consist of a plunger/barrel design with an open end. The device shown in Figure 25.3 was constructed of two cork borers, with the smaller cork borer having a solid tip for extruding the sample. Other designs include syringes with the tips removed and commercial sub-corers such as the one illustrated in Figure 25.4. The subcoring device is inserted into the bulk sample and an aliquot withdrawn. The aliquot is of a known volume and approximate weight. The aliquot can then be extruded into a tared container in such a way that it need not be disturbed until analysis. A subsample can be taken for percent moisture determination. A commonly used method would be to extrude the sample from the subcoring device into a 40-mL VOA vial sealed with a Teflon-lined septum cap. However, Teflon may be permeable to VOC; therefore an impermeable (aluminum-lined) cap is recommended, followed by purging of the entire contents of the vial (Lewis et al., 1991). A commercially available modified purge-and-trap cap (Figure 25.5) can be used to seal a 40-mL VOA vial. The modified cap seals the VOA vial in the field and is attached directly to a Tekmar purge-and-trap device without ever being opened to the ambient air.

Some researchers have used liners for shipping samples to the analytical laboratory. The ends of the liner or pre-cut sections of the liner are sealed with plastic endcaps, wrapped with aluminum foil, or sealed with paraffin. It is doubtful whether an airtight seal can be achieved with such materials. Additionally, plastic and paraffin are permeable to VOCs.

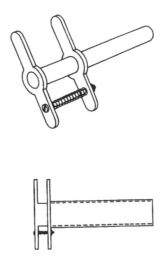

Figure 25.3.   Illustration of a hand-held sub-coring device made from two cork borers (taken from Lewis et al., 1991).

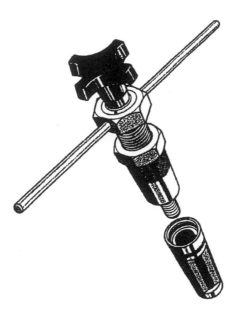

Figure 25.4. Small-diameter hand-held subcoring device made by Associated Design & Manufacturing Company (Alexandria, VA) (taken from Lewis et al., 1991).

Use of subcoring devices should produce analytical results of increased accuracy. In order to test this hypothesis, an experiment was conducted in which a bulk soil sample was spiked with 800 μg/kg of various VOCs (Maskarinec et al., 1988). Three aliquots were withdrawn by scooping, and three aliquots were withdrawn using a subcorer. The results are presented in Table 25.8. While neither method produced quantitative recovery, the subcorer approach produced results which were generally five times higher than the standard SW-846, Method 8240 approach (i.e., the contents of a 125-mL wide-mouth jar are poured into an aluminum tray, homogenized with a stainless steel spatula, and a 5-g sample placed in the sparger tube). Several compounds were problematic with both approaches, including styrene which polymerizes, bromoform which purges poorly, and 1,1,2,2-tetrachloroethane which degrades quickly (Maskarinec et al., 1988).

In a similar study (U.S. EPA, 1991), a large quantity of well characterized soil was spiked with 33 VOCs and homogenized. From the homogenized material 5-g aliquots of soil were placed in 40-mL VOA vials and sealed with modified purge-and-trap caps. The remaining soil was placed in 125-mL wide-mouth jars. To simulate actual sample handling procedures, each handling treatment was shipped via air carrier and analyzed by GC/MS with heated purge-and -trap. The 40-mL VOA vials were connected directly to a Tekmar sparger without exposure to the atmosphere. The wide-mouth jars were processed as per SW-846 Method 8240 specifications. Table 25.9 compares the results of the GC/MS analyses using the two pretreatment techniques. The modified method (40-mL VOA vial with modified cap) yielded consistently higher VOC concentrations than the traditional Method 8240 procedure (U.S. EPA, 1986).

Lewis et al. (1990) observed significant losses of VOCs when samples were homogenized as per Method 8240 specifications. Less loss was observed from small aliquots (1 to 5 g) placed in 40-mL VOA vials, sealed with a modified purge-and-trap

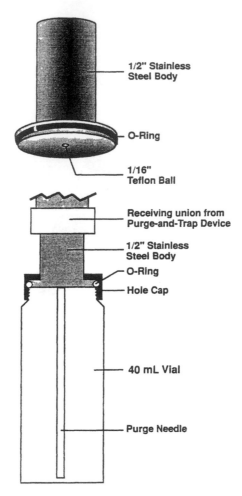

Figure 25.5. Modified purge-and-trap 40-mL VOA vial cap for containerizing samples in the field. Vial is attached directly to a Tekmar purge-and-trap system without exposure of sample to the atmosphere (taken from Lewis et al., 1991).

cap, and directly attached to a purge-and-trap device. In this case, a smaller sample size exhibited greater precision and accuracy. The MCL is dependent on precision and accuracy in sampling and analysis, and the relative analyte concentration. Relative error may be reduced by either increasing the number of samples or the mass of the sample. For instance, if 1- to 5-g aliquots collected in the field exhibit unacceptable errors, a 100-g sample will yield less error (Lewis et al., 1991). However, if the laboratory requires that a 1- to 5-g subsample be removed from the 100-g sample in a wide-mouth jar by opening the container and homogenizing the contents (as per SW-846, Method 8240), then the collection of a larger sample in the field will not reduce total measurement error because the subsampling procedures will compound these errors, as shown by Lewis et al. (1990).

Another option is to transfer the sample into a glass jar containing a known volume of methanol, usually in a ratio of one to one (volume/volume). This has the effect of preserving the volatile components of the sample at the time of containeri-

Table 25.8. Laboratory Comparison of Standard Method and Subcorer Method

Compound	Standard Method[a]	Subcorer Method[b]	Standard Method % Recovery of Spike	Subcorer % Recovery of Spike
Chloromethane	50	1225	6	153
Bromomethane	31	536	4	67
Chloroethane	78	946	10	118
1,1-Dichloroethene	82	655	10	82
1,1-Dichloroethane	171	739	21	92
Chloroform	158	534	20	67
Carbon tetrachloride	125	658	16	82
1,2-Dichloropropane	147	766	18	96
Trichloroethene	120	512	15	64
Benzene	170	636	21	80
1,1,2-Trichloroethane	78	477	10	60
Bromoform	30	170	4	21
1,1,2,2-Tetrachloroethane	46	271	6	34
Toluene	129	656	16	82
Chlorobenzene	57	298	7	37
Ethylbenzene	68	332	8	42
Styrene	30	191	4	24

[a] μg/kg (n = 3).
[b] μg/kg (n = 3).
Note: Standard method of sample transfer consists of scooping and subcorer method uses devices shown in Figure 25.3. Soil samples were spiked with 800 μg/kg of each VOC.

zation. Furthermore, surrogate compounds can be added at this time in order to identify possible changes in the sample during transport and storage. The addition of methanol to the sample raises the detection limits from 5–10 μg/kg to 500–1000 μg/kg, because of the attendant dilution. However, the resulting data have been shown to be more representative of the original content of the soil (Siegrist and Jennsen, 1990; Siegrist, 1990). In a comparison of transfer techniques, Siegrist and Jennsen (1990) demonstrated that minimum losses were obtained using an undisturbed sample followed by immediate immersion into methanol. The results for six VOCs are shown in Figure 25.6. At high VOC spike levels (mg/kg) the investigators found that headspace within the bottle caused a decrease in the concentration of VOCs in the sample. In another study, at lower spike levels, headspace did not seem to be a major contributor to VOC losses (Maskarinec et al., 1988). Lewis et al. (1990) found that a 5-g sample collected from a soil core and placed in a 40-mL VOA vial provided consistently higher VOC levels than a sample taken from the same core, placed in a 125-mL wide-mouth jar, homogenized in the laboratory, and a 5-g aliquot taken from the bulk as per Method 8240 specifications.

Also, the possibility exists of compositing soil samples in methanol in the field to improve precision and reduce variance.

The standard method for VOC analysis (SW-846, Method 8240) calls for the use of 125-mL wide-mouth jars. However, as previously described, these may not be the most appropriate containers. If, however, samples are collected in such containers, it is important to assure sample integrity as much as possible. Integrity is best

Table 25.9.   Mean and Statistical Analysis for Samples Analyzed by Method 8240 and Modified Method 8240 (Modified from U.S. EPA 1991)

VOC	Concentration (μg/kg)		
	Method 8240[a]	Modified Method 8240[b]	Difference
Bromomethane	9.30	44.4	35.1[e]
Vinyl Chloride	3.27	32.0	28.7[e]
Chloroethane	6.00	36.2	30.2[e]
Methylene Chloride	69.4	100.0	30.9[c]
Carbon Disulfide	32.3	82.7	50.4[c]
1,1-Dichloroethene	11.7	35.4	23.7[e]
1,1-Dichloroethane	34.2	82.9	48.7[c]
1,2-Dichloroethene	36.7	66.8	30.0[c]
Chloroform	54.6	95.6	41.1[c]
1,1,1-Trichloroethane	26.0	79.8	53.8[c]
Carbon Tetrachloride	18.4	61.1	42.7[c]
Vinyl Acetate	17.6	26.3	8.74
1,2-Dichloroethane	101.0	159.0	57.3[c]
1,2-Dichloropropane	79.7	152.0	72.0[c]
Cis-1,3-Dichloropropene	136.0	189.0	53.5[d]
Trichloroethene	48.2	87.4	39.2[c]
Benzene	55.6	114.0	55.8[d]
Bromodichloromethane	111.0	166.0	54.8[d]
Dibromochloromethane	121.0	159.0	37.8
1,1,2-Trichloroethane	142.0	193.0	51.8
Trans-1,3-Dichloropropene	154.0	203.0	48.6
Bromoform	116.0	140.0	24.6
Tetrachloroethene	61.5	124.0	62.5[c]
1,1,2,2-Tetrachloroethane	137.0	162.0	25.1
Toluene	85.4	161.0	75.7[d]
Chlorobenzene	91.1	132.0	41.3[c]
Ethylbenzene	85.2	135.0	49.7[c]
Styrene	86.4	114.0	28.1[d]
Total Xylenes	57.0	85.2	28.2[c]
Ketones			
Acetone	336.0	497.0	161.0[d]
2-Butanone	290.0	365.0	74.6
2-Hexanone	200.0	215.0	15.4
4-Methyl-2-Pentanone	264.0	288.0	23.5

[a]Method 8240 using 125-mL wide-mouth jar, mixing 5-g subsampling in laboratory, purge/trap analysis.
[b]Method 8240 using 40-mL VOA vial, 5-g sampled in the field, shipped to laboratory, purge/trap analysis.
[c]Difference is significantly greater than 0, with P-value <0.01.
[d]Difference is significantly greater than 0, with P-value between 0.01 and 0.05.
[e]Difference is significantly greater than 0, with P-value <0.01; however, data set contains zeros and makes results suspect.
[f]Difference is signficantly greater than 0, with P-value between 0.01 and 0.05; however, data set contains zeros and makes results suspect.
Note: Values based on duplicates.

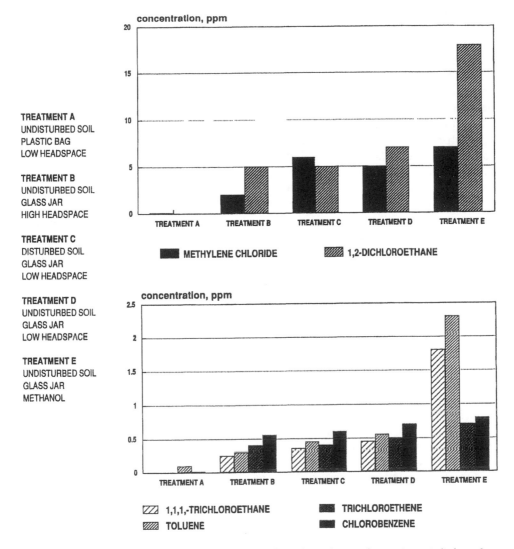

Figure 25.6. Volatile organic recovery as a function of sample treatment (taken from Lewis et al., 1991).

assured by using amber glass jars with phenolic resin caps with a foam-backed Teflon liner. Immersion into methanol will also improve VOC integrity in wide-mouth jars. Aluminum-lined caps are not available for the wide-mouth jars. Soil should be wiped from the threads of the jar to ensure a tight seal. A second recommended sample handling procedure is to collect a subsample from a coring device using a subcorer, and extrude the soil plug into a 40-mL VOA, and seal the vial with either an aluminum-lined cap or a modified purge-and-trap cap that allows direct connection to the purge-and-trap device (Lewis et al., 1991).

Field Storage and Preservation

It is important to remember that any material containing volatile organic compounds should be kept away from the sample and the sample container. Hand lotion, labeling tape, adhesives, and ink from waterproof pens contain VOCs which are often analytes of interest in the sample. Samples and storage containers should be kept away from vehicle exhaust. Any source of VOCs may cause contamination which may compromise the resulting data.

Two principal matrix-specific factors that can contribute to the loss of VOCs in soils are biodegradation and volatilization. Once samples are removed from the sampling device and placed in the appropriate storage container, appropriate preservation techniques are critical to control losses due to the aforementioned factors. Typically, the containers are placed in the dark at reduced temperatures (4°C). However, recent studies have shown that these preservation steps may not be adequate, at least for petroleum hydrocarbons. King (1992) evaluated 5 different preservation techniques for petroleum hydrocarbons in a sand matrix. The 5 hydrocarbon-sand mix preservation treatments were as follows:

1. held in brass tubes without refrigeration
2. held in brass tubes with refrigeration at 4°C
3. held in brass tubes and stored on dry ice
4. preserved in 40-mL VOA vials with 5-mL of methanol at room temperature
5. preserved in 40-mL VOA vials with 5-mL of methanol at 4°C

Six randomly selected samples from each of the 5 treatments were analyzed at 0, 3, 6, 10, and 14 days. Fourteen days is the widely accepted maximum allowable holding time for VOCs in soil. The results showed that in Treatment 1, gasoline concentrations decreased from 100 ppm to 20 ppm in 6 days (Figure 25.7). In Treatment 2, concentrations decreased from 135 ppm to 32 ppm within 6 days (Figure 25.8). Samples preserved using the remaining 3 treatments did not show any

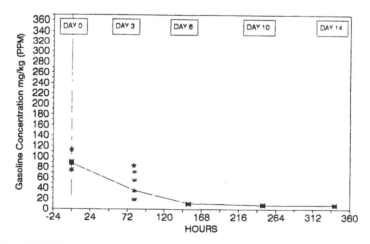

Figure 25.7.   Stability of gasoline concentrations in a sand matrix held in brass tubes at room temperature. Asterisks are sample results and triangles are batch medians (taken from King, 1992).

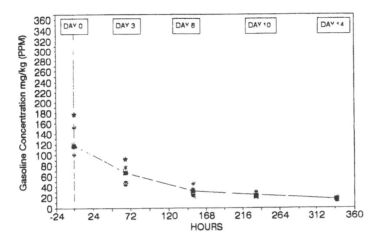

Figure 25.8. Stability of gasoline concentrations in a sand matrix held in brass tubes at 4°C. Asterisks are sample results and triangles are batch medians (taken from King, 1992).

deterioration exceeding the precision of the analytical method. King's (1992) results persuasively illustrate the inadequacy of EPA's preservation and holding time requirements for VOCs in soil.

In contrast to King's dry ice treatment, Maskarinec et al. (1988) state that excessively cold temperatures (<–10°C) should be avoided because greater losses of analytes may occur due to reduced pressure in the container, sublimation of water, and concomitant release of water-soluble VOCs into the headspace. Upon opening the container, the vacuum is quickly replaced with ambient air, thus purging out VOCs from the headspace (Maskarinec et al., 1988). However, at temperatures above freezing, bacterial action can have a significant impact on the observed soil VOC concentration.

Reduced temperature operates to control both biodegradation and volatilization. Some microorganisms, however, such as fungi, are biologically active even at 4°C. Wolf et al. (1989) investigated several methods (chemical and irradiation) for sterilizing soil. The researchers concluded that mercuric chloride is one of the most effective preservatives that cause minimal changes to the physical and chemical properties of the soil. Stuart et al. (1990) used mercuric chloride to preserve groundwater samples contaminated by petroleum hydrocarbons. In another study (U.S. EPA, 1991) researchers found that the amount of mercuric chloride needed to reduce biodegradation was directly related to the soil's organic carbon content. Typically, 2.5 mg of mercuric chloride were added per 5 g of soil. Addition of mercuric chloride may, however, introduce an additional hazardous waste disposal problem at the analytical laboratory.

Besides reducing the temperature of the sample, other preservation techniques have been evaluated for reducing the volatilization of VOCs from soil. Solid adsorbents, anhydrous salts, and water/methanol extraction mixtures have been evaluated (U.S. EPA, 1991). The most efficient preservatives for reducing volatilization were found to be two solid adsorbents, Molecular Sieve—5A™ (aluminum silicate desiccant) and Florasil™ (magnesium silicate desiccant). The addition of 0.2 mg per 5 g of soil greatly increased the recovery of spiked VOCs. The mechanism is believed to involve the displacement of water from binding sites on the soil particles, allowing VOCs to bind to the freed sites.

Other solid adsorbents that have been shown to be highly effective in binding VOCs are alkyammonium substituted montmorillonites and bentonites. These materials have been used as landfill liners for many years to reduce the migration of VOCs and other organics into ground water. Harper and Purnell (1989) have found that these materials were very effective in both thermal- and solvent-desorbable sampling systems. To date, these materials have not been evaluated as preservatives for VOCs in soils.

Shipping

Given the short holding times required for VOC analysis under SW-846 (U.S. EPA, 1986) (14 days from sample collection to analysis), samples are usually shipped via air carrier to the analytical laboratory. Samples should be well packed and padded to prevent breakage. Temperatures in cargo holds can increase to over 50°C during transit. The need for adequate cold storage is critical. Styrofoam coolers are commercially available to accommodate 40-mL and 125-mL size glass containers. Sufficient quantities of Blue Ice™ or Freeze-Gel™ packs should be placed in the shipping container to ensure that all samples are cooled for the duration of the shipment. Temperatures in ice chests are not uniform throughout; therefore, it is imperative that the samples are completely surrounded on all sides by cooling materials. A maximum-minimum thermometer (nonmercury) should be shipped with the samples. As mentioned previously, certain containers are more prone to leakage than others. This problem may be exacerbated by the reduced atmospheric pressures encountered in the cargo holds of various air carriers. High temperatures and low ambient pressures are the worst possible set of conditions for maintaining VOCs in containerized soil samples. Figure 25.9 illustrates the changes in temperature and pressure in the cargo hold of various air carriers. Lewis et al. (1990) noted decreases in VOC concentrations in soil samples that were shipped compared to samples that were analyzed in the field. This decrease may have been due in part to increased temperatures and decreased pressures in the cargo holds on the aircraft. If the container is of questionable or unknown integrity, it should either be evaluated prior to use or a previously characterized container should be used.

Samples that are immersed in methanol have special shipping requirements. These samples must be shipped as "Flammable Liquids" under the Department of Transportation (DOT) regulations (49 CFR, 1982). A secondary container is required for shipment of any item classified as a flammable liquid.

Blanks

For the most part, the reason for making a wrong decision based on soil VOC data stems from negative bias and/or false negatives. However, false positives may occur. Since soil is an efficient adsorbent for VOCs, it is possible under laboratory conditions for large concentrations to be reported. For example, a Contract Laboratory Program (CLP) laboratory reported as much as 0.3% of methylene chloride in soils collected at a Superfund site. The result was a laboratory artifact (Urban et al., 1989).

This example serves to stress the importance of trip blanks in VOC analyses. In

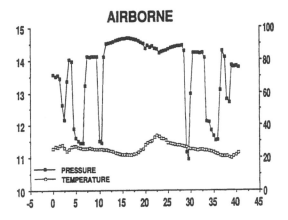

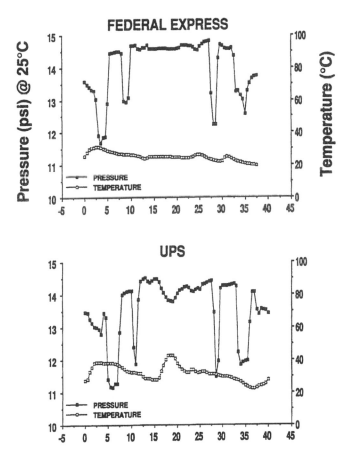

Figure 25.9.   Temperature and pressure fluctuations recorded in the cargo hold of various air carriers. Recording devices were shipped from Las Vegas, NV to Pearl River, NY and back (Lewis, unpublished data).

many cases, however, an inappropriate matrix is used as a trip blank. If a sample consists of a silty clay loam soil, a trip blank consisting of washed sand or distilled water would not be appropriate, because it would not retain VOCs in the event of cross contamination. The trip blank soil matrix should be similar in adsorption capacity to the actual sample. In addition, high- and low-level samples should be shipped in separate coolers in containers sealed with aluminum-lined caps (Supelco). Teflon-lined caps may permit permeation of VOCs through the Teflon (Lewis et al., 1991).

FIELD-SCREENING PROCEDURES

In the initial phases of a site investigation, field analytical techniques are useful for determining relative differences in VOC concentrations. Field screening procedures are gaining acceptance in the scientific community and are beginning to approach the precision and accuracy of laboratory-based methods. A variety of soil-VOC screening procedures have been used, including headspace analysis of soils using organic vapor analyzers, water (or NaCl-saturated water) extraction of soil followed by static headspace analysis using an organic vapor analyzer (OVA) or GC, colorimetric test kits, methanol extraction followed by headspace analysis or direct injection into a GC, and soil-gas sampling. Unfortunately, these procedures often are not well documented, and their effectiveness is not known. However, the procedures have been and continue to be used. The potential limitations and benefits must be carefully weighed and compared to the DQO.

Concerned about the large number of samples that are routinely shipped to CLP laboratories having below-detection levels of VOCs, Crockett and DeHaan (1991) evaluated six different soil extraction/headspace analysis treatments. They concluded that several procedures should be evaluated and compared using site-specific contaminated soils. A standard operating procedure should be defined that allows the field chemist to select the best screening procedure for its intended use. The authors pointed out that what is lacking is a good database showing the comparability between various field screening procedures and approved laboratory-based methods.

A recent study has provided these necessary data for comparing field screening and laboratory-based methods. Hewitt et al. (1992) conducted an experiment comparing aqueous extraction-headspace gas chromatography (HS/GC) and purge-and-trap gas chromatography mass spectroscopy (PT/GC/MS) measurements of VOCs in fortified soils. Soils (1 to 2 g) were placed in 40-mL VOA vials and fumigated in a vapor chamber for 4 and 39 days to generate a homogeneous starting material at low and high level concentrations, respectively. The aqueous headspace procedure consisted of addition of 30-mL of deionized water to the VOA vial, capping the container, shaking by hand for 2 min, and sampling the headspace with an airtight syringe. The purge-and-trap method followed EPA SW-846 Method 8240 (U.S. EPA, 1986). They found that the HS/GC method provided results that were not significantly different from the PT/GC/MS procedure. In conjunction with sound sampling and sample handling protocols, the HS/GC method may provide field-generated data that are as representative as laboratory-based PT/GC/MS results.

LABORATORY PROCEDURES

Sample Storage

Most regulatory procedures specify storage of samples for volatile organic analysis at 4°C in the dark. Sample coolers should be opened under chain-of-custody conditions and the temperature inside the cooler should be verified and recorded. Samples should be transferred to controlled-temperature (4°C) refrigerators until analysis. In many cases, insufficient cooling is provided during transport. In these cases, data quality may be compromised. However, as demonstrated by King (1992), soil samples immersed in methanol may be stable indefinitely, even at room temperatures.

Sample Preparation

The two most commonly used methods that satisfy regulatory requirements for the analysis of soil samples for VOCs are direct purge-and-trap and methanol extraction. As pointed out earlier, the use of a modified purge-and-trap cap is highly recommended if purge-and-trap procedures are to be used at the analytical laboratory. The advantages and disadvantages of the modified purge-and-trap procedure and the methanol immersion procedure are summarized in Table 25.10.

The analytical methods that can be used for the analysis of soils for VOCs are summarized in Table 25.11. An analytical method should be selected that is compatible with the recommended sample collection and containerizing procedures discussed above.

The efficiency of many methods for analyzing VOCs in soils is generally predicated on a high recovery of added compound (spike recovery) following short equilibration times (usually less than 24 hr). Few studies have examined recovery efficiency from real-world samples; i.e., where equilibration times could be on the order of years. With aging, hydrophobic organic compounds become increasingly harder to desorb (Di Toro and Horzempa, 1982; Karickhoff, 1980, 1984; Wu and Gschwend, 1986). Many methods, e.g., EPA Method 624, have shown nearly 100% recoveries of VOCs from waters. However, low recoveries in solid and semisolid matrices have been observed (Hiatt, 1981; Spraggins et al., 1981). Spraggins et al. (1981) used increasingly higher temperatures in a modified purge-and-trap system, and observed a progressive increase in the recovery of benzene, toluene, ethylbenzene, and xylene (BTEX) from 50° to 150°C. It is not known whether 100% recovery was achieved even at the highest temperature. The U.S. EPA recommended method for recovering volatiles from solid matrices calls for purging a heated (40°C) suspension of the sample for 11 min (US EPA, 1986). Based on the findings of Spraggins et al. (1981) the EPA-recommended temperature and purging times may be inadequate for complete recovery of VOCs from soils.

Sawhney et al. (1988) evaluated three common recovery methods using three different "real-world" soils contaminated with 1,2-dibromoethane (EDB). The three recovery methods evaluated were heated purge-and-trap, thermal desorption, and solvent extraction. Poor recoveries were seen for the heated purge-and-trap procedure, even at higher and longer purge times (Table 25.12). The thermal desorption method was performed by passing heated N_2 gas through soil columns up to 200°C. Less than 0.8% of the total EDB [total based on the recommended method (RM),

Table 25.10.   Advantages and Disadvantages of the Modified Purge-and-Trap and Methanol Immersion Procedures

Purge-and-Trap

Advantages

- Preserves VOC concentrations at levels comparable to those found in the field, i.e., no dilution.
- No special shipping precautions are required.
- Surrogate addition can be made to the soil in the field.
- Homogenization of the contents of the 40-mL VOA vial is not necessary as in the case of wide-mouth jars. The container is not opened to the atmosphere until it is attached to the purge-and-trap unit.
- The collection of a 5-g aliquot in the field can be performed. The field technician can more efficiently collect a representative sample than a laboratory analyst.

Disadvantages

- May be more susceptible to short-range spatial variability
- Multiple analyses of the same sample are not possible.
- Undiluted samples may be above the calibration range.

Methanol Immersion

Advantages

- Reduces losses of VOCs due to their solubility in methanol.
- Permits multiple analyses of the methanol extract.
- Surrogates can be spiked into the methanol prior to sample immersion.
- Permits compositing of several subsamples into one jar.
- Reduces biodegradation of VOCs.
- Larger sample may be less prone to short-range spatial variability.

Disadvantages

- Raises the detection limit due to attendant dilution in methanol.
- Methanol may be more susceptible to cross contamination.
- Requires special shipping procedures.

---

i.e., solvent extraction] was recovered, even at the highest temperatures. Post-thermally-desorbed soil was extracted with the RM, and 79% was found at the 100°C desorption temperature and 3.6% of the total at 200°C desorption temperature. The poorer recovery at the highest temperature suggests that EDB may have thermally decomposed. An earlier study by Steinberg et al. (1987) showed the slow release of EDB from soil into the aqueous phase, indicating a highly tortuous and sterically hindered path of diffusion. The retardation in the diffusion of EDB due to the aforementioned factors may have contributed to the thermal decomposition at the higher temperature. Perhaps the addition of a deflocculant would quicken the release of EDB, thereby reducing or preventing thermal decomposition at higher temperatures. Sawhney et al. (1988) defined the solvent extraction as the recommended method (i.e., 100% recovery) in their experiments because it consistently exhibited greater recoveries, especially when heated. In addition to ethanol, the researchers also evaluated some other commonly used extraction solvents; namely, hexane, acetonitrile, and acetone. Among the four solvents used, hexane was the least effective; in a 24-hr extraction at 20°C, only 3% EDB was recovered, as compared to 60%, 65%, and 78% recovered by methanol, acetone, and acetonitrile, respectively

Table 25.11. Methods for VOC Analysis of Soil

Method Extraction/Analysis	Sample Size (g)	Sample Preparation Procedure	Sensitivity (mg/kg)	Data Quality	Program	Comments
5030/8240 /8010 /8015 /8020 /8030 /8260 /3810	5	purge & trap	5–10	litigation	RCRA[a]	Sample transfer to purge & trap is critical.
3580/8240 /8010 /8015 /8020 /8030 /8260	5–100	methanol extraction (1:1 dilution)	500–1000	litigation	RCRA	Sensitivity loss but sample transfer facilitated.
5031/8240 /8010 /8015 /8020 /8260	5	field purge	5–10	semi-quantitative	RCRA	Sample can only be analyzed once, transfer & shipping facilitated.
3810/8240 /8010 /8015 /8020 /8260	5–100 (usu. 10)	Heat to 100°C in water and analyze headspace	1000	screening for purgeable organics	RCRA	Can be performed in the field.
3820	10	hexadecane extraction followed by GC/FID	500–1000	screening prior to GC or GC/MS analysis	RCRA	FID response varies with type of VOC.
624	5	purge & trap	5–10	litigation	CLP[b]	Similar to method 5030/8240 in RCRA SW–846.

Source: Taken from Lewis et al., 1991.
[a]U.S. EPA, 1986.
[b]U.S. EPA, 1982.

Table 25.12.   Release of EDB from Fumigated Soils by Purge-and-Trap Method

Temp. (°C)	Purge Time (min)	Purge No.	Amt. Released by Each Purge	
			(µg/kg)	(% of total[a])
		Cheshire fine sandy loam		
40	11[b]	1st	1.2	1.3
40	11[b]	2nd successive	0.9	1.0
40	30	3rd successive	3.2	3.6
80	30	4th successive	8.5	9.6
80	30	5th successive	9.1	10.2
		Agawam fine sandy loam		
80	11	1st	5.3	4.3
80	11	2nd successive	4.7	3.8

Source: Taken from Sawhney et al., 1988.
[a]Based on average values of EDB determined by recommended method.
[b]Conditions similar to EPA SW–846 Method 8240.

(Table 25.13). Sawhney et al. (1988) recommend methanol as the solvent of choice because of its broad applicability to the widely used purge-and-trap GC methods for analyses of VOCs, and its low hydrolyzation (3 ± 1%) under the conditions of their extraction [i.e., 24 hr at 75°C in 9:1 (v/v) methanol/water mixture].

Thus, the ASTM recommendation for using methanol immersion is justified. The

Table 25.13.   Extraction of EDB from the Cheshire Fine Sandy Loam Soil by Different Solvents

Treatment	Time (h)	Relative Recovery[a] (%)
1. Methanol, 75°C	2 × 24	(100)
	24	94
	16	93
	4	69
2. Methanol, 20°C	48	68
	24	60
	4	28
3. Hexane, 75°C	24	59
4. Hexane, 20°C	24	3
5. 1:1 Hexane/H$_2$O, reflux	4	56
6. Acetonitrile, 75°C	24	114
7. Acetonitrile, 20°C	24	78
8. Acetone, 75°C	24	113
9. Acetone, 20°C	24	65
	4	37

Source: Taken from Sawhney et al., 1988.
[a]Average of at least two determinations based on recovery by two 24 hr extractions with methanol at 75°C as 100%.

methanol not only retards biodegradation and reduces volatilization, it can also be heated to further improve recoveries.

CONCLUSIONS AND RECOMMENDATIONS

A suite of sampling approaches is currently used for collecting soil samples for volatile organic analyses. Sampling device selection will be site-specific, and no single device can be recommended for use at all sites. However, a number of different samplers which cover a broad range of sampling conditions and circumstances are recommended (Table 25.7). Proper use of a sampling device is an important factor for maintaining sample integrity. Transfer of the sample from the sampling device to the container is a critical step in the process. Liners are useful in making removal and sampling of soil from the collection device more efficient. The use of subcoring devices is recommended. Collection of the appropriate sample size in the field for a particular analytical procedure is optimal. Sample containers vary in terms of airtightness. Data are available to indicate that there is a decrease in pressure and an increase in temperature in the cargo holds of certain air carriers. This is the worst possible set of conditions for maintaining VOCs in containerized soil samples. Intact seals on storage containers and adequate cooling is thus critical for maintaining VOCs in soil samples. The 40-mL VOA vials seem to maintain the best seals. Commercially available shipping packages with built-in cooling materials (e.g., Freeze Gel packs or Blue Ice) are available. Based on current research data, the 125-mL wide-mouth jars commonly used and specified in Method 8240 are not recommended for use unless used in conjunction with the methanol extraction procedure. Whenever possible, an integrated sampling approach should be employed to obtain the most representative samples possible. Soil-gas surveying coupled with onsite soil sampling and analyses, followed by RCRA or CLP laboratory analyses, will provide data on the partitioning of VOC in three phases; gaseous, liquid, and solid. Knowledge of phase partitioning is critical for determining the extent and nature of migration and estimating the populations at risk.

REFERENCES

Acker, W.L., Basic Procedures for Soil Sampling and Core Drilling, Acker Drill Co., Inc., Scranton, PA, 1974.

Armson, K.A., *Forest Soils: Properties and Processes*, (Toronto: University of Toronto Press, 1977).

Arneth, J.-D., G. Milde, H. Kerndorff, and R. Schleyer, "Waste Deposit Influences on Groundwater Quality as a Tool for Waste Type and Site Selection for Final Storage Quality," in P. Baccini, Ed., *The Landfill: Reactor and Final Storage, Swiss Workshop on Land Disposal of Solid Wastes*, Gerzensee, Mar. 14–17, 1988, pp. 399–415.

Barcelona, M.J., "Overview of the Sampling Process," in Keith, L.H., Ed. *Principles of Environmental Sampling*, (Washington, DC: American Chemical Society, 1989), pp. 3–23.

Barth, D.S., B.J. Mason, T.H. Starks, and K.W. Brown, Soil Sampling Quality Assurance User's Guide (2nd Edition), EPA 600/8-89/046, U.S. EPA, EMSL-LV, Las Vegas, NV, March, 1989.

Black, C.A., Methods of Soil Analysis, American Society of Agronomy, Inc., Madison, WI, 1965, pp. 1–72.

Boucher, F.R., and G.F. Lee., "Adsorption of Lindane and Dieldrin Pesticides on Unconsolidated Aquifer Sands," *Env. Sci. Tech.*, 6:538–543 (1972).

Bouwer, E.J., "Biotransformations of Organic Micropollutants in the Subsurface," in *Petroleum Hydrocarbons and Organic Chemicals in Ground Water*, National Water Well Association, Dublin, OH, 1984.

Breckenridge, R.P., J.R. Williams, and J.F. Keck, Characterizing Soils for Hazardous Waste Site Assessments, EPA/540/4–91/001, Ground Water Forum Issue, U.S. EPA Environmental Monitoring Systems Laboratory, Las Vegas, NV, 1991.

Brusseau, M.L., R.E. Jessup, and P.S.C. Rao, "Nonequilibrium Sorption of Organic Chemicals: Elucidation of Rate-Limiting Processes," *Env. Sci. Tech.*, 25:134–142 (1991a).

Brusseau, M.L., T. Larsen, and T.H. Christensen, "Rate-Limited Sorption and Nonequilibrium Transport of Organic Chemicals in Low Organic Carbon Aquifer Materials," *Wat. Resour. Res.*, 27(6):1137–1145 (1991b).

Cameron, R.E., "Algae of Southern Arizona—Part 1, Introduction to Blue-Green Algae," *Rev. Alg. N.S.*, 6(4):282–318 (1963).

Campbell, G.S., *Soil Physics with BASIC*, (New York: Elsevier, 1985).

Chiou, C.T., "Partition and Adsorption on Soil and Mobility of Organic Pollutants and Pesticides," in Z. Gerstl, Y. Chen, U. Mingelgrin, and B. Yaron, Eds., *Toxic Organic Chemicals in Porous Media*, Chapter 7, (New York: Springer-Verlag, 1989), pp. 163–175.

Chiou, C.T., D.E. Kile, and R.L. Malcolm, "Sorption of Vapors of Some Organic Liquids on Soil Humic Acid and Its Relation to Partitioning of Organic Compounds in Soil Organic Matter," *Env. Sci. Tech.*, 22(3):298–303 (1988).

Chiou, C.T., and T.D. Shoup, "Soil Sorption of Organic Vapors and Effects of Humidity on Sorptive Mechanism and Capacity," *Env. Sci. Tech.*, 19(12):1196–1200 (1985).

Crockett, A.B., and M.S. DeHaan, "Field Screening Procedures for Determining the Presence of Volatile Organic Compounds in Soil," in *Proceedings of the Second International Symposium on Field Screening Methods for Hazardous Wastes and Toxic Chemicals*, February 12–14, 1991, Las Vegas, NV, 1991.

Dean, J.D., P.S. Huyakorn, A.S. Donigan, Jr., K.A. Voos, R.W. Schanz, Y.J. Meeks, and R.F. Carsel, Risk of Unsaturated/Saturated Transport and Transformation of Chemical Concentrations (RUSTIC)—Volume I: Theory and Code Verification, EPA/600/3–89/048a, Environmental Research Laboratory, U.S. EPA, Athens, GA, 1989.

Di Toro, D.M., and L.M. Horzempa, "Reversible and Resistant Components of PCB Adsorption-Desorption: Isotherms," *Env. Sci. Tech.*, 16:594–602 (1982).

Dorrance, D.W., L.G. Wilson, L.G. Everett, and S.J. Cullen, "A Compendium of Soil Samplers for the Vadose Zone," in *Handbook of Vadose Zone Characterization & Monitoring* (Chelsea, MI: Lewis Publishers, 1995), pp. 401–428.

Erb, T.L., W.R. Phillipson, W.L. Teng, and T. Liang, "Analysis of Historic Airphotos," *Photogrammetric Eng. Remote Sensing*, 47(9):1–12 (1981).

Farmer, W.J., et al., Land Disposal of Hexachlorobenzene Wastes: Controlling Vapor Movement in Soils, EPA-600/2–80–119, U.S. EPA, Environmental Research Laboratory, Cincinnati, OH, August, 1980.

Fuller, W.H., and A.W. Warrick, *Soils in Waste Treatment and Utilization. Vol. 2. Pollutant Contaminant Monitoring and Closure*, (Boca Raton, FL: CRC Press, 1985).

Gillman, R.W., and S.F. O'Hannesin, "Sorption of Aromatic Hydrocarbons by Materials Used in Construction of Ground-Water Sampling Wells," in D.M. Nielsen and A.I. Johnson, Eds., *Ground Water and Vadose Zone Monitoring*, ASTM STP 1053, American Society for Testing and Materials, Philadelphia, PA, 1990, pp. 108–122.

Harper, M., and C.J. Purnell, "Alkylammonium Montmorillonites as Adsorbents for Organic Vapors from Air," *Env. Sci. Tech.*, 24(1):55–62 (1989).

Hewitt, A.D., P.H. Miyares, D.C. Leggett, and T.F. Jenkins, Aqueous Extraction-Headspace/Gas Chromatographic Method for Determination of Volatile Organic Compounds in Soil, U.S. Army Corps of Engineers, Cold Regions Research & Engineering Laboratory, CRREL Report 92-6, April, 1992.

Hiatt, M.T., "Analysis of Fish and Sediments for Volatile Priority Pollutants," *Anal. Chem.*, 53:1541–1543 (1981).

Homsher, M.T., F. Haeberer, P.J. Marsden, R.K. Mitchum, D. Neptune, and J. Warren, "Performance Based Criteria: A Panel Discussion: Part II.," *Environ. Lab.* Dec./Jan. 1991/1992, pp. 10–15.

Jamison, V.W., R.L. Raymond, and J.O. Hudson, "Biodegradation of High-Octane Gasoline," in *Proceedings of the Third International Biodegradation Symposium*, (London: Applied Science Publishers Ltd., 1975).

Johnson, R.E., "Degradation of DDT by Fungi," *Residue Rev.*, 61:1–28 (1976).

Jury, W.A., A User's Manual for the Environmental Fate Screening Model Programs BAM and BCM, Dept. Soil and Environ. Sci., Univ. of California, Riverside, CA, Submitted to the California Department of Health Services, 1984.

Karickhoff, S.W., "Sorption Kinetics of Hydrophobic Pollutants in Natural Sediments," in R.A. Baker, Ed., *Contaminants and Sediments*, Vol. 2, (Ann Arbor, MI: Ann Arbor Science Publishers, Inc., 1980), pp. 193–204.

Karickhoff, S.W., "Organic Pollutant Sorption in Aquatic Systems," *J. Hydraul. Div. Am. Soc. Chem. Eng.*, 110:707–735, 1984.

Karickhoff, S.W., D.S. Brown, and T.A. Scott, "Sorption of Hydrophobic Pollutants on Natural Sediments," *Water Res.*, 13:241–248 (1979).

Kerfoot, H.B., "Subsurface Partitioning of Volatile Organic Compounds: Effects of Temperature and Pore-Water Content," *Ground Water*, 29(5):678–684 (1991).

King, P.H., Evaluation of Sample Holding Times and Preservation Methods for Gasoline in Fine-grained Sand, Master's Thesis, University of San Francisco, San Francisco, CA, August, 1992.

Kleopfer, R.D. et al., "Anaerobic Degradation of Trichloroethylene in Soil," *Env. Sci. Tech.*, 19:277–284 (1985).

Kobayashi, H., and B.E. Rittman, "Microbial Removal of Hazardous Organic Compounds," *Env. Sci. Tech.*, 16:170A-183A (1982).

Lechevalier, H.A., and M.P. Lechevalier, "Actinomycetes Found in Sewage Treatment Plants of the Activated Sludge Type," in *Actinomycetes: The Boundary Microorganisms*, (Tokyo: Toppen Co., Ltd., 1976).

Lewis, T.E., A.B. Crockett, R.L. Siegrist, and K. Zarrabi, Soil Sampling and Analysis for

Volatile Organic Compounds, EPA/540/4-91/001, Ground Water Forum Issue, U.S. EPA, Las Vegas, NV, 1991.

Lewis, T.E., B.A. Deason, C.L. Gerlach, and D.W. Bottrell, "Performance Evaluation Materials for the Analysis of Volatile Organic Compounds in Soil: A Preliminary Assessment," *J. Env. Sci. Health*, A25(5):505–531 (1990).

Lion, L.W., S.K. Ong, S.R. Linder, J.L. Swanger, S.J. Schwager, and T.B. Culver, Sorption Equilibria of Vapor Phase Organic Pollutants on Unsaturated Soils and Soil Minerals, ESL-TR-90-05, Environics Division, Air Force Engineering & Services Center, Engineering and Service Laboratory, Tyndall Air Force Base, FL, 1990.

Lotse, E.G., D.A. Graetz, G. Chesters, G.B. Lee, and L.W. Newland, "Lindane Adsorption by Lake Sediments," *Env. Sci. Tech.*, 2:353–357 (1968).

Maskarinec, M.P., personal communication to Tim Lewis, 1990.

Maskarinec, M.P., L.H. Johnson, S.K. Holladay, "Preanalytical Holding Times," paper presented at the Quality Assurance in Environmental Measurements Meeting, U.S. Army Toxic and Hazardous Materials Agency, Baltimore, MD, May 25–26, 1988.

Mason, B.J., Preparation of Soil Sampling Protocol: Techniques and Strategies, EPA/600/4-83-020, Environmental Monitoring Systems Laboratory, U.S. EPA, Las Vegas, NV, 1983.

McCoy, D.E., " '301' Studies Provide Insight into Future of CERCLA," *The Hazardous Waste Consultant*, 3/2:18–24 (1985).

Ong, S.K., and L.W. Lion, "Trichloroethylene Vapor Sorption onto Soil Minerals," *Soil Sci. Soc. Am. J.*, 55:1559–1568 (1991).

Pitard, F.F., *Pierre Gy's Sampling Theory and Sampling Practice. Volume I: Heterogeneity and Sampling*, (Boca Raton, FL: CRC Press, Inc., 1989).

Plumb, R.H., Jr., "A Practical Alternative to the RCRA Organic Indicator Parameters," in T. Bursztynsky, Ed., *Proceedings of Hazmacon 87*, Santa Clara, CA, Apr. 21-23, 1987, pp. 135-150.

Plumb, R.H., Jr., and A.M. Pitchford, "Volatile Organic Scans: Implications for Ground Water Monitoring," paper presented at the National Water Well Association/American Petroleum Institute Conference on Petroleum Hydrocarbons and Organic Chemicals in Ground Water, Houston, TX, Nov. 13-15, 1985.

Richardson, E.M., and E. Epstein, "Retention of Three Insecticides on Different Size Soil Particles Suspended in Water," *Soil Sci. Soc. Am. Proc.*, 35:884-887 (1971).

Roy, W.R., and R.A. Griffin, "Mobility of Organic Solvents in Water-Saturated Soil Materials," *Environ. Geol. Wat. Sci.*, 7(4):241-247 (1985).

Sawhney, B.L., J.J. Pignatello, and S.M. Steinberg, "Determination of 1,2-Dibromoethane (EDB) in Field Soils: Implications for Volatile Organic Compounds," *J. Env. Qual.*, 17(1):149-152 (1988).

Shen, T.T., and G.H. Sewell, "Air Pollution Problems of Uncontrolled Hazardous Waste Sites," in *Proceedings of 1982 Superfund Conference, Hazardous Materials Control Research Institute*, Washington, DC, 1982, pp. 76-80.

Siegrist, R.L., "Measurement Error Potential and Control When Quantifying Volatile Hydrocarbon Concentrations in Soils," paper presented at the Fifth Annual Conference on Hydrocarbon Contaminated Soils, Univ. of Massachusetts, Amherst, MA, September 24-27, 1990.

Siegrist, R.L. "Volatile Organic Compounds in Contaminated Soils: The Nature and Validity of the Measurement Process," *J. Haz. Mat.* 29:3–15 (1992).

Siegrist, R.L., and P.D. Jennsen, "Evaluation of Sampling Method Effects on Volatile Organic Compound Measurements in Contaminated Soils," *Env. Sci. Tech.*, 24:1387–1392 (1990).

Smith, J.A., C.T. Chiou, J.A. Kammer, and D.E. Kile, "Effect of Soil Moisture on the Sorption of Trichloroethene Vapor to Vadose-Zone Soil at Picatinny Arsenal, New Jersey," *Env. Sci. Tech.*, 24:6576–683 (1990).

Spencer, W.F., M.M. Cliath, W.A. Jury, and L.-Z. Zhang, "Volatilization of Organic Chemicals from Soil as Related to Their Henry's Law Constants," *J. Env. Qual.*, 17(3):504–509 (1988).

Spraggins, R.L., R.G. Oldham, C.L. Prescott, and K.J. Baughman, "Organic Analyses Using High-Temperature Purge and Trap Techniques," in L.H. Keith, Ed., *Advances in the Identification and Analysis of Organic Pollutants in Water*, Vol. 2. (Ann Arbor, MI: Ann Arbor Science Publishers, Inc., 1981), pp. 747–761.

Stanley, T.W., and S.S. Verner, "The U.S. Environmental Protection Agency's Quality Assurance Program," in *Quality Assurance for Environmental Measurements*, ASTM STP 867, American Society for Testing and Materials, Philadelphia, PA, 1985, pp. 12–19.

Steinberg, S.M., J.J. Pignatello, and B.L. Sawhney, "Persistence of 1,2-Dibromomethane in Soils: Entrapment in Intraparticle Micropores," *Environmental Science & Technology*, 21:1201–1208 (1987).

Stuart, J.D., V.D. Roe, W.M. Nash, and G.A. Robbins, Manual headspace method to analyze for gasoline contamination of ground water by capillary column gas chromatography, Personal communication to Tim Lewis, 1990.

Urban, M.J., J.S. Smith, E.K. Schultz, and R.K. Dickinson et al. "Volatile Organic Analysis for Soil, Sediment, or Waste," in *Fifth Annual Waste Testing and Quality Assurance Symposium*, U.S. Environmental Protection Agency, Washington, DC, July 24–28, 1989, pp. II-87 to II-101.

U.S. EPA, Test Method 624 (Purgeables), Methods for Organic Chemical Analysis of Municipal and Industrial Wastes, EPA-600/4–82–057, U.S. EPA Environmental Support Laboratory, Cincinnati, OH, 1982.

U.S. EPA, Test Methods for Evaluating Solid Waste (SW-846), Method 8240, Office of Solid Waste and Emergency Response, (3rd Edition), 1986.

U.S. EPA, Investigation of Sample and Sample Handling Techniques for the Measurement of Volatile Organic Compounds in Soil, University of Nevada, Las Vegas, Submitted to U.S. EPA, Environmental Monitoring Systems Laboratory, Las Vegas, NV (in preparation), 1991.

U.S. EPA, Characterizing Heterogeneous Wastes: Methods and Recommendations, EPA-600/R-92/033, U.S. EPA Environmental Monitoring Systems Laboratory, Las Vegas, NV, 1992.

Van Ee, J.J., L.J. Blume, and T.H. Starks, Rationale for the Assessment of Errors in the Sampling of Soils, EPA/600/4–90/013, Office of Research and Development, Environmental Monitoring Systems Laboratory, Las Vegas, NV, 1990.

Vogel, T.M., and P.L. McCarty, "Biotransformations of Tetrachloroethylene to Trichloroethylene, Dichloroethylene, Vinyl Chloride, and Carbon Dioxide under Methanogenic Conditions," *Appl. Environ. Microbiol.*, 49:1080–1084 (1985).

Voice, T.C., and W.J. Weber, Jr., "Sorption of Hydrophobic Compounds by Sediments, Soils, and Suspended Solids—I. Theory and Background," *Wat. Res.*, 17(10):1433–1441 (1983).

Wilson, J.T., and B.H. Wilson, "Biotransformation of Trichloroethylene in Soil," *Appl. Environ. Microbiol.*, 49(1):242–243 (1985).

Wolf, D.C., T.H. Dao, H.D. Scott, and T.L. Lavy, "Influence of Sterilization Methods on Selected Soil Microbiological, Physical, and Chemical Properties," *J. Env. Qual.*, 18:39–44 (1989).

Wu, S., and P.M. Gschwend, "Sorption Kinetics of Hydrophobic Organic Compounds to Natural Sediments and Soils," *Env. Sci. Tech.*, 20:717–725 (1986).

Yaron, B., and S. Saltzman, "Influence of Water and Temperature on Adsorption of Parathion by Soils," *Soil Sci. Soc. Am. Proc.*, 36:583–586 (1972).

26

In Situ Pore-Liquid Sampling in the Vadose Zone*,**

L.G. Wilson, D.W. Dorrance, W.R. Bond, L.G. Everett, and S. J. Cullen

INTRODUCTION

Reasons for sampling pore-liquids in the vadose zone include determining contaminant concentration/mass trends in time and depth, detection of contaminants, and tracing contaminant flow paths (Starr et al., 1991). Pore-liquid samples can be obtained either by direct, in situ methods (e.g., tile lines, pan lysimeters, and wells, for saturated regions, and suction samplers for unsaturated regions), or by indirect methods, e.g., soil core samples. A major difference between these two approaches is that core sampling methods are destructive, i.e., repetitive samples cannot be obtained from the same location. More importantly, as pointed out by Dorrance et al. (1991), the two techniques do not sample the same types of liquid. In situ samplers generally sample pore-liquids held under tensions below about 60 kilopascals (kPa)*** (Everett and McMillion, 1985). In contrast, pore-liquids extracted from soil cores include liquids held at tensions far in excess of 60 kPa. Extraction under several bars of (equivalent) tension may strip off cations preferentially sorbed in electrical double layers, sorbed organic compounds, and native soil components. These ions may or may not be present in the same concentrations in samples obtained by in situ pore-liquid samplers. Thus, Zabowski and Ugolini (1990) found solute concentrations in liquids obtained by centrifuging core samples were fre-

---

*This chapter is based in part on ASTM D 4696–92, Standard Guide for Pore-Liquid Sampling from the Vadose Zone. Copies may be purchased from ASTM, 1916 Race Street, Philadelphia, PA 19103.

**Abstracted with permission from D.W. Dorrance, L.G. Wilson, and S.J. Cullen, in *Groundwater Residue Sampling Design*, R.G. Nash and A.R. Leslie, Eds., ACS Symposium Series 465, American Chemical Society, Washington, DC, 1991, pp. 330–331.

***The appendix to this chapter lists common pressure-vacuum equivalents.

quently higher than those from suction samplers. They concluded that solutions from suction samplers are more representative of solute movement than core-derived samples.

Brown et al. (1990) evaluated the effectiveness of soil-core versus soil-pore water samplers for detecting the movement of organic constituents from land-treated industrial wastes. Based on this study, they advised using both soil-core sampling and soil-pore liquid sampling to effectively detect the movement of a wide spectrum of organic chemicals. If sampling for known organic chemicals, they suggested using values of the organic carbon partition coefficient (K_{oc}) of the chemicals as a means of choosing the sampling method. Specifically, soil pore-liquid sampling is the method of choice for detecting those N-alkanes with log K_{oc} values less than (<) 4.4; polynuclear aromatics <3; chlorophenols (mono-, di-, and tri-) <4; nitrophenols (mono-, di-) <2.3; and aromatics <3.3. N-alkanes with log K_{oc} values between 4.8 and 6.2 are detected equally well with either method. Otherwise, soil-core sampling is recommended for these chemicals when their log K_{oc} values are greater than those listed.

SELECTION OF DEVICES FOR A VADOSE ZONE MONITORING PROGRAM

The term "pore liquid" is applicable to any liquid within a soil, ranging from aqueous pore liquid to oil. In the context of this chapter, we refer to pore liquids as aqueous pore-liquids only. The ability of the described samplers to sample nonaqueous liquids may be quite different from the described methods.

By definition (see Bouwer, 1978, p. 2), vadose zones may contain both unsaturated and saturated regions. In the unsaturated regions, pore water exists at pressures less than atmospheric. Saturated regions may occur as local lenses of saturated soil (perched ground water), capillary fringes, and as water table mounds. Saturated flow also occurs in macropores. To sample from unsaturated regions, a suction is applied to pore liquids through a membrane (e.g., ceramic cup). Pore liquid is sucked into the device as a result of the pressure gradient. This technique also works in saturated regions. However, methods (e.g., monitor wells) generally used to sample from saturated regions cannot be used to sample unsaturated zones where pore pressures are less than atmospheric (Bouwer, 1978). The reason for this is that pore liquids cannot seep into a cavity (e.g., well perforations) unless there is sufficient pore pressure to overcome surface tension at the liquid-air interface.

Table 26.1 lists 12 important factors to consider when selecting pore-liquid samplers. Among these factors it is essential to consider the possible flow regimes within the soil mass. Thus, well-structured soils have two distinct flow regions, including macropores (e.g., interpedal openings, cracks, burrows, and root traces) and micropores (e.g., intrapedal openings between soil grains). During saturated flow (e.g., at the onset of infiltration), liquids move more rapidly through the macropores than micropores. Accordingly, contaminants transported by free drainage may bypass the finer pores. Thus, pore-liquids flowing in macropores have different chemistries than those in micropores (Beasley, 1976; Thomas and Phillips, 1979; Shaffer et al., 1979; Haines et al., 1982; Grossmann and Udluft, 1991). This is enhanced by the fact that the oxygen contents of macropores can change in a matter of hours during an infiltration event, whereas micropores may remain suboxic regardless of flow conditions (Anderson, 1986). In addition, micropores are less susceptible to leaching than macropores (Wilson et al., 1961; Morrison, 1983; Severson and Grigal, 1976;

Table 26.1.   Criteria for Selecting Pore-Liquid Samplers

1. Sampling depths
2. Sampling volumes
3. Soil characteristics
4. Chemistry and biology of the liquids to be sampled
5. Moisture flow regimes
6. Required durability of the samplers
7. Required reliability of the samplers
8. Climate
9. Installation requirements
10. Operational requirements
11. Commercial availability
12. Costs

Shuford et al., 1977; Tyler and Thomas, 1977). Given these differences, sample chemistry can vary from location to location and from time to time depending on the amount of liquid drawn from these two flow systems (Grossmann and Udluft, 1991). Accordingly, when dual-flow conditions are expected, it is wise to consider both unsaturated and saturated (free drainage) sampling devices. Inasmuch as macropores tend to diminish in width with depth (Beven and Germann, 1982), the choice of method is also depth dependent.

Elsewhere in this book, Wilson and Dorrance review samplers for saturated regions (i.e., perched ground water) of the vadose zone. In another chapter, they review methods for sampling macropore flow (i.e., free-drainage samplers). This chapter reviews common methods for obtaining in situ pore liquid samples from unsaturated regions. Particular emphasis is placed on samplers that are commercially available or easily fabricated. Experimental samplers are reviewed by Everett et al. (1984), Dorrance et al. (1991), Morrison (1983), and Eastern Research Group, 1993. The three types of porous samplers reviewed are suction-cup samplers (also called suction lysimeters), filter-tip samplers, and vacuum-plate samplers. Table 26.2 summarizes features of these samplers. Because of their widespread use, emphasis is placed on suction-cup samplers. For these units, detailed information is presented on construction materials, system design, and guidelines for their installation and use, sampling procedures, and constraints.

POROUS SUCTION SAMPLERS (LYSIMETERS)

Porous suction samplers are the dominant type of in situ pore liquid sampler. For example, suction sampling has been found to be an effective leaching monitoring tool at mining facilities when placed under the heap leach pads, below the pregnant, barren evaporation and overflow ponds. They are also used successfully as leakage monitoring devices around old and new solid-waste disposal areas. In the case of solid-waste disposal areas, they are useful in detecting the concentration of contaminants and evaluating the migration and size of plumes. In newly constructed waste-storage facilities they are being readily accepted by regulatory agencies as a first line detection system to monitor for leakage through synthetic liners. However, we must keep in mind that suction samplers are not designed to function as remedial tools and should not be considered as such. They are designed to detect leakage problems

Table 26.2. Suction Sampler Summary[a]

Sampler Type	Porous Section Material	Maximum Pore Diameter (μm)	Bubbling Pressure (kPa)	HB[b] HL[c]	Operational Suction Range (kPa)	Maximum Recommended Depth (m)
Single Chamber Vacuum Lysimeters	Ceramic	1.1–2.1	>100	HL	<60–80	<2
	PTFE	25–35	7–20	HB	< 7–20	<2
	Stainless Steel	6–14	20–50	HL	<20–50	<2
	Quartz	6–7	40–50	HL	<40–50	<2
Single Chamber Pressure Vacuum Lysimeters	Ceramic	1.1–2.1	>100	HL	<60–80	<15
	PTFE	25–35	7–20	HB	< 7–20	<2
	Stainless Steel	6–14	20–50	HL	<20–50	<15
	Quartz	6–7	40–50	HL	<40–50	<15
Dual Chamber Pressure Vacuum Lysimeters	Ceramic	1.1–2.1	>100	HL	<60–80	Unlimited
	PTFE	25–35	7–20	HB	< 7–20	Unlimited
	Stainless Steel	6–14	20–50	HL	<20–50	Unlimited
	Quartz	6–7	40–50	HL	<40–50	Unlimited
Vacuum Plate Samplers	Ceramic	2.1–2.5	>100	HL	<60–80	<7.5
	Stainless Steel	10–11	25–30	HL	<25	<7.5
	Fritted Glass	6	50	HL	<50	<7.5
Filter Tip Samplers	Ceramic	1.2–3.0	>100	HL	<100	Unlimited
	Stainless Steel	10–19	15–30	HL	<15–30	Unlimited

[a]Specifications and operational ranges subject to change following design modifications.
[b]HB = Hydrophobic.
[c]HL = Hydrophilic.

early enough to allow adequate planning before the contamination becomes a large vadose zone and ground-water pollution problem.

Suction samplers have been reviewed in detail by Litaor (1988), Grossmann and Udluft (1991), and Dorrance et al. (1991). The principles of suction-sampler operation are described by Everett and McMillion (1985) and Dorrance et al. (1991). Briefly, unsaturated portions of the vadose zone consist of interconnecting soil particles, interconnecting air spaces, and interconnecting liquid films. Attempting to sample pore liquids by applying a suction to an open-ended pipe driven into the ground will fail because only air will be drawn in. In contrast, when a porous sampler is installed, the liquid films in the soil are in hydraulic contact with liquid within the pores of the sampler (see Figure 26.1). When a vacuum greater than the pore-liquid tension is applied within the sampler, a hydraulic gradient is created toward the sampler. As demonstrated experimentally by Krone et al.(1952), this creates a potential field around the cup. Figure 26.2 shows an idealized potential field for a sampler in a homogeneous soil.

If menisci of the liquid in the soil or porous cup are able to withstand the applied

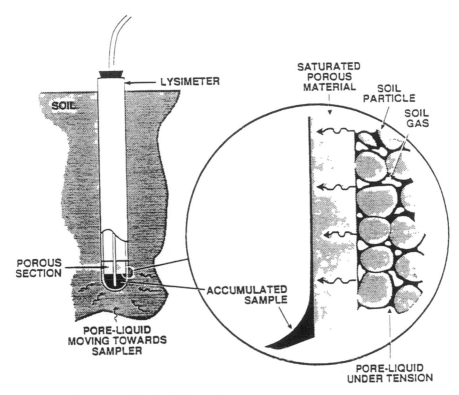

Figure 26.1.   Porous section/soil interactions.

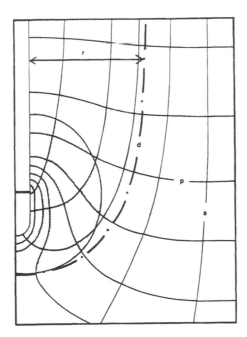

Figure 26.2.   Potential field around a suction cup in a homogeneous soil (r = radius of recharge area; d = dividing stream line; p = isopotential line; s = stream line; hatched field = space from which sample is taken) (after Grossmann and Udluft, 1991).

suction (depending on, among other factors, the maximum pore radii), liquid moves into the sampler. The ability of the menisci to withstand a suction decreases with increasing pore radii of the porous segment. This relationship is illustrated by the capillary rise equation (Hanks and Ashcroft, 1980):

$$\psi_m = -\frac{2\gamma \cos\alpha}{r}$$

where ψ_m = matric potential (dynes/cm^2)
          γ = surface tension (dynes/cm)
          α = contact angle
          r = pore radius

If the maximum pore radii are too large, the menisci are not able to withstand the applied suction. As a result, they break down, and air enters the sampler.

The hydrophilicity/hydrophobicity of the porous segment also affects the ability of menisci to withstand suctions. Ceramic, glass, and stainless steel samplers are hydrophilic, while PTFE samplers are hydrophobic (Table 26.2). As pointed out by Grossmann and Udluft (1991), the contact angle in the capillary rise equation is smaller than 90 degrees for hydrophilic materials (generally taken to be zero for water-glass interfaces), and between 90 and 180 degrees for hydrophobic materials. Based on the capillary rise equation, a smaller pore size is required to sustain a given matric potential when the material is hydrophobic.

The ability of a sampler to withstand applied vacuums is gauged by its bubbling pressure (Everett and McMillion, 1985; Soilmoisture Equipment Corp., 1988). The bubbling pressure is measured by saturating the porous segment, immersing it in water, and pressurizing the inside of the porous segment with air. The pressure at which air starts bubbling through the porous segment into the surrounding water is the bubbling pressure. The rationale for using bubbling pressure is to characterize the operating range of suction lysimeters and not their pore size. The magnitude of the bubbling pressure equals the magnitude of the maximum suction which can be applied to the sampler before air entry occurs (air entry value). This assumes that pneumatic pressure, ψ_p and the soil matric potential ψ_m, are equivalent, as stated by Hanks and Ashcroft (1980), as follows:

$$\psi_m = -\psi_p$$

(This relationship also assumes that the hydraulic potential and the gravitational potential are zero.) Thus, the bubbling pressure is equivalent to the maximum soil-water matrix potential that a suction lysimeter can theoretically sample before air enters the sampler. Given that the maximum vacuum that can be applied is one atmosphere, this means that suction samplers in theory could sample pore liquids up to a matrix potential of one bar. However, at one bar matrix potential the hydraulic conductivity of the soil is generally so low that the volume of liquid collected in a given time will be too small for laboratory analyses.

Table 26.2 summarizes bubbling pressure values of the samplers discussed in this review. Table 26.2 also shows the maximum pore sizes that are equivalent to these bubbling pressures, based on the capillary-rise equation. It should be noted that pore sizes specified by porous material manufacturers often do not translate into correct bubbling pressures. To be certain, one should always measure the bubbling pressure in water.

It is often surmised that when samplers fail to obtain samples, air is drawn through the pores of the porous segments. Indeed, this will occur when the matric suction of the soil exceeds the bubbling pressure of the sampler, a particular problem for cups with low bubbling pressures. For samplers with high bubbling pressures, the restricting factor is more likely to be related to the hydraulics of unsaturated flow. Thus, as soil pore-liquid tensions increase (low pore-liquid contents), pressure gradients toward the sampler decrease. Concurrently, the soil hydraulic conductivity decreases exponentially. These effects result in lower sampling rates by the sampler. Everett and McMillion (1985) found that at pore-liquid tensions above about 60 kPa the flow rates into ceramic suction samplers are effectively zero and samples cannot be collected.

Effect of Applying Suction When Sampling Organic Compounds in the Vadose Zone

Nightingale et al.(1985) indicated that normal suction sampling techniques are not suitable for sampling volatile organic compounds due to potential loss. Barbee and Brown (1986) also observed volatilization losses of xylene from ceramic suction-cup samplers. To minimize loss of volatile organics, Wood et al. (1981) devised a trapping device as part of the sampling assembly. Pettyjohn et al. (1991) also described a suitable system for sampling highly volatile organics. However, the reported system was limited to a maximum sampling depth of about 20 ft (6 m) and a small sample volume (5 to 10 mL). Smith et al. (1992) described a syringe method for sampling trichloroethene (TCE) in soil-pore liquids. This method involves inserting a 5-mL glass gas-tight syringe, sealed with Minnert valves, into the silicone discharge tubing of a suction sampler. As vacuum is applied to the sample line, water flowing past the tip of the needle is pulled into the syringe barrel by extending the syringe plunger. The method is similar to that employed for collecting soil gas samples from the vadose zone. Based on the results of statistical analysis of TCE concentrations collected during laboratory and field studies, Smith et al. (1992) found the syringe method to be superior to three other methods they evaluated. These methods included the standard method of sampling with suction samplers (vacuum collection into a vial), the filter-tip ("BAT®") method, and the methods of Wood et al. (1981) and Pettyjohn et al. (1981).

Factors in Selecting, Installing, and Operating Suction Samplers

Factors in selecting, installing, and operating suction samplers are as follows:

- materials for porous segments
- sampler type; maximum sampling depths
- monitoring depths, areal distribution, and required number of units
- preinstallation procedures
- assembly
- installation procedures
- sampling procedures

Materials for Porous Segments

A variety of materials have been used for the porous segment including ceramics, polytetrafluoroethylene (PTFE), stainless steel, fritted glass, sintered glass, and Alundum®. The sampler body tube has been made using PVC, ABS, acrylic, stainless steel, and PTFE. Selection from among the available porous segment materials is based on expected contaminants, vadose zone texture, expected pore-liquid suction range, and strength.

Expected Contaminants.   Suction samplers are frequently used in soil-nutrient studies (Angle et al., 1991). Zimmermann et al. (1978) found that significant differences were evident between nutrient concentrations before and after being filtered through ceramic samplers. Ammonia nitrogen and phosphate were particularly affected, with recoveries of only about 11% and 43%, respectively. Greater recoveries were found for nitrate nitrogen (96%) and nitrite nitrogen (85%). In contrast, percent recoveries using PTFE samplers ranged from 98% to 106% for all nutrients tested. During field comparisons using 10 pairs of ceramic and PTFE samplers, Beier and Hansen (1992) found that neither sampler contaminated soil-water samples nor retained any constituent (Na^+, K^+, Ca^{++}, Al^{+++}, NH_4^+, and H^+) except possibly magnesium. Nagpal (1982) found that ceramic cups had a greater affinity for phosphorus and potassium than nitrate and nitrite nitrogen. Angle et al. (1991) concluded that ceramic suction samplers are not recommended when phosphorus is the nutrient of interest. However, nitrate nitrogen concentrations in such samplers are representative of levels in the soil (Angle et al., 1991).

The use of suction samplers for sampling metals in pore liquids has been examined. McGuire et al. (1992) conducted a detailed study on the sorption of metals onto cups of various materials. They found that the general pattern of metal sorption on samplers was as follows:

$$\text{ceramic} > \text{stainless steel} \gg \text{fritted glass} = \text{PTFE}$$

They also found the general order of trace metal sorption onto samplers was:

$$Zn \gg Co > Cr > Cd$$

They concluded that adsorption-desorption processes can cause sampling errors when metals are present at microgram per liter concentrations. Sheppard et al. (1992) observed retention of both cations and anions on ceramic plates.

Given the role of organic compounds as a major pollutant, monitoring these pollutants in the vadose zone is of major interest. As pointed out by Finger and Hojaji (1991), the hydrophilic nature of ceramic suction samplers is an effective barrier to sampling nonpolar compounds. They report studies demonstrating that xylene and DDT were not effectively sampled by ceramic suction samplers. PTFE samplers were also found to undersample organic compounds (Finger and Hojaji, 1991). Gillham and O'Hannesin (1990) studied the sorption of six monoaromatic hydrocarbons onto seven materials. The hydrocarbons included benzene, toluene, ethylbenzene, and o-, m-, and p-xylene. The materials examined included stainless steel and PTFE, commonly used as porous segments in suction samplers. Stainless steel showed no significant sorption during the duration of their studies, while some sorption occurred on PTFE. Nielsen and Schalla (1991) reported results of studies by Reynolds and Gillham (1985) demonstrating that tetrachloroethene was strongly and rapidly absorbed by PTFE. Finger and Hojaji (1991) tested the ability of porous-glass elements to sample organic compounds. Analyte concentrations in the water sampled through the porous glass segments were within about 10% of test solution concentrations. Pore size of the elements was about 2 microns.

During long-term field studies, Beier et al. (1992) found that the concentrations of nonpurgeable organic carbon (NPOC) in samples from PTFE samplers were higher than NPOC concentrations in ceramic samplers. They suggested that differences may be attributable to filtration of organic molecules by the ceramic cups.

Dazzo and Rothwell (1974) concluded that ceramic suction-cup samplers do not yield valid water samples for fecal coliform analyses. Bell (1974) found a one-thousandfold reduction in *E. Coli* in a saline suspension passing through a 1-bar ceramic sampler. Powelson (1990) observed sorption of two bacteriophages (MS2 and PRD1) on ceramic samplers. In contrast, no loss of these viruses occurred when passing the solution through stainless-steel suction cups (Powelson et al., 1993)

Vadose Zone Textures and Expected Soil-Water Suctions. Soil textures and pore-liquid suctions control the amount of liquid that can be removed by a porous-cup sampler. At low pore-liquid suctions (see Figure 26.3), the slope on the pore-liquid release curve for a sand is steeper than that for finer-textured soils. This means there will be a larger quantity of pore-liquid released from a sand than from a silty clay soil for an equal change of pore-liquid suction at low suctions. Similarly, the sampling rate is higher in wet sandy soils because of their higher hydraulic conductivity values at low suctions.

At higher suctions (i.e., dryer soils), the slope of a liquid release curve for a fine-textured soil, such as a silty clay, is steeper than for a sand (see Figure 26.3). This indicates that more pore-liquid will be released from a clay than from a sand for an equal change in pore-liquid suction at the higher suctions. Inasmuch as the hydraulic conductivity of clay soils is greater than sandy soils at higher suctions (Jury et al.,

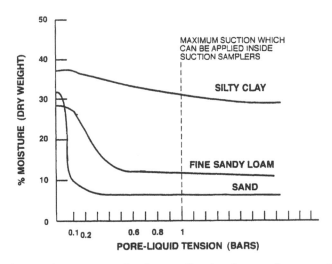

Figure 26.3. Water release curves for three soils, showing maximum suction which can be applied inside.

1991), the sampling rate will also be greater in clays. Consequently, sample recovery may be poor (or negligible) in coarse-grained soils at higher pore-liquid suctions.

Table 26.2 shows the operational ranges of the sampler materials discussed in this chapter. All of the samplers are capable of sampling either sandy or clayey textures under low-suction conditions. There may, however, be a difference in terms of sampling rate and yield. Thus, during laboratory studies in soils with high moisture content (i.e., low suctions), McGuire et al. (1992) found that fritted-glass and stainless steel samplers collected pore-liquid samples at a faster rate with more sample volume than ceramic samplers. They attributed this to the larger, more uniform pores in the fritted glass and stainless steel samplers. However, only ceramic-suction samplers were able to sample pore liquids from either a sandy or a clayey texture under high suctions (see Table 26.2). McGuire et al. (1992) also observed the failure of PTFE samplers to obtain samples when the soil-water suction was large. They concluded: "These samplers performed poorly due to large pore size (low air entry value) and the hydrophobic characteristic of the material."

Strength.   Resistance to damage is an important factor when handling and installing suction samplers. Of the materials currently available, stainless steel is the most rugged and least subject to damage.

Sampler Types; Maximum Sampling Depths

The selection of suitable sampler types at a particular location depends on expected sampling depths. (Factors involved in selecting monitoring depths and spatial distribution are reviewed in the section in this chapter entitled: Sampler depths, areal distribution, and required number of units.)

The three types are as follows:

- Single-chamber, vacuum operated samplers
- Single-chamber, vacuum-pressure operated samplers
- Dual-chamber, vacuum-pressure operated samplers.

Single-Chamber, Vacuum-Operated Samplers. Single-chamber, vacuum-operated samplers consist of a porous cup mounted on the end of a body tube (Figure 26.4). The body tube is commonly made from PVC. The cup is attached to the tube with adhesives or with "V" shaped flush threading sealed with an "O" ring. Commonly, rubber stoppers are inserted into the upper end of the body tube. A sample line, extending through an opening in the stopper to the bottom of the cup, is connected aboveground to a sample bottle. To obtain a pore-liquid sample, a sample collection bottle is connected to the sample line and a vacuum is then applied to the sample bottle, using a hand-operated vacuum pump or vacuum tank (Nelson, 1985). The suction caused by the applied vacuum draws the sample from the soil pores into the cup and sample line for delivery to the sample bottle.

The maximum operating depths of these samplers is a function of the bubbling pressures of the materials comprising the porous segments. These values are reported in Table 26.2. Inasmuch as commercial ceramic samplers have bubbling pressures of 200 kPa, they are theoretically capable of sampling to depths of 66 ft (20 m). However, a vacuum cannot be applied greater than the theoretical maximum suction lift of a column of water, i.e., about 33 ft (10 m). This limitation and losses in the system restrict ceramic samplers to recommended depths less than 1.8 m

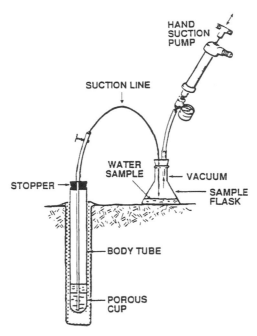

Figure 26.4. Single-chamber vacuum suction sampler (Soilmoisture Equipment Corp., 1981).

(personal communication, Soilmoisture Equipment Corp., 1993). For PTFE samplers the maximum bubbling pressures are about 10 kPa. Applying a vacuum in excess of 10 kPa will draw air into the sampler. Thus, the maximum sampling depths of these units are less than 3 ft (1 m). Stainless steel samplers are suitable to depths of about 10 ft (3 m), but the recommended maximum depth is 3.6 ft (1.5 m) (Soil Measurement Systems, 1993). For quartz-based samplers, the theoretical maximum sampling depth is 5 m; the practical depth is probably less than 1.8 m.

Single-Chamber, Pressure-Vacuum Operated Samplers. Figure 26.5 depicts single-chamber, pressure-vacuum operated samplers. These units allow sampling pore liquids beyond the reach of vacuum lysimeters. The most common design is a porous cup at the bottom of a body tube. However, commercial units are also available with the porous segment at the top of the sampler (Soil Measurement Systems, 1993). Two access tubes are forced through a two-hole stopper sealed into the upper end of the body casing. The discharge tube extends to the base of the sampler, and the pressure-vacuum tube terminates a short distance below the stopper. The body casing is commonly made from PVC, stainless steel or PTFE. Sample and pressure-vacuum tubes are constructed of PTFE, rubber, polyethylene, polypropylene, Tygon®, stainless steel, and nylon. Polyethylene tubing of dimensions 1/4 inches (0.63 cm) outside diameter, 1/16 inches (0.16 cm) wall thickness is commonly used for ceramic samplers. For stainless steel samplers, lengths of 3/16 inch (0.47 cm) inside diameter, stainless steel tubing are joined by stainless steel couplers. Stainless steel has an advantage over other tubing materials in being resistant to damage from burrowing animals. Above land surface, short lengths of neoprene tubing [3/16 inch (0.47 cm) inside diameter by 1/8 inch (0.32 cm) wall thickness] are

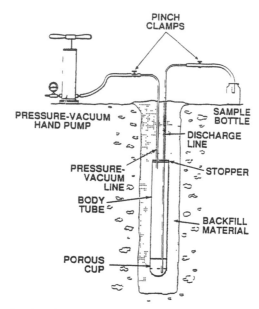

Figure 26.5. Single-chamber, pressure-vacuum suction sampler (Soilmoisture Equipment Corp., 1989).

forced over the ends of the sample and pressure-vacuum tubes to facilitate clamping. Pore-liquid is drawn into the body tube by application of vacuum through the pressure-vacuum line. The sample is then retrieved by pressurizing the sampler through the same line; this pushes the sample to the surface through the discharge line (Figure 26.5).

Inasmuch as samples are retrieved under pressure, the maximum depth of these samplers is restricted by the bubbling pressure of the porous segments. In other words, if the pressure applied to retrieve the sample exceeds the bubbling pressure of the porous segments, sample will be forced out through the cup. (This may not be a particular problem in units containing the porous segment on top of the body tube; however, pressures less than the bubbling pressure are recommended to avoid forcing air into the pores outside the sampler.) Table 26.2 shows the bubbling pressures of available sampler materials. For ceramic samplers with bubbling pressures of 200 kPa, depths up to about 66 ft (20 m) are theoretically possible. However, the maximum recommended depth is 50 ft (15 m) (personal communication, Soilmoisture Equipment Corp., 1993). For PTFE samplers, maximum depths appear to be about 3 ft (1 m) unless the porous segment is located at the top of the body tube. Stainless steel units are recommended for sampling down to 10 ft (3 m) (Soil Measurement Systems, 1993). The maximum sampling depths for quartz-based, cup samplers are probably less than 50 ft (15 m).

Dual-Chamber, Pressure-Vacuum Operated Samplers. Dual-chamber, high pressure-vacuum samplers operate in the same manner as single-chamber, pressure-vacuum samplers, and their dimensions are similar. As shown by Figure 26.6, the components of these units include a lower chamber, comprising the porous cup, and an upper, sample-collection chamber. Other designs include a transfer chamber between the sampler and the surface. The two chambers are connected through a

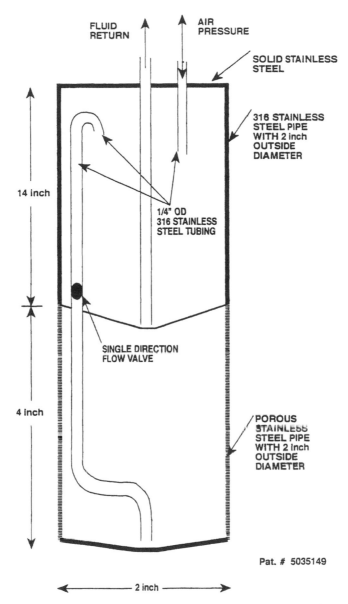

Figure 26.6. Double-chamber, pressure-vacuum suction sampler (Soil Measurement Systems, 1990).

one-way check valve constructed of stainless steel or brass. The valve opens during application of vacuum when sample that has collected in the cup is delivered to the upper chamber. The valve shuts during application of pressure as the sample is forced through the delivery tube. The check valve prevents sample loss through the porous section during pressurization. Thus, in contrast to the other two sampler types, the bubbling pressure is not a factor with these units. The internal sample and pressure-vacuum tubes terminate on the top of the sampler where they are attached to corresponding tubes extending to land surface. The samplers and their lines are generally manufactured from the same materials discussed for pressure-vacuum lysimeters. In deep installations, the sampler body is commonly connected to PVC riser tubing to facilitate installation and protect the sample-pressure-vacuum tubes from damage due to burrowing animals. Riser casing also prevents the samplers from "floating" in the bore hole, thus ensuring their proper depth placement.

The design of these samplers allows them to be installed at any depth in the vadose zone. In practice, however, the sampling depth will be restricted by head losses when applying pressure to the pressure-vacuum line. Installation of ceramic suction samplers as deep as 300 ft (91 m) have been reported (Bond and Rouse, 1985). This is also the recommended maximum sampling depth by the manufacturer (Soilmoisture Equipment Corp., personal communication, 1993). Stainless steel units are recommended for depths greater than 10 ft (3 m) (Soil Measurement Systems, Inc., 1993).

Sampler Depths, Areal Distribution, and Required Number of Units

The design of a suction sampler network for detecting pollutant movement at a waste disposal facility requires determining sampler depths, areal distribution of units, and the required numbers of units. Ideally, a network design should be an integral component of a comprehensive, subsurface-monitoring network design.

In general, the depths at which samplers are installed depends on the depth of the facility being monitored, depth to ground water, and subsurface geology. These factors should be determined during premonitoring activities. If the distance between the base of the facility and ground water is minimal, ground-water monitoring wells may be more suitable than suction cup samplers. For example, during a survey of 47 hazardous waste sites in 1984, Bumb et al. (1988) determined that most sites were located in areas with water tables that were too shallow to permit vadose zone detection monitoring. For deep profiles, suction cup samplers are warranted.

In their excellent review of factors associated with installing vadose zone monitoring systems, Robbins and Gemmell (1985) list three criteria for determining placement depth. *First*, samplers should be deep enough to intercept leakage. *Second*, samplers should be located above a depth where remedial action is possible if pollutant movement is detected. This critical depth is illustrated in Figure 26.7. Obviously, the critical depth should be shallow enough to minimize excavation costs, if excavation is the chosen remediation method. However, with the increasing availability of in situ vacuum extraction and bioremediation techniques, this crite-

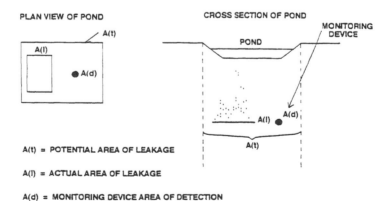

Figure 26.7. Monitoring depth criteria (Robbins and Gemmell, 1985).

rion is not as crucial as when soil excavation was the preferred method. This empha-sizes the importance of the first criterion, detecting leakage before a major portion of the vadose zone is contaminated.

Third, sampler depth should allow for a reaction time to intercept the pollutants before they reach the critical depth. Robbins and Gemmell (1985) point out that this criterion is important to avoid pollutant movement beyond the critical depth. Move-ment beyond the critical depth could require an expensive vadose-zone remediation program and possibly ground-water cleanup.

Selecting an appropriate sampling depth is intimately related to determining the areal distribution and density of sampling units. According to Robbins and Gem-mell (1985) and Bumb et al. (1988), there are three spatial regions that must be considered in selecting areal distributions and densities of units. These regions are: A(t), the total site area; A(l) actual area of leakage; and A(d), the detection area encompassed by monitoring units. Figure 26.8 depicts these areas for a lined pond. A(d) is not necessarily equal to the sampling area of a suction sampler; rather, it is the area encompassed by liquid as it spreads laterally during vertical movement beneath the leak. Lateral spreading is fortuitous, given that the sampling radius of suction samplers may be on the order of centimeters (Morrison and Lowery, 1990).

A(d) depends on soil textures beneath the leak and on textural layering in the vadose zone. Textural layering tends to increase the spreading area of a plume at the interface between different layers, particularly if coarse regions overlay finer re-gions. The greater the extent of lateral spreading the greater will be the opportunity for suction samplers to intercept contaminants. Additionally, sediments are more likely to approach saturation at such interfaces, enhancing the operation of suction samplers if placed immediately above the interface. The extent of lateral spreading is also a function of vadose-zone texture. Thus, during deep percolation, the widths of saturated plumes with depth are greater for finer than for coarser-textured sedi-ments. It should be noted, parenthetically, that most new ponds are located in fine sediments by choice, and underlined with compact clay to 10^{-7} cm/sec. In this case, lateral spreading would occur just below the liner.

McKee and Bumb (1988) describe an analytical-based model VADOSE for esti-

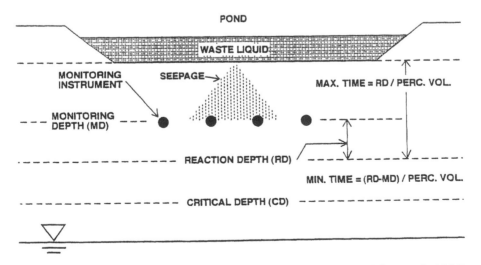

Figure 26.8. Conceptual approach for leak detection (Robbins and Gemmell, 1985).

mating the width/depth relationships of plumes in three homogeneous soil textures. Bumb et al. (1988) used the same model to determine the optimal depths and distance between units for three soils of progressively finer textures, i.e., loamy sand, silt loam, and silty clay. The model assumes steady state flow conditions and homogeneous soils. As demonstrated by Figure 26.9 for silt loam, the spacing of suction samplers increases to an optimum distance with depth, then decreases at greater depth. For this soil at residual saturation, the maximum lysimeter spacing was about 19 ft (5.7 m) at a depth of about 13 ft (3.9 m) from a 1 ft^2 (0.09 m^2) leak. For the loamy sand, Bumb et al. (1988) determined the maximum spacing of suction samplers was 15.5 ft (4.7 m) at a depth of 30 ft (9 m) to collect a 10 mL sample in one week from a 1 ft^2 (0.09 m^2) leak. For silt loam, the maximum lysimeter spacing was 17 ft (5.2 m) at a depth of 15 ft (4.5 m). For silty clays, the maximum spacing was 7 ft (2.1 m) at a depth of 2 ft (0.6 m). Another approach for simulating the distribution of plumes during deep percolation from a source is to use unsaturated flow models such as UNSAT2 (Davis and Neuman, 1983) or VST2D (Lappala et al., 1987). These models account for transient flow in layered conditions.

In the end, it may not be practical to install the units with complete areal coverage, due to economics or access. Therefore, the designer and regulatory agency must be comfortable with the spacing used and common sense.

Cup Conditioning

New samplers may be contaminated with dust and other contaminants from the manufacturing process (Neary and Tomassini, 1985; and Grossmann and Udluft, 1991). In order to reduce chemical interferences from substances within the porous sections, U.S. EPA (1986) recommended preparing ceramic units prior to installation following procedures originally developed by Wolff (1967), modified by Wood

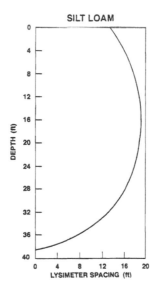

Figure 26.9. Suction sampler spacing corresponding to 60 cb of suction as a function of placement depth in silt loam initially at residual saturation (Bumb et al., 1988).

(1973) and recommended by Neary and Tomassini (1985). The process involves passing hydrochloric acid (HCl) through the porous sections and flushing with distilled water.

The maximum suction which can be applied is one atmosphere; therefore, flushing will be slow if suction is used to draw HCl through the porous segment. The flushing can be performed more rapidly if the porous segment is filled with HCl and pressurized to force the acid out of the porous segment, since more than one atmosphere of pressure can be applied. However, care must be taken to prevent overpressurization which might damage the porous section.

If it is not convenient to use the pressure method, an alternative approach is to prepare a deionized/acid mixture to a pH of about 2 in a clean bucket. A vacuum is then applied on the units overnight. This procedure is repeated twice with straight deionized water.

There does not seem to be a consensus of opinion on the concentration of acid and the extent of flushing for ceramic samplers. For example, Debyle et al.(1988) suggest leaching 1 L of 1 N HCl into ceramic samplers, followed by 1 L of distilled water. Maitre et al. (1991) flushed N/20 HCl through ceramic and PTFE cups overnight, followed by leaching with 500 mL 1 N HCl, then flushing with 5 to 6 L of distilled water. Hughes and Reynolds (1988) acid washed their ceramic samplers with 0.001 N HCl. Linden (1977) recommended applying a vacuum and pulling 250 to 500 mL of 0.1 N HCl through ceramic suction cup samplers, followed by an unspecified volume of deionized water. Grover and Lamborn (1970) recommend leaching with 50 to 60 pore volumes of 1 N HCl, then rinsing with 10 volumes of deionized water. Peters and Healy (1988) used H_2SO_4 rather than HCl.

One purpose of acid leaching is to remove metals and major ions from the porous segments. However, the value of this procedure is debatable, except when very low concentrations are important. Creasey and Dreiss (1986) conducted studies to determine cation releases from acid leached and unleached ceramic and PTFE samplers. They concluded that the potential is low for significant sample bias from cleaned and uncleaned samplers. Furthermore, cation concentrations released from both cleaned and uncleaned samplers were generally less than 10% of the concentrations found in soil water at pH 6–7 (Creasey and Dreiss, 1986).

Flushing with HCl strips cations off of the ceramic (Debyle et al. 1988; Wood, 1973; England, 1974). This results in an initial adsorption of cations from poreliquid onto the ceramic surface. (Sorption of contaminants of interest onto suction cup samplers is discussed in a previous section on selection of cup material.) This continues until the cation exchange capacity (CEC) of the ceramic has been satisfied. The effect is not reduced by distilled water flushing after the acid flushing. Therefore, Debyle et al. (1988) and Grossmann and Udluft (1991) recommend preconditioning the samplers, i.e., flushing them prior to installation, with a solution similar in composition to the expected soil solution. Alternately, the first sample after installation could be discarded. Bottcher et al. (1984) attributed increased adsorption of PO_4 to the acid leaching process. Therefore, they recommended a thorough flushing with a PO_4 solution of approximately the same concentration as that found in the soil solution, rather than the acid leaching procedure, when sampling for PO_4.

Porous segments may also be treated to inhibit microbial growth within the pores of suction samplers and in surrounding slurry material (Lewis et al., 1992). Such growth enhances the biodegradation of organic chemicals during the sampling process. This effectively reduces concentrations entering a sampler. Two mechanisms are postulated: (1) microbes colonizing the slurry and cups clog pores with biofilms that sorb biotransformable and nonbiotransformable chemicals, and (2) during the application of suction, the increase of flow velocities within the slurry/cup environment accelerates biodegradation rates. In contrast, the biodegradation rates of chemicals during background flow are mass-transport limited by the slow velocities within soil pores. To inhibit the colonization of microorganisms within the pore environment of suction samplers, Lewis et al. (1992) recommend treating slurry material and lysimeter cups with a 2% solution of copper sulfate (12.5 mmol/liter). During their studies, Lewis et al. (1992) also found that copper-treated samplers yielded more water than untreated samplers.

The flushing and preconditioning steps for ceramic samplers are also employed for samplers constructed from other materials. Timco (1988) described cleaning procedures for PTFE. The method includes passing 0.5 L of distilled water through the material or rinsing with HCl followed by a distilled water rinse. Soil Measurement Systems, Inc. (personal communication, 1993) recommends cleaning stainless-steel porous segments with isopropyl alcohol and 10% nitric acid. Stainless steel samplers used in virus studies are chlorinated to destroy resident microorganisms and then rinsed with a 10% solution of sodium thiosulfate to neutralize free chlorine in water sources.

Hydrochloric acid may corrode valves within dual chamber pressure-vacuum lysimeters; therefore, flushing these samplers should be performed prior to attaching the cup to the body, if possible.

Preinstallation Assembly

It is good practice to assemble the sampler unit aboveground prior to installation to check fittings and to look for leaks. The unit includes the following components: (1) porous cup plus body tube, (2) pressure/vacuum line, (3) sample delivery line, (4) riser tubing, and (5) short lengths of 3/8 inch (0.9 cm) neoprene tubing attached to the aboveground ends of the sample and pressure/vacuum lines (to seal in the vacuum). The riser tubing and the sample and pressure-vacuum lines are measured and cut to their correct lengths before installation. If the sampler is to be installed at great depths, it will be necessary to join lengths of the riser tubing. During installation, the sections of riser tubing should be threaded together, rather than glued, to prevent interferences when sampling for organic compounds. The sample and pressure-vacuum lines are cut to allow at least 5 ft (1.5 m) aboveground for access and handling. Different colored tubing for sample and pressure/vacuum lines is recommended to avoid confusion after installation. The neoprene tubes should be 8 inches (20 cm) long. A drop of silica sealer should be smeared on the ends of the sample and pressure/vacuum lines to facilitate sliding the neoprene tubing into place.

For single-chamber, vacuum-operated and single-chamber, pressure-vacuum operated units it is essential to correctly position the sampler delivery line within the inside base of the cup. An inner lip may be present at the juncture of the suction cup and the body tube (Everett and McMillion, 1985). Consequently, as the discharge line is pushed through the stopper at the top of the sampler, it may catch on this lip, preventing the line from reaching the bottom of the ceramic cup (see Figure 26.10). Consequently, an 80 mL error can occur in sampling rate determinations due to fluid remaining in the suction cup. This 80 mL of fluid accumulates in the cup, is not removed during sampling, and will cause cross contamination between sampling

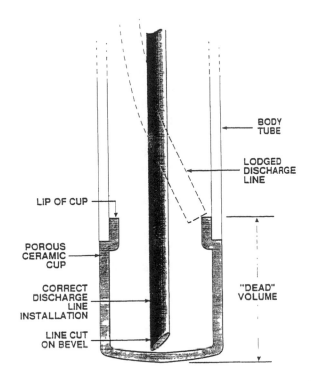

Figure 26.10. Location and potential dead volume in a suction sampler (Everett and McMillion, 1985).

events. The line can be kept from catching by cutting its tip at an angle (Figure 26.10) (Everett and McMillion, 1985). In all-PTFE and stainless-steel suction lysimeters, the discharge line is a rigid PTFE or stainless steel tube extending to the bottom of the cup. Thus, no fluid accumulates.

It may be very difficult to push the sample tubing through the openings in the neoprene plug/stopper in the top of the body tube. One approach is as follows:

- Remove the threaded ring clamp.
- Remove the plug by placing a screw driver through one of the hol·s and unscrewing or prying it out with a circular motion.
- After removing the plug, flush the interior of the sampler with deionized water to remove debris and dust.
- Apply a small drop of 100% silica sealer on the end of the tubing, and force the tubing through the openings in the stopper.
- Remove excess sealant. (The silica also helps seal the tubing for leaks.)
- Cut or adjust the tubing to fit within the bottom of the cup.
- Reinstall the plug (ensuring that the tubing does not hang up on the interior lip), and apply silica sealant to the region around the stopper/casing rim.
- Replace the threaded ring clamp.

To ensure that the pores remain saturated during the interim period before testing and final installation, the suction cup should be placed in a container filled with either deionized water or water representative of the expected sample quality.

Prior to installation, the sampler units should be checked to ensure that they are free of leaks by conducting either vacuum or bubbling tests. The porous cups are first saturated by inserting them in water and applying a vacuum. (Note: the cups should always be wet for testing, as dry units may lose vacuum quickly.) For the vacuum method, the porous segment of the cup is immersed in deionized water and the air inside of the porous segment is evacuated by clamping the outlet tubing (by using clamps on the neoprene tubing) and applying a vacuum to the pressure-vacuum line. For ceramic samplers this should be about 70 kPa. For other sampler types, the vacuum should approach the bubbling pressure. The vacuum, as read on a gauge, should remain constant. If the vacuum falls off, a leak exists in the system (see Figure 26.11). For the air pressure method, the entire unit, including tubing, is immersed in a tub of water and air pressure is applied. An air pressure gauge is included in the test system if bubbling pressures are also being determined (not applicable to dual-chamber, pressure-vacuum units because of the check valve between upper and lower chambers). As the air pressure increases, the assembly should be examined for air bubbles in the following locations: (1) the joint where the cup is attached to the body tube, (2) the neoprene plug at the top of the sampler body, (3) the openings where the pressure/vacuum and sampler lines exit the neoprene plug, and (4) around the neoprene tubing forced onto the upper ends of pressure-vacuum and sample lines. If there are no leaks, the pressure at which air starts bubbling through the porous segment into the surrounding water is the bubbling pressure.

Installation Procedures

A key, preliminary consideration in designing a suction-sampler system is site access for installation and future sampling of the samplers. Preferably, they should be installed prior to or during construction of the waste-disposal facilities. With some difficulty and less efficiency, they can be installed after construction is completed. In the case of new solution-storage ponds, it is preferable that the suction samplers be installed prior to completion of the subgrade and liner emplacement.

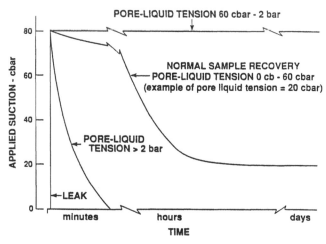

Figure 26.11.   Decay characteristics of suction applied to a two bar (bubbling pressure) ceramic suction sampler in equilibrium with soils in varying ranges of pore-liquid tension. Also shown is the almost instantaneous decay associated with an appreciable leak in the instrument.

When installing samplers around ponds, angle-drill holes can be used to place the units around the pond bottom perimeter. It is also important that the sample tubes be safely buried and terminated in a location that is out of danger for long-term sample station establishment. The construction contractor should be well informed of the installation procedures and equipment that will be needed during the construction phase. Trenching machines, road graders, bulldozers and other heavy equipment have been known to damage and cut sample lines after they have already been buried.

The two objectives of correct installation of samplers are (1) ensure a good contact between the suction cup and the soil to be sampled, and (2) prevent side leakage of liquid along the riser tubing. Good contact is required to ensure a hydraulic continuum between soil pores and pores in the porous cup. Side leakage must be prevented to avoid sample cross contamination.

In fine-grained deposits, such as clay, tailings, slimes, silt and fine sand, the porous segment may be placed in direct contact with the fine particles. The small size and hard surface area of the porous ceramic enables a direct contact with little effect of soil clogging the porous segment, which would deplete its available porosity.

For coarser material, such as coarse sand, alluvium, and loose fill material, contact is accomplished by placing a slurry of soil in the borehole covering the suction cup. The soil slurry is prepared by sieving the soil, to remove coarse particles, and adding deionized water until the mixture can be readily poured through a tremie pipe. In lieu of a slurry of native material, it has become a common practice to use 200 mesh silica flour. Surrounding the cup with silica flour not only ensures good hydraulic contact but also minimizes clogging of the porous cup by fine particulate matter transported by pore liquid (Everett and McMillion, 1985). Silica flour is recommended for extending the sampling range of PTFE suction cup samplers (Everett and McMillion, 1985). McGuire et al. (1992) do not recommend using silica flour if metals are the contaminant of concern. During their studies, at pH 8, they observed adsorption of trace metals on silica to be one order of magnitude larger than observed on ceramic samplers. Powelson (1990) also observed sorption of viruses on silica flour. A silica flour slurry is prepared by adding distilled water at the ratio of 0.45 kg silica flour to 150 mL distilled water.

Brose et al. (1986) described a method for freezing silica-flour slurry around the sampler prior to placement. The sampler and frozen pack are lowered to the sampling depth in the borehole. The cited advantages of this technique include assurance of proper sampler placement in the flour pack, and elimination of pack contamination by soils which may have caved into the borehole.

Side leakage is prevented by carefully backfilling the borehole following installation of the sampler assembly. If there is no contamination present, the material removed from the borehole can serve as the backfill material. It is important to try to reconstruct the borehole lithology when backfilling with the soil cuttings. In any case, the soil adjacent to the sample intake area should be void of contamination. Additionally, layers of mixed bentonite clay and soil are sometimes included, but not in contact with the water intake area. The bentonite clay swells in the borehole when hydrated with deionized water. However, bentonite may interfere with contaminant detection (Angle et al., 1991).

Construction of a borehole for installing samplers will require the use of drilling methods described by Dorrance et al. in this book. For shallow units in cohesive, nonstony material, hand augers may be appropriate. For deeper installation, it will be necessary to use power equipment. The preferred choice of power equipment is the hollow-stem auger.

Not all installations require extensive vertical drilling. Some installations use buried samplers in shallow boreholes or trenches with buried sample and pressure-vacuum lines extending laterally to a sampling station. The U.S. Environmental Protection Agency (1986) suggested installing suction cup samplers at an angle of 30 to 45 degrees from the vertical in trench walls constructed in land treatment areas. An advantage of this method is that an undisturbed column of soil remains above the cup. Thus, pore liquid samples will reflect flow through undisturbed pore sequences. Bond and Rouse (1985) describe situations where samplers are installed in trenches prior to constructing permanent leach pads, ponds, or tailing disposal structures (see Figure 26.12). They state that sample and pressure-vacuum lines may extend laterally up to 500 ft (165 m). For distances over 100 ft (30 m), they recommend using dual-chamber, pressure-vacuum units. To avoid damage to the lines, they are inserted through PVC conduit prior to installation, or buried in a sand backfill.

Selecting the least important or most shallow unit for installing the first suction sampler is highly recommended. It takes time for all concerned to become proficient in installing sampler assemblies and it is wiser to make mistakes on the least important unit, if there is one. However, this may not be possible when the borehole is also used for collecting lithological samples, aiding in determining sampler depths.

Ceramic and fritted glass samplers should be handled carefully because of their

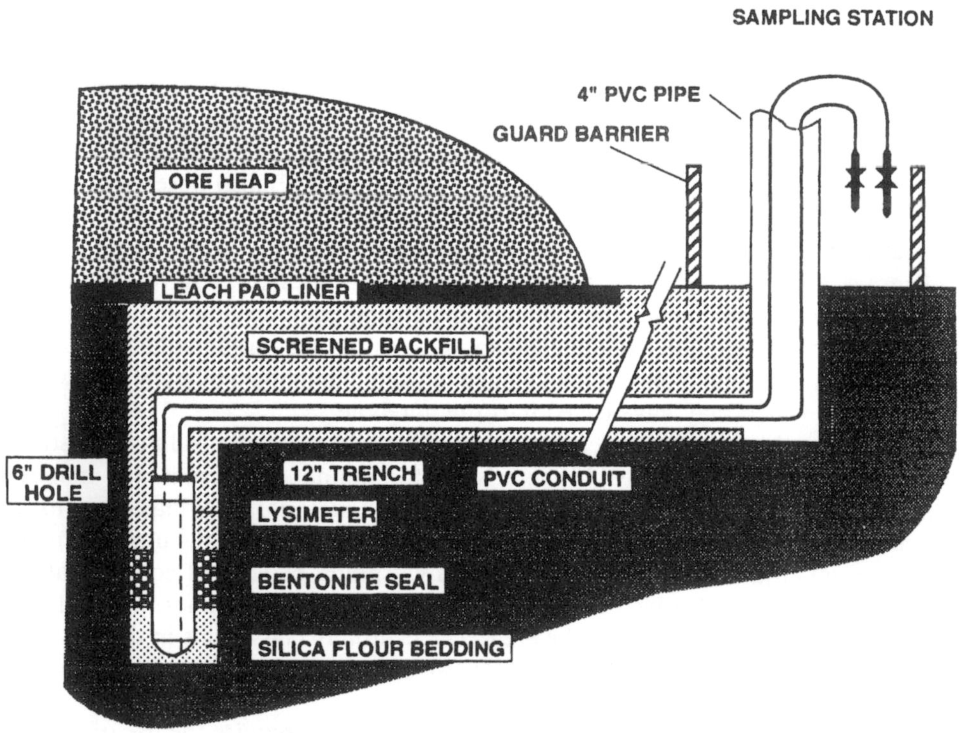

Figure 26.12.   Typical lysimeter installation in a heap leach mining operation (Bond and Rouse, 1985).

fragility. PTFE and especially stainless steel samplers have higher impact strengths. To minimize potential damage to ceramic units during transport to the site, it is suggested that the suction sampler be at least half full of deionized water. This also aids in checking the system continuity immediately after installation.

Procedure for Installing Shallow Units.   Linden (1977) described a step-by-step procedure for installing suction cup samplers in shallow boreholes constructed with hand augers. The method assumes that the borehole is constructed in cohesive material:

1. Following construction, pour in a volume of slurry (soil or silica flour) to a sufficient depth to completely surround the cup or intake section.
2. Press the sampler firmly into the slurry.
3. Slowly refill the borehole with soil collected during augering. Tamp the soil to eliminate large voids that could channel surface water down to the cup. If necessary, add bentonite clay.
4. Mound surface soil around the riser tube to minimize ponding and channel flow of surface water.

Procedure for Installing Deep Units.   Following is a suggested step-by-step procedure, modified from Graham's (1989) method, for installing suction-cup sampler assemblies within hollow-stem auger flights.

1. Auger a borehole to the desired depth. If possible, set aside the soil in the same sequence occurring in situ.
2. Use a tape measure inside the hollow stem auger flight to determine the distance to the bottom of the hole after the auger has reached total depth. Raise the hollow-stem auger flight about 1 ft (0.3 m) to ensure adequate clearance before installing the slurry.
3. Pour slurry through a tremie pipe [1-inch (0.4 cm) diameter or larger] to create a slurry bed that is no less than 2 inches (0.78 cm) thick after settling.
4. Lower the lysimeter assembly in the borehole, adding additional riser tubes as required and position the cup on top of the slurry. Ensure that the cup is entirely embedded in the slurry. Centralizers, attached to the riser tubing, may be required to center the sampler in the hole.
5. Pour additional slurry through the tremie pipe until it is about 1 ft (0.3m) above the top of the cup. This is determined by measuring inside the hollow-stem auger flight and calculating the borehole volume.
6. Check the integrity of the system by clamping the sample delivery line and applying a vacuum to the pressure-vacuum line. The vacuum should remain constant (see Figure 26.11). If a leak is detected, remove the assembly and determine the location of the leak.
7. Raise the auger flight about 2 ft (0.6 m). (If this is not done, backfill material may wedge the body tube and riser pipe inside the auger flight.)
8. After slurry has settled, add screened native material through the tremie pipe to a thickness of about 1 ft (0.3 m). Tamp the soil thoroughly by lowering a rod down the inside of the hollow stem.
9. Add about 1 ft (0.3 m) of dry bentonite [1/8 in (0.3 cm)] pellets through the tremie pipe. Shake the pipe while pouring the pellets to move the pellets

down. Also lift the pipe to assure that it is not plugged at the bottom. Pour distilled water through the tremie pipe to hydrate the pellets.

Note: The tremie pipe must be dry inside. Accordingly, two tremies may be required, one for dry material and one for slurry. Alternatively, the tremie pipe is used for wet slurry and deionized water and the inside of the hollow-stem auger flights for dry material.

10. Pour dry, screened native material through the tremie pipe and tamp thoroughly, ensuring that the material does not rise into the auger flight. In deep holes, i.e., 30 ft (9 m) or deeper, it is not critical to screen soil after placing 5 ft of material above the sampler.

11. Repeat the last step as the hole is filled, raising and removing auger flights as required. If possible, backfill the excavated soil above the bentonite in the order in which it was withdrawn. Additionally, try to compact the soil to its original bulk density.

12. Before the backfill reaches the land surface, place any security top riser into the borehole and cement the riser if necessary.

13. After backfilling to land surface, check the system for leaks by clamping off the sample line and applying a vacuum to the pressure-vacuum line. The vacuum should remain constant (see Figure 26.11).

14. Mound the backfill around the riser tubing to prevent ponded water from leaking along the side of the riser pipe.

15. Clamp lines and cover them with a plastic bag when not in use to prevent dust and insects from entering them. The lines should be protected from weather, sunlight exposure, and vandalism with a locked housing.

16. Carefully label the sampler with a permanent marking device. Use a simple mnemonic label to assist when collecting and analyzing data from the sampler. Additionally, the location of each sampler should be marked on a plan view of the facility being monitored.

17. Begin a logbook to keep records of the sampler's production and operation parameters.

Sampling Procedures

Following installation of samplers, the pores in the sampler and slurry should be purged of distilled or deionized water. Nitrogen or argon gas are recommended as purging agents. Debyle et al. (1988) suggested discarding the first one or two sample volumes when sampling dilute solutions with newly flushed (HCl method) and installed samplers. The purpose of this is to allow cation exchange between the porous segment and the pore-liquid (caused by the HCl flushing) to equilibrate. Bond and Rouse (1985) suggest discarding the first three samples. McGuire et al. (1992) recommend extracting several pore volumes of solution from a sampler for at least 24 hours before collecting a sample for metals analyses. However, in arid environments it is not unusual to have little or no sample until winter/summer rains, irrigation, or a leakage event from the facility being monitored.

It is important to calibrate the volume collected by the individual samplers as a check on their long-term performance and planning for laboratory analyses. Calibrated bottles are not typically available from laboratories; therefore, an alternative method must be developed to estimate the sample volume without transferring the sample from one vessel to another. Preferably, the sample bottles supplied by the laboratory are of the same size, shape, and volume from one sampling to another. If this is the case, a ruler or scale can be calibrated at the office to read the volume of water in the sample bottle. The precalibrated scale can be held on the side of the sample bottle and the water level read directly to indicate the corresponding volume on the scale. The calibrated scale should be marked in increments of 25 mL or less.

All samplers require a vacuum to draw soil pore liquids through the porous cup. The applied vacuum is a function of cup material, soil composition, and moisture content. For example, the upper operational range is 70 kPa for ceramic units, 20–30 kPa for stainless steel and fritted glass units, and less for PTFE units. Some sandy soils require a lower vacuum (30 to 40 kPa) application to induce steady flow and to avoid a dewatering halo effect in the immediate vicinity of the porous cup. Finer-grained sediments will require 50 to 60 kPa of vacuum.

Samples are collected under either constant or falling-head vacuum. Where electricity is available, constant vacuum is maintained by inexpensive, commercially-available vacuum pumps. Cole et al. (1961) constructed an array of samplers which was attached to a vacuum tank connected to an electric power source. This system allowed remote operation at a constant suction. Battery-operated vacuum pumps are also available for maintaining constant vacuum at remote stations (Soil Measurement Systems, 1993). Alternatively, Nelson (1985) suggested using portable vacuum tanks. The tanks are evacuated before they are transported to the field site. The desired vacuum in the samplers is maintained through a valve and vacuum gauge in the line connecting the tank and samplers. Nelson (1985) claims that the vacuum can be maintained over a long time. However, strictly speaking, this is a falling-head device. Vacuum for falling-head configurations may also be produced by simple hand pumps. Pressure-vacuum hand pumps are also available.

The selection of constant versus falling-head vacuums may affect the volume and rate of sample collection. Sample volume is important for analytical protocols. Collection rates may affect contaminant concentrations; Debyle et al. (1988) point out that longer collection times under unsaturated conditions may alter concentrations in collected samples. Sampling rates directly affect the sampling time; excessive collection times may exceed holding times for contaminants of concern. Morrison and Lowery (1990) conducted laboratory studies on sampling rates of ceramic porous-cup samplers under constant and falling vacuum conditions. When a constant vacuum is applied, they determined the decreasing order of importance of variables affecting sampling rate as follows:

vacuum > cup conductance > cup radius > cup surface area > cup length

For falling vacuum the order was:

vacuum > cup radius > cup conductance > internal sample volume
> cup surface area > cup length

The volume of sample collected by either the constant or falling-head methods depends, in part, on the type of cup material, hydraulic conductivity of the cup, soil texture, water content (soil suction), and applied vacuum. Morrison and Lowery (1990) determined that a constant vacuum increased the volume of sample collected by ceramic samplers by 1.65 more than a transient vacuum of 10, 30, 50, and 70 kPa. When collecting samples with a falling vacuum, they recommend providing a vacuum reservoir greater than the internal volume of the sampler and tubing to optimize sample collection rate, especially if the sampling period is greater than 8 hours. When a constant vacuum of 50 kPa was applied during their laboratory studies, McGuire and Lowery (1992) observed ceramic cups collected more sample in a sand at field moisture capacity than stainless steel or fritted glass plates. In contrast, stainless steel cups collected the largest samples from a silt loam at field capacity when a constant 25 kPa vacuum was applied. When a falling vacuum was applied, the fritted glass and stainless steel cups collected a larger sample volume at a greater rate than the ceramic cups in saturated sand. However, in sand at field capacity and silt loam at soil water suctions greater than 30 kPa, ceramic cups maintained their vacuum (70 kPa) and collected more samples during a 10 day test.

The effect of constant vacuum versus falling vacuum on sample characteristics is of concern. For their test conditions, Hansen and Harris (1975) found that a constant vacuum sampled more uniformly through time, producing a narrower range of sample concentrations compared with falling vacuum samplers. However, it does not necessarily follow that the data had less bias.

Based on their experimental observations, Morrison and Lowery (1990) concluded that the chemistry of initial pore-liquid samples from a constant vacuum sampler is most representative of water from larger pores adjacent to the sampler. Furthermore, the chemistry of later samples is more representative of water from large and small pore sequences near the sampler plus water from larger pores at distance (Morrison and Lowery, 1990). For falling-vacuum samplers, the initial high vacuums will draw water from both small and large pores. However, the predominant contribution will be from the larger pores because of the ease of fluid removal and greater velocities in these pores (Anderson, 1986). Later, as the vacuum declines, the contribution from the larger pores will dominate.

Air pressure is required to retrieve the sample. A hand-operated, pressure-vacuum hand pump is a simple means of obtaining a sample. Alternatively, compressed air in portable tanks is used. If there is concern that the sample composition may change during contact with compressed air, compressed nitrogen or argon gas is used. Nitrogen and argon gas are usually available from most air or welding supply companies.

Following is a suggested sampling procedure:

1. Clamp off the sample discharge line.
2. Apply a vacuum (either constant or falling) to the color-coded pressure-vacuum line. The degree of vacuum to apply depends on the type of suction cup. Thus, a vacuum of 70 kPa is commonly used for ceramic samplers, 20–30 kPa for stainless steel samplers, and less for PTFE samplers. (See the appendix to this chapter for conversions to other units, e.g., cm of mercury)
3. If vacuum is applied using a hand pump, check the vacuum gauge for 20–30 seconds to ensure that vacuum does not drop off, indicating a leak (Figure 26.11). If the vacuum holds, clamp off the pressure-vacuum line.
4. Allow a predetermined period of time to allow for sample collection. This is best determined experimentally for each sampler, type of soil, and depth as follows:
   • For constant vacuum systems, record the sample volume collected for various time increments. The time required to provide the necessary sample volume is selected.
   • For falling-head units, record the residual vacuum before extracting the

water and the sample volume for various time increments (Bond and Rouse, 1985). The time required to provide the necessary sample volume is selected. However, if the residual vacuum completely dissipates before an adequate sample is collected, it will be necessary to reapply the vacuum.

Note: The experimentally selected sample times will change with moisture status around the cup and recalibration will be required.

5. Upon arrival at the sample station, check the ends of the sample lines for dirt and dust accumulation. If necessary, flush the ends of the sample line using deionized water from a laboratory squirt bottle.

6. Connect the pressure-vacuum line to the vacuum gauge of the vacuum pump. Unclamp the pressure-vacuum lines and record the residual vacuum remaining on the unit; then release the vacuum from the unit.

7. Open the clamp on the discharge line.

8. Open the sample bottle provided by the laboratory. Insert the discharge line inside the top of the bottle.

9. Slowly apply an air (or nitrogen or argon) pressure to the pressure-vacuum line. The tank pressure valve should be operated at approximately 10 to 20 psi. If additional pressure is needed to retrieve the sample, the pressure may be increased in increments of 5 psi until adequate flow is achieved. For systems with long sample lines, record the pressure necessary for that particular unit for future reference. As pointed out by Bond and Rouse (1985), patience is required when collecting a sample from long sample lines. If no sample is collected, air will be felt exiting from the discharge line.

10. Discharge the sample into an appropriate collection bottle. Bond and Rouse (1985) suggest the following method if it is necessary to avoid contact of the sample with air:
    • insert a 50 mL (or larger) syringe head into the end of the sample tubing, without the needle
    • apply a pressure to the pressure-vacuum line
    • Withdraw the syringe plunger as the sample fills the chamber.

11. After collecting a sample, purge the sample line for about one minute by applying gas pressure to the pressure-vacuum line. When using inert gas, this will allow the gas to purge the sampler of residual atmospheric oxygen and soil air, and it will also purge the sample bottle head space. Place the cap on the sample bottle immediately.

12. Remove the gas source and clamp off the sample line. If necessary, place a vacuum on the unit at this time for the next sampling.

13. If the sample is to be split to other containers for duplicate analyses or acid preservation, it should be done at this time. In most cases. 0.45 micron filtering is not necessary.

14. If there is sufficient sample, obtain field measurements for pH, electrical conductivity, temperature, dissolved oxygen, or other parameters as needed.

15. Repeat Steps 1 through 14 for all samplers.

Collected samples should be stored and shipped using approved chain of custody procedures.

Field Monitoring and Recordkeeping

If sufficient sample volume is available, the measurement of field parameters such as pH, electrical conductivity, and temperature is very useful in evaluating the reproducible nature or viability of the samples from one sample collection episode to the next. Additional field monitoring should include the observation of the color, clarity, effervescence, and noticeable odor from the sample.

Maintaining a log or field record book is important to evaluate the performance of the individual sampling units. A page in the field logbook should be dedicated to each unit. Each page should provide ample space to record at least one year of data collection. Following is the recommended information to be included on each page:

- suction sampler identification number
- date of installation
- depth of porous segment
- sample date
- name of individual collecting the sample
- time of sample collection
- residual vacuum measured before sample collection
- volume of water sample retrieved
- sample description, such as odor, clarity, etc.
- field pH, electrical conductivity, and temperature
- vacuum applied, date, and time.

Laboratory Analyses and Priority Listing of Parameters

As suction samplers do not always yield enough sample volume for analysis of all desired constituents, an analytical priority list should be established. A full 22 inch (57 cm) long, 1.8 inch (4.5 cm) diameter suction sampler will provide about 0.9 liters of water. Customarily, the priority list is finalized after consulting with the regulatory personnel and the analytical laboratory. The laboratory will begin analyzing the high-priority parameters and continue down the list until the analysis is complete or the water sample has been depleted. The list should be altered to satisfy the analytical requirements as the project database develops.

FILTER TIP SAMPLERS

Filter tip samplers (commonly called BAT® samplers) are similar to suction cup samplers in that they collect pore liquids through a porous segment. At this time, these units are used for sampling saturated regions of the subsurface (e.g., Mines et al., 1993; Eckard et al., 1989; Hekma et al., 1990; Zemo et al., 1992; see also Wilson and Dorrance in this book). However, their design also accommodates sampling from unsaturated regions provided the porous segment has a high enough bubbling pressure. Unlike suction-cup samplers, with vacuum lines extending to the surface, a pore-liquid sample is drawn into the porous segment by connecting an evacuated glass-sample container to the tip. Figure 26.13 shows the three components of the sampler, including (1) a permanently-installed, sealed filter tip attached to an extension pipe; (2) an evacuated and sterilized, glass sample vial; and (3) a disposable double-ended hypodermic needle. The filter tip includes a pointed end to help with

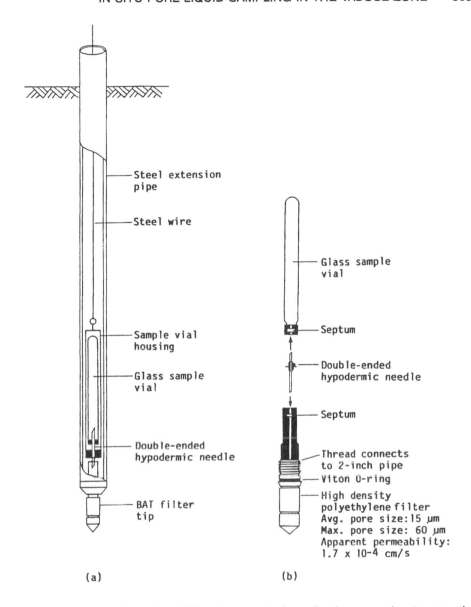

Steel extension
pipe

Steel wire

Sample vial
housing

Glass sample
vial

Double-ended
hypodermic needle

BAT filter
tip

Glass sample
vial

Septum

Double-ended
hypodermic needle

Septum

Thread connects
to 2-inch pipe

Viton O-ring

High density
polyethylene filter
Avg. pore size:15 µm
Max. pore size: 60 µm
Apparent permeability:
1.7 x 10^{-4} cm/s

(a)   (b)

Figure 26.13. (a) Configuration of filter tip sampler for collecting ground-water sample;
(b) detail of filter tip sampler and sampling components (Eckard et al.,
1989).

installation, a porous section, a cap, and a septum (Figure 26.13). The tip unit is
threaded onto riser pipes which terminate at the surface. Sample vials also contain a
resealable rubber septum. Mines et al. (1993) described a "cascading sampling technique" consisting of two hypodermic needles and two sample vials.

As described by Eckard et al. (1989), during sampling, a pre-sterilized, pre-
evacuated sample vial, sealed with a septum, is positioned within the housing (Figure
26.13). A spring-loaded guide sleeve, housing a double-ended hypodermic needle, is
threaded onto the lower end of the sample housing. The sample housing is lowered
down the extension pipe on a steel wire (Figure 26.13). A chain of weights on the

upper end of the sample housing aids the penetration of the double-ended needle through the septa. Thus, when the guide sleeve contacts the filter tip, the needle punctures both septa. The vacuum in the sample container causes pore liquid (or gas) to flow from the formation through the porous segment into the sample container. When the sample container is disconnected from the filter tip, the discs in the container and filter tip automatically reseal. In practice, the first sample, amounting to 8 to 30 mL, is purged prior to collecting samples for analyses. The sample container is then retrieved by reeling in the wire line.

The body of the filter tip is constructed from a variety of materials, including thermoplastic, stainless steel, or brass. The porous section is available in high density polyethylene (HDPE), porous ceramic, PTFE, or sintered stainless steel. The septum is made of natural rubber, nitrile rubber, or fluororubber (Haldorsen et al., 1985). Mines et al. (1993) found that much higher levels of VOCs were recovered when using stainless-steel filters rather than HDPE filters.

Two methods for installing filter-tip samplers are as follows (Eckard et al., 1989):

- Forcing the filter tip and extension pipe assembly into the ground to the desired sampling depth using a cone penetrometer rig or the hydraulic system of a conventional drill rig; and
- Driving the filter tip and assembly using a conventional 140-pound slide hammer either from land surface or in advance of a pre-drilled bore hole.

Forcing the filter-tip assembly into the ground using a cone-penetrometer rig, conventional drill rig, or by pounding, is limited to sandy soils and fine-grained materials, free of cobbles, coarse gravel, strongly-cemented layers, and otherwise consolidated soils (Hekma et al., 1990). For their studies in southern California, Eckard et al. (1989) drove the probes ahead of pre-drilled boreholes using a modified hammer. Hekma et al. (1990) installed filter-tip samplers using a 20-ton cone penetrometer.

In addition to permanently locating a sampler at a given depth, samplers are temporarily installed for reconnaissance studies. For this purpose, a sampler is available with a retractable sleeve to prevent cross contamination as the unit is positioned.

Information is not available on the use of a soil or silica flour slurry with these units. If the samplers are driven to final depth, a slurry may be unnecessary because a good soil/porous segment contact is probably achieved. Otherwise, a suitable slurry could be used.

Advantages and Limitations

Eckard et al. (1989) reviewed the advantages and limitations of filter-tip samplers. A principal advantage is the ability of these samplers to obtain samples in a hermetically-sealed filter tip without generating large volumes of purge water. Inasmuch as the sampling probe is sealed with a septum, only the fluid contained in the filter tip itself (i.e., 8 to 30 mL) needs to be purged prior to sampling. Additionally, since only a small borehole is required to install a filter-tip sampler, the mass of possibly contaminated vadose-zone sediments is reduced. Obviously, when installing units with a cone penetrometer rig, no deposits are generated. Another advantage is that pore-liquid samples can be taken in depth-wise increments.

The major limitation of filter-tip samplers for sampling pore-liquids in unsaturated regions of the vadose zone is probably the length of time that may be required to extract a sample, particularly under very dry conditions in fine material. Thus, a major concern is exceeding the holding time of constituents of interest. Research is needed to evaluate the utility of filter-tip samplers under various conditions in unsaturated regions.

VACUUM-PLATE SAMPLERS

A vacuum-plate sampler consists of a flat porous disk fitted with a nonporous backing attached to a suction line which leads to the surface (Figure 26.14). Plates are available in Alundum®, porous stainless steel, ceramic, or fritted glass. The nonpermeable backing can be a fiberglass resin, glass, plastic, or butyl rubber. Ceramic plates are available in diameters ranging from 0.6 to 10.9 inch (1.6 to 27.6 cm), and custom designs are easily arranged (Soilmoisture Equipment Corp., 1988; Morrison, 1983). An advantage of the larger plates is that they have large contact areas with the soil; therefore, larger sample volumes can be collected in shorter times than with vacuum lysimeters which have porous sections with smaller surface areas. They may also be in a position to intercept macropore fluids.

Ceramic plates are available in a variety of air entry values, ranging from 0.5 bar (high flow) to 5 bar (Soilmoisture Equipment Corp., 1988). For purposes of in situ pore-liquid extraction, one-bar plates are probably sufficient. One-bar, high flow plates have bubbling pressures of about 160 kPa, and pore sizes of 2.5 μm (Soilmoisture Equipment Corp., 1988).

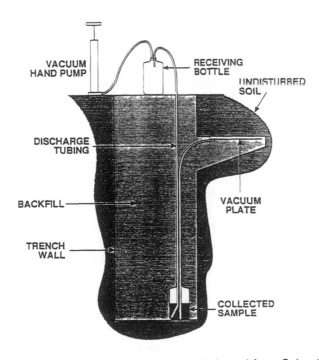

Figure 26.14. Vacuum plate sampler installation (Adapted from Cole, 1958).

Techniques for preparing porous plates are similar to those described for suction cup samplers, i.e., acid washing followed by thorough rinsing with distilled water. The plates are kept saturated until installation.

Installation techniques are described by Cole (1958) and Morrison (1983). Figure 26.14 shows that a trench is constructed with a horizontal tunnel. The tunnel should be about the same dimensions as the plate. The upper surface of the cavity should be scraped with a knife or wire brush to open pores that may have been smeared during excavation. Additionally, the upper surface should be completely flat to ensure an effective contact with the plate. After installation, pneumatic bladders or wedges are inserted beneath the plate to ensure a firm contact between the upper plate surface and the soil surface. Sample lines leading from the plate are protected from damage within a PVC manifold. The tunnel is backfilled firmly with soil.

Although only one unit is shown in Figure 26.14, conceivably, it is possible to install several units at different locations and depths in a trench.

The sampler tubing is inserted into a collection flask at the bottom of the trench. A second line is used to apply a vacuum to the flask and plate and to deliver the collected sample to the surface. Morrison (1983) and Cole et al. (1961) describe methods for applying a constant vacuum to the plates. Sampling by vacuum restricts the effective sampling depth. However, a system similar to the double-chamber, pressure-vacuum lysimeter design could be constructed to allow recovering samples by pressure. Thus, the effective sampling depth would be constrained only by installation equipment.

The advantages and limitations of this method are basically the same as for porous cup suction samplers. An added advantage is that a large volume can be obtained without disrupting the flow patterns (Eastern Research Group, 1993). A disadvantage is that trench installation procedures are more complicated; for example, compared to using a hollow-stem auger (Eastern Research Group, 1993).

CONSTRAINTS ON SUCTION SAMPLERS WHEN COLLECTING CHEMICAL DATA

Two field applications of pore-liquid suction samplers are (1) conducting field studies involving different soil or plant treatments (e.g., fertilizers), and (2) monitoring (detecting) contaminant transport. The major problem when collecting pore-liquid samples for either purpose is in interpreting the chemical data; i.e., do the chemical analyses truly represent the chemical quality of the in situ pore liquids? This problem is of greater concern for research studies attempting, for example, to quantify the mass flux of solutes beneath the root zone of crops. For monitoring studies, it is more important to detect the presence of a contaminant, and not necessarily the mass movement.

Factors affecting the interpretation of suction sampler data include those inherent with suction samplers and those associated with the physical nature of the vadose zone. Factors inherent with suction samplers are as follows:

- design and operational factors
- interactions between sampler material and contaminants of interest
    - inside the sample
    - within the cup pores
- effect of sampler on soil water flow regime
- limited area of influence

Vadose zone properties include:

- spatial variability of soil physical properties
- presence of macropores

Cup-Related Factors

According to Angle et al. (1991), factors inherent in the design and operation of
suction cup samplers affecting the chemical composition of collected sample include
the size of the cup, amount of vacuum applied, and volume of sample collected.
Regarding cup size, Silkworth and Grigal (1981) recommended using samplers with
large diameter porous sections since they showed less of a tendency to alter the pore-
liquid, they had lower failure rates, and they collected larger sample volumes. These
recommendations were supported by van der Ploeg and Beese (1977), who concluded that samplers with large cross sectional area porous sections used with low
extraction rates (suctions approaching those of the pore-liquid tensions) reduce the
effects of sampling on compositions of samples.

Hansen and Harris (1975) suggested the following guidelines to reduce sampler
data variability:

- select samplers with similar intake rates
- use a short sampling period
- use a similar sampler length (and diameter)
- use the same initial vacuum for all samplers
- do not composite samples as this may obscure biases associated with sampling technique

Chemical interactions between porous segments and the liquids which pass
through them affect the representativeness of the collected samples (Litaor, 1988;
Hansen and Harris, 1975; Grossmann and Udluft, 1991). Potential interactions
include sorption, desorption, cation exchange, precipitation, and screening (Hansen
and Harris, 1975). Biodegradation of chemicals is also a possibility (Lewis et al.,
1992). These interactions can also occur with other components of the samplers
contacted by liquids. However, the much higher surface area within the pores of
porous segments makes them the most critical component chemically. Table 26.3
summarizes the results of studies on interactions between selected inorganic and
organic chemical and porous sections (ceramic, fritted glass, PTFE, stainless steel,
and silica flour). Relevant features of the documented studies are listed. However,
the reader should refer to the original papers (Table 26.4) to determine if experimental techniques are applicable to the situation of concern. The absence of entries for a
constituent relative to a material does not infer absence of interactions. Additional
information on interactions of chemicals with different cup materials is included in
the section: Materials for Porous Segments, Expected Contaminants.

Changes in the composition of sampled liquids may also occur within the porous
cup because of gas exchange between the sample and the atmosphere within the cup
(Litaor, 1988; Grossmann and Udluft, 1991). This problem may cause volatilization
of organic compounds, changes in carbonate equilibrium, and precipitation of
metals. Short sampling times coupled with the use of inert gas to force the sample
into the sample lines will minimize some of these reactions. Thus, Litaor (1988)
observed that a short vacuum period (2-3 hours) followed by simultaneous evacua-

Table 26.3.   Porous Material Interactions[a-d]

	Material Absorbs Species	Material Desorbs or Releases Species	Material Screens Species	No Significant Interaction	No Interaction
Al		C(2,30)		C(16,27)	
Alkalinity				SF(11)	
Ca	C(20,28)	C(1,2,18) PTFE (26)		C(3,6,10,11,25) PTFE (3) FG(18,22)	PTFE(13)
C		FG(22)			
CO_3		C(2)			
HCO_3		C(2)			
Cd	C(9,11,29) SS(29) SF(29)			C(3) PTFE(3,29)	
Cl				C(11,25) SF(11)	PTFE(13)
Co	C(29) SS(29) PTFE (29) FG(29) SF(29)				
Cr	C(19) SS(29) PTFE (29) FG(29) SF(29)	C(3) PTFE (3)			
Cu	C(9,11)	C(3) PTFE(3)			
Fe	C(11)	PTFE(3,26)		C(3,25)	PTFE(13)
H				SF(11)	
K	C(5,6,15,28)	C(18)		C(1,25) FG(18,22)	
Mg	C(6,28)	C(2,3,11,18) PTFE(26)		C(10,25) PTFE(3) FG(18,22)	PTFE(13)
Mn	C(11)			C(3,9) PTFE(3)	PTFE(13)
Na	C(6,20)	C(2,18,28) FG(18,22) PTFE(26)		C(1,11,25)	PTFE(13)
NH_4	C(4,12,15)			PTFE(4)	
N		FG(22)			
NO_2				C(4,5) PTFE(4)	
$(NO_2$ + $NO_3)-N$				C(5)	
P	C(1,5,8, 15,18)			FG(18)	

Table 26.3. Porous Material Interactions[a–d] (Continued)

	Material Absorbs Species	Material Desorbs or Releases Species	Material Screens Species	No Significant Interaction	No Interaction
PO_4	C(4,5,7)			PTFE(4)	
PO_4-P				C(10)	
Pb	C(9)				PTFE(13)
SiO_2		C(2)			
Si		PTFE(26)		C(4) PTFE(4)	
SO_4				C(11)	
Sr		C(11)			
Zn	C(9)	C(11,29) SS(29) FG(29) SF(29)		PTFE(29)	PTFE(13)
High Molecular Weight Compounds			C(17,21)		
4-Nitro-phenol	PTFE (23)				
Chlorinated Hydrocarbons	PTFE(23,24)				
Diethyl-phthalate					PTFE(23)
Naphthalene	PTFE(23)				
Acenaphthene	PTFE(23)				
BTX	PTFE(30)				SS(30)
Bacteria	C(31)				
Virus	C(32)				SS(32)

[a]Comparisons of materials based on this table should be made cautiously. Differing experimental techniques should be considered as a source of differing conclusions. Undocumented factors often include material age and sampling history.
[b]Valence states are often not reported in studies.
[c]Abbreviations:
  1. C = porous ceramic
  2. PTFE = porous PTFE
  3. FG = fritted glass
  4. SS = porous stainless steel
  5. SF = silica flour
[d]Numbers in parentheses refer to references in Table 26.4.

Table 26.4. Porous Material Interaction Original Studies and Features

No. in Table	Reference	Porous Section Was Washed	Results are a Function of Several Factors	Dilute Solutions Were Tested	Experiments Were Performed on Non-Porous Materials
1	Grover and Lamborn, 1970	X			
2	Wolff, 1967	X			
3	Creasey and Dreiss, 1986	X			
4	Zimmermann et al., 1978	X			
5	Nagpal, 1982		X		
6	Debyle et al., 1988	X	X		
7	Bottcher et al., 1984	X			
8	Hansen and Harris, 1975	X			
9	Grossmann and Udluft, 1991	No information provided on experimental technique	X		
10	Levin and Jackson, 1977				
11	Peters and Healy, 1988	X			
12	Wagner, 1962	No information provided on experimental technique			
13	Morrison, 1983	X			
14	Neary and Tomassini, 1985	X	X	X	
15	Faber and Nelson, 1984	X			
16	Litaor, 1988		X		
17	Law, 1982	No information provided on experimental technique			
18	Silkworth and Grigal, 1981	X	X		
19	Anderson, 1986		X		
20	Barbarick et al., 1979	X	X		
21	Wagemann and Graham, 1974	X	X		
22	Jones and Miller, 1988		X	X	X
23	Barcelona et al., 1988		X	X	X
24	Johnson and Cartwright, 1980	X	X		
25	Maitre et al., 1991	X			
26	Hughes and Reynolds, 1988	Acid-washed and non-acid washed cuts were evaluated			
27	McGuire et al., 1992	X	X	X	
28	Hughes and Reynolds, 1990	X		X	
29	Dazzo and Rothwell, 1974				
30	Gillham and O'Hannesin, 1990				X
31	Dazzo and Rothwell, 1974	X			
32	Powelson, 1990	X			

tion, filtration, and acidification under an argon atmosphere prevented the oxidation of ferric to ferrous iron. Wood et al. (1981) describe a modified sampler which minimizes the loss of volatile organic compounds.

Figure 26.2 shows that application of a suction within a sampler creates a potential field around a suction cup, inducing flow toward the cup. According to Grossmann and Udluft (1991), the potential gradient acts on all pores, and pore flow velocity is a function of pore diameter. The relationship between flow to the cup and chemistry of the collected sample depends on the ambient flow conditions, i.e., stagnant, quasi-steady state, and transient. If the flow system is relatively stagnant, Grossmann and Udluft (1991) reason that the chemistry of collected samples should be independent of pore sizes, i.e., diffusion and dispersion will have promoted equilibration. For these conditions, they concluded that there is no reason to assume that suction cups extract liquid from pores of a certain size. In contrast, the work of Morrison and Lowery (1990), under quasi-static conditions in draining profiles, suggests that the chemistry of pore-liquid samples from a constant vacuum sampler initially represents contributions from larger pores adjacent to the sampler. Later samples represent the chemistry of macropore and micropore liquid near the sampler plus liquid from larger pores at distance.

During transient flow, the chemical composition of soil pore liquids is likely to vary spatially, given the highly variable nature of pore-liquid movement caused by preferential flow paths and other physical factors (Grossmann and Udluft, 1991). Litaor (1988), citing the work of Talsma et al. (1979), observed that in the presence of considerable flow, there is much less disturbance of the flow distribution around the porous cup than during unsaturated conditions. During saturated flow, the larger pore sizes are the major flow channels. Thus, the chemistry of collected samples reflects that in the larger pore sizes.

The effect of suction within a sampler on the flow regime is minimized by applying a vacuum similar to background pore-liquid suction. This is best determined by installing tensiometers or other units which measure the energy status of pore liquids.

The volume of soil pores affected by application of suction within a porous cup is very small compared with the total pore volume available for contaminant movement. As the work of Morrison and Lowery (1990) demonstrated, the sampling radius of a porous cup sampler may be on the order of centimeters. For example, Angle et al. (1991) estimated that under steady state conditions, a 0.78-inch (2-cm) cup would cause deflection of a uniform flow field within 3.93 inches (10 cm) of the cup. For research studies involving comparisons between treatments, many samplers will be required per plot to obtain the precision needed to estimate the mean and differences between means (Angle et al., 1991). The problem of replication is related to the spatial variability of soil properties, to be discussed subsequently. For monitoring programs, it is important for the samplers to detect contaminant flow. In this case, procedures described in the section dealing with sampler depths, areal distribution, and optimum number of units are necessary.

Vadose Zone Properties

Vadose zone properties governing the movement of pore liquids and entrained contaminants vary spatially in a geological profile. For example, according to Warrick and Amoozegar-Fard (1980), the following properties have coefficient of variation values greater than 100 percent:

- saturated hydraulic conductivity
- unsaturated hydraulic conductivity
- apparent diffusion coefficient
- pore water velocity.

The effect of spatial variability of soil properties on pore-liquid sampling was addressed during field studies by Biggar and Nielsen (1976), who found that " . . . point measurements such as the discharge from drainage tiles, solution samples from suction probes and piezometer wells, and excavated soil samples . . . provide good indications of relative changes in the amount of solute being transported but not quantitative estimates." An additional consequence of spatial variations in soil properties is that concentration data are lognormally distributed. Accordingly, data are skewed and more variable than normally-distributed parameters (Angle et al.,1991). Thus, a large number of samplers is required to collect a "representative" sample. For example, Alberts et al. (1977), as cited by Litaor (1988), determined that 246 replicates would be required to ensure locating the estimated mean concentration of nitrate in soil solution samples within 5% of the true mean. In practice, a greater number of samplers would be required to compensate for failure of some units. According to Angle et al. (1991), statistical methods are available to address the problem of spatial variability and improve experimental efficiency. Examples of such methods are incomplete block designs, nearest neighbor analyses, and trend analyses (Angle et al., 1991).

Macropore flow, a condition related to the spatial variability of soil physical properties, may affect the quality of samples collected by suction samplers. Specifically, during infiltration events, macropore flow may bypass suction lysimeters. The quality of fast-moving water in macropores will be different from the slower moving liquid in finer pore sequences. The goals of controlled experiments (estimating mass changes in solutes) and monitoring (detection of pollutants) will not be attained. For conditions favoring macropore flow, the use of free-drainage samplers together with suction samplers is recommended. Steenhuis et al., in this volume, present a case study comparing the effectiveness of suction samplers versus free drainage samplers in structured soils.

CONCLUDING REMARKS

This chapter provides a state-of-the-art review of pore liquid sampling and pore-liquid samplers in the vadose zone. Judging from the extensive literature review accompanying this section, and the thorough reviews by others (Litaor, 1988; Grossmann and Udluft, 1991), the research on pore liquid sampling technology is extensive. However, much work remains, particularly in light of options resulting from impending EPA regulations on vadose zone monitoring at alternative waste disposal facilities (Durant et al., 1993). Specifically, further studies are required to characterize interactions between chemicals of interest and the porous segments of samplers. The effect of sampling procedures on sample integrity also requires further study. The question of characterizing "sample representativeness" continues to be a challenge, particularly for practitioners conducting mass balances of nutrients and/or contaminants in the vadose zone.

Finally, despite many uncertainties in their use, pore-liquid sampling remains a valuable component of a comprehensive vadose zone monitoring system. We encourage practitioners to share their experiences, good and bad, with these devices by publishing in the technical literature and by giving oral/poster presentations at technical sessions.

ACKNOWLEDGMENT

Thanks to Helen Wilson for her help with some of the illustrations.

APPENDIX: PRESSURE-VACUUM EQUIVALENTS

1 bar is the equivalent of

- 100 centibars (cb)
- 100 kiloPascals (kPa)
- 0.1 MegaPascals (MPa)
- 0.987 atmospheres (atm)
- 10^6 dynes/cm^2
- 33.5 ft of water
- 401.6 inches of water
- 1020 cm of water
- 29.5 inches of Hg
- 75 cm of Hg
- 750 mm of Hg
- 14.5 pounds per square inch (psi)

REFERENCES

Alberts, E.E., R.E. Burwell, and G.E. Schuman, "Soil Nitrate-Nitrogen Determined by Coring and Solution Extraction Techniques," *Soil Science*, 41:90–92 (1977).

Anderson, L.D., "Problems Interpreting Samples Taken with Large-Volume, Falling Suction Soil-Water Samplers," *Ground Water*, 24:761–769 (1986).

Angle, J.S., M.S. McIntosh, and R.L. Hill, "Tension Lysimeters for Collecting Soil Percolate," in *Groundwater Residue Sampling*, R.G. Nash and A.R. Leslie, Eds., American Chemical Society, Washington, DC, 1991, pp. 290–299.

Barbarick, K.A., B.R. Sabey, and A. Klute, "Comparison of Various Methods for Sampling Soil Water for Determining Ionic Salts, Sodium, and Calcium Content in Soil Columns," *Soil Science Society of America Journal*, 43:1053–1055 (1979).

Barbee, G.C., and K.W. Brown, "Comparison Between Suction and Free-Drainage Soil Solution Samplers," *Soil Science*, 149–154, 1986.

Barcelona, M.J., J.A. Helfrich, and E.E. Garske, "Verification of Sampling Methods and Selection of Materials for Ground-Water Contamination Studies," *Ground-Water Contamination Field Methods*, American Society of Testing Materials, STP 963, Philadelphia, PA, 1988.

Beasley, R.S., "Contribution of Subsurface Flow from the Upper Slopes of Forested Watershed to Channel Flow," *Soil Science Society of America Journal*, 40:955–957 (1976).

Beier, C., K. Hansen, P. Gundersen, and B.R. Anderson, "Long-Term Field Comparison of

Ceramic and Poly(tetrafluoroethene) Porous Cup Soil Water Samplers," *Environmental Science and Technology*, 26:2005–2011 (1992).

Beier, C., and K. Hansen, "Evaluation of Porous Cup Soil-Water Samplers Under Controlled Field Conditions: Comparison of Ceramic and PTFE Cups," *Journal of Soil Science*, 43:261–271 (1992).

Bell. R.G., "Porous Ceramic Soil Moisture Samplers, An Application in Lysimeter Studies on Effluent Spray Irrigation," *New Zealand Journal of Experimental Agriculture*, 2:173–175 (1974).

Beven, K., and P. Germann, "Macropores and Water Flow in Soils," *Water Resources Research*, 18:1311–1325 (1982).

Biggar, J.W., and D.R. Nielsen, "Spatial Variability of the Leaching Characteristics of a Field Soil," *Water Resources Research*, 12:78–84 (1976).

Bond, W.R., and J.V. Rouse, "Lysimeters Allow Quicker Monitoring of Heap Leaching and Tailing Sites," *Mining Engineering*, 37:314–319 (1985).

Bottcher, A.B., L.W. Miller, and K.L. Campbell, "Phosphorous Adsorption in Various Soil-Water Extraction Cup Materials: Effect of Acid Wash," *Soil Science*, 137:239–244 (1984).

Bouwer, H., *Groundwater Hydrology*, (New York: McGraw-Hill Book Company, 1978).

Brose, R.J., R.W. Shatz, and T.M. Regan, "An Alternate Method of Lysimeter and Flour Pack Placement in Deep Boreholes," in *Proceedings of the Sixth National Symposium and Exposition on Aquifer Restoration and Ground Water Monitoring*, National Water Well Association, Water Well Publishing Co., Dublin, OH, 1986, pp. 88–95.

Brown, K.W., G.C. Barbee. J.C. Thomas, and H.E. Murray, "Detecting Organic Contaminants in the Unsaturated Zone Using Soil and Soil-Pore Water Samples," *Hazardous Waste and Hazardous Materials*, 7 (2):151–168 (1990).

Bumb, A.C., C.R. McKee, R.B. Evans, and L.A. Eccles, "Design of Lysimeter Leak Detector Networks for Surface Impoundments and Landfill," *Ground Water Monitoring Review Journal*, 8:102–114 (1988).

Cole, D.W., "Alundum Tension Lysimeter," *Soil Science*, 85:293–296 (1958).

Cole, D., S. Gessell, and E. Held, "Tension Lysimeter Studies of Ion and Moisture Movement in Glacial Till and Coral Atoll Sites," *Soil Science Society of America Proceedings*, 25:321–325 (1961).

Creasey, C. L. and S.J. Dreiss, "Soil Water Sampler: Do They Significantly Bias Concentrations in Water Samples?," *Proceedings of the NWWA Conference on Characterization and Monitoring of the Vadose Zone*, National Water Well Association, Water Well Publishing Co., Dublin, OH, 1986, pp. 173–181.

Davis, L.A., and S.P. Neuman, Documentation and User's Guide: UNSAT2-Variably Saturated Flow Model, Water, Waste and Land, Inc., Prepared for the U. S. Nuclear Regulatory Commission, NUREG/CR-3390, WWL/TM-1791-1, 1983.

Dazzo, F., and D. Rothwell, "Evaluation of Porcelain Cup Soil Water Samplers for Bacteriological Sampling," *Applied Microbiology*, 27:1172–1174 (1974).

Debyle, N.V., R.W. Hennes, and G.E. Hart, "Evaluation of Ceramic Cups for Determining Soil Solution Chemistry," *Soil Science*, 146:30–36 (1988).

Dorrance, D.W., L.G. Wilson, L.G. Everett, and S.J. Cullen, "Compendium of In Situ Pore-Liquid Samplers for Vadose Zone," in *Groundwater Residue Sampling*, R.G. Nash and A.R. Leslie, Eds., American Chemical Society, Washington, DC, 1991, pp. 300–331.

Durant, N.D., V.B. Myers, and L.A. Eccles, "EPA's Approach to Vadose Zone Monitoring at RCRA Facilities," *Ground Water Monitoring and Remediation*, 13-20, 1993.

Eastern Research Group, *Subsurface Characterization and Monitoring Techniques: A Desk Reference Guide, Volume II: The Vadose Zone, Field Screening and Analytical Methods*, United States Environmental Protection Agency, Office of Research and Development, EPA/625/R-93/003b, 1993.

Eckard, T.L., D. Millison, J. Muller, E. Vander Velde, and R.U. Bowallius, "Vertical Groundwater Quality and Pressure Profiling Using the BAT Groundwater Monitoring System," in *Proceedings of the Third National Outdoor Action Conference on Aquifer Restoration, Ground Water Monitoring, and Geophysical Methods*, National Ground Water Association, Orlando, FL, 1989.

England, C.B., "Comments on 'A Technique Using Porous Cups for Water Sampling at Any Depth in the Unsaturated Zone' by Warren W. Wood," *Water Resources Research*, 10:1049 (1974).

Everett, L.G., and L.G. McMillion, "Operational Ranges for Suction Lysimeters," *Ground Water Monitoring Review*, 51(5):51–60 (1985).

Everett, L.G., L.G. Wilson, and E.W. Hoylman, *Vadose Zone Monitoring Concepts for Hazardous Waste Sites*, (Park Ridge, NJ:Noyes Data Corporation, 1984).

Faber, W.R., and P.V. Nelson, "Evaluation of Methods for Bulk Solution Collection from Container Root Media," *Communications in Soil Science and Plant Analysis*, 15:1029–1040 (1984).

Finger, S. M. and H. Hojaji, "Effectiveness of Porous Glass Segments for Suction Lysimeters to Monitor Soil Water for Organic Contaminants," in *Field Screening Methods for Hazardous Wastes and Toxic Chemicals*, Second International Symposium, U.S. Environmental Protection Agency, U.S. Department of Energy, U.S. Army Toxic and Hazardous Materials Agency, U.S. Army Chemical Research, Development and Engineering Center, U.S. Air Force, Florida State University, National Environmental Technology Applications Corporation, and National Institute for Occupation Safety and Health, 1991, pp. 657–670.

Gillham, R.W., and S. F. O'Hannesin, "Sorption of Aromatic Hydrocarbons by Materials Used in Construction of Ground-Water Sampling Wells," in *Ground Water and Vadose Zone Monitoring*, D.M. Nielsen and A.I. Johnson, Eds., ASTM, STP 1053, American Society for Testing and Materials, Philadelphia, PA, 1990, pp. 108–122.

Graham, D.D., Methodology, Results, and Significance of an Unsaturated-Zone Tracer Test at an Artificial Recharge Facility, Tucson, AZ, U.S. Geological Survey Water Resources Investigation Report 89-4097, 1989.

Grossmann, J., and P. Udluft, "The Extraction of Soil Water by the Suction-Cup Method: A Review," *Journal of Soil Science*, 42:83–93 (1991).

Grover, B.L., and R.E. Lamborn, "Preparation of Porous Ceramic Cups To Be Used for Extraction of Soil Water Having Low Solute Concentrations," *Soil Science Society of America Proceedings*, 34:706–708 (1970).

Haines, B.L., J.B. Waide, and R.L. Todd, "Soil Solution Nutrient Concentrations Sampled with Tension and Zero-Tension Lysimeters: Report of Discrepancies," *Soil Science Society of America Journal*, 46:658–660 (1982).

Haldorsen, S., A.M. Petsonk, and B. Tortensson, "An Instrument for In Situ Monitoring of Water Quality and Movement in the Vadose Zone," in *Proceedings of the NWWA Conference on Characterization and Monitoring of the Vadose (Unsaturated) Zone*, Water Well Journal Publishing Co., Dublin, OH, 1985.

Hanks, R.J., and G.L. Ashcroft, *Applied Soil Physics*, (New York: Springer-Verlag, 1980).

Hansen, E.A., and A.R. Harris, "Validity of Soil-Water Samples Collected with Porous Ceramic Cups," *Soil Science Society of America Proceedings*, 39:528-536 (1975).

Hekma, L.K., K.K. Muraleetharan, and G.F. Boehm, "Application of the Electric Cone Penetration Test to Environmental Groundwater Investigations," unpublished paper presented at the Air and Waste Management Association, 83rd Annual Meeting and Exhibition, Pittsburgh, PA, 1990.

Hughes, S., and B. Reynolds, "Evaluation of Porous Ceramic Cups for Monitoring Soil-Water Aluminum in Acid Soils: Comment on a Paper by Raulund-Rasmussen (1989)," *Journal of Soil Science*, 41:325-328 (1988).

Johnson, T.M., and K. Cartwright, Monitoring of Leachate Migration in the Unsaturated Zone in the Vicinity of Sanitary Landfills, Illinois State Geological Survey Circular 514, Urbana, IL, 1980.

Jones, J.N., and G.D. Miller, "Adsorption of Selected Organic Contaminants onto Possible Well Casing Material," *Ground-Water Contamination Field Methods*, American Society of Testing Materials, STP 963, Philadelphia, PA, 1988.

Jury, W.A., W.R. Gardner, W.R., and W.H. Gardner, *Soil Physics*, (New York: John Wiley and Sons, Inc., 1991).

Krone, R.B., H.F. Ludwig, and J.F. Thomas, "Porous Tube Device for Sampling Soil Solutions During Water Spreading Operations," *Soil Science*, 73:211-219 (1952).

Lappala, E.G., R.W. Healy, and E.P. Weeks, Documentation of Computer Program VS2D to Solve the Equations for Fluid Flow in Variably-Saturated Porous Media, U.S. Geological Survey, Water Resources Investigations Report 83-4099, 1987.

Law Engineering Testing Company, *Lysimeter Evaluation Study*, American Petroleum Institute, 1982.

Levin, M.J., and D.R. Jackson, "A Comparison of In Situ Extracts for Sampling Soil Water," *Soil Science Society of America Journal*, 41:535-536 (1977).

Lewis, D.L., A.P. Simons, W.B. Moore, and D.K. Gattie, "Treating Soil Solution Samplers to Prevent Microbial Removal of Analytes," *Applied and Environmental Microbiology*, 58(1):1-5 (1992).

Linden, D.R., Design, Installation and Use of Porous Ceramic Samplers for Monitoring Soil-Water Quality, U.S. Department of Agriculture Technical Bulletin No. 1562, 1977.

Litaor, M.I., "Review of Soil Solution Samplers," *Water Resources Research*, 24:727-733 (1988).

Maitre, V., G. Bourrie, and P. Curmi, "Contamination of Collected Soil Water Samples by the Dissolution of the Mineral Constituents of Porous P.T.F.E. Cups," *Soil Science*, 152:289-293 (1991).

McGuire, P.E., B. Lowery, and P.A. Helmke, "Potential Sampling Error: Trace Metal Adsorption on Vacuum Porous Cup Samplers," *Soil Science Society of America Journal*, 56:74-82 (1992).

McKee, C.R., and A.C. Bumb, "A Three-Dimensional Analytical Model to Aid in Selecting Monitoring Locations in the Vadose Zone," *Ground Water Monitoring Review*, 8:124-136 (1988).

Mines, B.S., J.L. Davidson, D. Bloomquist, and T.B. Stauffer, "Sampling of VOCs with the

BAT Ground Water Sampling System," *Ground Water Monitoring and Remediation*, 13:115–120 (1993).

Morrison, R.D., and B. Lowery, "Effect of Cup Properties, Sampler Geometry, and Vacuum on the Sampling Rate of Porous Cup Samplers," *Soil Science*, 149:308–316 (1990).

Morrison, R.D., Ground Water Monitoring Technology, Timco MFG., Inc., Prairie du Sac, WI, 1983.

Nagpal, N.K., "Comparison Among and Evaluation of Ceramic Porous Cup Soil Water Samplers for Nutrient Transport Studies," *Canadian Journal of Soil Science*, 62:685–694 (1982).

Neary, A.J., and F. Tomassini, "Preparation of Alundum/Ceramic Plate Tension Lysimeters for Soil Water Collection," *Canadian Journal of Soil Science*, 65:169–177 (1985).

Nelson, A.B., "Use of Vacuum Tanks with Shallow Vacuum Lysimeters," *Ground Water*, 23:802–803 (1985).

Nielsen, D.M., and R. Schalla, "Design and Installation of Ground-Water Monitoring Wells," in *Practical Handbook of Ground-Water Monitoring*, D.M. Nielsen, Ed., (Chelsea, MI: Lewis Publishers, 1991), pp. 239–331.

Nightingale, H.I., D. Harrison, and J.E. Salo, "An Evaluation Technique for Ground Water Quality Beneath Urban Runoff Retention and Percolation Basins," *Ground Water Monitoring Review*, 5:43–50 (1985).

Peters, C.A., and R.W. Healy, "The Representativeness of Pore Water Samples Collected from the Unsaturated Zone Using Pressure-Vacuum Lysimeters," *Ground Water Monitoring Review*, 8:96–101 (1988).

Pettyjohn, W.A., W.J. Dunlap, R. Cosby, and J.W. Keeley, "Sampling Ground Water for Organic Contaminants," *Ground Water*, 19:180–189 (1981).

Powelson, D.K., C.P. Gerba, and M.T. Yahya, "Virus Transport and Removal in Wastewater During Aquifer Recharge," *Water Resources*, 27:583–590 (1993).

Powelson, D.K., personal communication, 1990.

Reynolds, G.W., and R.W. Gillham, "Adsorption of Halogenated Organic Compounds by Polymer Materials Commonly Used in Ground-Water Monitoring," in *Proceedings, Second Canadian/American Conference on Hydrogeology*, National Water Well Association, Dublin, OH, 1985, pp. 125–132.

Robbins, G.A., and M.M. Gemmell, "Factors Requiring Resolution in Installing Vadose Zone Monitoring Systems," *Ground Water Monitoring Review*, 5:75–80 (1985).

Severson, R.C., and D.F. Grigal, "Soil Solution Concentrations: Effect of Extraction Time Using Porous Ceramic Cups Under Constant Tension," *Water Resources Bulletin*, 12:1161–1169 (1976).

Shaffer, K.A., D.D. Fritton, and D.E. Baker, "Drainage Water Sampling in a Wet, Dual-Pore Soil System," *Journal of Environmental Quality*, 8:241–246 (1979).

Sheppard, M.I., D.H. Thibauld, and P.A. Smith, "Effect of Extraction Techniques on Soil Pore-Water Chemistry," *Communications in Soil Science and Plant Analyses*, 23:1643–1662 (1992).

Shuford, J.W., D.D. Fritton, and D.E. Baker, "Nitrate-Nitrogen and Chloride Movement Through Undisturbed Field Soil," *Journal of Environmental Quality*, 6:736–739 (1977).

Silkworth, D.R. and D.F. Grigal, "Field Comparison of Soil Solution Samplers," *Soil Science Society of America Journal*, 45:440–442 (1981).

Smith, J.A., H.J. Cho, P.R. Jaffe, C.L. Macleod, and S.A. Koehnlein, "Sampling Unsaturated-Zone water for Trichloroethene at Picatinny Arsenal, New Jersey," *Journal of Environmental Quality*, 21:264–271 (1992).

Soil Measurement Systems, *Product Documentation on Suction Lysimeters*, Tucson, AZ, 1993.

Soilmoisture Equipment Corp., Sales Division, Catalog of Products, Soilmoisture Equipment Corp., Santa Barbara, CA, 1988.

Starr, J.L., J.J. Meisenger, and T.B. Parkin, "Experiences and Knowledge Gained from Vadose Zone Sampling," in *Groundwater Residue Sampling*, R.G. Nash and A.R. Leslie, Eds., American Chemical Society, Washington, DC, 1991, pp. 279–289.

Talsma, T., P.M. Hallam, and R.S. Mansell, "Evaluation of Porous Cup Soil-Water Extractors, Physical Factors," *Australian Journal of Soil Resources*, 17:417–422 (1979).

Thomas, G.W., and R.E. Phillips, "Consequences of Water Movement in Macropores," *Journal of Environmental Quality*, 8:149–152 (1979).

Timco Mfg. Inc., Sales Division, Timco Lysimeters, Timco Mfg. Inc., Prairie du Sac, WI, 1988.

Tyler, D.D., and G.W. Thomas, "Lysimeter Measurements of Nitrate and Chloride Losses from Soil Under Conventional and No-Tillage Corn," *Journal of Environmental Quality*, 6:63–66 (1977).

U.S. Environmental Protection Agency Environmental Monitoring Systems Laboratory, *Permit Guidance Manual on Unsaturated Zone Monitoring for Hazardous Waste Land Treatment Units*, U.S. Environmental Protection Agency, Office of Solid Waste and Emergency Response, EPA/530-SW-86-040, 1986.

Van der Ploeg, R.R., and F. Beese, "Model Calculation for the Extraction of Soil Water by Ceramic Cups and Plates," *Soil Science Society of America Journal*, 41:466–470 (1977).

Wagemann, R., and B. Graham, "Membrane and Glass Fibre Filter Contamination in Chemical Analyses of Fresh Water," *Water Research*, 8:407–412 (1974).

Wagner, G.H., "Use of Porous Ceramic Cups to Sample Soil Water Within the Profile," *Soil Science*, 94:379–386 (1962).

Warrick, A.W., and A. Amoozegar-Fard, "Soil Water Regimes Near Porous Cup Water Samplers," *Water Resources Research*, 13:203–207 (1977).

Wilson, L.G., J.N. Luthin, and J.W. Biggar, Drainage Salinity Investigation of the Tulelake Lease Lands, California Agricultural Experiment Station, Bulletin 779, 1961.

Wolff, R.G., "Weathering Woodstock Granite, Near Baltimore, Maryland," *American Journal of Science*, 265:106–117 (1967).

Wood, W.W., "A Technique Using Porous Cups for Water Sampling at Any Depth in the Unsaturated Zone," *Water Resources Research*, 9:486–488 (1973).

Wood, A.L., J.T. Wilson, R.L. Cosby, A.G. Hornsby, and L.B. Baskin, "Apparatus and Procedure for Sampling Soil Profiles for Volatile Organic Compounds," *Soil Science Society of America Journal*, 45:442–444 (1981).

Zabowski, D., and F.C. Ugolini, "Lysimeter and Centrifuge Soil Solutions, Seasonal Differences Between Methods," *Soil Science Society of America Journal*, 54:1130–1135 (1990).

Zemo, D.A., T.A. Delfino, J.D. Gallinatti, V.A. Baker, and L.R. Hilpert, "Cone Penetrometer Testing and Discrete-Depth Groundwater Sampling Techniques: A Cost-Effective Method of Site Characterization in a Multiple-Aquifer Setting," *in Proceedings of the Sixth National Outdoor Action Conference on Aquifer Restoration, Ground Water Monitoring, and Geophysical Methods*, National Ground Water Association, Las Vegas, NV, 1992, pp. 341–355.

Zimmermann, C.F., M.T. Price, and J.R. Montgomery, "A Comparison of Ceramic and Teflon In Situ Samplers for Nutrient Pore Water Determinations," *Estuarine and Coastal Marine Science*, 7:93–97 (1978).

Case Studies of Vadose Zone Monitoring and Sampling Using Porous Suction Cup Samplers*

William R. Bond

INTRODUCTION

The porous suction cup sampler (PSCS) has proved to be a very useful tool for the monitoring and study of vadose zone water quality. The soil water PSCS was first used as a monitoring and research device for studies in agriculture. Since that time, they have been used for a wide variety of environmental monitoring purposes, from landfills to mine sites to hazardous waste facilities. They are capable of sampling soil moisture or ground water from remote areas that are not accessible by conventional water well installations. The most common installations are usually simple and cost-effective, and in many cases can be completed without the use of heavy drilling equipment. Perhaps most important, a PSCS properly placed and installed can detect contamination in the vadose zone within several days or weeks of a leak, whereas a conventional monitoring well may take months or years to detect a leak of the same magnitude after it has entered the ground-water regime.

The following case studies are selected examples of several systems that were designed and installed in part or in full by the author over the past 15 years. These installations have been chosen to illustrate the versatility of the porous suction cup sampler for vadose zone monitoring.

_____

*Portions of this chapter appeared previously in *Mining Engineering*, 37(4):314–319 (1985).

CASE STUDIES

Case Study 1:
Landfill Leachate Collection Pond Monitoring

As part of a comprehensive ground-water protection monitoring system at a landfill in Nevada County, California, the Central Valley Regional Water Quality Control Board (the Board) required a vadose zone monitoring system to be installed beneath a leachate collection pond. Because it was the first design of its kind in California in 1987, a negotiation phase was entered into with the Board to determine the acceptable spacing and depth of installation below the bottom of the pond liner. The lack of precedence for this type of vadose zone leak detection monitoring system required an innovative approach for the design and construction aspects.

The vadose zone monitoring tool of choice was the porous suction cup sampler. The primary question posed by the Board was at what spacing and what depth would the samplers be placed to monitor the leachate collection pond. After extensive negotiations with the Board, it was determined that an assumed horizontal to vertical seepage ratio of 10:1 would be utilized for determining the spacing between the sampling units. As a result, the units were buried three feet below the leachate recovery system liner, in the base of the clay liner, at a spacing of approximately 60 feet. Using a hypothetical leak scenario, calculations were provided to the Board to demonstrate the vertical to horizontal contamination seepage distribution, to justify that the spacing would provide adequate coverage. The final design included nine PSCS units in the pond bottom and ten units around the pond perimeter. The overall dimensions of the pond were 140 feet, by 450 feet, by 35 feet deep.

The pond bottom installations involved digging through the compacted clay liner for placement of the PSCS, and trenching up the sides of the pond to install the sampling tubes. It would have been preferable to install the units before the compacted clay liner was installed; however, the construction schedule did not allow for this. Of utmost importance was the need to rebuild the clay liner without damaging the PSCS units. To protect the PSCS ceramic cup during subsequent liner construction and compaction, the ceramic tip was firmly planted into a 6-inch silica flour slurry bed placed in the bottom of a 2-foot deep by 6-inch wide excavation (see Figure 27.1). The remainder of the trench was backfilled with six inches of hydrated granular bentonite and six inches of the excavated clay material to the subgrade surface. The polyethylene sampling tubes were protected by $1\frac{1}{2}$ inch schedule 40 PVC casing and buried in 8-inch deep trenches that ran up the side of the pond to the sampling station at the top. To minimize trenching, up to three different sampling tube sets were placed in the same PVC conduit. Because the tightly compacted clay would yield very little water, baseline sampling after installation proved fruitless. Consequently, a policy of "no water, no problem" was accepted by the Board. That is to say, the PSCS units would be sampled on a regular schedule, and the absence of a water sample would indicate that no leakage had passed through the synthetic liner and the leachate collection recovery system, and penetrated the clay liner.

Case Study 2:
Golf Green Design and Chemical Maintenance Evaluation

A $75 million resort development and golf course in an alpine meadow in the Sierra Nevada Mountains required the development of a "Chemical Application

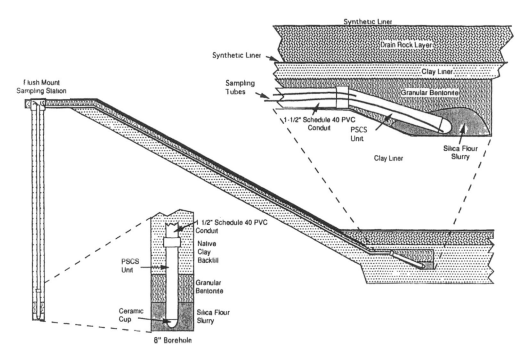

Figure 27.1. Landfill leachate collection pond monitoring.

Management Plan" for daily maintenance guidelines of fertilizer and pesticide application. To evaluate the application requirements, a vadose zone monitoring system was designed to monitor the downward migration or attenuation of fertilizer and pesticides that could potentially result from applications to the greens under normal maintenance and irrigation conditions. The focus of the testwork was to determine the optimal application of fertilizer and pesticides to maintain a suitable growth on a golf course putting green for professional play. A "test green" was constructed in the fall of 1985 and divided into five different cells using fiberglass panels. Four of the cells had similar construction that met golf course design criteria, and a fifth cell was constructed with a Hypelon™ liner, approximately three feet below ground surface. Each cell was instrumented with two sets of four PSCSs placed at depths of eight inches, two feet, four feet, and six feet below ground surface (see Figure 27.2). Three additional baseline/background monitoring points were also installed at similar depths in the surrounding native materials.

All of the PSCS units were installed at approximately 30 degree angles from horizontal to allow direct percolation of the chemicals through the soil material and into the ceramic cup without the influence of fluid migration down a borehole or the side of the PSCS. The setting hole for the PSCS was formed by driving a 2-inch split spoon sampler into the side of a pit excavated into the backfill. The soil core from the sampler was retained for geochemical analysis. Careful attention was paid to ensuring that the water discharge tube was placed on the bottom side of the ceramic cup to achieve maximum sample recovery. All the sample tubes from one pit were placed in a common trench and terminated in a converted "port-a-potty" building set up to handle up to 26 pairs of PSCS sample tubes.

The test green and PSCS monitoring system were operated continuously for more

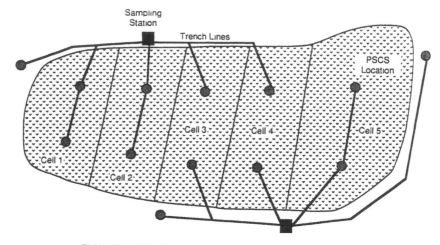

PLAN VIEW OF TEST GOLF GREEN AND PSCS LOCATIONS

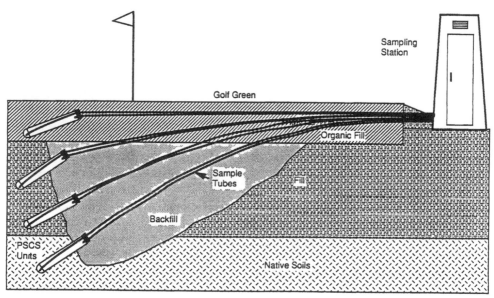

CROSS SECTION OF TYPICAL PSCS INSTALLATION IN TEST GOLF GREEN

Figure 27.2.   Golf green design and chemical maintenance evaluation.

than two years. Test green cells one through four had regularly scheduled fertilizer and pesticide applications at concentrations of 0, 33%, 66%, and full recommended strength, respectively. On the fifth cell with the Hypelon™ liner, the chemicals were applied at the manufacturer's recommended full strength. Complications in the sample program arose following the construction phase and several PSCSs ceased to function, most likely due to surface impact by construction equipment. Other PSCSs would not function during the extreme cold of winter months, most likely due to water freezing in the sample return line. Of the total 52 PSCSs installed, approximately 40 of the units continuously provided samples during the entire test procedure. Sample analysis included the nitrogen series (nitrate, nitrite, Kjeldahl nitrogen, ammonia) as well as total dissolved solids, phosphorous, potassium, and organophosphate compounds.

A typical sample sequence involved the application of a vacuum at one or two week intervals followed by sample collection within the next two days. The sample collection time was minimized to reduce the residence time of the water in the PSCS and decrease the potential for oxidation of the sample. As a further precaution, an inert gas (argon) was utilized as a pressure source to retrieve the water sample. All samples requiring preservation were immediately treated upon placement in the laboratory sample container.

The results showed that the golf green could be maintained in playing condition with only 66% of the recommended fertilizer strength, and the majority of the pesticides were attenuated in the upper two feet of the soil horizon. The Hypelon™ lined cell was found to be unnecessary and no further plans were made for this type of construction. The data collected during the study were used to prepare the "Chemical Application Management Plan" for optimum fertilizer application and minimal pesticide use to protect the shallow aquifer.

Case Study 3:
Vadose Zone Monitoring Around a Uranium Tailing Pond

This installation involved the emplacement of 15 high-pressure PSCS units around the perimeter of a newly constructed, membrane-lined uranium tailing disposal pond in April, 1981. The site is located in glacial outwash deposits in the state of Washington. The need for an early warning leak detection system became apparent when a contamination plume from an old pond at the site was detected in a nearby streambank. The site topography and stratigraphy allowed the plume to surface at the streambank prior to its detection in a nearby monitoring well. As a follow-up to a seepage plume investigation, the suggestion was advanced for the installation of PSCSs as monitoring devices for early warning leak detection around the new pond facility. At the time of installation, the use of the PSCS units as a tailings pond leak detection device was, to the author's knowledge, the first of its kind.

In selecting the depths and locations of the PSCS units, consideration was given to the nature of the glacial till material and the potential configuration of a leakage plume. A prior investigation of the earlier leakage plume, by means of resistivity survey and monitoring wells, gave a good indication of the dimensions and migration character of a potential contaminant plume. The pond held approximately 1200 acre feet, with an average depth of 70 feet. The final specifications included 15 PSCSs placed in auger boreholes located approximately 200 feet apart. The holes were drilled from a mid-depth construction bench on the side slope of the pond and terminated at approximately the 40 foot depth; therefore, most PSCSs were located about 5 to 10 feet below the pond bottom.

Timing of the PSCS system installation into the contractor's overall schedule was very important. The most critical aspect was the completion of all rough grade work for the pond base. It was emphasized that all rough grade work must be completed to avoid disturbing the PSCS units and buried sample lines with additional earth moving activities. The drill rig was then mobilized and installation began without concern for further disturbance. The installation of the PSCS units involved an in-situ placement technique. The in-situ placement was accomplished by using a 2-inch split spoon soil sampler, which is the same diameter as the PSCS unit. When the borehole reached the desired depth, the soil sampler was driven to form a "socket" hole in the soil (see Figure 27.3). The soil sample was retained for testing of ambient soil chemistry. The PSCS unit was then fed down the borehole by assembling 10-foot sections of 1 1/2 inch PVC pipe, feeding the sample tubes through each pipe length, and then pushing the unit into the 2-inch sampler hole, fitting it firmly in the undisturbed native soil. Following installation, the placement pipe was removed by sections, and the hole was then backfilled with a bentonite plug and screened native sand. The upper backfill surface material was tamped firmly to avoid additional settling and conduit of precipitation during completion of construction. (The PSCS placement pipe can also be left in place for protection and the borehole backfilled with bentonite-grout and/or drill cuttings, as the installation dictates. This type of installation has been preferred in rocky soils that could damage the sample tubes.) After installing the PSCS units and backfilling the boreholes, a small trenching machine was utilized to dig a trench from the top of the borehole to the top of the pond rim. Care was taken in backfilling the sample line trenches with clean fill to avoid pinching the lines with oversized cobbles. A final sample station for each PSCS was constructed on the pond rim.

The sampling of the PSCS began within one week of their installation. This early sampling was an effort to secure baseline vadose moisture samples prior to the completion of the pond liner and ultimate disposal of uranium tailings. The average volume recovery from the PSCSs was 150 to 200 milliliters of water collected over a one week period. At the time of sample collection, the samples were checked for field pH and conductivity. An ongoing record was kept of these parameters to note variations from the established baseline data. These two field parameters are indica-

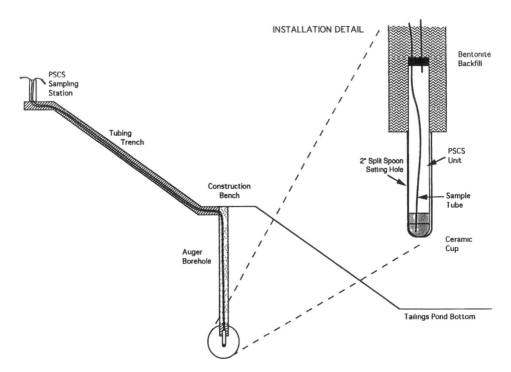

Figure 27.3.   Vadose zone monitoring around a uranium tailing pond.

tors of the possible migration of acid mill wastes from the pond through leaks in the liner. The baseline samples were fairly neutral pH and low conductivity; therefore, a drop in pH or increase in conductivity would be indicative of leakage.

A sufficient volume of water for radiochemical analysis could not be obtained with any regularity. For this reason, samples were analyzed for other, more mobile indicator parameters such as pH, specific conductance, TDS, sulfate, U^{308}, NO_3, manganese, chloride, NH_4, COD, and iron, which require less sample volume for analysis. The parameter list was agreed upon through negotiations with the Nuclear Regulatory Commission.

Case Study 4:
Vadose Zone Monitoring Below a Gold Heap Leach Facility

The selection of PSCS as a leakage monitoring device at a gold heap leach facility was based in part on satisfying concerns of state regulatory agencies that contaminants were not migrating into the subsurface, while assuring the mining personnel that pregnant solution containing precious metals was not being lost to the subsurface. The gold mine facility lies in a hilly/mountainous area of the northern Black Hills of South Dakota. Negotiations with state regulatory agencies in 1983 determined the acceptable PSCS spacing, depth, and analytical priority list of leak detection parameters. PSCS units were installed under both pad liners and solution holding ponds. The site is underlain primarily by weathered and fractured limestone, constituting an important regional aquifer. The heap leach pad construction removed all overburden organic materials, leaving behind as much clay soil as possible over the limestone.

In most instances, the PSCS installation depths were at or below the clay-limestone interface in the weathered limestone. The installation depths varied from 2 to 8 feet, spaced approximately 100 to 150 feet apart. Drilling was accomplished with an auger rig using a 4-inch auger. The shallow installation depths allowed for several holes to be drilled in rapid succession and left open for subsequent PSCS unit placement. Specifications for the installations of the units involved a dry bedding of silica flour around the porous ceramic tip. The silica flour bedding was then capped with a bentonite seal of approximately $1\frac{1}{2}$ foot thickness (see Figure 27.4).

The main difficulty of installation at this site was associated with trenching through the weathered limestone and clays for safe burial of the sample lines. The original specifications called for an 18-inch trench with the sample lines buried in clean bedding sand in the bottom one-third of the trench. The presence of cobble-size weathered bedrock material made trenching difficult and jeopardized the integrity of the sample lines, with sharp rocks collapsing on the tubing during backfill. To remedy this situation, the sample lines were fed through one-inch PVC pipe prior to burial. The trench was then filled with bedding sand to minimize the risk of line damage during subsequent construction backfilling and compaction of the leach pad. This same construction technique was used under the pregnant, barren and overflow ponds. At several locations, sample tubes up to 400 feet in length were installed from the sample station to the PSCS unit. Where possible, the long lines were run downgradient from the PSCS unit. Coordination and timing with the earth moving contractor was essential to minimize danger to PSCSs and sample tubes.

Samples were first collected one month after installation to establish baseline information prior to use of the leach pad. With time, all of the units eventually stopped producing water. This was attributed primarily to the isolation of the PSCS

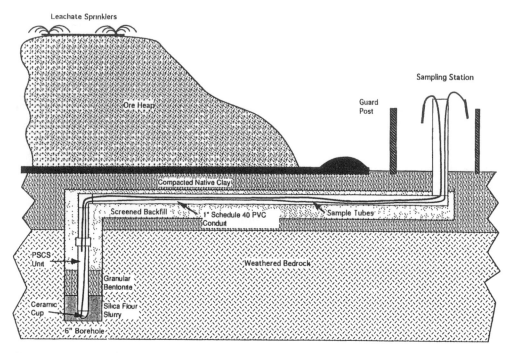

Figure 27.4. Vadose zone monitoring below a gold heap leach facility.

units below the pad with no available recharge. The nature of the installation and absence of recharge resulted in a "no water-no problem" approach to the monitoring system.

Case Study 5:
Investigation of Soil Water Movement in Old Tailing Deposits

This particular study in 1981 at the Whitewood Creek Superfund Site investigated the role of soil water migration in determining the distribution and movement of arsenic and other contaminants within old tailing deposits on a river floodplain in western South Dakota. PSCS units were used to recover soil water samples from the sand and slime tailings that were deposited in the meanders of the creek over a 96 year period. Saturated-zone sampling suggested that a complete understanding of arsenic migration in the tailings could not be determined solely by ground-water and soil sampling. The data collected from the PSCSs in this study proved valuable in understanding the movement of arsenic and other contaminants within the tailings.

A total of 24 PSCS units were installed at 11 different sites over a 9-mile stretch of the study area. At each different site, an array of ground-water wells and PSCSs were installed to obtain water samples from the various horizons of native material and tailings. All sites included one deep monitor well installed in the shallow bedrock, one or more monitoring wells in the stream alluvium and overlying tailings, and PSCS units at various depths within the unsaturated portion of the tailing material. Some areas involved nests of 3 wells and up to 5 PSCSs. With this type of installation array it was possible to correlate the analytical data to determine the relative movement of the elements of concern.

All PSCS units in this study were installed using the in-situ placement method (see Figure 27.5). The drill rig used 6-inch hollow stem augers, and soil samples were collected with a 2-inch split-spoon sampler. In shallow installations less than 6 feet deep, hand auguring was done because of the unconsolidated nature of tailing material. A 2-inch drive tube was used to form the setting hole for the PSCS tip. All holes were then backfilled with a bentonite seal and tailings material. The tailings surface was tamped to prevent excessive downward percolation of rainwater. Sample stations were constructed directly above each PSCS unit using 3-inch PVC casing supported with a steel "T" fence post.

All PSCS units installed in this study were purged with nitrogen gas prior to application of the first vacuum. This inert gas purge reduces the possibility of atmospheric oxygen influencing the redox conditions in the recovered fluid. Nitrogen gas pressure tanks were also utilized in all subsequent sampling of the PSCSs. In this manner, the PSCSs were constantly purged with nitrogen gas prior to the reapplication of vacuum to the unit.

Sampling of the PSCSs involved both the sample bottle collection method and syringe container method. The syringe sampling method was developed by the client during the study, when it was suspected that the samples had become altered by

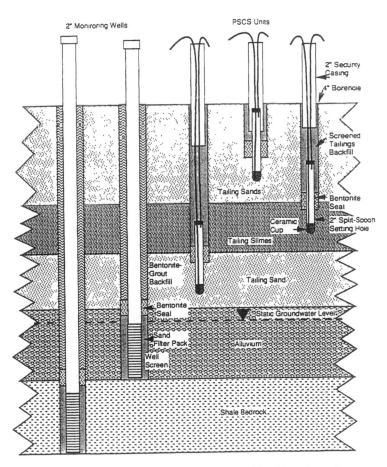

Figure 27.5. Investigation of soil water movement in old tailing deposits.

oxygen during sample collection and transport to the laboratory. During sampling, a 50-cc syringe was inserted directly into the neoprene sample tube of the PSCS, and the discharge pressure was used to fill the syringe. After collecting water samples in the syringes, they were quickly frozen in a dry ice-alcohol bath and maintained in a frozen state for shipment to the laboratory. This method provided a higher degree of confidence in the laboratory analyses of arsenic and iron speciation. Final results indicated that, in contrast to previous assumptions on the vadose water movement in the tailings, there was actually a net upward migration of the arsenic contaminants due to capillary rise.

28
_____

Special Problems in Sampling Fractured Consolidated Media

Daniel D. Evans and Todd C. Rasmussen

INTRODUCTION

The flow of water and the resulting transport of contaminants through unsaturated fractured media is a serious concern at many existing and potential waste disposal sites. Techniques for characterizing fractured media in the unsaturated zone are less established than techniques applied in saturated fractured rock or in unconsolidated media. This is due, in part, to difficulties associated with developing predictive models and devising monitoring methods suitable for characterizing contaminant migration in such media. The unique features of unsaturated fractured rock, such as the paucity of field data and the inability to identify networks of interconnected fractures over large distances, pose special problems when sampling for physical and chemical properties.

Fractures tend to increase the heterogeneity of hydraulic parameters relative to an unfractured medium, thus increasing the spatial variance of contaminant distributions. For infiltration into a dry soil with uniform hydraulic properties, a uniform contaminant distribution with depth is normally expected. As a contaminant infiltrates into a fractured rock, however, the contaminant often advances as fingers through the fracture system, bypassing much of the matrix, as seen in Figure 28.1A (Haldeman et al., 1991). Water and entrained contaminants also move into the matrix from the fracture due to concentration gradients associated with Fickian diffusion, and total head gradients associated with gravitational and pressure gradients (Figure 28.1B). Thus, there may exist a contaminant gradient both within and away from fractures. This spatial and temporal complexity in unsaturated fractured media adds greatly to the design of sampling protocols and the interpretation of observed contaminant concentrations.

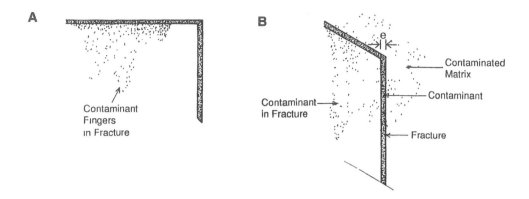

Figure 28.1. Spatial distribution of relative solute concentrations (A) within the fracture plane showing two preferential flow paths, and (B) in rock matrix away from fracture plane.

While it is often noted that open fractures provide preferential pathways for saturated fluid flow and, consequently, for the movement of contaminants (Figure 28.2A), fractures in the unsaturated zone can also form air-filled barriers to fluid flow under dry conditions at high suctions (Figure 28.2B). Another important transport possibility is that during and following infiltration events on the surface (e.g., rainfall, streamflow, snowmelt), fracture flow dominates even though the matrix rock remains unsaturated. These transient events, in which the suction in the fracture is not in equilibrium with the suction in the rock matrix, may result in significant fluid and contaminant flow within fractures, even under ostensibly unsaturated conditions. Therefore, contaminant movement and distribution within fractured rock depends on the relative fracture to matrix permeabilities and on the current and previous suction distributions.

Transport of contaminants in the vapor phase (e.g., BTX, TCE) is also strongly linked to the relative saturation of the fractures and matrix (Brooks, 1989). For fractured rock that is completely water saturated, the gas permeability is zero, but increases dramatically as the suction increases and the fracture system drains. Once

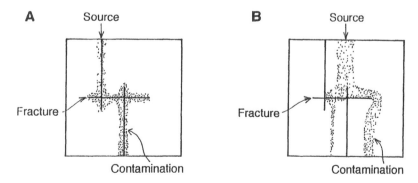

Figure 28.2. Contaminant transport may be (A) enhanced due to rapid transport through saturated fractures, or (B) reduced due to air-filled barriers associated with unsaturated fractures.

the fractures are air-filled, only an additional slight rise in gas permeability is observed as the matrix drains. Vapor transport of contaminants is governed by processes very similar to liquid transport, with the major difference being the substantially lower viscosity of gas as opposed to liquids. This viscosity contrast yields much greater vapor transport velocities, given similar total head gradients and saturations relative to the transport phase.

Contaminant sampling strategies in the consolidated and unconsolidated media are substantially different because of the significant differences in pore structure and pore continuity. Unconsolidated media transport properties are generally more time-dependent due to rapid changes in saturation and temperature near the surface. Flow fields in consolidated media are generally less time-dependent because they lie below unconsolidated media, thus reducing the influence of surface perturbations. While fractures in competent porous rock may respond to subsurface-imposed mechanical stresses and temperature perturbations (such as near a high-level nuclear waste repository), such changes are seldom encountered under natural conditions. In contrast, fractures in expanding and shrinking soils exhibit dynamic spatial and temporal behavior. Such soils often display cyclical fracture formation and filling with soil material on daily, seasonal, and annual intervals.

The material presented here is a summary of insights gained from over ten years of experimental field work by the University of Arizona at unsaturated field sites in fractured rock in Arizona. Characterization and model evaluation activities focusing on laboratory and field data at the Apache Leap Tuff Site, the Gringo Gulch Site, and the Santo Niño Mine Site have provided the opportunity to test alternate formulations for understanding the complex phenomena associated with unsaturated fractured flow and transport. It is hoped that the expertise developed and tested at these sites might be used at other sites to enhance the understanding of the complexity and the predictive capability for resolving the unique problems of flow and transport at each site.

DESCRIPTION OF APACHE LEAP TUFF SITE

The Apache Leap Tuff Site (ALTS) is located near Superior, Arizona, approximately 160 km north of Tucson, Arizona in the uppermost part of an approximately 20 million year old tuff formation. The formation varies from a slightly welded unit on top, through a moderately welded unit below, to a densely welded unit near the base of the formation. The base of the unit is composed of a nonwelded unit underlying a vitrophyre. The tuff covers an area of approximately 1000 km^2 and has a maximum depth of 600 m, with an average thickness of approximately 150 m. About one kilometer to the west of the Apache Leap Tuff Site is a 600-m escarpment, and immediately to the north there is a 200-m drop-off. The rest of the surface is dissected by ephemeral streams. The topography of the experimental area is nearly level compared to the surroundings.

Linear fracture traces are present, with some fractures containing a slight depression filled with unconsolidated material. These conditions provide an important water reservoir for water intake by the fractures. The annual average precipitation at the 1200-m elevation site is approximately 640 mm. Precipitation occurs in two seasons, from mid-July to late-September (characterized by high-intensity, short-duration thunderstorms during periods of high temperature and evapotranspiration demand), and from mid-November to late-March (characterized by longer duration and lower intensity storms during cooler periods with much lower evapotranspiration demands).

To obtain parameters for the rock matrix and the fractures at the site, three sets (named X-, Y-, and Z-series) of three boreholes (named 1, 2, and 3) each were installed with each borehole at a 45° angle from the horizontal. A diamond bit drill was used for installing the boreholes. The drill was also used to cut a 6.35-cm diameter core. A specially designed scribing technique provided an orientation mark every 1.52 or 3.05 m along the borehole. Nearly 100% core orientation for all boreholes was obtained by using the scribe marks. Approximately 270 m of oriented core were obtained. The 10-cm diameter boreholes of each set are in a vertical plane and offset by 10 m. The sets are parallel and offset by 5 m. Sets X and Y dip to the west, while set Z dips to the east. The variable lengths, approximately 15, 30, and 45 m, provide fracture-borehole intersections for individual fractures at different depths below the surface.

Characterization data sets for the tuff matrix were obtained from the oriented cores using sections cut from the core at approximately three-meter increments. Three segments are removed from each section. The segments are labeled 'large' (5 cm long by 6 cm diameter), 'medium' (2.5 cm long by 2.5 cm diameter), and 'small' (1 cm long by 1 cm diameter) to differentiate between the three segments. The nine boreholes provide ready access for various in situ measurements, while the cores provide information on fracture spacing and orientation as well as samples for measuring matrix properties in the laboratory. To control water inputs (e.g., precipitation) and outputs (e.g., evaporation) at the test area, the area was covered with plastic to a distance of 10 m in all directions beyond the boreholes, for a total area of 1500 m². Surface water is directed away from the site to prevent wetting of the region around the boreholes.

NATURAL WATER AND AIR FLOW IN FRACTURED MEDIA

The relationship between saturation and suction, called the moisture characteristic curve, in a fracture is a function of the distribution of fracture apertures and the nature of the interconnections between voids and contacts along the fracture walls. For parallel plates separated by a constant gap, or aperture, the moisture characteristic curve is a step function, with the critical suction at which the fracture changes between saturation and desaturation calculated using capillary theory:

$$\psi_c = \tau \cos(\alpha)/\gamma e \qquad (1)$$

where ψ_c = critical suction head, m;
  τ = liquid surface tension, Pa·m;
  α = solid-liquid contact angle;
  γ = liquid specific weight, Pa·m^{-1}; and
  e = fracture aperture, m.

Liquid flow along this idealized fracture, which by definition is nonexistent when the fracture is drained, is calculated using the following expression:

$$q = Q/w = T_f i \qquad (2a)$$

where

$$T_f = \begin{cases} 0 & \psi_c < \psi \\ (e^3\,\gamma)/(12\,\mu) & \psi_c > \psi \end{cases} \tag{2b}$$

and

q = liquid flux per unit width of fracture normal to the direction of flow, $m^2 \cdot s^{-1}$;

Q = total liquid flow through fracture, $m^3 \cdot s^{-1}$;

T_f = fracture transmissivity, $m^2 \cdot s^{-1}$;

i = total head gradient along fracture, dimensionless; and

μ = liquid dynamic viscosity, $Pa \cdot s$.

Equations 2a and 2b may also be used to find the flow rate of a gas through a fracture if appropriate physical constants are inserted. In this case, the flow of gas is nonexistent when the fracture is water-saturated. An additional feature of gas flow through porous media is the Klinkenberg slip-flow effect in which the gas tends to exhibit a higher flow rate than predicted due to nonzero velocities along the fracture walls. This effect only becomes significant in pores smaller than 1 μm, and generally does not affect gas flow in a fracture.

For geologic media that consist of significant nonfracture permeability, flow across the fracture, from one wall to the other, is also possible. In this case, the total head drop across the fracture from one wall to the other is zero if the fracture is saturated, and can increase to infinity for a drained fracture. This relationship can be formulated using the following expressions:

$$q' = Q/A = C_f\Delta H \tag{3a}$$

where

$$C_f = \begin{cases} 0 & \psi_c < \psi \\ \infty & \psi_c > \psi \end{cases} \tag{3b}$$

and

q' = flow rate across fracture per unit area of fracture, $m \cdot s^{-1}$;

Q = total flow across fracture, $m^3 \cdot s^{-1}$;

A = fracture area, m^2;

C_f = fracture conductance, s^{-1}; and

ΔH = total head drop across fracture, m.

For more realistic natural fractures in which the aperture is spatially variable, more complex relationships between suction, saturation, and fracture transmissivity are expected (Rasmussen, 1987; Vickers, 1990). Even with these more complex functions, however, the same general monotonic tendencies are expected (Myers, 1989). Experiments to identify these relationships and their trends are difficult to perform, however, and even slight changes in fracture normal and shear stresses may alter these relationships (Schrauf and Evans, 1986). These general relationships can be related to the direction and magnitude of water migration if the tensorial form of Darcy's law is written for unsaturated media:

$$q = -K(\psi)i \qquad (4)$$

where
q = vector of liquid flux, $m \cdot s^{-1}$;
$K(\psi)$ = suction-dependent hydraulic conductivity tensor, $m \cdot s^{-1}$; and
i = total fluid head gradient vector, dimensionless.

Spatial variations in liquid suction, ψ, affect both the total head gradient vector, as well as the hydraulic conductivity tensor.

A simple approach for obtaining the hydraulic conductivity tensor is to decompose the fracture and matrix hydraulic conductivities (K_f and K_m, respectively) into two, overlapping continua (Figure 28.4A), in which the effective permeability is the sum of the two permeabilities (see, e.g., Wang and Narasimhan, 1993):

$$K_e = K_f + K_m \qquad (5a)$$

where

$$K_f = d_f \, T_f$$

and

d_f = fracture density, m^{-1}; and
K_e = effective hydraulic conductivity for fractured medium, $m \cdot s^{-1}$.

Because this type of formulation assumes instantaneous equilibration of total head between the two media, essential features of the flow dynamics are overlooked. Such phenomena as capillary barriers (in which an unsaturated fracture impedes fluid flow) and rapid transient responses are, or should be, an important component of flow prediction and monitoring. This formulation also assumes that the flow direction is parallel to the plane of the fracture (Figure 28.3A).

Fractures may be open, partially filled, or completely filled with in-place weathered primary and secondary minerals, or with organic and inorganic materials that

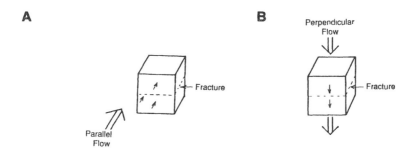

Figure 28.3.   The effective hydraulic conductivity can be calculated using fracture and matrix properties as the (A) arithmetic average for flow parallel to the fracture plane, or (B) harmonic average for flow perpendicular to the fracture plane.

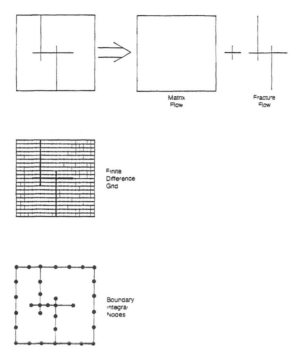

Figure 28.4. Model formulations for three-dimensional flow through fractured, porous rock: (top) Equivalent Porous Medium, (center) Dual Porosity, and (bottom) Boundary Integral.

have been transported from other locations. These fracture fillings and wall coatings may greatly alter physical and chemical interactions of fluids and contaminants within the fracture as well as in the host rock surrounding the fracture. As such, the effective permeability may be more appropriately calculated using the harmonic average of the fracture and matrix permeabilities (see, e.g., Rasmussen et al., 1989):

$$K_e^{-1} = d_f C_f^{-1} + K_m^{-1} \tag{5b}$$

where K_e is the effective hydraulic conductivity of the fractured medium. This formulation is more appropriate for flow perpendicular to the plane of the fracture (Figure 28.3B).

Fracture flow is often a transient hydraulic response during and immediately following precipitation events. At the Apache Leap Tuff Site, water velocities in fractures due to the infiltration of rainfall were estimated using known rainfall and runoff rates, and by accounting for losses due to evaporation and infiltration into the rock matrix (Rasmussen and Evans, 1993).

This water budget approach yielded a mean water intake rate of from 2 to over 7 $m \cdot hr^{-1}$. Water velocities in the fracture estimated from this intake rates and observed fracture densities ($0.77 m^{-1}$) were found to range from 122 to 293 $m \cdot hr^{-1}$, indicating the possibility for very rapid movement to great depth within the subsurface. Water flow in fractures propagates downward, either until the water table is reached, the water is imbibed into the porous matrix surrounding the fracture, or the water is discharged to the surface from the fracture system at a lower elevation.

Inclined fractures can intercept water percolating downward within a rock matrix. It can be shown (Rasmussen et al., 1989) that for inclined fractures lying at an oblique angle, α, to the vertically percolating water, the effective hydraulic conductivity is an average of the arithmetic and harmonic averages calculated above:

$$K_e = [\cos^2(\alpha) / K_h + \sin^2(\alpha) / K_a]^{-1} \tag{5c}$$

where K_h is the harmonic effective hydraulic conductivity calculated using Equation 5b and K_a is the arithmetic effective hydraulic conductivity calculated using Equation 5a. It is interesting to note that Equation 5c is the same result as would be obtained using a tensorial form, thus establishing the utility of Equation 4. Fracture flow in the vadose zone can also occur in zones where the rock matrix hydraulic conductivity falls to a value less than the downward water percolation rate. In this circumstance, fracture flow will increase as the matrix permeability decreases.

Air and vapor movement is generally less affected by gravitational forces than water transport, being instead more sensitive to air pressure, gas density, and temperature gradients. In areas with topographic relief, differential surface-subsurface temperatures cause a gradient in air density that results in forced convection of air through the subsurface (see, e.g., Weeks, 1987; Kipp, 1987). Barometric pressure changes also cause subsurface air movements. Air flow is more pronounced in fractures than in the rock matrix for both pressure and density driving forces. The higher air permeability associated with drained fractures results in preferential air flow that can penetrate to great depths (Smith, 1989). Also, vapor transport is enhanced by the natural air flow through the subsurface, induced by both pressure- and density-driven driving forces.

The prediction of gas flow requires that Equation 4 be rewritten in terms of the intrinsic permeability of the geologic medium:

$$q = k(\psi) \, f \, i \tag{6}$$

where $f = \gamma/\mu$
and k = saturation-dependent intrinsic permeability, m^2;
         f = fluidity of migrating fluid, m^{-1}·s^{-1};
         γ = specific weight of monitoring fluid, Pa·m^{-1}; and
         μ = viscosity of the migrating fluid, Pa·s.

The gas-phase total head gradient is a function of the gas pressure gradient and, to a minor extent, the gravitational gradient. The intrinsic permeability tensor is a function of the water potential, ψ.

More complex representations of the flow domain can be constructed by acknowledging the three-dimensional spatial complexity of fractured rock blocks separated by discrete fractures of finite areal extent. Dual porosity models are commonly used to simulate the interrelated flow between a fracture system and the surrounding rock matrix (see, e.g., Wang and Narasimhan, 1993; Updegraff et al., 1991). These models employ grids to discretize the flow domain into sets of cells representing both the fracture and rock matrix, with unique porosity and permeability distributions for each medium (Figure 28.4B). Such models become unwieldy, however, due to the large number of cells required to represent even a small rock volume.

Alternatively, boundary integral methods (see, e.g., Rasmussen et al., 1989; Rasmussen, 1991) can be used to reduce the dimensionality of the problem. The boundary integral method reduces a three-dimensional, spatially-variable head distribution within a matrix block, to a two-dimensional problem across the fracture surfaces surrounding the block (Figure 28.4). This improvement in efficiency can then be used to enlarge the domain of the problem, so that larger networks of models can be evaluated. One caveat of this method is the inability to track rapid changes in flux and transport.

CHARACTERIZATION OF FRACTURED MEDIA

Unfractured geologic media generally display a continuity at a pore scale that lends itself to characterization using core segments extracted from boreholes. The cores can be collected at various orientations and increasing scales and densities until a conceptual model is formed that adequately incorporates the geologic variation present at the site of interest. Undisturbed rock cores are normally used for obtaining laboratory characterization data for the matrix. Special efforts must be used to maintain the structural integrity of the matrix to assure representative properties are measured. Contacts between a rock core and a porous material, such as a porous plate or tensiometer cup, may be more difficult to maintain than for an unconsolidated core. A layer of deformable porous material at the interface is usually required. Procedures presented elsewhere in this volume that were developed for unconsolidated material are generally adaptable to rock cores.

Fractured media, on the other hand, display continuity on a scale larger than a

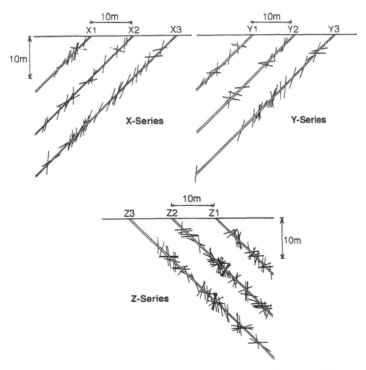

Figure 28.5.   Inclined boreholes are more appropriate for characterizing fractured rock where many fracture orientations are present.

pore or core. Fractures may connect voids on the order of a meter and even a kilometer apart. Such long-range connectivity makes laboratory and field-scale testing more difficult. Fractured porous media is even more difficult to characterize. The ability of water, air, and contaminants to migrate from a fracture into the rock matrix that bounds the fracture (and vice versa) can significantly affect their movement through the fracture. In situ characterization of a fractured medium is usually necessary to incorporate the scale of the heterogeneities inherent to the medium and to avoid the difficulty of removing and transporting a sample without major disturbance to the enclosed fractures. Because fracture transport properties may vary over large spatial scales (i.e., from tens to hundreds of meters or larger), field surface (Kilbury et al., 1986) and borehole tests (Rasmussen et al., 1993) are useful for obtaining appropriate material properties.

Because borehole tests can be interpreted in many ways, depending upon the assumed geometry of the flow field and the assumed boundary conditions, significant effort should be expended in trying to understand the physical system near the borehole. Fracture locations and orientations can be mapped using oriented core collected during borehole drilling, or from downhole televiewer logs. Inclined boreholes (Figure 28.5) are preferred in areas where vertical fractures are the dominant flow channels of interest because vertical boreholes may not intersect and adequately sample the near vertical fractures (Rasmussen et al., 1993).

In addition, sloping boreholes allow a simple procedure for obtaining oriented core. An unevenly weighted cylinder with a marker on the leading end is slid down the borehole. The marker makes a chip in the end of the core stub remaining after the removal of the last core segment. If the marker is on the lighter side of the cylinder, the orientation mark will be on the top side of the stub. The next core segment removed is oriented by the mark at the top of the segment. If the rock is sufficiently competent, the core for an entire borehole can be oriented and the strike and dip of any intersected fractures can be logged. Core segments without fractures can be used for matrix characterization and the boreholes are available for downhole measurements.

Auxiliary data collected using boreholes to provide access to the rock mass may include water potentials (Anderson, 1987), water content (Elder, 1988; Andrews, 1983), temperature (Davies, 1987), air pressure (Smith, 1989), and electrical resistivity (Andrews, 1983; Thornburg, 1990) along the boreholes. These measurements can be used to identify zones of constant or varying material properties, and the spatial and temporal variations in these measurements may provide useful information regarding liquid and solute movement through the matrix and, by inference, through fractures. Evaluation of fracture conditions per se is constrained because of the limited volume associated with fractures in relation to the volume of the surrounding matrix.

Borehole field permeability tests using water and air can yield information on potential flow and transport characteristics (Tidwell, 1988; Rasmussen et al., 1993). While laboratory tests using core samples collected from boreholes may be adequate for obtaining matrix permeabilities, the combined fracture-matrix permeabilities can only be obtained through field measurements. Field experiments are conducted by isolating borehole intervals using packers, and injecting or extracting water or air into the interval. The length of the packers must be sufficiently long to assure that conducting channels within fractures do not bypass the packer and reconnect with the test borehole, thus confusing the interpretation of the test. A string of four packers (two above the test interval and two below) are often recommended to form guard intervals above and below the test interval (Figure 28.6). Pressures in the guard intervals are used to test for leaks from the test interval.

Figure 28.7 presents plots of field borehole vs. laboratory core measurements of permeabilities using both air and water permeabilities. It can be noted that only a poor correlation exists between the two due to the failure to account for fracture flow during laboratory tests. The figure also demonstrates the good relationship between permeabilities obtained using air and water. Saturated water permeability tests obtained from an initially unsaturated zone can be conducted by maintaining a constant water level within a single borehole, or in a specified borehole interval, and measuring the resultant water outflow.

Figure 28.8 presents a plot of the laboratory core and field borehole permeability tests using both air and water. Core data were collected using permeameters designed to measure air and water flow rates along the long axis of 6-cm oriented cores extracted at 3-m intervals from boreholes at the Apache Leap Tuff Site. It can be noted that the field-estimated saturated permeability distribution obtained using water compares reasonably well to the estimated distribution obtained using laboratory core segments. The higher permeability observed in the field can be attributed to the inclusion of fracture flow in the test. The field-estimated permeability using air is plotted at an ambient suction in the field of approximately 1 bar, which is inferred from laboratory moisture release curves and field neutron probe logs. The observed permeability distribution compares favorably with the laboratory-estimated values obtained from unfractured cores.

Air permeability tests may be conducted in a similar manner using single-hole techniques (Figure 28.9A), but more useful information is obtained using either crosshole or dipole tests because of the larger effective volume (Trautz, 1984). In a crosshole test, one packed off borehole interval is used to inject (or extract) air, while other packed off intervals in the same or adjacent boreholes are used to monitor pressure response. Dipole tests are conducted by using two borehole intervals with one used for injection and the other for simultaneous extraction (Figure 28.9B). Crosshole and dipole flow tests using water are generally impractical in unsaturated fractured media because of the complex gravity force fields, as well as the relatively small spatial and large time scales of responses. Like field water injection tests, however, field air flow experiments are subject to ambiguity due to

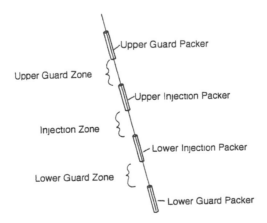

Figure 28.6. Packer string showing injection and guard intervals separated by packers.

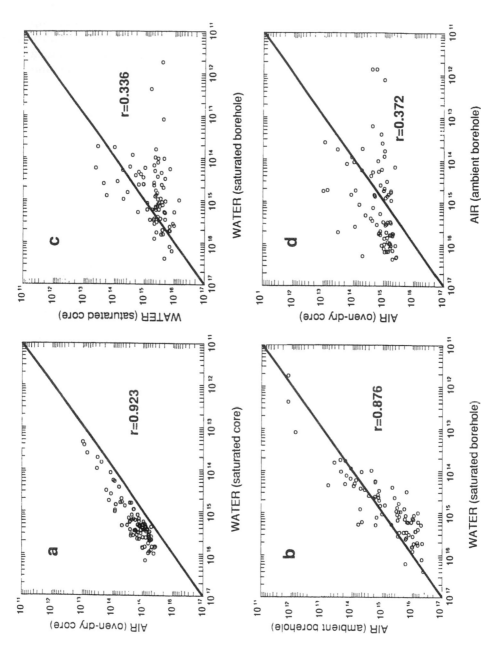

Figure 28.7. Scatterdiagrams including simple correlation coefficients and lines of equality for core and borehole permeabilities using air and water as the test fluids at the Apache Leap Tuff Site.

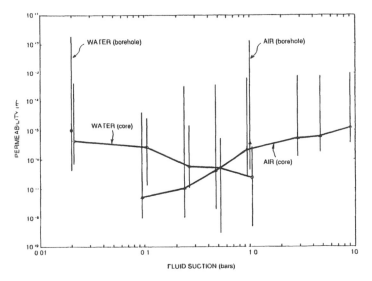

Figure 28.8. Apache Leap Tuff field and core permeability data using air and water as the test fluids.

the unknown geometry of the test region. For this reason, knowledge of the geologic conditions (e.g., fracture densities, locations, orientations, etc.) surrounding the borehole are critical for formulating the conceptual model.

If the principal conducting fractures are drained, the interpreted air permeability provides a good estimate of saturated water permeability. Because air flow tests are usually simpler to conduct and interpret than water flow tests in the unsaturated zone, this one-to-one relationship is extremely useful. No field method is currently available for measuring the water permeability of an unsaturated fractured rock, which is a serious constraint on site characterization. A method receiving increased attention (Tang, 1991) uses low-permeability geotextile membranes to impose a negative pressure on the rock surface. The method allows a positive pressure to be placed on the inside of a tubular membrane, which forces the membrane against the borehole wall. Because the head-drop across the membrane is larger in magnitude than the applied interior positive pressure, the resulting fluid pressure is negative on the face of the membrane in contact with the rock. The unsaturated permeability can be calculated by knowing the conductance and flow across the membrane, along with the injection pressure.

Initial results from the Apache Leap Tuff Site indicate that gas tracer tests, either

Figure 28.9. Test geometries for (A) single-hole injection/extraction and (B) dipole injection-extraction.

crosshole or dipole, can be used to provide additional characterization data about fracture interconnections between boreholes. Estimates of fracture dispersivities and matrix air-filled porosities can be obtained by using a suite of gas tracers. Helium and argon gas are inert, inexpensive, and insoluble air-phase tracers that can be used to determine fracture permeability variations as well as matrix porosities. Simultaneous injection of both gases and subsequent monitoring of their breakthrough curves can provide information on the location of fracture air-phase connections between boreholes. The difference in breakthrough times between the two gases provides information about the importance of matrix diffusion due to the difference in the molecular weights of the gases. The shape of the breakthrough curve also provides information about fracture diffusivities and channeling. Sulfur hexafluoride (SF_6) is an additional gas that is used as an air-phase tracer. The gas is suitable for measuring fracture diffusivities and channeling in the presence of large air-phase matrix porosities because of the large molecular weight of the gas, and, hence, lower molecular diffusivity.

SAMPLING OF CONTAMINANTS IN FRACTURED MEDIA

Contaminant sampling strategies should acknowledge the complex nature of flow through unsaturated fractured media. For rapid transient flow conditions, obtaining a representative liquid sample from a highly localized flow channel may require extensive arrays of sampling equipment (Figure 28.10A). For steady flow of contaminants through rock, contaminants may accumulate above horizontal, drained fractures due to blockage of fluid pathways by air. Sampling should then focus on collecting fluids from the rock matrix immediately above horizontal fractures (Figure 28.10B). In media with vertical fractures of finite length, a zone of saturation may form at the lowermost end of the fracture, extending upward to a point where the outflow from the fracture can either drain into another fracture or be absorbed by the porous rock matrix bounding the fracture (Rasmussen, 1991). Sampling in these circumstances should incorporate the physical properties of the contaminant, i.e., whether it is a DNAPL, LNAPL, etc. (Figure 28.10A).

If the objective is to ascertain the presence or absence of a foreign contaminant at a specific location in a fractured porous rock, sampling is much simpler than when

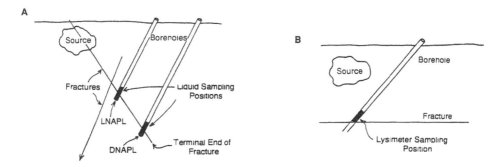

Figure 28.10.   Sampling strategies for unsaturated, fractured media: (A) Collecting a transient contaminant pulse in a fracture channel and a contaminant in a finite vertical fracture based on the physical property of the contaminant; (B) Collecting a contaminant trapped above a drained fracture.

attempting to define contaminant masses or concentrations (Roberts, 1987). Caution must be exercised regarding the size of sample taken because the first removed fluid would most readily come from fractures. Because the porosity of the fractures is relative small, part of the removed sample may come from a considerable distance from the point of extraction.

Contaminant transport is substantially different in saturated fractured media than in unsaturated media. The enhanced mobility of solutes in the liquid phase in saturated fractured media is contrasted with the greater mobility of vapor in the air phase of unsaturated fractured media. Sampling procedures are different for vapor and solute assessments in that liquid samples are more easily obtained from saturated fractured media while air samples are more easily obtained from unsaturated fractured media. For unsaturated consolidated media, ports can be installed in boreholes to sample gas composition at various depths. For some fracture systems, the boreholes should be inclined at an angle to intersect the fractures of primary interest. To isolate sampling intervals along a borehole, inflatable packers are placed on either end of the segment to be isolated.

Gas sampling is accomplished by placing a partial vacuum on the sampling port and measuring the presence or concentration of a contaminant in the extracted gas. The rate and total volume of extraction may affect the measured concentration, especially if the contaminant is found in the rock matrix as well as in fractures intersecting the port. The size of the sample interval may also affect the measured concentration, either through dilution of the sample or by enlarging the effective volume of the sampling. In situ sampling of liquids in fractured rock is constrained by high matrix suctions and the limited availability of fluids.

For saturated media, water samples may be obtained from installed sampling ports, as described above for gas sampling. At low ambient suctions, suction lysimeters installed at selected depths in a borehole may be applicable. However, hydraulic contact between the borehole wall and the suction cup must be assured. Also the applied suction may desaturate connecting fractures and greatly increase required sampling time. Suction lysimeters are particularly suitable for sampling just above a water table and especially for perched water table conditions. The collection volume for fracture-filling fluids may be augmented by placing the suction lysimeter within a large volume of silica sand that directly contacts a larger number of fractures. By maximizing the number of fractures connected to the lysimeter, larger volumes of fluids may be obtained.

The upper suction sampling limit for lysimeters may be extended by imposing an increased air pressure in the rock surrounding the lysimeter to force fluid into the lysimeter. For example, a suction lysimeter may be installed in a geologic medium where the ambient potential is 1.5 bars. A standard lysimeter does not provide sufficient suction to overcome the ambient matric suction of 1.5 bars. If air injection ports are installed above and below the lysimeter (Figure 28.11), and the air permeability of the geologic medium is sufficiently small, then air injection ports can be used to maintain a positive pressure of, say, 3 bars. The resulting gradient from the geologic medium to the suction lysimeter is sufficient to drive fluid from the geologic medium into the lysimeter.

One alternative for determining the concentration of a particular contaminant at an installed lysimeter location is to inject water into the media from the lysimeter (Amter, 1987). The injected water should contain a tracer that is not originally present at the site, and does not contain any of the target compound of concern. The injected liquid is allowed to equilibrate with the native liquid, and then extracted

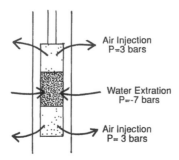

Figure 28.11.   Conceptual model of air injection-water extraction device to increase lysimeter fluid collection. Injection of air results in higher liquid pressure difference between geologic media and suction lysimeter. Total difference of 3.7 bars allows collection of fluids at matric suctions less than 3.7 bars.

using the lysimeter (Figure 28.12). The retrieved sample is then tested for both the tracer and the target compound. Quantitative estimates of the target compound concentration can be obtained if the tracer is used to determine the mixing ratio, obtained by comparing the injected concentration to the extracted concentration. This technique may be limited by the type of lysimeter employed. For example, dual chambered samplers are constructed with a check valve between chambers that prevent injection of fluids through the porous section.

Another approach for obtaining contaminant concentrations is to obtain rock cores and then mechanically squeeze the water from the pores for analysis (Mower et al., 1989; Yang et al., 1988). Drilling techniques have been developed using air as the cooling fluid which provide minimally altered water chemistry. Alternately, flushing of pore fluids from the core can be used to determine the presence of a target compound.

Still another promising technique being developed uses downhole laser-induced

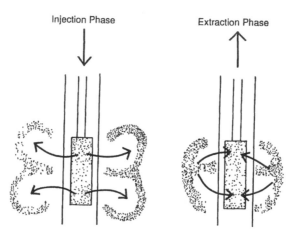

Figure 28.12.   Conceptual model of water injection-water extraction device to increase lysimeter fluid collection. During the first phase, water is injected into the geologic medium and allowed to equilibrate with native pore fluids. During second phase, the water is extracted and collected to determine native pore fluid composition.

fluorescence. Many chemicals can be shown to fluoresce when excited with light (Goering, 1988). This phenomenon can be used to advantage by remotely monitoring downhole fluorescence. A laser generator is positioned on the surface, the laser beam is transmitted to any downhole location using a fiber optic cable, focused onto the borehole wall, and the amount of fluorescence is monitored using a downhole sensor (Figure 28.13). The intensity of the resulting fluorescence can indicate the presence of a specific compound on the borehole wall. An example of its application is determining the arrival time of an injected tracer at different depths along a borehole. The device can be left in the field at a specific location, or can be raised and lowered within the borehole to map specific tracer locations.

CALIBRATION AND EVALUATION OF UNSATURATED, FRACTURED MEDIA MODELS

Water and air characterization data and chemical sample analyses are commonly employed to construct a conceptual or computer model of the flow regime. Inputs for matrix properties are usually obtained from laboratory tests, while fracture properties are determined using field tests. Calibration of the flow model can be accomplished using laboratory and field characterization data, reserving the chemical analysis data for model evaluation. In many circumstances, however, all available data must be used for model calibration due to the complexity of the geologic environment. Model evaluation should then be accomplished by conducting an independent field test that consists of flow conditions dissimilar to previous tests. The field test should be modeled a priori using calibration data sets, and a prediction forecast should be generated. The field test should be designed to test all components of the conceptual model in such a way as to assure that the results of the test

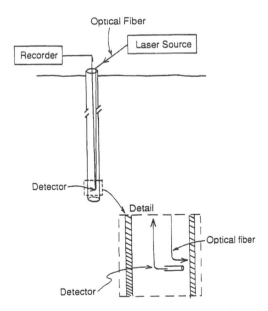

Figure 28.13. Laser-induced fluorescent technique for in situ monitoring of contaminants in unsaturated fractured rock.

unambiguously resolve assumptions central to the model. Computer modeling can be performed prior to conducting the test for the purpose of identifying field tests that provide the most useful information. Especially useful tests generally involve coupled flow and tracer tests in the vertical direction that incorporate both matrix and fracture flow components.

This evaluation exercise requires that parameter, model, and experimental uncertainties be quantified prior to the design of the experiment. Parameter uncertainties should be estimated based on repetitive measurements of rock fracture and matrix properties. The spatial variation in material properties can be determined by assessing the variation of samples collected over a range of distances. Quantification of experimental uncertainties requires that instrument and monitoring uncertainties be determined to assess the importance of measurement errors. Propagation of parameter and experimental uncertainties can be used to provide simulation forecast confidence intervals. The result of the field experiment can then be compared to the prediction interval to provide a rigorous test of model performance. Figure 28.14 summarizes the use of uncertainties in the model evaluation process.

FURTHER INFORMATION

Essentially all of the technical papers on flow and transport through unsaturated fractured rock have been published during the last ten years as the result of interest in the potential emplacement of high-level radioactive waste in a geologic repository located in unsaturated fractured tuff. An additional source of information is also

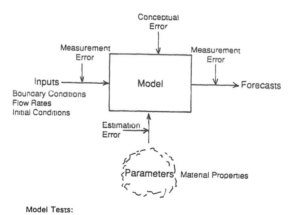

Model Tests:

1. Do observed experimental results fall within model forecast confidence intervals?

2. If yes, are model forecast confidence intervals unacceptably large?

3. If yes, what additional data can be collected to minimize forecast confidence intervals?

Figure 28.14. Propagation of uncertainties from model parameterization and field measurements. Increasing uncertainties result in inability to provide narrow forecast limits for model evaluation and the possible acceptance of an incorrect model.

available in the petroleum literature. Also, there has been an increased realization that many contaminated sites and proposed waste disposal sites involve flow and transport through unsaturated fractured rock and that improved understanding and assessment techniques are required. For a general coverage of the topic, see the American Geophysical Union monograph edited by Evans and Nicholson (1987). For a description of unsaturated fractured rock testing procedures and data sets, see Rasmussen et al. (1990) and Rasmussen et al. (1993). Evans (1983) and Rasmussen and Evans (1986) provide a general discussion of unsaturated flow and transport monitoring techniques. A summary of fracture flow models can also be found in Mercer et al. (1983) and Bear et al. (1993).

REFERENCES

Amter, S., "Injection-Recovery Techniques for Vacuum Lysimeter Sampling in Highly Unsaturated Media," unpublished MS thesis, Department of Hydrology and Water Resources, The University of Arizona, Tucson, AZ, 1987.

Anderson, I.C. "Measurement of Unsaturated Rock-Water Potential In Situ," unpublished MS thesis, Department of Hydrology and Water Resources, The University of Arizona, Tucson, AZ, 1987.

Andrews, J.W., "Water Content of Unsaturated, Fractured, Crystalline Rocks from Electrical Resistivity and Neutron Logging," unpublished MS thesis, Department of Hydrology and Water Resources, The University of Arizona, Tucson, AZ, 1983.

Bear, J., C.F. Tsang, and G. de Marsily, Eds., *Flow and Contaminant Transport in Fractured Rock*, (San Diego, CA: Academic Press, 1993).

Brooks, G.P., "Forced Ventilation of Chlorinated Hydrocarbons in Layered, Unsaturated Soil Material: A Laboratory Evaluation," unpublished MS thesis, Department of Hydrology and Water Resources, The University of Arizona, Tucson, AZ, 1989.

Davies, B., "Measurement of Thermal Conductivity and Diffusivity in an Unsaturated, Welded Tuff," unpublished MS thesis, Department of Hydrology and Water Resources, The University of Arizona, Tucson, AZ, 1987.

Elder, A.N., "Theoretical Calibration of Neutron Gauge," unpublished MS thesis, Department of Hydrology and Water Resources, The University of Arizona, Tucson, AZ, 1988.

Evans, D.D., Unsaturated Flow and Transport Through Fractured Rock — Related to High-Level Waste Repositories, Phase I, NUREG/CR-3206, U.S. Nuclear Regulatory Commission, Washington, DC, 1983.

Evans, D.D., and T.J. Nicholson, Eds., *Flow and Transport Through Unsaturated Fractured Rock*, AGU Geophysical Monograph 42, 187 pp., American Geophysical Union, Washington, DC, 1987.

Goering, T., "Use of Fluorescent Dyes as Tracers in Unsaturated, Fractured Rock," unpublished MS thesis, Department of Hydrology and Water Resources, The University of Arizona, Tucson, AZ, 1988.

Haldeman, W.R., Y. Chuang, T.C. Rasmussen, and D.D. Evans, "Laboratory Analysis of Fluid Flow and Solute Transport Through a Fracture Embedded in Porous Tuff," *Water Resour. Res.*, 27(1):53–66 (1991).

Kilbury, R.K., T.C. Rasmussen, D.D. Evans, and A.W. Warrick, "Water and Air Intake of Surface-Exposed Rock Fractures in Situ," *Water Resour. Res.*, 22(10):1431–1443 (1986).

Kipp, K.L., Jr., "Effect of Topography on Gas Flow in Unsaturated Fractured Rock: Numerical Simulation," in D.D. Evans and T.J. Nicholson, Eds., *Flow and Transport Through Unsaturated Fractured Rock*, AGU Geophysical Monograph 42, 187 pp., American Geophysical Union, Washington, DC, 1987, pp. 171–176.

Mercer, J.W., P.S.C. Rao and I.W. Marine, Eds., *Role of the Unsaturated Zone in Radioactive and Hazardous Waste Disposal*, (Ann Arbor, MI: Ann Arbor Science, 1983).

Mower, T.E., I.C. Yang and J.D. Higgins, "Triaxial- and Uniaxial-Compression Testing Methods Developed for Extraction of Pore Water from Unsaturated Tuff, Yucca Mountain, Nevada," in *Focus 89*, Proceedings of Nuclear Waste Isolation in the Unsaturated Zone, September 17–21, 1989, Las Vegas, NV, p. 426, American Nuclear Society, La Grange Park, IL, 1989.

Myers, K.C., "Water Flow and Transport Through Unsaturated Discrete Fractures in Welded Tuff," unpublished MS thesis, Department of Hydrology and Water Resources, The University of Arizona, Tucson, AZ, 1989.

Rasmussen, T.C. and D.D. Evans, Unsaturated Flow and Transport Through Fractured Rock — Related to High-Level Waste Repositories, Phase II, NUREG/CR-4655, U.S. Nuclear Regulatory Commission, Washington, DC, 1986.

Rasmussen, T.C., "Computer Simulation of Steady Fluid Flow and Solute Transport Through Three-Dimensional Networks of Variably Saturated, Discrete Fractures," in D.D. Evans and T.J. Nicholson, Eds., *Flow and Transport Through Unsaturated Fractured Rock*, AGU Geophysical Monograph 42, 1987, American Geophysical Union, Washington, DC, 1987, pp. 107–114.

Rasmussen, T.C., T.C.J. Yeh and D.D. Evans, "Effect of Variable Matrix-Permeability/Fracture-Permeability Ratios on Three Dimensional Fractured Rock Hydraulic Conductivity," in Geostatistical, Sensitivity and Uncertainty Methods for Ground-Water Flow and Radionuclide Transport Modeling, Proceedings of DOE/AECL Conference, Sept. 15–17, 1987, San Francisco, CA, 1989.

Rasmussen, T.C., D.D. Evans, P.J. Sheets, and J.H. Blanford, Unsaturated Fractured Rock Characterization Methods and Data Sets at the Apache Leap Tuff Site, NUREG/CR-5596, U.S. Nuclear Regulatory Commission, Washington, DC, 1990.

Rasmussen, T.C., "Steady Fluid Flow and Travel Times in Partially Saturated Fractures Using a Discrete Air-Water Interface," *Water Resour. Res.*, 27(1):67–77 (1991).

Rasmussen, T.C., and D.D. Evans, "Water Infiltration into Exposed Fractured Rock Surfaces," *Soil Sci. Soc. Am. J.*, 57:324–329 (1993).

Rasmussen, T.C., D.D. Evans, P.J. Sheets and J.H. Blanford, "Permeability of Apache Leap Tuff: Borehole and Core Measurements Using Water and Air," *Water Resour. Res.*, 29(7):1997–2006 (1993).

Roberts, M., "Volatile Fluorocarbon Tracers for Monitoring Water Movement in the Unsaturated Zone," unpublished MS thesis, Department of Hydrology and Water Resources, The University of Arizona, Tucson, AZ, 1987.

Schrauf, T.W. and D.D. Evans, "Laboratory Studies of Gas Flow Through a Single Natural Fracture," *Water Resour. Res.*, 22(7):1038–1050 (1986).

Smith, S.J., "Natural Airflow through the Apache Leap Tuff near Superior, Arizona," unpublished MS thesis, Department of Hydrology and Water Resources, The University of Arizona, Tucson, AZ, 1989.

Tang, J., "Hydraulic Impedance Technique for the Characterization of Unsaturated Fractured Rock," unpublished MS thesis, Department of Hydrology and Water Resources, The University of Arizona, Tucson, AZ, 1991.

Thornburg, T.M., "Electrical Resistivity of Unsaturated Fractured Tuff: Influence of Moisture Content and Geologic Structure," unpublished MS thesis, Department of Hydrology and Water Resources, The University of Arizona, Tucson, AZ, 1990.

Tidwell, V.C., "Determination of the Equivalent Saturated Hydraulic Conductivity of Fractured Rock Located in the Vadose Zone," unpublished MS thesis, Department of Hydrology and Water Resources, The University of Arizona, Tucson, AZ, 1988.

Trautz, R.C., "Rock Fracture Aperture and Gas Conductivity Measurements In Situ," unpublished MS thesis, Department of Hydrology and Water Resources, The University of Arizona, Tucson, AZ, 1984.

Updegraff, C.D., C.E. Lee, and D.P. Gallegos, "DCM3D: A Dual-Continuum, Three-Dimensional, Ground-Water Flow Code for Unsaturated, Fractured, Porous Media," NUREG/CR-5536, SAND90-7015, U.S. Nuclear Regulatory Commission, Washington, DC, 1991.

Vickers, B.C., "Aperture Configuration of a Natural Fracture in Welded Tuff," unpublished MS thesis, Department of Hydrology and Water Resources, The University of Arizona, Tucson, AZ, 1990.

Wang, J.S.Y. and T.N. Narasimhan, "Unsaturated Flow in Fractured Porous Media," in J. Bear, C.F. Tsang, and G. de Marsily, Eds., *Flow and Contaminant Transport in Fractured Rock*, (San Diego, CA: Academic Press, 1993).

Weeks, E.P., "Effect of Topography on Gas Flow in Unsaturated Fractured Rock: Concepts and Observations," in D.D. Evans and T.J. Nicholson, Eds., *Flow and Transport Through Unsaturated Fractured Rock*, AGU Geophysical Monograph 42, American Geophysical Union, Washington, DC, 1987, pp. 165–170.

Yang, I.C., A.K. Turner, T.M. Sayre and P. Montazer, "Triaxial Compression Extraction of Pore Water From Unsaturated Tuff, Yucca Mountain, Nevada," USGS Water Resources Investigation 88–4189, Denver, CO, 1988.

29

Soil Gas Sampling

William L. Ullom

INTRODUCTION

Soil gas sampling is a method used to support measurement of soil atmosphere characteristics which may indicate processes occurring in and below the sampling horizon. Soil gas sampling can suggest the presence, composition, and origin of pollutants in and below the vadose zone; however, the method is somewhat qualitative, and conclusions based upon soil gas data must take this into account. Soil gas monitoring is a valuable screening method for detection of volatile organic contaminants, the most abundant analytical group of ground-water pollutants (Plumb and Pitchford, 1985).

PRINCIPLES

Soil gas surveys largely measure the effects of movement of contaminants in the vadose zone. Four processes control this movement. These are partitioning, migration, emplacement and degradation.

Partitioning represents a group of processes which control contaminant movement from one physical phase to another. These phases are:

- liquid
- free vapor, i.e., through-flowing air (Stonestrom and Rubin, 1989)
- occluded vapor, i.e., locally accessible air and trapped air (Stonestrom and Rubin, 1989)
- solute
- sorbed.

Partitioning is the initial step by which contaminants begin to move away from their source. Partitioning occurs in water saturated and unsaturated environments, and is difficult to quantify due to the unique makeup of the vadose matrix, i.e., air-filled porosity (microporous and macroporous), pore water, free product, solid-phase soil organic matter, clay, and discrete inorganic soil particles. Important individual processes of partitioning are dissolution, volatilization, air-water partitioning, soil-water partitioning, and soil-air partitioning (Lyman, 1987). Models of contaminant movement in the vadose zone consider these processes individually, albeit incorporating numerous assumptions in practice.

Migration refers to contaminant movement over distance with any vertical, horizontal, or temporal component. Unsaturated flow of gas or liquid is highly complex and is controlled by soil characteristics, contaminant composition, and contaminant phase (American Petroleum Institute, 1972). Vapor migration through unsaturated matrix can occur through a variety of diffusion, dispersion, and mass transport mechanisms which operate differently under saturated and unsaturated conditions.

One major division in the migration action of contaminants is defined by their solubility or immiscibility in water. Pollutants are often introduced into the soil as liquid mixtures and the components of these mixtures immediately begin to partition into other phases. The makeup of a contaminant chemical mixture that establishes soil residence behind a migratory front can change in relative composition with distance from a point of entry to the soil. As contaminant migration continues, pathways for individual components can become divergent. This is largely driven by the magnitude of water solubility and vapor pressure of each component of the chemical mixture, so that the composition of the mixture continues to change as migration proceeds. Transport of contaminants by downward percolation of meteoric waters and upward movement of ground water accelerate the contact of contaminants with ground water. Although gaseous migration is often attenuated in soils with high clay or organic matter contents due to sorption and tortuosity, care must be exercised not to take this action for granted. For example, chlorinated aliphatic hydrocarbons may pool on low-permeability soils causing shrinkage and cracking of clay minerals, opening new macroporous flowpaths available for pollutant migration (Barbee, 1994).

The impact of gaseous and liquid migration processes on soil gas measurement is significant. It is impractical to estimate actual migration mechanisms by modeling prior to most soil gas surveys. A thorough knowledge of relevant site characteristics, such as the presence or absence of barriers to vertical or horizontal migration, i.e., moist clay layers, foundations, buried pavement, or perched ground water, as well as preferential pathways for contaminant migration, i.e., backfill rubble, utility vaults, storm sewers, or soil cracks, can assist investigators to assess the migration impact on soil gas survey design.

Emplacement refers to establishment of contaminant residence in any phase within any residence opportunity. This dynamic process has been described as an in-situ chromatographic-like separation of contaminants (Kerfoot and Sanford, 1986). Contaminants partition from one phase to another as changes occur in both chemical and physical equilibria. Important changes impacting phase residence change include temporal variations in moisture content, barometric pressure, soil temperature, and level of microbial activity. In natural systems, temporal increases in soil moisture cause gradual increases in solute phase emplacement at the expense of other phases; increases in soil moisture can decrease the amount of pore space available for gaseous flow. Of changes in moisture content, barometric pressure, soil temperature, and level of microbial activity, change in soil moisture has the greatest overall effect on phase residence. In very permeable soils, barometric pressure changes have a large impact on soil gas survey data. Soil temperature changes can have measurable impact on relatively shallow investigations, but the most notable are the obvious influences of frost.

Degradation is the process whereby contaminants are attenuated by oxidation or reduction in the vadose zone, either through biogenic or abiogenic processes. Soil gas monitoring measures the result of the interaction of these processes in a dynamic equilibrium. Degradation is most often recognized in shallow, permeable soils where favorable conditions exist for oxidation of labile compounds; however, other vadose environments can be conducive to degradation. Specific environmental conditions are required for degradation processes to occur. Most soils contain naturally occurring populations of various microorganisms that can degrade petroleum products (Dragun, 1985). Contaminant biodegradation is known to occur in ground water and in soils prior to contaminant partitioning into a vapor phase (Barker et al., 1987; Davis, 1969). Contaminant biodegradation rates for some compounds are highly variable and are controlled by a number of kinetic factors influencing the distribution of microorganisms responsible for degradation. These include soil moisture content, aerobic versus anaerobic environments, the availability of nutrient salts, contaminant type, and soil temperature (Jensen et al., 1985; White et al., 1985).

SUMMARY PROCEDURE FOR SOIL GAS SAMPLING

Soil gas sampling programs commonly involve six primary procedures. These are:

- A planning and preparation step including definition of data quality objectives;
- The act of sampling soil gas in the field;
- Handling and transporting the sample;
- Sample analysis;
- Interpretation of the results of analysis; and
- Preparation of a report of findings.

Planning and Preparation

The planning and preparation step begins with the formulation of project objectives, including purpose of the survey, appropriate application of the data to be collected, and data quality objectives. The planning and preparation step continues with the evaluation of available information already gathered for the project area. Project planning culminates in the selection of an appropriate soil gas monitoring method and a survey design which best fits the project objectives within budgetary constraints.

Soil Gas Sampling: A Variety of Methods

Soil gas sampling methodology has evolved over time. The equipment available is highly varied but is categorized into basic types. The application of any method must be controlled by strict adherence to a standard operating procedure. Occasional deviations as dictated by unusual field conditions should be recorded in the project field notebook. Inadvertent minor deviations in field procedure can result in changed data results, which can lead to possible misinterpretation of the data acquired.

Methods for sampling soil gas can be described based upon the collection of soil gas by a whole-air or sorbent method in an active or passive approach. Most soil gas surveys are accomplished by some variant of an active whole-air approach. Passive sorbent methods are also used.

A unique method loosely categorized as a "soil gas sampling method" involves the measurement of constituents in contained headspace above a soil (or water) sample. However, contained atmosphere methods do not yield samples representative of in situ soil atmospheres and should be used for rudimentary screening purposes only.

Soil Gas Sampling Using Active Methods

This method of soil gas sample collection involves the forced movement of bulk soil atmosphere from the sampling horizon to a collection or containment device through a probe or other similar apparatus (Jensen et al., 1985; Kerfoot, 1987). Figure 29.1 (ASTM D 5314–93 and Devitt et al., 1987) is an example of a whole-air active sampling system. Contained samples of soil atmosphere are then transported to a laboratory for analysis, or the sampling device is directly coupled to an analytical system. Whole air-active sampling is best suited to soil gas monitoring efforts where contaminant concentrations are expected to be high and the vadose zone is highly permeable to vapor. Probes can be very simple and light-weight for low cost mobilization or they can be affixed to vehicle mounted drills or hammers useful for larger, more complex surveys at a higher cost of mobilization (Kerfoot, 1984). Ground probes can be of small to large internal volume. The development of sampling devices with smaller internal volumes equating to smaller purge volumes is a significant improvement, providing samples which are more representative of soil atmosphere, and a greater ease of equipment decontamination between usages.

The success of the active approach is strongly dependent upon soil clay, organic matter, and moisture content. Driven probes tend to destroy natural soil permeability around the body of the probe due to soil compaction concurrent with insertion. This can be a severe limitation in moist, heavy clay soils. In very dry or cemented soils, driven probes can create radial fractures which can enhance soil permeability to vapor concurrent with insertion. These fractures can communicate atmospheric air with soil atmosphere, a limiting factor for obtaining representative, large-volume soil gas samples. The effect can be so severe as to lower recovered contaminant concentrations in the soil gas sample below the limits of analytical detection. Some investigators have attributed the poor recoveries of some compounds exclusively to degradation (Nadeau et al., 1985; Thompson and Marrin, 1987).

Methods requiring a pre-existing hole for probe insertion made with a commercially available "slam bar" can provide supportable contaminant data where contaminant concentrations and soil permeability to vapor are high (U.S. EPA, 1988). However, the act of making a hole with a "slam bar" and subsequent removal of the "slam bar" can encourage soil venting. Insertion of the sampling probe into this hole further degrades representativeness by additional venting of contaminants as the probe displaces the atmosphere in the hole upon insertion. Purging of the probe prior to sampling under conditions of low soil permeability and low contaminant concentration may lower contaminant levels below the limits of analytical detection.

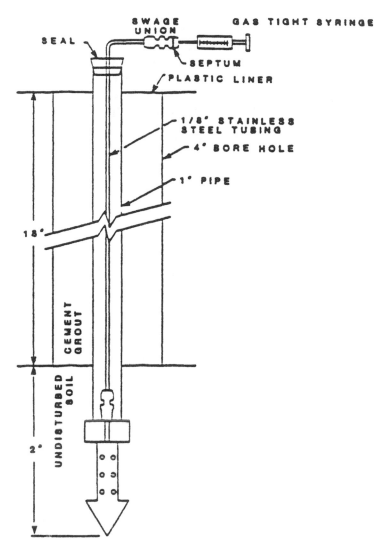

Figure 29.1.   Example of a whole-air active sampling system (after Devitt et al., 1987).

Methods requiring a pre-existing hole for probe insertion are not recommended for soil gas sampling from soils with high clay and moisture contents.

Caveats to active methods are often forgotten in the field. Soil gas sampling on steep slopes, near animal burrows, or over utility conduits can lead to falsely negative results. Unseen barriers to migration of vapors can provide opportunity for accumulation of contaminant vapors and may cause an inexperienced soil gas investigator to believe that he or she is near a source area.

Soil Gas Sampling Using Passive Methods

In contrast to active methods, passive methods are based upon the passive movement of contaminants in soil to a sorbent collection device or flux chamber over time. Sorbent samplers have market prevalence over flux chambers in most field applications. Passive samplers that have been applied to sampling soil gases of environmental concern include occupational health volatile organic compound monitors and a sampler originally developed for detecting the presence of hydrocarbons in petroleum exploration (Kerfoot and Mayer, 1987; Mayer, 1989; Voorhees et al., 1984). Both devices use charcoal as a sorbent; the former as a flat film and the latter coated on a wire. Other sorbent materials can be selected by the user as governed by the chemicals of interest and their physiochemical properties.

Passive samplers are housed in containers up to several inches in diameter, depending upon the design. Figure 29.2 (ASTM D 5314-93 and Devitt et al., 1987) is an example of a sorbed contaminant passive system. Passive samplers are placed open end down in holes that are usually less than five feet deep, which are then backfilled (Mayer, 1989). These monitors are generally left in place from two to ten

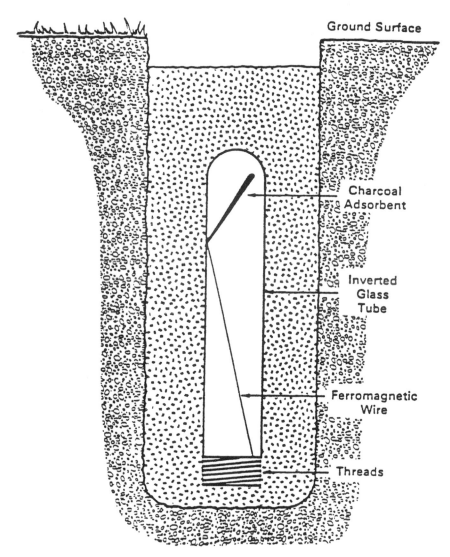

Figure 29.2. Schematic diagram of emplacement of a sorbed contaminant passive system (after Devitt et al., 1987).

days, although certain passive collectors can be left in place for a period of 30 days or more for certain applications.

The sorbed contaminants-passive approach can be employed in a wide range of geological conditions. Frozen ground and high water saturation may not limit the ability of the monitors to collect contaminants (Wesson and Armstrong, 1975). The composition of the contaminant suite may be impacted by related alterations in partitioning equilibria.

Most passive approaches depend upon the ability of contaminants to move through the vadose zone to the passive collection device. The principle of passive-sorbent monitors relies on adsorbent reduction of the equilibrium concentration of contaminants around the monitor over time, therefore creating a concentration sink in the vicinity of the monitor. This can encourage continued migration of contaminants toward the monitor when conditions for contaminant partitioning into the vapor phase are favorable. Migration of contaminants in the vadose zone toward a passive-sorbent device is strongly controlled by vadose zone character and the chemical and physical properties of the subject contaminants. Contaminants may move from a few feet to thousands of feet, or not at all. In highly permeable soils, this movement can be impacted, if not driven, by changes in barometric pressure.

As with active sampling protocols, specific issues exist affecting the function and calibration of passive monitors. Soil gas, even in the drier climates, will be at a relatively high humidity condition, approaching 100%. This humidity can affect the collection efficiency of the adsorbent media. In soils of low permeability, contaminants commonly move very slowly. This can create a condition of near-zero contaminant concentration in the soils immediately adjacent to the monitor. When soil contaminant concentrations are rapidly depleted due to invasion of the sampling horizon by meteoric water, the passive monitor can source contaminants back to the soil in this condition of dynamic equilibrium.

Use of passive systems does not allow collection of real-time data. Passive systems can provide measurements of flux rather than soil contaminant concentration data. It is not possible to measure the efficiency of passive monitoring devices because the bulk volume of soil gas affected by the sorbent trap cannot be measured. Care must be taken not to contaminate the sorbent samples during installation or by backfilling with contaminated soil. Such care is comparable to potential problems for any measurement method in which a contaminated layer is penetrated.

Sample Handling and Transport

Consideration of sample handling and transport are not trivial to monitoring gas phase contaminants at very low levels.

The period of sample handling and transport represents the greatest opportunity for loss or gain of contaminants from or to sample containers. Loss can occur by contaminant condensation within the sampling system, sorption, solution into condensed water in the sampling system, thermal degradation, or leakage to the atmosphere. Gain of contaminants can be attributable to "memory" from insufficient decontamination procedure. In general, the time between sample collection and analysis should be minimized. Practitioners should protect samples against light, heat, and leakage.

Investigators are responsible for selecting materials for soil gas sampling, transfer, and containment that will not impact sample integrity. Acceptable materials can be conveniently decontaminated prior to soil gas recovery. Materials that cannot be decontaminated effectively between samples must either be replaced between samples, considered in QA/QC planning as a survey limitation, or abandoned in favor of more suitable materials.

Problems of sample handling and transport are minimized by integration of the sampling and analytical system. Cross-contamination is a concern with integral systems and is often attributable to tubing, flow meters, or analyzer components.

The method of transfer of samples from sampling device to containers, if used, is largely dependent upon the volume of soil gas recovered. Small volume samples are commonly recovered by syringe for immediate injection into an analyzer or small volume container. Small volume sampling is quite sensitive to variations in sample transfer technique. Large volumes of soil gas are commonly recovered by hand or mechanical pumps installed at the end of the sampling system. Tubing is commonly used in large volume sampling. For low level detection, tubing can present a cross contamination problem if not replaced in the sampling train prior to sampling at a new location.

A wide variety of sample containers is employed by field investigators. Container selection is based upon the physical properties of the contaminants sampled, the volume of the sample recovered, the physical properties of suspected contaminants, the sampling system employed, the anticipated sample holding time prior to analysis, and the analytical method chosen. Container type for a soil gas survey should be held constant within the survey. A change in container type can impart bias to a portion of the data, due to sorptive or desorptive processes related to container type. Whole air samples can be contained in any device made of suitable materials that conveniently satisfy survey, handling, transport, and analytical requirements. Atmospheric sampling bags can present difficulties due to leakage after improper or repeated use. Sorbent traps are commonly self-contained. Care must be exercised to select a trapping device that is compatible with the properties of the target compounds and the technique of desorption chosen.

Some investigators process soil vapor samples prior to analysis. Extraction is a sample processing step used to remove soil contaminants from soil cores or other similar samples. This technique can efficiently recover contaminants from all residence phases, not just the vapor phase. As a result, the technique yields samples which are not representative of soil atmosphere contaminant suites. Refrigeration is the only sample "processing" step which is nearly universally justifiable and widely used.

If samples are to be transported to an offsite laboratory for analysis, they must be properly packaged to avoid damage to sample containers. Care must be taken to keep samples from becoming overly warm or agitated during transport. Overnight air express is highly convenient if samples are properly contained, but air freight is not recommended if samples are held in containers such as gastight syringes or bags. Soil gas samples have limited shelf life even in the most effective containers. Soil gas sample life is strongly container-dependent. The safest practice is to minimize sample storage time. This problem is greatest when offsite laboratories are engaged to analyze the samples.

Sample Analysis

Soil gas analysis procedure is based upon pre-existing protocol established for the analysis of contaminants in ambient air. Some analytical methods are highly general, satisfying only the most rudimentary requirements of contaminant screening and quality assurance objectives. Others are sophisticated, providing identification and relative concentration information for numerous chemical compounds determined to be present in a soil gas sample. The choice of basic analytical approach in soil gas analysis is driven by the purpose of the soil gas survey, quality assurance objectives, and budgetary constraints placed upon investigators.

Soil gas surveying as a field screening technique can often be effective without the commitment of expenditure for highly sophisticated techniques. This survey purpose is merely to locate other, more direct, techniques. Caution is suggested when choosing highly sophisticated analytical methods for field screening by soil gas monitoring. This selection is controlled largely by the need for the analytical method chosen to be cost-effective.

Other applications of soil gas monitoring require more thorough analytical protocols. The independent monitoring of multiple classes of contaminants in soil gas normally requires analytical systems with multiple detectors.

Contaminant concentrations in soil gas can vary from levels below the detection limit of the most sophisticated equipment to percent of a whole-air sample. Ideally, the analytical system chosen has enough flexibility to determine contaminants in a wide range of concentrations. Care should be taken to select an analytical system sensitive enough to avoid false negative results which can lead to invalid conclusions. Many analytical systems are not designed to perform to specifications in very high concentration environments, requiring sample dilution prior to analysis or selection of a less sensitive method.

Of primary importance to the successful analysis of soil gas is the familiarity and experience of the analyst with the analytical system chosen. The analyst must be able to independently care for and maintain the equipment as well as recognize symptoms of procedural error. The success of an analytical effort lies wholly with operator ability and experience. Excessive machine capability cannot compensate for operator inexperience.

Soil gas may be analyzed by a number of methods, including portable VOC analyzers, gas elution chromatography, and colorometric detector tubes. Gas chromatography (GC) or GC-based hand-held detectors are the most widely used analytical instruments for soil gas analysis.

Portable VOC analyzers used for fugitive emission screening and industrial hygiene monitoring have been adopted for soil gas analytical purposes by numerous investigators. These devices are easily transported to and from the field, require minimal operator skill, provide immediate data, and serve to eliminate many sample handling and transport steps which can result in uncertainty. Portable VOC analyzers are limited in application to very low level detection due to the absence of a concentration step. They exhibit limited selectivity and do not have the ability to separate contaminant compounds, leading to potential interference. These devices also are limited in accuracy, due to the inability to calibrate for the wide variety of contaminant compounds encountered in soil gas, each compound having its own character of detector response.

Soil gas analysis by GC is by far the most versatile and the most costly soil gas analytical method. Instrumentation can be varied to accommodate field mobility; however, this is not always required. The technique provides separation of compounds in a chromatographic column, tentative identification of compounds determined to be present, and a relative quantitation of compound concentration based upon comparison to a known standard. Soil gas is introduced into the GC and conveyed through a chromatographic column by a carrier gas, separating the soil gas constituents as they pass through the column. The separation is obtained when the sample mixture in the vapor phase passes through a column containing a stationary phase possessing special adsorptive properties. As the gas stream emerges from the column, it passes through a detector, providing for measurement of a specific sample property through the recording of detector electrical response. These responses, or peaks, are recorded as a function of time. Comparison of known standard compound response time with the response time of an unknown represented by a peak results in the tentative identification of the unknown. Comparison of the magnitude of detector response to the newly identified compound versus detector response to the same compound of known concentration, a laboratory standard, results in a relative quantitation of a subject compound concentration in the sample.

Detector tubes have been applied to safety and health atmospheric monitoring, agriculture, and the chemical industry. These devices are designed to be compound-specific, although this characteristic is dependent upon the contaminant compounds present in the sample drawn through the tube. Detector tubes may be used for short-term sampling (grab sampling; 1 to 10 minutes) or long-term sampling (dosimeter sampling; 1 to 8 hours). Short-term sampling involves the movement of a given volume of gas through the tube by a mechanical pump. If the substance for which the detector tube was designed is present, the indicator chemical in the tube will change color (stain). The concentration of the gas component may be estimated by either the length of the stain compared to a calibration chart or by the intensity of the color change. Detector tubes are relatively inexpensive and provide immediate results. Their use is restricted to applications with few interfering compounds.

Data Interpretation

Soil gas surveys not specified to only provide raw data require some measure of interpretation and formulation of conclusions based upon that raw data. Data interpretation is largely an iterative process of review of the raw soil gas data out of context, a review of the soil gas data in context of other site characteristics, and the formulation of conclusions based upon all known information.

The tool most often used to aid in data interpretation is contour mapping of soil gas concentration data. Areas of highest soil gas constituent concentration termed "hot spots" can be interpreted to be potential source areas or release areas. Such interpretations can be tenuous at best when impacted by site characteristics including barriers to or conduits for migration, the presence of clay soils, or lateral soil moisture changes in the sampling horizon. However, in cases where tests are to be made to determine the effectiveness of remediation, this technique can be very beneficial to locate poorly affected areas.

Other interpretive tools include cross sections through data derived from three-dimensional soil gas surveys, cross-plots of soil gas constituent concentrations, and comparison of soil gas data to data derived from other monitoring methods.

It is important to remember that soil gas surveying, as generally practiced in the field, is a more qualitative method than many other site characterization methods used. Bearing this in mind, it is prudent to caution against overinterpretation of soil gas data.

Report Preparation

A soil gas survey report should include the information necessary to describe the results of that survey performed for a particular application. Certain applications require a thorough treatment of a significant number of factors impacting the meaning and usefulness of soil gas data interpretations. Examples of such applications include field screening in support of environmental assessments, contaminant source identification or tests of the effectiveness of remediation.

Quality assurance data elements are necessary without regard to data application. These include precision, accuracy, data comparability, representativeness, completeness, and analytical detection limits whenever possible. At a minimum, a general discussion of the reliability of results and analytical detection limits is warranted. Efforts to limit reporting requirements for the sake of short-term time and money cost savings usually result in low confidence level treatment of the report and/or an ultimate time and money cost gain.

Every subject of every vadose zone monitoring effort has unique characteristics. Those characteristics which could impact the results of the soil gas monitoring effort should be described to provide a meaningful context in which to interpret the soil gas data. Local conditions should be described with regard to soil type(s), moisture content in the vadose zone, soil/bedrock interface, stratigraphy and lithology, ground-water bearing zones, flow directions and gradients, potentiometric levels, aquifer characteristics and ground-water quality. If known and appropriate, the characteristics of a contaminant source or spill should be addressed. Examples of such characteristics are contaminant composition, the likelihood of single or multiple contamination events, or the reaction potential (above, within, and beneath the vadose zone) of multiple contaminant mixtures.

A detailed description should be given of the type of soil gas survey conducted. Details should include selection of active or passive method, whole air or passive sample collection method, sampling array, background sampling, equipment decontamination procedure employed prior to the survey, field or laboratory analytical methods, and QA/QC procedures. Any unusual conditions should be noted, such as rainfall events during the course of the survey (especially when moveable soil gas chiefly originates from vadose zone microporosity), high pressure or low pressure front movement across the survey area during the course of the survey (especially when moveable soil gas chiefly originates from vadose zone macroporosity), or visual observations of contamination at sampling points.

Data collected during the field sampling and field or laboratory analyses should be compiled in table form and be included in a preliminary or final report, preferably as appendices. Such data should include a listing of sampling and analysis dates, soil/rock description at each sampling point, depth and diameter of sampling point, quantity of soil gas purged prior to sampling, quantity of sample extracted, chromatogram and/or mass spectra for each sample, and a tabulation of QA/QC samples recovered.

Whenever possible, discussion should be provided that correlates soil gas data to ground truth. The most common and widely accepted form of ground truth is data from ground-water monitoring wells.

The report should contain a section which discusses the conclusions drawn from the results of the soil gas study and any recommendations which seem appropriate to enhance the value of conducting such a soil gas study. Conclusions should include identification of the compounds detected, if any, an assessment of the appropriateness of the soil gas study method used, and any circumstances which may have significantly impacted the results of the investigation, such as weather conditions or equipment calibration. Recommendations should address need for establishing ground truth, extension of the study to adjacent areas of interest, the need for a different soil gas study method, actions to resolve questionable QA/QC results, or need for additional chemical analyses for contaminant identification.

REFERENCES

American Petroleum Institute, *The Migration of Petroleum Products in Soil and Ground Water: Principles and Countermeasures*, API Publication No. 4149, Washington, DC, 1972.

ASTM D 5314-93, *Standard Guide for Soil Gas Monitoring in the Vadose Zone*, American Society for Testing and Materials, Philadelphia, PA.

Barbee, G.C., "Fate of Chlorinated Aliphatic Hydrocarbons in the Vadose Zone and Ground Water," *Ground Water Monitoring Review*, Winter, 1994, pp. 129–140.

Barker, J.F., G.C. Patrick, and D. Major, "Natural Attenuation of Aromatic Hydrocarbons in a Shallow Sand Aquifer," *Ground Water Monitoring Review*, 7:66–71 (1987).

Davis, J.B., "Microbiology in Petroleum Exploration," *Unconventional Methods in Exploration for Petroleum and Natural Gas*, W. B. Heroy, Ed., Southern Methodist University, Institute for the Study of Earth and Man, SMU Press, 1969, pp. 139–157.

Devitt, D.A., R.B. Evans, and W.A. Jury, et al., *Soil Gas Sensing for the Detection and Mapping of Volatile Organics*, National Water Well Association, Dublin, OH, 1987.

Dragun, J., "Microbial Degradation of Petroleum Products in Soil," *Proceedings, Conference on the Environmental and Public Health Effects of Soils Contaminated with Petroleum Products*, Amherst, MA, October 30–31, 1985.

Jensen, B., E. Arvin, and A.T. Gundersen, "The Degradation of Aromatic Hydrocarbons with Bacteria from Oil Contaminated Aquifers," *Proceedings, NWWA/API Conference on Petroleum Hydrocarbons and Organic Chemicals in Groundwater*, Houston, TX, November 13–15, 1985, pp. 421–435.

Kerfoot, H.B., "Shallow-Probe Soil-Gas Sampling for Indication of Ground Water Contamination by Chloroform," *International Journal of Environmental and Analytical Chemistry*, 1987, No. 30, pp. 167–181.

Kerfoot, H.B., and C.L. Mayer, *The Use of Industrial Hygiene Samplers for Soil-Gas Measurement*, U. S. EPA Environmental Monitoring Systems Laboratory, Advanced Monitoring Division, Contract No. 68-03-3249, September, 1987.

Kerfoot, W.B., "A Portable Well Point Sampler for Plume Tracking," *Ground Water Monitoring Review*, 4:38–42 (1984).

Kerfoot, W.B., and W. Sanford, "Four-Dimensional Perspective of an Underground Fuel Oil Tank Leakage," *Proceedings, NWWA/API Conference on Petroleum Hydrocarbons and Organic Chemicals in Ground Water*, Houston, TX, November 12–14, 1986, National Water Well Association, Dublin, OH, pp. 383–403.

Lyman, W.J., "Environmental Partitioning of Gasoline in Soil/Groundwater Compartments," Seminar on the Subsurface Movement of Gasoline, Edison, NJ, 1987.

Mayer, C.L., *Draft Interim Guidance Document for Soil-Gas Surveying*, U. S. EPA Environmental Monitoring Systems Laboratory, Office of Research and Development, Contract No. 68–03–3245, September, 1989.

Nadeau, R.J., T.S. Stone, and G.S. Clinger, "Sampling Soil Vapors to Detect Subsurface Contamination: A Technique and Case Study," *Proceedings, NWWA Conference on Characterization and Monitoring of the Vadose (Unsaturated) Zone*, November 19–21, 1985, Denver, CO, National Water Well Association, Dublin, OH, pp. 215–226.

Plumb, R.H., Jr., and A.M. Pitchford, "Volatile Organic Scans: Implications for Ground Water Monitoring," *Proceedings, NWWA/API Conference on Petroleum Hydrocarbons and Organic Chemicals in Ground Water-Prevention, Detection and Restoration*, Houston, TX, November 13–15, 1985.

Stonestrom, D.A., and J. Rubin, "Air Permeability and Trapped-Air Content in Two Soils," *Water Resources Research*, 25:1959–1969 (1989).

Thompson, G.M., and D.L. Marrin, "Soil Gas Contaminant Investigations: A Dynamic Approach," *Ground Water Monitoring Review*, 7:88–93 (1987).

U. S. EPA. Environmental Response Team, Standard Operating Procedure 2149: Soil Gas Sampling, September 30, 1988.

Voorhees, K.J., J.C. Hickey, and R.W. Klusman, "Analysis of Ground Water Contamination by a New Surface Static Trapping/Mass Spectrometry Technique," *Analytical Chemistry*, 56:2602–2604 (1984).

Wesson, T.C., and F.E. Armstrong, The Determination of $C_1 - C_4$ Hydrocarbons Adsorbed on Soils, Bartlesville Energy Research Center Report of Investigations BERC/RI-75/13, U. S. Energy Research and Development Administration, Office of Public Affairs, Technology Information Center, Bartlesville, OK, December, 1975.

White, K.D., J.T. Novak, C.D. Goldsmith, and S. Bevan, "Microbial Degradation Kinetics of Alcohols in Subsurface Systems," *Proceedings, NWWA/API Conference on Petroleum Hydrocarbons and Organic Chemicals in Ground Water-Prevention, Detection and Restoration*, Houston, TX, November 13–15, 1985.

30

Case Studies of Soil Gas Sampling

Kirk A. Thomson

INTRODUCTION

The technical acceptance, popularity, accuracy, and use of soil gas survey techniques have increased tremendously over the past five years. Due to the volatile nature of many solvents and fuel hydrocarbons, soil sample analytical results are often inaccurate (not representative) due to loss of volatile organic compounds (VOCs) from samples during sample collection, transport, and analysis. In-situ measurement of soil gas samples is often a more accurate means of assessing the presence and extent of VOCs in unsaturated medium-to coarse-grain geologic materials. The volatile nature of many solvents and fuel hydrocarbons has made vapor extraction the preferred remediation technique for soil impacted by these chemicals and, therefore, it is prudent to sample the gas or vapor phase to evaluate remediation progress. This chapter describes the objectives and limitations of soil gas surveying, provides specific investigative methods and procedures, and presents two case studies conducted in Southern California.

OBJECTIVES

The objectives of soil gas surveys typically include:

- Identification of possible vadose zone source areas of VOCs.
- Assessment of the lateral and vertical extent of soil impacted by VOCs.
- Aiding in the effective placement of the soil borings and/or ground-water monitoring wells.
- Assessment of the presence and extent of VOCs in underlying ground water (plume mapping).

LIMITATIONS

Applications of soil gas sampling and analysis are limited by site-specific conditions which can influence VOC distributions in the vadose zone. Some factors affecting the gas-phase distribution of VOCs in the subsurface include:

- The location of VOC source areas (e.g., surface spills, subsurface leaks, atmospheric, and ground-water contamination).
- The liquid-gas partitioning coefficient of the compounds of interest (the "volatility" of the compound).
- The vapor diffusivity (a measure of how quickly a volatile compound "spreads out" within a volume of gas).
- Retardation of the individual compounds as they migrate in the soil gas. Retardation may be due to degradation, adsorption on the soil matrix, tortuosity of the soil profile, or entrapment in unconnected soil pores.
- The presence of impeding layers, wetting fronts of freshwater, or perched water tables, between the regional water table and ground surface.
- Movement of soil gas in response to barometric pressure changes.
- The preferential migration of gas through zones of greater permeability (e.g., natural lithologic variation or backfill).

Owing to these limitations, soil gas sampling and analytical data should be used in conjunction with other site-specific data, whenever possible.

SOIL GAS SAMPLING METHODS AND PROCEDURES

Introduction

In Southern California, the Regional Water Quality Control Board has developed soil gas sampling and analysis protocols for use in Environmental Protection Agency programs. Basin-wide ground-water contamination by VOCs, primarily chlorinated solvents, has resulted in the investigation of many sites in the San Gabriel basin. Site investigations are often designed to assess the presence, concentration, and extent of VOCs in soil gas, soil, and ground water. Site assessment requirements may include the following work elements:

Task 1: Shallow soil gas survey
Task 2: Drilling, soil sampling, and laboratory analysis of soil matrix samples
Task 3: Installation of Nested Soil Gas Probes or Ground-Water Monitoring Wells with Nested Probes

The lateral and vertical extent of contamination in soil gas samples, soil samples, and ground-water samples is used to evaluate whether or not a site requires remediation and whether the site has contributed to regional ground-water contamination. Sites requiring remediation are typically treated using vapor extraction techniques. Sites found to have contributed to ground-water contamination may be forced to pay settlements for ground-water remediation.

This section describes specific methods and procedures used to conduct two soil gas surveys, and presents the lateral and vertical distributions of VOCs at each study site.

Shallow Soil Gas Surveys

Soil gas samples were obtained using a perforated soil probe, Teflon™ or Nylaflow™ tubing, a vacuum pump, and an instrumentation assembly. Analysis of the soil gas samples was accomplished using a laboratory-grade, field-operable gas chromatograph.

Initially, probes were installed to depths of approximately 5 to 10 feet below grade. Soil gas probes were located using a horizontal grid system. Based on results of the shallow survey, additional probes were installed deeper using the same methods. A typical soil gas sampling probe is shown in Figure 30.1. Probes were installed using a percussion hammer or hydraulic ram. Once the probe was installed to the desired depth, the probe shaft was withdrawn, leaving the probe point and sampling tube in the soil. A small amount of silica sand was poured into the probe hole. The remaining open probe-hole was backfilled with hydrated bentonite grout to the ground surface. The probe point and sampling tube assembly remained in place to serve as a long-term soil gas monitoring point. This allows for subsequent soil gas sampling and analysis, if desired.

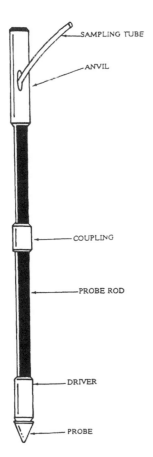

Figure 30.1. Typical construction of a soil gas sampling probe.

Sample Collection and Handling

Soil gas samples were collected using a soil gas sampling system as shown in Figure 30.2. Sample collection, handling, and analysis was the same for samples collected from deeper clustered probes and nested probe installations. Initially, site-specific probe purging and sample volume calibrations were performed to evaluate the appropriate volume of gas to be purged from each probe prior to sample collec-

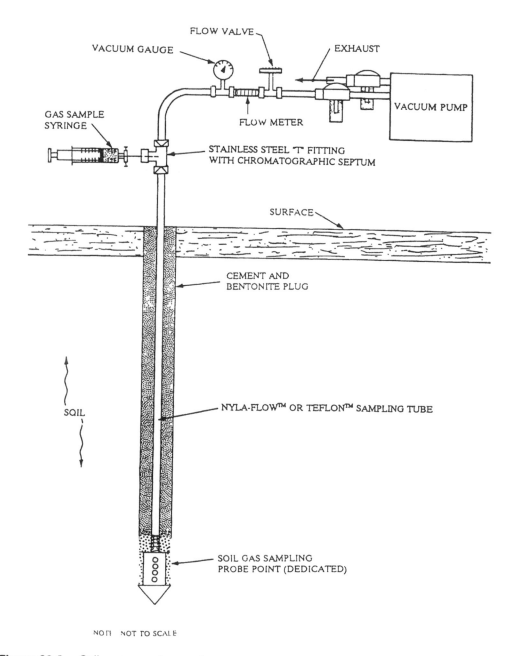

Figure 30.2. Soil gas sampling system.

tion. This was done by performing time-series sampling of at least one probe at each site to evaluate trends in soil gas concentrations as a function of purge volume. Soil gas samples were analyzed in the field immediately following collection.

Sample Analysis

Soil gas samples were analyzed in the field using a field-operable gas chromatograph equipped with a photoionization detector (PID) and an electrolytic conductivity detector (ELCD). The PID and ELCD were used in-series to analyze for EPA Method 8010/8020 compounds, including halogenated hydrocarbons and benzene, toluene, ethylbenzene, and total xylene (BTEX). Detection limits for the EPA Method 8010/8020 compounds were about 1.0 microgram per liter (μg/L) of gas. Soil gas samples may be analyzed for various other constituents, including volatile hydrocarbon mixtures such as gasoline, mineral spirits, or jet fuel using other detectors [e.g., flame-ionization detector (FID), electron capture detector (ECD), thermal conductivity detector (TCD), etc.]. Lower detection limits for selected compounds may be achieved using an ECD; however, greater sensitivity causes increased concern regarding cross-contamination and accuracy of calibration.

Equipment Calibration

The chromatographic equipment used for soil gas analyses was calibrated using high-purity solvent-based standards obtained from certified vendors. Calibration using solvent-based standards was performed using varying injection volumes of the stock solvent-based standard without dilution. If necessary, stock solvent-based standards were diluted to an appropriate concentration. Standards prepared by dilution were prepared by introducing a known volume of stock solvent-based standard into a known volume of high-purity solvent.

Initial field calibration was performed for halogenated and aromatic compounds (EPA Methods 8010/8020 compounds). The gas chromatograph was calibrated using three standard injections to establish a three-point calibration curve. The lowest standard was not higher than ten times the Method Detection Limit (or 10 μg/L). Identification and quantitation of compounds in the field were based on calibration under the same analytical conditions as for the three-point calibration.

Once in the field, daily calibration of the gas chromatograph consisted of a standard injection containing a minimum of nine compounds, including three aromatic compounds and six halogenated compounds representing short, medium, and long retention time groups. Daily calibration was performed prior to the first sample analysis of the day. One-point calibration was performed for all compounds detected at each site to ensure accurate quantitation. The response factor for each compound was within 15% of the corresponding value from the three-point calibration, or the GC was re-calibrated. Subsequent calibration episodes, if deemed necessary, consisted of at least one injection of the standard exhibiting a similar detector response as that of samples encountered in the field.

Equipment Blanks

The sample collection system consisted of stainless-steel, Teflon™, or Nylaflow™. Portions of the system in contact with the soil gas sample stream were decontaminated prior to sampling each probe. Each soil gas sampling syringe was decontaminated prior to use. Syringes and adapters were baked in the gas chromatograph oven at a minimum temperature of 180°C.

Prior to sampling each day, a syringe used for soil gas sample collection was filled with ambient air or ultra-high-purity carrier-grade gas from a compressed gas cylinder. The ambient air or high-purity gas was then injected directly into the gas chromatograph. This sample injection served as a blank to detect contamination of the syringe used for sampling and to verify the effectiveness of decontamination procedures.

Equipment Decontamination

Probes and equipment in contact with the soil gas sample stream were decontaminated prior to initiation of sampling. Decontamination of soil gas sampling equipment was conducted by repeated washing and/or by baking in the gas chromatograph oven. Washing included the use of a phosphate-free detergent wash, tap water rinse, organic-free water rinse, and was followed by air drying.

Quality Control (QC) Check Samples

Two QC check samples (obtained from a source different from the Daily Calibration Standard) were analyzed each working day (one at the beginning and one at the end of the day), bracketing the analysis of environmental samples. A minimum of nine compounds, as described above, were checked. The acceptance criteria was to obtain a response factor for each compound that was within 20% of the corresponding true value. If the beginning QC check sample did not meet the acceptance criteria, the problem was resolved before proceeding with sample analysis. If the end of the day QC check sample did not meet the acceptance criteria, additional QC check samples were run to meet the QC criteria, if possible.

Vertical Soil Gas Profiling

Volatile compounds partition from liquid to gas phase and move, by molecular diffusion and advective transport, away from source areas toward regions of lower concentrations. Provided sufficient time, a gas-phase concentration gradient may be established from the source to adjacent areas of lower concentration.

The vertical distribution of gas-phase VOCs was evaluated using multi-depth nested soil gas probe installations. A typical nested soil gas probe installation is shown in Figure 30.3. Nested probes were installed in conjunction with a ground-water monitoring well, as shown in Figure 30.4. Ground-water monitoring wells with associated nested soil gas probes allow the evaluation of soil gas concentrations with depth and assessment of underlying ground-water conditions. Data of this kind were used to assess the nature of the VOC source, whether vadose zone or underlying contaminated ground water.

Theoretically, if a VOC source exists near the ground surface, the vertical distribution of VOCs would likely indicate a general decreasing trend with increasing depth. A theoretical vertical soil gas concentration profile for a shallow vadose zone source is shown in Figure 30.5. Research has shown that VOCs may partition from a contaminated ground-water source and result in gas phase concentrations of VOCs in the vadose zone (Thomson, 1985). A theoretical vertical soil gas concentration profile for a contaminated ground-water source is shown in Figure 30.6. Variations

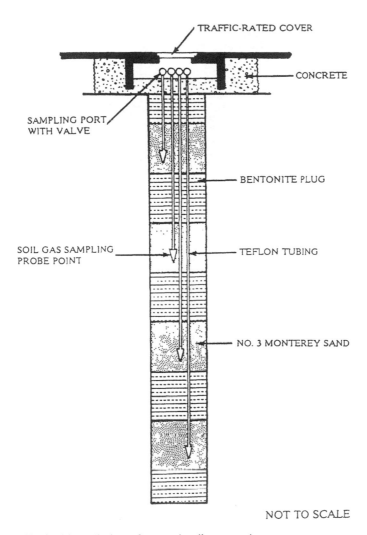

TRAFFIC-RATED COVER

CONCRETE

SAMPLING PORT
WITH VALVE

BENTONITE PLUG

SOIL GAS SAMPLING
PROBE POINT

TEFLON TUBING

NO. 3 MONTEREY SAND

NOT TO SCALE

Figure 30.3. Typical installation of nested soil gas probes.

in these ideal vertical profiles exist due to site-specific conditions, and may cause difficulties when attempting to identify VOC sources.

CASE STUDIES

This section describes two case studies performed in Southern California. Each study involved shallow soil gas surveys and subsequent drilling, soil sampling, and installation, sampling, and analysis of ground-water monitoring wells with nested soil gas probes.

Site 1: Southern California (Possible Vadose Zone Source)

Subsurface geologic materials encountered at Site 1 consisted primarily of fine-to medium-grain sands to a depth of about 50 feet below grade. The sands were light

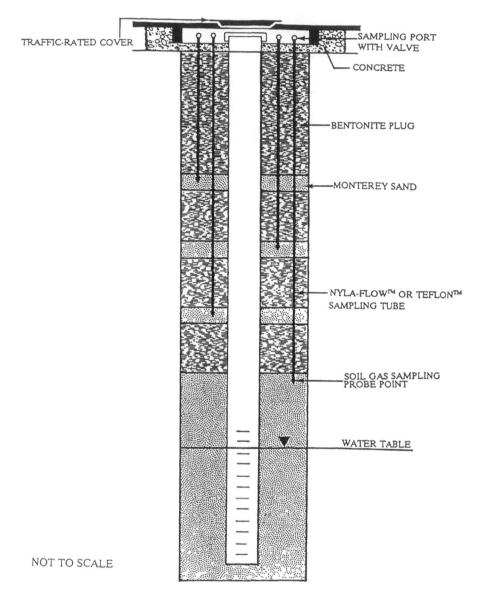

Figure 30.4. Typical construction of a ground-water monitoring well with nested soil gas probes.

gray to dark gray, well-sorted, medium-dense, and moist. Very coarse sands with minor gravels were encountered from 50 feet and persisted to 70 feet. Depth to ground water was about 55 feet below grade.

Shallow Soil Gas Survey

A total of 12 shallow (5-foot) soil gas probes were installed at Site 1 during the initial soil gas survey. The appropriate locations of the probes are shown in Figure 30.7. Results of field analyses for shallow soil gas samples collected at Site 1 are summarized in Table 30.1. Soil gas samples collected during the soil gas survey

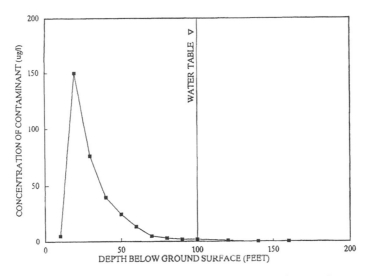

Figure 30.5. Theoretical vertical soil gas concentration profile for a vadose zone source.

contained elevated concentrations of 1,1,1-trichloroethane (TCA) and perchloroethene (PCE). Lower levels of 1,1-dichloroethene (DCE), 1,1-dichloroethane (DCA), and trichloroethene (TCE) were also detected at the site. Soil gas isoconcentration contour maps for TCA and PCE are provided in Figures 30.8 and 30.9.

Concentrations of TCA were detected in all 14 of the soil gas probes with concentrations ranging from 126 micrograms per liter (μg/L) to 2,765 μg/L. The maximum concentration of 2,765 μg/L TCA was detected in Probe SG-2. PCE was detected in 13 of the 14 soil gas probes. Concentrations ranged from 16 μg/L to 4,362 μg/L. The maximum concentration of 4,362 μg/L PCE was also detected in Probe SG-2.

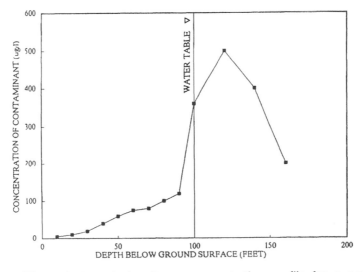

Figure 30.6. Theoretical vertical soil gas concentration profile for a contaminated ground-water source.

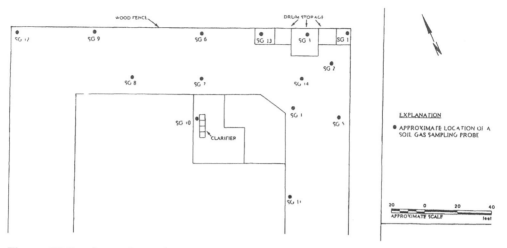

Figure 30.7. Approximate locations of shallow soil gas probes and the ground-water monitoring well with nested probes (Site 1).

Lower concentrations of PCE were detected in the remaining probes, except for Probe SG-12, in which PCE was not detected.

Soil gas concentrations of VOCs indicate the presence of a vadose zone source area in the northeast corner of the site. Deeper subsurface investigation was necessary to evaluate the vertical distribution of VOCs and to assess the impact on underlying ground water.

Table 30.1. Field Analyses Results for Shallow Soil Gas Samples (Site 1) [Concentrations in micrograms per liter (mg/L) of gas]

Sample I.D.	Depth (Feet)	1,1-DCE	1,1-DCA	1,1,1-TCA	TCE	PCE
SG-1	5	ND	ND	321	ND	2,066
SG-2	5	ND	23	2,765	9	4,362
SG-3	5	ND	10	1.966	6	3,379
SG-4	5	15	167	1,402	11	1,774
SG-5	5	ND	66	1,459	ND	1,528
SG-6	5	ND	3	265	2	358
SG-7	5	ND	72	1,012	5	1,600
SG-8	5	ND	42	1,275	4	740
SG-9	5	ND	ND	211	ND	59
SG-10	5	ND	54	706	ND	554
SG-11	5	ND	7	440	ND	16
SG-12	5	ND	ND	126	ND	ND
SG-13	5	ND	5	544	ND	1,088
SG-14	5	ND	77	2,009	14	3,525

1,1-DCE = 1,1-Dichloroethene
1,1-DCA = 1,1-Dichloroethane
1,1,1-TCA = 1,1,1-Trichloroethane
TCE = Trichloroethene
PCE = Perchloroethene
ND = Not detected above reportable limits of quantitation

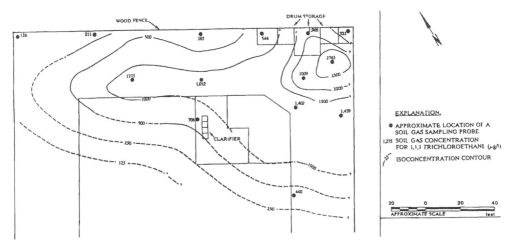

Figure 30.8.   Soil gas concentration contours for 1,1,1-trichloroethane (Site 1).

Vertical Soil and Soil Gas Profiling

A soil boring was drilled to a total depth of 65 feet for installation of a ground-water monitoring well with nested soil gas sampling probes (refer to Figure 30.4). During drilling, soil samples were collected for laboratory analyses of VOCs. The ground-water monitoring well was developed and sampled, and the ground-water sample was analyzed for VOCs. About one month following installation (to allow installation to reach equilibrium with native subsurface material), soil gas samples were collected from the nested soil gas probes and analyzed onsite for EPA Method 8010/8020 compounds. The vertical distribution of VOCs detected in soil, soil gas, and ground water is summarized in Table 30.2.

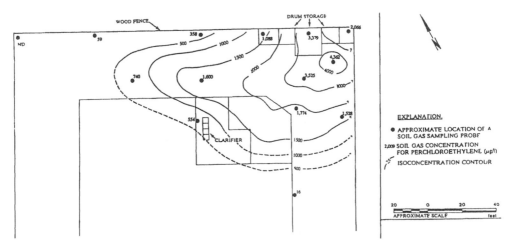

Figure 30.9.   Soil gas concentration contours for perchloroethene (Site 1).

Table 30.2. Laboratory Analyses Results for Soil Samples, Soil Gas Samples from Nested Probes, and a Ground-Water Sample (Site 1)

Sample I.D.	Depth (Feet)	1,1-DCE	1,1-DCA	C 1,2-DCE	TCE	1,1,1-TCA	PCE
NP1–20′	20	23	64	ND	4	981	740
SB1–20′	20	ND	ND	ND	ND	340	1,400
NP1–30′	30	46	112	ND	1	930	130
SB1–30′	30	NA	NA	NA	NA	NA	NA
NP1–40′	40	45	54	ND	2	821	106
SB1–40′	40	ND	ND	ND	ND	ND	ND
NP1–45′	45	37	30	ND	ND	680	53
SB1–45′	45	ND	ND	ND	ND	ND	ND
Ground water	NA	ND	ND	ND	ND	120	16

1,1-DCE = 1,1-Dichloroethene
1,1-DCA = 1,1-Dichloroethane
C 1,2-DCE = Cis-1,2-Dichloroethene
TCE = Trichloroethene
1,1,1-TCA = 1,1,1-Trichloroethane
PCE = Perchloroethene
NP = Nested Probe (soil gas sample), concentrations in micrograms per liter
SB = Soil boring (soil sample), concentrations in micrograms per kilogram
Ground-water concentrations in micrograms per liter
ND = Not detected above reportable limit of quantitation
NA = Not analyzed

Analytical Results for Soil Samples. Soil samples were analyzed from depths of 20, 40, and 45 feet below grade. Samples contained concentrations of TCA and PCE. In many instances, VOCs were detected in soil gas samples and not in soil samples. This is probably due to volatilization of VOCs from soil samples during collection, handling, transport, and laboratory analysis. Tabular data comparing soil gas and soil matrix concentrations of VOCs in the same boring, at the same depths, indicate the utility of in-situ soil gas sampling and analyses to detect VOCs in the vadose zone.

Field Analytical Results for Soil Gas Samples from Nested Probes. Soil gas samples were collected from depths of 20, 30, 40, and 45 feet below grade. Samples contained concentrations of 1,1-DCE, 1,1-DCA, TCE, 1,1,1-TCA, and PCE. Concentrations of 1,1-DCE, 1,1-DCA, and TCE may be present due to the biodegradation of TCA and PCE. The vertical distribution of TCA and PCE in the vadose zone at Site 1 is shown in Figures 30.10 and 30.11, respectively. Soil gas data indicate a decreasing trend with increasing depth to ground water, suggestive of a likely vadose zone source.

Laboratory Analyses Results for Ground-Water Samples. The ground-water sample contained concentrations of TCA and PCE. Ground-water contamination may be derived from overlying soil gas contamination as a result of VOCs partitioning from the vadose zone gas into ground water.

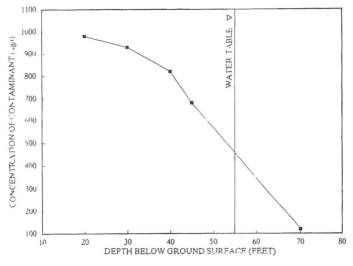

Figure 30.10. Vertical distribution of 1,1,1-trichloroethane (Site 1).

Site 2: Southern California (Possible Contaminated Ground-Water Source)

The Site 2 investigation included a shallow depth soil gas survey, drilling, soil sampling and analyses, and installation and sampling of a ground-water well with nested soil gas probes.

Shallow Soil Gas Survey

The shallow soil gas survey included the installation and sampling of 48 5-foot-deep soil gas probes and two 15-foot-deep soil gas probes. A general grid-spacing of 10 feet was used for the soil gas survey. Approximate locations of soil gas sampling

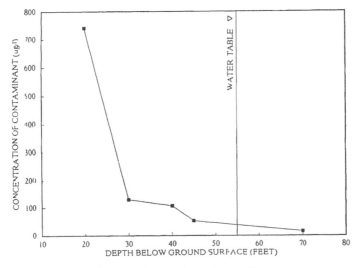

Figure 30.11. Vertical distribution of perchloroethene (Site 1).

sites are shown in Figure 30.12. Suspected contaminant source areas investigated included areas near former underground storage tanks, and adjacent to an existing industrial waste clarifier.

Results of field analyses for soil gas samples collected at Site 2 are summarized in Table 30.3. Soil gas samples collected during the soil gas survey at Site 2 contained low concentrations of TCA, PCE, and TCE. Soil gas isoconcentration contours for TCA and PCE are shown in Figures 30.13 and 30.14, respectively. Contours suggest the presence of a vadose zone source of VOCs in the vicinity of a hazardous materials storage room and a dip-dye tank.

Concentrations of PCE detected above reportable limits ranged from 1 μg/L to 45 μg/L. The maximum concentration of PCE (45 μg/L) was detected in Probe SG-47. Lower concentrations above 1 μg/L of PCE were detected in approximately 25% of all probes sampled. Concentrations of this magnitude may be derived from an

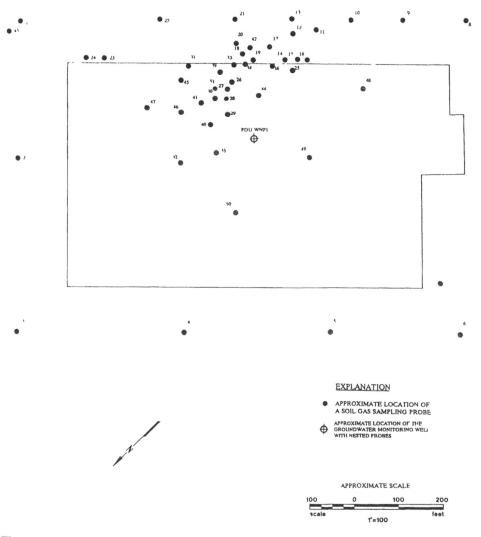

Figure 30.12.   Approximate locations of shallow soil gas probes and the ground-water monitoring well with nested probes (Site 2).

Table 30.3. **Field Analyses Results for Shallow Soil Gas Samples (Site 2) (Concentrations in micrograms per liter (mg/L) of gas)**

Sample I.D.	Depth (Feet)	1,1-DCE	1,1-DCA	1,1,1-TCA	TCE	PCE
SG-1	5	1	ND	62	ND	ND
SG-2	5	ND	ND	ND	ND	ND
SG-3	5	ND	ND	ND	ND	ND
SG-4	5	ND	ND	ND	ND	ND
SG-5	5	ND	ND	ND	ND	ND
SG-6	5	ND	ND	ND	ND	ND
SG-7	5	ND	ND	ND	ND	ND
SG-8	5	1	ND	2	ND	ND
SG-9	5	29	ND	19	ND	3
SG-10	5	42	2	159	ND	8
SG-11	5	42	8	126	7	8
SG-12	5	18	8	137	4	4
SG-13	5	29	11	137	6	11
SG-14	5	2	1	74	ND	ND
SG-15	5	2	ND	79	ND	ND
SG-16	5	4	4	95	ND	ND
SG-17	5	3	3	89	ND	2
SG-18	5	ND	1	ND	ND	ND
SG-19	5	ND	ND	2	ND	ND
SG-20	5	ND	ND	15	ND	ND
SG-21	5	1	ND	42	ND	2
SG-22	5	ND	ND	13	ND	2
SG-23	5	ND	ND	ND	ND	4
SG-24	5	ND	ND	3	ND	3
SG-25	5	2	6	74	ND	ND
SG-26	5	ND	26	52	ND	2
SG-27	5	ND	39	95	ND	ND
SG-28	5	ND	1	86	ND	1
SG-29	5	ND	ND	26	ND	ND
SG-30	5	9.87	96	153	ND	1
SG-31	5	ND	ND	14	ND	ND
SG-32	5	ND	ND	2	ND	2
SG-33	5	ND	ND	1	2	ND
SG-34	5	ND	ND	6	ND	ND
SG-35	5	ND	ND	15	ND	ND
SG-36	5	1	2	74	ND	1
SG-37	5	5	11	84	ND	1
SG-37D	15	ND	9	123	ND	ND
SG-38	5	ND	4	1	ND	ND
SG-39	5	ND	ND	31	ND	ND
SG-40	5	ND	ND	14	ND	ND
SG-41	5	ND	1	123	ND	ND
SG-42	5	ND	3	5	ND	ND
SG-43	5	10	ND	93	ND	ND
SG-44	5	ND	16	ND	1	ND
SG-45	5	ND	ND	40	ND	ND
SG-46	5	ND	ND	14	ND	3
SG-47	5	ND	ND	3	ND	45

1,1-DCE = 1,1-Dichloroethene
1,1-DCA = 1,1-Dichloroethane
1,1,1-TCA = 1,1,1-Trichloroethane
TCE = Trichloroethene
PCE = Perchloroethene
ND = Not detected above reportable limits of quantitation.

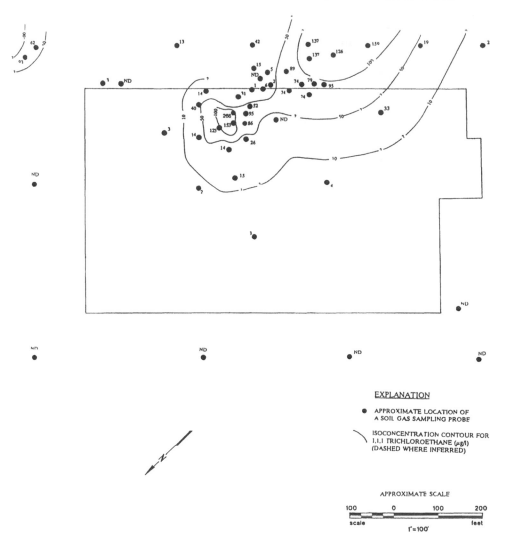

Figure 30.13.   Soil gas concentration contours for 1,1,1-trichloroethane (Site 2).

underlying ground-water source, or may be residual levels remaining from a small quantity release.

Vertical Soil and Soil Gas Profiling

The location of a soil boring and the installation of a ground-water monitoring well with nested soil gas probes was selected based on the results of the shallow soil gas survey. One 65-foot ground-water monitoring well was installed at Site 2. Nested soil gas probes were installed in the annulus of the 65-foot boring at depths of 10-, 20-, 30-, and 40-feet below grade. The location of the ground-water monitoring well with multi-depth nested soil gas probes is shown in Figure 30.12. Previous soil gas survey data indicated the presence of VOCs in the vicinity of the dip-dye unit. These data, in conjunction with the apparent ground-water flow direction, were used to select the location of the ground-water monitoring well with nested soil gas probes.

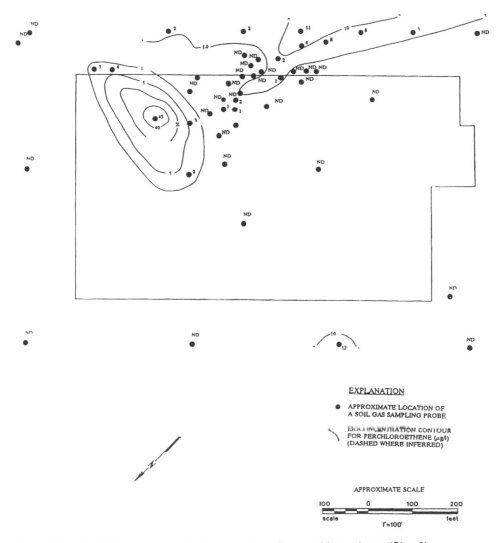

Figure 30.14.   Soil gas concentration contours for perchloroethene (Site 2).

Laboratory analytical results for soil samples, soil gas samples from nested probes, and a ground-water sample are shown in Table 30.4.

Soil Sample Analytical Results

Soil samples for laboratory analyses were collected from depths of 15, 30, and 40 feet below grade. EPA Method 8260 compounds were not detected in the three soil samples collected from the soil boring.

Field Analytical Results for Soil Gas Samples from Nested Probes

Field analytical results for soil gas samples collected from nested soil gas probes are summarized in Table 30.4. Soil gas samples collected from the nested soil gas probes at Site 2 contained PCE, 1,1,1-TCA, TCE, cis-1,2-DCE, and 1,1-DCA. The

Table 30.4. Laboratory Analyses Results for Soil Samples, Soil Gas Samples from Nested Probes, and a Ground-Water Sample (Site 2)

Sample I.D.	Depth (Feet)	1,1-DCE	1,1-DCA	C 1,2-DCE	TCE	1,1,1-TCA	PCE
NP1–10′	10	ND	6	5	3	ND	17
SB1–15′	15	ND	ND	ND	ND	ND	ND
NP1–20′	20	ND	5	ND	ND	2	14
SB1–20′	20	ND	NA	NA	NA	NA	NA
NP1–30′	30	ND	ND	ND	ND	ND	176
SB1–30′	30	ND	ND	ND	ND	ND	ND
NP1–40′	40	ND	ND	ND	7	10	3,928
SB1–40′	40	ND	ND	ND	ND	ND	ND
Groundwater	NA	ND	ND	ND	ND	6	10

1,1-DCE = 1,1-Dichloroethene
1,1-DCA = 1,1-Dichloroethane
C 1,2-DCE = Cis-1,2-Dichloroethene
TCE = Trichloroethene
1,1,1-TCA = 1,1,1-Trichloroethane
PCE = Perchloroethene
NP = Nested Probe (soil gas sample), concentrations in micrograms per liter
SB = Soil boring (soil sample), concentrations in micrograms per kilogram
Ground-water concentrations in micrograms per liter
ND = Not detected above reportable limit of quantitation
NA = Not analyzed

vertical distributions of 1,1,1-TCA and PCE are illustrated in Figures 30.15 and 30.16, respectively.

Concentrations of PCE were detected in each nested probe with concentrations generally increasing with increasing depth below grade. Concentrations of PCE in

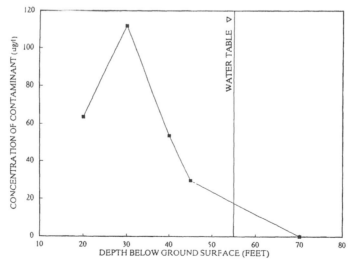

Figure 30.15. Vertical distribution of 1,1,1-trichloroethane (Site 2).

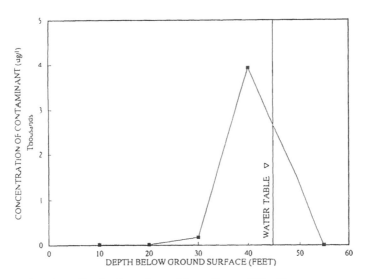

Figure 30.16. Vertical distribution of perchloroethene (Site 2).

10 and 20 foot probes were 16.8 $\mu g/L$ and 13.9 $\mu g/L$, respectively. At 30 feet below grade, concentrations of PCE increased to 175.9 $\mu g/L$, and a maximum concentration of 3,928.2 $\mu g/L$ PCE was detected at 40 feet below grade.

Concentrations of TCA were detected in nested probes installed at 20 and 40 feet below grade. A concentration of 1.8 $\mu g/L$ TCA was present at 20 feet. Concentrations of TCA were also observed to increase with depth where a maximum of 10.1 $\mu g/L$ was detected in the 40-foot probe.

Analytical Results for Ground-Water Samples. The ground-water sample collected from the monitoring well on Site 2 was analyzed for VOCs using EPA Method 502.2. Concentrations of TCA and PCE were detected in the ground-water sample at concentrations of 5.7 $\mu g/L$ and 10 $\mu g/L$, respectively. Maximum contaminant levels (as established by California Department of Health Services) for these compounds in drinking water are 200 $\mu g/L$ for TCA and 5 $\mu g/L$ for PCE. Soil gas data for these compounds indicate an increasing concentration trend with increasing depth to ground water, suggestive of a possible up-gradient ground-water source.

CONCLUSIONS

Soil gas surveying techniques have proved to be very effective for assessing the presence, concentration, and extent of soil contamination by VOCs. Due to the potential for site-specific interferences that can cause data anomalies, soil gas sampling and analysis data should be used in conjunction with soil matrix and ground-water data when possible. Vertical soil and soil gas profiling aids in evaluating the persistence of VOCs with depth, and may be used to assess whether soil or soil gas contamination in the vadose zone has impacted underlying ground water.

In some instances, soil gas surveying may be used to map the general extent of ground water contaminated by VOCs. The effectiveness of plume mapping is site-specific and should be used only following the collection and evaluation of vertical soil and soil gas profiling data. Soil gas surveying has become a fast, cost-effective means of site assessment and can be used to reduce the amount of subsequent subsurface investigation necessary to evaluate the nature and extent of soil and ground-water contamination by VOCs.

REFERENCES

Thomson, K.A., 1985. "Vertical Diffusion of Volatile Organic Contaminants from a Water Table Aquifer; Field and Laboratory Studies," unpublished MA thesis, Department of Hydrology and Water Resources, University of Arizona, Tucson, AZ.

31

In Situ Pore-Liquid Sampling in Saturated Regions of the Vadose Zone[*],[**]

L.G. Wilson and D.W. Dorrance

INTRODUCTION

Saturated regions may occur in the vadose zone as perched ground water and possibly as water-table mounds. Perched ground water develops at the interface between regions of varying hydraulic conductivity, e.g., a coarse zone overlying a finer-textured zone (Everett et al., 1984, Wilson and Schmidt, 1979), while water-table mounds develop above a regional water table. Perched systems are underlain by unsaturated sediments. Figure 31.1 shows the growth of perched ground water and a water-table mound in stratified alluvium during a recharge event in the Santa Cruz River, an ephemeral stream near Tucson, Arizona (Wilson and Schmidt, 1979). The two regions were separated by an unsaturated transmission zone which remained at a water content sufficient to transmit vertical leakage from the perched system into the water table mound. Piezometers in both regions manifested positive pressures, yielding water samples. Elsewhere in this book, Schmidt discusses perched ground-water systems in more detail.

Sampling liquid from perched systems or mounds is attractive because liquid is collected over a large area. Such integrated samples are more representative of areal

---

[*]This chapter is based in part on ASTM 4696–92, Standard Guide for Pore Liquid Sampling from the Vadose Zone. Copies may be purchased from ASTM, 1916 Race Street, Philadelphia, PA 19103.

[**]Abstracted with permission from Dorrance, D.W., L.G. Wilson, L.G. Everett, and S.J. Cullen, in *Groundwater Residue Sampling Design*, R.G. Nash and A.R. Leslie, Eds., ACS Symposium Series 465, American Chemical Society, Washington, DC, 1991, pp. 300–331.

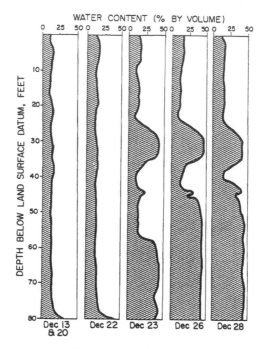

Figure 31.1.   Water content profiles obtained in a neutron access tube during flow in the Santa Cruz River, December, 1967, showing perched water and a water table mound (after Wilson and Schmidt, 1979; Courtesy of the American Water Resources Association).

conditions than suction samples (Wilson and Schmidt, 1979). In addition, larger sample volumes can be collected than with suction samplers.

Perched water systems and mounds can be difficult to find and delineate. Possible methods for locating saturated regions in the vadose zone include nuclear logging (Poeter, 1988) and video logging of existing wells. Perched systems tend to be ephemeral; therefore, suction samplers are sometimes required as a backup. As with all samplers, potential chemical interactions between sampler materials and the constituents of interest should be considered (Dunlap, 1977).

Following are the methods for sampling perched ground water discussed in this section:

- point samplers
- multi-level samplers
- wells
- cascading water samplers
- drainage systems.

Except for cascading water samplers and possibly drainage samplers, these methods are commonly used for monitoring ground water in aquifers.

Point Samplers

Point samplers are open-ended pipes or tubes, such as piezometers or wells with screened points, installed to collect samples from a discrete location in saturated material. Water samples from a point sampler array with individual well points terminating at different depths will indicate the vertical displacement of indigenous ground water with the recharge source. Similarly, if a number of arrays are distributed in a horizontal transect, samples from individual units terminating at the same elevation will manifest quality changes during the lateral spread of a pollution plume.

Maintaining a tight fit between the casing and surrounding soil is critical. Accordingly, these samplers are commonly driven into place by sledge hammers for shallow units, or by mechanical devices for deeper units; for example, using cone penetrometer rigs or the hydraulic system of a conventional drill rig. Samplers are also driven in place using conventional slide hammers either from land surface or in advance of a pre-drilled borehole (Eckard et al., 1989). Alternatively, they may be installed in constructed boreholes with bentonite or grout seals between the borehole wall and casing to prevent side leakage. Clustering of units within a common borehole is possible, provided the borehole is sealed between units to prevent cross-contamination. Point samplers are made from a variety of materials including stainless steel, PVC, PTFE, ABS, and fiberglass. Samples are collected by bringing liquid which has seeped by hydrostatic pressure into the well point to the surface by a bailer or pump. Figure 31.2 shows an alternative sampling technique consisting of a well point containing a porous segment, a pressure line with check valve, and an internal sample-retrieval line. Sample collected through the porous segment is forced to the surface via the sample line by pressurizing the pressure line.

Two other point-sampling devices for collecting samples from saturated regions of the vadose zone are the HydroPunch® and BAT® (filter tip) units. When equipped with shielded screens, both of these devices are capable of discrete depth sampling (Zemo et al., 1992). After driving the points to the sampling depth (e.g., by drive hammer or cone penetrometer), the screen is retracted and a sample is collected. With the HydroPunch method, a sample is collected in a sample chamber by hydrostatic pressure (see Figure 31.3). The sample is retained in the chamber between two check valves. The sample chamber is brought to the surface and the sample is decanted into a sample bottle. Samples are collected in the BAT sampler by retracting the shield and applying a vacuum through a sample vial (see Figure 31.4). The probe is sealed at the top by a septum, which prevents the entry of ground water (Zemo et al., 1992). The evacuated vial, also sealed by a septum, is lowered into the drive casing on a weighted wire line, together with a spring-loaded, double-ended hypodermic needle. The needle pierces both septa, causing water to enter the sampler and sample vial. The vial and needle unit is withdrawn to the surface. The septum on the sampler reseals. In contrast to the other methods, samples collected by the BAT sampler are not exposed to the atmosphere. During a field comparison of the HydroPunch and BAT methods, Zemo et al. (1992) determined that the two devices may be used interchangeably during screening investigations of chlorinated aliphatic compounds. However, the two methods cannot be used interchangeably when sampling for chlorinated benzenes. Mines et al. (1993) found that BAT samplers recovered higher levels of VOCs than bailer samples from adjacent monitoring wells screened over large intervals.

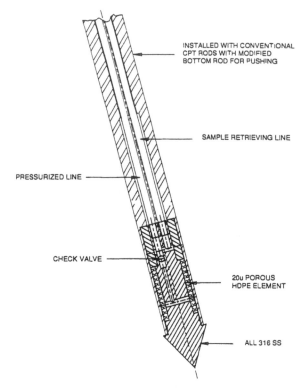

INSTALLED WITH CONVENTIONAL
CPT RODS WITH MODIFIED
BOTTOM ROD FOR PUSHING

SAMPLE RETRIEVING LINE

PRESSURIZED LINE

CHECK VALVE

20u POROUS
HDPE ELEMENT

ALL 316 SS

Figure 31.2.   Well point with internal sample retrieval line (Courtesy of Hogentogler and Co., Inc.; patent pending).

Advantages and Limitations

An advantage of point samplers is their use in obtaining depth-wise samples from perched zones. Additionally, they may be used to measure hydraulic gradients, indicating the vertical direction of water movement in perched regions. Hekma et al. (1990) and Eckard et al. (1989) used BAT probes to measure pore-water pressures and hydraulic conductivity values during their studies in California.

Locating point samplers in fractured zones is difficult. Proper placement of the units requires knowledge of the hydrogeologic framework; e.g., hydraulic gradients, flow paths, and geologic structure. Flow trajectories are not necessarily perpendicular to piezometer contours. In addition, sampling may affect the flow regime around the well point. Perched ground water may be ephemeral, with sampling units failing during the dry periods. Conversely, rising water tables may require additional wells for sampling new perched ground-water lenses. Chemical reactions between the sampler and pollutants may occur, releasing nontypical substances. Oxidative environments may occur near the water table, requiring prolonged pumping to remove nonrepresentative ground water before sampling. Dilution of pollutant plumes with surface infiltration from heavy rains may also produce nonrepresentative samples. Without careful construction, freezing soils may breach the borehole-casing seal, allowing surface fluids to penetrate. These fluids could contaminate borehole samples and then move laterally away from the piezometer. In addition, above-ground parts of piezometers completed with PVC should be protected from sun-

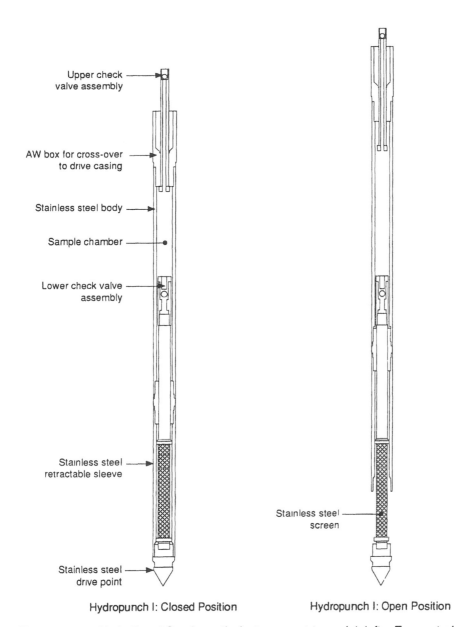

Figure 31.3. HydroPunch® schematic (cutaway not to scale) (after Zemo et al., 1992).

light. These wells should be thoroughly developed following installation. Special drilling techniques are required to install piezometers near existing landfills and waste disposal areas.

Installation of point samplers is difficult in compacted, consolidated media. Thus, forcing the BAT assembly into the ground using a cone-penetrometer rig, conventional drill rig, or by pounding, is limited to sandy soils and fine-grained materials, free of cobbles and coarse gravel, strongly-cemented layers, and otherwise consolidated soils (Hekma et al., 1990).

Eckard et al. (1989) reviewed the advantages and limitations of BAT samplers.

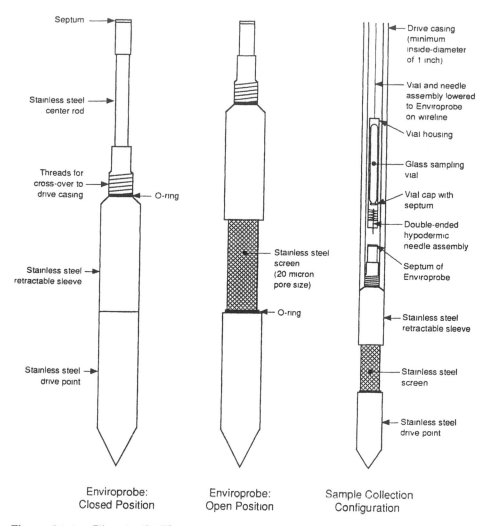

Figure 31.4. Filter-tip (BAT®) sampler (after Zemo et al., 1992).

Obtaining samples in a hermetically-sealed filter tip without generating large volumes of purge water is a major feature of these units. Inasmuch as the sampling probe is sealed with a septum, only the fluid contained in the filter tip itself (i.e., 8 to 30 mL) needs to be purged prior to sampling. Thus, costs associated with removal of purged, contaminated ground water are eliminated. Additionally, since a smaller borehole is required to install a filter-tip sampler than when constructing a conventional monitor well, the mass of possibly contaminated vadose zone sediments is reduced. Obviously, when installing units with a cone penetrometer rig, no deposits are generated.

Eckard et al. (1989) cite the following limitations of filter-tip samplers: (1) the fill rate of the sample container is limited by the diameter of the double-ended hypodermic needle and the permeability of the filter tip and the formation material, i.e., impractically-long sampling times are required to extract large samples from low permeability formations; (2) the length of the filter area limits the sampling interval of the probe; and (3) data collected are representative of only a small area.

Zemo et al. (1992) indicate that the HydroPunch sampler provides water samples rather quickly in coarse-grained units containing fines compared to the BAT sampler because of a coarser screen and larger intake ports. However, the HydroPunch may produce turbid samples. The HydroPunch is also more prone to damage when forced through tight sediments because of its greater length. Additionally, it must be located at least five feet beneath the water table to provide sufficient hydrostatic pressure to fill the sampler (Zemo et al., 1992).

Multilevel Samplers

Pickens et al. (1978) described a multilevel sampler, a type of point sampling device, particularly suited for sampling in cohesionless deposits with shallow perched ground water in which flow occurs predominantly in the horizontal direction. The sampler consists of a PVC pipe, ports or openings at desired incremental depths, screened coverings on the openings, and polypropylene tubing sealed into the openings within the pipe. The polypropylene tubing extends to the surface. Units may be designed in the field by locating the position of openings using stratigraphic information from drilling.

The tubing is installed by common drilling techniques such as the hollow-stem auger. To collect samples, a vacuum is applied to the polypropylene tubing. The water is pumped into an air-free cell for measurement of pH and redox potential (Eh). Quick delivery from the perched zone to the cell probably minimizes degassing of the water (Pickens et al., 1978).

Advantages and Limitations

An advantage of the multilevel sampler is that concentration profiles can be obtained with a single unit at a lower cost than by installing piezometer nests (Pickens and Grisak, 1979). Sampling ports located in regions with fluctuating water tables can be used to sample recharge waves, providing valuable information on layering characteristics of pollutant plumes.

A disadvantage of these samplers is they are limited to shallow depths, less than the suction lift of water (about 25 feet). Additionally, unlike wells or piezometers, there is no effective method for developing multilevel samplers following installation. Thus, because the water yielding formation is disturbed during drilling, a prolonged time period is required to ensure that samples are representative of the undisturbed conditions (Pickens et al., 1981). Sample degassing may occur as a result of the sampling technique. Residual fluids in delivery tubes of multilevel samplers may freeze during winter months. Soil heaving from freezing and thawing may loosen soil around the casing, introducing the possibility of cross-contamination of samples and short circuiting of surface fluids to depth. Since sunlight will render PVC brittle, aboveground parts of the casing must be protected. Multilevel samplers may be unsuitable in fractured material because of side leakage and difficulty in locating sampling ports opposite fractures. They are difficult to install in well-indurated formations. Installation processes may affect permeability near the sampling ports, resulting in long periods of time for return of "natural" conditions.

Wells

A monitoring well is similar to a point sampler except the screened interval is longer. Therefore, samples are averaged over the screened length (Pickens and Grisak, 1979). Components of a well generally include the well casing, well screens, filter packs, annular seals, and surface protection. If casing size permits, dedicated submersible pumps may also be installed, which minimizes labor costs. Methods for constructing, developing, and equipping monitor wells in perched zones and regional ground water are identical. The most comprehensive text available on the subject was prepared by Nielsen and Schalla (1991). This reference presents guidelines for designing and constructing monitor wells, including selecting casing and screen material, determining casing diameter, designing well screens, and selecting drilling methods, filter packs, and annular seals. Details are also presented on post-installation development and completion techniques (Kraemer et al., 1991). Herzog et al. (1991) review common drilling methods for monitor wells (fluid rotary drilling, auger drilling, and percussion drilling).

According to Nielsen and Schalla (1991), casing materials for monitor wells can be categorized into four types:

- thermoplastic material (e.g., PVC)
- fluoropolymer materials [e.g., polytetrafluoroethylene (PTFE)]
- metallic material (e.g, galvanized steel, stainless steel)
- fiberglass-reinforced material

Selection of an appropriate material is critical to ensure strength and to minimize interactions between the well casing and water being sampled.

Samples are brought to the surface by a variety of techniques. Herzog et al. (1991) group samplers into three broad categories: grab techniques (bailers and syringe samplers), suction-lift samplers (e.g., centrifugal and peristaltic pumps), and positive displacement samplers (e.g., gas-drive devices, gas-operated bladder pumps, electric submersible pumps, and gas-driven piston pumps). Herzog et al. (1991) summarize the advantages and disadvantages of each of these techniques. These authors also review procedures for purging wells, filtering, selecting sample containers, and preserving and storing samples. Samples from discrete depths along the screened interval can also be obtained using packer-pump setups such as those described by Fenn et al. (1977).

Advantages and Limitations

The major advantage of monitor wells is yielding large volumes of perched ground water for analysis. As with the other methods described in this section, a major disadvantage is that perched lenses may be ephemeral. Thus, a well may become dry and unavailable for sampling part of the year. Sampling schedules should be adjusted to meet these conditions. Proper placement of monitor wells requires information on the hydraulic gradient and flow direction of the perched ground-water body. In areas where flow is vertically downward in wells, surface pollutants may be short circuited into deeper zones. In these areas, caution must be used in developing monitoring wells to minimize side leakage through the well annulus. Special well development techniques discussed by Kaufmann et al. (1981) are required to develop wells in existing waste disposal sites or at landfills.

Cascading Water Samplers

Cascading water occurs when a well is screened throughout a perched layer or water table mound, or when water leaks through casing joints at the perched layer. Figure 31.5 illustrates cascading water from a perched system. Water flowing into the well in the portion open to the perched layer cascades downward to the water table. This situation is common in some areas where the practice has been to install water wells with large screened intervals. As discussed by Everett et al. (1984), cascading water samples may reflect the integrated quality of water draining from an overlying perched zone. This will be particularly true when the overlying source is diffuse, such as irrigation return flow. Because of the integrated nature of the water in these perched regions, samples may be obtained using one or more wells at a fraction of the cost of installing batteries of suction cups.

The data in Table 31.1 illustrate the contrast in water chemistry between cascading water samples and pumped ground water in two wells sampled by the senior author in 1973. Cascading water was obtained from an abandoned irrigation well and pumped ground water was obtained from an irrigation well within a mile of the abandoned well. The abandoned well was located on the boundary of a field used at the time for disposal of treated sewage effluent from a nearby community. Note the marked contrast in quality between the two samples; for example, the total dissolved salts in cascading water was about seven orders of magnitude greater than the ground-water sample. This reflects the large increase in sodium, chloride, and sulfate concentrations. Nitrate concentrations increased by almost nine orders of magnitude.

Wilson and Schmidt (1979) described techniques for sampling from cascading wells. The simplest method uses a bailer or bucket, decontaminated and lowered to a position in the well below the cascading water but above the water table. When the sampler is full, it is pulled back to the surface. In sampling for chemical constitu-

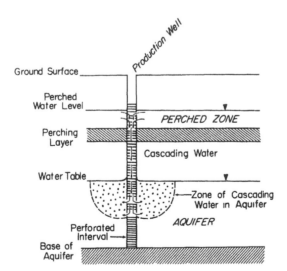

Figure 31.5. Conceptualized cross section of a well showing cascading water from a perched zone (after Wilson and Schmidt, 1979; Courtesy of the American Water Resources Association).

Table 31.1.   Chemical Analyses of Water from Wells at Coolidge, Arizona (values are in mg/L)

Parameter	Well Discharge	Cascading Water
Date	10/22/73	10/20/73
TDS	1,059	7,020
pH	7.7	7.5
Calcium	94	652
Magnesium	27	131
Sodium	230	1,550
Potassium	NA	NA
Chloride	396	2,224
Sulfate	189	2,109
Carbonate	0	0
Bicarbonate	112	263
Fluoride	1	5
Nitrate	10	88

ents, the bailer and sample bottles should be rinsed several times with cascading water before collecting a sample. To sample for microorganisms, a sterilized container should be used. Because cascading water is exposed to the atmosphere, some problems may be experienced in determining pH, dissolved oxygen, and oxidation-reduction potential.

An alternative sampling method is shown in Figure 31.5. For the situation shown, the chemistry of the water table immediately surrounding a well which has been shut down for an extended period will be dominated by cascading water. Therefore, a sample of cascading water is collected from the water table during the initial stages of pumping.

Access for sampling cascading water in production wells may be restricted by pump assemblies; however, when the pumps are removed for maintenance, sampling can be coordinated with pump maintenance personnel.

Elsewhere in this book, Schmidt discusses criteria for determining perched ground water at greater length, and presents case studies involving field testing to determine whether or not perched water is actually present.

Advantages and Limitations

The main advantage of sampling cascading wells is the availability of large sample volumes for chemical or microbial analyses. Sampling is relatively easy if access to the casing is readily available.

Cascading water may enter a well from several distinct perched systems (Wilson and Schmidt, 1979). Thus, it may not be possible to determine the horizon or horizons affected by detected pollutants. Cascading water is most often sampled from pre-existing wells used for other purposes. As a result, materials used in the well construction may alter those chemical constituents of interest. Wells used for irrigation and water supply often have lubricant oils from the pump floating in them. With fluctuating water levels, these oils become smeared along the casing, and may even move out into the surrounding soils. Therefore, traces of these oils may appear in samples.

Sampling Drainage Flows

Shallow perched systems may spread contamination, cause problems with structures, or interfere with agriculture. Drainage systems are installed to alleviate these problems. These systems cause gravity flow of perched ground water to a ditch or sump from which it is pumped out. This outflow can be sampled. Typical drainage systems include open ditches, tile lines, perforated pipes, synthetic sheeting, or even layers of gravel and sand. Depending on the design of the system, it may be possible to sample outflows which drain different land-use sites such as agricultural areas (Eccles and Gruenberg, 1978) and sanitary landfills (Wilson and Small, 1973).

The most widespread use of drainage systems is in the agricultural sector. Jury (1975) pointed out that approximately 53 million hectares of cropland in the United States are artificially drained. Thus, the composition of tile-drainage water is an index of contaminant movement from agricultural areas. The contaminants of concern include pesticides, boron, tds, nutrients (N,K,P), and trace elements such as Se. Hallberg et al. (1986) reviewed the advantages of using tile lines to measure water quality. However, Jury (1975) concluded from a literature review of case studies of tile drainage monitoring that no simple correspondence between surface practices and drainage water quality exists unless other factors are considered. Among these factors is preferential flow through macropores. For example, based on field studies, Czapar and Kanwar (1991) concluded that the rapid disappearance of herbicide soon after rainfall suggested that preferential flow is an important transport mechanism in structured soils. They recommended continuous monitoring of drain tiles following heavy rains to identify chemical peaks from preferential transport.

Richard and Steenhuis (1988) also recommend tile drain sampling to characterize preferential flow on a field scale. They recognized that tile drains measure solute transport on a scale which integrates spatial variability within the sampling volume (e.g., an agricultural field). This approach is an alternative to geostatistical approaches which aggregate many small-scale samples. During field studies, they observed that solute transport measured in tile drain outflow was more rapid than predicted by a homogeneous flow equation developed by Jury (1975). Specifically, the influence of macropores was observed by solute travel times of 3 to 4 hours compared with calculated values of 3 to 4 weeks. They concluded the following:

> The results are consistent with a two-domain concept of water movement in which rapid flux occurs through macropores while solutes equilibrate slowly with the less mobile water in the soil matrix. Mass balance accounting of chloride movement over the course of the experiment demonstrated that tile drains could be used to sample solute transport through the unsaturated zone . . . In heterogeneous soils with a water table close enough to the surface to be tapped by tile drains, this method has excellent potential for large-scale sampling of solute movement through the unsaturated zone.

Describing design and construction techniques for drainage systems is beyond the purview of this book. Interested readers should consult a text on land drainage edited by Schilfgaarde (1974) containing several papers on the design and construction of drainage systems. Willis et al. (1991) used the water management model DRAINMOD (Skaggs, 1980) to aid in designing an experimental draintube system. Gilliam et al. (1974), Gambrell et al. (1975), Eccles and Gruenberg (1978), and Gilliam et al. (1979) described sampling from tile drains. Gilliam et al. (1974) and

Jacobs and Gilliam (1985) described sampling from drainage ditches. Generally, sampling drainage effluent is a simple matter. Grab samples may be collected where tile lines or drainage pipes discharge to ditches or sumps. Time-sequencing of sampling can become tedious with the grab sampling method, and automatic samplers are available. For their field experiments, Richard and Steenhuis (1988) located access chambers on each tile line and sampled tile drain effluent using ISCO 1680 automatic samplers. Willardson et al. (1973) described a flow-path, ground-water sampler for collecting water following different flowpaths along a tile drainage system.

Measuring drainage-system discharge rates during a sampling program facilitates calculating a mass balance of chemicals leaving a field. For open ditches, flumes (e.g., Parshall flumes) and weirs (e.g., V-notch weirs, broad crested weirs, etc.) are suitable for measuring discharge. Automatic water-stage recorders mounted on stilling wells are used for continuous recording of discharge. Tile drain discharge is easily measured by the trajectory or bucket methods. Wilson et al. (1961) describe a slotted tube device mounted on the end of a tile line for measuring discharge. The device is connected to a stilling well and an automatic stage recorder for continuous measurement of discharge. Flow measuring devices may also be installed within the tile line system. For example, during their field studies, Richard and Steenhuis (1988) measured drain tile flow rates every 30 minutes using a 30-degree V-notch weir with stilling well and a float activated level recorder. Palmer-Bowlus critical depth meters may also be installed within tile lines.

Advantages and Limitations

A major advantage of sampling drains is the availability of large sample volumes for chemical analyses. As indicated, they may also account for transport in a dual-flow system. As discussed in the last paragraph, the ability to easily measure discharge rates facilitates conducting chemical mass balances.

Because of the limitations of excavation equipment, drainage samplers are limited to shallow depths. In addition, the systems are difficult to install in rocky or steep terrains. In areas which experience freeze-thaw cycles, they may be damaged by soil heaving. Unless covered with an appropriate filter, tile drains may eventually clog as fine particles and chemically precipitated material accumulate on the drain openings. Pollutants that are heavier than water may move below the drain if it is not located at the bottom of the perched zone. As with all sampling techniques, there is the possibility of chemical interactions between the sampling system (e.g., tile line) and the chemical constituents of interest. In the case of drainage sampling systems, this effect is amplified because contaminants may have to travel considerable distances through drains before being sampled. In addition, normally nonaerated solutions may be aerated and chemically altered as they travel through drains.

Thomas and Barfield (1974) conducted nitrate-N balance studies in an area drained by an open ditch and by tile lines. For their conditions, they concluded that tile effluent gave an unreliable picture of nitrate-N drainage from farm land for two reasons: (1) the proportion of total flow contributed from tile lines was rather small, and (2) the concentration of nitrate-N in the tile effluent was higher than that in the nontile flow. Apparently, the oxidized region in the vicinity of tile openings precluded denitrification, producing higher nitrate-N levels in tile-line outflow. It should also be noted that, depending on the distribution of soil textures, tile drains may sample ground-water discharge predominantly from more distant areas. Thus, results should be interpreted with caution.

REFERENCES

Czapar, G.F., and R.S. Kanwar, "Field Measurement of Preferential Flow Using Subsurface Drainage Tiles," in *Preferential Flow*, T.J. Gish and A. Shirmohammadi, Eds. Proceedings of the National Symposium, Chicago, IL, American Society of Agricultural Engineers, 1991, pp. 122–128.

Dunlap, W.J., J.F. McNabb, M.R. Scalf, and R.L. Cosby, Sampling for Organic Chemicals and Microorganisms in the Subsurface, EPA-600/2-77-176, U.S. Environmental Protection Agency, 1977.

Eccles, L.A., and P.A. Gruenberg, "Monitoring Agricultural Waste Water Discharges for Pesticide Residues," *Proceedings: Establishment of Water Quality Monitoring Programs*, American Water Resources Association, 1978, pp. 319–327.

Eckard, T.L., D. Millison, J. Muller, E. Vander Velde, and R.U. Bowallius, "Vertical Groundwater Quality and Pressure Profiling Using the BAT Groundwater Monitoring System," in *Proceedings of the Third National Outdoor Action Conference on Aquifer Restoration, Ground Water Monitoring, and Geophysical Methods*, National Ground Water Association, Orlando, FL, 1989.

Everett, L.G., E.W. Hoylman, L.G. Wilson, L.G. McMillion, "Constraints and Categories of Vadose-Zone Monitoring Devices," *Ground Water Monitoring Review*, 4:26–32 (1984).

Fenn, D., E. Cocozza, J. Isbister, O. Briads, B. Yare, and P. Roux, Procedures Manual for Ground Water Monitoring at Solid Waste Disposal Facilities, EPA/530/SW-611, U.S. Environmental Protection Agency, 1977.

Gambrell, R.P., J.W. Gilliam, and S.B. Weed, "Denitrification in Subsoils of the North Carolina Coastal Plain as Affected by Soil Drainage," *Journal of Environmental Quality*, 4:311–316 (1975).

Gilliam, J.W., R.D. Daniels, and J.F. Lutz, "Nitrogen Content of Shallow Ground Water in the North Carolina Coastal Plain," *Journal of Environmental Quality*, 2:147–151 (1974).

Gilliam, J.W., R.W. Skaggs, and S.B. Weed, "Drainage Control to Diminish Nitrate Loss from Agricultural Fields," *Journal of Environmental Quality*, 8:137–142 (1979).

Hallberg, G.R., J.L. Baker, and G.W. Randall, "Utility of Tile-Line Effluent Studies to Evaluate the Impact of Agricultural Practices on Groundwater," in *Proceedings of the Agricultural Impacts on Ground Water*, August 11–13, 1986, Omaha, NE, National Water Well Assoc., Dublin, OH, 1986, pp. 298–325.

Hekma, L.K., K.K. Muraleetharan, and G.F. Boehm, "Application of the Electric Cone Penetrometer Test to Environmental Groundwater Investigations," Presented at the Air and Waste Management Association 83rd Annual Meeting and Exhibition, Pittsburgh, PA, 1990.

Herzog, B.L., J.D. Pennino, and G.L. Nielsen, "Ground-Water Sampling," in *Practical Handbook of Ground-Water Monitoring*, D.L. Nielsen, Ed. (Chelsea, MI: Lewis Publishers, 1991), pp. 449–499.

Jacobs, T.C., and J.W. Gilliam, "Riparian Losses of Nitrate from Agricultural Drainage Waters," *J. Environ Qual.*, 14:472–478 (1985).

Jury, W.A., "Solute Travel Estimates for Tile-Drained Fields: I. Theory," *Soil Science Society of America Proceedings*, 39:1020–1024 (1975).

Kaufmann, R.F., T.A. Gleason, R.B. Ellwood, and G.P. Lindsey, "Ground Water Monitoring Techniques for Arid Zone Hazardous Waste Disposal Sites," *Ground Water Monitoring Rev.*, 1(3):47–54 (1981).

Kraemer, C., J.A. Shultz, and J.W. Ashley, "Monitoring Well Post-Installation Considerations," in *Practical Handbook of Ground-Water Monitoring*, D.L. Nielsen, Ed. (Chelsea, MI: Lewis Publishers, 1991, pp. 333–365.

Mines, B.S., J.L. Davidson, D. Bloomquist, and T.B. Stauffer, "Sampling of VOCs with BAT Ground Water Sampling System," *Ground Water Monitoring and Remediation*, 13(1):115–120 (1993).

Nielsen, D.M., and R. Schalla, "Design and Installation of Ground-Water Monitoring Wells," in *Practical Handbook of Ground-Water Monitoring*, D.L. Nielsen, Ed. (Chelsea, MI: Lewis Publishers, 1991, pp. 239–331.

Pickens, J.F., and G.E. Grisak, "Reply to the Preceding Discussion of Vanhof et al. on 'A Multilevel Device for Groundwater Sampling and Piezometric Monitoring,'" *Ground Water*, 17:393–397 (1979).

Pickens, J.F., J.A. Cherry, G.E. Grisak, W.F. Merritt, and G.A. Risto, "A Multilevel Device for Ground-Water Sampling and Piezometric Monitoring," *Ground Water*, 16:322–327 (1978).

Pickens, J.F., J.A. Cherry, R.M. Coupland, G.E. Grisak, W.F. Merritt, and B.A. Risto, "A Multi-Level Sampling Device for Ground-Water Sampling," *Ground Water Monitoring Review*, 1(1):48–51 (1981).

Poeter, E.P., "Perched Water Identification with Nuclear Logs," *Ground Water*, 26:15–21 (1988).

Richard T.L., and T.S. Steenhuis, "Tile Drain Sampling of Preferential Flow on a Field Scale," *Journal of Contaminant Hydrology*, 3:307–325 (1988).

Schilfgaarde, J. van, Ed., *Drainage for Agriculture*, No. 17 in Agronomy Series, American Society of Agronomy, Madison, WI, 1974.

Skaggs, R.W., *Drainmod-Reference Report; Methods for Design and Evaluation of Drainage-Water Management Systems for Soils with High Water Tables*. USDA-SCS National Technical Center, Fort Worth, TX, 1980.

Thomas, G.W., and B.J. Barfield, "The Unreliability of Tile Effluent for Monitoring Subsurface Nitrate-Nitrogen Losses from Soil," *J. Environmental Quality*, 3:183–185 (1974).

Willardson, L.S., B.D. Meek, and M.J. Huber, "A Flow Path Ground Water Sampler," *Soil Sci. Soc. Amer. Proc.*, 36:965–966 (1973).

Willis, G. H, J.L. Fouss, J.S. Rogers, C.E. Carter, and L.M. Southwick, "System Design for Evaluation and Control of Agrichemical Movement in Soils Above Shallow Water Tables," in *Groundwater Residue Sampling Design*, R.G. Nash and A.R. Leslie, Eds., ACS Symposium Series 465, American Chemical Society, Washington, DC, 1991, pp. 195–211.

Wilson, L.G., and K.D. Schmidt, "Monitoring Perched Ground Water in Vadose Zone in *Establishment of Water Quality Monitoring Programs*, American Water Resources Association, 1979, pp. 134–149.

Wilson, L.G., and G.G. Small, "Pollution potential of a sanitary landfill near Tucson," in

Hydraulic Engineering and the Environment, Proc. 21st Annual Hydraulics Division Specialty Conference, ASCE, 1973, pp. 427–436.

Wilson, L.G., J.N. Luthin, and J.W. Biggar, *Drainage Salinity Investigation of the Tulelake Lease Lands*, University of California, California Agricultural Experiment Station, Bulletin 779, 1961.

Zemo, D.A., T.A. Delfino, J.D. Gallinatti, V.A. Baker, and L.R. Hilpert, "Cone Penetrometer Testing and Discrete-Depth Groundwater Sampling Techniques: A Cost-Effective Method of Site Characterization in a Multiple-Aquifer Setting," in *Proceedings of the Sixth National Outdoor Action Conference on Aquifer Restoration, Ground Water Monitoring, and Geophysical Methods*, National Ground Water Association, Las Vegas, NV, 1992, pp. 341–355.

32

Sampling from Macropores with Free-Drainage Samplers*,**

L.G. Wilson and D.W. Dorrance

INTRODUCTION

Luxmoore (1991) defines macropore flow as water movement through holes or channels that are 1 mm or more in width or diameter. In contrast, mesopore flow occurs in pores less than 1 mm in equivalent diameter. Micropores are those mesopores capable of retaining water by surface tension at field capacity. Macropores have lower surface areas than mesopores, with little opportunity for affecting water quality during drainage. In contrast, mesopores have greater surface areas, affecting chemical reactions between the soil and pore liquids. Although the percentage of the total pore space occupied by macropores may be small (e.g., on the order of 0.5% to 5%), in many soils they may account for the bulk of the flow (Richard and Steenhuis, 1988). Water moving in macropores is generally termed "new water," while water retained in micropores is called "old water" (Luxmoore, 1991). During studies in a steep humid catchment in New Zealand, McDonnell (1990) determined that macropore flow contained both "old water" as well as "new water". Additional aspects of macroscopic flow as a factor in preferential flow are reviewed by Steenhuis and Parlange (1991).

The presence of macropore flow (also called bypass flow) creates uncertainties when designing a vadose-zone monitoring system. For example, such flow may bypass suction samplers (Barbee and Brown, 1986; Shaffer et al., 1979). For vadose

---

*This chapter is based in part on ASTM D 4626-92, Standard Guide for Pore-Liquid Sampling from the Vadose Zone. Copies may be purchased from ASTM, 1916 Race Street, Philadelphia, PA 19103.

**Abstracted with permission from Dorrance, D.W., L.G. Wilson, L.G. Everett, and S.J. Cullen, in *Groundwater Residue Sampling Design*, R.G. Nash and A.R. Leslie, Eds., ACS Symposium Series 465, American Chemical Society, Washington, DC, 1991, pp. 300–331.

zone profiles where both macropore flow and flow in unsaturated regions occur (i.e., dual-flow systems), free-drainage samplers are recommended to supplement suction samplers. A free-drainage sampler consists of a collection chamber placed in the soil to collect liquid from macropore regions of the vadose zone which are intermittently saturated because of rainfall, flooding, or irrigation. Gravity drainage creates a slightly positive pressure at the soil-sampler interface, causing fluid to drip into the sampler. In contrast to suction samplers, which require application of a vacuum, free-drainage samplers are generally passive collectors, automatically collecting the percolating liquids. However, a small suction is applied to some free-drainage samplers in order to break the initial surface tension at the soil-sampler interface. Liquid samples are generally retrieved by applying a vacuum to a suction line within a collection bottle (Figure 32.1).

FREE DRAINAGE SAMPLERS

Free-drainage samplers include the following units:

- sand-filled funnel samplers
- pan lysimeter
- glass-block lysimeters
- caisson lysimeters
- trough lysimeters
- capillary wicking-type samplers.

The first five of these samplers are also classified in the literature as zero-tension lysimeters. Not included in the list are drainage systems (e.g., drain tiles) which also sample macropore flow. Wilson and Dorrance describe such systems in Chapter 31.

Sand-Filled Funnel Samplers

Figure 32.1 shows a sand-filled funnel installation for collecting freely-draining liquid in cohesive soils. The funnel is filled with clean sand and inserted through the sidewall of a trench into a tunnel extending beneath undisturbed soil. The funnel is forced up against the roof of the tunnel to ensure a good contact. It is good practice to scratch the roof of the tunnel with a wire brush or other scraping device to eliminate smearing. The funnel is connected through tubing to a collection bottle. Application of suction to a separate tube within the collection bottle pulls the sample to land surface. As shown on Figure 32.1, the trench may be backfilled. Alternatively, the trench walls may be supported with a wooden framework. Such operations will require filing an OSHA Approved Safety Plan.

Shaffer et al. (1979) designed a funnel extractor system employing an inflated airplane tire to force a collection funnel against the tunnel ceiling, and a vacuum pump for applying a vacuum to the sample line and sample flask to enhance sampling.

Pan Lysimeters

A pan lysimeter generally consists of a metal (e.g., galvanized steel, aluminum) pan of varying dimensions. Figure 32.2 shows a pan with a copper tube soldered to a

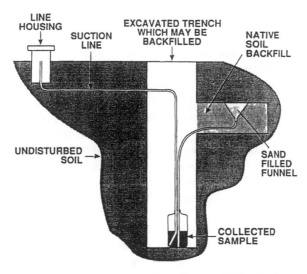

Figure 32.1. Sand-filled funnel sampler installation (U.S. Environmental Protection Agency, 1986).

raised edge of the pan. Plastic or Tygon^R tubing connects the copper tube to a collection vessel. Any liquid that accumulates on the pan drains through the tubing into the vessel. Jemison and Fox (1992) described the installation of pan lysimeters for their experimental studies.

A pan lysimeter can often be pushed or driven directly into the side wall of a trench. However, if the soil is resistant, an opening for the sampler can be created by hammering a sheet metal blade into the soil profile with a sledge hammer. The pan is placed in the side wall so that it slopes gently toward the trench. Any voids above or below the pan are filled with soil. The end of the copper tubing is connected to plastic tubing and a sample bottle.

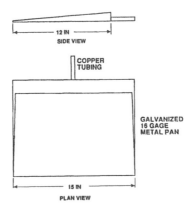

Figure 32.2. Example of pan lysimeter (U.S. Environmental Protection Agency, 1986).

Glass-Block Lysimeters

Figure 32.3 shows a free-drainage sampler made from a hollow glass brick. These glass bricks, which are produced as ornamental masonry, have dimensions of 12 inches by 12 inches by 4 inches (i.e., about 30 by 30 by 10 cm), and have a capacity of 5.5 L. Holes are drilled around the perimeter of one of the square surfaces of a brick to allow pore-liquid entry into the block. For the samplers used in field studies by Barbee and Brown (1986), nine holes, e.g., each 3/16th inch (i.e., 0.47 cm) in diameter, were drilled into the blocks. Appropriate-size nylon tubing is inserted into one of the holes to allow for sample removal. The collecting surface is fitted with a fiberglass sheet to improve contact with the soil and to prevent soil from entering the block. Pore-liquid collection is enhanced by a raised lip along the edge of the surface.

A glass block lysimeter is installed in a tunnel excavated in the side of a trench. Barbee and Brown (1986) used a wooden model of the sampler in order to achieve the correct cavity size during excavation. They used a small knife to score the ceiling of the cavity in order to expose any pores which may have been smeared shut during excavation. Care should be taken to keep the ceiling of the cavity smooth and level to ensure good contact with the upper surface of the glass block. The edges of the sampler should be in contact with the soil for the entire perimeter of the sampler in order to prevent liquid from running out through spaces between the soil and the sampler. Level blocks are important so that the majority of the collected sample can be retrieved. However, the inside glass surface of some blocks may be uneven, with "dead spots" where residual sample could collect between sampling cycles. This leads to cross-contamination of samples. Such blocks should be rejected. The glass block is pushed to the end of the cavity and wedges are used to hold its collecting surface firmly against the ceiling of the tunnel. Both the cavity and trench are partially backfilled. Barbee and Brown (1986) recommend pressing a sheet of aluminum foil against the wall of the trench, extending below the top of the brick, before final backfilling, in order to minimize any lateral migration of liquid from the disturbed portion of the soil profile to the undisturbed portion.

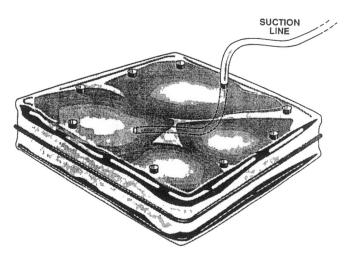

Figure 32.3.   Glass block lysimeter (U.S.Environmental Protection Agency, 1986).

In Chapter 33 of this book, Artiola presents a case study comparing sample collection with glass blocks and suction samplers.

Trough Lysimeters

Trough lysimeters, also known as Ebermayer lysimeters, use troughs or pails to collect pore-liquid (Morrison, 1983). The basic design of this sampler comprises a semicircular trough, e.g., created by cutting a PVC pipe lengthwise down the middle. Both ends are capped, but a tube is inserted into one end for sample collection. The unit is forced against the ceiling of a tunnel constructed within a trench. A problem with this basic design is that surface tension occurring at the soil-air interface prevents some of the soil water from dripping into the sampler (Jordan, 1968). To overcome this problem, Jordan (1968) designed a sampler with a fiberglass screen suspended inside the trough to maintain a firm contact with the edges of the sampler and the soil. The screen is lined with glass wool and covered with soil until the soil is even with the top of the trough (Jordan, 1968). Additionally, two parallel metal rods run along the bottom of the fiberglass cradle, just touching the screen. These rods terminate within the sample collection tube. These rods cause liquid that collects at the base of the fiberglass cradle to be pulled by capillary action away from the screen. By maintaining a slope to the rods, liquid migrates along these rods toward the collection tube. A modification of this design consists of a metal trough with a length of perforated PVC pipe mounted inside (Moore et al., 1981; Morrison, 1983). The trough is filled with graded gravel so that coarse material is immediately adjacent to the PVC pipe and fine sand is at the edges and the top of the trough.

Trough lysimeters are installed in the same manner as other free-drainage samplers.

Caisson Lysimeters

A caisson lysimeter consists of collector pipes, radiating from a vertical chamber (Morrison, 1983). Figure 32.4 illustrates a caisson lysimeter proposed by Aulenbach et al. (1980) for collecting percolating water beneath sand beds used during the land application of wastewater. As shown, a buried, 10 ft diameter steel culvert is used as

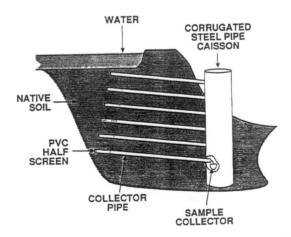

Figure 32.4. Example of a caisson lysimeter (after Aulenbach and Clesceri, 1980).

a central collection chamber, with collector pipes extending into the soil. A section at the end of the pipes is cut in half lengthwise and replaced with a screen. Percolating water entering the collector pipes is conducted to collection bottles within the caisson. An advantage of this design is that the caisson is installed outside the sand beds, thus eliminating any disturbance of the sand beds. McMichael and McKee (1966) also used a large diameter culvert as the central chamber during their studies on wastewater reclamation at a water spreading site at Whittier Narrows, California. Pan lysimeters were installed at various depths to collect percolating water in a 10 ft deep profile during flooding periods. Individual sampling pans were filled with gravel to limit clogging. Each sampler was connected to the central chamber via tubing.

Another caisson lysimeter design reported by Schneider et al. (1984), consists of the following components: (1) stainless steel tubing extending diagonally upward through the caisson wall into the native soil, (2) a screened plate assembly within each tube to retain the soil, (3) a purging system used to redevelop each sampler when it becomes clogged, (4) an airtight cap that prevents exchange between the air in the caisson and the soil air.

Caissons for housing free-drainage samplers are constructed with corrugated culverts or concrete drainage pipes. Schneider and Oaksford (1986) installed caissons by excavating soil from within a concrete pipe using a crane operated shovel and manual labor. Each concrete pipe section, weighing 222.5 kN (25 tons), was set in place with a crane. As excavation inside the pipe progressed, the pipe advanced downward under its own weight. Lateral collectors or free-drainage samplers are installed in cavities augered by hand or by power-driven equipment through holes in the caisson walls. This is another constructed device that will probably require an OSHA Approved Safety Plan.

Wicking Soil Pore-Liquid Samplers

Figure 32.5 shows a wicking sampler which combines the attributes of free-drainage samplers and pressure-vacuum lysimeters (Hornby et al., 1986). The sampler collects both free-drainage liquid and liquid held at tensions to about 4 kilopascals (kPa). A hanging "Hurculon" fibrous column acts as a wick to exert a tension on the soil pores in contact with a geotextile fiber which serves as a plate covering a 12 inch by 12 inch by 0.5 inch (30.5 by 30.5 by 1.3 cm) pan. The terminus of the fibrous column is sealed into the cap of a tubular chamber containing an inlet pressure-vacuum line and a sample collection tube. Materials for the sample collection tube depend on the constituents being sampled. Glass and polytetrafluoroethylene (PTFE) are recommended when sampling for organics (Hornby et al., 1988). In Chapter 34 of this book, Steenhuis et al. compared the collection effectiveness of wicking-type samplers with gravity pan samplers, suction samplers, and piezometers.

Holder et al. (1991) described a capillary-wick sampler which uses glass wicks in a hanging column encased in a Pyrex or PVC pipe (Figure 32.6). Glass wicks are particularly desirable because they are nonreactive with solutions containing inorganic and organic constituents. The upper ends of the wicks are frayed and spread evenly across the surface of a glass plate. Woven glass cloth is placed on top of the frayed wick material to help collect sample from across the entire surface of the plate (Holder et al., 1991). The basal end of the wicks is inserted into a collection flask. The wicks terminate 54 cm below the glass plate, corresponding to an air entry

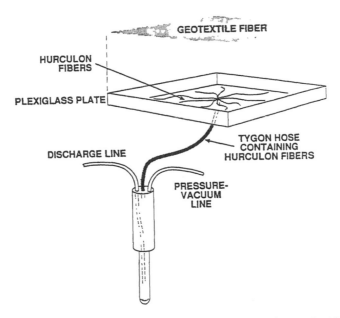

Figure 32.5. Wicking-type soil-pore liquid sampler (after Hornby et al., 1986).

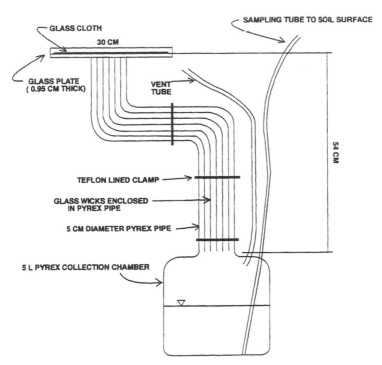

Figure 32.6. Capillary-wick pore-liquid sampler (after Holder et al., 1991).

value of –5.4 kPa of the wick material. The number of wicks required is based on the hydraulic conductivities of the soil and a single wick. As described by Holder et al. (1991), under saturated conditions, water flowing through the soil is intercepted by the sampler plate, drawn into the wick and conducted to the sampling flask. Accumulated water in the flask is withdrawn by applying a vacuum to a nylon tube terminating at the base of the flask.

During field studies, the sampler collected pore-liquid samples from soils with water potentials ranging from 0 to –6 kPa. However, some of the samplers continued to collect samples to a potential of –8.5 kPa (Holder et al., 1991). The suction produced by the hanging column collected samples continuously across this range without the need for an auxiliary vacuum source.

Samplers are installed within a tunnel extending from a trench. The glass plate must be level and firmly wedged against the ceiling of the tunnel.

Pan Collection Efficiency; Number of Units Required

Radulovich and Sollins (1987) define pan collection efficiency (PCE) as the volume of water collected divided by water flux, as calculated by a water balance. Factors tending to lower PCE include water flow being diverted away from the pan because of textural changes and lithological discontinuities in the soil profile, and capillary attraction of unsaturated soil above the pan causing flow around the pan (Jemison and Fox, 1992). Russell and Ewel (1985) found PCE values of less than 10% using the 0.015 m^2 pans designed by Jordan (1968). Radulovich and Sollins (1987) found that by increasing the pan size to 0.25 m^2 and forcing the pan rim upward into the soil, the PCE was increased to 36% under grass and 17% under forest soil. Using pans of 0.46 m^2 area to evaluate PCE in a well-structured Hagerstown silt loam, Jemison and Fox (1992) found PCE values ranging from 45% to 58%. They concluded that in well-structured soils, larger pans are the most efficient. Thus, with larger pans and good contact, there is a better opportunity for intercepting macropore flow and less water movement around the pans.

The studies of Jemison and Fox (1992) are also noteworthy in comparing bromide tracer, the Mather water balance method, and the model LEACHM as a means of estimating PCE. They concluded that using a tracer, such as bromide, is a practical means of estimating PCE.

Holder et al. (1991) used the following equation to determine the number of samplers for an accurate estimate of the chemical concentration of percolating liquid:

$$n = \sigma^2/L^2$$

where n = number of samplers required
σ = population standard deviation
L = allowable error in sample mean

Holder et al. (1991) conducted a field study using a plot in each of three soils representing sand, silt loam, and clay textures. Each plot was about 8 ft by 12 ft (2.4 m by 3.7 m). They found that the number of samplers to estimate the concentration of a given constituent in the soil water with 95% confidence was 31 for sand, 6 for silt loam, and 2 for clay soils.

Advantages and/or Limitations of Free-Drainage Flow Samplers

Physical Advantages and Limitations

A major advantage of free-drainage samplers is that they are essentially passive (aside from the wicking-type units), thus they do not alter pore-liquid flow paths. Additionally, infiltration events can be sampled without the necessity of immediately traveling to the field, unless there is concern the collection bottles may overflow.

Unlike suction samplers, macropore flow samplers can be used to estimate the volume of liquid moving through macropores in the vadose zone, particularly when samplers with large collection areas are used (Jemison and Fox, 1992). For example, Wilson et al. (1991) installed six stainless-steel samplers in the seepage face of a small forested subcatchment and monitored subsurface discharges using tipping buckets. These units monitored subsurface flows in macropores that bypassed soil matrix water.

Samplers collect liquid from undisturbed overlying soils. Thus, the opportunity for cross contamination is minimal.

Excluding wicking-type samplers, the major disadvantage of free-drainage samplers is that samples can only be obtained when soil moisture conditions are in excess of field capacity. Saturated conditions usually occur at the onset of an infiltration event, e.g., during rainfall events and irrigation, or require constant application of surface liquid. Under drier conditions, free-drainage samplers fail to yield any liquid and suction samplers are required.

In contrast to suction samplers (particularly dual chamber, pressure vacuum units) which can be installed at great depths, free-drainage samplers are generally limited to shallow depths; e.g., the operating depths of trenching equipment. However, given that structural cracks and openings generally diminish in width in deeper reaches of the profile (Beven and Germann, 1982), the effectiveness of these samplers is naturally restricted to the near-surface.

Additional limitations of free-drainage samplers, cited by Jemison and Fox (1992), include cost of installation, particularly when construction of trenches or caissons is required, and low PCE when pans of small collection areas are involved.

Advantages for Chemical Sampling

A major advantage of macropore samplers for sampling chemical transport is that the samplers do not distort natural flow patterns as do suction samplers. Inasmuch as samples are collected over known areas, quantitative mass balance estimates are possible.

Larger samples are collected with macropore samplers than generally available with suction samplers, facilitating laboratory analyses of constituents requiring large volumes. Additionally, inasmuch as macropore samplers are generally constructed of inert material, there is less likelihood of reactions with pore liquids (e.g, ammonia nitrogen).

Free-drainage samplers with large cross-sectional areas are cumulative collectors. As a result, they collect samples which average soil heterogeneities, and therefore give a more representative picture than suction samplers of chemical movement through wet soil, particularly through well-structured soils (Barbee and Brown,

1986). In addition, the samplers use suction only to retrieve samples from a collection bottle. As a result there is less potential for loss of volatile compounds than with suction samplers (Barbee and Brown, 1986).

The holding time of samples collected is minimal with free-drainage samplers compared to suction samplers, which may require days or weeks to collect a useable sample.

Inasmuch as free-drainage samplers collect macropore liquids, they are more likely to detect pulses of soluble, nonreactive contaminants escaping below the root or treatment zone of a soil used for the treatment of wastes or wastewaters.

Advantages for Sampling Microorganisms

Inasmuch as free-drainage samplers do not have the minute openings that porous ceramic suction samplers have, they do not screen out colloidal-sized particles and soil bacteria. Consequently, they yield more representative values for suspended solids or Biochemical Oxygen Demand (BOD) measurements (Parizek and Lane, 1970).

Limitations for Modeling

Zhang et al. (1992) modeled the performance of free-drainage samplers using an approximation to the Richards equation. They concluded that these samplers may in some cases collect water from mesopores of unsaturated soils by creating a saturated zone at the soil-sampler interface. However, the numerical model tested by these researchers did not account for preferential flow. There is strong evidence that preferential flow contributes significantly to both volume and quality of the pore liquid collected by these devices. To date there are no predictive models that can separate and quantify Darcian flow from preferential flow. Thus, pan sampler soil-pore data is difficult to use in flow and solute-transport models.

ACKNOWLEDGMENT

The senior author expresses his appreciation to his daughter, Helen Wilson, who helped with some of the illustrations.

REFERENCES

Aulenbach, D., and N. Clesceri, "Monitoring for Land Application of Wastewater," *Water, Air, and Soil Pollution*, 14:81–94 (1980).

Barbee, G.C., and K.W. Brown, "Comparison Between Suction and Free-Drainage Soil Solution Samplers," *Soil Science*, 141:149–154 (1986).

Beven, K., and P. Germann, "Macropores and Water Flow in Soils," *Water Resources Research*, 18:1311–1325 (1982).

Holder, M., K.W. Brown, J.C. Thomas, D. Zabcik, and H.E. Murray, "Capillary-Wick Unsaturated Zone Soil-Pore Water Sampler," *Soil Science Society of America Journal*, 55:1195–1202 (1991).

Hornby, W.J., J.D. Zabcik, and W. Crawley, "Factors Which Affect Soil-Pore Liquid: A

Comparison of Currently Available Samplers with Two New Designs," *Ground Water Monitoring Review*, 6:61–66 (1986)

Jemison, J.M., and R.H. Fox, "Estimation of Zero-Tension Pan Lysimeter Collection Efficiency," *Soil Science*, 154:85–94 (1992).

Jordan, C. F., "A Simple, Tension-Free Lysimeter," *Soil Science*, 105:81–86 (1968).

Luxmoore, R.J., "On Preferential Flow and Its Measurement," in *Preferential Flow*, T.J. Gish and A. Shirmohammadi, Eds., Proceedings of the National Symposium, Chicago IL, American Society of Agricultural Engineers, 1991, pp. 113–121.

McDonnell, J.J. "A Rationale for Old Water Discharge Through Macropores in a Steep, Humid Catchment," *Water Resources Research*, 26:2821–2832 (1990)

McMichael, F.C., and J.E. McKee, Wastewater Reclamation at Whittier Narrows, Water Quality Publication No. 33, State of California, 1966.

Moore, B.B., B. Sagik, and C. Sorber, "Viral Transport to Ground Water at a Wastewater Land Application Site," *Journal Water Pollution Control Federation*, 53:1492–1501 (1981).

Morrison, R. D., *Ground Water Monitoring Technology*, Timco MFG., Inc., Prairie du Sac, WI, 1983.

Parizek, R.R., and B.E. Lane, "Soil-Water Sampling Using Pan and Deep Pressure-Vacuum Lysimeters," *J. Hydrology*, 1: 1–21 (1970).

Radulovich R., and P. Sollins, "Improved Performance of Zero-Tension Lysimeters," *Soil Science Society of America Journal*, 51:1386–1388 (1987).

Richard T.L., and T.S. Steenhuis, "Tile Drain Sampling of Preferential Flow on a Field Scale," *Journal of Contaminant Hydrology*, 3:307–325 (1988).

Russell, A.E., and J.J. Ewel, "Leaching from a Tropical Andept During Big Storms: A Comparison of Three Methods," *Soil Science*, 139:181–189 (1985).

Schmidt, C., and E. Clements, Reuse of Municipal Wastewater for Groundwater Recharge, U.S. Environmental Protection Agency, 68-03-2140, Cincinnati, OH, 1978, pp. 110–125.

Schneider, B.J., J. Oliva, H.F.H. Ku, and E.T. Oaksford, "Monitoring the Movement and Chemical Quality of Artificial-Recharge Water in the Unsaturated Zone on Long Island, New York," *Proceedings of the Characterization and Monitoring of the Vadose (Unsaturated) Zone*, National Water Well Association, Las Vegas, NV, 1984, pp. 383–410.

Schneider, B.J., and E.T. Oaksford, Design and Monitoring Capability of an Experimental Artificial-Recharge Facility at East Meadow, Long Island, New York, U.S. Geological Survey Open-File Report 84-070, 1986.

Shaffer, K.A., D.D. Fritton, and D.E. Baker, "Drainage Water Sampling in a Wet, Dual-Pore Soil System," *Journal of Environmental Quality*, 8:241–246 (1979).

Steenhuis, T.S, and J. -Y, Parlange, "Preferential Flow in Structured and Sandy Soils," in *Preferential Flow*, T.J. Gish and A. Shirmohammadi, Eds., Proceedings of the National Symposium, Chicago, IL, American Society of Agricultural Engineers, 1991, pp. 12–21.

U.S. Environmental Protection Agency, Environmental Monitoring Systems Laboratory, Permit Guidance Manual on Unsaturated Zone Monitoring for Hazardous Waste Land Treatment Units, U.S. Environmental Protection Agency, Office of Solid Waste and Emergency Response, EPA/530-SW-86-040, October 1986.

Wilson, G.V., P.M. Jardine, R.J. Luxmoore, L.W. Zelazny, D.A. Lietzke, and D.E. Todd, "Hydrogeochemical Processes Controlling Subsurface Transport from an Upper Catchment of Walker Branch Watershed During Storm Events. 1. Hydrologic Transport Processes," *Journal of Hydrology*, 123:297–316 (1991).

Zhang, R., A.W. Warrick, and J.F. Artiola, "Numerical Modeling of Free-Drainage Water Samplers in the Shallow Vadose Zone," *Advances in Water Resources*, 15:215–258 (1992).

33

Long-Term Use of Glass Brick Lysimeters and Ceramic Porous Cups to Monitor Soil-Pore Water Quality in a Nonhazardous Waste Land Treatment Case Study

Janick F. Artiola and Wayne Crawley

INTRODUCTION

Land treatment of nonhazardous secondary wastewater treatment sludges is a safe method for recycling wastes high in water and macronutrients. However, all secondary wastewater treatment wastes contain significant fractions of suspended and dissolved solids which can affect the soil-pore water quality, and even ground-water resources over time. Land treatment facilities are required by federal and state regulations to monitor any changes on the upper vadose zone resulting from repeated waste applications. This case study describes the use of glass brick lysimeters and ceramic porous cups to monitor the soil-pore water quality during the initial seven years of operation of a nonhazardous land treatment facility located on the Gulf Coast. Background information will be provided on the waste quality, and management practices, as well as soil and climate conditions at this location. Information on the installation of the glass brick lysimeters, their operation and performance will be presented and discussed. Finally, soil-pore water data collected from glass brick lysimeters, ceramic porous cups and soil core samples will be presented and compared.

0–87371–610–8/95/$0.00 + $.50

LAND TREATMENT DESCRIPTION AND LOCATION

This land treatment facility has been in operation since 1984 and has a total combined surface area of about 75 acres. This facility was designed with 11 treatment cells or management units, and a central run-off collection impoundment. The annual rainfall in the region typically ranges from 48 to 53 inches per year. The upper vadose zone or soil treatment zone (1.5 m) of the land treatment facility is located on a moderately heavy clay soil (Houston Clay series) which has a slow to very slow permeability in the top 3m of the profile. The ground-water table is known to be deeper than 10 m, and no perched water tables are known to exist within the soil treatment zone (1.5 m).

WASTE CHARACTERISTICS

The land treatment facility treats biocake waste generated by dewatered sludge from the flocculator and gravity thickener underflows from two industrial wastewater treatment plants. The chemical and physical characteristics of this biocake are listed in Table 33.1. The biocake waste contains about 15% solids, and must be applied on the land treatment plots with farming equipment. Subsequently, the biocake is incorporated (plowed) into the top 20–30 cm of the soil profile with a tractor and disk implement.

Table 33.1 also shows that the biocake waste applied to the land treatment facility has high concentrations of soluble constituents. Sodium, potassium, calcium, bicarbonate, sulfate, and chloride ions are found at high concentrations in the water phase of the biocake waste. While soluble salts are a management concern, nitrogen is the rate limiting constituent. Ammonia-N and TKN are more than 5,000 mg/L in the waste. Metals (Ni, Cu, Cr, and Zn) are present in moderate concentrations and represent the capacity limiting constituents. However, the information presented in this case study will be limited to Electrical Conductivity (EC) and Sodium Adsorption Ratios (SAR), estimated using the concentrations of the major cations. The EC and soluble Na, Ca, and Mg ion concentrations needed to compute SAR were measured directly from the soil-pore water samples. Note: Sodium Adsorption Ratio (SAR) values are not adjusted for bicarbonates because the soil-pore water data did not include bicarbonate measurements during the first two years of monitoring.

GLASS BRICKS AND CERAMIC POROUS CUPS

The following sections present a brief description of the installation of these two soil-pore water sampling devices. Subsequent sections deal with statistical and comparative evaluations of the glass brick and porous cup data sets.

Table 33.1. Major Characteristics of Sludge Applied to Landfarm

	EC (mmhos/cm)	Water (%)	SAR	Na (mg/kg)	Ca (mg/kg)	Mg (mg/kg)	K (mg/kg)	HCO$_3$ (mg/kg)
Mean	1.77	84.4	17.3	672	98.0	17.3	63.8	1080
Variance	0.729	2.83	9.11	428	30.8	2.87	8.96	374

Note: These values are for samples (n = 60) collected during 1988–1990.

Location and Installation of Glass Brick Lysimeters and Porous Cups

Soil-pore water monitoring equipment was installed at random locations within the landfarm subplots. A total of 7 glass brick lysimeters and 17 ceramic porous cups were installed at a depth of 1.3 to 1.5 m during the spring of 1984. These devices were installed using access trenches, as shown in Figure 33.1. (Note: horizontal distances between porous cups and glass brick devices shown in Figure 33.1 are for illustration purposes only). The glass brick lysimeters where installed against the exposed upper soil surface of an access cavity dug down along the side of the access trench (see Figure 33.1). Using other access trenches, porous cups were installed about 20–30 meters away from the glass brick. This was accomplished using access holes which were dug into the side of the trench using a 5 cm in diameter hand auger

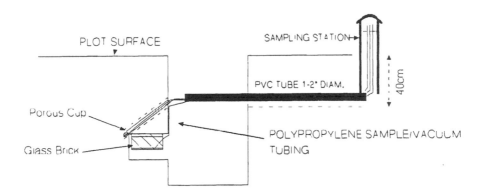

TRENCH CROSS-SECTION

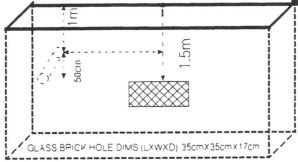

TRENCH 3-D VIEW

Figure 33.1.   Glass brick and ceramic porous cup field installation: cross section (top) and 3-D view (bottom).

(see Figure 33.1). The porous cups were fitted into the holes using a native soil paste. Subsequently, the top of the porous cup access holes and the side of the glass brick access cavity where sealed with a Na-bentonite paste, and the trenches buried. The sampling tube lines were also buried below the plow layer to a depth of about 60 cm, and sampling stations were located along the edges of the landfarm cells. The sampling tubing was protected with PVC pipe along the buried path. The method of installation reduces the risk of damage to the sampling devices and sampling tubing during farming activities. The glass brick lysimeters are made using construction glass bricks (30cm × 30cm × 10cm). Five holes, 10 cm in diameter, were drilled on one of the two faces. A geotextile fiber was placed over the glass brick to filter out soil particles. The access tube was inserted into the glass brick cavity using a sixth hole drilled on one of the four sides. Standard 24-inch long ceramic porous cups (1 bar bubbling pressure) were obtained from Soilmoisture Equipment Corp., Santa Barbara, CA. Polypropylene 1/4-inch O.D. with 1/16-inch walls tubing was used for all sampling and vacuum lines (one for each glass brick and two for each porous cup), as shown in Figure 33.1.

Performance Evaluation

The general long-term performance of these two devices for quarterly monitoring of soil-pore water in this land treatment facility has been very good. Since their installation in 1984, about 80% of the porous cups and glass brick are still in use. Records indicate that the volumes of soil-pore liquid collected have been very consistent in both types of devices. Soil-pore liquid volumes from glass bricks typically collect 200–800 mL per sampling event, while the porous cups typically average 30–250 mL per sampling event (vacuum at 60 centibars for 24 hours).

Data Statistical Analysis

Soil cores were collected beginning in 1985 at two depth intervals, 0–30 cm and 30–60 cm, once a year. Unfortunately, no soil core data are available from the same depth as the location of the sampling devices (130 cm). Therefore, comparisons between the soil-pore and the soil core data will not be made. The glass brick lysimeter and porous cup data collection started in 1984. Although data on more than 20 parameters are available, the following data evaluation and discussion will be limited to soil and soil-pore water Electrical Conductivity (EC mmhos/cm) and Sodium Adsorption Ratio (SAR). The data presented here were generated from soil cores and soil-pore water monitoring equipment installed in 3 of the 11 land treatment cells. The data presented in Figures 33.2 to 33.9 were plotted and a linear regression was fitted on the means of the three cells and the background data. Subsequently, a trend analysis of each data set using the Mann-Kendall trend test (non-parametric) for slope analysis (Gilbert,1987) was done. This procedure is designed to test whether or not the slope of the line (shown in Figures 33.2 to 33.9) is significantly different from zero. Thus, acceptance of the null hypothesis would indicate that there is no significant trend (upward or downward) in the data with time. All of the results discussed here assume significance at the 95% confidence level (P = 0.05).

Soil-Pore Water Data

Figures 33.2 and 33.3 present the soil-pore water data collected with glass brick lysimeters and ceramic porous cups from 1984 to 1990. The regression lines show an upward trend in the slope, indicating an increase in solutes over time. The upward trends are statistically significant as tested by the M-K procedure described above. However, the background data, which is limited to 4 years, did not show any significant increases with time (see Figures 33.2 and 33.3).

The sodium adsorption ratio was also computed from the Na, Ca, and Mg values obtained from soil-pore analyses. Since this ratio uses one mobile ion (Na) and two moderately immobile ions (Ca and Mg), any change in this ratio indicates changes in the concentrations of these three ions. Figures 33.4 and 33.5 present the SAR data computed from the glass brick and porous cup water. No significant trend change in the SAR of the waste treatment cells is detected with either type of equipment with time. Thus, the data collected from glass bricks and porous cups suggest that although the EC is increasing in the soil-pore water, the SAR is not. However, the figures show much lower SAR values in the glass bricks liquid than in the porous cups.

Soil Core Data

Figures 33.6 and 33.7, and 33.8 and 33.9 present soil core data collected from the sample plots during the same time intervals, at 0–30 cm and 30–60 cm depth intervals, respectively. The EC and SAR values were obtained from $1+1$ soil-water extracts using dry soil. Figures 33.6 and 33.7 show that the solute concentrations decrease dramatically from the 0–30 cm interval to the next. Also, the EC increases significantly with time at 0–30 cm, but at 30–60 cm, the M-K test barely failed ($P = 0.057$, indicating no significant changes in slope with time. The SAR data presented in Figures 33.8 and 33.9 also have the same trends as the EC. As expected, the EC and SAR soil core data did not change significantly with time in the background soil, as shown in Figures 33.6 to 33.9. The soil core data at 30–60 cm depth show lower EC values than those observed in the soil-pore water collected at a 130 cm depth. However, as previously stated, direct comparison of this soil core data with the soil-pore data sets is not possible, since data did not come from the same soil profile depth. It is apparent from data presented in Figures 33.6 to 33.9 that the soil core data have much smaller variances, and thus better regression fits.

Statistical Comparisons of Data Sets

The glass brick and porous cup data have similar (relative) trends in EC and SAR in the soil-pore water. However, this does mean that the data of the two soil-pore water collection devices are comparable in absolute terms. Student-t and F test analyses of the means and variances of the EC data set indicate that the overall means of the two sets of data are equal ($P = 0.05$). However, the variances of the same two sets of data are not equal. The data distributions of both data sets do not conform well to the goodness of fit tests for normality. A natural log transformation of the data did improve the goodness of fit test results for the porous cup data, making it normally distributed. However, this transformation was not sufficient to improve the glass brick data behavior and did not change the mean and variance

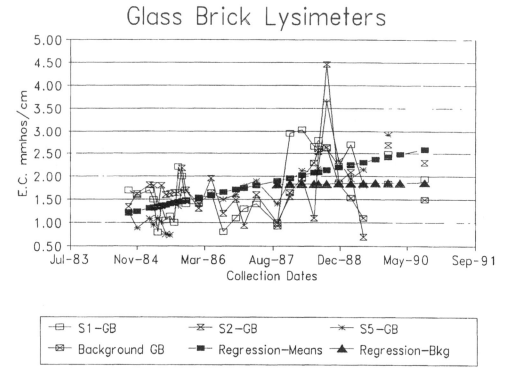

Figure 33.2. Glass brick lysimeter EC data and linear regression fits.

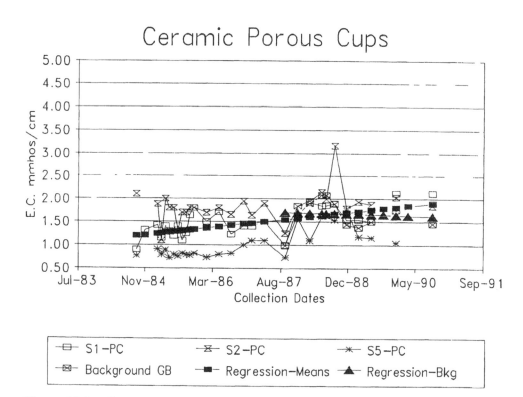

Figure 33.3. Ceramic porous cup EC data and linear regression fits.

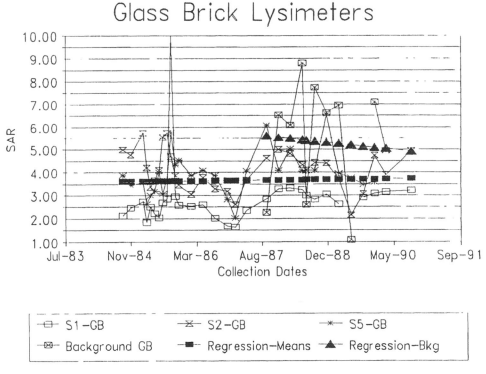

Figure 33.4. Glass brick lysimeter SAR data and linear regression fits.

comparisons (see Table 33.2). A linear correlation of the two EC data sets indicates that these are significantly correlated at $P = 0.05$ [$n = 29$, Corr. Coef.(r) = 0.837]. However, this correlation is only an indication that the means of the data sets have the same relative upward trend. The SAR data sets are also not normally distributed, and the ln transformation did not improve their fit to a normal distribution significantly. At the same time, mean and variance comparisons indicate that the means are not the same, but the variances appear to be equal. A linear regression of the two SAR data sets did not reveal a significant correlation between the two variables (SAR-PC and SAR-GB) at $P = 0.05$ [$n = 29$, Corr. Coef.(r) = 0.146]. These comparisons point out two major differences in the glass brick and porous cup data sets. First, glass bricks show much higher EC variances in the data than porous cups. Second, the SAR values are much lower in the glass bricks than the porous cups.

CONCLUSIONS

Glass bricks and porous cups are useful for long-term monitoring of soil-pore water quality in active land treatment units. Both types of devices are able to show similar long-term trends in overall soil-pore water quality. However, in this case study, glass bricks usually have higher variances in the data. Most importantly, the means of the data sets obtained with either of these two devices may not be comparable. In the case discussed above, the EC data sets had similar means and well correlated upward trends with time. However, the SAR data sets were not comparable nor well correlated. The fact that the SAR mean of the porous cups was signifi-

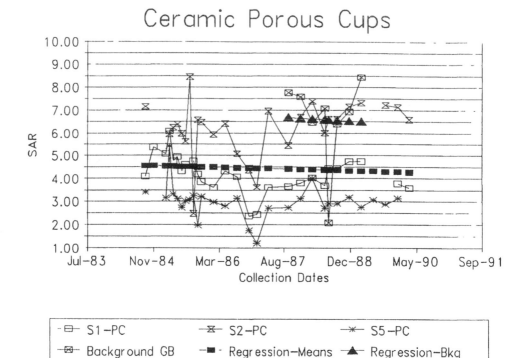

Figure 33.5.   Ceramic porous cup SAR data and linear regression fits.

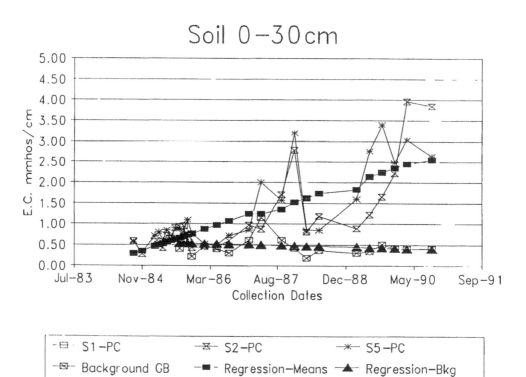

Figure 33.6.   Soil core (0–30cm) EC data and linear regression fits.

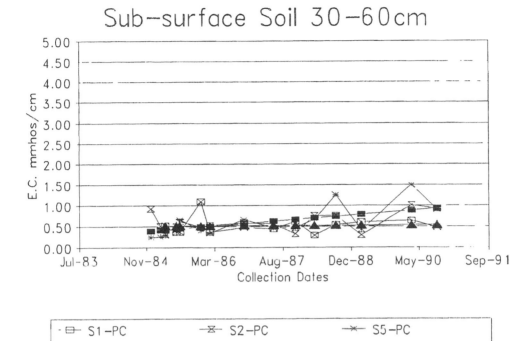

Figure 33.7. Soil core (30–60cm) EC data and linear regression fits.

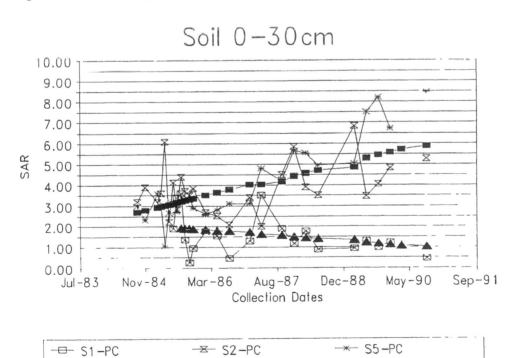

Figure 33.8. Soil core (0–30cm) SAR data and linear regression fits.

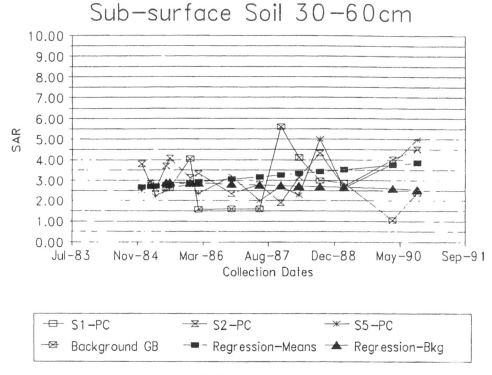

Figure 33.9. Soil core (30–60cm) SAR data and linear regression fits.

cantly higher than that of the glass bricks may indicate that the two devices do not sample the same type of soil-pore water. Calcium and Mg ions necessary to lower the SAR may not move as fast in micropores. This is consistent with previous observations made by Hornsby et al. (1985) using these two sampling devices. These study results indicated that TDS in porous cups liquids were much higher than in glass bricks. Furthermore, the same study (Hornsby et al., 1985) pointed out that the ratios of Ca, Mg, Na, and K in porous cups were different from the ones in glass bricks. Silkworth and Grigal (1981) also reported alteration of the major cation concentrations in soil solutions sampled with small and large ceramic porous cups. Concerns about the possible alteration of ion concentrations in soil solutions by ceramic porous cups have been documented, as reviewed by Litaor (1988). Thus, it is likely porous cups may exclude Ca and Mg ions through the ceramic membrane,

Table 33.2. Comparison of Means and Variances Between Glass Brick and Porous Cup Data Sets

	Glass Brick	Porous Cup	Glass Brick Ln Transforms	Porous Cup Ln Transforms
Means-EC	1.728	1.454	.501 (1.650)	0.358 (1.431)
Variances-EC	0.323	.0699	.089 (1.093)	0.0319 (1.023)
Means-SAR	3.651	4.409	1.273 (3.568)	1.466 (4.332)
Variances-SAR	0.681	0.624	0.043 (1.044)	0.040 (1.041)

Note: n = 29, numbers in parentheses are the antilog of the means and variances.

either because they are not there or because of surface precipitation/sorption reactions. Also, ceramic porous cups may create nonequilibrium conditions in moderately soluble minerals such as Ca and Mg carbonates during sampling. These conditions may exist when high water gradient conditions are imposed in the surrounding soil as a result of excessive vacuum. This being the case, porous cups may tend to overestimate SAR. On the other hand, as previously reported by Barbee and Brown, glass bricks are more likely to collect soil-pore water from macropores which carry water that may or may not be in equilibrium with the bulk of the soil. In a study by Brown and Barbee (1986) these researchers concluded that glass bricks are better suited to collect fast-moving macropore water from well-structured clay soils than porous cups are. This observation is consistent with data presented here, and may explain the higher variances and different SAR observed in the data sets from the glass bricks.

REFERENCES

Barbee, G.C, and K.W. Brown. "Comparison Between Suction and Free-Drainage Soil Solution Samplers," *Soil Science*. 141(2):149–154 (1986).

Gilbert, R.O., *Statistical Methods for Environmental Pollution Monitoring*. (New York: Van Nostrand Reinhold, 1987), Chapter 16.

Hornsby, W.J., J.D. Zabcik, and W. Crawley. "Factors Which Affect Soil-Pore Liquid Sample Quality: A Comparison of Currently Available Samplers with Two New Designs," *Fifth National Symposium Exposition on Groundwater Monitoring*. Torrance, CA, Sept. 30 to Oct. 3. National Water Well Association, 1985.

Litaor, M.I. "Review of Soil Solution Samplers," *Water Resources Research*. 24(5):727–733 (1988).

Silkworth, D.R., and D.F. Grigal. "Field Comparison of Soil Solution Samplers," *Soil Sci. Soc. Am. J.*, 45:440–442 (1981).

34

Field Evaluation of Wick and Gravity Pan Samplers

Tammo S. Steenhuis, Jan Boll, Eric Jolles, John S. Selker

INTRODUCTION

Protection of ground water quality requires accurate knowledge of chemical transport through the vadose zone, which is, in large part, limited by the accuracy of available sampling techniques. Methods for continuous sampling of soil solutes generally involve the acquisition of water drained either by the force of gravity (e.g., gravity pan samplers, agricultural tile lines, and shallow wells), or by applying a "capillary" suction (e.g., porous cup samplers and wick pan samplers). While all these sampling techniques result in a solute concentration for the vadose zone, only tile lines and pan samplers define the sampling volume and, hence, allow a mass balance and prediction of solutes moving to the aquifer. In this chapter, the effectiveness of wick and gravity pan samplers in assessing solute transport in the vadose zone is discussed. Concentrations measured with the pan samplers are compared to those obtained from porous cup samplers.

Pan Samplers

Gravity pan samplers have been used, among others, by Jordan (1968), Parizek and Lane (1970), Shaffer et al. (1979), Brinsfield et al. (1987), Barbee and Brown (1986), Steenhuis et al. (1989), Boll et al. (1991), and Jamison and Fox (1992). In essence, gravity pan samplers are designed to collect water from saturated regions in the vadose zone. However, in many cases, such regions are not present naturally, and the sampler induces a pressure increase in soil above the pan. This artifact often leads to bypassing of the pan samplers (Kung, 1988; Jamison and Fox, 1992).

A modification to the gravity pan was the alundum tension plate sampler, in

© 1995 by Lewis Publishers   629

which samples were drawn from unsaturated soil by applying a suction across alundum filter discs (Tanner et al., 1954; Cole, 1958). Pan samplers with fiberglass wicks are based on the same principle. The wicks are self-priming and act as a hanging water column, providing a suction in the soil above the pan. Wick pan samplers were used by Brown et al. (1986) and improved by Boll et al. (1991, 1992). The design of Brown et al. (1986) consisted of a 30 by 30 cm pan with one large wick in the center. The drawback to this design is that chemicals entering the sampler at the sides need to travel a considerable distance to the center, while solutes near the middle flow without delay, giving rise to a large instrument dispersion coefficient. Boll et al. (1991) decreased the pan-induced dispersion by dividing the 30 by 30 cm plate into a 5 by 5 grid, and sampling each 6 by 6 cm cell separately with a fiberglass wick.

TWO CASE STUDIES

Gravity and wick pan samplers were tested in conjunction with porous cup samplers on Cornell University's Thompson Vegetable Crops Farm near Freeville, NY (the "Freeville Site") and in the Cornell University Apple Orchard in Ithaca, NY (the "Orchard Site"). Gravity and wick pan samplers were installed 1 m apart, 0.6 m below the soil surface in 0.9-m long, horizontal tunnels excavated in the side of a trench. The pan consisted of a 5 by 5 grid of individually sampled compartments, and was pressed against the native soil above using screw-jack supports (Steenhuis et al., 1989). The difference between the wick and gravity pan samplers was the way effluent was collected. In the 320 by 320 mm gravity pan, the 25-mm high sampling compartments were filled with pea gravel so that water dripped into the bottles from saturated soil. The undisturbed soil above each cell of the wick pan was sampled under continuous suction applied by a 0.4-m long, 95-mm diameter fiberglass wick. The upper end of the wick was spread over an acrylic plate, and each plate was seated on a compression spring (65 mm tall, 24 mm in diameter), assuring good contact with the soil. Each wick was encased in 13 mm (i.d.) tygon tubing and was suspended well above a collection bottle to prevent upward movement in the wick (Boll et al., 1991). A schematic drawing of one cell in the wick pan sampler is given in Figure 34.1.

Porous cup samplers consisted of a ceramic cup having an outside diameter of 48 mm in the Freeville Site and 24 mm in the Orchard Site, and an overall length of 53 mm. The cup was cemented to a PVC pipe plugged with a two-hole stopper. Plastic tubing was put through the stopper holes; a short length to apply the vacuum (around 0.2 bars daily), and a longer section reaching into the cup for sample retrieval. Each sampler was installed through vertical augered holes backfilled with a slurry of the original soil mixed with a small amount of bentonite. All holes were sealed with a 250-mm layer of bentonite to prevent leakage from the surface. All porous cup samplers were installed at a depth of 0.6 m.

Freeville Site

The soil was a well-drained Genesee silt loam characterized by a root zone of 60 cm of dark brown sandy to loamy soil containing worm and root holes, overlaying a dark grayish-brown silt loam to very fine sandy loam from 60 to 200 cm, and a substratum of layers of gravel and sand at a depth of >200 cm. Two gravity and two

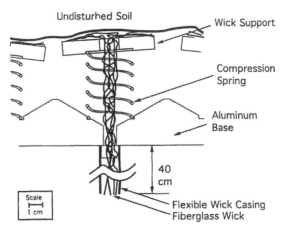

Figure 34.1.   Schematic diagram of one 6 by 6 cm cell in the wick pan sampler by Boll et al. (1991).

wick pan samplers, and four porous cup samplers were installed according to above descriptions. The grass-covered plot (6 by 6 m) received 2 cm of irrigation twice a day. After a steady-state outflow pattern had been reached, a 0.1 M bromide solution was applied for one day in 4 cm of rain, followed by 22 days of irrigation.

Orchard Site

The soil is a Rhinebeck clay loam in which structural cracks are a distinct feature. One gravity and one wick pan sampler, and 24 porous cup samplers were installed in each of two 2 by 6 m plots. Each plot was subjected to a different management practice to suppress weeds under the apple trees: a mowed sod and a moss cover resistant to roundup (glyphosate) (sprayed yearly). Both practices had been in place for five years prior to the experiment. In the plot with mowed sod cover, many very fine roots were evident in the upper 30 cm, with no cracks larger than 1 mm observed to this depth. Below 30 cm, the original soil structure was maintained and water could flow through structural cracks between hexagonally shaped peds 20 to 30 cm in diameter. Large surface-connected cracks (sometimes as large as 1 to 2 cm wide) were visible in the moss-covered plot. These cracks remained open to a depth of at least one meter. Water was applied to both plots daily at a rate of 1 cm/hr for three to four hours. The duration of the irrigation period was three weeks on the mowed grass plots and slightly less than two weeks on the moss-covered plots. A 0.1 M bromide solution was applied with the first irrigation.

RESULTS

Freeville Site

Figure 34.2 shows breakthrough curves of bromide for each pan sampler. Each concentration value was obtained from the total mass and volume collected in all 25 cells. The concentration peak for both sampling methods occurred 9 or 10 days after

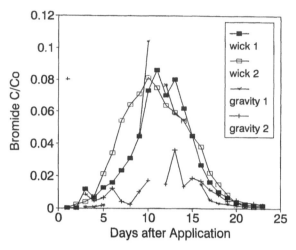

Figure 34.2. Spatially averaged breakthrough curves for two wick and two gravity pan samplers at the Freeville Site. (The missing line segments indicate that no water was collected.)

application, with a peak concentration of approximately 650 mg/L for both wick samplers, and 813 and 280 mg/L for the gravity samplers. Although initially, a few cells in wick sampler 1 and gravity sampler 2 collected high bromide concentrations as a consequence of preferential flow, experimental data were fitted closely using the standard convective-dispersive (Gaussian) curve. This indicates that matrix flow was the predominant component for transport of the bulk of the bromide in the upper 60 cm of the soil.

Wick pan samplers collected nearly 100% of the applied water and, on average, 63% of the total bromide, whereas the gravity pan samplers intercepted, on average, 28% of the applied water and only 7% of the total bromide (Table 34.1). Water and solutes clearly bypassed the gravity pan samplers and, to a much lesser degree (if at all), the wick pan samplers. Evidence of bypass flow also is derived from the distribution of the cumulative loss of bromide in the pan for each cell during the experimental period. Figure 34.3 shows the total percentage of bromide intercepted by each cell in one gravity and one wick pan sampler. As observed, the highest percentage of bromide was collected in the center of the gravity pan sampler, whereas in the wick pan sampler, a more random distribution was apparent. Water

Table 34.1.   Average Water and Total Bromide Collected by Wick and Gravity Pan Samplers Installed at 60 cm Depth at Freeville Site

Sampler Type	Average Water Collected/Day		Total Bromide Collected	
	(cm)	(%)[a]	(g)	(%)
Wick 1	4.3	108	19.68	67
Wick 2	3.9	98	17.43	59
Gravity 1	0.5	13	1.59	5
Gravity 2	1.7	42	2.33	8

[a]Percentage of applied after correction for evaporation.

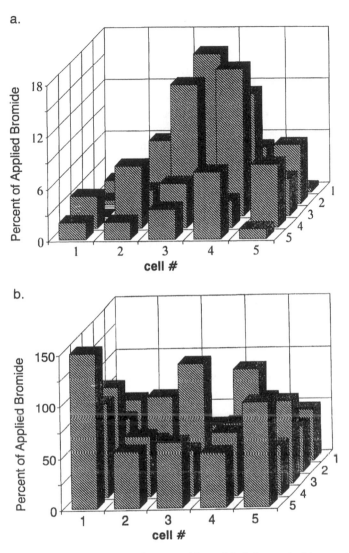

Figure 34.3.   a. Spatial distribution of mass of bromide intercepted by each cell during the experimental period as a percentage of the amount applied for one gravity pan sampler at the Freeville Site. b. Spatial distribution of mass of bromide intercepted by each cell during the experimental period as a percentage of the amount applied for each cell of one wick pan sampler at the Freeville Site.

and bromide bypassed the outer edges of the gravity pan sampler due to the hydraulic potential gradient between saturated soil above the sampler and unsaturated soil in its surroundings. The soil above the wick pan sampler was unsaturated and the potential gradient did not manifest itself under the imposed flux, resulting in a more random pattern of high and low fluxes. Accordingly, the distribution of fluxes in this soil was much better represented in the wick pan sampler than in the gravity sampler. Under very low flow conditions (not tested here), water may also bypass the wick sampler, as noted recently by some researchers (Daliparthy et al., 1993).

Knutson and Selker (1994) and Rimmer et al. (1994) found that when the soil and wicks have similarly shaped soil conductivity curves, bypass flow is small.

As shown in Figure 5.10 of Steenhuis et al. (1994), the four porous cup samplers at the 60 cm depth differed not only from the pan samplers, but also from each other. Porous cup 1 had a peak concentration of 968 mg/L which occurred on day 15. Porous cups 2 and 3 had high peak concentrations, 1792 and 1304 mg/L, respectively, which occurred on day 4, three days after application. Porous cup 4 did not collect except on the first day. It is noteworthy that only three of the 100 pan sampling cells provided concentration data similar to the patterns of porous cups 2 and 3, strongly suggesting that the porous cup sampling method had distorted the flow pattern. This was confirmed at the end of the experiment when blue dye was applied shortly before the samplers were removed. Blue dye had moved along thin sampling tubes (despite careful installation), directly to the porous cup samplers at the 60 cm depth.

Orchard Site

Breakthrough curves for the weighted average of all cells show that, in both plots, the highest concentrations arrived with the wetting front, due to preferential flow, while the matrix flow contribution was insignificant (Figure 34.4). Concentrations in the moss-covered plot almost reached the pulse concentration and less than half that in the mowed sod plot. So, in this clay loam soil, the gravity and wick pan samplers gave almost identical results because most of the water flowed through cracks, and sideways matrix flow movement was minimal due to the dense and almost impermeable soil matrix.

In both plots, the pan samplers collected approximately all of the applied bromide, but not always all of the applied water (Table 34.2). Bromide was collected during the first five days of the experiment, when most cracks were still largely open at the surface. Subsequently, overland flow increased as some of the surface cracks closed, resulting in reduced infiltration. Furthermore, the water collection pattern was a function of the microtopography and contributing area of the cracks. The

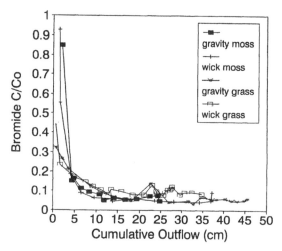

Figure 34.4.   Spatially averaged breakthrough curves for wick and gravity pan samplers for the mowed-grass and moss-covered plots at the Orchard Site.

Table 34.2.   Total Water and Bromide Collected by Wick and Gravity Pan Samplers Installed at 60 cm Depth at Orchard Site

Sampler Type	Total Water Collected		Total Bromide Collected	
	(cm)	(%)[a]	(g)	(%)
Grass Covered Plot				
Wick	34.8	70	21.6	115
Gravity	45.6	93	20.1	107
Moss Covered Plot				
Wick	37.2	106	19.6	92
Gravity	24.1	69	22.1	104

[a]Percentage of applied after correction for evaporation.

latter is illustrated in Figure 34.5, which again shows the cumulative percentage of bromide collected in each cell in the gravity and wick pan samplers during the experimental period. Cells with great losses of bromide were observed adjacent to cells with very little loss of bromide. Obviously, the large amounts of bromide were carried preferentially through the cracks and neighboring cells were not affected as lateral movement in the region above the samplers was insignificant due to the extreme low permeability of the soil matrix. It was coincidental that the highest amounts of bromide were collected in the center of the gravity pan sampler. Under the experimental conditions tested, the data from the gravity and wick pan samplers appear to have accurately captured the characteristics of the soil and the difference between the treatments.

The average concentration of the 24 porous cups is depicted in Figures 34.6a and 34.6b for the mowed-sod plot and the moss-covered plot, respectively. The most striking feature is that the early bromide breakthrough, as seen in the pan samplers, was not observed with the porous cups. Especially in the moss-covered plots, all of the applied solutes bypassed the porous cup samplers. Bypass of the cups in the mowed-sod also was evident, however, less pronounced.

CONCLUSIONS

Two case studies showed the relative importance of matrix and preferential flow inferred from the appearance of breakthrough curves obtained with wick and gravity pan samplers. The effectiveness of the wick and gravity pan samplers depended on the type of soil. Wick pan samplers represented flow through the vadose zone much more accurately than gravity samplers when water and solutes were transported by matrix flow in the loamy soil. The bulk of transport in the clay loam soil followed macropore flow paths, in which case both wick and gravity pan samplers performed equally well. Porous cup samplers spuriously indicated preferential solute movement in the loamy soil, where pan samplers indicated a predominance of matrix flow. In the clay loam soil, the porous cup samplers failed to register flow through existing cracks, which was unmistakably observed with the pan samplers. Hence, under conditions similar to those presented here, use of porous cups in the upper part of the vadose zone is not recommended.

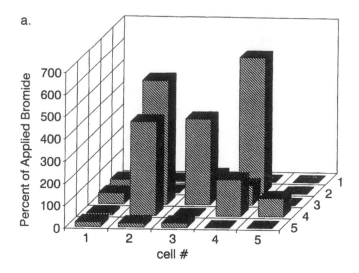

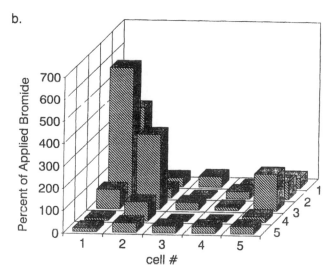

Figure 34.5.   a. Spatial distribution of mass of bromide intercepted by each cell during the period as a percentage of the amount applied for each cell of the gravity pan sampler for the moss-covered plot at the Orchard Site. b. Spatial distribution of mass of bromide intercepted by each cell during the period as a percentage of the amount applied for each cell of the wick pan sampler for the moss-covered plot at the Orchard Site.

ACKNOWLEDGMENTS

We thank the USDA CSRS Special Water Quality Grant Research Program for the funding of the research on which this chapter is based, and Ms. Betty Czarniecki for the editing.

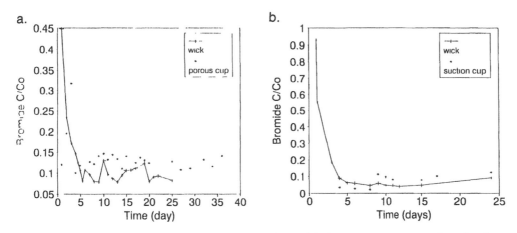

Figure 34.6. a. Comparison of wick pan samplers with the porous cup samplers for the mowed-grass plot at the Orchard Site. b. Comparison of wick pan samplers with the porous cup samplers for the moss-covered plot at the Orchard Site.

REFERENCES

Barbee, G.C., and K.W. Brown, "Comparison Between Suction and Free-Drainage Soil Solution," *Soil Sci.*, 141:149 (1986).

Boll, J., J.S. Selker, B.M. Nijssen, T.S. Steenhuis, J. Van Winkle, and E. Jolles, "Water Quality Sampling Under Preferential Flow Conditions," in *Proc. International Symposium on Lysimetry: Lysimeters for Evapotranspiration and Environmental Measurements*, R.G. Allen, T.A. Howell, W.O. Pruitt, I.A. Walter, and M.E. Jensen, Eds., American Society of Civil Engineers, New York, NY, 1991, p. 290.

Boll, J., T.S. Steenhuis, and J.S. Selker, "Fiberglass Wicks for Sampling Water and Solutes in the Vadose Zone," *Soil Sci. Soc. Am. J.*, 56:701 (1992).

Brinsfield, R.B., K.W. Staver, and W.L. Magette, "Impact of Tillage Practices on Pesticide Leaching in Coastal Plain Soils," Paper No. 87-2631, American Society of Agricultural Engineers, St. Joseph, MI, 1987, p. 20.

Brown, K.W., J.C. Thomas, and M.W. Holder, Development of a Capillary Wick Unsaturated Zone Water Sampler, Cooperative Agreement CR812316-01-0, U.S. Environmental Protection Agency, Environmental Monitoring Systems Laboratory, Las Vegas, NV, 1986.

Cole, D.W., "Alundum Tension Lysimeter," *Soil Sci.*, 85:293 (1958).

Daliparthy, J., S.J. Herbert, P.L.M. Veneman, G.V. Litchfield, and F.X. Mangan, "Monitoring Nitrate Leaching in Flood Plain Soils Under Alfalfa-Corn Rotation," in *Ground Water Management Book 16 of the Series*, National Ground Water Association, Columbus, OH, 1993, p. 901. Proc. Focus Conference on Eastern Regional Groundwater Issues, Burlington, VT, September 27-29, 1993.

Jamison, J.M. and R.H. Fox, "Estimation of Zero-Tension Pan Lysimeter Collection Efficiency," *Soil Sci.*, 154:85 (1992).

Jordan, C.F., "A Simple Tension Free Lysimeter," *Soil Sci.*, 105:81 (1968).

Knutson, J.H., and J.S. Selker, "Unsaturated Hydraulic Conductivities of Fiberglass Wicks in Designing Capillary Wick Pore-Water Samplers," *Soil Sci. Soc. Am. J.*, 1994 (in press).

Kung, K.-J.S., "Ground Truth About Water Flow Pattern in a Sandy Soil and Its Influence on Solute Sampling and Transport Modeling," in *International Conference and Workshop Proc.: Validation of Flow and Transport Models for the Unsaturated Zone*, P.J. Wierenga and D. Bachelet, Eds., Ruidoso, NM, May 23–26, 1998, p. 224. (Also Research Report 88-SS-04, Department of Agronomy and Horticulture, New Mexico State University, Las Cruces, NM, 1988.)

Parizek, R.R., and B.E. Lane, "Soil-Water Sampling Using Pan and Deep Pressure Vacuum Lysimeters," *J. Hydrology*, 11:1 (1970).

Rimmer, A., T.S. Steenhuis, and G. Albrecht, "Wick Samplers: An Evaluation of Solute Travel Times," 1994 (submitted).

Shaffer, K.A., D.D. Fritton, and D.E. Baker, "Drainage Water Sampling in a Wet Dual Pore System," *J. Env. Quality*, 8:241 (1979).

Steenhuis, T.S., J.R. Hagerman, N.B. Pickering, and W.F. Ritter, "Flow Path of Pesticides in the Delaware and Maryland Portion of the Chesapeake Bay Region," in *Proc. Ground Water Issues and Solutions in the Potomac River Basin/Chesapeake Bay Region Conference*, Washington, DC, March 14–16, 1989, p. 397.

Steenhuis, T.S., J.-Y. Parlange, and S.A. Aburime, "Preferential Flow in Structured and Sandy Soils: Consequences for Modeling and Monitoring," Chapter 5 in *Handbook of Vadose Zone Characterization & Monitoring*, L. Wilson, L. Everett, and S. Cullen, Eds., [Chelsea, MI: Lewis Publishers, 1994 (this book)].

Tanner, C.B., S.J. Bourget, and W.E. Holmes, "Moisture Tension Plates Constructed from Alundum Filter Discs," *Soil Sci. Soc. Am. Proc.*, 18:222 (1954).

35

Monitoring Perched Ground Water in Arid Lands

K. D. Schmidt

INTRODUCTION

Perched ground water has turned out to be one of the more elusive phenomena to hydrogeologists evaluating alluvial basins of the southwestern U.S. This chapter covers the following aspects, including:

- the historical and present hydrogeologic uses of the term "perched water"
- methods of testing for perched water
- well monitoring and sampling for perched water
- case studies, based on the results of site-specific monitoring programs designed to evaluate perched water
- summary and conclusions.

HISTORICAL USES OF THE TERM PERCHED WATER

Perched water is generally referenced more in the literature that concerns arid areas, where water levels are often deep (50 to 500 or more feet), than in humid areas, where water levels are often relatively shallow (a few tens of feet or less). Perched water in Arizona and California has been mentioned by numerous investigators, including Haskell and Bianchi (1967), Wilson, Herbert, and Ramsey (1975), and Wilson and Schmidt (1978). When ground water was encountered at relatively shallow depth (i.e., the upper 20 to 30 feet or so), and if a low permeability layer (i.e., a clay layer or hardpan) was found beneath it, it was commonly assumed that the shallow ground water was perched. The water level in a nearby supply well would often range from several hundred to more than 500 feet deep, considerably deeper than for the "perched" water. This caused many investigators to speculate that perched water must be present.

Another common observation in arid alluvial basins was the phenomenon of cascading water in deep supply wells. In this instance, water entering the well casing above the water level falls or cascades down the well inside the casing. This cascading of the water down the well can often be heard from near the top of the well at the land surface. This is a fairly common situation in arid alluvial basins where there has been significant ground-water overdraft, and water levels have fallen several hundred feet. In such situations, perforations in the well casing that were below the water level several decades ago, may now be located substantially above the water level as a result of the declining water levels. This falling or cascading water has also been interpreted by many investigators as evidence of perched water.

Farmers in many arid areas where surface water has been used for several decades as the sole or primary source of irrigation water, have also frequently encountered shallow ground water. In the absence of ground-water pumping and conjunctive use of surface water and ground-water supplies, shallow ground-water levels have often risen in such areas, due to a hydrologic imbalance between the applied water and the crop consumptive use. In most irrigated areas where shallow ground water was encountered (i.e., eastern part of the Westlands Water District in the western San Joaquin Valley), a perched layer was assumed to be present.

An exception to the previously discussed cases, was the work of L.G. Wilson (1971) at The University of Arizona Water Resources Research Center in Tucson. In Wilson's studies, the existence of a true perched zone was documented adjacent to the Santa Cruz River following significant streamflow through the extensive use of neutron probe measurements. Specially installed access wells were used to obtain these measurements, which are indicative of moisture contents. The presence of an intervening unsaturated zone was documented by these measurements. Even though this approach was successfully used in 1967, it has not been widely used at many sites since that time in determining the presence of perched water.

PRESENT-DAY HYDROGEOLOGIC TERMINOLOGY

In this chapter, the term "confining bed" is used for low permeability layers below the water table. The occurrence of perched water can be verified by two different approaches. The first is when unsaturated deposits can be documented below the perched water and above the regional aquifer. This is normally done at the time of drilling by utilizing methods that do not require drilling fluids (i.e., the hollow-stem auger, cable-tool, and casing hammer methods). The second approach is based on water-level elevations in wells that tap strata in discrete depth intervals. If the water level in the strata below the perching layer is below the base of this layer, it is inferred that an unsaturated zone is present below the layer. Thus the present hydrogeologic definition of perched water is for a saturated zone to be present, underlain by an unsaturated zone above the regional aquifer. The perching layer is the low permeability layer above which the perched water is present.

Some historical evidence in the western San Joaquin Valley indicates that perched water may have developed above a well-known, regional confining clay layer (the Corcoran Clay or E-Clay), after decades of ground-water overdrafting. The Corcoran Clay has an average thickness of less than 100 feet, and water-level differences in strata above and below the clay were often as great as several hundred feet in the mid-1950s. Considering the thickness of the confining bed, this would require a vertical gradient much greater than unity. However, a problem in evaluating such head differences is that wells tapping strata above and below the confining layer were often perforated over hundreds of feet of alluvial deposits, and other clay layers or confining beds are present at many locations. In some cases, when the total thickness of all of these confining beds is considered, the vertical hydraulic gradient did not exceed unity. In order to appropriately use this second approach, good quality geologic logs and electric logs and specially designed observation wells or monitor wells are essential. This approach is more readily applied at point-source contamination sites (i.e., plumes), where substantial hydrogeologic data are available from numerous monitor wells that tap discrete intervals at different depths.

Experience in the past two decades (primarily at contamination sites) indicates that the more thoroughly an alluvial ground-water basin is examined, the more complicated it appears to be. For example, whereas much of the San Joaquin Valley was once envisioned as a two-layer system several decades ago, more recent investigators (Williamson, Prudic, and Swain, 1985, and Belitz, 1988) have used a layered approach, sometimes with five or more individual layers. At a specific site in developed alluvial basins, the depth to the water level in an individual water-producing stratum is commonly different than in other strata. This is because of the layered nature of the alluvium, and the resulting low vertical permeability compared to the horizontal permeability.

One of the most surprising findings from detailed evaluations of hydrogeologic conditions in alluvial basins of the southwest in recent years has been the discovery that perched water is much less common than previously believed. That is, once water is encountered at a particular depth, the underlying strata are often found to be saturated. In general, true perched water (according to the present hydrogeologic definition) has probably been found in less than 10% of the cases where it was previously thought to be present. One of the reasons for this is that deep percolation beneath irrigated land is relatively small compared to infiltration from streams and canals. Deep percolation from agricultural land usually ranges from about one-half foot to several feet per year, whereas infiltration from unlined canals or streams can range from ten to hundreds of feet per year. Near canals or streams, perched water may only be present for short periods of time, such as during and following surface flow. The creation of perched water beneath irrigated land would require a perching layer with a very low vertical permeability. Experience has shown that most clay layers in alluvial deposits of the southwest have a vertical permeability that is large enough to readily transmit several feet of water per year downward. Such a permeability is approximately 10^{-6} cm per second.

Although intensive, detailed monitoring at many contamination sites has documented the general absence of perched ground water, an interesting phenomenon has been observed at some sites in some deposits far below the "regional water table." This phenomenon has been observed mostly in continuous core samples that contain unsaturated deposits, sometimes 50 or more feet below the regional water table. These findings, although not well understood, have been corroborated by occasional observations of cable-tool drillers and others of "dry" strata at various depths in alluvial aquifers.

METHODS OF DRILLING

Some of the most commonly used drilling methods in alluvial basins of the southwest are: bucket augers, hollow-stem augers, cable-tools, casing hammer (air percussion casing advance), air rotary, direct (mud) rotary, and reverse rotary. The first five of these methods have considerable potential to allow the detection of perched water, whereas the last two have much less potential. This is because drilling fluid is used during drilling by the latter two methods, either mud or water, both of which can obscure the detection of perched water. Since the last two methods are now the most common ones used for drilling new domestic, irrigation, and public-supply wells in these alluvial basins, this means that perched water often won't be detected during the drilling of water-supply wells. Thus, most of the valid information on perched water will be derived from the drilling of monitor wells, observation wells, test holes, and test wells.

For the bucket auger, cable-tool, and casing hammer methods of drilling, water may occasionally need to be added to stabilize the hole above the water level, and this may complicate the detection of perched zones. However, once such a perching zone is encountered, the cable-tool and casing hammer methods have the advantage that this zone can be cased off before drilling deeper. These methods are therefore preferable to the auger and rotary methods for detecting perched ground water.

Although perched ground water usually cannot be directly observed when drilling by the mud rotary or reverse rotary methods, down-the-hole geophysical logs may indicate such perched zones. The neutron log and the sonic or velocity log can be useful in this regard. However, it is not presently feasible to use the neutron log in most situations in Arizona and California, because of regulatory constraints and a perceived environmental threat. The sonic log has been used, with varying degrees of success, to delineate both the top of the saturated zone and the overlying perched zone. If the salinity in the perched zone is high enough, it may be possible to detect the perched zone by sampling and testing of the drilling fluid itself. This is because ground water encountered during drilling comes in contact with the drilling fluid. Past experience has shown that high salinity ground water can noticeably increase the salinity of the drilling fluid.

The air-rotary method can often be used to depths of about 50 to 70 feet at many locations, and a perched zone sometimes can be detected. However, confirmation of this zone is difficult because deeper drilling usually requires the addition of drilling fluid.

In summary, methods where the use of drilling fluid is minimized and casing is driven during drilling (such as the cable-tool and casing hammer) appear to be preferable for clearly delineating and confirming the presence of perched water. Interestingly enough, neither of these methods has been used very much for this purpose in the southwest.

Costs of the various methods differ, and in general, drilling with a casing is more expensive than merely drilling a hole. If geophysical logging is conducted, the costs must be added to the drilling cost. Collecting better data often entails incurring more costs than for methods where little data are needed or collected. Costs for drilling range from as low as about ten dollars per foot of depth for test holes alone, to as much as hundreds of dollars per foot for deeper holes and cased wells.

TEST HOLES, MONITOR WELLS, AND CASCADING WELLS

Samples of perched water can be collected from test holes or wells, from specially designed monitor wells, and from existing supply wells with cascading water. For test holes, advantages are that the costs of well casing and annular seal (normally cement) may be avoided. Disadvantages are that this provides a one-time sample, routine monitoring is not possible, and obtaining non-turbid water samples can be difficult. This is because the hole will not normally remain open in unconsolidated deposits, and it is difficult to control fine sand migration with the water. Also, the commonly observed problem of "well trauma" (i.e., the disturbance of ground-water quality due to the drilling process itself) is unavoidable in test holes. Collecting samples by this approach is normally done as part of a reconnaissance exploration program, as opposed to routine ground-water monitoring.

Figure 35.1 is a schematic diagram showing the occurrence of cascading water from a perched zone. Water from a cascading supply well is normally sampled when the permanent pump is pulled, such as for routine maintenance. A bucket or other type of bailing device is lowered to just below where the water enters the well, and a water sample is then bailed from the well. This type of sampling may not be entirely valid for some chemical constituents, such as volatile halocarbons, because of volatilization losses during sampling. However, such sampling has been shown to be useful for characterizing concentrations of many inorganic chemical constituents in the ground water. A case history of the use of this approach is presented later in this chapter.

Monitor wells that tap perched layers are most useful if they are equipped with a well casing that is at least four inches in diameter. Four-inch diameter Schedule 40, five-inch diameter 160 psi, or six-inch diameter Schedule 80 PVC casing has proved to be highly effective in virtually all perched water zones in alluvial deposits that are less than about 400 feet deep. Such monitor wells can be readily developed, even if drilling mud is used, and representative water samples can be collected from them by a variety of methods. In order to obtain meaningful values of aquifer transmissivity and permeability for coarse-grained permeable alluvial deposits tapped by such

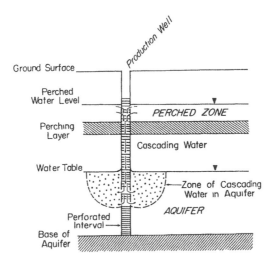

Figure 35.1. Schematic diagram for cascading water.

monitor wells, a pumping rate of at least 100 gpm is often necessary. In order to achieve this rate, an inside diameter of the casing that is large enough to accommodate a four-inch diameter pump is necessary.

A key factor is to drill a large enough hole (i.e., 10 to 14 inches in diameter) to allow the proper placement of gravel and the annular seal between the casing and the wall of the hole. The adequacy of auger methods for this purpose has been highly over-stated in the literature. Rather, normal well drilling methods for water-supply wells and experienced water-well drilling contractors are highly preferable. This is because such contractors are readily available at economic costs, and most are highly familiar with developing adequate amounts of sediment free water. Many auguring firms aren't aware of proper well development procedures, or do not have the equipment to properly develop monitor wells.

Many ground-water monitoring debacles in the southwest have resulted from inadequate well construction and development, inadequate water sample removal prior to sampling, and inadequate determination of aquifer parameters. Most of these problems stem from the use of monitor wells that are too small, and the overuse of augers in trying to construct monitor wells. This has frequently resulted in a wide discrepancy between the apparent extent of a plume based on data from such wells, and the results of remedial extraction projects, wherein data are obtained from extraction or supply wells.

WATER SAMPLING METHODS

Because in most cases perched water is present at a depth of less than about 100 feet, a variety of methods can be used to obtain water samples from test holes, test wells, monitor wells, and cascading wells that tap the perched water. Included in these methods are air-lifting, bailing, hand pumps, and submersible pumps.

Air-lifting is not normally used to retrieve water samples from perched zones. However, if an air rotary or casing hammer rig is being used for drilling, this method can be successfully used for sampling for analysis of many chemical constituents. Experience has shown that air-lifting provides adequate results for most inorganic chemical constituents, except for iron and hydrogen sulfide. In the case of iron, the added air can change the oxidation state and a loss of iron in the sampled water usually results. For gases such as hydrogen sulfide, the air-lifting acts as an air stripping process, and the hydrogen sulfide may no longer be present in the water sample. This means that air-lifting can be used successfully for monitoring situations that do not require measuring concentrations of constituents such as volatile trace organics, or gases such as radon.

Bailing can be grouped into several categories: (1) water from cascading wells, (2) large bailers that are used in the construction of water-supply wells, and (3) small bailers that are used in monitor wells to attempt to obtain "in-situ" samples. Although bailing is probably necessary for cascading wells in many cases, care needs to be taken for constituents such as volatile constituents, in order to minimize volatilization losses. The use of bailers at adequate rates of bailing (i.e., up to 100 gpm, or as much as the perched zone can produce on a sustained basis) is useful for many chemical constituents. Experience in the Salt River Valley of Arizona with ground-water levels less than about 150 feet in depth has shown that even volatile halocarbons can be measured on a reconnaissance level by such bailing. Although losses of volatile halocarbons during sampling by this approach have approached 50%

(Schmidt, 1986), this method is still feasible for some purposes. The use of small well bailers to collect very small "in-situ" samples at specific points in the aquifer has been demonstrated to be of limited value in heterogenous alluvial deposits, particularly when large volumes of contaminated ground water are being monitored. That is, larger samples are necessary to represent volumes of ground water that are often in the range of thousands of acre-feet. One application where bailing is useful is where there is a free product (such as gasoline) floating on the water table.

Pumped samples are superior in most cases, for evaluating dissolved chemical constituents in ground water, including volatile halocarbons and hydrogen sulfide. For depths to water of less than about 50 feet, hand pumps can be used. Submersible pumps used in water-supply wells are highly efficient for deeper applications. Pumping can be conducted to minimize turbulence, and to avoid significant losses of volatile constituents prior to chemical analyses. This has been demonstrated at hundreds of contamination sites. At a number of these sites, the results of initial ground-water monitoring (utilizing properly sized and sampled monitor wells) during the remedial investigation phase are consistent with results obtained during the subsequent remediation phase. Permanent pumps are often the most cost-effective to use, because they minimize manpower requirements, equipment cleaning, and potential cross contamination problems, compared to other devices.

CASE HISTORIES

Two case histories are described, both of which concern the effect of irrigation on ground water in alluvial basins. The first of these is in the central San Joaquin Valley of California, in the well-known agricultural drainage problem area. In this case, a supposed shallow "perched zone" and "perching layer" were evaluated through installation of monitor wells and observation wells and subsequent aquifer testing. The second case is located in the southeastern part of the Salt River Valley in Arizona. In this case, numerous cascading water-supply wells were sampled, and the quality of the supposed "perched water" and the source of this water were both investigated. In both cases, the impact of irrigation on ground-water quality was well documented.

Central San Joaquin Valley

Figure 35.2 shows the locations of three sites on the north part of the Panoche alluvial fan, about 40 miles west of Fresno. The sites are west of the San Joaquin River near Firebaugh. At each site, a hand auger was used to auger down to the shallowest ground water (in fine-grained deposits), and thence to an underlying "perching layer" (normally clay). At each site, three small-diameter observation wells were installed, perforated within the "perched zone." The observation wells were about 10 feet in average depth. Three-quarter inch diameter PVC casing was used, perforated opposite the lower 4 to 5 feet. A rotary drilling rig was used to install three monitor wells at each site, to tap the shallowest underlying coarse-grained, permeable strata (in the depth interval from about 30 to 50 feet). In general, from 5 to 10 feet of coarse-grained, permeable deposits (sand or gravel) were encountered, and the remainder of the deposits were fine-grained (clay or silt). The holes for the monitor wells tapping this deeper zone were drilled by the air rotary method, and no unsaturated zone was observed below the "perching layer." These

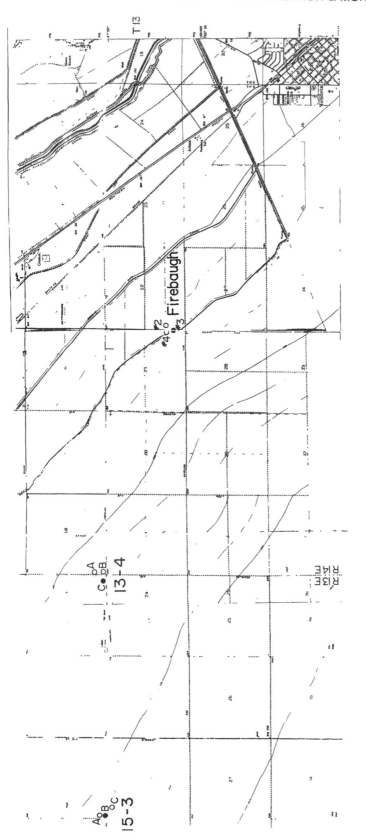

Figure 35.2. Locations of monitored sites on the Panoche Fan.

wells ranged from about 40 to 60 feet in depth and were equipped with five-inch diameter 160 psi PVC casing. The over-lying fine-grained deposits were sealed off in these wells. Precise elevations of the water-level measuring points for the observation wells and the monitor wells were determined. Water levels were then measured and water-level elevations determined. Water-level elevations under nonpumping conditions in strata above the "perching layer" were approximately the same as those in strata below the layer. This and other information indicated that horizontal flow of the shallow ground water was predominant at the site.

Aquifer tests averaging several days in duration were subsequently conducted at each of the three sites. One deep monitor well at each of the three sites was used as the pumped well, and the other two deep monitor wells and the shallow observation wells at each site were used as observation wells. At one site, several other more distant observation wells were present. Past experience with aquifer tests in alluvial deposits in Arizona and California has shown that when a well tapping a confined aquifer is pumped, water levels normally rise in the overlying unconfined aquifer during pumping, and fall during the recovery period after pumping has stopped. This has been attributed to a temporary compaction of clay strata comprising the confining bed, and a subsequent settling of the overlying unconsolidated deposits. However, in this case, water levels in the shallow observation wells declined at all three sites during pumping, and rose during recovery. Figure 35.3 shows the water-level drawdown in some observation wells for the test at the easterly of these sites. Drawdowns of more than 3 feet were observed in three observation wells within 15 feet of the pumped well. Drawdowns of almost 1 foot were observed in two observation wells from 180 to 250 feet from the pumped well.

The water-level declines and the apparent transmissivities determined from the observation well and monitor well measurements indicated that there was excellent

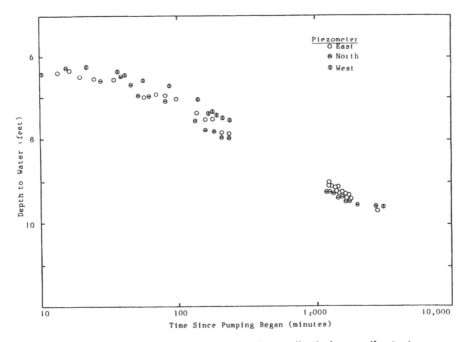

Figure 35.3. Water-level drawdown in observation wells during aquifer test.

hydraulic communication between the supposed "perched zone" and the ground water in the underlying permeable deposits. These tests proved that the clay layer was not a perching layer and was also not an effective confining bed. Regional studies in the area by the U.S. Geological Survey (Belitz, 1988) indicated that at specific sites, no unsaturated zone was present beneath the shallowest saturated zone. These results, based on site-specific monitoring and aquifer testing, showed that perched water was not present in most of the area. Instead, the shallow ground water is coincident with the top of a regional aquifer that is several hundred feet thick in the area. The shallow ground water is not used in most of the area because of high salinity and boron contents.

Southeastern Salt River Valley

Schmidt (1983) reported on results of an extensive program of sampling cascading water from wells in the southeastern part of the Salt River Valley. The study area for this evaluation was primarily south of the Salt River and east of Kyrene Road. Records on cascading wells that had been previously inventoried in this area were obtained from the U.S. Geological Survey and U.S. Bureau of Reclamation. During November 1981-February 1982, a field program was undertaken to measure the depth to the cascading water and the water level in the wells. Water samples were collected from cascading wells whenever possible during the early 1980s. This work was done as part of the Maricopa Association of Governments (MAG) 208 Water Quality Management Program, in cooperation with the Salt River Project (SRP).

Water-Level Measurements

Depth to the top of the cascading water was measured in 17 of the wells shown in Figure 35.4. It was not possible to measure the other wells shown. Depth to the top of the cascading water in most wells was between 120 and 220 feet in 1981 to 1982. The distance between the top of the cascading water and the water level in the wells was usually between 15 and 40 feet.

For decades, cascading water in wells in the Salt River Valley and elsewhere in Arizona was believed to be indicative of perched water. At that time, a one or two aquifer system was envisioned. However, detailed ground-water monitoring at numerous contamination sites in the Salt River Valley in recent years has confirmed the presence of three major geologic units. These are now termed the Upper Alluvial Unit (UAU), Middle Alluvial Unit (MAU), and Lower Alluvial Unit (LAU), following the terminology of Laney and Hahn (1986). All of these units contain saturated deposits in the study area. No significant unsaturated zone below the shallowest saturated strata has been found, based on monitor well drilling. Many of the cascading wells are perforated over hundreds of feet and are composite, in that they tap more than one geologic unit. The MAU is a major confining bed in much of the southeastern Salt River Valley. Depth to water measured in recent years in monitor wells tapping solely the UAU agree well with the depths to cascading water that were measured in 1981 to 1982. This indicates that the cascading water was not from perched water, but rather ground water in the regional aquifer in the UAU, which had a higher water level than in the composite well. Because of this, water from the UAU entered the well, fell down the inside of the well casing, and then exited in one or more of the underlying units.

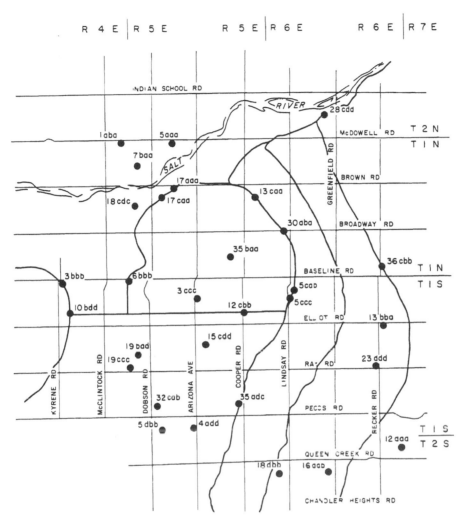

Figure 35.4. Location of cascading wells measured in the southeastern Salt River Valley.

The Ground Water Planning Division of the SRP has conducted television surveys down some of these cascading wells (Small, 1982). Most of the cascading wells were near major canals, laterals, or the Salt River, where high-rate infiltration (tens or hundreds of feet of water per year) could occur. In some cases, cascading water was observed entering SRP wells at several different depths. In addition, television surveys indicated that perched water was falling down the well in the annular space on the outside of the casing. Such annular spaces are common for cable-tool drilled wells, where the drive shoe was slightly larger than the outside diameter of the casing. The consolidated nature of some of these coarse-grained deposits enabled the annular space in at least part of the well to be maintained for decades. At some depths, (i.e., where a clay stratum is present) no such annular space around the well casing is present, and cascading water falling on the outside of the casing is forced to enter the well at that point, if openings are present. In some cases, the composite well is not open or perforated opposite all of the UAU. Because of this, the static water level in the UAU at the well may be higher than the top of the cascading water in a specific well.

Water Sampling

Between March 1980 and January 1983, water samples were bailed from nine cascading wells, as part of the MAG 208 program. The major inorganic chemical constituents were analyzed for all of these samples, and the stable isotopes deuterium and oxygen-18 ratios were determined in some samples. Stable isotope ratios are useful in differentiating between the two major sources of recharge in the area: canal seepage and deep percolation of irrigation return flow. SRP personnel also collected samples of cascading water from five other SRP wells between June 1981 and March 1982 (Small, 1982). These samples were analyzed by the SRP laboratory for major inorganic chemical constituents. Thirteen samples of canal water were collected by Salt River Project personnel and made available for stable isotope analyses. These samples were also analyzed for major inorganic chemical constituents. These samples were primarily collected from the Tempe Canal, Consolidated Canal, and South Canal during March-July, 1983. Locations of sampled wells and nearby sites where canal water was sampled by the SRP are show in Figure 35.5.

Based on typical total dissolved solids (TDS) and nitrate contents, the samples of cascading water were grouped into three categories:

Type	No. of Wells	Nitrate (mg/L)	Total Dissolved Solids (mg/L)
A	4	<10	400–500
B	5	15–20	830–940
C	6	40–52	1,010–1,220

On a long-term basis, the TDS content of canal water in the study area averages between about 400 and 500 mg/L. Smith, et al. (1982) showed that the inorganic chemical composition of cascading water of the Type A was identical to that of canal water. Thus cascading water of Type A is derived from recharge due to canal seepage.

Table 35.1 shows three examples of the chemical composition of cascading water of Type A. For comparison, an analysis is provided for a composite of samples of canal water from the sampling point nearest to the cascading wells. All of the wells with this type of cascading water were located adjacent to major canals. The inorganic chemical analyses of cascading water and canal water were virtually identical, indicating little change in ground-water quality during percolation. The cascading waters were of the sodium bicarbonate chloride type, and chloride contents ranged from about 100 to 150 mg/L. Nitrate contents were 6 mg/L or less.

Evidence indicated that Type C cascading water, characterized by relatively high salinity and nitrate content, is derived primarily from deep percolation of irrigation return flow. Table 35.2 shows four examples of the chemical composition of cascading water of Type C. All of the wells with this type of cascading water were located at some distance from major canals. These waters were of the sodium chloride or sodium chloride bicarbonate type, and chloride contents ranged from about 320 to 440 mg/L. Nitrate contents ranged from 40 to 50 mg/L. The salinity of irrigation return flow is higher than that of the applied water, most of which is from canals in

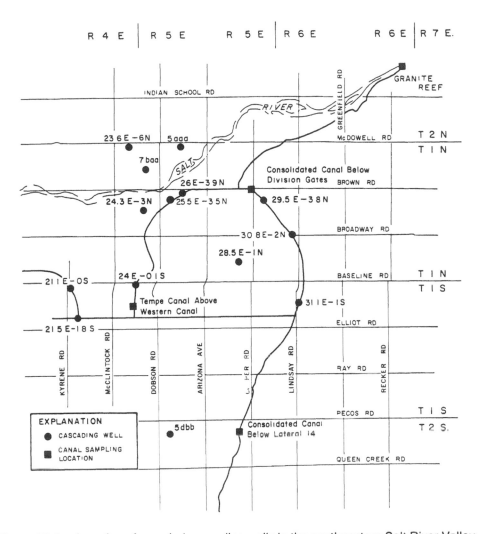

Figure 35.5. Location of sampled cascading wells in the southeastern Salt River Valley.

this area, due to crop evapotranspiration. The high nitrate contents are likely due to the use of nitrogen fertilizers and concentration by evapotranspiration. Comparing the TDS contents in canal water and cascading water for Type C, a concentration factor of about 2.4 is apparent. This would be expected with an irrigation efficiency of about 60%, which is in agreement with previous estimates made by the water-budget approach for the area (Schmidt, 1983).

Water of Type B is of intermediate composition between that of the other two types. Table 35.3 shows five examples of the chemical composition of cascading water of Type B. All of the wells with this type of cascading water were also located along major canals. These waters are of the sodium chloride or sodium bicarbonate type, and chloride contents ranged from about 250 to 340 mg/L. Nitrate contents ranged from 16 to 29 mg/L. This water appears to be derived from a mixture of canal seepage and deep percolation of irrigation return flow.

Table 35.1. Chemical Composition of Canal Water and Cascading Water Derived from Canal Seepage

Constituent (mg/L)	25.5E–3.5N	26E–3.9N	30.8E–2N	Consolidated Canal
Calcium	42	47	44	59
Magnesium	17	16	17	5
Sodium	120	103	119	115
Potassium	4	3	2	4
Carbonate	0	0	0	0
Bicarbonate	178	196	213	159
Sulfate	65	70	53	51
Chloride	154	138	147	173
Nitrate	5	0	6	0
Boron	0.1	0.1	0.2	–
pH	7.3	8.4	8.3	8.1
Electrical Conductivity (micromhos/cm @ 25°C)	880	890	915	920
Total Dissolved Solids	497	477	498	485
Depth (feet)	145	109	219	–
Date	11/5/81	9/28/81	6/30/81	Composite
Laboratory	BC Labs	SRP	SRP	SRP

Table 35.2. Chemical Composition of Cascading Water Derived from Irrigation Return Flow

Constituent (mg/L)	(A–1–5) 5aaa	(A–1–5) 7baa	24.3E–3N	28.5E–1N
Calcium	124	68	87	115
Magnesium	42	35	34	46
Sodium	301	315	247	280
Potassium	8	8	6	8
Carbonate	0	0	12	0
Bicarbonate	407	434	347	460
Sulfate	144	133	91	130
Chloride	440	341	319	394
Nitrate	46	50	41	40
Boron	–	0.9	0.5	0.5
pH	7.4	7.9	8.2	7.5
Electrical Conductivity (micromhos/cm @ 25°C)	1,820	1,250	1,780	2,200
Total Dissolved Solids	1,187	1,047	1,011	1,217
Depth (feet)	140	102	142	194
Date	3/7/80	11/23/81	3/12/82	1/18/83
Laboratory	BC Labs	BC Labs	SRP	BC Labs

Table 35.3. Chemical Composition of Cascading Water of the Intermediate Type

Constituent (mg/L)	21.1E–OS	21.5E–1.8S	24E–0.1S	29.5E–3.8N	31.1E–1S
Calcium	28	64	70	295	84
Magnesium	37	24	32	37	27
Sodium	248	249	255	237	190
Potassium	6	4	7	7	10
Carbonate	12	0	0	0	0
Bicarbonate	158	378	445	188	272
Sulfate	153	126	100	108	100
Chloride	297	249	266	343	275
Nitrate	16	29	21	23	18
Boron	0.5	0.5	0.8	0.1	–
pH	8.5	8.1	8.0	8.4	7.5
Electrical Conductivity (micromhos/cm @ 25°C)	1,610	1,610	1,725	1,630	1,400
Total Dissolved Solids	880	938	887	882	830
Depth (feet)	132	144	159	280	216
Date	12/21/81	2/9/82	11/23/81	12/15/81	1/7/82
Laboratory	SRP	SRP	BC Labs	SRP	BC Labs

Stable isotope ratios were determined on 13 samples of canal water and 2 samples of cascading water. Values are referenced to Standard Mean Ocean Water (SMOW). Most of the samples of canal water are from the Tempe Canal above the Western Canal, the Consolidated Canal below Lateral 14, and the South Canal at Granite Reef Diversion Dam. Results indicate that $\delta^{18}O$ values of canal water were usually less than -9.1 0/00. Other evidence, based on sampling of monitor wells in irrigated areas (Schmidt, 1983), indicates that $\delta^{18}O$ values for irrigation return flow in this area are normally –8.6 0/00 or greater. Stable isotopes of oxygen and hydrogen would not be expected to be significantly fractionated during seepage from canals, because evaporation losses are relatively insignificant in relation to the amount of seepage. However, in the case of irrigation return flow, evapotranspiration consumes the majority of the applied water, which should result in a fractionation which increases the relative contents of the heavier isotopes. The sample of cascading water from Well 6.5E-16.4N had a $\delta^{18}O$ value of –9.3 0/00, indicative of canal seepage. This well is north of the study area in Scottsdale, and the inorganic quality of the cascading water was similar to that of canal seepage. The sample of cascading water from Well 28.5E-1N had a $\delta^{18}O$ value of -8.6 0/00, indicative of irrigation return flow. The results thus confirm the interpretations previously made based on the inorganic chemical composition of the cascading water. Deuterium analyses did not allow differentiating canal seepage from irrigation return flow as well as did oxygen-18 values, although the same general trends were present as for the oxygen-18 determinations.

CONCLUSIONS

The results of site-specific monitoring in alluvial basins in Arizona and California in recent years have indicated that perched ground water is not as common as was once believed. The existence of perched water has commonly been inferred based on shallow ground water underlain by a clay or other low permeability layer, deeper water levels in adjacent or nearby supply wells, or cascading or falling water in supply wells. Modern-day hydrogeologic criteria for the existence of perched water mandate that an unsaturated zone must be present beneath the perched zone and above the regional aquifer.

When perched water is present, it often cannot be detected by the most common methods now being used to drill supply wells in the southwest (rotary methods). However, the cable-tool and casing hammer methods, whereby casing is driven during drilling, often provide the means to readily detect such zones. Down-the-hole geophysical logs such as the velocity or sonic log and the neutron log offer a means to detect perched zones in holes drilled by many methods.

Water samples can be collected from monitor wells, test holes, and test wells tapping perched zones by a variety of methods. Properly drilled and developed monitor wells with at least a four inch diameter PVC casing are preferred, and normal submersible pumps have proved highly effective in characterizing the quality of water in perched zones less than about 400 feet deep. Steel casing may be advisable for greater depths. Pumped samples are suitable for all constituents, including volatile ones, except for separate phases floating on the water (such as free-product gasoline or oil).

REFERENCES

Belitz, K., Character and Evolution of the Ground-Water Flow Systems in the Central Part of the Western San Joaquin Valley, California, U.S. Geological Survey, Open-File Report 85–573, 1988.

Haskell, E.E., and W.C. Bianchi, "The Hydrologic and Geologic Aspects of a Perching Layer-San Joaquin Valley, Western Fresno County, California," *Ground Water*, 5:4,12 (1967).

Laney, R.L., and M.E. Hahn, Hydrogeology of the Eastern Part of the Salt River Valley Area, Maricopa and Pinal Counties, Arizona, U.S. Geological Survey Water-Resources Investigations Report 86–4147, 1986.

Schmidt, K.D., "Management of Ground Water Quality Beneath Irrigated Arid Lands," *Proceedings of the Western Regional Conference on Groundwater Management*, San Diego, CA, October, 1983, National Water Well Association, 1983, p. 74.

Schmidt, K.D., Results of the Final Groundwater Quality Monitoring Phase, November 1981-September 1983, prepared for Maricopa Association of Governments 208 Water Quality Management Program, Phoenix, AZ, 1983.

Schmidt, K.D., "Monitor Well Drilling and Sampling in Alluvial Basins in Arid Lands," *Proceedings of the FOCUS Conference on Southwestern Groundwater Issues*, National Water Well Association, October 20–22, Tempe, AZ, 1986.

Small, G.G., "Groundwater Quality Impacts of Cascading Water in the Salt River Project Area," in *Proceedings of the Deep Percolation Symposium*, October 26, 1982, Scottsdale, AZ, Arizona Dept. of Water Resources Report No. 4, 1982.

Smith, S.A., G.G. Small, T.S. Phillips, and M. Clester, Water Quality in the Salt River Project, Salt River Project, Groundwater Planning Section, Phoenix, AZ, 1982.

Williamson, A.K., D.E. Prudic, and L.A. Swain, Ground Water Flow in the Central Valley, California, U.S. Geological Survey Open-File Report 85–345, 1985.

Wilson, L.G., *Investigations on the Subsurface Disposal of Waste Effluents at Inland Sites*, Water Resources Research Center, The University of Arizona, 1971.

Wilson, L.G., R.A. Herbert, and R.C. Ramsey, "Transformations in Quality of Recharging Effluent in the Santa Cruz River," in *Proceedings of the 1975 Meetings of the Arizona Section AWRA*, 5:169 (1975).

Wilson, L.G., and K.D. Schmidt, "Monitoring Perched Ground Water in the Vadose Zone," in *Proceedings of the Symposium on Establishment of Water Quality Monitoring Programs*, American Water Resources Association, San Francisco, CA, June 12–14, 1978, p. 134.

36

Emerging Technologies for Detecting and Measuring Contaminants in the Vadose Zone*

Eric N. Koglin, Edward J. Poziomek, and Mark L. Kram

INTRODUCTION

This chapter focuses on the innovative and emerging technologies that can be used to detect and measure contaminants in the vadose zone. It is meant to serve as an adjunct to discussions of the state-of-the-art technologies for monitoring and measuring the *chemical and physical characteristics of the vadose zone*. These technologies often yield superior results with respect to time, cost, and accuracy. Of course, in many cases the technologies can be used in other environmental and nonenvironmental applications (for example, ground-water monitoring and process analytical chemistry). Our objective is to provide the interested user with a brief description of the basic operating principles of the different technologies, where the technologies stand with regard to commercial availability, the intended applications, and a list of contaminants detected. For more in-depth technical and application information, the reader is referred to the list of references at the end of the chapter.

What is driving innovation in environmental monitoring technology development and commercialization? The most significant driving force in the emergence of innovative technology is the ever-increasing cost associated with environmental compliance. Large portions of the costs of compliance are due to monitoring requirements specified in the various pieces of environmental regulation. Most, and in many cases all, of the monitoring data are generated through the use of fixed analytical laboratories. The "tried and true" methods have been able to stand the test of time in satisfying regulators and the courts. However, it is quickly becoming

*Portions of this chapter were presented at the Conference on Groundwater Contamination, The Danish Academy of Technical Sciences, Denmark, March 1994.

apparent to those responsible for paying for environmental monitoring and cleanup that more cost-effective tools need to be employed. For example, as much as 80% of the costs of site characterization during the cleanup of a Superfund site can be attributed to laboratory analyses, many of which render "nondetectable" results. In addition, sample manipulation during collection, transport, and storage can affect sample integrity and thus analyte concentration. A typical approach to characterizing a site often requires iterative mobilization periods to collect chemical and physical data. This iterative process may not be completed for as much as three years. Without development and commercialization of deployable sensors and field screening/analytical techniques, it is going to become more difficult, and therefore more costly, to comply with existing and new pieces of environmental regulation. Therefore, the user community needs more cost-effective alternatives.

The emphasis in the present review is on deployable sensors and field screening/ analytical techniques. An appreciation of the state of the technology can be gained by scanning the proceedings of a recent symposium on field screening methods for hazardous wastes and toxic chemicals (Air & Waste Management Association, 1993) and an overview by Koglin and Poziomek (1993). A discussion of trends in field methods development and use was given by Poziomek (1993). An alternative to field screening can involve the use of rapid laboratory screening. The U.S. Environmental Protection Agency (EPA) recently initiated a quick turnaround method (QTM) project. This project has been developing through Special Analytical Services (SAS) activities under the sponsorship of the EPA Sample Management Office. Meierer et al. (1993) reported on recent progress. In the QTM, program laboratories are required to process, analyze, and report analytical results from a specified number of samples per day. Up to five analytical fractions may be required; i.e., volatile organics, phenolic compounds, pesticides, polynuclear aromatic hydrocarbons, and PCBs as Aroclors. For a maximum of three different fractions, summary data are required to be delivered through electronic means within 48 hours. The QTM project has important implications for the future in cost-savings using laboratory analysis. However, use of deployable sensors and field methods remain attractive to the user community, and will continue to gain attention.

Another driving force for innovation can be attributed to the Department of Energy (DOE) and the Department of Defense (DOD). With the passage of the Federal Facilities Compliance Act of 1992, federal agencies are required to comply with the myriad of environmental legislation. The estimated costs associated with remediating DOD and DOE facilities are in excess of $200 billion dollars. DOE alone has over 4,000 contaminated sites covering tens of thousands of acres. The sites are contaminated with organics, inorganics, metals, and radionuclides. The contaminants are in soils, ground water, surface water, and structures. The central challenge to DOE is to develop new technologies that can be used to clean up its sites faster, more effectively, and cheaper. In addition, DOE needs to develop better monitoring tools for managing its on-going waste management operations (U.S. DOE, 1993a). DOD is faced with similar problems. Both agencies are investing heavily in innovative environmental technologies by leveraging the capabilities of the federal laboratory system and the private and academic sectors.

This chapter focuses on the vadose zone; however, the technologies described should be adaptable to other applications as well.

CHEMICAL DETECTION TECHNOLOGIES

Chemical Sensors

One of the most exciting areas of environmental monitoring technology research and development is in chemical sensors. A chemical sensor has been defined by Janata and Bezegh (1988) as a ". . .transducer which provides direct information about the chemical composition of its environment. It consists of a physical layer and a chemically selective layer" (Figure 36.1). Chemical sensor development is occurring in laboratories across all sectors, public and private. Much of that activity is focused on developing sensors that can be used for continuous, in-situ monitoring or for remote sensing. Biennial reviews on chemical sensors based on a search of the Chemical Abstracts database have been published by Janata and Bezegh (1988) and Janata (1990 and 1992). These provide an appreciation of the recent progress that has been made in chemical sensor research and development.

A recent series of reports by the U.S. Department of Energy (U.S. DOE, 1993b; 1993c; 1993d; and 1993e) provides a snapshot of the state of the art in chemical and radiochemical sensors. The intent of these reports was to canvass the academic and industrial groups developing, demonstrating, and commercializing chemical sensors, particularly those sensors important to DOE in meeting its environmental cleanup responsibilities. These reports also identify the gaps in technology where DOE may choose to focus some of its resources to accelerate development.

An important role of chemical sensors is to provide real-time monitoring and rapid screening capabilities necessary to support remediation and hazard assessment. The development of chemical sensors represents a significant departure from the classical laboratory analytical methods, in that the analytical tool can be taken to the sample. Sophisticated fixed laboratory analysis will continue to be necessary to provide confirmation, identification, and characterization of unknown and unquantified contaminants.

Chemical sensors can be broadly classed into five categories:

- mass sensors
- optical sensors
- electrochemical sensors
- radiochemical sensors
- thermal sensors

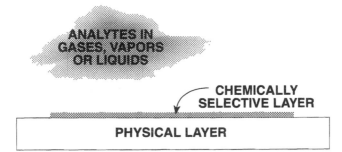

Figure 36.1. A schematic illustrating a chemical sensor. The chemical layer provides selectivity. The physical layer serves to sense changes in the chemically selective layer as the result of interaction/reaction with analytes.

Most of the interest in sensors for environmental applications lies in the first four categories. Electrochemical sensors, by and large, have been associated with the detection of contaminants in water or liquids (for example, ion selective electrodes). However, there are a variety of electrochemical sensors for gases and vapors such as CO_2, NO_x, CO, H_2, Cl_2, NH_3, chlorinated hydrocarbons and combustible organic vapors (Janata, 1992). The development of real-time radiochemical sensors is important to characterize sites contaminated with mixed and radioactive wastes. The demand for these types of sensors rests principally with the Departments of Energy and Defense; however, there will be applications beyond their needs alone. Thermal sensors represent the smallest group among chemical sensors (Janata, 1992). These sensors will undoubtedly receive more attention in the future for environmental applications. The discussion which follows in this section deals with mass, optical, electrochemical, and radiochemical sensors.

Mass Sensors

Surface acoustic wave (SAW) probes and bulk acoustic wave [quartz crystal microbalance (QCM)] sensors represent a class of chemical microsensors which measure changes in mass when analytes sorb and/or react with the surface of the device or a coating on the device. Referring to Figure 36.1, the chemical layer would be a coating for specific interaction or reaction with the analyte of interest. The physical layer in this case would be either a SAW or a QCM device. Such mass measurement devices are conceptually simple, rugged, and low cost. Many investigators have been exploring the potential applications of mass sensors for environmental purposes. QCM sorption detectors have been described for three decades (King, 1964). Most of the research and applications to date have focused on the detection of vapors (Ward and Buttry, 1990; Katritzky and Offerman, 1989; Mierzwinski and Witkiewicz, 1989; Ballantine and Wohltjen, 1989; and Guilbault and Jordan, 1988), which is an attractive feature for monitoring in the vadose zone. The use of mass sensors has much appeal for in situ applications, as well as for integration with one of the emerging push technologies (e.g., cone penetrometry and Geoprobe®) for real-time monitoring in the vadose zone.

Mass microsensors have the potential for achieving sensitivities at the parts-per-billion (ppb) level. SAW and QCM devices are conceptually similar. A chemically selective coating (which could be chemical or biochemical in nature) is coupled to a physical probe (SAW or QCM) that acts to sense mass changes as the result of analyte sorption and/or reaction. SAW microsensors possess advantages over the QCM ones, including higher sensitivity (Ballantine and Wohltjen, 1989). SAW devices can be fabricated in several configurations (Grate and Klusty, 1991).

An excellent case study in development of a mass sensor system as opposed to an individual sensor has been published by Grate et al. (1993c). A review of this paper should be useful to practitioners designing QCM or SAW systems for various applications, including vadose zone monitoring. The sensor system was developed for the detection of highly toxic organophosphorus and organosulfur vapors at trace concentrations. The features of the system are an array of four SAW vapor sensors, temperature control on the sensors, use of pattern recognition to analyze the sensor data, and an automated sampling system including thermally-desorbed preconcentration tubes. The effort is an excellent example of a systems approach to mass sensor development. The system was designated "smart," since it can discriminate between different classes of vapors, can identify more than one class of vapor, and can provide answers about the presence or absence of a hazard, rather than simply providing raw sensor data.

A portable SAW system for volatile organic compounds has been described (Cernosek et al.,1993). The system consists of two SAW devices (a sensor and a reference), radio frequency (RF) oscillator electronics to drive the devices, digital interface/communications electronics, gas handling hardware for sampling, and a notebook computer to control the hardware and display responses. The system has been demonstrated for trichloroethylene (TCE) and perchloroethylene (PCE) using headspace in monitoring wells and from off-gas lines from the vacuum extraction system at the U.S. DOE's Savannah River Site (SRS). It has also been demonstrated at the U.S. DOE's Hanford Reservation for downhole monitoring of carbon tetrachloride (CCl_4) in the vadose zone (Figure 36.2). The TCE/PCE mixture measured at the SRS ranged between 260–280 ppm, and the SAW data agreed well with gas chromatography analysis performed onsite. The CCl_4 concentrations measured at the Hanford Reservation were made using the SAW system deployed in a downhole probe configuration. CCl_4 was measured at 265 and 1450 ppm in two different wells that were open to the vadose zone.

One important objective of mass sensor development is to meet a real-world need such as detection of pollutants in the vadose zone. This demands that selectivity be achieved. A critical point is to find selective sorbents which can be used as recognition coatings with mass sensors. Obviously, this applies to other classes of sensors as well. Improvements continue to be made in chemical sensor microelectronics and

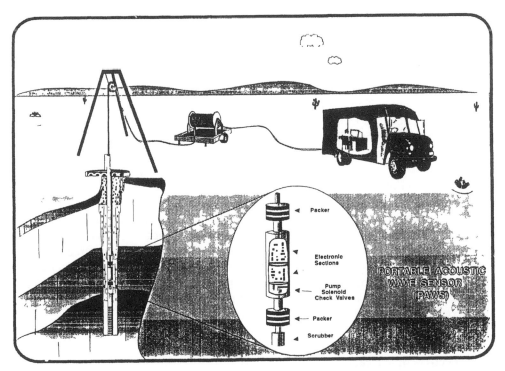

Figure 36.2. Downhole SAW system related and related onsite hardware used to determine chemical contaminant concentrations in vadose zone monitoring wells at the Hanford Site (from Cernosek et al., 1993).

device miniaturization. Excellent concepts exist for systems approaches to the development of chemical sensors. However, the results in applying these advances to environmental monitoring have been mixed (Poziomek, 1992a). The major technology barrier to the development of mass-sensitive and other chemical sensors continues to be the proper selection of the sensor coating materials. For example, there is little specific information on chemical reactions and molecular association effects in vapor-solid and liquid-solid phases which can be drawn from to develop chemical sensor coatings. In addition, few guidelines describe the selection of solid phase coatings for use in conjunction with chemical sensors. Standard methods for screening and evaluating candidate solid phase coatings in sensor applications are not readily available (Poziomek, 1992a and 1992b). An important guide to the design of reversible sorbent coatings for chemical sensors is a publication by Grate and Abraham (1991). A systematic variation of the solubility properties of recognition coatings is suggested to achieve selectivity. Another problem that affects sensor performance, regardless of coating, is water vapor condensation. Water will absorb on almost any surface up to a relatively high temperature and over a wide range of pressures. Condensation of water vapor on the crystal will affect the mass and, therefore, decrease the signal to noise ratio. An appreciation of the recognition coatings that have been utilized with SAW and QCM sensors for specific test vapors can be gained by scanning Tables 36.1 and 36.2 (Poziomek, 1989).

Optical Sensors

At face value, the term "optical sensing" can include any type of sensing technology that measures the behavior of analytes when subjected to electromagnetic radiation. Although laboratory-based systems can be and are used to support the characterization of the vadose zone, this discussion exclusively focuses on the environmental application of optical fibers to field analysis of contaminants in soil, soil gas, and water. There are a number of reviews available on fiber optic sensors and their applications (Angel, 1987; Wolfbeis, 1991; Poziomek, 1992a and 1992b; National Research Council, 1993; Gillispie, 1992; Seitz, 1988; Arnold, 1992; Hirschfeld, 1981; and Walt, 1993a and 1993b).

As pointed out by Walt (1993a), the introduction of optical fiber technology revolutionized the telecommunications industry and may produce "a subsequent and equal impact on chemical sensor technology." Fiber optic chemical sensors are a subclass of chemical sensors in which the optical fiber is used to transmit light (Arnold, 1992 and Seitz, 1988). The intended application is to obtain quantitative information from a spectroscopic measurement performed in situ or directly from a sample that is collected and brought to the device. Research in fiber optic chemical sensors has focused on detecting and measuring organic, inorganic, and radioactive contaminants in all types of environmental media.

Current research and development on fiber optic chemical sensors for environmental applications can be placed into three categories:

- spectroscopic measurements of the pollutants themselves
- molecular association effects in which the pollutants participate in reversible binding with indicator molecules, and
- chemical and biochemical reactions of pollutants.

In the latter two categories, *changes* are required which can be measured through some form of spectroscopy. These categories are depicted in Figure 36.3.

An excellent summary of the principal molecular spectroscopic techniques for hazardous waste site techniques is available (Eastwood and Vo-Dinh, 1991). An emphasis is placed on fluorescence and phosphorescence for direct spectroscopic measurements. A major application to date of direct spectroscopy has been for in situ measurement of fuels in soil (Lieberman et al., 1991 and 1993; Bratton et al., 1993; Gillispie et al., 1993; and St. Germain et al., 1993). This involves laser-induced fluorescence spectroscopy using fiber optics in conjunction with cone penetrometry. A Tinker Air Force Base demonstration showed that a prototype laser-induced fluorescence/electronic cone penetrometer could detect at least 100 mg total petroleum hydrocarbons per kilogram of soil (Bratton et al., 1993). The fluorescence properties of more than 20 aromatic hydrocarbons in aqueous solution have been characterized (Meidinger et al., 1993). The list includes benzene, toluene, ethylbenzene and xylenes (BTEX), and naphthalene and its 1- and 2-methyl derivatives. These are found in most fuels. In situ analysis of BTEX using laser induced fluorescence and fiber optics was also discussed by Chudyk et al. (1990).

A fiber optic refractive index sensor has been reported for the detection of dense nonaqueous phase liquids (DNAPLs) (Ewing et al., 1993). The authors indicate that the sensor is capable of discriminating between the gaseous, aqueous, and organic phases, with the organic phases being TCE, CCl_4, and chloroform ($CHCl_3$).

A field portable ultraviolet absorption instrument has been demonstrated in the analysis of benzene in ground water up to 50 m depths (Haas et al., 1991). There would be difficulty in using the technique in the vadose zone, but it should be applicable to perched ground water.

A second category of fiber optic chemical sensor involves some type of molecular association in which the pollutant binds reversibly to a recognition coating (indicator). This coating is usually contained at the tip of the optical fiber. The molecular association produces a change (for example, refractive index, color, or fluorescence) which can be measured and related to the concentration of the pollutant. A fiber optic chemical sensor system, based on the refractive index change of a recognition coating upon sorption of a pollutant in a vapor, has been described (Arman et al., 1993). Parts-per-million sensitivity is claimed for hydrocarbons (for example, toluene).

Tetracyanoethylene (TCNE) has been used as a reversible indicator for aromatic hydrocarbons (Shahriari et al., 1993). The TCNE was immobilized in a porous glass fiber. Exposure to an aromatic hydrocarbon such as benzene led to the appearance of color. Low ppm sensitivity was observed for benzene in water. Another approach is to coat the end of an optical fiber with fluorescent reagents. Changes in light emission in the presence of pollutant vapors is used for monitoring. Such systems have been described for p-xylene vapor (DeFilipppi and Cody, 1993) and organic halogen compounds (Butler et al., 1993). Yet another approach using indicators with fiber optic chemical sensors is fluorescence quenching. For example, fluorescence quenching of a charge-transfer pair (perylene and anthracene) has been used to construct a fiber optic sensor for halothane (bromochlorotrifluoroethane) (Sharma and Fasihi, 1993). Systems responding to vapors should be adaptable for use in the vadose zone.

A classical example in the use of chemical reactions in conjunction with fiber optics for monitoring analytes of environmental interest is the irreversible development of color in the reaction of certain organochloro compounds (for example, TCE

Table 36.1.   Recognition Coatings Used with Surface Acoustic Wave Sensors

Coating	Test Vapor
Metallophthalocyanines	Nitrogen dioxide Other electron acceptors
Phthalocyanines Poly(2-vinylpyridine) copper complexes	Iodine
Tetra-4-*tert*-butyl silicon phthalocyanine dichloride Copper (II) phthalocyanine/ aminopropyltriethoxysilane polymer	Nitrogen dioxide
Pyridinium compounds	Organophosphorus compounds Halocarbons
Pyridinium oximes Copper (II) complex of tetramethylethyl enediamine in a polyvinyl pyrrolidone matrix 2-(Diphenylacetyl)-1,3-indanedione hydrazone Poly(acrolein oxime) Poly(*N*-vinylimidazole) Hexafluorodimethylcarbinol functionalized polystyrenes, polyacrylates and poly(3,4-isoprenes)	Dimethyl methylphosphonate
Tungsten oxide	Hydrogen sulfide
Triethanolamine	Sulfur dioxide
Acrylamidoxime Acrylamidoxime/acrylonitrile/butadiene copolymer	Methanesulfonyl fluoride Dimethyl methylphosphonate
Pyridinium tetracyanoquinodimethane	Nitrogen dioxide
Palladium	Hydrogen
Platinum	Ammonia
Polyimide cellulose acetate butyrate Hygroscopic polymer Halogenated polymers	Water
PtCl$_2$(ethylene) pyridine	Vinyl acetate
Poly(ethylene maleate)	Cyclopentadiene Dimethyl methylphosphonate
Poly(4(5)-vinyl-imidazole)	Diisopropyl methylphosphonate
Fluoropolyol	*N,N*-Dimethyl acetamide Tributyl phosphate
	Organophosphorus compounds
	N,N-Dimethyl acetamide Dimethyl methylphosphonate

Table 36.1.   Continued

Coating	Test Vapor
Fluoropolyol (continued)	Dimethyl methylphosphonate N,N-Dimethyl acetamide Isopropyl methylphosphorofluoridate
Ethyl cellulose	Aromatic hydrocarbons (e.g., toluene) Halocarbons (e.g., dichloromethane)
Poly(isobutlyene)	Aromatic hydrocarbons
Tenax GC	Various organic solvents
Poly(ethyleneimine)	Alcohols (e.g., methanol) Water
N,N,N',N' Tetrakis(2-hydroxyethyl)- ethylenediamine	Sulfur dioxide
Lead acetate	Hydrogen sulfide

Source: Adapted from Poziomek, 1989.

and chloroform) with pyridine in the presence of hydroxide ion (Milanovich et al., 1991). The research has taken place for a number of years at the Lawrence Livermore National Laboratory (LLNL) under U.S. DOE sponsorship. It is applicable to vadose zone monitoring and was demonstrated successfully in the field at LLNL at a location where estimates of TCE concentration were less than 10 ppb (Milanovich et al., 1991). The samples were pumped at 450 cc/min to a remote mobile laboratory. More recent work involved a demonstration of the fiber optic TCE system at the Savannah River Site (Rossabi et al., 1993a). There were two configurations, both applicable to vadose zone monitoring. The first incorporated the sensor into a down-well instrument bounded by two inflatable packers capable of sealing an area for a discrete depth analysis. The second involved an integration of the sensor into the probe tip of a cone penetrometry system (Figure 36.4). The authors cite the capability of the system to measure TCE concentrations below 200 ppbv. Commercialization of the sensor system for TCE is underway (Wells et al., 1993). Detection limits for TCE are given as a few micrograms per liter in water and 0.1 ppmv in air.

Another area which is drawing a great deal of attention is the use of biologically-derived materials as recognition coatings with fiber optics for environmental monitoring applications. The term "biosensor" has been defined in a number of ways; for example, "an analytical device composed of a biological sensing element (i.e., enzyme, receptor, microbe, or antibody) in intimate contact with a physical transducer (i.e., electrochemical, mass, or optical) which together relate the concentration of an analyte to a measurable electrical signal" (Rogers and Poziomek, 1993). For example, a fiber optic biosensor is under development which measures the formation of fluorescent complexes (Anderson et al., 1993). Biosensors generally require that the analyte be in the aqueous phase. The method seems most suitable for ground water but would be applicable to perched ground water. A reference text is available on biosensors utilizing fiber optics (Wise and Wingard, 1991).

Fiber optic chemical sensors do offer certain advantages. In particular, they can be used in continuous monitoring applications; they can be internally calibrated; no deleterious effects are expected from electromagnetic sources; the signal can be

Table 36.2. Recognition Coatings Used with Quartz Crystal Microbalance Sensors

Coating	Test Vapor
Parathion antibodies	Parathion Malathion Methyl parathion p-Nitrophenol Disulfoton Ethion
Cholinesterases (horse serum, eel and bovine)	Malathion
Gold	Mercury
Mercuric oxide and gold	Carbon monoxide
Triphenyl amine Quaternary amines Trimethylamine hydrochloride	Hydrogen chloride
Acetone extract from carbon blacks	Hydrogen sulfide

Metal halides
   Iron (III) chloride
   Zinc chloride
   Cobalt (II) chloride
   Mercury (II) bromide
   Zinc iodide
	Amines

Didodecylamine Dioctadecylamine Tetrakis(hydroxyethyl)ethylenediamine	Carbon dioxide
Ucon-LB-300X Ucon-75-H-900000 Paprika extracts Ascorbic acid (and their mixtures with AgNO$_3$) L-Glutamic acid hydrochloride Pyridoxine hydrochloride Nonylphenoxypolyethoxylate and pyridoxine hydrochloride Polyvinylpyrrolidone	Ammonia
Sodium tetrachloromercuriate Carbowax 20M Styrene-maleic dimethylaminopropylimide copolymer p-Toluidine Triethanolamine Triethanolamine/triisopropanolamine Quadrol Amine 220 Armeen 2S Ethylenedinitrilotetraethanol Alkyl derivatives of amines Gold amalgam	Sulfur dioxide

Table 36.2.   Continued

Coating	Test Vapor
1,4-Polybutadiene	Ozone

Liquid crystals
    4-pentyl-4'-cyanodiphenyl
    4-pentyl-4'-propylazobenzene
    4-propyl-4'-methylazoxybenzene
    cholesteryl oleylcarbonate
	Benzene
	Toluene
	Chlorobenzene
	o- and m-Dichlorobenzene
	Nitrobenzene
	o-, m- and p-Diethylbenzene
	DDVP
	Phosdin

Cu(butyrate)$_2$
Cu(butyrate)$_2$ ethylenediamine
Cu(butyrate)$_2$ diethylenediamine
Cu(butyrate)$_2$ (ethylenediamine)$_2$
Polymer bonded copper complexes
Copper (II) chloride
Metal salts (e.g., mercury (II) bromide,
    magnesium bromide)
Nickel (II) chloride
Cadmium chloride
Iron (III) chloride complex with diisopropyl
    methylphosphonate
Pyridinium oximes (e.g.,
    1-n-dodecyl-3-hydroximinomethyl
    pyridinium iodide and its mixtures with
    sodium hydroxide and surfactants)
Succinyl choline chloride
Polyvinylpyrrolidone/
    tetramethylenediamine
Polyvinylbenzyl chloride/
    tetramethylenediamine
Gold
Silver
Nickel
	Diisopropyl methylphosphonate

Iron (III) chloride	Paraoxon Diisopropyl methylphosphonate
L-Histidine hydrochloride	Diisopropyl methylphosphonate Parathion Malathion
Iron (III) chloride complex with paraoxon	Paraoxon
Styrene-4-vinylhexafluorocumyl alcohol copolymer	Dimethyl methylphosphonate
D,L-Histidine hydrochloride	Parathion Malathion

Cobalt complex of isonitrilobenzoyl
    acetate
Cobalt complex of isonitrilobenzoyl
    acetate and paraoxon
Sodium salt of isonitrilobenzoyl acetate
	DDVP

Table 36.2.   Continued

Coating	Test Vapor
Succinyl choline iodide	Diisopropyl methylphosphonate Malathion Parathion
Methyltrioctylphosphonium dimethyl phosphate	Phosgene
Glutathione, NAD	Formaldehyde
OV-105 OV-225	Propylene glycol dinitrate
Silica	Thiophene
Carbowax 1000	Trinitrotoluene Mononitrotoluene
Charcoal Quadrol	Nitrobenzene
Carbowax 400 Silicone fluid DC high vacuum silicone grease Sillastic LS-420	Toluene diisocyanate
Carbowax 550	Toluene
Carbowax 20M	Chloroform Acetone
Pluronic L-64	Ethylbenzene o-Xylene Hexane
Squalane	Cyclohexane
[6.6.6]Cyclophane hexalactam trimer	Chloroform Methylene chloride
Zinc oxide	Alcohols
Tetrabutylphosphonium chloride Semicarbazide	Acetoin
Pyridinium tetracyanoquinodimethane	Nitrogen dioxide
trans-Chlorocarbonylbis-(triphenylphosphine) iridium	Aromatics (e.g., xylenes, benzaldehyde, 1,3,5-trimethylbenzene, anisole, and butylbenzene)
Squalane Silicone oil Apiezon grease	Nonselective hydrocarbon detection
Polyethylene glycol Sulfolane Dinonyl phthalate Aldol-40 Tide (alkyl sulfonate)	Selective detection of polar molecules (e.g., aromatic, oxygenated unsaturated)

Table 36.2. Continued

Coating	Test Vapor
Silica gel Molecular sieve Alumina Hygroscopic polymers Silicon oxides Lithium and calcium halides Gelatin	Water
Lead acetate Metallic silver Metallic copper Anthraquinonedisulfonic acid	Hydrogen sulfide
Silicone oil (DC 190)	Halogenated hydrocarbons
Palladium	Hydrogen Deuterium

Source: Adapted from Poziomek, 1989.

multiplexed (spectroscopic techniques provide information-rich signals); and they can be made very small. However, as pointed out by Walt (1993b), there are a number of challenges in designing and deploying fiber optic sensors for use in environmental applications. These include interferences in the sample matrix, precision, sensitivity, selectivity, and power requirements. Critical issues in using an optical fiber as a light pipe in the direct measurement of pollutants are the light source, the detection sensitivity, and the data processing algorithms (Gillispie, 1992). There are a number of important considerations including spectral dimensionality

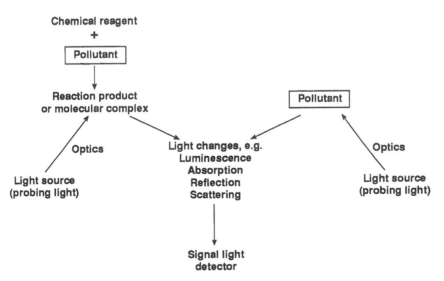

Figure 36.3. Categories of fiber optic chemical sensor research and development activities.

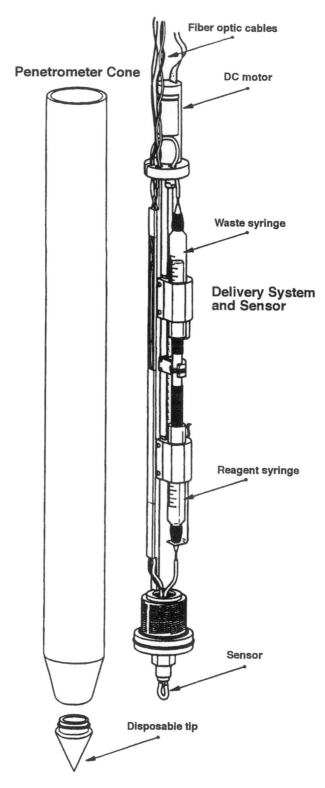

Figure 36.4.   Integration of a TCE (trichloroethylene) sensor system into the probe tip of a cone penetrometer (from Rossabi et al., 1993).

(absorption versus fluorescence), spectral structure, light source intensity, signal/noise ratio, appropriate standards, calibration criteria, etc. Despite these considerations, fiber optic sensors have been designed to monitor threshold levels.

Poziomek (1992b) pointed out that there are some other challenges to the use of fiber optics in environmental applications. Some examples include:

- From a practical point of view, transmission of light across long distances is limited to the visible region, thus severely limiting the options for remote spectroscopy.
- Organic pollutants most frequently found at hazardous waste sites such as chloroaliphatic, chloroalkene, and low molecular weight aromatic compounds are not easily measured and/or differentiated using UV-visible spectroscopy.
- Use of fiber optic IR and normal Raman spectroscopy provides opportunities for high selectivity; however, there is a severe tradeoff because of low sensitivity.
- The dynamic concentration range may be limited, thereby reducing flexibility of use.
- Indicator systems used in the chemically-selective coatings tend to degrade with time.

The continuous monitoring capability of these sensors makes them ideal candidates for post-closure monitoring or as an alarm sensor. They can be easily deployed using either push or drilling technology. As the technical challenges are overcome, fiber optic chemical sensors will undoubtedly prove to be valuable components in vadose zone monitoring systems.

Electrochemical Sensors

Electrochemical sensors fall into three basic categories, potentiometric (measurement of voltage), amperometric (measurement of current), and conductometric (measurement of conductivity). There are a number of subclasses in each category. Gas and vapor electrochemical sensors are relevant to vadose zone monitoring. Table 36.3 contains a list of analytes that have been detected using electrochemical sensors. The table was compiled from a review of recent literature (Janata, 1992). Chlorinated hydrocarbons were detected with all three types of electrochemical sensors. A key to development of electrochemical sensors is that the analyte of interest be electroactive, i.e., easily reduced or oxidized.

Publications on potentiometric sensors, including ion-selective electrodes (ISEs) remain the strongest group among chemical sensors relative to numbers of papers (Janata, 1992). Classical examples of potentiometric measurements include pH measurements with glass electrodes and redox measurements with metal electrodes. New directions are reflected by a recent review of polymer membrane-based ion-, gas-, and bio-selective potentiometric sensors (Yim et al., 1993).

Amperometry is a classical electroanalytical tool for a variety of electroactive analytes in the liquid phase. For gas analysis, the Teflon-bonded diffusion electrode made a convenient triple phase (solid/liquid/gas) boundary possible (Stetter et al., 1991). This development increased the potential use of amperometric gas sensors for pollutants in the vadose zone. The sensors operate by reducing or oxidizing the analyte and using the resultant cell current as the detection signal. Sensitivity and

Table 36.3.   Electrochemical Sensors

Electrochemical Sensor Category	Gas and Vapor Phase Analyte
Potentiometric	CO_2
	NO_X
	O_2
	CO
	H_2
	Cl_2
	Arsenic oxides
	Azides
	Chloroform
	Oxidizable pollutants
Amperometric	O_2
	H_2
	NH_3
	NO_2
	NO
	CO
	CO_2
	Water
	Ethanol
	Volatile organic solvents
	Chlorinated hydrocarbons
	Combustible organic vapors
Conductometric	H_2
	NH_3
	NO_2
	CO
	CO_2
	Cl_2
	AsH_3
	I_2
	PH_3
	Alcohols
	Organic solvents
	Anesthetics
	Chlorinated hydrocarbons
	Organophosphates
	Hydrazine
	Water

Source: Compiled from Janata, 1992.

selectivity can be improved by suitable choice of the electrode material for electrocatalysis, the electrolyte, the electrochemical potential, and the geometry. A commercially available amperometric sensor for organic vapors is briefly described in U.S. DOE (1993d). Depending on the type of porous membrane being used and the particular sensing electrode, the sensor was reported to have useful sensitivity for 50 different organic compound vapors. The range of sensitivity for ethylene oxide was given as 0–300 ppm with a response time of less than 360 seconds and a recovery time of less than 360 seconds.

An interesting method involves catalytic pyrolysis of chemical vapors combined with electrochemical detection (Penrose et al., 1991). Compounds that are not normally thought of as electrochemical analytes, such as chloroform or cyclohexane, can be partially oxidized on a hot platinum surface (Stetter et al., 1984). The resulting volatile products give a response on a porous-electrode electrochemical sensor. More recent developments involve the use of an array of such sensors. Based on an assessment of current technology, vapor samples would need to be pumped to the detection system from the vadose zone. Figure 36.5 shows a configuration containing two catalyst filaments and four electrochemical sensors.

Two types of conductometric sensors seem readily amenable to field screening and monitoring applications, i.e., chemiresistors and semiconducting oxides (Poziomek, 1992a). The principle of chemiresistors involves conductivity change of a recognition sensor coating as a result of analyte vapor sorption. The coatings are usually deposited over interdigital electrodes microfabricated on quartz substrates. Metallophthalocyanines have been most widely investigated as coatings for chemiresistors (Snow and Barger, 1989). Conducting polymers (e.g., polypyrrole, poly-N-methylpyrrole, poly-5-carboxyindole and polyaniline) do respond to organic vapors such as methanol, ethanol, acetone, and toluene, but at high concentrations (Bartlett and Ling-Chung, 1989). Several analyte gases have been detected using metallophthalocyanine at concentrations below 1 ppm (Wohltjen et al., 1985). Chemiresistors are normally used for strong electron-donor and electron-acceptor gases/vapors (Poziomek, 1989). Table 36.4 includes a few examples.

Sorption electron-acceptor gases on most phthalocyanine coatings leads to an increase of sensor film conductivity, whereas electron-donors cause conductivity to decrease. Gases such as CO, CO_2, CH_3, NO, SO_2, and benzene, which are neither strong electron acceptors nor donors give low response (Snow and Barger, 1989). However, as reported by DOE(1993d), chemiresistors have been used for underground storage tank monitoring.

The challenge to develop recognition coatings for chemiresistors is more formidable than for the SAW and QCM mass sensors described earlier, since appropriate

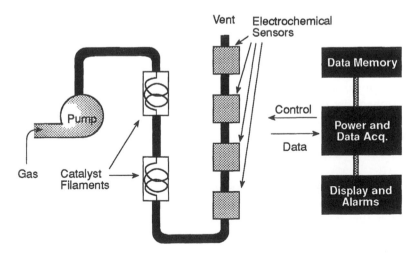

Figure 36.5. Configuration of a monitor fitted with two catalyst filaments and four electrochemical sensors (from Penrose et al., 1991).

Table 36.4.   Electron-Donor and Electron-Acceptor Gases and Vapors

Electron-Donors	Electron-Acceptors
Ammonia	Nitrogen dioxide
Hydrazine	Boron trifluoride
Dimethyl methylphosphonate	Halogens
Phosphine	Nitric oxide
Water	Hydrogen chloride
	Oxygen

conductivity properties need to be met in addition to sorption ones. Combinations of SAW and chemiresistor devices using the same coatings have been explored (Snow et al., 1986).

Semiconducting oxides [also called metal-oxide semiconductors (MOS)] have been used for many years to detect combustible gases. The most commonly used material for the substrate is tin oxide; however, many other oxides have been used (Azad et al., 1993). In recent years, semiconducting oxide sensors have been used for the detection of NH_3, H_2S, CO, thiols, ethanol, hydrogen, arsine, and acetic acid, to name a few. The mechanism of detection in air is usually a catalytic oxidation at the surface of the oxide, inducing an increase in conductivity. For example, a semiconducting oxide has been developed for organochloro and organobromo compounds (Penrose et al., 1991). The sensor consists of a coil of platinum wire heat-treated with a mixture of lanthanum oxide, lanthanum fluoride, and a binder. It is used as a sensor by heating the coil to 550°C with an electric current. Conductivity is measured between the heated coil and a separate platinum electrode. When a vapor containing an organochloro compound comes in contact with the sensor, the conductivity increases. Exposure to 100 ppm concentrations of chlorobenzene, benzene, and n-hexane only gave a response to chlorobenzene.

A survey of 25 vendors of toxic gas sensors and monitoring systems for hazardous gases and chemicals used by the semiconductor industry included electrochemical sensors (Korolkoff, 1989). Though the target list of gases contained mostly inorganic compounds (other than Freon™), the survey reflected the state of commercial electrochemical technology. The minimum limit of detection was in the low-ppm range, generally less than 1 ppm. Electrochemical gas sensors were cited as having excellent response times but lacking the sensitivity required for some gases. It was also mentioned that most electrochemical sensors are affected by humidity. Manufacturers have taken steps to minimize the problems with water vapor. However, the potential impact of water vapor on electrochemical sensor performance in the vadose zone and headspace analysis of ground water will have to be taken into account.

A very attractive feature of electrochemical sensors is their small size. Miniaturization continues. Small electrochemical sensors may find a number of in situ applications including monitoring bioremediation (for measuring and monitoring biogenic gases) and soil gas monitoring for volatile and semivolatile organic compounds. They may also serve as hand-held sensors, where the sample is brought to the device for contaminant detection. Major challenges relate to increasing sensitivity and selectivity.

Radiochemical Sensors

A final category of sensors includes those designed to measure radionuclides. The DOE is faced with remediating a staggering number of waste sites scattered across the Weapons Complex. The above-discussed technologies focused on the analysis of organic compounds; however, heavy metals contamination, particularly radionuclides, need to be addressed. Soil, sediments, surface water and ground-water contamination by heavy metals and radionuclides is a widespread problem (Office of Technology Assessment, 1991). A necessary component of the DOE's site characterization programs is to identify the nature and extent of radionuclide contamination, and, clearly, the vadose zone at many of these locations will have been substantially contaminated.

There are three approaches that can be applied in detecting and quantifying radionuclides. The first approach, and probably the most common, is to measure the presence and quantity of a radionuclide based on its characteristic energy emissions (alpha, beta, and gamma rays). This usually requires some form of sample preparation to concentrate the radionuclides prior to counting to achieve a reasonable degree of sensitivity, especially for radionuclides with short half-lives. For more abundant radionuclides (for example, uranium, thorium, and plutonium), direct counting of samples may be possible. Another means to analyze a sample for radionuclides is through the use of wet chemical or instrumental techniques. Inductively-coupled plasma/mass spectrometry (ICP/MS) and inductively coupled plasma/atomic emission spectroscopy (ICP/AES) have been used. Both of these methods are relatively expensive. Zamzow et al. (1994) reported on the in situ determination of uranium in soil by laser ablation/ICP. Laser ablation enables sampling and analysis to be performed for solids without acid dissolution of the samples. Laser radiation is focused into the sample, and causes vaporization and ejection of material from the surface. The laser ablated vapor and particulates are transported to a secondary plasma source such as ICP. Zamzow et al. (1994) utilized a 25 m long silica core fiber optic cable routed out of the mobile laboratory to the soil site to be analyzed. The material ablated from the soil surface is swept out of the ablation cell by argon gas through a length of polyvinyl tubing and into an ICP torch mounted in an instrumentation platform connected to the rear of a mobile lab trailer. This is not applicable to in situ monitoring in the vadose zone at the present time, but illustrates progress being made in field measurements; it may be applicable to the vadose zone in the future.

Another approach relies on energy-dispersive techniques, such as neutron activation analysis (NAA) or x-ray fluorescence spectrometry (XRF). Neither of these techniques is feasible for radionuclides with short half-lives. Field portable XRFs have been used for the detection of metals and elements in soils and water; however, their sensitivity is typically in the parts per million range. For additional information on field portable XRF, the reader is referred to Raab et al. (1991).

There has been progress in the development of field portable radiochemical sensors. However, much of the current analytical work is still done in fixed laboratories using conventional radiochemical analyses. Radiochemical sensors have been defined as those "devices that use the detection of atomic or nuclear reactions to identify chemical processes or constituents" (DOE, 1993b). DOE (1993c) provides an excellent review of the currently available radiochemical detection technologies. These technologies rely on conventional detection techniques—such as scintillation detectors, solid state detectors, high purity germanium detectors, and multichannel analyzers—and are confined to fixed or mobile laboratories. DOE has supported some research and testing of novel approaches. Gammage et al. (1993) reported on the pilot testing of electrets and alpha-track detectors (ATDs) for inexpensive passive monitoring of alpha contamination on man-made surfaces or in soil. They cited having successfully monitored soil containing only a few pCi/g of plutonium, exposing the ATD to the soil surface for 24 hours. Plans for vertical profiling of soil contaminants were mentioned.

Most likely, reliance on fixed laboratories will continue in the immediate future for analyzing radioactive samples drawn from hazardous waste sites, including those samples collected from the vadose zone.

Field Test Kits

As discussed previously, there is tremendous appeal in being able to conduct near-real-time or real-time characterization and profiling of contaminants in the subsurface. It is desirable to do as much noninvasive characterization as possible by using, for example, geophysical methods. However, noninvasive techniques alone are not adequate. It is desirable to supplement the information with results of chemical analysis. Earlier we discussed the range of applications that chemical sensors have or may have in real-time monitoring. Field test kits provide an alternative for generating qualitative and quantitative information about contaminants in environmental matrices.

A number of companies have been offering field test kits and a variety of detector tubes for many years. These typically rely on classical colorimetric techniques. There are many known and well documented chemical reactions that generate specific colors in the presence of inorganic and organic substances. The kits and detector tubes provide the necessary reagents in convenient and easy to use forms. Battery-operated spectrophotometers are available for more quantitative work. There has been a great deal written on the use of kits and detector tubes for environmental applications. The technology is considered mature but improvements continue to be sought. For example, there is a new class of field kit emerging that is based on biochemical reactions—immunoassays.

Very simply, immunoassay techniques rely on antibodies that have a high degree of affinity to a target analyte. Quantitation of target analytes can be accomplished by monitoring a color change or by measuring radioactivity or fluorescence (Van Emon and Lopez-Avila, 1992). Kits have been designed to measure contaminants in all types of matrices. Immunochemical techniques are also utilized in devices to monitor breathing-zone air for specific compounds. Van Emon and Lopez-Avila (1992) provide a table that contains a list of the immunoassay methods that have been reported in the literature (Table 36.5). Other compounds for which kits have been or are being designed include: bioresmethrin, carbofuran, chlorpyrifos, chlorpyrifos-methyl, cyclodienes (e.g., chlordane), DDT, diazinon, fenitrothion, isoproturan, metalaxyl, methoprene, nicotine, pirimiphos, procymidone, trisulfuron, and urea herbicides.

Immunoassay kits have a number of features that make them particularly useful in field screening and field analytical applications. These include: (1) a high degree of analyte selectivity, (2) sensitivity of parts per billion to parts per million concentration in most matrices, (3) quick sample turnaround, (4) relatively low cost per analysis, (5) size and ruggedness well suited for field use, (6) relatively small sample size (as low as 5 mL), and (7) ease of use. The utility of these kits has been recog-

Table 36.5.   Immunochemical Methods Reported in the Literature for Environmental Analysis

Alachlor	Isoproturon
Aldicarb	Lindane & BHC
Atrazine	Metolachlor
Benomyl	Metalaxyt
Bionesmethrin	Methoprene
BTEX (total)	PAH (total)
Captan	Paraquat
Carbofuran	PCB
Carboryl	Pentachlorophenol
Chloropyrifos	Petroleum Hydrocarbons
Chlorothalonil	Pirimiphos
Cyanzine	Procymidone
Cyclodienes	Simazine
2,4-D	Triasulfuron
DDT	Triazine
Fenitrithion	Trifluralin
Imazaquin	Urea Herbicide

Source: Van Emon, 1993, cited in Poziomek et al., 1993.

nized by the EPA. Recently, draft methods for the use of immunoassays for screening soil for petroleum hydrocarbons (BTEX) and PCBs and a screening method for pentachlorophenol have been proposed for inclusion into the Resource Conservation and Recovery Act (RCRA) methods manual, *Test Methods for Evaluating Solid Waste* (SW-846). The EPA has also tested a number of kits under the Superfund Innovative Technology Evaluation (SITE) program (Gerlach et al., 1993). [Additional information on other technologies demonstrated under the SITE program can be found in a U.S. government reference (U.S. EPA, 1993a)]

All of the techniques mentioned above would require samples to be brought to the surface from the vadose zone; e.g., vapors and pore liquids. Sample preparation for immunoassays would require the use of water for the tests to be performed.

Miniaturized Laboratory Instrumentation

Clearly, regulators and the regulated community are very (and probably overly) confident with data that are generated in fixed analytical laboratories using conventional instrumentation. It is understandable then that there has been much activity on miniaturizing laboratory instrumentation and re-engineering instrumentation for use in mobile laboratories. These emerging applications of conventional technologies are valuable for those responsible for environmental monitoring and environmental management. Although these technologies, for the most part, cannot be miniaturized to the point where they can be deployed down a borehole or emplaced in the subsurface (the exception is ion mobility spectrometry), they can provide practitioners with equipment that can generate quantitative and qualitative data in near-real time. This capability can provide a more cost-effective alternative for environmental monitoring and hazardous waste site cleanup.

This section provides highlights of current developments and innovations in gas chromatography (GC), gas chromatography/mass spectrometry (GC/MS), and ion mobility spectrometry (IMS). The reader is referred to the references for greater technical detail.

Field Portable Gas Chromatographs

Field portable gas chromatography is one of the most heavily relied upon tools used in environmental monitoring and environmental problem-solving. Case studies are frequently described in the literature (Driscoll, 1993 and Stout et al., 1993). There have also been many studies done comparing the performance of field GCs for VOC analysis to laboratory and other methods (Jenkins et al., 1988; Hewitt, 1993; Simmons and Knollmeyer, 1993; and Zemansky et al., 1993). GC represents the most popular and mature field screening technology available. Chappell and Simmons (1993) provide an excellent primer on the performance requirements for portable GCs in the field.

Recent efforts have been focused in five main areas: (1) method development, (2) sample preparation/introduction methods, (3) new and improved column materials and designs, (4) instrument re-engineering, and (5) detectors. Levine et al. (1993) reported on the development and demonstration of a "Fast-GC" that has been successfully used to measure mixtures of compounds in air. Total analysis times were up to 100 seconds (depending on the number of compounds in the mixture) and the limit of detection for single compounds of approximately 1 ppb(v/v). Spittler (1993) reported on the utility of two-column chromatography in the field. He identified three applications. The first is to use one column as a total detector for rapid screening of soil or water samples. The limits of detection will vary depending on the detector selected. A second relies on a short column (a second column) for rapid screening of samples once the composition of the contaminant mix has been determined. This can provide the field analyst with the capability to rapidly screen large numbers of samples without dealing with the detailed data provided by a longer, slower column. Finally, two-column chromatography can be used to make qualitative measurements of nonpolar solvents. This application requires two columns that have different retention times for the commonly encountered nonpolar solvents. The retention time differences can be used to identify compounds with picogram sensitivity, rivaling the performance of portable GC/MS equipment. In a final example of recent developments, Carney et al. (1993) described their experience in exploring the limits of detection and chromatographic capabilities of a microchip GC. They successfully improved the detection limits of a commercially available instrument by using an automatic, integral sorbent tube concentrator. This improved the performance of the instrument by 1 to 2 orders of magnitude.

Transportable Gas Chromatography/Mass Spectrometry (GC/MS)

Another important step in bringing a powerful analytical tool to the field has been the miniaturization of mass spectrometers for use in conjunction with gas chromatographs. Meuzelaar (1993) and Arnold et al. (1993) provide very insightful discussions on the present capabilities of transportable GC/MS systems. A number of research institutions are actively involved in the development and testing of "portable," "transportable," and "wearable" instrumentation. Table 36.6 lists devices considered to be "man-portable." Many of these instruments exist as prototypes but are likely to appear commercially within the next few years. Most have undergone preliminary field testing and evaluation.

Table 36.6. Engineering Design and Performance Specifications of Man-Portable MS Systems Currently Under Development

System Parameter	University of Utah Micro Analysis & Reaction Chemistry	Jet Propulsion Laboratory	Lawrence Livermore National Library	Massachusetts Institute of Technology
(Sample Inlet)				
–automated/ manual injection	automated	automated	GC/MS hand injection	continuous
–inlet valve type	virtual/fluidic	mechanical	split/splitless	membrane
–minimum pulse duration	20 msec	28 msec	variable(0–60sec)	n/a
–enrichment option	trap-and-desorb	trap-and-desorb	concentrated liquids	n/a
(Gas Chromatography)				
–column dia/length	100 μm i.d./6ft	50 μm i.d./10ft	25 μm/15 meter	n/a
–carrier gas/flow	H_2or N_2/.5 or.15 ml min^{-1} RT–	He/.05 ml min^{-1}	H_2/.5 ml min^{-1}	n/a
–column temperature	200°C isothermal	RT-200°C isothermal?	RT - 280°C programmed	n/a
–typical analysis time	10-60 sec	10–60 sec	5–45 minutes	n/a
(Mass Spectrometry)				
–MS type	quadrupole (HP-MSD)	mini. Mattauch Herzog	modif. (HP 5971A MSD)	cycloidal magnetic
–mass range	10–650 amu	40-250 amu	10–650 amu	2–150 amu
–maximum acquisition speed	2000 amu sec^{-1}	4000 amu sec^{-1}	2000 amu sec^{-1}	manual, 1 amu/s typical
–minimum detectable quantity	2.5×10^{-12} g benzene (SIM)	7.5×10^{-14} g benzene	2.5×10^{-12} g benzene	50 pg Argon
(Vacuum System)				
–pump type	hybrid getter	getter/ion pump	hybrid getter (custom)	ion getter
–pumping speed	10–50- l sec^{-1}	(5–10 l min^{-1})	10-30 L/sec	8 l/s
–operating pressure	5×10^{-5} torr	5×10^{-5} torr	5×10^{-5} torr	5×10^{-11} torr
(Engineering Parameters)				
–weight	60 lbs incl.power source	(75 lb)	75 lbs excl. power source	70 lbs incl. power source
–size	5.5 × 19 × 24 in	10 × 18 × 28 in	9.5 × 18 × 27 in	8 × 14 × 24 in
–power requirements	90 W (24 Vdc)	?	560 W pk (110 VAC)	30 W est.(12 VDC)
–portability mode	backpack + flexible probe	briefcase	briefcase	backpack
(Data Processing Module)				
–computer type	PC386 notebook	Lap-top PC	graphics LC + keyboard	micro w/RS232 interface
–software	HP MSD software	custom + commercial	HP MSDsoftware + LLNL custom tattletail software	custom

Source: Arnold et al., 1993.

In addition to these man-portable systems, the DOE has been sponsoring research and development on transportable GC/MS systems. Leibman et al. (1991) and Roberts et al. (1993) report on a transportable GC/MS system developed and tested at Los Alamos National Laboratory. Over the last five years, Wise et al. (1991 and 1993) have been developing a transportable direct sampling ion trap mass spectrometer (DSITMS) at Oak Ridge National Laboratory. The DOE has deployed units at the Savannah River National Laboratory and the Hanford Reservation. Henriksen and Booth (1992) prepared an evaluation of the cost-effectiveness of the DSITMS. They concluded that "one sample analysis using the DSITMS costs about 18% of that using the conventional methodology."

There are at least two commercial, transportable GC/MS systems that have been used in environmental applications. Kowalski et al. (1993) and Robbat et al. (1992a and 1992b) provide a description of the performance of the the Bruker Mobile Environmental Monitor (also called Bruker Mobile Mass Spectrometer MM-1). The MM-1 was successfully used to analyze soils for PAHs and PCBs, and for volatile organic compounds (VOCs) in water. Three different methods were used in the field, which were variations of standard EPA methods. The demonstrations illustrate that a high level of data quality was achievable using a transportable GC/MS and that the field-generated results were in good agreement with the confirmatory analyses.

Matz and Schröder (1993) developed and tested field analytical methods for analyzing soil contaminated with organic compounds using a transportable GC/MS system. Their work was supported by the German Ministry of Research and Technology and the city of Hamburg. They report that the methods are being used routinely for site characterization because of the cost-effectiveness of the technology and the immediate identification and semi-quantitation of classes and individual organic compounds.

Transportable GC/MS systems have been demonstrated to be cost-effective and efficient tools for environmental problem-solving. This technology is likely to become, in the not too distant future, an integral component of most subsurface characterization programs. However, most of the above-mentioned systems will require operators who have considerable experience with analytical instrumentation. We anticipate that as the technology evolves, so too will the ease of the operation, but for the time being, the instrumentation is not very user-friendly.

Ion Mobility Spectrometry

Ion mobility spectrometry (IMS) is a technique for the detection and characterization of organic compounds as vapors at parts-per-billion concentrations in air (Eiceman, 1991). IMSs, ranging from hand-held to laboratory bench models, are available on the commercial market. IMS has been known by two other names; gaseous electrophoresis and plasma chromatography. The term "ion mobility spectrometry" adequately represents the basic principle of the technology, though it is sometimes confused with mass spectrometry.

IMS is conceptually simple and relies on the drift time of molecular or cluster ions. It differs from mass spectrometry in that there is little, if any, fragmentation and the ions are not mass analyzed. The ions are analyzed instead by charge and by mobility. IMS involves the gentle ionization of molecules and the analysis of subsequent ions using ionic mobilities.

An IMS is operated at atmospheric pressure. Most commonly, ambient air is pumped into the IMS through semipermeable membranes. Ions are formed from air by using an ionization source such as Nickel [63]. These ions then react with analyte molecules to form ion clusters which are subject to atmospheric pressure "time of flight" measurements. The ions are allowed to enter a drift region, where they move under the influence of an applied electric field to a collector electrode. The electrode current is monitored continuously, thus allowing a mobility spectrum to be measured. Separations of ion clusters are a function of ion size (Figure 36.6).

Selectivity of IMS for different analytes is based on the atmospheric pressure ionization events themselves (which relate to the proton and electron affinities of the analytes), the polarity of the products (i.e., positive versus negative ions), and the mobility of the ions. Target analytes with electron or proton affinities higher than those of other chemicals in the ambient environment can be differentiated and detected readily. Analytes with low affinities can be measured as long as chemicals with strong affinities are absent. The sensitivity is generally in the low ppm to low ppb range with the possibility of parts-per-trillion (ppt), depending on the analyte and instrument parameters.

Poziomek and Eiceman (1992a and 1992b) have developed a method for measuring dimethyl phthalate in water using solid-phase extraction membranes and a thermal desorption device. Bell and Eiceman (1991) and Snyder et al. [as reported by Eiceman (1993)] have successfully incorporated a GC column onto a hand-held IMS and used the device to detect organics in air. Table 36.7 identifies the compounds used by Bell and Eiceman to study the capability of the GC/IMS.

Eiceman (1993) concluded that "The potential for details [sic] chemical information with instrumentation that is small, low in weight, and miserly in power needs portends a future for IMS or some configuration of IMS in environmental field screening." He also stated that commercialization of the technology is the biggest barrier to practical application. It appears likely that GC/IMS can be made to be placed down a well or a borehole. It holds great promise for vadose zone monitoring and subsurface characterization. Readers are referred to Eiceman (1993) and Hill et al. (1990) for more in-depth historical and technical discussions on IMS.

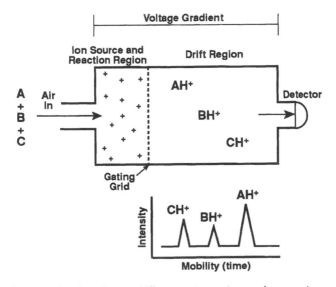

Figure 36.6.   Schematic of an ion mobility spectrometer and a spectrum.

Table 36.7. Listing of Analytes Studied Using GC/IMS

Positive Mode

Alcohols	Ketones	Esters
Methanol	Acetone	Methyl methanoate
Ethanol	2-Butanone	Methyl ethanoate
n-Propanol	3-Methyl-2-butanone	Methyl propanoate
i-Propanol	2-Penanone	Methyl butanoate
n-Butanol	3-Pentanone	Methyl pentanoate
i-Butanol		Ethyl methanoate
s-Butanol		Ethyl ethanoate
t-Butanol		

Aromatics	Aldehydes	
Benzene	Propanal	
Toluene	Butanal	
Ethylbenzene	3-Methylbutanal	
o-Xylene	Pentanal	
m-Xylene	Hexanal	
p-Xylene		
Styrene		

Negative Mode

Halocarbons	Chlorinated Aromatics
Methylene chloride	Chlorobenzene
Chloroform	o-Dichlorobenzene
Carbon tetrachloride	2-Chlorotoluene
Trichloroethene	
1,1,1-Trichloroethane	
Tetrachlorethane	
1,2-Dichloroethane	
1,1,2,2-Tetrachlorethane	

Source: Bell and Eiceman, 1991.

SOIL AND SOIL GAS SAMPLING TECHNOLOGIES

An issue of immense concern to those faced with characterizing the extent of chemical contamination in the subsurface is collecting samples with representative concentrations of semivolatile and volatile organic compounds. Indeed, this was the theme of a recent symposium (U.S. EPA, 1993b). This is not to imply that there are no concerns about sample collection and analysis for extractable organics, metals, or radionuclides. However, the following discussion predominantly focuses on the emerging techniques for sampling semivolatile and volatile organic compounds in soil and soil gas.

It has been recognized for a long time that the VOC results from samples sent to fixed laboratories may not be truly indicative of the extent of contamination. The inherent volatility of light, aromatic organics makes them extremely unstable, despite best efforts to preserve soil or soil gas samples. Many investigators have chosen to move some of the instrumentation described in the preceding sections into the field to at least minimize the time from sample collection to analysis. However, this is not always practical. New sampling and sample handling techniques are needed if sample integrity and data quality are to be maintained.

Sample Collection Techniques

In many instances, soil sampling is done through the use of core barrels or split spoon samplers. Often, an entire core is shipped to the laboratory for subsampling and analysis. Alternatively, the core can be subsampled in the field. This typically requires that the split spoon or core barrel be opened (to the atmosphere), and the sample collected with a small spade or spoon and placed in a sampling container. Hewitt (1992 and 1993) reports on the use of a novel subcoring device that uses a 10-cm^3 syringe with the needle end removed [the technique is called the limited-exposure handling (LDE) method]. This syringe size fits nicely into the mouth of a 40-mL VOA vial. A subsample can be collected and containerized within 10 seconds after the barrel is opened, thereby minimizing the loss of volatiles. It was demonstrated that the LDE method was an important first step in ensuring the quality of data, and that the cleanliness of the sealing surfaces of the vials plays an important role as well.

Triegel (1993) reports on the development and promulgation of standard methods for sampling waste and soils for volatile organics. The American Society for Testing and Materials (ASTM) recently approved the "Standard Practice for Sampling Waste and Soils for Volatile Organics" (D 4547-91). This method describes two methods for sampling loose granular materials: (1) the use of metal rings inserted in split spoons, and (2) subsampling using a metal coring cylinder [similar to that described by Hewitt (1992 and 1993)]. The standard also describes three sample handling methods. One involves the handling and shipment of a core to a fixed laboratory, where it is subsampled and analyzed. The other two concern the handling of subsamples collected in the field. The first method involves the extrusion of the sample into a wide-mouth jar that contains methanol. The work of Siegrist and Jenssen (1989 and 1990) was referenced because they demonstrated that the use of methanol significantly reduced the loss of VOCs prior to analysis. The second method involves the placement of the sample in a glass vial with a modified cap. The modified cap allows the sample vial to be directly connected with a purge-and-trap device to do a headspace analysis.

Sampling soil gas has also proved to be difficult, and the analytical results are sometimes questionable. There are a number of traditional soil-gas sampling methods, some of which are described previously in this volume. They include, for example, driving a tube into the subsurface to withdraw soil gas for analysis at the surface and burying passive sorbent devices. Jenkins et al. (1993) have developed and tested a multisorbent array for collecting soil gas samples from their point of "generation" in the subsurface (Figure 36.7). The sampler is designed for two applications. The first application is in situations where relatively high concentrations (ppm level) of contaminants exist in a low porosity medium. The sampler can be used to collect very small volume samples (10–20 mL), minimizing the disturbance to the subsurface equilibrium. The other application focuses on high porosity media, where higher flow rates can be used to improve the sensitivity.

Stutman (1993) reports the development of a passive sorbent collection device for the collection of volatile and semivolatile organic compounds. The design goals were to improve sensitivity; provide rapid and unobtrusive installation and retrieval; be able to deploy the device anywhere in the soil profile (above or below the water

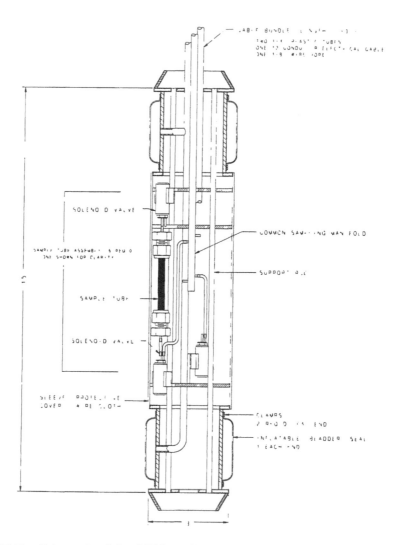

Figure 36.7. Schematic of the ORNL multi-sorbent array (from Rossabi et al., 1993b).

table); archive samples; and eliminate analytical interferences from construction materials. The sorbent containers and insertion/retrieval cords are constructed of inert, hydrophobic, microporous GORE-TEX™ expanded polytetrafluoroethylene (ePTFE). This material allows efficient vapor transfer, which is critical because the container has to be filled with a suitable granular sorbent material or resin. The collection device is placed in a narrow hole and left for a few days or a few weeks. Once retrieved, it is resealed in a glass vial for transport to the laboratory. The device has successfully been used to measure many semivolatile compounds including, pyrene, diakyl phenols, dibenzofuran, fluorene, naphthalene, and nitrobenzene. It was concluded that these devices are particularly useful at chemical plants and refineries, where it has proved difficult in the past to collect soil gas samples from these types of sites using conventional methods.

The most common means to collect soil and/or soil-gas samples is to use one of the many drilling methods to install a monitoring well. Boreholes are typically cased with polyvinyl chloride (PVC), Teflon®, stainless steel, fiberglass composites, and even aluminum. The casing materials can be configured to allow for soil gas sampling by selecting fixed sampling locations and mounting collection ports. Once installed, it is virtually impossible to change the sample port geometry or to go back into the hole to collect additional soil samples. Keller and Lowry (1990) introduced an innovative approach that can function as a means to case a borehole and to serve as a support and deployment platform for sampling devices and instrumentation. The technology is called SEAMIST™.* SEAMIST™ is comprised of an impermeable liner attached to a tether that is inserted into the subsurface using air (Figure 36.8). Once emplaced, positive pressure is applied to maintain the integrity of the borehole and to keep the membrane firmly pressed against the subsurface materials. SEAMIST™ has many applications for collecting samples, making measurements, and transporting sensors, all without allowing the instruments to come into contact with the contaminated media.

Keller and Travis (1993) recently reported on the use of SEAMIST™ to install and retrieve absorbent collectors used for sampling pore fluids in the vadose zone. Absorbent pads were placed at specific locations along the length of the membrane. As the membrane was inflated, the absorbent pads were pressed against the borehole wall. The intent was to wick moisture from the hole wall to collect samples for VOC analysis. They concluded that fluid samples can be collected in reasonable volume and over a short period of time for many, but not all, soil types. They also found that the selection of the absorber type is important, in that some absorbers can be used in places where soils are too dry for suction lysimeters. There remains some additional research and development necessary to fully capitalize on this technique, but it already has shown some promising and valuable attributes.

The DOE has recognized the capabilities of SEAMIST™ and has demonstrated it a number of times as part of the Mixed Waste Landfill Integrated Demonstration (DOE, 1993f). Henriksen and Booth (1993) conducted an evaluation of the cost-effectiveness of the SEAMIST™ system at the request of the DOE. In their study, they used four hypothetical scenarios based on the expressed needs of the DOE. Two of the scenarios involved vertical boreholes and two involved horizontal boreholes. The purpose was to focus on contamination posed by VOCs, SVOCs, other water soluble substances, and low level radioactive waste. In all scenarios, SEAMIST™ demonstrated significant cost savings compared to traditional methods. They concluded that "The most cost- and performance-effective applications of SEAMIST™ occur when the requirements for more than one characterization technology are combined. Whereas several different characterization applications can be integrated into one SEAMIST™ system, the same is not usually true of the corresponding baseline technologies." Obviously, SEAMIST™ has shown great promise for vadose zone monitoring applications.

Another approach to collecting soil gas, soil, and ground-water samples from the vadose zone is through the use of cone penetrometer technology (CPT) and the various attachments available. The cone penetrometer has been in use for many years, primarily to collect continuous lithologic profiles, estimate changes in water quality using electrical conductivity, and to evaluate stability of soils for buildings, dams, etc. Cone penetrometer testing is considered one of the most useful in situ soil test methods available. Large truck-mounted cone penetrometer systems, capable of

---

*The SEAMIST™ system (Pat. No. 5176207) was invented by C. Keller, and the patent and trademark are currently owned by Eastman Cherrington Environmental, Houston, TX.

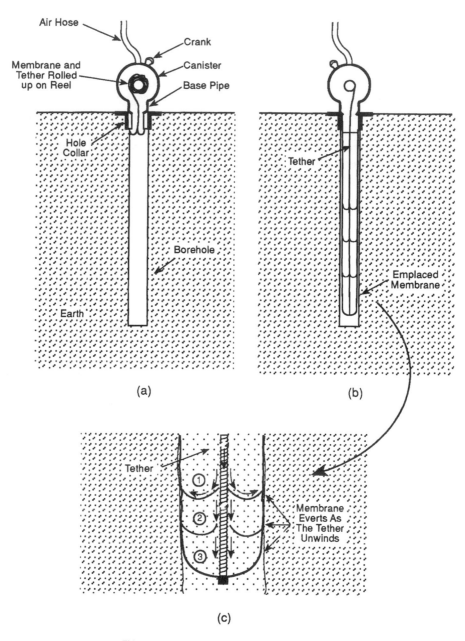

Figure 36.8. SEAMIST™ system deployment sequence: (a) canister placed on surface casing, (b) membrane inverting into borehole under air pressure, and (c) detail of the membrane inversion process (adapted from Henriksen and Booth, 1993).

pushing a penetrometer rod to depths of approximately 150 feet in unconsolidated soils, have been used for investigating strength properties in foundations and road subgrades for over 40 years.

It has only been within the last 5 years that it has been used in environmental applications, and only the last few where it has been used as a means to collect soil,

soil gas, and ground-water samples. In the recent Seventh National Outdoor Action Conference and Exhibition (National Ground Water Association, 1993) there were five papers presented on the application of CPT to environmental problem-solving. The applications described included traditional uses of CPT for collecting stratigraphic, soil strength, and pore pressure data (Williamson and Thomson, 1993), CPT to collect ground-water samples using HydroPunch[TM]* (Cronk and Vovk, 1993), CPT to collect soil-gas samples using a hydrocone (Fierro and Mizerany, 1993), and CPT to collect soil and ground-water samples (Kimball and Tardona, 1993 and Varljen, 1993).

Recently, the Department of Defense (DOD) and DOE have used the penetrometer as an access device for the application of geophysical and chemical methodologies under the aegis of the Site Characterization and Analysis Penetrometer System (SCAPS) program. One of the important aspects of the SCAPS program is the integration of cone penetrometer technology with real-time, downhole sensing capabilities (for example, fiber optic chemical sensors and gamma ray counters)(Figure 36.9). SCAPS sensor-derived data yields real-time information regarding lithology, ground-water depth, ground-water quality, and contaminant plume delineation. In

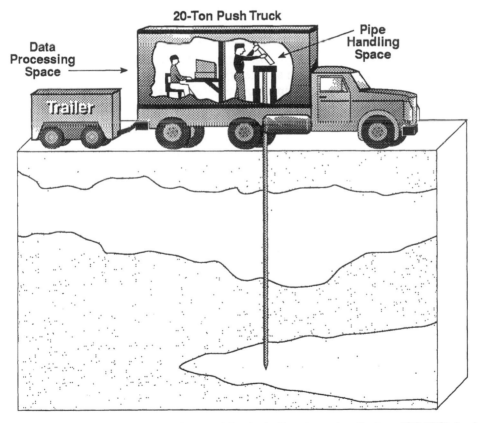

Figure 36.9.   Site Characterization and Analysis Penetrometer System (SCAPS) for in situ monitoring and characterization of contamination.

---

*HydroPunch[TM] is a trademark of QED Groundwater Specialists, Ann Arbor, MI.

addition, the SCAPS can be equipped with soil, soil gas, and ground-water depth sampling capabilities.

Although many sensors are presently being evaluated, the SCAPS penetrometers are equipped with three types. Soil type is determined using strain gauges measuring the force exerted on the penetrometer tip and the sidewall (or sleeve) friction developed on the rod directly above the tip. Estimates of inorganic contamination of soil and ground water, as well as location of lithologic boundaries, are determined using an electrical resistivity sensor. Detection and delineation of plumes of contaminants which fluoresce under ultraviolet light are accomplished using a fiber optic-based laser-induced fluorometer system. Presently the SCAPS fluorometric capabilities are best suited for detection of "heavy" fuel fractions, such as diesel fuel, marine, and JP-5.

SCAPS data processing capabilities allow for the real-time visualization of lithologic properties, as well as the depiction of the location and relative concentrations of contaminants present at a site. This processed data output provides for decision-making capabilities in the field, and can serve as a guide for subsequent characterization and remediation activities. When the SCAPS is equipped with a location/positioning system, the combination allows for real-time, onsite mapping of contamination in the subsurface.

The SCAPS technology allows for cost-effective, real-time, onsite, three-dimensional contaminant plume mapping. In addition, the system can be used for determining optimum placement location for wells and remediation equipment desired. When one compares the SCAPS approach to a conventional investigation utilizing soil gas surveys to screen and determine well emplacement locations at fuel spill sites, the SCAPS approach does not yield false negatives or false positives due to horizontal migration of the soil gas. Rather, the SCAPS approach is a direct measurement versus the soil gas method, which is an indirect assessment of the location of liquid fuel plumes. In addition, the conventional approach generally requires more than one well installation phase, whereas the SCAPS approach typically requires only one well installation phase. When operating in a geophysical and chemical detection mode (not in a sampling mode), the SCAPS is very safe to operate, since workers are not exposed to contaminated materials.

The SCAPS cannot generally be deployed at sites which consist of volcanic, igneous, limestone, cemented, or conglomerate subsurface lithology. Also, the SCAPS fluorometer does not presently detect relatively light molecular weight organic compounds (including benzene, toluene, and xylene). In addition, the fluorometer signal is not analyte-specific nor quantifiable. However, semiquantification is achievable, provided fresh standards are used. Factors which affect the fluorometer sensor response include: grain size distribution, mineralogy, moisture content, and degree of soil aggregation of the substrate.

In terms of deployment logistics, standards for calibration and use of the SCAPS fluorometer method have not been prepared to date. In addition, comparability and validation studies have yet to be completed. Therefore, regulatory acceptance, although expected in the near future, is not ubiquitous.

It appears that cone penetrometer technology will play an increasing role in hazardous waste site characterization, and it will be particularly useful in monitoring and characterizing the vadose zone. Additional information on the application cone penetrometry for environmental applications can be found in Chiang et al. (1992), Chudyk et al. (1985), Cooper et al. (1988), Hirshfield et al. (1981), Lieberman et al. (1991), Malone and Lee (1992), and Robertson and Campanella (1989).

Sample Preparation Techniques

Recently there has been a great deal of attention paid to the use of supercritical fluid extraction (SFE) for the extraction of total petroleum hydrocarbons (TPHs) and other organics (e.g., PCBs) from contaminated soil. The EPA has recently announced that a laboratory SFE-based method is expected to be approved for routine use in the near future. SFE is similar to conventional solvent extraction techniques except that the solvent used is a supercritical fluid (SCF) instead of a liquid. An SCF is a substance in a condition above its critical temperature and pressure. A pump and heater arrangement is used to maintain the solvent in a supercritical state. At these conditions, the physical properties of the SCF are between those of a gas and those of a liquid. Density and solvent strength (the ability to become miscible with a solute) are like those of a liquid, whereas transport properties and compressibility are like those of a gas. During the extraction, the soluble analytes are partitioned from the bulk sample matrix in the SCF, then swept through a flow restrictor into a collection device which is normally at ambient conditions. The fluids used for SFE are generally gases at ambient conditions and can be vented from the collection device, while the extracted analytes are left behind. The extract can then be analyzed by GC, GC/MS, or other techniques.

Common SCFs include carbon dioxide, ammonia, ethane, water, and nitrous oxide. SFCs are attractive solvents for organic compounds because of their characteristically low viscosity, high diffusion coefficient, low toxicity, low flammability, and low surface tension. They retain mass transfer properties similar to a gas, which enables them to penetrate the innermost recesses of a sample matrix. Solvent strength and compound specificity can be controlled by modifying the pressure. For example, using carbon dioxide, extraction at relatively lower pressures will target less polar analytes, while higher pressures will target more polar and higher molecular weight analytes. Addition of modifiers such as acetone and methanol can also accentuate solvent strength of SCFs.

Hawthorne et al. (1993) have evaluated a portable SFE instrument for analyzing soil samples contaminated with TPHs in the field. Included in their study was a portable infrared spectrometer (IR) to make the TPH determinations after extraction. They also compared the analyses made after extraction by SFE with those made after using conventional Soxhlet extraction. Their study demonstrated that field analyses of gasoline, diesel, motor oil, and crude oil contaminated soils using the SFE-IR method compared nicely (less than 10% variation) with samples extracted and analyzed in the laboratory. The investigators were able to have the equipment up and running within about 20 minutes. They did point out that there was a problem with icing of the outlet restrictor; however, it was easily remedied. They also mentioned that the weight (38 kg) and power requirements of the tested instrument would probably restrict its use to those sites with easy access to sampling locations and available power. This success in a field application will likely stimulate the development of more compact designs.

SFE can be used to separate target compounds from mixed organic waste streams and contaminated soils. Solvent efficiency (in terms of percent leached) is comparable to commonly used methods for contaminated soils. In addition, solvents are recovered for re-use once they have been returned to ambient conditions. Another advantage of SFE is its ability to directly couple the extraction effluent from a sample matrix to an analytical instrument for quantification. If so desired, SFE can also be used in an off-line collection mode. This is typically done for the purpose of method development and analytical characterization. In the off-line mode, extraction effluent can be collected and analyzed using a number of devices (i.e., gas chromatograph, liquid chromatograph, mass spectrometer, infrared spectrometer, ultraviolet spectrometer, nuclear magnetic resonance spectrometer, etc.).

Present limitations include: handling of a fluid in a supercritical state can pose a health and safety problem; capital costs are relatively high; and regulatory hurdles need to be overcome. Recent breakthroughs include: use of carbon dioxide which requires relatively lower temperatures (typically less than 100°C) and is nontoxic; lower costs due to commercialization of SFE units; and moderate regulatory acceptance due to successful peer-reviewed demonstrations. Additional information concerning the environmental applications of SFE can be found in Dial (1985), Fan et al. (1994), Freeman and Rowlinson (1960), and Hogen (1985).

Thermal desorption has recently been applied in investigations using field GC and field GC/MS methodologies (Robbat et al., 1992a and 1992b). Thermal desorption is not a new concept; however, it has only recently been used in field analytical applications aimed at screening for known and unknown volatile and semivolatile compounds.

Essentially, a sample is placed in a chamber or tube and heated in a stepwise manner, generally ramped up to approximately 300°C (depending on the goal of the investigation). Temperature can be computer-controlled. The heated effluent is injected into the head of a GC column for separation and identification. Some models are equipped with an injector/desorber unit, which alleviates the need for cumbersome transfer lines or improperly matched flow rates.

The most obvious advantage to thermal desorption techniques is that they alleviate the need for solvents when investigating the presence of volatiles and semivolatiles in liquid and solid samples. This not only reduces time requirements, but also alleviates potential health and safety considerations. In addition, less sample handling is required, thereby reducing the potential for contamination, loss of analyte, and personnel exposure.

One limitation to thermal desorption applications is that the compounds entrained in the vapor stream via heating are not necessarily bioavailable at the site. This is because the heat disrupts the representative sample in a way that does not occur in the "real world." Essentially, one measures the amount and type of constituents released given the conditions the sample is exposed to. This is also true for solvent extraction applications. In addition, the amount of a compound released by heating is a function of soil and liquid matrix characteristics. Compound-specific desorption kinetics and partitioning characteristics are not yet fully understood. Also, although not conclusive, there is a possibility for alteration of some analytes by heating. Because of these reasons, thermal desorption applications may only yield qualitative and semiquantitative results.

Poziomek and Eiceman (1992a and 1992b) report sorption of organic analytes from water onto solid phase extraction disks, followed by thermal desorption and use of IMS as a potential field screening method. The use of solid-phase extraction disks should also be applicable to sampling semivolatile organic vapors in the vadose zone. They would be installed as passive samples. An alternative would be to install micro tubes containing solid phase extraction material and draw vapors through them.

SUMMARY

Much of the current effort on emerging technologies for detecting and measuring contaminants in the vadose zone is focused on developing tools that can be used to generate data and information in near-real time. There is considerable activity in the development of chemical sensors, miniaturized or field portable laboratory instrumentation, noninvasive characterization techniques, minimally-invasive techniques, and data and information management tools. Effective, rapid, onsite data acquisition, data management and evaluation with capabilities for onsite, real-time visualization of the data could significantly improve overall decision-making and reduce the associated costs and schedule requirements. In addition, the tools and techniques being developed will support monitoring during remediation to optimize the treatment technology and to measure the progress and effectiveness of remediation; hazardous waste site characterization; hazardous waste treatment facility monitoring; and post-closure monitoring.

NOTICE

The U.S. Environmental Protection Agency (EPA), through its Office of Research and Development (ORD), partially funded and managed the compilation of information included in this chapter. It has been subjected to the Agency's peer review process and has been approved as an EPA publication. Mention of trade names or commercial products does not constitute endorsement or recommendation for use. The U.S. government retains a nonexclusive, royalty-free license to publish or reproduce the published form of this contribution, or to allow others to do so for U.S. government purposes.

REFERENCES

Air & Waste Management Association, *Proceedings of the Third International Symposium on Field Screening Methods for Hazardous Wastes and Toxic Chemicals*, Air & Waste Management Association, Pittsburgh, PA, 1993, p. 1323.

Anderson, G.P., R.A. Ogert, L.C. Shriver-Lake, D.C. Wijesuriya, J.P. Golden, and F.S. Ligler, "Fiber Optic-Based Biosensor for Environmental Monitoring," in *Proceedings of the Third International Symposium on Field Screening Methods for Hazardous Wastes and Toxic Chemicals*, Air and Waste Management Association, Pittsburgh, PA, 1993, p. 737.

Angel, S.M., "Chemically Selective Fiber-Optic Sensors," *Spectroscopy*, 2:38 (1987).

Arman, H., S.M. Klainer, and J.R. Thomas, "An On-Line Fiber Optic Chemical Sensor (FOCS™) System for Monitoring Above and Below-Ground Hydrocarbon Storage Tanks," in *Proceedings of the Third International Symposium on Field Screening Methods for Hazardous Wastes and Toxic Chemicals*, Air and Waste Management Association, Pittsburgh, PA, 1993, p. 352.

Arnold, M.A., "Fiber-Optic Chemical Sensors," *Analytical Chemistry*, 64:1015 (1992).

Arnold, N.S., P.A. Cole, D.W. Hu, B.Watteyne, D.T. Urban, and H.L.C. Meuzelaar, "The Next Horizon in Portable GC/MS for Field Air Monitoring Applications," in *Proceedings of the Third International Symposium on Field Screening Methods for Hazardous Wastes and Toxic Chemicals*, Air and Waste Management Association, Pittsburgh, PA, 1993, p. 915.

Azad, A., L. Younkman, and S. Akbar, "Semiconducting Oxides for Carbon Monoxide Detection," in *Proceedings of the Third International Symposium on Field Screening Methods for Hazardous Wastes and Toxic Chemicals*, Air and Waste Management Association, Pittsburgh, PA, 1993, p. 1323.

Ballantine, D.S., and H. Wohltjen, "Surface Acoustic Wave Devices for Chemical Analysis," *Analytical Chemistry*, 61:704A (1989).

Bartlett, P.N., and S.K. Ling-Chung, "Conducting Polymer Gas Sensors, Part III: Results for Four Different Polymers and Five Different Vapours," *Sensors and Actuators*, 20:287 (1989).

Bell, S.E., and G.A. Eiceman, "Hand-Held GC-Ion Mobility Spectrometry for On-Site Analysis of Complex Organic Mixtures in Air or Vapors Over Waste Sites," in *Proceedings of the Second International Symposium on Field Screening Methods for Hazardous Wastes and Toxic Chemicals*, National Ground Water Association, Dublin, OH, 1991, p. 153.

Bratton, W.L., J.D. Shinn III, and J.L. Bratton, "Air Force Site Characterization and Analysis Penetrometer System for Fuel-Contaminated Soils," in *Proceedings of the Third International Symposium on Field Screening Methods for Hazardous Wastes and Toxic Chemicals*, Air and Waste Management Association, Pittsburgh, PA, 1993, p. 431.

Butler, M., S.M. Klainer, J. Tussey, and J. Thomas, "A Two Channel Portable Fluorimeter for Use with Fiber Optic Chemical Sensors (FOCS[TM])," in *Proceedings of the Third International Symposium on Field Screening Methods for Hazardous Wastes and Toxic Chemicals*, Air and Waste Management Association, Pittsburgh, PA, 1993, p. 183.

Carney, K.R., E.B. Overton, A.M. Mainga, and C.F. Steel, "Detection Limits and Chromatographic Capabilities of an Improved Microchip Gas Chromatograph," in *Proceedings of the Third International Symposium on Field Screening Methods for Hazardous Wastes and Toxic Chemicals*, Air and Waste Management Association, Pittsburgh, PA, 1993, p. 149 (abstract only).

Cernosek, R.W., G.C. Frye, and D.W. Gilbert, "Portable Acoustic Wave Sensor Systems for Real-Time Monitoring of Volatile Organic Compounds," in *Proceedings Ideas in Science & Electronics '93*, The Institute of Electrical and Electronics Engineers, Inc., Albuquerque, NM, 1993, pp. 44–51.

Chappell, S.E., and M.S. Simmons, "Performance Requirements for Field Applications of Portable Gas Chromatographs," in *Proceedings of the Third International Symposium on Field Screening Methods for Hazardous Wastes and Toxic Chemicals*, Air and Waste Management Association, Pittsburgh, PA, 1993, p. 899.

Chiang, C.W., K.R. Loos, and R.A. Klopp, "Field Determination of Geological/Chemical Properties of an Aquifer by Cone Penetrometry and Headspace Analysis," *Ground Water*, 30(3):428 (1992).

Chudyk, W.A., M.M. Carrabba, and J.E. Kenny, "Remote Detection of Ground Water Contaminants using Far-Ultraviolet, Laser-Induced Fluorescence," *Analytical Chemistry*, 57:1237 (1985).

Chudyk, W., K. Pohlig, K. Exarhoulakos, J. Holsinger, and N. Rico, "In Situ Analysis of Benzene, Ethylbenzene, Toluene and Xylenes (BTEX) Using Fiber Optics," in *Ground Water and Vadose Zone Monitoring*, D.M. Nielsen and A.I. Johnson, Eds., American Society for Testing and Materials, ASTM STP 1053, Philadelphia, PA, 1990, pp. 266–271.

Cooper, S.S., P.G. Malone, R.S. Olsen, and D.H. Douglas, Development of a Computerized Penetrometer System for Hazardous Waste Site Soils Investigations, U.S. Army Toxic and Hazardous Materials Agency, Aberdeen Proving Ground, Report No. AMXTH-TR-TE-882452, 1988.

Cronk, G.D., and M.A. Vovk, "Conjunctive Use of Cone Penetrometer Testing and Hydro-Punch™ Sampling to Evaluate Migration of VOCs in Groundwater," in *Proceedings of the Seventh National Outdoor Action Conference and Exposition*, National Ground Water Association, Dublin, OH, 1993, p. 459.

deFilippi, R.P., and T.J. Cody Jr., "A Monitoring System for Hydrocarbon Leakage from Petroleum and Chemical Product Sites," in *Proceedings of the Third International Symposium on Field Screening Methods for Hazardous Wastes and Toxic Chemicals*, Air and Waste Management Association, Pittsburgh, PA, 1993, p. 1241.

Dial, C.J., Hazardous Waste Treatment Research, U.S. Environmental Protection Agency, EPA/600/M-85/013, 1985.

Driscoll, J.N., "Hydrocarbon Screening Methods," *Environmental Testing & Analysis*, 2:23 (1993).

Eastwood, D., and T. Vo-Dinh, Molecular Optical Spectroscopic Techniques for Hazardous Waste Site Screening, U.S. Environmental Protection Agency, Project Summary, EPA/600/S4-91/011, 1991.

Eiceman, G.A., "Ion Mobility Spectrometry: Opportunities, Challenges, and Future Directions," in *Proceedings of the Third International Symposium on Field Screening Methods for Hazardous Wastes and Toxic Chemicals*, Air and Waste Management Association, Pittsburgh, PA, 1993, p. 55.

Eiceman, G.A., "Advances in Ion Mobility Spectrometry: 1980–1990," *Critical Reviews in Analytical Chemistry*, 22:471 (1991).

Ewing, K.J, T.G. Bilodeau, G.M. Nau, and I.D. Aggarwal, "Fiber Optic Sensor for the Detection of Dense Non-Aqueous Phase Liquids," in *Proceedings of the Third International Symposium on Field Screening Methods for Hazardous Wastes and Toxic Chemicals*, Air and Waste Management Association, Pittsburgh, PA, 1993, p. 163.

Fan, C., S. Krishnamurthy, and C.T. Chen, "A Critical Review of Analytical Approaches for Petroleum Contaminated Soil," in *Analysis of Soil Contaminated with Petroleum Constituents*, T. A. O'Shay and K.B. Hoddinott, Eds., ASTM STP 1221, American Society for Testing and Materials, Philadelphia, PA, 1994.

Fernandez, C., C.P. Lee, J. Guyton, and S.S. Parmar, "Field Evaluation of Photovac 10S Plus Portable GC for Analysis of VOCs in Ambient Air," in *Proceedings of the Third International Symposium on Field Screening Methods for Hazardous Wastes and Toxic Chemicals*, Air and Waste Management Association, Pittsburgh, PA, 1993, p. 214.

Fierro, P., and J.E. Mizerany, "Utilization of Cone Penetrometer Technology as a Rapid, Cost-Effective Investigative Technique, in Groundwater," in *Proceedings of the Seventh National Outdoor Action Conference and Exposition*, National Ground Water Association, Dublin, OH, 1993, p. 473.

Freeman, P.I., and J.S. Rowlinson, "Lower Critical Points in Polymer Solutions," *Polymer*, 1:20 (1960).

Gammage, R.B., J.C. DePriest, M.E. Murray, R.V. Wheeler, M.R. Salasky, J.C. Dempsey, and P. Kotrappa, "In-Situ Passive Monitoring of Alpha-Emitting Radionuclides," in *Proceedings of the Third International Symposium on Field Screening Methods for Hazardous Wastes and Toxic Chemicals*, Air and Waste Management Association, Pittsburgh, PA, 1993, p. 1323.

Gerlach, R.W., R.J. White, N.F.D. O'Leary, and J.M. Van Emon, *Superfund Innovative Technology Evaluation (SITE) Program Evaluation Report for Antox BTX Water Screen*

(BTX Immunoassay), prepared for the U.S. Environmental Protection Agency, EPA/540/R-93/518, 1993, p. 91.

Gillispie, G.D., R.W. St. Germain, and J.L. Klingfus, "Subsurface Optical Probes: Current Status and Future Prospects," in *Proceedings of the Third International Symposium on Field Screening Methods for Hazardous Wastes and Toxic Chemicals*, Air and Waste Management Association, Pittsburgh, PA, 1993, p. 793.

Gillispie, G.D., "Remote Fiber Optic Spectroscopy," in *Proceedings DOD Fiber Optics '93, 3rd Biennial Department of Defense Fiber Optics Conference*, McLean, Virginia, 1992, p. 115.

Grate, J.W., and M. Klusty, "Surface Acoustic Wave Vapor Sensors Based on Resonator Devices," *Analytical Chemistry*, 63:1719 (1991).

Grate, J.W., S.J. Martin, and R.M. White, "Acoustic Wave Microsensors, Part I," *Analytical Chemistry*, 65:940A (1993a).

Grate, J.W., S.J. Martin, and R.M. White, "Acoustic Wave Microsensors, Part II," *Analytical Chemistry* 65:987A (1993b).

Grate, J.W., S.L. Rose-Pehrsson, B.L. Venezky, M. Klusty, and H. Wohltjen, "Smart Sensor System for Trace Organophosphorus and Organosulfur Vapor Detection Employing a Temperature-Controlled Array of Surface Acoustic Wave Sensors, Automated Sample Preconcentration, and Pattern Recognition," *Analytical Chemistry*, 65:1868 (1993).

Grate, J.W., and M.H. Abraham, "Solubility Interactions in the Design of Chemically Selective Sorbent Coatings for Chemical Sensors and Arrays," *Sensors and Actuators*, B, 3:85 (1991).

Guilbault, G.G. and J.M. Jordan, "Analytical Uses of Piezoelectric Crystals," *Critical Reviews in Analytical Chemistry*, 19:1 (1988).

Haas, J.T. III, T.G. Matthews, and R.B. Gammage, "In Situ Detection of Toxic Aromatic Compounds in Groundwater Using Fiberoptic UV Spectroscopy," in *Field Screening Methods for Hazardous Wastes and Toxic Chemicals Second International Symposium*, PB92-125764, EPA 600/9-91/028,1991, p. 677.

Hawthorne, S.B., D.J. Miller, and K.M. Hegvik, "Field Evaluation of the SFE-Infrared Method for Total Petroleum Hydrocarbon (TPH) Determinations," *Journal of Chromatographic Science*, 31:26 (1993).

Henriksen, A.D., and S.R. Booth, "Cost Effectiveness of an Innovative Technology for VOC Detection: The Direct Sampling Ion Trap Mass Spectrometer," Los Alamos National Laboratory, LA-UR 92-3527, 1992, p. 43.

Henriksen, A.D., and S.R. Booth, "Cost-Effectiveness Analysis of the SEAMIST™ Membrane System Technology," Los Alamos National Laboratory, Los Alamos, NM, LA-UR-93-3750, 1993, p. 49.

Hewitt, A.D., "Comparison of Site Characterization for Trichloroethylene by Headspace Gas Chromatography and Purge-and-Trap Gas Chromatography Mass Spectrometry," in *Proceedings of the Third International Symposium on Field Screening Methods for Hazardous Wastes and Toxic Chemicals*, Air and Waste Management Association, Pittsburgh, PA, 1993, p. 1323.

Hewitt, A., "Review of Current and Potential Future Sampling Practices for Volatile Organic Compounds," in *Proceedings of the 16th Annual Army Environmental R&D Symposium*, Williamsburg, VA, 1992, p. 75.

Hewitt, A.D., "Comparison of Sample Collection and Handling Practices for the Analysis of Volatile Organic Compounds in Soils," in *U.S. EPA* (1993b).

Hill, H.H., W.F. Siems, R.H. St. Louis, and D.G. McMinn, "Ion Mobility Spectrometry," *Analytical Chemistry*, 62:1201A (1990).

Hirschfeld, T., T. Deaton, F.P. Milanovich, and S. Klainer, "The Feasibility of Using Fiber Optics for Monitoring Ground Water Contaminants," *Optical Engineering*, 22:527 (1981).

Hogen, G.G., "Extraction with Supercritical Fluids - Why, How, and So What?," *U.S. Chemtech*, 15(7):440 (1985).

Janata, J., and A. Bezegh, "Chemical Sensors," *Analytical Chemistry*, 60:62R (1988).

Janata, J., "Chemical Sensors," *Analytical Chemistry*, 62:33R (1990).

Janata, J., "Chemical Sensors," *Analytical Chemistry*, 64:196R (1992).

Jenkins, R.A., W.H. Griest, R.L. Moody, M.V. Buchanan, J.P. Maskarinec, F.F. Dryer, and C.-h Ho, Technology Assessment of Field Portable Instrumentation for Use at Rocky Mountain Arsenal. Final Report, Oak Ridge National Laboratory, ORNL/TM-10542, 1988, p. 105.

Jenkins, R.A., C.E. Higgins, T.M. Gayle, G.W. Allin, and R.R. Smith, "Field Experiences with a Multisorbent Arrayed Sampler for In-Situ Collection of Vadose Zone Volatile Organic Compounds," in *Proceedings of the Third International Symposium on Field Screening Methods for Hazardous Wastes and Toxic Chemicals*, Air and Waste Management Association, Pittsburgh, PA, 1993, p. 1018.

Katritzky, A.R., and R.J. Offerman, "The Development of New Microsensor Coatings and Short Survey of Microsensor Technology," *Critical Reviews in Analytical Chemistry*, 21:83 (1989).

Keller, C., and W. Lowry, "A New Vadose Zone Fluid Sampling System for Uncased Holes," in *Proceedings of the Fourth National Outdoor Action Conference*, National Groundwater Association, Dublin, OH, 1990, pp. 3–10.

Keller, C., and B. Travis, "Evaluation of the Potential of Fluid Absorber Mapping of Contaminants, in Groundwater," in *Proceedings of the Seventh National Outdoor Action Conference and Exposition*, National Groundwater Association, Dublin, OH, 1993, p. 421.

Kimball, C.E., and P.M. Tardona, "A Case History of the Use of a Cone Penetrometer to Assess an UST Release That Occurred on a Property That is Adjacent to a DNAPL Release Site," in *Proceedings of the Seventh National Outdoor Action Conference and Exposition*, National Ground Water Association, Dublin, OH, 1993, p. 487.

King, W. Jr., "Piezoelectric Sorption Detector," *Analytical Chemistry*, 36:1735 (1964).

Koglin, E.N., and E.J. Poziomek, "Field Screening and Analysis Technologies: An Overview of the State of the Art," in *Monitoring and Measurement Technologies for Hazardous Substances*, Water Environment Federation, Alexandria, VA, 1993, p. 7.

Korolkoff, N.O., "Survey of Toxic Gas Sensors and Monitoring Systems," *Solid State Technology*, 6:4 (1989).

Kowalski, P., J. Wronka, and F. Laukien, "Multiple Uses and Applications of the Bruker Mobile Mass Spectrometer MM-1 for the Detection of Chemical Warfare Agents and Other Hazardous Substances," in *Proceedings of the Third International Symposium on Field Screening Methods for Hazardous Wastes and Toxic Chemicals*, Air and Waste Management Association, Pittsburgh, PA, 1993, p. 624.

Leibman, C.P., D. Dogruel, and E.P. Vanderveer, "Transportable GC/Ion Trap Mass Spectrometry for Trace Field Analysis of Organic Compounds," in *Proceedings of the Second International Symposium on Field Screening Methods for Hazardous Wastes and Toxic Chemicals*, National Ground Water Association, Dublin, OH, 1991, p. 367.

Levine, S., J. Gonzalez, and R. Berkley, "Advances in High Speed Gas Chromatography for Monitoring Gas and Vapor Contaminants in Workplace and Ambient Air," in *Proceedings of the Third International Symposium on Field Screening Methods for Hazardous Wastes and Toxic Chemicals*, Air and Waste Management Association, Pittsburgh, PA, 1993, p. 909.

Lieberman, S.H., G.A. Theriault, S.S. Cooper, P.G. Malone, R.S. Olsen, and P.W. Lurk, "Rapid Subsurface In-Situ Screening of Petroleum Hydrocarbon Contamination Using Laser Induced Fluorescence Over Optical Fibers," in *Field Screening Methods for Hazardous Wastes and Toxic Chemicals Second International Symposium*, PB92-125764, EPA 600/9-91/028, 1991, p. 57.

Lieberman, S.H., S.E. Apitz, P.G. Malone, and S.S. Cooper, "Real Time In Situ Measurements of Fuels in Soil: Comparison of Fluorescence and Soil Gas Measurements," in *Proceedings of the Third International Symposium on Field Screening Methods for Hazardous Wastes and Toxic Chemicals*, Air and Waste Management Association, Pittsburgh, PA, 1993, p. 1123.

Malone, P.G., and L.T. Lee, The Site Characterization and Analysis Penetrometer System, a Breakthrough in Hazardous Waste Site Investigations, Army Research, Development & Acquisition Bulletin, July-August 1992, p. 34.

Matz, G., and W. Schröder, "Fast GC-MS Analysis of Contaminated Soil: Routine Field Screening in Hamburg," in *Proceedings of the Third International Symposium on Field Screening Methods for Hazardous Wastes and Toxic Chemicals*, Air and Waste Management Association, Pittsburgh, PA, 1993, p. 963.

Meidinger, R.F., R.W. St. Germain, V. Dohotariu, and G.D. Gillispie, "Fluorescence of Aromatic Hydrocarbons in Aqueous Solutions," in *Proceedings of the Third International Symposium on Field Screening Methods for Hazardous Wastes and Toxic Chemicals*, Air and Waste Management Association, Pittsburgh, PA, 1993, p. 395.

Meierer, R.E., D. McCormack, J. Jennings, and R. Sutton, Abstract Guide, *SUPERFUND XIV Conference*, Hazardous Materials Control Resources Institute, Rockville, MD, 1993, p. 2.

Meuzelaar. H.L.C., "Man-Portable GC/MS: Opportunities, Challenges and Future Directions," in *Proceedings of the Third International Symposium on Field Screening Methods for Hazardous Wastes and Toxic Chemicals*, Air and Waste Management Association, Pittsburgh, PA, 1993, p. 35.

Mierzwinski, A., and Z. Witkiewicz, "The Application of Piezoelectric Detectors for Investigations of Environmental Pollution," *Environmental Pollution*, 57:181 (1989).

Milanovich, F.P. P.F. Daley, K. Langry, B.W. Colston Jr., S.B. Brown, and S.M. Angel, "A Fiber Optic Sensor for the Continuous Monitoring of Chlorinated Hydrocarbons," in *Proceedings of the Second International Symposium on Field Screening Methods for Hazardous Wastes and Toxic Chemicals*, U.S. Environmental Protection Agency, Las Vegas, NV: 1991, p. 43.

National Ground Water Association, *Proceedings of the Seventh National Outdoor Action Conference and Exposition*, Dublin, OH, 1993, p. 746.

National Research Council, *Application of Analytical Chemistry to Oceanic Carbon Cycle Studies*, Washington, DC: National Academy Press, 1993.

Office of Technology Assessment, Complex Cleanup: The Environmental Legacy of Nuclear Weapons Production, Washington, D.C., OTA-O-484, 1991.

Penrose, W.R., J.R. Stetter, M.W. Findlay, W.J. Buttner, and Z. Cao, "Arrays of Sensors and Microsensors for Field Screening of Unknown Chemical Agents," in *Proceedings of the Second International Symposium on Field Screening Methods for Hazardous Wastes and Toxic Chemicals*, U.S. Environmental Protection Agency, Las Vegas, NV, 1991, p. 85.

Poziomek, E.J., "Detection and Analysis of Emissions at Hazardous Waste Sites: Part 2 - Coatings for Microsensors in the Detection and Analysis of Hazardous Material," Environmental Science and Engineering Fellows Program Report, American Association for the Advancement of Science, Washington, DC, 1989.

Poziomek, E.J., "Trends in Field Methods Development and Use," in *Proceedings of the International Symposium on Field Screening Methods for Hazardous Wastes and Toxic Chemicals*, Air and Waste Management Association, Pittsburgh, PA, 1993, p. 1307.

Poziomek, E.J., "Fiber-Optic and Other Chemical Sensors for Field Screening: Technology Barriers," in *Hazardous Waste Site Investigations: Toward Better Decisions*, R.B. Gammage, and B.A. Berven, Eds., (Boca Raton, FL: Lewis Publishers, 1992a), p. 143.

Poziomek, E.J., "Fiber Optic Chemical Sensors. A Review," in *Proceedings DOD Fiber Optics '92, 3rd Biennial Department of Defense Fiber Optics Conference*, McLean, VA, 1992b, p. 115.

Poziomek, E.J., W.H. Engelmann, and E.N. Koglin, "Coating Selection and Evaluation: Weak Links in the Development of In-Situ Reactive Sensors," in *Proceedings of Real-Time In Situ Monitoring of Groundwater Workshop*, Office of Technology Development, Department of Energy, Dallas, TX, 1990.

Poziomek, E.J., and G.A. Eiceman, "Use of Ion Mobility Spectrometry in Determining Organic Pollutants in Water," in *Proceedings of the 1992 Federal Environmental Restoration Conference, Hazardous Materials Control Research Institute*, Vienna, VA, 1992a, p. 252.

Poziomek, E.J., and G.A. Eiceman, "Solid Phase-Enrichment, Thermal Desorption, and Ion Mobility Spectrometry for Field Screening of Organic Pollutants in Water," *Environmental Science & Technology*, 26:1318 (1992b).

Poziomek, E.J., W.H. Engelmann, E.N. Koglin, and B. Nielsen, "Field Screening Technologies for Hazardous Waste Sites," in *Abstract Book, Superfund XIV*, Hazardous Materials Control Resources Institute, Silver Spring, MD, 1993, p. 1.

Raab, G.A., R.E. Enwall, W.H. Cole, M.L. Faber, and L.A. Eccles, "X-Ray Fluorescence Field Method for Screening of Inorganic Contaminants at Hazardous Waste Sites," in *Hazardous Waste Measurements*, M.S. Simmons, Ed., (Chelsea, MI: Lewis Publishers, 1991), pp. 159–192.

Robbat, A., T. Liu, and B. Abraham, "Evaluation of a Thermal Desorption GC/MS: On Site Detection of PCBs at a Hazardous Waste Site," *Analytical Chemistry*, 64(4):358–364 (1992a).

Robbat, A., T. Liu, and B. Abraham, "On Site Determination of PAHs in Contaminated Soils by Thermal Desorption GC/MS," *Analytical Chemistry*, 64(13):1477–1483 (1992b).

Roberts, J., F. Trujillo, and C.P. Leibman, "Transportable GC/MS for Volatile Organic Analysis—The Sequel: A Design Derived from Field Experience," in *Proceedings of the Third International Symposium on Field Screening Methods for Hazardous Wastes and Toxic Chemicals*, Air and Waste Management Association, Pittsburgh, PA, 1993, p. 947.

Robertson, P.K., and R.G. Campanella, *Guidelines for Geotechnical Design Using the Cone Penetrometer Test and CPT with Pore Pressure Measurement*, Hogentogler & Co., Inc., Columbia, MD, 1989, p. 185.

Rogers, K.M., and E.J. Poziomek, "Environmental Applications of Biosensors: Opportunities and Future Directions," in *Proceedings of the Third International Symposium on Field Screening Methods for Hazardous Wastes and Toxic Chemicals*, Air and Waste Management Association, Pittsburgh, PA, 1993, p. 26.

Rossabi, J., B. Colston Jr., S. Brown, F. Milanovich, and L.T. Lee Jr., "In Situ Subsurface Monitoring of Vapor-Phase TCE Using Fiber Optics," in *Proceedings of the Third International Symposium on Field Screening Methods for Hazardous Wastes and Toxic Chemicals*, Air and Waste Management Association, Pittsburgh, PA, 1993a, p. 1165.

Rossabi, J.B., R.A. Jenkins, M.B. Wise, M.R. Guerin, P.M. Kearl, S. Ballard, G.J. Elbring, G. Frye, F.P. Milanovich, P.F. Daley, A.L. Ramirez, W.D. Daily, E. Kaplan, B. Neilsen, J. Evans, and K.B. Olsen, Demonstration of Innovative Monitoring Technologies at the Savannah River Integrated Demonstration Site[U], Prepared for the U.S. Department of Energy under Contract No. DE-AC09-89SR18035, Report Number WSRC-TR-93-671, 1993b.

St. Germain, R.W., G.D. Gillispie, and J.L. Klingfus, "Variable Wavelength Laser System for Field Fluorescence Measurements," in *Proceedings of the Third International Symposium on Field Screening Methods for Hazardous Wastes and Toxic Chemicals*, Air and Waste Management Association, Pittsburgh, PA, 1993, p. 1113.

Seitz, W.R., "Chemical Sensors Based on Immobilized Indicators and Fiber Optics," *CRC Critical Reviews in Analytical Chemistry*, 19:135 (1988).

Shahriari, M.R., J. Liu, G.H. Sigel Jr., and M. Liva, "Ruggedized Fiber Optic Environmental Sensors Based on Porous Optical Materials," in *Proceedings of the Third International Symposium on Field Screening Methods for Hazardous Wastes and Toxic Chemicals*, Air and Waste Management Association, Pittsburgh, PA, 1993, p. 175.

Sharma, A. and A.Z. Fasihi, "Optical Sensor for Halothane Based on Energy Transfer," in *Proceedings of the Third International Symposium on Field Screening Methods for Hazardous Wastes and Toxic Chemicals*, Air and Waste Management Association, Pittsburgh, PA, 1993, p. 419.

Siegrist, R.L., and P.D. Jenssen, "Evaluation of Sampling Method Effects on Volatile Organic Compound Measurements in Contaminated Soils," *Environmental Science and Technology*, 24:1387 (1990).

Siegrist, R.L., and P.D. Jenssen, *Sampling Method Effects on Volatile Organic Compound Measurements in Solvent Contaminated Soil*, Institute for Georesources and Pollution Research, Norway, 1989.

Simmons, K.E., and S.L. Knollmeyer, "Field Analysis of Volatile Organics Using Triple Detector Gas Chromatography," in *Proceedings of the Third International Symposium on Field Screening Methods for Hazardous Wastes and Toxic Chemicals*, Air and Waste Management Association, Pittsburgh, PA, 1993, p. 203.

Snow, A.W., W.R. Barger, M. Klusty, H. Wohltjen, and N. L. Jarvis, "Simultaneous Electrical Conductivity and Piezoelectric Mass Measurement on Iodine-Doped Phthalocyanine Langmuir-Blodgett Films," *Langmuir*, 2:513 (1986).

Snow, A.W., and W. Barger, "Phthalocyanine Films in Chemical Sensors," in *Phthalocyanines: Principles and Applications*, A.P.B. Lever and C.C. Len, Eds., (New York: VCH, 1989).

Snyder, A.P., C.S. Harden, A.H. Brittain, M.-G. Kim, N.S. Arnold, and H.L.C. Meuzelaar, "Portable, Hand-Held Gas Chromatograph-ion Mobility Spectrometer," *International Laboratory*, 1993, pp. 30–33.

Spittler, T., "Two Column Chromatography Using a Portable GC," in *Proceedings of the Third International Symposium on Field Screening Methods for Hazardous Wastes and Toxic Chemicals*, Air and Waste Management Association, Pittsburgh, PA, 1993, p. 148 (abstract only).

Stetter, J.R., S. Zaromb, and M.W. Findlay, "Monitoring Electrochemically Inactive Compounds by Amperometric Gas Sensors," *Sensors and Actuators*, 6:269 (1984).

Stetter, J.R., W.R. Penrose, G.J. Maclay, Z. Cao, L.J. Luskus, and J.D. Mulik, "The Amperometric Gas Sensor: One Electroanalytical Solution for Pollution," Preprint extended abstract, Division of Environmental Chemistry, American Chemical Society, Atlanta, GA, 1991, p. 333.

Stout, R.J., P.J. Mulligan, and D.D. Nelson, "Cost-Effective & Reliable Field Analysis of TPH Saves Federal Dollars," *Hazardous Materials Control*, 6:46 (1993).

Stutman, M., "A Novel Passive Sorbent Collection Apparatus for Site Screening of Semi-volatile Compounds," in *Proceedings of the Third International Symposium on Field Screening Methods for Hazardous Wastes and Toxic Chemicals*, Air and Waste Management Association, Pittsburgh, PA, 1993, p. 579.

Takahata, K., "Tin Dioxide Sensors—Development and Applications," in *Chemical Sensor Technology*, Vol. 1, T. Seiyama Ed., (New York, NY: Elsevier, 1988), pp. 39–55.

Triegel, E.K., "Development of an ASTM Standard for Sampling Soils for VOCs," in *U.S. EPA* (1993).

U.S. DOE, Environmental Restoration and Waste Management Five-Year Plan, Fiscal Years 1994–1998, Volume 1, DOE/S-00097P Vol. 1, Washington, DC, 1993a.

U.S. DOE, Literature Search, Review, and Compilation of Data for Chemical and Radio-chemical Sensors: Task 1 Report, prepared by the Hazardous Waste Remedial Actions Program, Oak Ridge, TN, DOE/HWP-130, 1993b.

U.S. DOE, Literature Search, Review, and Compilation of Data for Chemical and Radio-chemical Sensors: Task 2 Report, prepared by the Hazardous Waste Remedial Actions Program, Oak Ridge, TN, DOE/HWP-133, 1993c.

U.S. DOE, Chemical Sensor R&D Status Based on Research and Industrial Interviews: Task 3 Report, prepared by the Hazardous Waste Remedial Actions Program, Oak Ridge, TN, DOE/HWP-138, 1993d.

U.S. DOE, Literature Search, Review, and Compilation of Data for Biosensors and Thermal Sensors: Task 4 Report, prepared by the Hazardous Waste Remedial Actions Program, Oak Ridge, TN, DOE/HWP-144, 1993e.

U.S. DOE, Office of Technology Development FY 1993 Program Mid-Year Summaries Research, Development, Demonstration, Testing and Evaluation, Washington, DC, DOE/EM- 0110P, 1993f.

U.S. EPA, The Superfund Innovative Technology Evaluation Program: Technology Profiles, Sixth Edition, EPA/540/R-93/526, 1993a.

U.S. EPA, Measuring and Interpreting VOCs in Soils: State of the Art and Research Needs: A Symposium Summary, EPA 540/R-94/506, 1993b.

Van Emon, J.M., and V. Lopez-Avila, "Immunochemical Methods for Environmental Analysis," *Analytical Chemistry*, 64:79 (1992).

Varljen, M.D., "Combined Soil Gas and Ground Water Field Screening Using the Hydro-Punch™ and Portable Gas Chromatography," in *Proceedings of the Seventh National Outdoor Action Conference and Exposition*, National Ground Water Association, Dublin, OH, May 1993, p. 499.

Walt, D.R., "Continuous Monitoring of Groundwater Using Fiber Optic Chemical Sensors," The Center for Environmental Management, Tufts University, Medford, MA, 1993a, p. 29.

Walt, D.R., "Fiber Optic Sensors," in *Proceedings of the Third International Symposium on Field Screening Methods for Hazardous Wastes and Toxic Chemicals*, Air and Waste Management Association, Pittsburgh, PA, 1993b, p. 1323.

Ward, M.D., and D.A. Buttry, "In Situ Mass Detection with Piezoelectric Transducers," *Science*, 249:1000 (1990).

Wells, J.C., P.G. Blystone, M.D. Johnson, W.R. Haag, and B.W. Colston Jr., "A Commercialized Fiber Optic Sensor for the Detection of Volatile Chlorinated Hydrocarbons," in *Proceedings of the Third International Symposium on Field Screening Methods for Hazardous Wastes and Toxic Chemicals*, Air and Waste Management Association, Pittsburgh, PA, 1993, p. 345.

Williamson, P., and J. Thomson, "A Technique for Double-Cased Cone Penetrometer Testing Beneath a Leachate-Saturated Landfill," in *Proceedings of the Seventh National Outdoor Action Conference and Exposition*, National Ground Water Association, Dublin, OH, 1993, p. 451.

Wise, D.L., and L.B. Wingard, Eds., *Biosensors with Fiberoptics* (Clifton, NJ: Humana Press, 1991).

Wise, M.B., R.H. Ilgner, M.V. Buchanan, and M.R. Guerin, "Rapid Determination of Drugs and Semivolatile Organics by Direct Thermal Desorption Ion Trap Mass Spectrometry," in *Proceedings of the Second International Symposium on Field Screening Methods for Hazardous Wastes and Toxic Chemicals*, National Ground Water Association, Dublin, OH, 1991, p. 823.

Wise, M.B., C.V. Thompson, M.R. Guerin, and R.A. Jenkins, "Development and Testing of a Field Transportable Direct Sampling Ion Trap Mass Spectrometer," in *Proceedings of the Third International Symposium on Field Screening Methods for Hazardous Wastes and Toxic Chemicals*, Air and Waste Management Association, Pittsburgh, PA, 1993, p. 543.

Wohltjen, H., W.R. Barger, A.W. Snow, and N.L. Jarvis, "A Vapor Sensitive Chemiresistor Fabricated with Planar Microelectrodes and a Langmuir-Blodgett Organic Semiconductor Film," *IEEE Trans. Elec Dev.*, ED-32(7): 1170 (1985).

Wolfbeis, O.S., Ed., *Fiber Optic Chemical Sensors and Biosensors*, Vols. 1 and 2, (Boca Raton, FL: CRC Press, 1991).

Yim, H.S., C.E. Kibbey, S.C. Ma, D.M. Kliza, D. Liu, S.B. Park, C. Espados Torre, and M.E. Meyerhoff, "Polymer Membrane-Based Ion-, Gas- and Bio-Selective Potentiometric Sensors," *Biosensors and Bioelectronics*, 8:1 (1993).

Zamzow, D.S., D.P. Baldwin, S.J. Weeks, S. J. Bajic, and A.P. D'Silva, "In Situ Determinisation of Uranium in Soil by Laser Ablation-Indirectively Coupled Atomic Emission Spectrometry," *Environmental Science & Technology*, 28:352 (1994).

Zemansky, G.M., R.W. Wood, S.R. Alewine, and J.H. Frerichs, "Field Screening Soil/Sediment Samples for PCBs," in *Proceedings of the Third International Symposium on Field Screening Methods for Hazardous Wastes and Toxic Chemicals*, Air and Waste Management Association, Pittsburgh, PA, 1993, p. 229.

Glossary of Acronyms

AES	Atomic Emission Spectroscoypy
ALTS	Apache Leap Tuff Site
AMZ	aggregated mixing zone
API	American Petroleum Institute
ASAE	American Society of Agricultural Engineers
ASTM	American Society for Testing and Materials
ATDs	alpha-track detectors
bls	below land surface
BOD	biochemical oxygen demand
BTC	breakthrough curve
BTEX	benzene, toluene, ethylbenzene, and xylene
CCR	California Code of Regulations
CDE	convection-dispersion equation
CEC	cation exchange capacity
CLP	Contract Laboratory Program
CPT	cone penetrometer technology
C.V.	coefficient of variation
DCA	1,1-dichloroethane
DCE	1,1-dichloroethene
DNAPLs	dense nonaqueous phase liquids
DOD	Department of Defense
DOE	Department of Energy
DOT	Department of Transportation
DPPS	drive point/piston sampler
DQOs	data quality objectives
DSITMS	direct sampling ion trap mass spectrometer
EC	exchange capacity; electrical conductivity
ECD	electron capture detector
Eh	redox potential
ELCD	electrolytic conductivity detector
EM	electromagnetic induction
EMSL	Environmental Monitoring Systems Laboratory
EPA	Environmental Protection Agency
FC	field capacity
FEM	frequency domain measurement

FID	flame ionization detector
GC	gas chromatography
GC/MS	gas chromatography/mass spectrometry
GPR	Ground Penetrating Radar
HDPE	high density polyethylene
HS/GC	headspace gas chromatography
ICP/AES	inductively coupled plasma/atomic emission spectrometry
ICP/MS	inductively-coupled plasma/mass spectrometry
ID	inside diameter
IMS	ion mobility spectrometry
IR	infrared
ISEs	ion-selective electrodes
LAU	Lower Alluvial Unit
LDE	limited-exposure handling (method)
LEAP	liquid extracting air permeameter
LLNL	Lawrence Livermore National Laboratory
MAU	Middle Alluvial Unit
MCL	minimum confidence level
MMO	methane monooxygenase
MOS	metal-oxide semiconductors
MSE	mean square error
NA	not analyzed
NAA	neutron activation analysis
NAPL	nonaqueous phase liquid
ND	not detected
NPOC	nonpurgeable organic carbon
OD	outside diameter
OSHA	Occupational Safety and Health Administration
OVA	organic vapor analyzer
PAHs	polyaromatic hydrocarbons
PCBs	polychlorinated biphenyls
PCE	pan collection efficiency; perchlorethylene
pdf	probability density function
PDP	portable dielectric probe
PEM	performance evaluation material
PID	photoionization detector
PSCS	porous suction cup sampler
PTFE	polytetrafluoroethylene
PT/GC/MS	purge-and-trap gas chromatography mass spectroscopy
PVC	polyvinyl chloride
PWP	permanent wilting point
QA	quality assurance
QC	quality control
QCM	quartz crystal microbalance
QTM	quick turnaround method
RCRA	Resource Conservation and Recovery Act
REV	representative elementary volume
RF	radio frequency
RI/FS	remedial investigation/feasibility study
RM	recommended method

%RSD	percent relative standard deviation
SAR	sodium adsorption ratio
SAS	Special Analytical Services
SAW	surface acoustic wave
SCAPS	Site Characterization and Analysis Penetrometer System
SCF	supercritical fluid
SFE	supercritical fluid extraction
SITE	Superfund Innovative Technology Evaluation
SMOW	Standard Mean Ocean Water
SRP	Salt River Project
SRS	Savannah River Site
STP	standard temperature and pressure
SVE	soil vapor extraction
SWMUs	solid waste management units
TCA	1,1,1-trichloroethane, tricarboxylic acid
TCD	thermal conductivity detector
TCE	trichloroethene
TCNE	tetracyanoethylene
TDR	time domain reflectometry
TDS	total dissolved solids
TEM	transverse electric and magnetic (mode)
TFM	transfer function model
TKN	total Kjeldahl nitrogen
TLC	thin layer chromatography
TNE	transport nonequilibrium
TPHs	total petroleum hydrocarbons
TREM	transient electromagnetic induction
UAU	Upper Alluvial Unit
USCS	United Soil Classification System
USGS	United States Geological Survey
UST	Underground Storage Tanks
VOA	volatile organic analysis
VOC	volatile organic compound
XRF	x-ray fluorescence spectrometry

List of Contributors

Sunnie A. Aburime
Department of Plant and Soil Science
Alabama A&M University
P.O. Box 1208
Normal, AL 35762

Janick F. Artiola
Soil and Water Science Department
429 Shantz Building, #38
University of Arizona
Tucson, AZ 85721

Jan Boll
Department of Agricultural and Biological Engineering
Cornell University
Riley-Robb Hall
Ithaca, NY 14853

William R. Bond
Terra Vac, Inc.
12596 West Bayaud Avenue—Suite 205
Lakewood, CO 80228

Herman Bouwer
U.S. Water Conservation Laboratory
USDA ARS
4331 East Broadway Road
Phoenix, AZ 85040

Mark L. Brusseau
Department of Soil and Water Science and
Department of Hydrology and Water Resources
University of Arizona
Tucson, AZ 85721

Wayne Crawley
K.W. Brown Environmental
500 Graham Road
College Station, TX 77841

Stephen J. Cullen
Vadose Zone Monitoring Laboratory
Institute for Crustal Studies
University of California
Santa Barbara, CA 93106-1100
and
Geraghty & Miller
5425 Hollister Avenue, Suite 100
Santa Barbara, CA 93111

Stanley N. Davis
Department of Hydrology and Water Resources
University of Arizona
Tucson, AZ 85721

D.W. Dorrance
ENSR
2700 Wycliff Road, Suite 300
Raleigh, NC 27607

Neal D. Durant
Department of Geography and Environmental Engineering
Johns Hopkins University
313 Ames Hall
3400 N. Charles St.
Baltimore, MD 21218

Lawrence A. Eccles
Environmental Monitoring Systems Laboratory
U.S. Environmental Protection Agency
P.O. Box 93478
Las Vegas, NV 89193

Daniel D. Evans
Department of Hydrology and Water Resources
University of Arizona
Tucson, AZ 85721

Lorne G. Everett
Vadose Zone Monitoring Laboratory
Institute for Crustal Studies
University of California
Santa Barbara, CA 93106-1100
and
Geraghty and Miller
5425 Hollister Avenue, Suite 100
Santa Barbara, CA 93111

Graham E. Fogg
Hydrologic Science
University of California, Davis
225 Veihmeyer Hall
Davis, CA 95616

Charles P. Gerba
Department of Soil and Water Science
University of Arizona
Tucson, AZ 85721

Amado Guzman-Guzman
Department of Hydrology and Water Resources
College of Engineering and Mines
Building 11
University of Arizona
Tucson, AZ 85721

Joseph P. Hayes
Weber, Hayes and Associates
120 Westgate Drive
Watsonville, CA 95076

A. A. Hussen
8426 Timber Mill Road
San Antonio, TX 78250

Eric Jolles
Department of Agricultural and Biological Engineering
Cornell University
Riley-Robb Hall
Ithaca, NY 14853

Barry Keller
Metcalf and Eddy, Inc.
104 West Anapamu Street, Suite L
Santa Barbara, CA 93101

Eric N. Koglin
U.S. EPA
Environmental Monitoring Systems Laboratory
P.O. Box 93478
Las Vegas, NV 89193–3478

Mark L. Kram
P.O. Box 2586
Santa Barbara, CA 93120

John H. Kramer
Vadose Zone Monitoring Laboratory
Institute for Crustal Studies
University of California
Santa Barbara, CA 93106

and
Condor Earth Technologies, Inc.
P.O. Box 3905
Sonora, CA 95370

Timothy E. Lewis
Bureau of Land Management
Forest Health Monitoring
Forestry Sciences Laboratory
3041 Cornwallis Road
Research Triangle Park, NC 27709

Raina M. Miller
Department of Soil and Water Science
University of Arizona
429 Shantz Building
Tucson, AZ 85721

Vernon B. Myers
Office of Solid Waste
U.S. Environmental Protection Agency
401 M Street, SW
Washington, DC 20460

Donald R. Nielsen
Hydrologic Science
University of California, Davis
123 Veihmeyer Hall
Davis, CA 95616

J.-Yves Parlange
Department of Agricultural and Biological Engineering
Cornell University
Riley-Robb Hall
Ithaca, NY 14853–5701

David K. Powelson
Department of Environmental Sciences
University of Virginia
Clark Hall
Charlottesville, VA 22903

Edward J. Poziomek
Harry Reid Center for Environmental Studies,
University of Nevada-Las Vegas,
Las Vegas, NV 89154–4009;
currently at Department of Chemistry and Biochemistry,
Old Dominion University, Norfolk, VA 23529–0126.

Todd C. Rasmussen
Warnell School of Forest Resources
University of Georgia
Athens, GA 30602

Shirlee C. Rhodes
Bellatrix, Inc.
8040 E. Morgan Trail, Suite 5
Scottsdale, AZ 85258

Donald D. Runnells
Shepherd Miller, Inc.
1600 Specht Point Drive, Suite F
Fort Collins, CO 80525

Kenneth D. Schmidt
Kenneth D. Schmidt and Associates
1540 East Maryland, Suite 100
Phoenix, AZ 85014

John S. Selker
Department of Bioresource Engineering
Oregon State University
Gilmore Hall
Corvallis, OR 97331

D. Shibberu
ICF Kaiser Engineers Inc.
Suite 200
11290 Point East Drive
Rancho Cordova, CA 95742

David S. Springer
Geraghty & Miller
5425 Hollister Avenue, Suite 100
Santa Barbara CA 93111

Tammo S. Steenhuis
Department of Agricultural and Biological Engineering
Cornell University
Riley-Robb Hall
Ithaca, NY 14853–5701

Kirk A. Thomson
Environmental Support Technologies, Inc.
23011 Moulton Parkway, Suite E-6
Laguna Hills, CA 92653

David C. Tight
Burlington Environmental, Inc.
5901 Christie Avenue
Emeryville, CA 94608

William L. Ullom
Vadose Research, Inc.
1645 Cleveland Avenue, NW
Canton, OH 44703

A.W. Warrick
Department of Soil and Water Science
University of Arizona
Tucson, AZ 85721

Ian White
CSIRO Centre for Environmental Mechanics
Institute of Natural Resources and Environment
GPO Box 821, Canberra, AUSTRALIA ACT 2601

Peter J. Wierenga
Department of Soil and Water Science
Shantz Building, Room 429
University of Arizona
Tucson, AZ 85721

Michael A. Williams
Black & Veatch Waste Science, Inc.
4717 Grand Avenue—Suite 500
Kansas City, MO 64112

L.G. Wilson
Department of Hydrology and Water Resources
University of Arizona
Tucson, AZ 85721

T.-C. Jim Yeh
Department of Hydrology and Water Resources
College of Engineering and Mines
Building 11
University of Arizona
Tucson, AZ 85721

S.J. Zegelin
CSIRO Centre for Environmental Mechanics
Institute of Natural Resources and Environment
GPO Box 821, Canberra, AUSTRALIA, ACT 2601

Index